USING AND UNDERSTANDING

Mathematics

A QUANTITATIVE
REASONING APPROACH

Dedication

To those who want to understand the world, and especially to those who have struggled with mathematics in the past, we hope this book will help guide you in the new millennium.

And, of course, to those who have supported us as we've worked on this book during the past decade, especially Julie, Lisa, Katie, and Grant (whose birth will coincide with publication), thank you for your love and patience.

USING AND UNDERSTANDING
Mathematics
A QUANTITATIVE
REASONING APPROACH

Jeffrey O. Bennett
University of Colorado at Boulder

William L. Briggs
University of Colorado at Denver

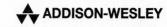
ADDISON-WESLEY
An imprint of Addison Wesley Longman, Inc.

Reading, Massachusetts • Menlo Park, California • New York • Harlow, England
Don Mills, Ontario • Sydney • Mexico City • Madrid • Amsterdam

Sponsoring Editor	Bill Poole
Project Manager	Elka Block
Managing Editor	Karen Guardino
Associate Editor	Rachel Reeve
Senior Production Supervisor	Peggy McMahon
Design Supervisor	Barbara T. Atkinson
Art Editor and Designer	The Davis Group, Inc.
Marketing Supervisor	Michael Boezi
Senior Manufacturing Manager	Ralph Mattivello
Cover Design	Barbara T. Atkinson
Cover Art	Anne Burns
Composition	Black Dot Graphics
Technical Art Illustration	George Nichols, John Goshorn
Situational Art Illustration	Network Graphics

About the Cover:

The landscape on the front cover may look like a painting, but it actually is a mathematically-generated image created by mathematician Anne Burns at Long Island University, C. W. Post Campus. Dr. Burns studies the relationships between mathematics and art. She is particularly interested in using computers to create landscapes and intricate geometric designs. Her landscapes are generated using fractal geometry, which can be used to describe irregular shapes and forms found in nature. Fractal geometry, along with other aspects of mathematics and art, is discussed in Chapter 10.

Credits appear on page C-1.

Library of Congress Cataloging-in-Publication Data
Bennett, Jeffrey O.
    Using and understanding mathematics : a quantitative reasoning approach / Jeffrey O.
    Bennett, William L. Briggs.
        p.   cm.
    Includes bibliographical references and index.
    ISBN 0-201-65642-6
    1. Mathematics.   I. Briggs, William L. II. Title.
QA39.2.B4766   1998
510—dc21   98-3885
                                                                   CIP
Reprinted with corrections, March 1999

4 5 6 7 8 9 10 CRW 010099

Contents

Preface

Human history becomes more and more a race between education and catastrophe.

—H. G. Wells, in The Outline of History, 1920

What is This Book About?

There is no escaping the importance of mathematics in the modern world. However, for most people, the importance of mathematics lies not in the abstract ideas often presented in school, but in how it applies to personal and social issues. Thus everything in this book is designed with a very practical goal in mind: helping students prepare for the challenges they will face in all future endeavors. In particular, we've designed this book with three specific purposes:

- To prepare students for the mathematics they will encounter in other **college** courses, particularly core courses in social and natural sciences.

- To help students develop the ability to reason with quantitative information in a way that will help them achieve success in their **careers.**

- To provide students with both the specific background and the critical-thinking skills they need to understand the major issues they will face in **life,** both on a personal level and as citizens in a modern democracy.

Who Is This Book For?

We hope this book will be useful to everyone, but it is designed primarily for students who are *not* planning additional course work in mathematics. In particular, we've written this book with students in mind who have ever felt fear or anxiety when dealing with mathematics. Rather than a set of abstract or theoretical ideas, this book will give students exactly what they need to succeed mathematically in college, careers, and life.

Throughout this book, students will find issues and applications that directly affect their lives. Wherever students' interests lie—social sciences, environmental issues, politics, business and economics, art and music, or any of many other topics—they will find them covered in this book. Perhaps they will also find that mathematics is much more important to their lives than they had guessed, especially if they have not been fond of mathematics in the past.

The most important idea for students to take away from this book is that mathematics can help anyone understand a variety of topics and issues. Despite the many topics and examples covered in the book, there are many more that we could not cover in a limited space. But, once students have studied this book, they should be prepared to understand almost any quantitative issue they will encounter.

Structure of the Book

This book has a modular structure designed to allow instructors maximum flexibility in creating a course that meets particular needs. The twelve chapters are organized broadly by mathematical topics. Each chapter, in turn, is divided into a set of self-contained *units* that focus on particular issues or applications. Because the units are self-contained, instructors can pick and choose among them in designing a course. Some of the later units build upon material covered in earlier units, but instructors can find several model curricula in the *Instructor's Guide and Solutions Manual* to help them shape their course.

Prerequisite Mathematical Background

Because of its modular structure, this book can be used by students with a wide range of mathematical backgrounds. Many of the units require nothing more than arithmetic and a willingness to think about quantitative issues in new ways. Others use techniques of algebra or geometry, which are reviewed as they arise. In particular, Chapter 6 provides a review of key algebraic ideas that are used in many subsequent units. Note, however, that *nothing in this book is remedial*: While we provide reviews of mathematical techniques, the techniques always arise in the context of issues and applications that require students to demonstrate a high level of critical thinking.

SUPPLEMENTS FOR INSTRUCTORS

Instructor's Guide and Solutions Manual

ISBN 0-201-59083-2
Prepared by the authors, this supplement contains model curricula, unit-by-unit teaching hints, and technical notes. It also contains solutions to all text exercises. See your Addison-Wesley sales representative.

Instructor's Testing Manual

ISBN 0-201-38401-9
Prepared by Laurel Technical Services, this printed supplement contains four alternate tests for each unit utilizing both free-response and multiple-choice formats. Instructors can select questions from the units covered to create tests suitable for their classes. Answer keys for all tests are provided. See your Addison-Wesley sales representative.

TestGen-EQ with QuizMaster-EQ

Windows: ISBN 0-201-38377-2
Macintosh: ISBN 0-201-38378-0

The test items of the testing manual are available in software. TestGen-EQ's friendly graphical interface enables instructors to easily view, edit, and add questions, transfer questions to tests, and print tests in a variety of fonts and forms. A built-in question editor gives the user the ability to create graphs, import graphics, insert mathematical symbols and templates, and insert variable numbers or text. QuizMaster-EQ enables instructors to create and save tests using TestGen-EQ so students can take them for practice or a grade on a computer network. Instructors can set preferences for how and when tests are administered. See your Addison-Wesley sales representative.

Videotape Series

Qualified adopters of this text may be eligible to receive a classroom copy of the exciting new PBS video series, *Life By the Numbers*, produced by WQED, Pittsburgh. See your Addison-Wesley sales representative.

SUPPLEMENTS FOR STUDENTS

Student's Study Guide and Solutions Manual

ISBN 0-201-59084-0
Prepared by William Briggs, this supplement contains hints for learning and studying. It also contains complete solutions to exercises that are answered in the back of the text. Ask your bookstore about ordering.

SUPPLEMENTS FOR INSTRUCTORS AND STUDENTS

Using and Understanding Mathematics Web Site

Prepared by Jeffrey Bennett and William Briggs. The textbook is really the tip of an iceberg: It covers many topics and issues, but many of the issues covered in this text are constantly developing. To help everyone find the rest of the iceberg, the authors have a supplemental web site to the textbook at

http://hepg.awl.com (keyword:Bennett)

This Web site features chapter-by-chapter updates along with new examples and applications, as well as links to useful sites. Some of these links are connected to the Web Watch feature in the text.

Acknowledgments

This textbook has been developing for more than a decade, ever since the authors first began teaching a quantitative reasoning course in 1987 at the University of Colorado. Along the way, drafts of this book were published in several preliminary forms from which we have received invaluable feedback from many instructors and students. Because of the changes we have made in response to this feedback, the first edition looks unlike any previous draft. We would particularly like to thank the individuals who carefully reviewed the manuscript for this first edition:

W. Wayne Bosché, Jr., Dalton College
Michael Bradshaw, Caldwell Community College and Technical Institute
Kellie Evans, York College of Pennsylvania
Pat Foard, South Plains College
Barbara Grover, Salt Lake City Community College
Bonnie Kelly, University of South Carolina
Norbert Kuenzi, University of Wisconsin, Oshkosh
Jay Malmstrom, Oklahoma City Community College
Paul O'Heron, Broome Community College
Evelyn Pupplo-Cody, Marshall University
Scott Reed, College of Lake County
Hugo Rossi, University of Utah
Judith Silver, Marshall University
Terry Tolle, Southwestern Community College

We would also like to thank those who reviewed earlier drafts of the text, including:

Bob Bernhardt, East Carolina University
W. E. Briggs, University of Colorado at Boulder
Walter Czarnec, Framingham State College
Marsha J. Driskill, Aims Community College
John Emert, Ball State University
Lynn R. Hun, Dixie College
Jim Koehler, University of Colorado at Denver
Timothy C. Swyter, Frederick Community College
Emily Whaley, DeKalb College
Donald J. Zielke, Concordia Lutheran College

Throughout the development process, we have received many excellent ideas from former graduate students and other instructors teaching the course. We particularly would like to thank Hal Huntsman, John Supra, David Wilson, and David Theobald. Others who have made important contributions to this book include our accuracy checker, Paul Lorczak; Richard Blumenthal, a public accountant who reviewed Chapter 5 on financial management; Anne Burns, who graciously provided fractal imagery; and John Goshorn, who drafted most of the illustrations appearing in this first edition.

We especially appreciate the faith that all our editors and friends at Addison Wesley Longman have placed in this project, including Bill Poole, Greg Tobin, Peggy McMahon, Rachel Reeve, Barbara Atkinson, Michael Boezi, Brenda Bravener, Kim Ellwood, Joe Vetere, Andy Fisher, Christine O'Brien, Mary Clare McEwing, and Linda Davis. Extra special thanks go to Elka Block, who meticulously read every page of this book multiple times, making countless suggestions for improvement both editorially and pedagogically. Without Elka, this project would never have succeeded.

Finally, we deeply appreciate the efforts of Cherilynn Morrow, who helped us frame the concept of quantitative reasoning upon which this book is based, and worked with us as a co-author on the preliminary edition published in 1995.

Jeffrey O. Bennett
William L. Briggs

Prologue
LITERACY FOR
THE MODERN WORLD

Mathematics is the key to opportunity. No longer just the language of science, mathematics now contributes in direct and fundamental ways to business, finance, health, and defense. For students, it opens doors to careers. For citizens, it enables informed decisions. For nations, it provides knowledge to compete in a technological economy. To participate fully in the world of the future, America must tap the power of mathematics.

FROM *EVERYBODY COUNTS*

WHAT IS QUANTITATIVE LITERACY?

Literacy is the ability to read and write, and it comes in varying degrees. Some people can recognize only a few words and write only their names; others read and write in many languages. A primary goal of our educational system is to provide citizens with a level of literacy sufficient to read and write about the important issues of our time.

Today, the abilities to interpret and reason with **quantitative** information—information that involves mathematical ideas or numbers—are crucial aspects of this literacy. This so-called **quantitative literacy** is essential to understanding modern issues that appear in the news everyday. The process of interpreting and reasoning with quantitative information is called **quantitative reasoning**. The purpose of this book is to help you gain the skills of quantitative reasoning as it applies to issues you will encounter in

- your subsequent course work,
- your career, and
- your daily life.

QUANTITATIVE REASONING AND CULTURE

Quantitative reasoning enriches the appreciation of both ancient and modern culture. The historical record shows that all great cultures devoted substantial energy to mathematics and to science (or to observational studies that predated modern science). Without a sense of how quantitative concepts are used in art, architecture, and science, you cannot fully appreciate the incredible achievements of the Mayan civilization of Central America, the builders of the great city of Zimbabwe in Africa, the ancient Egyptian and Greek civilizations, the ancient Polynesians with their navigational expertise, or countless other cultures.

A ROCK, A RIVER, A TREE

Hosts to species long since departed

Marked the mastodon,

The dinosaur, who left dry tokens

Of their sojourn here

On our planet floor,

Any broad alarm of their hastening doom

Is lost in the gloom of dust and ages.

—*FROM* ON THE PULSE OF THE MORNING, *BY MAYA ANGELOU*

Similarly, quantitative concepts can help you understand and appreciate the great works of modern artists. The excerpt from the Maya Angelou poem can be more fully appreciated by knowing that dinosaurs have been extinct for 65 million years, and by being able to put this amount of time into perspective. The ties between mathematics and music can be found in modern and classical music, as well as in the digital production of music. Indeed, it is hard to find popular works of art, film, or literature that do not make at least some indirect reference to mathematics.

QUANTITATIVE REASONING IN THE WORK FORCE

Quantitative reasoning is important in the work force. A lack of adequate quantitative skills closes off many of the most challenging and highest paying jobs. Table 1 defines skill levels in language and mathematics on a scale of 1 to 6, and Table 2 shows the typical levels needed in many jobs. (Tables 1 and 2 are adapted from a chart in *Education: The Knowledge Gap*, supplement to *The Wall Street Journal*, February 9, 1990.)

Table 1   Skill Levels

Level	Language Skills	Math Skills
1	Recognizes 2500 two- or three-syllable words. Reads at a rate of 95–120 words per minute. Writes and speaks simple sentences.	Adds and subtracts two-digit numbers. Does simple calculations with money, volume, length, and weight.
2	Recognizes 5000–6000 words. Reads 190–215 words per minute. Reads adventure stories and comic books, as well as instructions for assembling model cars. Writes compound and complex sentences with proper grammar and punctuation.	Adds, subtracts, multiplies, and divides all units of measure. Computes ratio, rate, and percentage. Draws and interprets bar graphs.
3	Reads novels and magazines, as well as safety rules and equipment instructions. Writes reports with proper format and punctuation. Speaks well before an audience.	Understands basic geometry and algebra. Calculates discount, interest, profit and loss, markup, and commissions.
4	Reads novels, poems, newspapers, and manuals. Prepares business letters, summaries, and reports. Participates in panel discussions and debates. Speaks extemporaneously on a variety of subjects.	Deals with complex algebra and geometry, including linear and quadratic equations, logarithmic functions, and axiomatic geometry.
5	Reads literature, book and play reviews, scientific and technical journals, financial reports, and legal documents. Can write editorials, speeches, and critiques.	Knows calculus and statistics; able to deal with econometrics.
6	Same types of skills as level 5, but more advanced.	Works with advanced calculus, modern algebra, and statistics.

Table 2   Skill-Level Requirements

Occupation	Language Level	Math Level	Occupation	Language Level	Math Level
Biochemist	6	6	Corporate executive	4	5
Computer engineer	6	6	Computer sales agent	4	4
Mathematician	6	6	Management trainee	4	4
Cardiologist	6	5	Insurance sales agent	3	4
Social psychologist	6	5	Retail store manager	3	4
Lawyer	6	4	Cement mason	3	3
Tax attorney	6	4	Dairy farm manager	3	3
Newspaper editor	6	4	Poultry farmer	3	3
Accountant	5	5	Tile setter	3	3
Personnel manager	5	5	Travel agent	3	3
Corporate president	5	5	Telephone operator	3	2
Weather forecaster	5	5	Janitor	3	2
Secondary teacher	5	5	Short-order cook	3	2
Elementary teacher	5	4	Assembly-line worker	2	2
Disc jockey	5	3	Toll collector	2	2
Financial analyst	4	5	Laundry worker	1	1

Note that the occupations requiring higher skill levels are generally the most prestigious and highest paying. Note also that most of these occupations require high skill levels in *both* language and math, refuting the myth that if you're good at language you don't have to be good at mathematics, and vice versa.

Calvin and Hobbes
by Bill Watterson

Jobs are becoming more demanding, more complex. But our schools don't seem up to the task. They are producing students who lack the skills that business so desperately needs to compete in today's global economy. And in doing so, they are condemning students to a life devoid of meaningful employment.

—*FROM* THE WALL STREET JOURNAL

𝒯HINKING ABOUT…
People Who Studied Mathematics.

Critical thinking provided by the study of mathematics is valuable in many careers. The following is only a small sample of people who studied mathematics, but became famous for work in other fields. (Most of the names on this list are from an article by Steven G. Buyske ("Famous Nonmathematicians," *American Mathematical Monthly*, November 1993)).

Corazon Aquino, former president of the Philippines. A mathematics minor; **Harry Blackmun**, Supreme Court justice. Summa cum laude in mathematics, Harvard University; **Lewis Carroll** (Charles Dodgson), author of *Alice in Wonderland*. A mathematician; **David Dinkins**, former mayor of New York City. BA in mathematics, Howard University; **Alberto Fujimori**, president of Peru. MS in mathematics, University of Wisconsin; **Art Garfunkel**, musician. MA in mathematics, Columbia University; **Mae Jemison**, first African-American woman in space. Studied mathematics as part of her degree in chemical engineering from Stanford University; **John Maynard Keynes**, economist. MA in mathematics, Cambridge University; **Lee Hsien Loong**, politician in Singapore. BA in mathematics, Cambridge University; **Edwin Moses**, three-time Olympic Champion in the 400-meter hurdles. Studied mathematics as part of his degree in physics from Morehouse College; **Florence Nightingale**, pioneer in nursing. Studied mathematics and applied it to her work; **William Perry**, former secretary of defense. Ph.D. in mathematics, Pennsylvania State University; **Sally Ride**, first American woman in space. Studied mathematics as part of her Ph.D. in physics from Stanford University; **David Robinson**, basketball star. Bachelor's degree in mathematics, U.S. Naval Academy; **Alexander Solzhenitsyn**, Nobel prize-winning Russian author. Degrees in mathematics and physics from the University of Rostov; **Bram Stoker**, author of Dracula. Studied mathematics at Trinity University, Dublin; **Laurence Tribe**, Harvard law professor. Summa cum laude in mathematics, Harvard University; **Virginia Wade**, Wimbledon champion. Bachelor's degree in mathematics, Sussex University.

MISCONCEPTIONS ABOUT MATHEMATICS

Do you consider yourself to be either "math phobic" (fear of mathematics) or "math loathing" (dislike of mathematics)? We hope not; but if you do, you aren't alone. Many adults harbor fear or loathing of mathematics and, unfortunately, these attitudes are often reinforced by classes that present mathematics as an obscure and sterile subject.

In reality, mathematics is not nearly so dry as it often seems in school. Indeed, attitudes toward mathematics often are directed not at what mathematics really is, but at some common misconceptions about mathematics. Let's investigate a few of these misconceptions and the reality behind them.

Misconception One: Math Requires a Special Brain

One of the most pervasive misconceptions is that some people just aren't good at mathematics because learning mathematics requires special and rare abilities. The reality is that nearly everyone can do mathematics, although it requires self-confidence and hard work.

Why should anyone think it otherwise? Years of work are required to learn to read, to master a musical instrument, or to become skilled at a sport. Indeed, the belief that mathematics requires special talent found in a few elite people is peculiar to the United States. In other countries, particularly in Europe and Asia, *all* students are expected to become proficient in mathematics.

Of course, different people learn mathematics at different rates and in different ways. For example, some people learn by concentrating on concrete problems, others by thinking visually, and still others by thinking abstractly. No matter what type of thinking style you prefer, you can succeed in mathematics.

Misconception Two: Math in Modern Issues Is Too Complex

Some people claim that the advanced mathematical concepts underlying many modern issues are too complex for the average person to understand. However, while only a few people receive the training needed to work with or discover advanced mathematical concepts, *most people* are capable of understanding enough about the mathematical basis of important issues to develop informed and reasoned opinions.

You may already recognize a similar idea in other fields. For example, years of study and practice are required to become a proficient, professional writer, but most people can read a book. It takes hard work and law school to become a lawyer, but most people can understand how the law affects them in a particular situation. And though few have the musical talent of Mozart, anyone can learn to appreciate his music. Mathematics is no different. If you've made it this far in school, you can understand enough mathematics to succeed as an individual and a concerned citizen. Don't let anyone tell you otherwise!

THE FAR SIDE   By GARY LARSON

Math phobic's nightmare

Equity for all requires excellence for all; both thrive when expectations are high.
—FROM EVERYBODY COUNTS

Misconception Three: Math Makes You Less Sensitive

Some people believe that learning mathematics will somehow make them less sensitive to the romantic and aesthetic aspects of life. In fact, understanding the underlying mathematics that produces the colors of the sunset or the geometric beauty in a work of art can actually enhance aesthetic appreciation. Furthermore, many people find beauty and elegance in mathematics itself. It is no accident that many people trained in mathematics have made important contributions to art, music, and many other fields.

Misconception Four:
Math Makes No Allowance for Creativity

The "turn the crank" nature of the problems in many textbooks may give the impression that mathematics stifles creativity. Some of the facts, formalisms, and skills required for mathematical proficiency are fairly cut and dried; but *using* these mathematical tools demands creativity. Consider designing and building a home. The task demands specific skills to lay the foundation, frame in the structure, install plumbing and wiring, and paint walls. But building the home involves much more: It requires creativity to develop the architectural design, to respond to on-the-spot problems during construction, and to factor in constraints based on budgets and building codes. The mathematical skills you've learned in school are like the skills of carpentry or plumbing. But the essence of applying mathematics is like the creative process of building a home.

Probably the most harmful misconception is that mathematics is essentially a matter of computation. Believing this is roughly equivalent to believing that writing essays is the same as typing them.

—JOHN ALLEN PAULOS

Misconception Five: Math Provides Exact Answers

A mathematical formula may yield a specific result, and in school that result may be marked right or wrong. But when you use mathematics in real-life situations, answers are never so clear-cut. For example:

> *A bank offers simple interest of 5%, paid at the end of one year (that is, after one year the bank pays you 5% of your account balance). If you deposit $1000 today and make no further deposits or withdrawals, how much will you have in your account after one year?*

A straight mathematical calculation *seems* simple enough: 5% of $1000 is $50; so you should have $1050 at the end of a year. But will you? How will your balance be affected by service charges or taxes on interest earned? What if the bank fails? What if the bank is located in a country in which the currency collapses during the year? Choosing a bank in which to invest your money is a *real* mathematics problem. It is a problem that doesn't necessarily have a simple or definitive solution.

Misconception Six: Math Is Irrelevant to My Life

No matter what your path in college, career, and life, you will find mathematics involved in many ways. A major goal of this text is to show you hundreds of examples in which mathematics applies to everyone's life. We hope you will find that mathematics is not only relevant, but also interesting and enjoyable.

WHAT *IS* MATHEMATICS?

In discussing misconceptions, we identified what mathematics is *not*; now, let's describe what mathematics *is*. The word *mathematics* is derived from the Greek word *mathematikos*, which means "inclined to learn." Thus, literally speaking, to be mathematical is to be curious, open-minded, and interested in always learning more! Today, we tend to look at mathematics in three different ways: as the sum of its branches, as a way to model the world, and as a language.

Mathematics As the Sum of Its Branches

As you progressed through school, you undoubtedly learned to associate mathematics with some of its branches. Among the better known branches of mathematics are:

- **logic**—the study of principles of reasoning;
- **arithmetic**—methods for operating on numbers;
- **algebra**—methods for working with unknown quantities;
- **geometry**—the study of size and shape;
- **trigonometry**—the study of triangles and their uses;
- **probability**—the study of chance;
- **statistics**—methods for analyzing data; and
- **calculus**—the study of quantities that change.

One way to view mathematics is as the sum of its branches. Indeed, most "traditional" mathematics books focus on one branch of mathematics at a time. In this book, we emphasize the connections between the branches rather than their differences. Nevertheless, familiarity with the various branches of mathematics is helpful, even if only to know when you are using them.

Mathematics As a Way to Model the World

Mathematics also may be viewed as a tool for creating **models**, or representations that allow us to study real phenomena. Modeling is not unique to mathematics. For example, a road map is a model that represents the roads in some region.

Mathematical models can be as simple as a single equation that predicts how the money in your bank account will grow or as complex as a set of thousands of interrelated equations and parameters used to represent the global climate. By studying models, we gain insight into otherwise unmanageable problems. A global climate model, for example, can help us understand weather systems, and allows us to ask "what if" questions about how human activity may affect the climate. Further, when a model is used to make a prediction that does *not* come true, it can point to areas where further research is needed. Today, mathematical modeling is used in nearly every field of study. The diagram shown on the next page indicates some of the various fields of study that use mathematical modeling to solve problems.

Skills are to mathematics what scales are to music or spelling is to writing. The objective of learning is to write, to play music, or to solve problems—not just to master skills.

—*FROM* EVERYBODY COUNTS

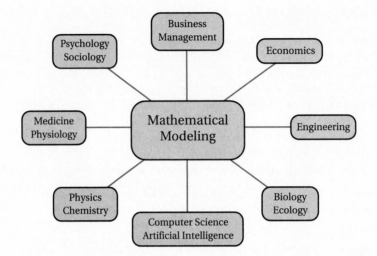

Mathematics As a Language

A third way to look at mathematics is as a language with its own vocabulary and grammar. Indeed, mathematics often is called "the language of nature" because it is so useful for modeling the natural world. Like any language, different degrees of fluency are possible. From this point of view, quantitative literacy is the level of fluency required for success in today's world.

The idea of mathematics as a language also is useful in thinking about how to *learn* mathematics. Table 3 shows a useful analogy between learning a language and learning mathematics, along with a comparison to learning art.

Table 3 Learning Mathematics: An Analogy to Language and Art		
Learning a Language	**Learning the Language of Art**	**Learning the Language of Mathematics**
Become familiar with many styles of speaking and writing such as essays, poetry, and drama.	Become familiar with many styles of art such as classical, renaissance, impressionist, and modern.	Become familiar with techniques from many branches of mathematics such as arithmetic, algebra, and geometry.
Place literature in context by studying its history and the social conditions in which it was created.	Place art in context by studying its history and the social conditions under which it was created.	Place mathematics in context by studying its history, purposes, and applications.
Learn the elements of language—such as words, parts of speech (nouns, verbs, etc.), and rules of grammar—and practice their proper use.	Learn the elements of visual form—such as lines, shapes, colors, and textures—and practice using them in your own art work.	Learn the elements of mathematics—such as numbers, variables, and operations—and practice using them to solve simple problems.
Develop the ability to analyze language in complex forms critically—novels, short stories, essays, poems, speeches, debates, and similar works.	Develop the ability to analyze works of art critically—painting, sculpture, architecture, photography, and similar works.	Develop the ability to analyze quantitative information critically—mathematical models, statistical studies, economic forecasts, investment strategies, and similar works.
Use language creatively for your own purposes, such as writing a term paper or story, or engaging in debate.	Use your sense of art creatively, such as in designing your house, taking a photograph, or making a sculpture.	Use mathematics creatively to solve problems you encounter and to help you understand important issues in the world around you.

HOW TO SUCCEED IN MATHEMATICS

If you are reading this book, you probably are enrolled in a mathematics course of some type at a college or university. The keys to success in your course include approaching the material with an open and optimistic frame of mind, paying close attention to how useful and enjoyable mathematics can be in *your* life, and studying effectively and efficiently. The following sections offer a few more specific hints that may be of use as you study.

Using *This* Book

Before we get into more general strategies for studying, here are a few guidelines that will help you use *this* book most effectively.

- Before doing any assigned problems, read assigned material *twice*:
  - On the first pass, read quickly to gain a "feel" for the material and concepts presented.
  - On the second pass, read the material in more depth, and work through the examples carefully.

- During the second reading, take notes that will help you when you go back to study later. In particular:
  - *Use the margins!* The wide margins in this textbook are designed to give you plenty of room for making notes as you study.
  - Don't highlight—underline! Using a pen or pencil to underline material requires greater care than highlighting, and therefore helps to keep you alert as you study.

- After you complete the reading, and again when studying for exams, make sure you can answer the *review questions* at the end of each unit.

- You'll learn best by *doing*, so do plenty of the end-of-unit problems. In particular, try some of the problems that have answers in the back of the book to supplement the problems assigned by your instructor.

Budgeting Your Time

A general rule of thumb for college classes is that you should expect to study about 2 to 3 hours per week *outside* class for each unit of credit. For example, a student taking 15 credit hours should spend 30–45 hours each week studying outside of class. Combined with time in class, this works out to a total of 45–60 hours per week—not much more than the time required of a typical job. Moreover, except for class time, you get to choose your own hours. Of course, if you are working while you attend school, you will need to budget your time carefully.

As a rough guideline, your studying time in a mathematics course might be divided as shown in the table on page 10.

If Your Course Is:	Time for Reading the Assigned Text (Per Week)	Time for Homework Assignments (Per Week)	Time for Review and Test Preparation (Average Per Week)	Total Study Time (Per Week)
3 credits	1 to 2 hours	3 to 5 hours	2 hours	6 to 9 hours
4 credits	2 to 3 hours	3 to 6 hours	3 hours	8 to 12 hours
5 credits	2 to 4 hours	4 to 7 hours	4 hours	10 to 15 hours

If you find that you are spending fewer hours than these guidelines suggest, you can probably improve your grade by studying more. If you are spending more hours than these guidelines suggest, you may be studying inefficiently; in that case, you should talk to your instructor about how to study more effectively for a mathematics class.

General Strategies for Studying

- Don't miss class. Listening to lectures and participating in discussions is much more effective than reading someone else's notes. Active participation will help you retain what you are learning.

- Budget your time effectively. An hour or two each day is more effective, and far less painful, than studying all night before homework is due or before exams.

- If a concept gives you trouble, do additional reading or problem solving beyond what has been assigned. If you still have trouble, ask for help: you surely can find friends, colleagues, or teachers who will be glad to help you learn.

- Working together with friends can be valuable, as you will solidify and enhance your own understanding when discussing concepts with others. However, be sure that you learn *with* your friends and do not become dependent on them.

Preparing for Exams

- Rework problems and other assignments; try additional problems to be sure you understand the concepts. Study your performance on assignments, quizzes, or exams from earlier in the semester.

- Study your notes from lectures and discussions. Pay attention to what your instructor expects you to know for an exam.

- Reread the relevant sections in the textbook, paying special attention to notes you have made in the margins.

- Study individually before joining a study group with friends. Study groups are effective only if *every* individual comes prepared to contribute.

- Don't stay up too late before an exam. Don't eat a big meal within an hour of the exam (thinking is more difficult when blood is being diverted to the digestive system).

- Try to relax before and during the exam. If you have studied effectively, you are capable of doing well. Staying relaxed will help you think clearly.

QUESTIONS FOR DISCUSSION OR STUDY

1. Mathematics in Modern Issues. Describe at least one way that mathematics is involved in each issue below.

Example: The spread of AIDS: Mathematics is used to study the probability of contracting AIDS.

a. The long-term viability of the Social Security system.

b. The appropriate level for the federal gasoline tax.

c. The fairness of medical insurance rates.

d. Job discrimination against women or ethnic minorities.

e. Effects of population growth (or decline) on your community.

f. Possible bias in standardized tests (e.g., the SAT).

g. The degree of risk posed by carbon dioxide emissions.

h. Immigration policy of the United States.

i. Violence in public schools.

j. Pick an issue of your choice from today's newspaper.

2. Quantitative Concepts in the News. Identify the major unresolved issue discussed on the front page of today's newspaper. List at least three areas in which quantitative concepts play a role in the policy considerations of this issue.

3. Mathematics and the Arts. Choose a well-known, historical figure in a field of art in which you have a personal interest (e.g., a painter, sculptor, musician, or architect). Briefly describe how mathematics played a role in or influenced that person's work.

4. Quantitative Literature. Choose a favorite work of literature (poem, play, short story, or novel). Describe one or more instances in which quantitative reasoning is helpful to understanding the subtleties intended by the author.

5. Your Quantitative Major. What is your major field of study? Identify ways in which quantitative reasoning is important within that field. (If you haven't yet chosen a major, pick a field that you are considering for your major.)

6. Career Preparation. Realizing that most Americans change careers several times during their lives, identify at least three occupations in Table 2 that interest you. Do you have the necessary skills for them at this time? If not, how can you acquire these skills?

7. Attitudes Toward Mathematics. Most children have a natural affinity for mathematics; they take pride in their counting skills and enjoy puzzles, building blocks, and computers. Unfortunately, this natural interest seems to be snuffed out in most people by the time they reach adulthood. What is your attitude toward mathematics? If you have a negative attitude, can you identify when in your childhood that attitude developed? If you have a positive attitude, can you explain why? How might you encourage someone with a negative attitude to become more positive?

SUGGESTED READING

The following popular and nontechnical books give good introductions to mathematics from historical and contemporary points of view. Books are listed alphabetically by title.

Innumeracy, J. A. Paulos (New York: Vintage Press, 1988).

Islands of Truth: A Mathematical Mystery Cruise, I. Peterson (New York: W. H. Freeman, 1990).

Mathematics: The New Golden Age, K. Devlin (New York: Penguin Books, 1988).

Mathematics: The Science of Patterns, K. Devlin (New York: Scientific American Library, 1994).

Mathematics and the Search for Knowledge, M. Kline (New York: Oxford University Press, 1985).

Mathematics in Western Culture, M. Kline (New York: Oxford University Press, 1953).

The Mathematical Tourist: Snapshots of Modern Mathematics, I. Peterson (New York: W. H. Freeman, 1988).

The Universe and the Teacup: The Mathematics of Truth and Beauty, K. C. Cole, Harcourt Brace, New York, 1998.

The World of Mathematics (4 Volumes), Ed. J.R. Newman (New York: Simon and Schuster, 1956).

What Is Mathematics Really?, R. Hersh (New York: Oxford University Press, 1997).

Chapter 1
PRINCIPLES OF REASONING

Skills of reasoning lie at the heart of problem solving and the interpretation of quantitative information. Although much of your everyday reasoning probably is intuitive, it still can be useful to break down the process of reasoning into its most basic components. Once you understand these components, you may find your skills of reasoning and problem solving greatly enhanced.

Civilized life depends upon the success of reason in social intercourse, the prevalence of logic over violence in interpersonal conflict.
JULIANA GERAN PILON

In a republican nation, whose citizens are to be led by reason and persuasion and not by force, the art of reasoning becomes of first importance.
THOMAS JEFFERSON

To be able to be caught up into the world of thought—that is being educated.
EDITH HAMILTON

*W*e use arguments nearly everyday to try to persuade other people of our views. However, arguments can take many different forms, some of which are more useful than others. Imagine, for example, that you overhear the following argument between two classmates.

Bob:	"The death penalty is immoral."
Susan:	"No it isn't."
Bob:	"Yes it is! Judges who give the death penalty should be impeached."
Susan:	"You don't even know how the death penalty is decided."
Bob:	"I know a lot more than you know!"
Susan:	"I can't talk to you, you're an idiot!"

This heated conversation may be typical of many arguments heard daily, but it accomplishes little. Rather than helping Bob or Susan to sway the other, this argument is likely to leave both of them upset and angry. Fortunately, there is a better way for them to argue: with **logic**, the study of the methods and principles involved in reasoning. Although arguing logically may not cause either of them to change their positions, it can at least help them understand each other. Then, perhaps, they'll discover common ground that will allow them to work together despite their differences.

People generally quarrel because they cannot argue.

—G. K. Chesterton (1874–1936), English author

An argument over the death penalty may seem far removed from mathematics, but the same principles of logic that could improve Bob and Susan's debate can also help you approach and solve mathematical problems. Thus we begin this book with a brief study of logic.

UNIT 1A

THE FORCES OF PERSUASION— AN OVERVIEW OF COMMON FALLACIES

CASE STUDY   *Joe Camel*

Studies published in the *Journal of the American Medical Association* showed that illegal sales of Camel brand cigarettes to minors skyrocketed from $6 million to $476 million per year between 1988 and 1991. This rise coincided with the beginning of an advertising campaign by the R. J. Reynolds Tobacco Company that featured a cartoon character named Joe Camel. A 1991 study of 229 preschool children by researchers at the Medical College of Georgia concluded that, by age 6, children recognized Joe Camel as readily as Mickey Mouse. A separate study of 131 teenaged smokers found that the popularity of Camel cigarettes had increased in parallel with the buildup of the Joe Camel advertising campaign.

Did the Joe Camel advertising campaign deliberately target children? R. J. Reynolds' executives denied it, but their opponents called them "merchants of death" seeking to hook young children on cigarettes to replace the customers killed by their product. What is the truth? And how can a cartoon character persuade anyone—child or adult—to buy a product?   ■

By the Way ············
R. J. Reynolds dropped the Joe Camel advertising campaign in 1997.

The Joe Camel case study shows the dramatic effects that are possible from advertising. The success of such advertisements is not accidental. More than $100 billion per year is spent on advertising in the United States, and the best advertising firms do extensive research before embarking on new campaigns.

We are surrounded by the powers of persuasion. Just as businesses use sophisticated advertising, politicians use public relations consultants to target their messages to sway voter opinion. Even on an individual level, we are constantly subjected to arguments from friends, relatives, and teachers who hope to influence our opinions or behavior.

Faced with such a barrage of persuasive powers, how can we formulate rational decisions and beliefs? One of the simplest ways is to distinguish between logical arguments and deceptive arguments, or **fallacies**. There's no magic to the process, but it takes a lot of practice—after all, public relations specialists have spent billions of dollars to discover what types of fallacies are most likely to fool us! The examples that follow illustrate some of the most common among the more than one hundred different types of fallacies.

By the Way ············
The word *fallacy* comes from the Latin for *deceit* or *trick.*

TIME-OUT TO THINK: In *Star Trek* (original cast), Mr. Spock believed that all decisions should be based on logic. Captain Kirk argued that logic should be only one ingredient in decision making. Which side do you take? When should logic be used, and when should other factors be considered?

EXAMPLE 1   *Appeal to Popularity*

"Ford makes the best automobile in America; after all, more people drive Fords than any other American car."

Analysis: The fact that more people drive Fords does not, in itself, mean that they are the best cars. This argument suffers the fallacy of *appeal to popularity* (or *appeal to majority*), in which the fact that large numbers of people believe a proposition is used inappropriately as evidence of its truth.

We can represent the general form of a fallacy with a simple diagram in which we use the letters *p* or *q* to stand for particular statements. In this case, illustrated in Figure 1.1, *p* stands for the statement that "Ford makes the best automobile in America."   ■

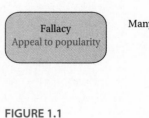

Fallacy
Appeal to popularity

Many people believe *p* is true, therefore…

p is true.

FIGURE 1.1

EXAMPLE 2   *False Cause*

"I placed the quartz crystal on my forehead and in five minutes my headache was gone. The crystal alleviated my pain."

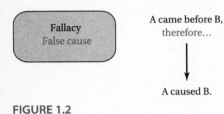

Fallacy
False cause

A came before B,
   therefore...

↓

A caused B.

FIGURE 1.2

Analysis: The fact that the crystal was on the forehead before the pain went away certainly does not mean the crystal *caused* the headache to go away. This argument suffers from the fallacy of *false cause*, in which the mere fact that one event came before another is incorrectly taken as evidence that the first event *caused* the second event (Figure 1.2).   ■

EXAMPLE 3   *Hasty Generalization*

"Three cases of childhood leukemia have occurred along the street where the high voltage power lines run. The power lines must be the cause of these illnesses."

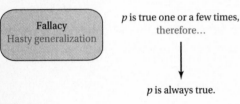

Fallacy
Hasty generalization

p is true one or a few times,
   therefore...

↓

p is always true.

FIGURE 1.3

Analysis: Three supporting cases certainly are not enough to establish a pattern, let alone to prove, that the power lines are the cause of the illnesses. This argument illustrates the fallacy of *hasty generalization*, in which a conclusion is drawn from an inadequate number of cases or cases that have not been sufficiently analyzed (Figure 1.3). If any connection between power lines and leukemia exists, it will have to be established with far more evidence than provided in this argument.   ■

EXAMPLE 4   *Appeal to Ignorance*

"Scientists have not proven that global warming will have any dire consequences for the human race. Therefore the global warming catastrophes that environmentalists shout about are bunk."

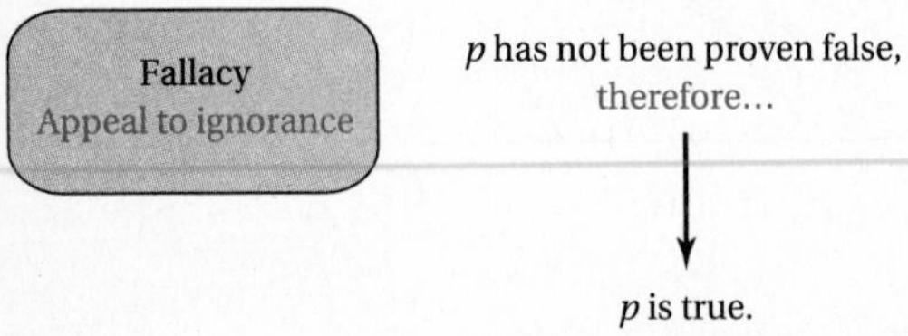

Fallacy
Appeal to ignorance

p has not been proven false,
   therefore...

↓

p is true.

FIGURE 1.4

Analysis: A *lack* of proof that global warming will have dire consequences does not mean that it won't have such consequences. The argument suffers from the fallacy of *appeal to ignorance*, in which lack of knowledge about the truth of a proposition is taken to imply the truth of the opposing proposition (Figure 1.4).   ■

TIME-OUT TO THINK: Suppose that a person is tried for a crime and found *not* guilty. Can you conclude that the person is innocent? Why or why not? Based on your answer, why do you think our legal system demands that prosecutors prove guilt, rather than demanding that defendants (suspects) prove innocence?

EXAMPLE 5   *Limited Choice*

"If you don't support the President then you are not a patriotic American."

Analysis: This argument suggests that there are only two types of Americans: patriotic ones who support the President and unpatriotic ones who don't. But there are many other possibilities, such as being patriotic while disliking a particular president. The fallacy is called *limited choice* (or *false choice*) because it artificially precludes choices that ought to be considered (Figure 1.5). Another simple and common form of this fallacy is "You're wrong, so I must be right." ■

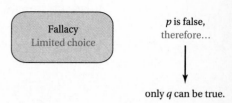

FIGURE 1.5

EXAMPLE 6   *Appeal to Emotion*

In a television commercial for Coors beer, a group of attractive men and women in bathing suits are playing volleyball on the beach and passing around cans of beer.

Analysis: The commercial offers no logical reason for drinking beer, let alone a particular brand of beer. The advertisers simply hope that the images will evoke positive emotions that you will associate with Coors beer. This attempt to evoke an emotional response as a tool of persuasion represents the fallacy of *appeal to emotion* (Figure 1.6). ■

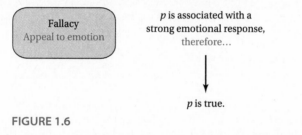

FIGURE 1.6

> **TIME-OUT TO THINK:**   Turn on the television. How long do you have to wait before you find a commercial that makes an appeal to emotion?

EXAMPLE 7   *Appeal to Force*

"If my opponent is elected, your tax burden will rise."

Analysis: This argument is only slightly more subtle than outright blackmail. Rather than saying "vote for me or I'll break your leg," it is essentially saying "vote for me or it's going to cost you money." Any such case in which persuasion is based on intimidation, pressure, or threats represents the fallacy of *appeal to force* (Figure 1.7). As the example shows, appeals to force are common in negative political advertisements. ■

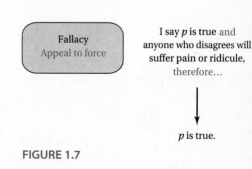

FIGURE 1.7

EXAMPLE 8   *Personal Attack (Ad Hominem)*

Gwen: "You should stop drinking because it is hurting your grades, endangering people when you drink and drive, and destroying your relationship with your family."
Merle: "I've seen you drink a few too many on occasion yourself!"

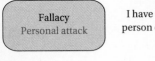

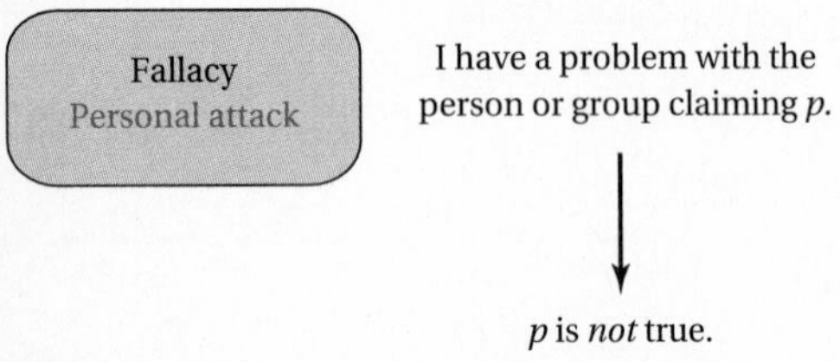

Fallacy
Personal attack

I have a problem with the person or group claiming *p*.

↓

p is *not* true.

FIGURE 1.8

Analysis: Gwen's argument is well-reasoned, with premises offering strong support for her conclusion that Merle should stop drinking. Merle rejects this argument by noting that Gwen sometimes drinks too much herself. Even if Merle's claim is true, it is irrelevant to Gwen's point. Merle has resorted to attacking Gwen personally rather than arguing logically, so we call this fallacy *personal attack* (Figure 1.8); this fallacy is also called *ad hominem*, Latin words meaning "to the person."

The fallacy of personal attack can also apply to groups. For example, you might hear someone say, "This new bill will be an environmental disaster because its sponsors received large campaign contributions from heavy industrial polluters." This argument is fallacious because it doesn't challenge the bill; instead, it questions the motives of the sponsors. ■

TIME-OUT TO THINK: A person's (or group's) character, circumstances, and motives occasionally *are* logically relevant to an argument. That is why, for example, witnesses in criminal cases often are asked questions about their personal lives. If you were a judge, how would you decide when to allow such questions?

EXAMPLE 9   *Circular Reasoning (Begging the Question)*

"Society has an obligation to shelter the homeless because the needy have a right to the resources of the community."

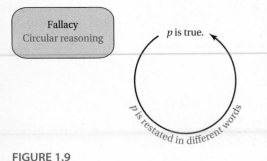

Fallacy
Circular reasoning

p is true.

p is restated in different words

FIGURE 1.9

Analysis: First, note that this argument states the conclusion (society has an obligation to shelter the homeless) before the premise (the needy have a right to resources of the community). Such "backwards" structures are common in everyday speech, and are perfectly legitimate as long as the argument is well-reasoned. However, in this case the premise and the conclusion both say essentially the same thing, though in subtly different ways. Thus this argument suffers from *circular reasoning* (Figure 1.9). This fallacy is also called *begging the question* because the premises "beg" the listener to accept the conclusion by asserting it, though in a somewhat disguised form. ■

EXAMPLE 10   *Subjectivism*

"I don't care what the Supreme Court says. I was brought up to believe that prayer is an important part of every day, so I am sure that our Constitution cannot prohibit prayer in the public schools."

Analysis: The conclusion of this argument is that prayer in public schools is constitutional. However, the premises offered in support are personal opinions. The speaker is certainly entitled to believe that prayer in the schools is a good idea, but this belief is irrelevant to the question of constitutionality. The argument suffers from the fallacy of *subjectivism*, in which the only support for the conclusion comes from *subjective* (personal) beliefs or desires (Figure 1.10). ■

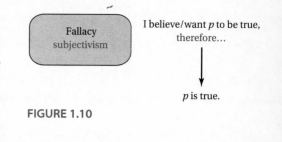

FIGURE 1.10

TIME-OUT TO THINK: Explain why celebrity endorsements of products essentially represent examples of subjectivism.

EXAMPLE 11   *Diversion (Red Herring)*

"We should not continue to fund cloning research because there are so many ethical issues involved. Ethics is at the heart of our society, and we cannot afford to have too many ethical loose ends."

Analysis: The argument begins with its conclusion: that we should stop funding cloning research. However, the discussion is all about ethics. This argument represents the fallacy of *diversion* (Figure 1.11) because it attempts to divert attention from its real issue (funding for cloning) by instead focusing on some other issue (ethics). The issue to which attention is diverted is sometimes called a *red herring* (herring is a fish that turns red when rotten). ■

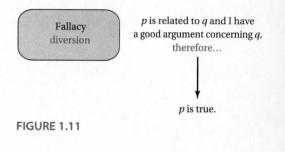

FIGURE 1.11

EXAMPLE 12   *Straw Man*

"We know that there is no such thing as absolute truth because Einstein's theory of relativity proved that everything is relative."

Analysis: This argument may sound reasonable at first, but it suffers from a major flaw: Einstein's theory of relativity does not say that *everything* is relative; it is a theory that applies only to the physics of motion, light, and gravity. The speaker either does not understand Einstein's theory, or is deliberately distorting Einstein's conclusions. Any argument that is based on a distortion of someone's ideas or beliefs is called a *straw man*, a term that comes from the idea that the speaker is arguing against something of his or her own invention (i.e., "build a man made out of straw"), rather than against a real person or issue (Figure 1.12). ■

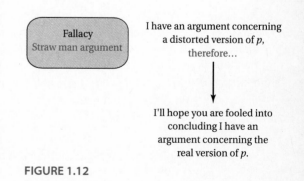

FIGURE 1.12

REVIEW QUESTIONS ∙∙

1. What is *logic*? Briefly explain how logic can be useful both to everyday arguments and to mathematical problem solving.

2. What is a *fallacy*? Why is it important to be able to recognize fallacies?

3. Why are fallacies so common in advertising and in political campaigns?

4. Make a list of the fallacies described in the twelve examples given. For each one, come up with your own example.

PROBLEMS ∙∙∙

Finding Fallacies. *For each type of fallacy listed in Problems 1–12, find and describe at least one example from an advertisement, a news report, or a political campaign.*

1. Appeal to popularity
2. False cause
3. Hasty generalization
4. Appeal to ignorance
5. Limited choice
6. Appeal to emotion
7. Appeal to force
8. Personal attack
9. Circular reasoning
10. Subjectivism
11. Diversion
12. Straw man argument

Identifying Fallacies. *For each of the arguments in Problems 13–23, identify one or more of the twelve fallacies described in this unit. Explain how the fallacy is involved.*

13. Following President Reagan's defense buildup, the Soviet Union began the process of democratization that ultimately led to its breakup. Therefore Reagan is responsible for the changes that led to the demise of the Soviet Union.

14. The Golden Rule is a sound ethical principle because it is basic to every system of ethics in every culture.

15. Lawyer to defendant: "Have you stopped abusing your wife?"

16. Of the four candidates for mayor, the former dog catcher is leading in the polls, so I will vote for her.

17. My mom says that I should never smoke, but I'm not going to pay any attention because she smokes at least a pack a day.

18. If you love me, you'll loan me the money.

19. I believe in telepathy, because no one has ever proven that it doesn't exist.

20. The preponderance of male doctors and lawyers tells me that males have an aptitude for these careers.

21. There's no way that Senator Smith's bill can help the cause of gun control because he is one of the biggest recipients of campaign contributions from the National Rifle Association.

22. Violent crime by youth has risen in virtual lockstep with increased violence on television. Can there be any doubt that television violence leads to real violence?

23. Since 1960, the percentage of the population over 18 that smokes has decreased from 40% to about 20%. During the same period, the percentage of overweight people has increased from 25% to 35%. Clearly, quitting smoking leads to overeating.

24. **Mathematical Battle of the Sexes.** Often a claim is made that boys inherently are more capable at mathematics than girls. The following arguments are commonly used to make this point:

    - Most of the great mathematicians in history have been men.

    - Males outnumber females in fields requiring mathematical expertise, such as the physical sciences and engineering.

    - Boys score higher, on average, than girls on standardized mathematics tests such as the GRE and SAT.

- Even in preschool and kindergarten, boys show more interest in mechanical things than girls do.

a. Identify and analyze as many fallacies as you can find in these arguments.

b. A conclusion can be true even if the argument supporting it is fallacious. Do you think that boys inherently are more capable at mathematics than girls? Defend your opinion.

c. Essay: Study the following quotation. Then write a short essay describing what you think the writer means in this quote, and whether the "prevailing opinion" she describes could have an influence on observed differences between the sexes in mathematical performance. Be sure to defend your opinions well.

It would be an endless task to trace the variety of meanness, cares, and sorrows into which women are plunged by the prevailing opinion that they were created rather to feel than to reason, and that all the power they obtain must be obtained by their charms and weakness.

—*MARY WOLLSTONECRAFT (1759–1797)*

25. Johnson Versus Goldwater. In 1964, President Lyndon Johnson ran for reelection against Barry Goldwater. The following passage describes a television ad made by the Johnson campaign:

> A little girl, strolling through a field, picks a daisy and plucks its petals while counting, "one, two, three. . . ." In the background, a gruff male voice begins a countdown, "ten, nine, eight . . . ," slowly drowning out the little girl's voice. As the countdown ends, the screen lights up with an atomic explosion. It is followed by the voice of President Johnson, who begins, "These are the stakes: to make a world in which all of God's children can live, or go into the dark."

a. What implication does the ad make about Goldwater? What did Johnson's campaign hope to accomplish with the ad?

b. What fallacy is committed in Johnson's narration at the end (these are the stakes . . .)? Discuss.

c. Identify and discuss any other fallacies committed in the ad's attempt to get people to vote for Johnson.

26. Project: Political Ads. Refer to Problem 25. Briefly research the 1964 presidential campaign. How significant was this ad to the outcome of the election?

27. Project: Fallacies in Advertising. Pick a single night and a single television channel and record the advertisements that are shown over a two-hour (or longer) period. Describe each advertisement and identify any fallacies involved in it. Write a summary of your findings and include answers to the following questions. What fraction of the advertisements involve fallacies? Which fallacies are the most common? Why do you think that fallacies are so common in advertising?

28. Project: Fallacies in Politics. Research a recent major election (e.g., for president, governor, mayor) and describe the role that fallacies played in the campaign. Were particular types of fallacies more common than others? Do you believe that fallacies influenced the outcome of the vote? Write a short essay summarizing your findings, and defend your opinions clearly.

29. Topics for Discussion or Essay. Write about or discuss each of the following questions.

a. Relate a personal situation in which your logic told you to make one choice, but other factors (emotions, etc.) told you to make another. Which choice did you make (the logical or the nonlogical)? In retrospect, was it the right decision? Why or why not?

b. Some believe that applying logic can reveal unquestionably which side of a debate has the better argument. Do you agree? Explain.

c. Identify and describe an instance in which you were persuaded of something that you later decided was untrue. Explain how you were persuaded, and why you later changed your mind.

d. What do you think about the preponderance of fallacious logic in our society? Do you believe that it is harmful? Can anything be done to reduce the incidence of fallacies? Explain.

UNIT 1B

PROPOSITIONS—BUILDING BLOCKS OF LOGIC

Having discussed common fallacies in Unit 1A, we next move toward studying proper logical arguments. The building blocks of logical arguments are called **propositions**—statements that *propose* something to be either true or false. A proposition must have the structure of a complete sentence and must make a distinct assertion or denial. For example:

- *Joan is sitting in the chair* is a proposition because it is a complete sentence that makes an assertion.

- *I did not take the pen* is a proposition because it is a complete sentence that makes a denial.

- *Are you going to the store?* is not a proposition because it is a question; it does not assert or deny anything.

- *Three miles south of here* is not a proposition both because it does not make any claim and because it does not have the structure of a complete sentence.

- $3y \times 4y = 12y^2$ is a proposition: it can be read as a complete sentence, and it makes a distinct claim.

- $7 + 9 = 2$ also is a proposition, although it is false.

"Contrariwise," continued Tweedledee, "if it was so, it might be; and if it were so, it would be; but as it isn't, it ain't. That's logic."

—LEWIS CARROLL, FROM THROUGH THE LOOKING GLASS.

NEGATION (OPPOSITES)

The opposite of a proposition is called its **negation**. For example, the negation of *Joan is sitting in the chair* is *Joan is not sitting in the chair*. The negation of $7 + 9 = 2$ is $7 + 9 \neq 2$. If we represent a proposition with a letter such as p or q, we represent its negation by placing the symbol "$\sim$" in front of it. Thus $\sim p$ is read as "not p" and $\sim q$ is read as "not q."

The *claim* made by a proposition may or may not be true. Thus we say that any proposition has a **truth value** of either true (T) or false (F). Clearly, if a particular proposition is true, its negation must be false, and vice versa. We can represent these facts with a simple **truth table**.

p	$\sim p$
T	F
F	T

The first column shows the two possible truth values of p: it is either true or false. The second column shows the corresponding truth values for $\sim p$: it is false when p is true, and true when p is false.

EXAMPLE 1   *Negation*

Consider the proposition *Betsy is the fastest runner on the team*. Make a truth table for the proposition and its negation.

Solution: Because the proposition is rather long, let's use *q* to represent the proposition.

$$q = Betsy\ is\ the\ fastest\ runner\ on\ the\ team.$$

The negation of *q* is *Betsy is not the fastest runner on the team.* If *q* is true, then its negation, $\sim q$, must be false. If *q* is false, then $\sim q$ must be true. The truth table below summarizes these facts.   ∎

q	$\sim q$
T	F
F	T

EXAMPLE 2   *Double Negative*

Consider the double negative statement, *the deal is not not good.* Assuming this statement is true, is the deal good? Answer with a truth table.

I cannot say that I do not disagree with you.

—GROUCHO MARX

Solution: Let's use *p* to represent the statement *the deal is good.* Then the given statement, *the deal is not not good*, is represented $\sim\sim p$. We make our truth table by starting with the two possible truth values of *p*, then adding columns for the truth values of $\sim p$ and $\sim\sim p$:

p	$\sim p$	$\sim\sim p$
T	F	T
F	T	F

The table shows that the truth values of *p* and $\sim\sim p$ are always the same. Thus if $\sim\sim p$ is true, *p* also is true; translating back to words, a deal that is "not not good" is a good deal.   ∎

> **TIME-OUT TO THINK:**   From a recent newspaper: "The Senator opposes the ban on the demonstrations." If the Senator has her way, will there be demonstrations? Explain how the sentence from the paper involves a double negative, despite the fact that the word *not* does not appear.

LOGICAL CONNECTORS: AND, OR, IF . . . THEN

Propositions can be joined into *compound statements* by using **logical connectors** such as *and*, *or*, and *if . . . then*. The truth value of a compound statement depends on the truth values of its individual propositions.

And Statements (Conjunctions)

Consider the following two propositions:

$$p = The\ test\ was\ hard.$$
$$q = I\ got\ a\ good\ grade.$$

If we join the two propositions with *and*, we get their **conjunction**: *The test was hard and I got a good grade*. In logic, the connector *and* is denoted by the symbol $\wedge$, so the conjunction *p and q* is written symbolically as $p \wedge q$.

Each individual proposition can be either true or false, so the compound statement can be built from four possible pairs of truth values: (1) p and q both true; (2) p true and q false; (3) p false and q true; and (4) p and q both false. A truth table for $p \wedge q$ has four rows giving these four possibilities.

p	q	$p \wedge q$
T	T	T
T	F	F
F	T	F
F	F	F

As the truth table shows, $p \wedge q$ is true only if both of its individual propositions are true.

EXAMPLE 3   *And Statements*

Evaluate the truth value of the following two statements:

a) The capital of France is Paris and Antarctica is cold.

b) The capital of France is Paris and the capital of America is Madrid.

Solution:

a) The two distinct propositions are *The capital of France is Paris* and *Antarctica is cold*. Because both propositions individually are true, their conjunction also is true.

b) The two distinct propositions are *The capital of France is Paris* and *The capital of America is Madrid*. Although the first proposition is true, the second is false. Thus their conjunction is false. ■

EXAMPLE 4   *Triple Conjunction*

Suppose that you are given three individual propositions p, q, and r. Make a truth table for the conjunction $p \wedge q \wedge r$.

Solution: Each of the three individual propositions can be either true or false, so together the three propositions allow $2 \times 2 \times 2 = 8$ different sets of truth values. Thus the truth table should have 8 rows. As the table shows, $p \wedge q \wedge r$ is true only if all three individual propositions are true. ■

p	q	r	$p \wedge q \wedge r$
T	T	T	T
T	T	F	F
T	F	T	F
T	F	F	F
F	T	T	F
F	T	F	F
F	F	T	F
F	F	F	F

> **TIME-OUT TO THINK:**   As shown previously, the truth table involving two propositions has four rows, and the truth table involving three propositions has eight rows. How many rows would there be in a truth table involving four propositions? Verify your answer by creating a truth table for the conjunction $p \wedge q \wedge r \wedge s$.

Or Statements (Disjunctions)

The connector *or* can have two different meanings. If a health insurance policy covers hospitalization in cases of illness *or* injury, it probably means that it covers either illness or injury, *or both*. This is an example of the **inclusive or** that means "either or both." In contrast, when a restaurant offers a choice of soup *or* salad, you probably are not supposed to choose both. This is an example of the **exclusive or** that means "one or the other but not both."

In general, the only way to determine whether an *or* statement is inclusive or exclusive is by its context. However, in logic we always assume that *or* is *inclusive* unless stated otherwise.

> **TIME-OUT TO THINK:**   Kevin's insurance policy states that his house is insured for earthquakes, fire, *or* robbery. Suppose that a major earthquake levels much of his house, and the rest burns in a fire caused by earthquake-damaged gas lines. Then his few remaining valuables are looted in the aftermath. Explain why Kevin hopes that the *or* in his insurance policy is *inclusive* rather than exclusive.

A compound statement made with *or* is called a **disjunction**. In logic, the connector *or* is denoted by the symbol $\vee$. Using our earlier propositions, p = *the test was hard* and q = *I got a good grade*, the disjunction $p \vee q$ is the statement: *the test was hard or I got a good grade*. Because we assume the *or* is inclusive, this statement is true if either p or q is true or if both are true. Thus a disjunction $p \vee q$ is false only if both individual propositions are false. The truth table for $p \vee q$ is as follows.

p	q	$p \vee q$
T	T	T
T	F	T
F	T	T
F	F	F

EXAMPLE 5   *Smart Cows?*

Consider the compound statement, *airplanes can fly or cows can read*. Is it true?

Solution: Let's identify the two individual propositions in the statement as p = *airplanes can fly* and q = *cows can read*. Proposition p is clearly true, while q is clearly false. As shown in the preceding truth table, the disjunction $p \vee q$ is true unless both propositions are false. Thus the statement *airplanes can fly or cows can read* is true. ∎

EXAMPLE 6   *Truth Table for p or not q*

Make a truth table for the statement $p \vee \sim q$. For what truth values of p and q is this statement *false*?

Solution: We begin by putting columns in our truth table for p, q, and $\sim q$. Then we add the truth values for the compound statement $p \vee \sim q$.

p	q	$\sim q$	$p \vee \sim q$
T	T	F	T
T	F	T	T
F	T	F	F
F	F	T	T

Looking at the table, we see that $p \vee \sim q$ is false only in the third row: when p is false and q is true.   ∎

EXAMPLE 7   *Key Word Search*

You are doing a research report for which you need to find articles about the federal deficit that appeared in either *Time* or *Newsweek* between 1995 and 1997. You have access to a computerized database that allows you to search on date, magazine, and title. How should you structure the search?

Solution: In general, using OR in a key word search widens the search while using AND narrows the search.

For the date, the key word search

1995 OR *1996* OR *1997*

will generate a list with articles published in any of the three years (because *or* is assumed to be inclusive). Similarly, you would search on

Time OR *Newsweek*

for the magazine to ensure that both magazines are included on the list.

The title requires a bit more thought. For example, searching on *federal* OR *deficit* would lead to a list with many articles unrelated to the federal deficit, such as articles on federal crimes or attention deficit disorder. Searching on *federal* AND *deficit* would limit the search to articles containing both words in the title, but might still miss relevant articles with different key words in their titles; for example, it would miss an article with the title "the United States budget deficit." One way to solve this problem is to use a more complex set of search criteria, such as:

(*federal* OR *United States* OR *government*) AND (*deficit* OR *budget*)

This search would find articles cataloged with any of the following word combinations in their titles: United States deficit; government deficit; federal budget; United States budget; or government budget, as well as federal deficit.   ∎

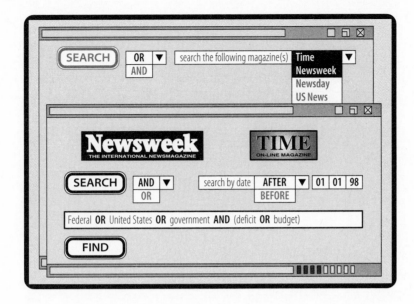

If . . . Then **Statements (Conditionals)**

Another common way to connect propositions is with the words "if . . . then," such as in the statement, "If all politicians are liars, then Representative Smith is a liar." Statements of this type are called **conditional propositions** (or *implications*) because they propose something to be true (the *then* part of the statement) on the *condition* that something else is true (the *if* part of the statement).

We can represent a conditional proposition as *if p, then q.* In logic, a right arrow ($\Rightarrow$) is also used to represent a conditional by writing $p \Rightarrow q$, which is read "*p* implies *q*." Proposition *p* is called the **antecedent** (because it comes before *q*) and the proposition *q* is called the **consequent** (because it is a consequence of *p*).

Making a truth table for a conditional is a bit subtle, so consider the following statement as an example.

If it is Tuesday, then this is Belgium.

This proposition has the standard form *if p, then q*, where *p = it is Tuesday* and *q = this is Belgium*. Now let's consider the four possible cases for the truth values of *p* and *q*.

1. *p* and *q* both true: it *is* Tuesday and this *is* Belgium. Thus the entire statement, "If it is Tuesday, then this is Belgium," is true.

2. *p* true and *q* false: it *is* Tuesday but this is *not* Belgium. In this case, the conditional statement is false; if it had been true, then this would have been Belgium.

3. *p* false and *q* true: it is *not* Tuesday, but this *is* Belgium. The conditional statement tells us what happens only if it *is* Tuesday. Because it is not Tuesday, the conditional statement gives us no basis for saying whether or not this is Belgium. Thus we have no basis for declaring the entire conditional statement false. By convention, the conditional statement is therefore considered true. Although this may seem a stretch, it may help if you think of it as a sort of logical equivalent of the principle of "innocent until proven guilty."

By the Way ··········
If It's Tuesday, It Must Be Belgium was the title of a 1969 movie starring Ian McShane and Suzanne Pleshette.

4. *p* and *q* both false: it is *not* Tuesday and this is *not* Belgium. As in Case (3), the fact that *p* is false means we have no basis to declare $p \Rightarrow q$ to be false. Therefore we say that it is true.

Summarizing, the statement $p \Rightarrow q$ is true in all cases except when the antecedent *p* is true and the consequent *q* is false. Here is the truth table:

p	*q*	$p \Rightarrow q$
T	T	T
T	F	F
F	T	T
F	F	T

EXAMPLE 8   *Conditional Truths*

Following the rules of logic summarized in the preceding truth table, evaluate the truth of the statement, "If $4 \times 5 = 19$, then $3 \times 3 = 10$."

Solution: The statement has the form $p \Rightarrow q$, where $p = $ "$4 \times 5 = 19$," and $q = $ "$3 \times 3 = 10$." Both *p* and *q* are clearly false. However, according to the rules of logic, the conditional $p \Rightarrow q$ is true any time the antecedent *p* is false, no matter what *q* says. Thus the statement "If $4 \times 5 = 19$, then $3 \times 3 = 10$" is true!   ■

EXAMPLE 9   *Rephrasing Conditional Propositions into Standard Form*

Conditional propositions often are "hidden" in everyday speech. Rephrase each of the following statements into the standard conditional form *if p, then q*. Identify the statements *p* and *q* in each case.

a) I'm not coming back if I leave.

b) More rain will cause a flood.

c) Your ad says that the price should be only $20.

Solution:

a) *I'm not coming back if I leave* is equivalent to *If I leave, then I am not coming back.* In the standard form *if p, then q*, we have *p = I leave* and *q = I am not coming back.*

b) *More rain will cause a flood* is equivalent to *If there is more rain, then there will be a flood.* In the standard form *if p, then q*, we have *p = there is more rain* and *q = there will be a flood.*

c) *Your ad says that the price should be $20* is equivalent to *If your advertisement is true, then the price is $20.* In the standard form *if p, then q*, we have *p = your advertisement is true* and *q = the price is $20.*   ■

Converse, Inverse, and Contrapositive

The order of the propositions does not matter in conjunctions or disjunctions: $p \wedge q$ is the same as $q \wedge p$; and $p \vee q$ is the same as $q \vee p$. However, $p \Rightarrow q$ is not the same as $q \Rightarrow p$. For example, the statement *if you are sleeping then you are breathing* is not the same as *if you are breathing, then you are sleeping*. When we switch the order of the propositions in a conditional, we create the **converse** of the conditional. Thus $q \Rightarrow p$ is the converse of $p \Rightarrow q$. Using the same logic we used for $p \Rightarrow q$, the converse $q \Rightarrow p$ must be true unless the antecedent (q in this case) is true and the consequent (p in this case) is false. Here is the truth table with a column added for the converse:

p	q	$p \Rightarrow q$	$q \Rightarrow p$ *(converse)*
T	T	T	T
T	F	F	T
F	T	T	F
F	F	T	T

There are two other common variations on a conditional of the form $p \Rightarrow q$. The **inverse** of the conditional $p \Rightarrow q$ is the statement $\sim p \Rightarrow \sim q$. The **contrapositive** of $p \Rightarrow q$ is $\sim q \Rightarrow \sim p$.

> **TIME-OUT TO THINK:**   What is the inverse of $q \Rightarrow p$? Based on your answer, explain why making the contrapositive of $p \Rightarrow q$ is equivalent to first making its converse and then making the inverse of the converse.

To extend the truth table to include the inverse and contrapositive, we add columns for the negations $\sim p$ and $\sim q$, and then use the rule that a conditional is false only when the antecedent is true and the consequent is false. Here are the results:

p	q	$\sim p$	$\sim q$	$p \Rightarrow q$	$q \Rightarrow p$ *(converse)*	$\sim p \Rightarrow \sim q$ *(inverse)*	$\sim q \Rightarrow \sim p$ *(contrapositive)*
T	T	F	F	T	T	T	T
T	F	F	T	F	T	T	F
F	T	T	F	T	F	F	T
F	F	T	T	T	T	T	T

Note that the truth table for the conditional $p \Rightarrow q$ is the same as the truth table for its contrapositive. We therefore say that a conditional and its contrapositive are **logically equivalent**: if one is true, so is the other, and if one is false, so is the other. Similarly, the table shows that the converse and inverse are logically equivalent.

EXAMPLE 10   *Logical Equivalence*

Consider the statement: *if you are sleeping, then you are breathing*. Write its converse, inverse, and contrapositive. Which statements are logically equivalent?

Solution: The statement has the form $p \Rightarrow q$, where $p = $ *you are sleeping* and $q = $ *you are breathing.* Thus we find:

converse ($q \Rightarrow p$): *if you are breathing, then you are sleeping.*

inverse ($\sim p \Rightarrow \sim q$): *if you are not sleeping, then you are not breathing.*

contrapositive ($\sim q \Rightarrow \sim p$): *if you are not breathing, then you are not sleeping.*

If you read carefully, you'll see that the original conditional and its contrapositive have the same meaning and therefore are logically equivalent. Similarly, the converse and inverse have the same meaning and are logically equivalent. ∎

CATEGORICAL PROPOSITIONS

A particularly important and simple type of proposition is one that expresses a relationship between two categories, or **sets**, in a simple sentence. One set appears as the *subject* of the sentence and the other appears in the *predicate*. For example, the proposition

All politicians are liars.

compares the *subject set* (politicians) to the *predicate set* (liars). Propositions of this type are called **categorical propositions** because they compare two categories or sets. If we use the letter S for the subject set and the letter P for the predicate set, we can rewrite the preceding proposition as

All S are P, where $S = $ politicians and $P = $ liars.

The statement *all S are P* is one of four standard forms for a categorical proposition.

The Four Standard Categorical Propositions			
Form	Example	Subject Set (*S*)	Predicate Set (*P*)
All *S* are *P.*	All apes are mammals.	apes	mammals
No *S* are *P.*	No fish are mammals.	fish	mammals
Some *S* are *P.*	Some doctors are women.	doctors	women
Some *S* are not *P.*	Some teachers are not men.	teachers	men

The English logician John Venn (1834–1923) invented a simple, visual way of describing relationships between sets. His diagrams, now called **Venn diagrams**, use circles to represent sets. Venn diagrams are fairly intuitive once you get the basic idea, so let's look at the Venn diagrams for each of the four standard categorical propositions.

All S are P. It's easiest to construct the Venn diagram with an example, such as *all apes are mammals.* Because all members of the subject set ($S = $ apes) also are members of the predicate set ($P = $ mammals), the subject circle is drawn *inside* the predicate circle (Figure 1.13). The fact that one circle is inside the other is the *only* important feature of the Venn diagram; the sizes of the circles do not matter.

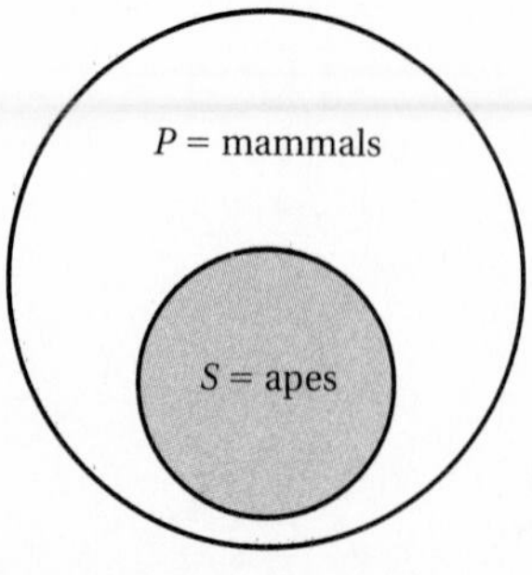

All *S* are *P*

FIGURE 1.13

No *S* are *P*. Let's use the example *no fish are mammals*. In this case, the subject set (*S* = fish) and predicate set (*P* = mammals) are completely distinct. We therefore draw two circles that don't intersect (Figure 1.14).

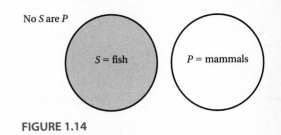

No *S* are *P*

FIGURE 1.14

Some *S* are *P*. Consider the example *some politicians are liars*. We can draw its Venn diagram with two overlapping circles, one for *S* = politicians, and one for *P* = liars. However, this proposition introduces some ambiguity: It makes clear that at least *some* politicians are liars, but it doesn't say whether any politicians are *not* liars. We can avoid confusion by adding labels to identify what is stated and what is unstated (Figure 1.15).

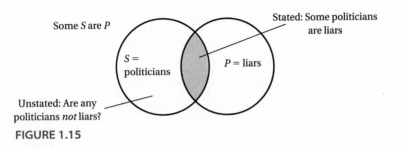

FIGURE 1.15

Some *S* are not *P*. This time we use the example *some princes are not charming*, which in standard form reads *some princes are not charming people*. The sets *S* = princes and *P* = charming people can be drawn as overlapping circles, but ambiguity arises because the proposition doesn't state whether any princes *are* charming. Again, we avoid confusion by adding labels to indicate what is stated and unstated (Figure 1.16).

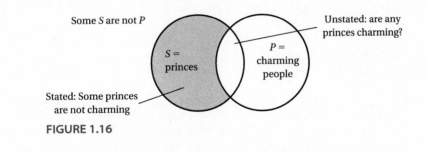

FIGURE 1.16

EXAMPLE 11   *Rephrasing Categorical Propositions into Standard Form*

Many statements in everyday speech are equivalent to one of the four standard categorical propositions, even when expressed somewhat differently. Convert each of the following propositions into one of the four standard forms of categorical propositions. In each case, state which of the four Venn diagrams applies (Figures 1.13–1.16).

a) All diamonds are valuable.

b) Some birds do not fly.

c) Some people never learn.

d) Elephants never forget.

Solution:

a) *All diamonds are valuable* can be rephrased as *All diamonds are things of value.* We identify *S* = *diamonds* and *P* = *things of value*; the proposition has the form *all S are P*, with the Venn diagram shown in Figure 1.13.

b) *Some birds do not fly* can be rephrased as *Some birds are not flying animals.* We identify *S* = *birds* and *P* = *flying animals.* The proposition has the form *some S are not P*, shown in Figure 1.16.

c) *Some people never learn* can be rephrased as *Some people are people who never learn.* This proposition has the form *some S are P*, shown in Figure 1.15, where *S* = *people* and *P* = *people who never learn.*

d) *Elephants never forget* can be rephrased as *No elephants are creatures that forget.* This proposition has the form *No S are P*, shown in Figure 1.14, where *S* = *elephants* and *P* = *creatures that forget.* ∎

EXAMPLE 12 *Categorical Propositions with Singular Sets*

Convert each of the following propositions into one of the four standard forms for categorical propositions. Identify the subject and predicate sets in each case.

a) Jane is a doctor.   b) My dinner is good.

Solution: Both of these propositions involve sets that contain only one member, or **singular sets**. We can translate them into standard form by recognizing that the single member represents *all* members of a singular set.

a) In the proposition *Jane is a doctor*, we identify *S* = *the set consisting of Jane* and *P* = *doctors.* Because Jane represents all members of *S*, this proposition has the form *all S are P.*

b) In the proposition *My dinner is good*, we identify *S* = *the set of all dinners that I am eating right now* and *P* = *good dinners.* Because *my dinner* is the only member of *S*, this proposition also has the form *all S are P.* ∎

REVIEW QUESTIONS

1. What is a proposition? Give a few examples, and explain why each is a proposition.

2. What do we mean by the negation of a proposition? How do we show negation in symbols? Make a simple truth table for a proposition and its negation.

3. Define conjunction, disjunction, and conditional, and give an example of each in words.

4. Suppose that a true proposition and a false proposition are connected by the word *and* to form a compound statement. Is the compound statement true or false? Explain. What if the same two propositions are connected by the word *or*?

5. Make a truth table for each of the following: $p \wedge q$, $p \vee q$, $p \Rightarrow q$. Explain all of the truth values in your tables.

6. What is the difference between an inclusive *or* and an exclusive *or*? Give an example of each.

7. What is a *key word search*? What would you use one for?

8. Explain why, in logic, a conditional is considered true except in the case where the antecedent is true and the consequent is false. Is this idea of "truth" the same as the idea used in everyday life? Explain.

9. Describe how you make the converse, inverse, and contrapositive of a conditional proposition. Make a truth table for each. What is logical equivalence?

10. State the general form for each of the four standard categorical propositions, and draw a Venn diagram for each one.

11. What is a *singular set*? How can we put categorical propositions with singular sets into standard form?

PROBLEMS

Identifying Propositions. *For Problems 1–5, determine whether each statement is a proposition and give your reasons.*

1. High taxes penalize the wealthy.

2. Fly me to the moon.

3. Every monkey is a mammal.

4. She talks in her sleep.

5. Go west, young man, go west.

Negating Propositions. *For Problems 6–13, write the negation of each proposition. Then state the truth value of the original proposition and its negation.*

6. London is the capital of England.

7. The Gettysburg Address was given by George Washington.

8. Caesar was a Roman.

9. Mark Twain wrote *Tom Sawyer*.

10. Paris is not the capital of France.

11. New York is not the capital of the United States.

12. Some U.S. Senators are Republican. (Hint: the negation must be false if the proposition is true, and vice versa.)

13. All snakes are mammals. (Hint: See hint for Problem 12.)

Multiple Negatives. *For Problems 14–19, analyze each statement and answer the question that follows with an explanation.*

14. "The city council did not approve the no-confidence vote for the police chief." Did the city council oppose or support the police chief?

15. "The Senate voted to override the President's veto on the disaster relief bill." Did the Senate oppose or support the bill?

16. "The ban on anti-nuclear demonstrations was lifted." Are anti-nuclear demonstrations now allowed?

17. "The Dean opposes the repeal of laws supporting affirmative action." Does the Dean support affirmative action?

18. "The overturning of the anti-discrimination policy violates the Constitution." Is the anti-discrimination policy consistent with the Constitution?

19. "The Congressman opposes the ban on anti-war demonstrations." In general terms, where does the Congressman stand on anti-war demonstrations?

Truth of "And" Statements. *For Problems 20–24, state whether the individual propositions of each "and" statement are true or false. Then determine whether the entire statement is true or false. Explain your reasoning.*

20. London is the capital of England and Paris is the capital of China.

21. Quebec is the capital of Canada and Moscow is the capital of Russia.

22. Emily Dickinson is an American poet and Emily Dickinson is from Costa Rica.

23. Ben is married and Ben is a bachelor.

24. $8 \times 4 = 32$ and $7 \times 3 = 22$.

Interpreting "Or". *For each statement of Problems 25–30, interpret the use of the connector "or". Is it used in the exclusive or inclusive sense? Should the statement be further qualified to clarify it?*

25. The menu offers a choice of appetizer or dessert.

26. The car warranty covers parts for three years or 36,000 miles.

27. If I win the lottery, I will go to Brazil or Nepal.

28. I will wear either boots or shoes.

29. The road will be made of asphalt or concrete.

30. The insurance policy covers fire or theft.

Truth of "Or" Statements. *For Problems 31–35, state whether the individual propositions of each "or" statement are true or false. Then determine whether the entire statement is true or false. Explain your reasoning.*

31. Washington, D.C., is the capital of the United States or Paris is the capital of France.

32. William Shakespeare was an Englishman or Mark Twain was an American.

33. Either Benjamin Franklin or Martin Luther King was a U.S. President.

34. Ben is married or Ben is a bachelor.

35. $8 \times 4 = 32$ or $7 \times 3 = 22$.

Truth of Compound Statements. *For Problems 36–41, determine the truth value of each compound statement when p is true, q is false, and r is true.*

36. $p \wedge q$ **37.** $p \wedge \sim q$

38. $p \vee q$ **39.** $p \vee \sim q$

40. $p \vee q \vee r$ **41.** $p \wedge q \wedge r$

Truth Tables for Compound Statements. *For Problems 42–49, create a truth table for each proposition. Determine the truth values of the individual propositions that make the entire statement true. (Evaluate expressions within the parentheses first.)*

42. $p \wedge \sim q$ **46.** $(p \wedge q) \vee r$

43. $p \vee \sim q$ **47.** $(p \vee q) \wedge r$

44. $\sim p \wedge q$ **48.** $p \wedge (q \vee \sim r)$

45. $\sim p \vee q$ **49.** $(p \vee q) \wedge \sim r$

50. Library Search. Suppose that you wanted to find the call number of the book *Finnegan's Wake* by James Joyce. For each of the following key word search combinations, state whether the book would be on the search list. Explain.

a. Finnegan

b. Finnegan and Joyce

c. Joyce or Yeats

d. Joyce and James

e. Joyce and Yeats

f. Finnegan and (Joyce or Yeats)

51. Interpreting Connectors. A consumer survey of popcorn preferences found 24 people who eat WonderCorn only, 32 people who eat PrimePop only, and 18 people who eat both brands.

a. How many people eat WonderCorn?

b. How many people eat PrimePop?

c. How many people eat WonderCorn and PrimePop?

d. How many people eat WonderCorn or PrimePop?

Truth of "If ... Then" Statements. *For Problems 52–57, identify the antecedent and consequent in each statement, and state their truth values. Then determine whether the entire statement is true or false. Explain your reasoning.*

52. If a trout is a fish, then a trout can swim.

53. If London is the capital of England, then Paris is the capital of China.

54. If Boston is the capital of Colorado, then Moscow is the capital of Russia.

55. If Boston is the capital of Colorado, then Moscow is the capital of France.

56. If the Pope is unmarried, then the Pope is a bachelor.

57. If $8 \times 4 = 32$, then $8 \times 8 = 64$.

Symbolic "If ... Then" Statements. *For Problems 58–62, assume that p is true, q is false, and r is true. Then determine whether each compound statement is true or false. Explain why.*

58. $p \Rightarrow \sim q$ **61.** $(p \vee q) \Rightarrow r$

59. $\sim p \Rightarrow q$ **62.** $(p \wedge q) \Rightarrow r$

60. $p \Rightarrow (q \vee r)$

Truth Tables for "If ... Then" Statements. *For Problems 63–66, create a truth table for each statement. Determine the truth values of p, q, and r that make the entire proposition true. (Evaluate expressions within the parentheses first.)*

63. $p \Rightarrow (q \wedge r)$ **65.** $(p \wedge q) \Rightarrow r$

64. $p \Rightarrow (q \vee r)$ **66.** $p \vee (q \Rightarrow r)$

Converse, Inverse, and Contrapositive. *For Problems 67–71, write the converse, inverse, and contrapositive of each statement.*

67. If Marco lives in Chicago, then he lives in the United States.

68. If Maria lives in Nevada, then she lives in Germany.

69. If I go swimming, then I will get wet.

70. If it's a penguin, then it can't fly.

71. If it's a reptile, then it's cold-blooded.

Conditional Propositions. *For each statement in Problems 72–75, (i) rephrase it in the standard form "if p, then q"; (ii) identify p and q; and (iii) explain why you think the conditional proposition is true or false.*

72. A resident of Miami is a resident of the United States.

73. A member of Congress must be a lawyer.

74. A person who lives in the Pacific Time Zone lives west of the Mississippi River.

75. All musicians can play the saxophone.

Working with Categorical Propositions. *For each of the propositions in Problems 76–83, (i) rephrase the proposition as one of the four standard forms for categorical propositions, if necessary; (ii) identify the subject set (S) and the predicate set (P); and (iii) draw a Venn diagram for the proposition.*

76. All biology courses are science courses.

77. Some police officers are women.

78. Some police officers are not tall people.

79. No Republican is a socialist.

80. Some states have no coastlines.

81. No bachelors are married.

82. Every nurse knows CPR.

83. Ronald Reagan was a great president.

84. Ambiguity in the News. The following are direct quotes from news sources. In each case, describe if and how it is ambiguous. What further information would you need to remove the ambiguity?

   a. "LOS ANGELES—Two jumbo jets with more than 350 people aboard nearly collided over the city during a landing attempt . . ."

   b. "There has been an alarming 25% rise in [child abuse] cases between 1988 and 1993, a year when 2.9 million incidents were reported to child welfare agencies."

   c. "France must decrease its public deficit from 6% of its gross domestic product, or \$83 billion, to 3%."

85. Another Use of Venn Diagrams. Venn diagrams can be used to describe relationships between different sets even when no argument is involved. Consider the following Venn diagram that represents people in a room. The number in each region represents the number of people matching the characteristics of that region.

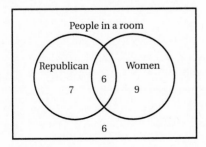

   a. How many people are in the room?

   b. How many women are in the room?

   c. How many men are in the room?

   d. How many Republicans are in the room?

   e. How many men are Republicans?

   f. How many men are not Republicans?

   g. How many women are Republicans?

   h. How many women are not Republicans?

86. Venn Diagrams for Sets. Draw a Venn diagram to represent each of the following situations. Your diagrams should look similar to the one shown for Problem 85.

   a. There are 24 girls on a school bus of which 12 play soccer only, 6 play softball only, 2 play soccer and softball, and 4 play neither softball or soccer.

   b. A movie critic rated 15 feature movies and 10 documentaries. He gave 7 feature movies a "thumbs up" and the other feature movies a "thumbs down." He gave 9 of the documentaries a "thumbs up." (Hint: label one circle "feature movies" and the other "thumbs up.")

   c. A doctor testing a new antibiotic puts 15 patients on a daily dose of the drug and 15 patients on a daily dose of a placebo (a substance that looks like the actual drug, but has no active ingredients). Twelve of the patients on the actual drug improve after a week while 6 of the patients on the placebo improve. (Hint: Think carefully about how to label the two circles.)

87. Interpreting a Survey. A survey of newspaper readers asked which of the following three newspapers they read daily: *New York Times (NYT), Washington Post (WP)*, and *The Wall Street Journal (WSJ)*. The results are shown below. Interpret the connecting words carefully and answer the following questions. A Venn diagram may also be useful.

Paper(s)	Readers	Paper(s)	Readers
NYT only	24	*NYT* and *WSJ* only	14
WSJ only	27	*NYT* and *WP* only	16
WP only	26	*WP* and *WSJ* only	13
None	15	All three	8

   a. How many people read *NYT*?

   b. How many people read the *WP* and the *WSJ*?

c. How many people read the *NYT* and *WSJ* and *WP*?

d. How many people read the *NYT* or *WSJ* or *WP*?

e. How many people read the *NYT*, but not the *WP*?

88. Interpreting the Internal Revenue Service. The following are IRS instructions describing who must make estimated tax payments during the course of a year.

"You must make estimated tax payments if you expect to owe ... at least $500 in tax and you expect your withholding and credits to be less than the *smaller* of:

1. 90% of the tax shown on next year's tax return, or

2. The tax shown on last year's tax return (110% of that amount if you are not a farmer or fisherman and the adjusted gross income shown on that return is more than $150,000 or, if filing separately next year, more than $75,000)."

Based on these rules, decide whether you would need to pay estimated taxes in each of the following cases. Explain.

a. You expect to pay $450 in taxes next year.

b. You are a teacher and expect to pay more than $500 in taxes. Your adjusted gross income is $35,000. You expect your withholding and credits to be 85% of your tax for next year and you expect next year's taxes to equal last year's taxes.

c. You are a teacher and expect to pay more than $500 in taxes. Your adjusted gross income is $35,000. You expect your withholding and credits to be 95% of your tax for next year and 120% of your tax for last year.

d. You are a lawyer and expect to pay more than $500 in taxes. Your adjusted gross income is $185,000. You expect your withholding and credits to be 80% of your tax for next year and 115% of your tax for last year.

e. You are a farmer and expect to pay more than $500 in taxes. Your adjusted gross income is $40,000. You expect your withholding and credits to be 95% of your tax for next year and 95% of your tax for last year.

UNIT 1C

LOGICAL ARGUMENTS: DEDUCTIVE AND INDUCTIVE

An argument consists of a collection of propositions. The **premises** of the argument are propositions representing the ideas or assumptions upon which the argument is based. The **conclusions** are propositions that, if the argument is well-constructed, follow from the premises in a compelling way. Arguments come in two basic types, illustrated by the following two arguments. As you study them, ask yourself whether the premises lead to the conclusions in a compelling way.

Argument 1 (Deductive)

Premise:	All politicians are crooks.
Premise:	All crooks are liars.
Conclusion:	All politicians are liars.

Argument 2 (Inductive)

Premise:	Birds fly up into the air but eventually come back down.
Premise:	People or animals that jump into the air fall back down.
Premise:	Rocks thrown into the air come back down.
Premise:	Balls thrown into the air come back down.
Conclusion:	What goes up must come down.

Let's first examine Argument 1. Either of its two premises might provoke a long and heated debate. However, *if* these premises are true, then the conclusion *necessarily* follows. A Venn diagram makes this point clear (Figure 1.17). We can represent the first premise (*all politicians are crooks*) by drawing the circle for *politicians* inside the circle for *crooks*. The second premise (*all crooks are liars*) then tells us to put the circle for *crooks* inside the circle for *liars*. The resulting Venn diagram has the circle for *politicians* inside the circle for *liars*, which is a representation of the conclusion (*all politicians are liars*). Thus the conclusion follows from the premises.

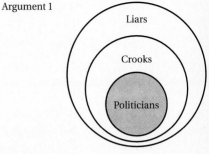

FIGURE 1.17

Argument 1 is an example of a **deductive argument** because it allows us to *deduce* a specific conclusion from a set of more general premises. As long as a deductive argument is properly structured, or **valid**, its conclusion follows necessarily from its premises. Note that validity is concerned only with the *logical structure* of an argument, and hence is very different from the issue of truth. For example, Argument 1 is valid even though the truth of both of its premises may be matters of opinion. If an argument is valid *and* has true premises, we say that argument is **sound**—the highest test of the reliability of an argument. Argument 1 is not sound because its premises are not true.

Now let's turn to Argument 2. Its premises are clearly true, and each premise lends support to the statement in the conclusion. Indeed, this argument would have seemed quite compelling for most of human history. However, we now know that the conclusion is false: a rocket launched with sufficient speed can leave the Earth permanently, going up without ever coming back down.

Argument 2 is an example of an **inductive argument** because a conclusion is formed by *generalizing* from a set of more specific premises. The concepts of validity and soundness do not apply to inductive arguments. Instead, we discuss the **strength** of inductive arguments; that is, we make a subjective judgment about how well the premises support the generalization in the conclusion. Thus an inductive argument never *proves* its conclusion to be true; at best, a strong inductive argument shows that its conclusion *probably* is true. Note that the conclusion of an inductive argument can be proven false by just a single counterexample. Such is the case with Argument 2, where a single rocket that never returns to Earth proves that the conclusion *what goes up must come down* is false.

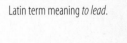

By the Way
The word *induce* comes from a Latin term meaning *to lead*.

I think, therefore I am.
—*RENÉ DESCARTES
(1596–1650), FRENCH
PHILOSOPHER AND
MATHEMATICIAN*

We use both inductive and deductive logic in everyday life. Inductive processes probably are more common because we often rely on patterns and routines that we assume will continue to be useful in the future. We use deductive processes when we need *proof*, such as in proving that a suspect could not have committed a crime because he was somewhere else when the crime occurred.

TIME-OUT TO THINK: Consider several decisions you made recently. For each case, decide whether the reasoning you used to reach the decision was inductive or deductive, and explain your reasoning process.

A Comparison of Deductive and Inductive Arguments	
Deductive Arguments	**Inductive Arguments**
A specific conclusion is deduced from a set of more general (or equally general) premises.	A conclusion is formed by *generalizing* from a set of more specific premises.
Deductive arguments can be analyzed in terms of their *validity*. In a valid deductive argument, the conclusion *necessarily* follows from the premises.	Inductive arguments can be analyzed only in terms of their *strength*. The strength of an inductive argument depends on a *subjective* judgment about how well the premises support the generalization in the conclusion.
Validity concerns only logical structure—a deductive argument can be valid even when its conclusion is blatantly false. A deductive argument is *sound* if it is valid, *and* its premises (and hence its conclusion) are true.	At best, a strong inductive argument shows that its conclusion *probably* is true. However, the conclusion can be shown to be *false* by a single fact that contradicts it.

THE VALIDITY OF DEDUCTIVE ARGUMENTS

Logically speaking, a deductive argument either is valid or is not valid—there are no other possibilities. However, it is not always easy to determine validity. Some very complex deductive arguments, such as legal analyses or mathematical proofs, may go on for hundreds of pages. Even experts may need years for a complete analysis of such an argument. Fortunately, the principles needed for analyzing a complex deductive argument are essentially the same as those needed for simple ones.

Argument 1 discussed earlier represents one of the simplest forms of deductive argument: It has just two premises and one conclusion, each of which is a categorical proposition. As we did with Argument 1, we can test the validity of any such argument with Venn diagrams. The test proceeds in three steps.

By the Way ············
A deductive argument consisting of three propositions—two premises followed by a conclusion—is called a *syllogism*.

Venn Diagram Test for a Categorical Deductive Argument

Step 1. Make sure the premises and conclusion are in standard form for categorical propositions, and clearly identify the three sets that appear in the argument.

Step 2. Draw a single Venn diagram for both premises.

Step 3. Think about the requirement of the argument's conclusion. If this requirement is clearly shown by the diagram for the premises, the argument is valid. Otherwise, the argument is invalid.

The examples that follow illustrate the four basic cases that can arise in the analysis of deductive arguments.

Case 1. If an argument is valid *and* its premises are true, its conclusion also must be true. Such an argument is *sound*. Of course, disagreement over whether a premise is true will lead to disagreement over whether the argument is sound.

Case 2. If an argument is *valid* but its premises are false, its conclusion may be either true or false. The argument is valid but not sound.

Case 3. If an argument's premises are true but its conclusion is false, a flaw in its logical structure must exist; that is, it is an **invalid argument**.

Case 4. If an argument has a true conclusion but is invalid, the conclusion does not follow from the premises; the truth of the conclusion might be attributed to a "lucky guess."

EXAMPLE 1   *A Valid, Sound Argument*

Premise: All narcotics induce drowsiness.
Premise: Some drugs are narcotics.
Conclusion: Some drugs induce drowsiness.

Analysis: We apply the Venn diagram test of validity.

Step 1. We put the propositions in standard form by replacing the words "induce drowsiness" with "are drowsiness-inducing substances":

Premise: All narcotics are drowsiness-inducing substances.
Premise: Some drugs are narcotics.
Conclusion: Some drugs are drowsiness-inducing substances.

The three sets in the argument are: *narcotics*, *drugs*, and *drowsiness-inducing substances*.

Step 2. The first premise tells us to draw the circle for *narcotics* inside the circle for *drowsiness-inducing substances*. The second premise tells us that the circle for *drugs* overlaps the circle for *narcotics*. However, the premises say nothing about whether any drugs are *not* narcotics. We include all possible cases by drawing the *drugs* circle so that it overlaps both of the other circles and then indicate what is stated and what is unstated by the premises (Figure 1.18).

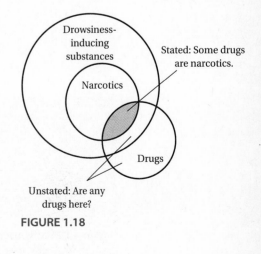

FIGURE 1.18

Step 3. The conclusion of the argument requires that some *drugs* be inside the circle for *drowsiness-inducing substances*. The Venn diagram for the premises clearly shows that to be the case, so the conclusion follows from the premises. Therefore the argument is valid. In this case, the argument also is sound because the premises are true. ■

EXAMPLE 2   *Valid But Not Sound*

Premise: All fish are mammals.
Premise: All mammals are human beings.
Conclusion: All fish are human beings.

Analysis: We apply the Venn diagram test of validity.

Step 1.   The three propositions are already in standard form; the
three sets involved are: *fish*, *mammals*, and *human beings*.

Step 2.   The first premise tells us that the circle for *fish* belongs
inside the circle for *mammals*. The second premise tells us
that the circle for *mammals* should be inside the circle for
human beings (Figure 1.19).

Step 3.   The conclusion requires the circle for *fish* to be inside the
circle for *human beings*—and it is. The argument is therefore valid, even though both its premises and conclusion
are blatantly false. The argument is not sound.   ■

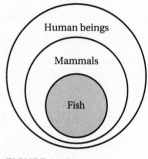

FIGURE 1.19

EXAMPLE 3   *True Premises, but Invalid*

Premise: Fish live in the water.
Premise: Whales are not fish.
Conclusion: Whales do not live in the water.

Analysis: There is a quick way to evaluate this argument. Both of its premises are true, yet
its conclusion is false, so it must be invalid. We can also eval-
uate this argument with a Venn diagram test.

Step 1.   A bit of rephrasing puts the propositions in standard form, showing that the three sets in the argument are *fish*, *whales*, and *water dwellers* (creatures
that live in the water).

Premise: All fish are water dwellers.
Premise: No whale is a fish.
Conclusion: No whale is a water dweller.

Step 2.   The first premise tells us to draw the circle for *fish*
inside the circle for *water dwellers*. The second
premise tells us that the circle for *whales* must be
separate from the circle for *fish*. However, it does
not indicate whether the *whales* circle is inside, outside, or overlapping the *water dwellers* circle. We
therefore draw all three possibilities, using dotted
lines to show that they are *possible* locations for the *whales* circle (Figure 1.20).

Step 3.   The conclusion requires the *whales* circle to be outside the *water dwellers* circle—but that is only one of three possibilities shown in the Venn diagram. Thus
the conclusion does *not* necessarily follow from the premises, so the argument
is invalid. Because it is invalid, it cannot be sound.   ■

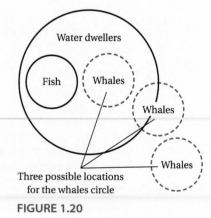

Three possible locations
for the whales circle

FIGURE 1.20

EXAMPLE 4   *Invalid, But True Conclusion*

Premise: All twentieth-century U.S. presidents were male.
Premise: Harry Truman was a male.
Conclusion: Harry Truman was a twentieth-century U.S. president.

Analysis: Once again, we apply the Venn diagram test.

Step 1.   Rephrasing this argument into standard form requires identifying Harry Truman as the single member of *the set consisting of Harry Truman*. The other two sets involved are *twentieth-century U.S. presidents* and *males*.

Premise: All twentieth-century U.S. presidents are males.
Premise: All members of the set consisting of Harry Truman are males.
Conclusion: All members of the set consisting of Harry Truman are twentieth-century U.S. presidents.

Step 2.   The first premise puts the circle for *twentieth-century U.S. Presidents* (*USP* for short) inside the circle for *males*. The second premise tells us that the circle for *the set consisting of Harry Truman* (*HT* for short) also is inside the circle for *males*. However, the premises say nothing about whether or how the *USP* circle and the *HT* circle overlap, so we must show both general possibilities (Figure 1.21).

Step 3.   The conclusion requires the *HT* circle to be inside the *USP* circle, but the diagram shows that this is not necessarily the case. Thus the argument is invalid. You can also spot the invalidity by substituting another name (e.g., Mahatma Ghandi) for Harry Truman. Structurally the argument remains the same, but the conclusion is then clearly false.   ■

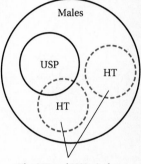

The HT and USP circles may or may not overlap

FIGURE 1.21

Deductive Arguments with One Conditional Premise

Another important type of deductive argument begins with a conditional proposition. Consider the following example:

Premise: If it is Tuesday, then this is Belgium.
Premise: It is Tuesday.
Conclusion: This is Belgium.

This argument is clearly valid: if we accept the first premise and it *is* Tuesday, then we can definitively conclude that *this is Belgium*. If we let p = *it is Tuesday* and q = *this is Belgium*, this argument has the following general form.

Premise: If p, then q.
Premise: p is true.
Conclusion: q is true.

This form of argument is called *affirming the antecedent* because it begins with a conditional proposition *if p, then q*, then affirms the truth of the antecedent p. (Recall from Unit 1B that p is the antecedent and q is the consequent in a conditional proposition *if p, then q*.) We can show that this form of argument is always valid with a modification

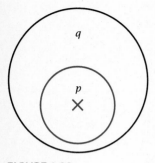

FIGURE 1.22

of the Venn diagram technique. We use circles to represent the *p* and *q* statements (rather than sets *S* and *P*). The conditional proposition *if p, then q* is represented by placing the *p* circle inside the *q* circle. Next, we place an X in the *p* circle to indicate the assertion that *p* is true (Figure 1.22). Because the X also lies within the *q* circle, we can validly conclude that *q* is true.

Three other types of argument can be formed from one conditional proposition: denying the antecedent, affirming the consequent, and denying the consequent. Conditional arguments are summarized in Table 1.1.

Table 1.1 Four Basic Conditional Arguments				
Structure of Argument	If *p*, then *q*. *p* is true. *q* is true.	If *p*, then *q*. *p* is not true. *q* is not true.	If *p*, then *q*. *q* is true. *p* is true.	If *p*, then *q*. *q* is not true. *p* is not true.
Validity	Valid	Invalid	Invalid	Valid
Name of Argument	Affirming the Antecedent (also called *modus ponens*)	Denying the Antecedent	Affirming the Consequent	Denying the Consequent (also called *modus tollens*)

EXAMPLE 5   *The Fallacy of Affirming the Consequent*

Evaluate the argument: "I knew that, if I got a B on the final, then I'd pass the course. Well, the professor told me that I passed! Therefore I must have gotten a B on the final."

Solution: We first put this statement into the standard form of a conditional argument.

Premise:	If I got a B on the final, then I passed the course.
Premise:	I passed the course.
Conclusion:	I got a B on the final.

The first statement has the form *if p, then q*, where *p = I got a B on the final* and *q = I passed the course*. The second premise asserts the truth of *q*, so this argument has the form of *affirming the consequent*. Common sense tells you that the argument is invalid—the student might have passed by getting an A on the final, rather than a B.

We can see why affirming the consequent is invalid by drawing a Venn diagram. The general form of this fallacy is

If *p*, then *q*.
 q is true.
 p is true.

As before, we place the *p* circle inside the *q* circle to represent the conditional proposition *if p, then q*. The premise *q is true* requires an X inside the *q* circle. But there are two possible locations for the X inside the *q* circle: one that is also within the *p* circle and one that is not (Figure 1.23). The conclusion, *p is true*, requires an X in the *p* circle—which is only one of the two possibilities. Thus the conclusion does not necessarily follow from the premises and the argument is invalid. ■

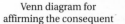

Venn diagram for
affirming the consequent

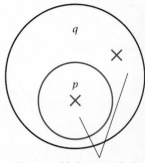

Two possible locations for
"X" consistent with being
inside *q*

FIGURE 1.23

EXAMPLE 6   *The Fallacy of Denying the Antecedent*

Evaluate the following argument.

> If you liked the book, then you'll love the movie.
> You did *not* like the book.
> You will *not* love the movie.

Solution: This argument begins with a conditional *if p, then q*, where *p = you liked the book* and *q = you'll love the movie*. The second premise says that you did *not* like the book, which means that *p* is not true. Hence the argument has the form of denying the antecedent.

> If *p*, then *q*.
> *p* is not true.
> *q* is not true.

In drawing a Venn diagram, we begin with the *p* circle inside the *q* circle to represent the conditional *if p, then q*. Because the second premise tells us that *p* is *not* true, we must place an X *outside* the *p* circle. However, the argument says nothing about whether this X should also be outside the *q* circle, so we must draw both possibilities (Figure 1.24). The conclusion that *q is not true* requires that the X be outside the *q* circle—which is only one of the two possibilities. Hence the conclusion does not necessarily follow from the premises, and the argument is invalid. In this particular example, you might have liked the movie (*q* is true) even if you did not like the book (*p* is not true). ■

Venn diagram for
denying the antecedent

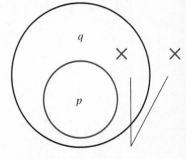

Two possible locations for "X"
consistent with being outside *p*

FIGURE 1.24

EXAMPLE 7   *The Validity of Denying the Consequent*

Using *p = we are eating at home* and *q = I need to go shopping*, construct and evaluate a conditional argument that denies the consequent.

Solution: We can use the argument form of denying the consequent and then substitute the specific premises *p* and *q* that are given.

> If *p*, then *q*.   ⇒   If we are eating at home, then I need to go shopping.
> *q* is not true.   I do *not* need to go shopping.
> *p* is not true.   We are *not* eating at home.

We can show that this argument is valid with another Venn diagram test. The conditional is represented by drawing the *p* circle inside the *q* circle. The premise that *q is not true* requires an X *outside* the *q* circle (Figure 1.25). Because this X also is outside the *p* circle, there is no doubt about the conclusion that *p is not true*. Thus the argument is valid. ■

Venn diagram for
denying the consequent

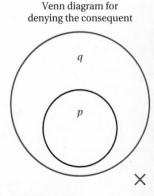

FIGURE 1.25

Deductive Arguments with a Chain of Conditionals

Another type of deductive argument involves a chain of three or more conditionals. Such arguments have the following form.

> Premise: If p, then q.
> Premise: If q, then r.
> Conclusion: If p, then r.

This particular chain of conditionals is clearly valid: if p implies q and q implies r, it must be true that p implies r.

EXAMPLE 8   *A Chain of Conditionals*

Analyze the argument: "If elected to the school board, Maria Lopez will force the school district to raise academic standards, which will benefit my children's education. Therefore, my children will benefit if Maria Lopez is elected."

Solution: This argument can be rephrased as a chain of conditionals.

> Premise: If Maria Lopez is elected to the school board, then the school district will raise academic standards.
> Premise: If the school district raises academic standards, then my children will benefit.
> Conclusion: If Maria Lopez is elected to the school board, then my children will benefit.

Cast in this form, we see a clear chain of conditional propositions from p = *Maria Lopez is elected* to q = *the district will raise academic standards* to r = *my children will benefit*. Therefore, the argument is valid.   ∎

EXAMPLE 9   *Invalid Chain of Conditionals*

Analyze the following argument. "We agreed that if you shop, I make dinner. We also agreed that if you take out the trash, I make dinner. Therefore if you shop, you should take out the trash."

Solution: Let's assign p = *you shop*, q = *I make dinner*, r = *you take out the trash*. Then this argument has the following form.

> Premise: If p, then q.
> Premise: If r, then q.
> Conclusion: If p, then r.

The conclusion is invalid because there is no chain from p to r.   ∎

INDUCTION AND DEDUCTION IN MATHEMATICS

Perhaps more so than any other subject, mathematics relies on the idea of proof. A mathematical proof is a deductive argument in which a set of basic starting "premises," or **axioms**, is used to establish a conclusion, or **theorem**. Proofs must be conclusive, so a theorem is considered proven only when its entire chain of logic is deductive.

Although mathematical proof relies on deduction, induction may give reason to believe that a proposed theorem is true even before it is proved. Consider the **Pythagorean theorem**, that applies to a *right triangle* (a triangle with one 90° angle):

$$a^2 + b^2 = c^2,$$

where c is the length of the longest side, or **hypotenuse**, and a and b are the lengths of the other two sides (Figure 1.26a). A geometric construction shows how the *squares* of the sides are related (Figure 1.26b).

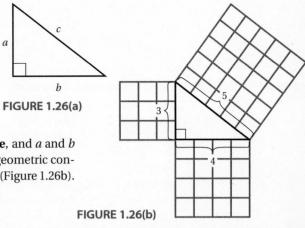

FIGURE 1.26(a)

FIGURE 1.26(b)

$\mathcal{T}$*HINKING ABOUT* ...

Proof of the Pythagorean Theorem

Over the centuries, mathematicians found many different ways to prove the Pythagorean theorem deductively. One of the simplest proofs is attributed to a twelfth-century Hindu mathematician named Bhaskara. His proof begins with a diagram of a large square, inside of which is a smaller square surrounded by four right triangles (Figure 1.27). Note that the diagram simply divides the large square into five separate regions (the small square and the four triangles) so that the

area of *large* square = (area of *small* square)
         + (area of the four right triangles).

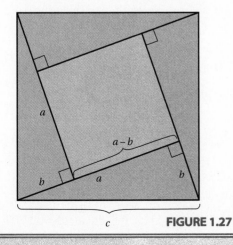

c **FIGURE 1.27**

Now we can write each of these areas as follows:

- The side length of the *large* square is c, so its area is c^2.
- The side length of the *small* square is $a - b$, so its area is $(a - b)^2$.
- The area of any triangle is given by the formula $\frac{1}{2} \times$ base $\times$ height. Each of the four triangles has base length a and height b, so each has area $\frac{ab}{2}$.

Substituting these areas into the preceding equation gives

$$c^2 = \underbrace{(a - b)^2}_{\substack{\text{area of} \\ \text{small square}}} + 4 \times \underbrace{\frac{ab}{2}}_{\substack{\text{area of} \\ \text{triangle}}} = (a - b)^2 + 2ab.$$

$\underbrace{\quad}_{\substack{\text{area of} \\ \text{large square}}}$

We continue by using a bit of algebra to show that $(a - b)^2 = a^2 - 2ab + b^2$. Substituting this expression into the previous equation we get

$$c^2 = (a - b)^2 + 2ab = a^2 - 2ab + b^2 + 2ab = a^2 + b^2.$$

We have arrived at the Pythagorean theorem by following a deductive chain of logic from the starting diagram. Legend has it that, when Bhaskara showed his proof to others, he accompanied it with just a single word: "Behold!"

The Pythagorean theorem is named for the Greek philosopher Pythagoras (c. 580–500 B.C.) because he was the first person known to have proved it deductively. However, the theorem was known to many ancient cultures and already had been used for at least a thousand years before the time of Pythagoras. These ancient cultures probably never bothered to prove it. Instead, they simply noticed that it was true every time they tested it and that it was very useful in art, architecture, and construction. The many test cases essentially formed an inductive argument, each one lending more support to the conclusion that the theorem is true.

The principle of seeking inductive evidence can be very useful whenever you are having difficulty remembering whether a particular theorem or mathematical rule applies: Try a few test cases, and see if the rule works. Although test cases can never constitute a proof, they often are enough to satisfy yourself of a rule's truth. However, the "rule" cannot be true if even one test case fails.

EXAMPLE 10   *Inductively Testing a Mathematical Rule*

Test the following rule. For all numbers a and b, $a \times b = b \times a$.

Solution: We begin with some test cases, using a calculator as needed.

Does $7 \times 6 = 6 \times 7$? $\Rightarrow$ Yes!

Does $(-23.8) \times 9.2 = 9.2 \times (-23.8)$? $\Rightarrow$ Yes!

Does $4.33 \times \left(-\dfrac{1}{3}\right) = \left(-\dfrac{1}{3}\right) \times 4.33$? $\Rightarrow$ Yes!

The three test cases are somewhat different (mixing fractions, decimals, and negative numbers), yet the rule works for all three. This outcome offers a strong inductive argument in favor of the rule. Although we have not proved the rule $a \times b = b \times a$, we have good reason to believe that it is true. Our belief would be strengthened by additional test cases that confirm the rule.   ∎

EXAMPLE 11   *Invalidating a Proposed Rule*

Suppose that you cannot recall whether adding the same amount to both the numerator and denominator (top and bottom) of a fraction such as $\dfrac{2}{3}$ is legitimate. That is, you are wondering whether it is true that, for any number a,

$$\frac{2}{3} = \frac{2+a}{3+a}.$$

Solution: Again, we can check the rule with test cases.

Suppose that $a = 0$. Is it true that $\dfrac{2}{3} = \dfrac{2+0}{3+0}$? $\Rightarrow$ Yes!

Suppose that $a = 1$. Is it true that $\dfrac{2}{3} = \dfrac{2+1}{3+1}$? $\Rightarrow$ No!

Although the rule worked in the first test case, it failed in the second. Thus it is *not* generally legitimate to add the same value to the top and bottom of the fraction $\dfrac{2}{3}$. Generalizing (inductively), we conclude that it is not legitimate to add the same value to the top and bottom of *any* fraction.   ∎

REVIEW QUESTIONS ••

1. Summarize the differences between deductive and inductive arguments. Which tends to go from general premises to a specific conclusion? Which goes from specific premises to a general conclusion?

2. Does the validity of a deductive argument depend on whether its premises are true or false? If a deductive argument has a false premise, can it be valid? Explain.

3. Can an inductive argument be sound? Explain.

4. Explain the procedure used to test the validity of a deductive argument with a Venn diagram. Create your own deductive argument with three categorical propositions, and test its validity with a Venn diagram.

5. Give an example of a simple argument that affirms the consequent. Then use the same propositions to create an example of each of the other three forms of conditional argument. Which forms are valid? Why?

6. What is a *chain of conditionals*? Give an example of a valid argument made from such a chain.

7. Can inductive logic be used to prove a mathematical theorem? Explain.

8. How can inductive testing of a mathematical rule be useful? Give an example.

PROBLEMS ••

Deductive Arguments with Categorical Propositions. *For Problems 1–11, (i) rephrase the argument so that all propositions are in a standard categorical form, if necessary; (ii) draw a Venn diagram for the premises and use it to determine whether the argument is valid; and (iii) discuss the truth of the premises and whether the argument is sound.*

1. Premise: All islands are tropical.
   Premise: All tropical lands have jungles.
   Conclusion: All islands have jungles.

2. Premise: All horses are mammals.
   Premise: All horses have long tails.
   Conclusion: All mammals have long tails.

3. Premise: All dairy products contain protein.
   Premise: No soft drinks contain protein.
   Conclusion: No soft drinks are dairy products.

4. Premise: All salty foods cause high blood pressure.
   Premise: Some snack foods are salty.
   Conclusion: All snack foods cause high blood pressure.

5. Premise: No women are NFL quarterbacks.
   Premise: Some NFL quarterbacks are tall.
   Conclusion: Some tall people are not women.

6. Premise: Some reptiles are snakes.
   Premise: All snakes can fly.
   Conclusion: Some flying animals are reptiles.

7. Premise: Some lobbyists work for the oil industry.
   Premise: All lobbyists are persuasive people.
   Conclusion: Some persuasive people work for the oil industry.

8. Premise: No horses have wings.
   Premise: All animals with wings can breathe.
   Conclusion: No horses can breathe.

9. Premise: No one can get medical treatment without health insurance.
   Premise: Some people do not have health insurance.
   Conclusion: Some people cannot get medical treatment.

10. Premise: All U.S. presidents have been men.
    Premise: George Washington was a man.
    Conclusion: George Washington was a U.S. president.

11. Premise: States in the Eastern Standard Time Zone are east of the Mississippi River.
    Premise: Maine is in the Eastern Standard Time Zone.
    Conclusion: Maine is east of the Mississippi River.

Valid and Sound. *For Problems 12–16, state whether a three-line argument with categorical propositions (two premises and a conclusion) can have the given properties. If so, make up an example.*

12. Valid and sound.

13. Not valid and sound.

14. Valid and not sound.

15. Valid with false premises and a true conclusion.

16. Not valid with true premises and a true conclusion.

Deductive Arguments with One Conditional. *For Problems 17–24, (i) rephrase the argument so that it fits a standard form, if necessary; (ii) name the standard form; (iii) draw a Venn diagram from the premises and use it to determine whether the argument is valid; and (iv) if the argument is invalid, explain the fallacy.*

17. Premise: If I don't eat breakfast, then I eat lunch.
    Premise: I didn't eat breakfast.
    Conclusion: I ate lunch.

18. Premise: If I don't eat breakfast, then I eat lunch.
    Premise: I didn't eat lunch.
    Conclusion: I ate breakfast.

19. Premise: If I don't eat breakfast, then I eat lunch.
    Premise: I ate breakfast.
    Conclusion: I didn't eat lunch.

20. Premise: If we can put a man on the Moon, we can build a VCR that works.
    Premise: We can build a VCR that works.
    Conclusion: We can put a man on the Moon.

21. Premise: When interest rates decline, the bond market improves.
    Premise: Last week the bond market improved.
    Conclusion: Interest rates must have declined.

22. Premise: When it rains, it pours.
    Premise: It is pouring.
    Conclusion: It is raining.

23. Premise: Nurses must know CPR.
    Premise: Tom is a nurse.
    Conclusion: Tom knows CPR.

24. Premise: Knowledge implies power.
    Premise: The president is powerful.
    Conclusion: The president is knowledgeable.

Make Your Own. *For Problems 25–28, create a simple three-line argument of the given form. Choose your example so it illustrates clearly whether or not the argument is valid.*

25. Affirming the antecedent.

26. Affirming the consequent.

27. Denying the antecedent.

28. Denying the consequent.

Chains of Conditionals. *For Problems 29–33, (i) rephrase the argument as a chain of conditionals, if necessary; (ii) evaluate its validity; and (iii) discuss whether the argument is sound.*

29. Premise: If you shop, I make dinner.
    Premise: If I make dinner, you take out the trash.
    Conclusion: If you shop, you take out the trash.

30. Premise: If fish have wings, then fish can fly.
    Premise: If dogs can bark, then fish can fly.
    Conclusion: If fish have wings, then dogs can bark.

31. Premise: In the United States, we have the right to say anything at any time.
    Premise: Yelling "fire!" in a theater is saying something.
    Conclusion: In the United States, we have the right to yell "fire!" in a theater.

32. Premise: If taxes are cut, then taxpayers will have more disposable income.
    Premise: With more disposable income, taxpayer spending will fuel the economy.
    Conclusion: A tax cut will fuel the economy.

33. Premise: If taxes are cut, the U.S. government will have less revenue.
    Premise: If there is less revenue, then the deficit will be larger.
    Conclusion: Tax cuts will lead to a larger deficit.

Analyzing Inductive Arguments. *For Problems 34–38, (i) determine the truth of the premises; (ii) discuss the strength of the argument; and (iii) assess the truth of the conclusion.*

34. Premise: Cows have four limbs, and they are mammals.
    Premise: Monkeys have four limbs, and they are mammals.
    Premise: Lions have four limbs, and they are mammals.
    Conclusion: All animals with four limbs are mammals.

35. Premise: $(-6) \times (-4) = 24$
Premise: $(-2) \times (-1) = 2$
Premise: $(-27) \times (-3) = 81$

Conclusion: Whenever we multiply two negative numbers, the result is a positive number.

36. Premise: Michael Jordan wears Nike shoes, and he is a great basketball player.
Premise: Charles Barkley wears Nike shoes, and he is a great basketball player.
Premise: Shaquille O'Neal wears Nike shoes, and he is a great basketball player.

Conclusion: All people who wear Nike shoes are great basketball players.

37. Premise: Bach, Buxtehude, Beethoven, Brahms, Berlioz, and Britten are great composers.

Conclusion: All great composers have names that begin with B.

38. Premise: Sparrows are birds and they fly.
Premise: Eagles are birds and they fly.
Premise: Hawks are birds and they fly.
Premise: Larks are birds and they fly.

Conclusion: All birds fly.

Testing Mathematical Rules with Inductive Arguments. *For Problems 39–42, test the statement with several different numbers and determine whether you think the statement is true.*

39. Is it true for all numbers a and b that $a + b = b + a$?

40. Is it true for all numbers a and b that $a \div b = b \div a$?

41. Is it true for all numbers a and b that $a^2 + b^2 = (a + b)^2$?

42. Is it true for all positive integers n that

$$1 + 2 + 3 + \ldots + n = \frac{n \times (n + 1)}{2}?$$

Everyday Logic. *For Problems 43–45, explain whether the argument is inductive or deductive.*

43. I saw the same tall man leaving that office building every day at 5:00 P.M. He must work in the building.

44. If I carry a heavy load in the trunk of my car, the rear wheels squeak. If the rear wheels squeak, then I get a headache. Therefore if I carry a heavy load in the trunk of my car, I get a headache.

45. The last four times I went skiing, the traffic was light on Tuesdays and heavy on Saturdays. Weekdays must have lighter traffic than weekends.

46. **Project: The Goldbach Conjecture.** Recall that a prime number is any number whose only factors are itself and 1 (for example, 2, 3, 5, 7, 11, . . .). The Goldbach conjecture, posed in 1742, claims that every even number can be expressed as the sum of two primes. For example, $4 = 2 + 2$, $6 = 3 + 3$, and $8 = 5 + 3$. A *deductive* proof of this conjecture has never been found. Test the conjecture for at least 10 even numbers and present an *inductive* argument for its truth. Do you think the conjecture is *really* true? Why or why not?

· ·

ANALYZING REAL ARGUMENTS

**UNIT
1D**

Arguments in everyday life rarely are as clean and simple as the arguments presented so far. Real arguments may follow a chain of logic that includes both inductive and deductive portions, and often leave some premises unstated.

A useful technique for examining an argument's construction is to make a *flow chart* that maps its logical chain from premises to conclusions. We already used simple flow charts in the examples of fallacies in Unit 1A. Now we examine the technique more carefully.

INDEPENDENT AND ADDITIVE PREMISES

Consider the following inductive argument, in which Propositions (1)–(4) are premises and Proposition (5) is the conclusion.

(1) Birds are animals and they are mortal.

(2) Fish are animals and they are mortal.

(3) Spiders are animals and they are mortal.

(4) Human beings are animals and they are mortal.

(5) All animals are mortal.

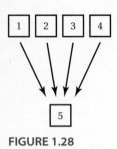

FIGURE 1.28

No single premise in this argument is essential to reaching the conclusion, but each premise lends the conclusion more strength. We therefore say that each premise *independently* supports the conclusion. In a flow chart, we show that each premise flows independently toward the conclusion by drawing an arrow pointing from *each* premise to the conclusion (Figure 1.28).

Now consider the following simple deductive argument.

(1) All carcinogens (cancer-causing materials) are dangerous.

(2) Tobacco smoke is a carcinogen.

(3) Tobacco smoke is dangerous.

In this case, neither premise supports the conclusion by itself. The premise *all carcinogens are dangerous* does *not* suggest that tobacco smoke is dangerous—unless we also know that tobacco smoke is a carcinogen. Similarly, the premise that *tobacco smoke is a carcinogen* says nothing about danger—unless we also know that carcinogens are dangerous. The two premises support the conclusion *additively*, but not independently. In a flow chart, we link the two premises with a "+" sign and draw only one arrow from the combined premises to the conclusion (Figure 1.29).

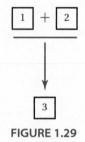

FIGURE 1.29

SUMMARY OF FLOW CHARTS FOR INDUCTIVE AND DEDUCTIVE ARGUMENTS

- Inductive arguments involve premises that *independently* support the conclusion.
- Deductive arguments involve premises that must be combined *additively* to support the conclusion.

ASSUMED PREMISES

Consider the argument:

"We should support the new Forest Service regulations because they will protect old-growth forests."

We can identify and number two propositions in this argument.

(1) We should support the new Forest Service regulations.

(2) The new Forest Service regulations will protect old-growth forests.

Clearly, Proposition (1) is the conclusion and Proposition (2) is a premise. However, this premise *by itself* doesn't support the conclusion. It supports the conclusion only if it is combined with two unstated assumptions: that protecting old-growth forests is a good idea and that we should support good ideas. We call these assumptions the **assumed premises** of the argument, and label them (A1) and (A2) ("A" for "assumed"):

(A1) Protecting old-growth forests is a good idea.

(A2) We should support good ideas.

Figure 1.30 shows that Propositions (2), (A1), and (A2) must all be combined additively to support the conclusion, which is Proposition (1). *With* the assumed premises, the argument is deductively valid because the conclusion follows necessarily. *Without* the assumed premises, the conclusion seems to come out of nowhere.

As this example suggests, assumed premises usually are additive. We look for assumed premises when an argument is not valid with the stated premises alone. Assumed premises usually are statements that the speaker (or writer) considers so obvious that they need not be stated.

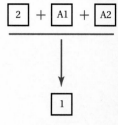

FIGURE 1.30

> **TIME-OUT TO THINK:**   Do you think that everyone would agree that the assumed premises in the forest service argument are "obvious"? Why or why not?

INTERMEDIATE CONCLUSIONS

So far, we have dealt with arguments in which a set of premises leads directly to a conclusion. But real arguments often contain many layers of premises and conclusions. Consider the following argument.

> *"I was driving 55 mph (miles per hour) in a 40 mph zone when the cop gave me a ticket. The penalty is $10 for each mile per hour over the speed limit, so I had to pay $150."*

We can identify and number five distinct propositions in this argument.

(1) I was driving 55 mph.

(2) The speed limit was 40 mph.

(3) The cop gave me a ticket.

(4) The state penalty is $10 for each mile per hour over the speed limit.

(5) I had to pay $150.

Proposition (5) is the conclusion, and it is based on three facts: that the driver got a ticket (Proposition 3); that the penalty was $10 for each mile per hour over the speed limit (Proposition 4); and that the driver was going 15 mph over the speed limit (to make a penalty of $15 \times \$10 = \$150$). However, the latter fact is *not* explicitly stated in the argument. Instead, we deduced it from the first two propositions: that the driver was going 55 mph and that the speed limit was 40 mph. This fact therefore represents an unstated **intermediate conclusion:** We label intermediate conclusions with an I and a number, so this intermediate conclusion is denoted by (I1):

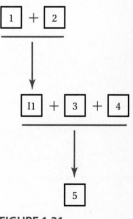

(I1) I was driving 15 mph over the speed limit.

The complete flow chart for this argument is shown in Figure 1.31. Note that Propositions (1) and (2) additively make a deductive argument for the intermediate conclusion (I1). It then combines with Propositions (3) and (4) to make a deductive argument for the final conclusion (Proposition 5).

FIGURE 1.31

> **TIME-OUT TO THINK:**   In the preceding argument, we called the unstated fact that the driver was going 15 mph over the speed limit an intermediate conclusion. Why didn't we call it an assumed premise? How do assumed premises differ from unstated intermediate conclusions?

PUTTING IT ALL TOGETHER

The tools we have discussed in this chapter should be enough to help you evaluate almost any argument, whether it is a carefully considered editorial or a spur-of-the-moment claim made during a heated discussion. If you find an argument difficult to follow, it may help to analyze it in four basic steps.

> ### Four-Step Argument Analysis
>
> **Step 1.**   Identify all the stated propositions and determine which are premises and which are conclusions.
> **Step 2.**   Look for and identify any assumed premises or intermediate conclusions.
> **Step 3.**   Analyze the overall flow of the argument; a flow chart can be a useful tool for this purpose but is not required.
> **Step 4.**   Evaluate the argument. If it is deductive, is it valid? If it is inductive, how strong is it?

EXAMPLE 1   *Argument Analysis with the Four-Step Process*

Apply the four-step analysis to the following argument: "We should buy a house, as interest rates have declined to reasonable levels and we have enough money for a down payment. Also houses in this part of town have appreciated in value during the last two years, and we could use another bedroom."

Analysis:

Step 1.   We identify and number the stated propositions.

(1) We should buy a house.

(2) Interest rates are affordable at our income level.

(3) We have enough money for a down payment.

(4) Houses are appreciating in value.

(5) We could use another bedroom.

Proposition (1) is the conclusion, and the other propositions are premises.

Step 2.   We look for assumed premises and intermediate conclusions. Premises (2) and (3) together seem to imply an intermediate conclusion—that *we can afford to buy a house.* However, this implication is based on the assumption that the combination of money for a down payment and reasonable interest rates makes a house affordable. With this assumed premise, a deductive argument leads to the intermediate conclusion:

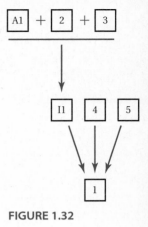

(A1) If interest rates are affordable *and* we have enough money for a down payment, then we can afford to buy a house.

(2) Interest rates are affordable.

(3) We have enough money for a down payment.

(I1) We can afford to buy a house.

Step 3. The intermediate conclusion *we can afford to buy a house* offers one reason to support the final conclusion *we should buy a house*. Propositions (4) and (5) independently support the same conclusion. The overall flow of the argument is shown in Figure 1.32.

FIGURE 1.32

Step 4. None of the statements in (I1), (4), or (5) alone is a very strong reason to buy a house. But together they provide a fairly strong inductive argument. ∎

EXAMPLE 2   *Hidden Complexity in Arguments*

Analyze the following argument: "We should build more prisons because incarcerating more criminals will reduce the crime rate."

Analysis: This argument looks deceptively simple, as it consists of only two propositions:

(1) We should build more prisons.

(2) If we incarcerate more criminals, then the crime rate will be reduced.

Proposition (1) is the conclusion and Proposition (2) is a premise. However, Proposition (2) by itself says nothing about building prisons. Clearly, there are hidden assumptions. We can begin to make sense of the argument by identifying two assumed premises that support an intermediate conclusion additively with Proposition (2).

(A1) If we build more prisons, then more criminals can be incarcerated.

(2) If we incarcerate more criminals, then the crime rate will be reduced.

(A2) If the crime rate is reduced, then we will have a more desirable society.

(I1) If we build more prisons, then we will have a more desirable society.

This intermediate argument is deductive because it is a chain of conditionals. It can be combined with a third assumed premise to reach the final conclusion.

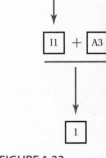

(I1) If we build more prisons, then we will have a more desirable society.

(A3) All policies that lead to a more desirable society should be enacted.

(1) We should build more prisons.

Figure 1.33 shows the complete flow chart. The argument is deductively valid with the inclusion of the assumed premises and intermediate conclusion. However, of the four premises and one intermediate conclusion necessary to reach the final conclusion, the original argument stated only one! Moreover, some of these premises are highly debatable. For example, many people would not agree with assumed premise (A3), arguing instead

FIGURE 1.33

that a good policy should *not* be enacted if it has a particularly high cost or if another policy could achieve the same goal at lower cost.

In summary, the original argument is difficult to follow because it left so much unstated. Moreover, at least one of the unstated assumptions is highly debatable. Thus the argument is quite weak. ◼

ONWARD TO QUANTITATIVE REASONING

The only difference between the reasoning we've discussed so far and *quantitative* reasoning is that the latter adds mathematical or quantitative information. The same principles of logical analysis apply to all reasoning. Moreover, these principles can also be used to "pick apart" a complex problem, as the final example in this chapter shows.

EXAMPLE 3   *Which Ticket to Buy?*

Airlines typically offer many different prices for the same trip. Analyze the following situation. You are planning a trip six months in advance and discover that you have two choices in purchasing an airline ticket:

(A) the lowest fare is $1100, but 25% of the fare is nonrefundable if you change or cancel the ticket; or

(B) a fully refundable ticket is available for $1900.

Analysis: Because we are analyzing options, rather than a specific argument, we need to modify our strategy for making flow charts. (One of the keys to problem solving is being flexible in choosing strategies!) We can think of each of the two options as a pair of conditional propositions. Under Option (A), you will lose 25% of $1100, or $275, if you cancel your trip. Thus Option (A) represents the following pair of conditional propositions.

(1A) If you purchase ticket (A) and *go* on the trip, then you will pay $1100.

(2A) If you purchase ticket (A) and *cancel* the trip, then you will pay $275.

Similarly, Option (B) represents the following pair of conditional propositions.

(1B) If you purchase ticket (B) and *go* on the trip, then you will pay $1900.

(2B) If you purchase ticket (B) and *cancel* the trip, then you will pay $0.

Figure 1.34 shows a flow chart representing the four possibilities. Clearly, option (A) is the better buy if you go on the trip, and option (B) is the better buy if you end up canceling your trip. However, because you are planning six months in advance, it's impossible to foresee all the circumstances that might lead you to cancel your trip. Therefore you might want to analyze the *difference* between the two tickets under the two possibilities (going on the trip or canceling).

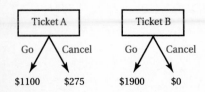

FIGURE 1.34

If you go on the trip: ticket (B) costs $800 more than ticket (A).

If you cancel the trip: ticket (A) costs $275 more than ticket (B).

In effect, you must decide which ticket to purchase by balancing the risk of spending an extra $800 if you go on the trip against spending an extra $275 if you cancel. How would *you* decide?   ■

> **TIME-OUT TO THINK:**   Consider the two ticket options in Example 3 from the point of view of the airline. How does offering the two options help the airline maximize its revenue?

REVIEW QUESTIONS

1. Explain why deductive arguments involve *additive* premises and inductive arguments involve *independent* premises.

2. Identify an assumed premise in the argument *we should support the new Forest Service regulations because they will protect old-growth forests*. How is this assumed premise important to the argument?

3. Explain the term *intermediate conclusion*. How does an intermediate conclusion differ from an assumed pre-

mise? How are assumed premises and intermediate conclusions similar?

4. Describe the four-step process of argument analysis presented in this unit. Give an example of how you could apply it.

5. In Example 3, briefly explain what determines the better option in a particular situation.

PROBLEMS

Flow Charts for Arguments. *For Problems 1–8, evaluate the argument according to the four-step process described in the text. Specifically, try to express propositions in standard form; identify stated, unstated, and assumed premises; and make a flow chart for the argument.*

1. This school should be closed because its walls contain asbestos, a known cancer-causing material.

2. Charlie obviously has something to hide. He pleaded the Fifth Amendment in court last week; only people with things to hide plead the Fifth.

3. I saw Jenny in a limousine, so she must be rich.

4. The national inspiration afforded by the Moon landings in 1969–1972 was justification enough for the cost of the program.

5. The Soviet Union lost the war in Afghanistan because the United States provided weapons to the Afghani rebels.

6. Every American has a right to adequate medical treatment. Therefore I can only support a health care reform package that guarantees insurance coverage for all U.S. citizens.

7. Statistics show that in the United States a criminal offense occurs every 2 seconds, violent crimes occur every 16 seconds, and robberies occur every 48 seconds. Clearly we need to increase the conviction rate of offenders and strengthen police forces.

8. Good advice in investing is to "buy low and sell high." By the same reasoning, as the number of business degrees awarded by universities has been increasing for twenty years and the number of mathematics degrees has been decreasing for twenty years, becoming a mathematics major would be wise.

Quantitative Reasoning. *For Problems 9–13, analyze each situation and describe how you would go about making a decision in the situation.*

9. You are planning a trip to visit your family two months in advance. The airline offers you two ticket choices: (1) you may purchase a ticket for $350, and the ticket price can be refunded for a $75 penalty; or (2) you may purchase a fully refundable ticket for $600.

10. You need three sticks of butter for baking. Your local store sells individual sticks for 40¢ each and packages of four sticks for $1.25.

11. You own a small business and need to be in another city on Monday for a meeting. You may fly to the meeting on Monday morning and return Monday evening on a ticket priced at $750. However, if you include a Saturday night stay in your trip, the ticket price is only $335. A hotel will cost $105 per night, and you estimate that meals away from home will cost you $55 per day.

12. You are married and expecting a baby. Your current health insurance costs $115 per month, but doesn't cover prenatal care or delivery. You can upgrade to a policy that covers prenatal care and delivery, but your new premium will be $275 per month. The cost of prenatal care and delivery is approximately $4000.

13. You fly frequently between two cities 1500 miles apart. Airline A offers the trip for an average round-trip cost of $350. Airline B offers the same trip for only $325. However, Airline A has a frequent flyer program by which you earn a free round-trip ticket after you fly 25,000 miles. Airline B does not have a frequent flyer program.

Complete Argument Analysis. *For Problems 14–24, use any of the logical tools at your disposal to evaluate each argument. Explain the details of your evaluation clearly and be sure to identify assumed premises and intermediate conclusions. Discuss the validity and soundness of any deductive arguments (or parts of arguments) and the strength of any inductive arguments.*

14. Amoebas are not plants because they are capable of locomotion, and no plant has that capacity.

15. The athletic program is given more money than any academic department, so this university must value athletics over academics. Also, the football coach is the highest paid university employee.

16. Dependency on foreign oil has put the U.S. economy at great risk, yet the nation has at least a decade's worth of oil untouched in wilderness areas and offshore. The federal government should immediately open all potential oil fields to drilling, regardless of the environmental consequences.

17. During the next decade the number of Americans aged 5–15 will increase. Furthermore, the number of Americans aged 25–35 will decline. Therefore the demand for teachers will increase during the next decade.

18. On average, Americans save 4.6% of their disposable incomes, whereas Japanese people save 14.6% of their incomes. Moreover, the personal debt of Americans has increased sixfold in 20 years, and the consumer confidence index has dropped since 1990. These trends show that, overall, Americans have adopted an increasingly "live for the moment" attitude.

19. That we have a health care crisis is clear: Large numbers of Americans are uninsured and the cost of health care to our nation is skyrocketing. I believe that only the government has the power to step in and solve this crisis, so I urge you to support candidates who will make health care reform their top priority.

20. The Federal Reserve Board should lower interest rates because the economy is slowing down, and when the economy slows down there is less investment.

21. The populations of China and India dwarf that of the United States, yet they use far less energy. In the future, the people of China and India no doubt will demand a standard of living comparable to that in the United States. To achieve it, the governments of China and India are likely to exploit their enormous reserves of coal as an energy source. Their burning of so much fossil fuel undoubtedly will cause a worldwide environmental catastrophe.

22. Overpopulation is not a real problem. Modern technology, especially in bioengineering, will enable scientists to develop far more efficient agriculture. In addition, advances in irrigation technology, along with the development of crops that can grow in salt water, will enable the conversion of much of the world's desert wastelands into productive farms. As a result, agribusinesses will be able to produce enough food for at least 50 billion people, or some eight times the current world population. Some people claim that such large-scale agriculture would cause environmental damage, but this result is unlikely. The human ingenuity of a much larger population undoubtedly will prevent such damage.

23. Fourth-grade students should use pencil and paper to learn to add fractions and strengthen their mental skills. Drill problems also induce good discipline. Therefore fourth-grade students should not be allowed to use calculators. Anyway, their teachers and parents never used calculators when they were learning.

24. Word processors with spell checkers allow students to grow up without any spelling skills. Now, word processors have grammar checkers, so students will not learn grammar. What's next? Word processors should not be allowed in the schools.

25. **A Financial Decision.** Consider the following situation: "I need a special computer for a project I will be working on for the next three months; after that, I will no longer need this computer. I can lease the computer for $350 per month, or I can buy it for $2100. If I resell it after the three months, I can expect to get $1200. Oops, I almost forgot, sales tax is 5%."

    a. Identify all the propositions and make a flow chart for the argument.

    b. What should you do if you want to make the most economical choice for acquiring a computer?

    c. Identify those premises that are "hard facts" and those that are estimates that could affect the outcome of the argument.

    d. What would you do in this situation? Why?

26. **Editorial Analysis.** Find a newspaper editorial on a topic of interest to you. Attach a copy of the editorial. Analyze the arguments made in the editorial, and identify and describe any fallacies you find. Overall, do you agree with the editorial? Defend your opinion.

27. **Poetry and Mathematics.** Consider the following poem by the English classical scholar and poet A. E. Housman (1859–1936).

    Loveliest of Trees

    Loveliest of trees, the cherry now
    *   Is hung with bloom along the bough*
    And stands about the woodland ride
    *   Wearing white for Eastertide.*
    Now, of my threescore years and ten,
    *   Twenty will not come again,*
    And take from seventy springs a score,
    *   It only leaves me fifty more.*

    And since to look at things in bloom
    *   Fifty springs are little room,*
    About the woodlands I will go
    *   To see the cherry hung with snow.*

    a. How old was the poet at the time he wrote this poem? (Hint: a *score* is 20.)

    b. Based on your reading of this poem, how much longer does the poet expect to live? Explain.

28. **Project: Analysis of College Education.** The following statement is a collection of observations concerning American higher education. Analyze and discuss the argument. Can you bring other facts to the argument that might alter the conclusion?

    American public universities are approaching a crisis. State funding is declining as priorities shift to public school systems and prisons. At the same time, population projections indicate a 30% increase in high school graduates in the next 15 years. Many universities are at capacity and cannot expand even if funds were available. The solution is to raise both admission standards and tuition at four-year colleges and universities, thereby limiting the number of students admitted to these institutions. Even if the number of university graduates remains level, the U.S. would have one of the highest university graduation rates of any country in the world.

29. **Project: Fuzzy Logic.** The system of logic that you studied in this chapter is inherited from the ancient Greeks and is sometimes called *two-valued*, or *binary* logic. It is based on the assumption that a proposition must be capable of being either true or false, but not both and certainly nothing in-between. Recently, other systems of logic

have been devised in which other "truth values" are possible; some of these systems allow for shades of doubt and uncertainty. One form of logic, called *fuzzy logic*, allows for a continuous range of values between absolutely true and absolutely false.

a. Discuss situations in which these multiple-valued systems of logic would be useful and more realistic than the traditional two-valued system.

b. Many engineers are attempting to improve the performance of modern computers and appliances by incorporating fuzzy logic in their designs. Investigate and report on at least one use of fuzzy logic in modern technology. Be sure to explain the advantages offered by the use of fuzzy logic instead of two-valued logic.

CHAPTER 1 •

SUMMARY

*I*n this chapter we introduced the basic principles of reasoning, which are needed to reason quantitatively. Key ideas to keep in mind from this chapter include the following.

- The forces of persuasion in modern society can be very strong. Nevertheless, you can avoid being misled by learning to recognize and challenge common fallacies.

- Validity and truth are very different. Validity concerns only the logical structure of an argument. Truth of a proposition is nearly always subjective.

- Arguments may be either deductive or inductive. Deductive arguments can be analyzed for validity. Inductive arguments involve generalization, and can be analyzed only in terms of their strength.

- Real-world arguments rarely are clear and simple. Nevertheless, the principles developed in this chapter can help you analyze arguments in depth, and thereby better your understanding of their strengths and weaknesses.

- The principles that underlie the logical techniques presented in this chapter are crucial for all types of reasoning—including quantitative reasoning.

SUGGESTED READING

Concise Introduction to Logic, P. J. Hurley (Belmont, California: Wadsworth Publishing, 1988).

Goodbye, Descartes: The End of Logic and the Search for a New Cosmology of the Mind, K. Devlin (New York: Wiley, 1997).

Introduction to Logic, I. Copi (New York: Macmillan, 1968).

Logic and Knowledge; Essays, 1901–1950, Bertrand Russell, edited by Robert Charles Marsh (London: G. Allen & Unwin, 1956).

Chapter 2
STATISTICAL REASONING

*I*s your drinking water safe? What is the latest on the President's approval rating? How much is the cost of health care rising? You can hardly pick up a newspaper today without being inundated with statistical results addressing questions such as these. The ability to understand and reason with statistics is crucial to understanding modern society.

There are three kinds of lies: lies, damned lies, and statistics.
BENJAMIN DISRAELI

Statistical thinking will one day be as necessary for efficient citizenship as the ability to read and write.
H. G. WELLS

To understand God's thoughts, we must study statistics, for these are the measure of His purpose.
FLORENCE NIGHTINGALE

*I*n 1990, medical doctors revised the standard tables used to determine a healthy weight for adults. The 1990 tables were based on the latest research, and suggested that middle-aged adults face no medical danger in adding 20 to 30 pounds. These guidelines offered a great sense of relief to the many Americans whose weight increases in middle age.

Only five years later, a new report directly contradicted the recommendations in the 1990 tables. This report was based on the analysis of a study that followed the health of 115,000 nurses for 16 years. Among many other findings, the analysis showed that any weight gain at all led to a greater risk of early death from cancer or heart disease. Instead of relief for people who gain weight in middle age, the new report suggested that any weight gain is unhealthy.

How could two sets of findings about weight, published only five years apart, directly contradict each other? Each finding was based on **statistics**—numbers that are used to describe large sets of data. Understanding the contradiction in the reports and forming an opinion about either study requires investigating the *science* of statistics—the science of collecting, organizing, evaluating, and interpreting data.

If you follow the news, you know it's difficult to find a lead story that is *not* based on an analysis or interpretation of statistical data. In this chapter, we'll discuss how you can interpret the kinds of statistical reasoning that you see regularly in the media. Later, in Chapter 9, we'll look into the methods of calculation used in statistics.

By the Way ⋯⋯⋯⋯
The word *statistics* has two different meanings: (1) the *numbers* that describe data (e.g., an average); (2) the *science* of collecting, organizing, and interpreting data.

UNIT 2A

FUNDAMENTALS OF STATISTICS

Consider each of the following types of study.

- The U.S. Labor Department publishes a monthly unemployment rate, determined by surveying 60,000 households out of the entire U.S. population.
- Meat inspectors working for the government randomly select samples of meat from markets to be tested for contamination.
- Astronomers study a relatively small number of stars in great detail and use the results to learn about the general properties of stars.

Despite their apparent differences, these studies share at least one important characteristic: each study draws conclusions about some large **population** by studying a much smaller **sample**.

- In the unemployment survey, the population is *all* households in the United States and the sample is the 60,000 households surveyed.
- In the meat inspection, the population is all meat products sold in the United States and the sample is those few chosen for testing.

• Web • Watch •
Complete reports on the U.S. labor force can be found at the Bureau of Labor Statistics World Wide Web site.

- For astronomers, the population is all the stars in the universe and the sample comprises the relatively few stars studied in detail.

In each case a sample is studied because studying the entire population would be impractical. Finding the exact unemployment rate would require surveying every household in the United States every month—a task that would cost hundreds of millions of dollars each time, if it were possible at all. In the case of meat inspection, a sample that is tested cannot be sold; if all meat were tested, nothing would be left to sell. Astronomers face an even more daunting problem: the number of stars in the universe is greater than the number of grains of sand on all the beaches of the world. Clearly, astronomers will never be able to study more than a tiny fraction of all stars.

These examples illustrate that a primary purpose of statistics is *generalization*. As with all processes of generalization, uncertainties are necessarily involved at almost every step of statistical generalization.

Get the facts first and then you can distort them as much as you please.

—MARK TWAIN

Because statistical generalization involves so many uncertainties, some people say that statistics can be used to support *any* conclusion. Although this claim has a grain of truth, most statistical studies are carefully researched and provide valuable insights into important issues. The key to dealing with statistics lies in evaluating the degree to which a statistical analysis is reasonable and whether it should affect your opinions or actions.

THE PROCESS OF STATISTICAL STUDY

Consider the study that tracked 115,000 nurses over a 16-year period. The purpose of the study was to learn about health issues affecting the *population* of all Americans. The 115,000 nurses were the sample selected for study so that the researchers could learn about the much larger population.

The researchers collected an enormous amount of information about their sample: the height and weight changes of each nurse during the 16 years, the illnesses suffered by the nurses, and much more. This huge base of information is called the **raw data** of the study. The researchers then summarized the raw data in the form of **sample statistics**—numbers that describe the sample, such as the average weight of nurses of a particular height and age, or the percentage of nurses who suffered heart attacks during the study.

If a study has been done well, the sample statistics should accurately reflect the characteristics of the sample. However, the goal of a study is to learn about **population parameters**—the characteristics of the population itself. This is done by generalizing the results from the sample to the population. Overall, the

Elements of a statistical study

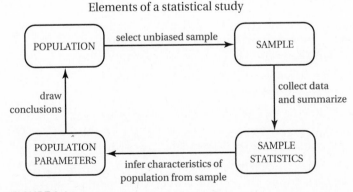

FIGURE 2.1

By the Way ·············

Statistics often is divided into two major branches. *Descriptive statistics* is the branch that deals with *describing* the raw data in the form of sample statistics. *Inferential statistics* is the branch that deals with *inferring* characteristics of the population from the sample.

process of statistical study is an interplay between describing the sample and inferring characteristics of the population (Figure 2.1).

You may recognize a parallel between statistics and logic (as discussed in Chapter 1). The sample statistics are *deduced* from the raw data, so this process involves deductive logic. The population parameters are *inferred* from the sample statistics by generalizing, making this a process of inductive logic. Recall that an inductive argument cannot *prove* the truth of its conclusion; at best, the conclusion of a strong inductive argument is *probably* true. Similarly, in a statistical study, we can never be certain that the conclusions drawn about a population are true. Thus the crucial final step in a statistical study is to evaluate the likelihood that its conclusions are accurate.

The following five steps summarize the process of conducting a statistical study.

Basic Steps in a Statistical Study

1. Determine the goal of the study (e.g., evaluating the health risk due to weight gain) and identify the *population* to which the study applies.

2. Choose an appropriately sized *sample* from the population and collect raw data from the sample.

3. Consolidate the data collected from the sample into a set of *sample statistics*: numbers, such as averages, that summarize the characteristics of the sample.

4. Infer the characteristics of the population, called the *population parameters*, based on the sample results.

5. Evaluate the likelihood that the inferred characteristics of the population reflect the true population parameters, and draw appropriate conclusions.

EXAMPLE 1 *Statistical Terminology*

Suppose that you want to learn about the heights and weights of students at your school. You select 100 students randomly and measure their heights and weights. Describe the population, sample, raw data, sample statistics, and population parameters for your study.

Solution:

- The *population* is the group you want to learn about; in this case, the entire student body at your school.

- The *sample* is the 100 students selected for measurement.

- The *raw data* is all the data you collect: the heights and weights of each of the 100 students.

- The *sample statistics* are numbers that summarize the raw data, such as the average height and average weight for the 100 students.

- The *population parameters* are the corresponding characteristics of the entire population, such as the average height and average weight for *all* students at your school. Because you are studying only 100 students out of a larger student body, you will not actually measure the population parameters. However, if the study is conducted well, there's a good chance that your sample statistics will accurately reflect the true population parameters. ■

> **TIME-OUT TO THINK:**   The study concluding that weight gain represents a health risk was based on a sample made up entirely of nurses. Do you think that conclusions drawn from this sample can be reasonably used to make inferences about the health effects of weight gain in all Americans? Why or why not? Can you think of any reasons why researchers might have chosen to use nurses for their sample?

STATISTICAL STUDIES AND RANDOM SAMPLING

Statistical methods . . . are necessary in reporting the mass data of social and economic trends. . . . But without writers who use [statistics] with honesty and understanding, and readers who know what they mean, the result can only be semantic nonsense.

—DARRELL HUFF,
FROM HOW TO LIE
WITH STATISTICS

The only way to determine true population parameters is by studying *every* member of a population. For example, you can calculate the true average height of all students at your school only by measuring the height of every student. In a sense, a statistical study is a substitute for going to the trouble of studying every member of a population. The study focuses only on a sample drawn from the population, and the sample statistics provide a means of making an educated guess about the population parameters.

If this educated guess about population parameters is to be close to the truth, the statistical study must be carried out with great care. Critical errors can be introduced at any stage of the five-step process. Statisticians have identified many common errors (some of which we will discuss shortly), and have developed guidelines for preventing such errors.

Perhaps the most important step in conducting a statistical study is the selection of the sample. If the sample is not *representative* of the population, there is little hope that generalizations from sample statistics to population parameters will be accurate. For example, you wouldn't try to learn the average height of all students at your school by choosing only members of the basketball team for your sample!

In most cases, the best way to obtain a representative sample is by choosing *randomly* from the population. There are many forms of random sampling, but the simplest, called **simple random sampling**, ensures that every member of the population has an equal chance of being selected for the sample. For example, you could assign each student a number and draw numbers randomly from a hat. The students whose numbers are drawn would become the members of your sample. Computers and some calculators can select random numbers very quickly using a built-in random number generator as shown in the illustration on the next page.

> **TIME-OUT TO THINK:**   Look for a random number key on your calculator. What happens when you push it? How would you use this key to generate a list of random numbers?

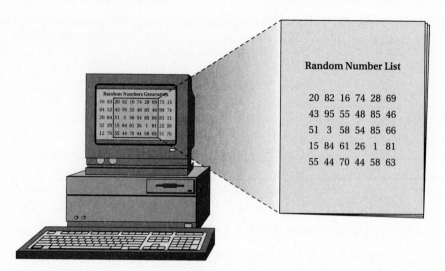

Random Number List

20 82 16 74 28 69
43 95 55 48 85 46
51   3 58 54 85 66
15 84 61 26   1 81
55 44 70 44 58 63

By the Way ....
In 1996, 194 million people were eligible to vote in the United States; of these, 65.9% registered to vote and 54.2% actually voted in the presidential election. These percentages vary greatly with age: citizens over age 65 are more than twice as likely to vote as those age 18–20.

EXAMPLE 2   *Preelection Survey*

Suppose that you are conducting a poll of 1,500 people to predict the outcome of the next presidential election. What is the population under study? Suggest a method for choosing your sample. Discuss potential problems with your sampling method and potential solutions.

Solution: Because you are trying to predict the outcome of an election, the population consists of all people *who will vote* in the election; people who will *not* vote in the election are not part of this population because they will not affect the outcome of the election. How can you sample the population of people who will vote? Because citizens must register to vote, one approach might be to select individuals randomly from a list of all registered voters in the United States. However, this method presents at least two problems.

- Only a fraction of registered voters actually *will* vote on election day, so the sample drawn from registered voters may not be truly representative of the population of actual voters.

- Because presidential elections are decided state by state by *electoral* votes (see Unit 11D), the popular vote need not reflect the outcome of the election. Thus, even if the sample accurately reflects the popular vote, it might not predict the winner of the election.

You can alleviate the first problem by including responses only from people who actually *will* vote. Unfortunately, this isn't easy to do. One possibility is simply to ask "do you plan to vote?," but many people with good intentions of voting don't ultimately get around to it. Another approach is to ask a question such as "when did you vote last?" and include responses only from people who voted in recent elections.

Alleviating the second problem requires sampling voters within each state, rather than nationally. You could predict which candidate will win in each state and then project the winner of the election from the predicted total electoral vote count. ∎

> **TIME-OUT TO THINK:**   In Example 2 suppose that you chose your sample by randomly selecting names in telephone books, rather than from lists of registered voters. Do you think this sample would be representative of voters? Why or why not?

A BRIEF REVIEW

Percentages.

Percentages have been widely used since ancient times, when their earliest use probably was for levying taxes. For example, the Roman Emperor Augustus (27 B.C.– A.D. 14) is said to have levied a 1% tax on proceeds from goods sold at auctions. Fortunately, percentages are easy to work with because they are just another way to express fractions. The word *percent* means *per hundred* or *divided by one hundred*. For example, 19% means

$$19\% = \frac{19}{100} = 0.19.$$

Converting Percentage to Decimals or Fractions

We can convert a percentage to a fraction or decimal simply by carrying out the division by 100 implied by the % sign. Some examples are shown below.

$$25\% = 25 \div 100 = \frac{25}{100} = 0.25$$

$$86.2\% = 86.2 \div 100 = \frac{86.2}{100} = 0.862$$

$$724\% = 724 \div 100 = \frac{724}{100} = 7.24$$

$$0.1\% = 0.1 \div 100 = \frac{0.1}{100} = \frac{1}{1000} = 0.001$$

Note that we can convert a percentage to decimal form with an even simpler rule: *Drop the % symbol and move the decimal point two places to the left.*

Converting Decimals to Percentages

Note that 100% is just a fancy way of writing 1:

$$100\% = \frac{100}{100} = 1$$

We can therefore convert a decimal to a percentage by multiplying by 100%; because we really are multiplying by 1, we are not changing the value of the original number. Some examples are shown below.

$$0.43 = 0.43 \times 100\% = 43\%$$

$$0.003 = 0.003 \times 100\% = 0.3\%$$

$$0.07 = 0.07 \times 100\% = 7\%$$

$$25 = 25 \times 100\% = 2{,}500\%$$

$$1.02 = 1.02 \times 100\% = 102\%$$

$$-0.1 = -0.1 \times 100\% = -10\%$$

Converting Fractions to Percentages

Converting a fraction to a percentage requires two steps.

1. Convert the fraction to a decimal.
2. Convert the decimal to a percentage.

Here are a few examples.

$$\frac{1}{2} = 0.5 = 50\%$$

$$\frac{5}{6} = 0.8333\ldots = 83.33\% \text{ (rounded)}$$

$$\frac{3}{4} = 0.75 = 75\%$$

$$\frac{5}{3} = 1.666\ldots = 166.67\% \text{ (rounded)}$$

By the Way ···········
Arthur C. Nielsen founded his company and invented market research in 1923. He introduced the Nielsen Radio Index to rate radio programs in 1942 and extended his methods to TV programming in the 1960s. Today, the Nielsen TV ratings are based on a sample of 5000 households all across America.

CONFIDENCE INTERVAL AND MARGIN OF ERROR

Suppose that a random sample of 1000 Americans reveals that 330, or 33%, had watched the show *Friends* on a particular evening. Can we conclude that 33% of *all* Americans watched *Friends* that night? Certainly not! It's highly unlikely that the choice made by the 1000 people in the sample will *perfectly* mirror the choice of the more than 260 million Americans in the population. Using our statistical terminology, the 33% of sample viewers watching *Friends* is a *sample statistic*. The actual percentage of all Americans watching the show is a *population parameter*. Even with random sampling, we shouldn't expect the sample statistic and population parameter to be identical—but we might hope that they are *close*.

In general, the more people included in the sample, the greater the likelihood that the sample statistic and population parameter will be close. If you survey only ten people, it's unlikely that the percentage watching a particular show will be the same as the overall percentage in the population. But if you survey 1 million people, there's a good chance that the result from the sample will accurately reflect the true characteristics in the population. To quantify this idea, statisticians define a **margin of error** in any survey. We'll discuss how the margin of error is calculated in Chapter 9, but for now it is enough to know that the margin of error depends on the size of the sample: a large sample means a relatively small margin of error, and a small sample means a relatively large margin of error.

For the sample of 1000 people in the *Friends* survey, the margin of error turns out to be about 3%. Adding and subtracting the margin of error from the sample statistic of 33% gives a range, or *interval*, from

$$33\% - 3\% = 30\% \qquad \text{to} \qquad 33\% + 3\% = 36\%.$$

Statistical science tells us that it is very likely that the population parameter (the actual percentage of Americans watching *Friends*) lies in the range of 30% to 36% defined by the sample statistic and the margin of error. In fact, statisticians have found that this range represents a **95% confidence interval**: we can be 95% confident that the population parameter lies in this range. The precise meaning of a confidence interval is a bit subtle and will be discussed in Unit 9C. But, roughly speaking, it means there is a 95% chance that the population parameter lies in this range and a 5% chance that it does *not* lie in this range.

Mathematical Note:
In this text, we assume that the *margin of error* corresponds to a 95% confidence interval, which is the practice followed by most polling organizations. However, some researchers define their margin of error to represent a different confidence range.

EXAMPLE 3   *Interpreting the Margin of Error*

A marketing survey finds that 65% of consumers in a sample prefer the taste of Clear Cola to that of Cloudy Cola, with a margin of error of 5%. Interpret the survey results.

Solution: We add and subtract the margin of error from the sample result to find the 95% confidence interval:

$$65\% - 5\% = 60\% \qquad \text{to} \qquad 65\% + 5\% = 70\%.$$

We can be 95% confident that between 60% and 70% of *all* consumers prefer Clear Cola over Cloudy Cola. There is roughly a 5% chance that the percentage of consumers who prefer Clear Cola is *not* between 60% and 70%. ∎

EXAMPLE 4   *Who Will Win the Election?*

An election eve poll finds that 52% of surveyed voters plan to vote for Smith and 48% plan to vote for Jones. The margin of error in the poll is 3%. Who will win the election?

Solution: Based on the poll results and the margin of error, we can be 95% confident that the actual percentage of voters planning to vote for Smith is between

$$52\% - 3\% = 49\% \quad \text{and} \quad 52\% + 3\% = 55\%.$$

Similarly, we can be 95% confident that the actual percentage of voters planning to vote for Jones is between

$$48\% - 3\% = 45\% \quad \text{and} \quad 48\% + 3\% = 51\%.$$

Note that these ranges leave open the possibility of either candidate winning. Based on this poll, the election is too close to call.   ■

TYPES OF STATISTICAL STUDY: OBSERVATIONS AND EXPERIMENTS

Broadly speaking, statistical studies can be divided into two major categories.

- In **observational studies**, researchers observe or measure characteristics of the sample members, but do not attempt to influence these characteristics. Thus all members of the sample are treated the same way. For example, a study to measure heights of students is observational because we measure all student heights in the same way and the measurements do not influence student height.

- In **controlled experiments**, researchers deliberately create two (or more) groups of individuals or objects. One group receives special treatment and the other, called the **control group**, does not; researchers then look for differences between the two groups. For example, researchers might test a new weed killer by applying it to 100 lawns and comparing the results to a control group of another 100 lawns that do *not* receive the weed killer.

Special care is required when experiments involve people. For example, suppose that you are testing the effectiveness of a new anti-depression drug. You find 500 people who are suffering from depression, and divide them into a group that receives the new drug and a control group that does not. You then ask people in each group about their mood, and compare the groups to learn whether the drug alleviates depression. Will your results be valid?

Unfortunately, it's quite possible that the mood of people receiving the drug will improve simply because they're glad to be getting some kind of treatment. One way to prevent this problem is by giving the members of the control group a **placebo**—something that looks just like the new drug, but lacks its active ingredients. In that case, the participants won't know whether they are members of the control group or the group receiving the drug. Such a study, in which participants don't know which group they belong to, is called a **single-blind experiment** (the participants are "blind" to their category).

By the Way ·············
Many people treated with a placebo actually improve, even though it has no active ingredients. When this happens, it is called a *placebo effect* (see Chapter 9).

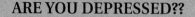

ARE YOU DEPRESSED??

Do you have depressed moods, loss of energy, change in sleep or weight? Volunteers, 18-49 yrs. old, needed to participate in a research study. Study includes an 8-week trial of an experimental medication for the treatment of depression and two MRI scans free of charge.

Depression Research Laboratory
County Medical School
1-800-555-5555

There may still be problems even if the participants don't know their group. For example, suppose that the researchers inadvertently smile more at the people who receive the drug than those in the control group. The smile might improve those participants' moods, leading to invalid results. Therefore it may be useful to make a **double-blind experiment** in which neither the participants *nor the researchers* know who received the drug at the time of the experiment.

Sometimes it may be impractical or unethical to create a controlled experiment. For example, suppose that you want to study how alcohol consumed during pregnancy affects newborn babies. It would be unethical to divide a sample of pregnant mothers randomly into two groups—one that receives alcohol and one that doesn't. However, you may be able to conduct a **case-control study** in which the participants naturally form groups by choice. In this example, the **cases** consist of mothers who consume alcohol during pregnancy *by choice*, and the **controls** consist of mothers who choose *not* to consume alcohol. Note that a case-control study is *observational* because the researchers do not change the behavior of the participants. However, it also resembles a controlled experiment because it allows comparison between different groups.

TIME-OUT TO THINK: Suppose that you were designing a double-blind study to test a particular drug. How would you set up the study so that neither the researchers nor the subjects know whether a particular subject receives the drug or a placebo? Remember that you must have some way to determine who got what in the end, so that you can analyze the differences between the two groups.

EXAMPLE 5   *Which Type of Study?*

For each of the following questions, what type of statistical study is most likely to lead to an answer? Why?

   a)   What is the average income of stockbrokers?

   b)   Do seat belts save lives?

   c)   Can lifting weights improve runners' times in a 10-kilometer race?

   d)   Does skin contact with a particular glue cause a rash?

   e)   Can a new herbal remedy reduce the severity of colds?

Solution:

a) An *observational study* can answer the question of the average income of stock-brokers. We need only to survey the brokers, and the survey itself will not change their incomes.

b) It would be unethical to do an experiment in which some people are told to wear seat belts and others are told *not* to wear them. Thus a study to determine whether seat belts save lives must be *observational.* However, this observational study can take the form of a case-control study. Some people choose to wear seat belts (the cases) and others choose not to (the controls). By comparing the death rates in accidents between cases and controls, we can learn whether seat belts save lives.

c) We need a *controlled experiment* to determine whether lifting weights can improve runners' 10K times. One group of runners will be put on a weight-lifting program, and a control group will be asked to stay away from weights. We must try to ensure that all other aspects of their training are similar. Then we can see whether the runners in the weight-lifting group improve their times more than the control group. Note that this experiment cannot be blind because there is no way to prevent participants from knowing whether they are lifting weights!

d) An experiment can help us determine whether skin contact with a particular glue causes a rash. In this case, we are best off with a *single-blind experiment* in which we apply the actual glue to participants in one group, and apply a placebo that looks the same, but lacks the active ingredient, to members of the control group. There is no need for a double-blind experiment because there does not seem to be any way that the researchers can inadvertently influence whether a person gets a rash.

e) We should use a *double-blind experiment* to determine whether a new herbal remedy can reduce the severity of colds. Some participants get the actual remedy, while others get a placebo. We need the double-blind conditions because the severity of a cold may be affected by mood or other factors that researchers might inadvertently influence. In the double-blind experiment the researchers do not know which participants belong to which group, and thus cannot treat the two groups differently. ∎

REVIEW QUESTIONS

1. Why do statisticians study a *sample*, rather than an entire *population*?

2. Describe each of the five basic steps in a statistical study. Where does the process of generalization come in?

3. What is *random sampling*, and why is it so important in statistical research?

4. What do we mean by the *margin of error* of a survey? How is it related to the idea of a *95% confidence interval*?

5. How do observational studies differ from controlled experiments? How is a *case-control* observational study similar to a controlled experiment?

6. What is a *placebo*, and when is it useful in statistical studies?

7. What is the difference between a *single-blind* and a *double-blind* experiment? Under what conditions is each type of experiment useful?

PROBLEMS

1. The Word Statistics. The word *statistics* has the same root as the word *status*. Briefly explain how the two words are related.

2. Statistics in the News. Identify at least three stories reported in today's news that involve statistics in some way. For each case, briefly describe in one or two paragraphs the role of statistics in the story.

3. Statistics in Sports. Most sports keep track of many statistics such as win–loss percentages, or batting or shooting averages. Choose a sport and describe at least three different statistics commonly tracked by participants or spectators of the sport. For each one, briefly describe the importance of the statistic to the sport.

Identifying Statistical Terms. *For the statistical studies described in Problems 4 and 5, answer the following questions.*

  a. Identify the population and sample for the study. What is the population parameter?

  b. What is the raw data for the study?

  c. What is the sample statistic for the study?

  d. What is your estimate of the population parameter? Comment on the likelihood that your estimate is close to the true value of the population parameter.

4. Suppose that you would like to know the average amount of sleep for the 3500 students at your college. One day you ask each of the 35 students in your mathematics class to report their average amount of sleep. You then average the sleep time for these 35 students and find that the average is 7.3 hours. Your conclusion is that, on average, the 3500 students at your college sleep 7.3 hours.

5. Suppose you want to know which of two lawn fertilizers gives greener lawns. You choose 30 friends and tell them to apply More-Grow Fertilizer to half of their lawn and Go-Green Fertilizer to the other half of their lawn. You also tell them to give both halves of their lawn the same care. After six weeks each of your 30 friends reports which half of their lawn is greener: 18 lawns were greener with More-Grow Fertilizer and 12 lawns were greener with Go-Green Fertilizer. You conclude that 18/30 = 60% of all lawns do better with More-Grow Fertilizer.

6. Drinking Survey. Suppose that you want to determine the (alcoholic) drinking habits of students at your college by asking one hundred students to fill out a survey about their alcohol consumption. What is the population you are studying? Suggest a method for surveying a random sample of students that is likely to be representative of the population.

7. Preelection Survey. Suppose that you are conducting a poll to determine who is most likely to win an upcoming student government election. The poll will be conducted a week before the election. What is the population? Suggest a method for choosing a random sample that is likely to be representative of the population.

8. Football Player Body Fat. You decide to make a study of the average percentage body fat for collegiate football players. Which of the following would be the best sample for your study, and why? Also explain why each of the other choices would not make a good sample for your study.
- The linebackers at your college.
- The freshmen football recruits at your college.
- The quarterbacks for each of the top twenty teams in the nation.
- The entire team at your college.
- The entire team at the college with the longest losing streak in the nation.

Samples and Populations. *For Problems 9–14, state the population under study and discuss factors in choosing a random sample.*

9. You want to determine the percentage of people in this country with each of the four major blood types (A, B, AB, and O).

10. You want to determine the number of smokers in this country who die from lung cancer each year.

11. You want to determine the number of lung cancer victims (each year) who smoked.

12. You want to determine the average mercury content of the tuna fish consumed by U.S. residents.

13. You want to determine whether drinking three cups of herbal tea per day reduces the chances of getting a cold.

14. You want to determine the percentage of students in a large high school who listen to classical music.

Margin of Error in Polls. *Problems 15–17 describe survey results. For each one, state the 95% confidence interval for the results and interpret the results in words.*

15. In a 1997 TIME/CNN poll, 748 adults were asked whether they believed their children would have a higher standard of living than they have. With a margin of error of 3.6%, 63% of those polled said "yes."

16. A 1997 Yankelovich poll determined that 70% of 4000 people surveyed agreed with the statement. "People have to realize that they can only count on their own skills and abilities if they're going to win in this world." The margin of error in the poll was ±1.6%.

17. The U.S. Census Bureau estimates the unemployment rate in the United States monthly by surveying 60,000 households. Suppose that for a particular month the unemployment rate is 5.6% with a margin of error of 0.4%.

Types of Studies. *For Problems 18–24, determine whether the study is observational or a controlled experiment. If it is observational, say whether it is a case-control study. If it is an experiment, indicate if a control group is used.*

18. In a study of hundreds of Swedish twins, it was determined that among identical twins (twins coming from a single egg) the level of mental skills was more similar than in fraternal twins (twins coming from two separate eggs). (*Science*, June 6, 1997)

19. A National Cancer Institute study of 716 melanoma patients and 1014 cancer-free patients matched by age, sex, and race found that those having a single large mole had twice the risk of melanoma; having 10 or more moles was associated with a 12-times greater risk of melanoma. (*Journal of the American Medical Association*, September 1997)

20. A European study of 1500 men and women with exceptionally high levels of the amino acid homocysteine had double the risk of heart disease. However, the risk was substantially lower for those in the study who took Vitamin B supplements. (*Journal of the American Medical Association*, June 1997)

21. A 1996 Gallup survey done for CNN/USA TODAY found that 80% of Americans think we're less civil than ten years ago, and 67% think that Americans are more likely to use vulgar language than ten years ago.

22. In a study of the effect of pulsing magnetic fields on multiple sclerosis (MS) patients, those equipped with an active magnetic pulsing device showed improvement in motor and language skills compared to patients equipped with an identical device that was inactive. (*Journal of Alternative and Complementary Medicine*, November 1, 1997)

23. A June 1997 Gallup poll sponsored by the National Sleep Foundation discovered that one-third of adult Americans are "excessively sleepy during the daytime."

24. A breast cancer study began by asking questions of 25,624 women about how they spent their leisure time. The health of these women was tracked over the next fifteen years. Those women who said they exercise regularly were found to have a lower incidence of breast cancer. (*New England Journal of Medicine*, May 1, 1997)

Types of Statistical Studies. *For Problems 25–30, state what type of statistical study is most likely to lead to an answer. Explain your answer. Comment on whether a placebo could be used and whether the study could be single- or double-blind.*

25. What is the average income of public school teachers?

26. What effect does jazz have on studying and grades?

27. Is red dye #40 harmful to humans when eaten in foods?

28. Does playing soccer help swimmers improve their times?

29. Does an aspirin a day reduce the incidence of heart attacks?

30. Does a self-proclaimed mind reader really have supernatural abilities?

UNIT 2B

SHOULD YOU BELIEVE A STATISTICAL STUDY?

The focus in this chapter is not on *doing* statistical studies, but rather on learning how to *evaluate* statistical results that you may see in the news. Most researchers presumably conduct their statistical studies with integrity and carry out their statistical research with care. Nevertheless, conducting and interpreting statistical research is sufficiently complex that errors can easily arise. In particular, statistical research is prone to many different forms of **bias**—a term that covers a variety of errors, often unintentional, that can prejudice a statistical study.

News reports do not always provide all the information you need to answer the question, "should I believe a statistical study?" However, you can generally find or infer a fair amount of information about how and why a study was conducted. In this unit, we discuss eight guidelines that you can use to help you evaluate statistical reports in the news.

GUIDELINE 1: IDENTIFY THE GOAL, TYPE, AND POPULATION

Before you can decide whether to believe a statistical study, you must first know what was being studied. Ask yourself a few basic questions such as:

- What was the goal of the study?
- Was the study observational or experimental? If it was observational, was it a case-control study? If it was experimental, was it single- or double-blind? Considering the goal, was the type of study appropriate?
- What was the population under study? Was the population clearly and appropriately defined?

If you can't find reasonable answers to these questions, it will be difficult to evaluate any further aspects of the study. In that case, your only choices are to look for a better article that might contain more information about the study, or to dismiss it until you are given better information.

EXAMPLE 1   *Appropriate Type of Study?*

A newspaper reports the following: "Researchers gave each of the one hundred participants their astrological horoscopes, and asked them whether the horoscopes appeared to be accurate. Eighty-five percent of the participants reported that the horoscopes were accurate. The researchers concluded that horoscopes are valid."

Analysis: Clearly, the goal of the study was to determine the validity of horoscopes. Based on the news report, it appears that the study was essentially *observational*: The researchers simply asked the participants about the accuracy of the horoscopes. However, because the accuracy of a horoscope is somewhat subjective, this study should have been a controlled experiment in which some people were given an actual horoscope and others were given a fake horoscope. Then the researchers could have looked for differences between the two groups. Moreover, because researchers could easily influence the results by how they

By the Way ··········
Different news reports on the same study may give vastly different amounts of information. Local newspapers often truncate articles from national sources; look at the by-line to find the source of the complete article (e.g.,"adapted from *The Wall Street Journal*"). News articles may also tell you if and where the original research was published (e.g.,"published in the *New England Journal of Medicine*").

By the Way ··········
Surveys show that nearly half of Americans believe their horoscopes. However, in controlled experiments, the predictions of horoscopes come true no more often than would be expected by chance.

questioned the participants, the experiment should have been double-blind. In summary, the type of study was inappropriate to the goal, and its results are therefore meaningless.   ■

EXAMPLE 2   *Appropriate Population?*

A news report states that: "Researchers tracked 1000 men for fifteen years. They found that the men who exercised regularly had lower mortality rates than those who did not. They concluded that regular exercise leads to longer lives."

Analysis: This study is observational, but takes the form of a case-control study because the men naturally divided into two groups: a case group that exercised regularly and a control group that did not. This type of study seems appropriate, which should give us some confidence in its conclusions. However, the population is not clearly defined—at least not in the news report. Were *all* men included in the population? In that case, several factors may complicate the interpretation of the results. For example, it may not have been a good idea to include smokers in the population, because exercise might affect smokers and nonsmokers differently. Because the news report does not clearly state how the population was defined, we should withhold final judgment on the conclusions until we learn more.   ■

GUIDELINE 2: CONSIDER THE SOURCE

Statistical studies are supposed to be objective, but the people who carry them out and fund them may be biased. Thus it is important to consider the source of a study and evaluate the potential for biases that might invalidate its conclusions.

Bias may be obvious in cases where a statistical study is carried out for marketing, promotional, or other commercial purposes. For example, a car advertisement that claims "97% of Prizm owners would recommend Prizm to a friend" appears to be statistically based, but we are given no details about how the survey was conducted. Because the advertisers obviously want to say good things about Prizms, it's difficult to take the statistical claim seriously without much more information about how the result was obtained.

Other cases of bias may be more subtle. For example, suppose that a carefully conducted study concludes that a new drug helps cure cancer. On the surface, the study might seem quite believable. But what if the study was funded by a drug company that stands to gain billions of dollars in sales if the drug is proven effective? The researchers may well have carried out their work with great integrity despite the source of funding, but it might be worth a bit of extra investigation to be sure.

Major statistical studies are usually evaluated by unbiased experts. For example, the process by which scientists examine each others' research is called **peer review**. Reputable scientific journals require all research reports to be peer reviewed before the research is accepted for publication. Peer review does not guarantee that a study is valid, but it certainly lends credibility.

EXAMPLE 3   *Is Smoking Healthy?*

By 1963, enough research on the health dangers of smoking had accumulated so that the Surgeon General of the United States publicly announced that smoking is bad for health. The evidence has continued to accumulate ever since. However, while the vast majority of studies show that smoking is unhealthy, a few studies have shown no dangers from smoking, and perhaps even health *benefits*. These studies generally were carried out by the Tobacco Research Institute, funded by the tobacco companies.

Analysis: Even in a case like this, it can be difficult to decide whom to believe. However, the studies showing smoking to be unhealthy come primarily from peer reviewed research. In contrast, the studies carried out at the Tobacco Research Institute have a clear potential for bias. The *potential* for bias does not mean their research *is* biased, but the fact that it contradicts virtually all other research on the subject should be cause for concern.   ■

EXAMPLE 4   *Press Conference Science*

The nightly TV news shows scientists at a press conference announcing that they've discovered evidence that people with particular personality types are more prone to common colds.

Analysis: Scientists often announce the results of their research at a press conference so that the public may hear about their work as soon as possible. However, it may require a great deal of expertise to evaluate their study for possible biases or other errors—which is the point of the peer review process. Until the work is peer reviewed and published in a reputable journal, any findings should be considered preliminary.   ■

GUIDELINE 3: LOOK FOR BIAS IN THE SAMPLE

If a sample is biased, there is little hope that it will be representative of the population under study. Unfortunately, statistical studies are particularly prone to bias in their samples. There are two particularly common forms of bias that can affect sample selection.

- **Selection bias** occurs when researchers select their sample in such a way that it is unlikely to be representative of the population. For example, a preelection poll that surveys only registered Republicans has selection bias because it is unlikely to reflect the opinions of voters of all parties (and independents).
- **Participation bias** occurs any time participation in a study is voluntary. In general, people who feel strongly about an issue are more likely to volunteer. Hence the opinions or actions of the volunteers may not reflect the larger population that is less emotionally attached to the issue.

By the Way ....
Participation bias generally is a consideration only when the population involves people—meat inspectors, for example, don't have to worry about a steak refusing to be part of a sample!

EXAMPLE 5   *The 1936 Literary Digest Poll*

The *Literary Digest*, a popular magazine of the 1930s, successfully predicted the outcomes of several elections through large polls. In 1936, editors of the *Literary Digest* conducted a particularly large poll in advance of the presidential election. They randomly chose a sample of 10 million people from lists of names in telephone books, rosters of clubs and associations, and similar sources. They mailed a postcard "ballot" to each of these 10 million people. About 2.4 million people returned the postcard ballots. Based on the preferences indicated on the returned postcard ballots, the editors of the *Literary Digest* predicted that Alf Landon would win the presidency by a margin of 57% to 43% over Franklin Roosevelt. Instead, Roosevelt won with 62% of the popular vote.

Analysis: The *Literary Digest* poll suffered from both selection bias and participation bias. The selection bias arose because only relatively affluent people could afford telephones in 1936, so selecting names from telephone books tended to draw people who were wealthier than average. Drawing names from club rosters and similar sources also was biased toward more affluent voters. The participation bias arose because return of the postcard ballots was voluntary and only 2.4 million people returned the ballots out of the 10 million who received them.

Both forms of bias favored Landon over Roosevelt. The selection bias favored Landon because he was the Republican, and studies have shown that affluent voters tend to vote for Republican candidates more often than for Democratic candidates. The participant bias favored Landon because he was the challenger. People who were dissatisfied with President Roosevelt could express their desire for change by indicating their preference for Landon in the survey. In contrast, Roosevelt's supporters weren't seeking any major changes and thus were less likely to return the survey cards. Together, the two forms of bias made the poll sample unrepresentative of the population that voted in the election. ■

By the Way ....
A young pollster named George Gallup conducted his own survey prior to the 1936 election. Sending postcards to only 3000 randomly selected people, he correctly predicted not only the outcome of the election, but also the outcome of the *Literary Digest* poll to within 1%! Gallup went on to establish a very successful polling organization.

EXAMPLE 6   *Call-In Poll*

A television talk show host asks viewers to respond to a poll asking, "Should immigration be further restricted?" Viewers are given two 900 numbers to call: one for "yes" answers and one for "no" answers. Each call costs the viewer 50¢. At the end of the show, the host announces that 62% of Americans believe immigration should be restricted further.

Analysis: This survey suffers severely from both selection bias and participation bias. The announced results suggest that the host considered the *population* under study to be all Americans, but the sample was selected only from the show's viewers—a group unlikely to be representative of all Americans. Participation bias occurs not only because participation was strictly voluntary, but also because viewers had to *pay* the 50¢ cost of the phone call to participate. This cost makes it even more likely that only those viewers who feel very strongly will participate. All in all, the biases involved in this survey are so severe that its results should be considered meaningless. ■

GUIDELINE 4: LOOK FOR DIFFICULTIES IN *DEFINING* QUANTITIES OF INTEREST

Statistical studies always attempt to measure *something*. The things being measured in a statistical study are its **quantities of interest**. If these quantities are difficult to define, results may be especially difficult to interpret. For example, consider a survey designed to determine the number of unemployed people. What qualifies a person as "unemployed?" Should "unemployed" include people who quit their jobs voluntarily or who are retired? Should it include teenagers who are full-time students? Clearly, the definition of "unemployed" will have a major impact on the survey results. Similarly, who is included in a

By the Way ....
The U.S. Department of Labor says that: "persons are classified as unemployed if they do not have a job, have actively looked for work in the prior four weeks, and are currently available for work."

$\mathcal{T}$HINKING ABOUT...

Have You Been Sugged *Lately?*

In early 1990, the Nielsen television rating system reported an 8% drop in the number of people watching television. Such a drop would make television advertising time less valuable and might have a severe impact on the economics of the television industry. The industry claimed that the Nielsen survey results were misleading. Nielsen's data was based on a sample of homes in which a meter monitors when the television is on and what is being watched. By 1990, only 47% of the people that Nielsen contacted agreed to use the meter—down from 68% in 1980. The television industry argued that the supposed decline in television viewing reflected *participation bias* in the Nielsen surveys, rather than a real change in television viewing habits. In the words of NBC's head of research at the time, Robert Niles, "We don't know whether the people who turned down the meter are like those who agreed to take it."

The Nielsen example shows that unbiased sampling is increasingly difficult to achieve. Today, more than a third of all Americans routinely shut the door or hang up the phone when contacted for a survey.

Why? Researchers suspect several factors: First, many people are contacted repeatedly for different surveys and simply tire of the inconvenience. Second, people concerned about privacy are fearful that giving information may put them on yet another mailing list or find its way into a credit report. Finally, researchers have found that many people resent a practice known in the marketing industry as "selling under the guise" of market research, or by its acronym of "*sug*ging." In this practice, a telemarketer tells you that you are part of a survey, but really is trying to get you to buy something. This practice has become so common that even people who are willing to participate in a legitimate survey may be uncooperative because they suspect "sugging" is taking place.

survey of "all Americans?" Does it include children or citizens living abroad? Does it include permanent residents or citizens in territories?

The problem of defining measurement quantities becomes even more difficult in cases of intangible quantities. Surveys designed to gauge feelings such as satisfaction with a product, level of approval of a political official, or whether a new drug is improving one's mood are particularly prone to difficulties.

EXAMPLE 7   *Life Is Swell*

A TIME/CNN survey (*Time*, May 19, 1997) concluded that "by almost any measure, life [in America] is swell." The survey was conducted by asking the following questions in telephone interviews with 1,017 adult Americans. Some of the results are shown in the table on the next page.

Analysis: The goal of the study apparently was to measure satisfaction with various aspects of life in America. However, the questions are rather subjective, which makes it difficult to interpret the responses. The first two questions depend on respondents' interpretation of *good times* and *bad times*. The wording of the third and fourth questions is curious: it could be a good time to make a decision about buying an appliance or going on vacation, even if the decision turns out to be "no." Regarding the fifth question, some people may worry even when their job is secure, while others may not worry even when their job is at risk.

An even greater concern may be that the interpretation of the study is subjective. Even if we take the survey results at face value, only 63% of respondents said that they were going through good times. Yet, the article concluded that "life [in America] is swell." Do you agree that 63% is a large enough positive response to conclude that life is swell?   ■

Is the U.S. going through a period of good times or bad times right now?	Good times 54%	Bad times 33%	Both or neither 12%
Are you going through a period of good times or bad times right now?	Good times 63%	Bad times 24%	Both or neither 12%
Is this a good time for you to make a decision about buying a major appliance?	Yes 62%	No or don't know 38%	—
Is this a good time for you to make a decision about taking an expensive vacation?	Yes 28%	No or don't know 72%	—
Do you worry about losing your job?	Yes 19%	No 80%	—

Mathematical Note: The total of the responses to each question in the TIME/CNN poll is not always 100%. For example, the yes and no responses to the last question total to only 19% + 80% = 99%. This can happen because of rounding the survey results to the nearest percent, or because some people may not respond to a question.

TIME-OUT TO THINK: Because the preceding example used a telephone survey, it was certainly subject to participation bias: some people undoubtedly chose not to answer the survey questions. Could this bias have affected the results? For example, could unhappy people be more likely to hang up the phone on the survey takers?

GUIDELINE 5: LOOK FOR DIFFICULTIES IN *MEASURING* QUANTITIES OF INTEREST

Even when quantities of interest are clearly defined, they may not be easy to measure. For example, imagine trying to conduct a study of how exercise affects resting heart rates. It is relatively easy to measure a person's resting heart rate at some particular moment (although even this measurement may be affected by such factors as stress, medication, room temperature, or the time of the person's last meal). It is much more difficult to quantify how much a person exercises: it depends not only how much time a person spends exercising, but also on a subjective evaluation of how vigorously the person exercises.

EXAMPLE 8   *Illegal Drug Supply*

Law enforcement authorities try to stop illegal drugs from entering the country. A commonly quoted statistic is that they succeed in stopping only about 10% to 20% of the drugs entering the United States.

Analysis: It should be relatively easy to measure the quantity of illegal drugs that law enforcement officials intercept. However, because the drugs are illegal, it's unlikely that anyone is reporting the quantity of drugs that are *not* intercepted. How, then, can anyone know that the intercepted drugs are 10% to 20% of the total? In a *New York Times* analysis, a police officer was quoted as saying that his colleagues refer to this type of statistic as "P.F.A.," for "pulled from the air."   ■

GUIDELINE 6: IF A SURVEY IS INVOLVED, CONSIDER ITS SETTING AND WORDING

The setting in which a survey is administered can have a significant effect on its outcome. Telephone surveys can introduce bias because they may produce a high nonresponse rate and often sample a disproportionate number of women and older people. Mail surveys often produce very low response rates because they allow participants to procrastinate. Personal interview surveys depend critically on the time and place at which they are given. A survey given at a shopping mall during a working day is likely to survey a very different group of people than the same survey on a weekend. If a survey concerns sensitive subjects, such as personal habits or income, people may be unwilling to answer honestly if the setting does not assure them of complete confidentiality.

The wording of a survey is also crucial. Poorly worded questions may lead to results that are misleading or meaningless. It is particularly important to watch out for what psychologists call the **availability error**: the tendency to make judgments based on what is *available* in the mind. In its simplest form, the availability error explains why pictures of tacos, burritos, and enchiladas make many people want Mexican food—the pictures make Mexican food "available" to your mind.

> **TIME-OUT TO THINK:**   Consider the following two survey questions. Question 1: *What is your favorite cola?* Question 2: *Is Pepsi your favorite cola?* Using the idea of the availability error, explain why Question 2 is likely to find more people who say that Pepsi is their favorite.

EXAMPLE 9   *Do You Favor Campaign Finance Reform?*

A professional polling organization asked the following question, taken from a survey produced by Ross Perot's 1992 presidential campaign team:

> *Should laws be passed to eliminate all possibilities of special interests giving huge sums of money to candidates?*

By the Way ············
This example is adapted from John Allen Paulos's book *A Mathematician Reads the Newspaper*.

Of the people polled, 80% responded *yes* and 18% responded *no*. Then the question was rephrased and asked in a different way (to a different sample of people):

> *Should laws be passed to prohibit interest groups from contributing to campaigns, or do groups have a right to contribute to the candidate they support?*

The results were dramatically different: only 40% in favor of new laws, and 55% against.

Analysis: Both questions appear to be asking whether respondents favor campaign finance reform, yet they obtained dramatically different results. The explanation lies in the different ways the questions were worded and how they led to the availability error. The original question was worded in such a way that it made arguments in favor of campaign finance reform *available* to respondents, but did not make arguments against reform available. Thus the question was biased toward a *yes* answer. The second question associated the word *prohibit* with limits on contributions, and the word *rights* with allowing contributions. Because most people enjoy their rights and dislike prohibitions, the second question probably was biased toward answers *opposing* campaign finance reform.   ∎

GUIDELINE 7: CHECK FOR CONSISTENCY BETWEEN RESULTS AND INTERPRETATION

Even when a statistical study is done well, it may be misinterpreted. Researchers may occasionally misinterpret the results of their own studies, particularly when the results are unexpected. In other cases, news reporters or others may misinterpret a survey—and then pass their misinterpretation on to you! Such misinterpretations can sometimes be discovered simply by looking for inconsistencies between the interpretation and any actual data given along with it.

EXAMPLE 10   *Does the School Board Need a Statistics Lesson?*

The School Board in Boulder, Colorado, created a hubbub when it announced that 28% of Boulder school children were reading "below grade level," and hence concluded that methods of teaching reading needed to be changed. The announcement was based on reading tests on which 28% of Boulder school children scored below the national average for their grade.

Analysis: By definition, half of all children must score below average and the other half score above average. Thus, if Boulder were representative of the nation at large, we should expect 50% of Boulder children to score below average on their reading tests. The fact that only 28% scored below average clearly implies that Boulder children, on the whole, are doing *better* than the national average. The School Board's ominous statement about students reading "below grade level" comes about only because they have chosen to interpret "grade level" as average. This interpretation is quite subjective. Others might look at the same data, and claim that reading is being taught very successfully because 72% of Boulder children score above the national average.   ∎

> **Mathematical Note:** The analysis in Example 10 applies only if the average refers to a *median* (the middle value in a set of scores) rather than a *mean*. In this case, because national test scores follow a *normal distribution*, the mean and median are the same. We'll explore means, medians, and normal distributions in Chapter 9.

GUIDELINE 8: STAND BACK AND CONSIDER THE CONCLUSIONS: WHAT DOES IT MEAN TO YOU?

Suppose that a study reports that wearing a gold chain makes people more likely to survive automobile accidents. Even if the study seems to have been conducted properly, you might be quite skeptical of its conclusion. After all, how could a thin chain around your neck help you in a high speed collision? Of course, it's *possible* that the conclusion is correct, but you'd probably want to understand much more before accepting it.

This example illustrates the importance of stepping back and considering the conclusions of a study, even when it checks out in every other way. Ask yourself questions like:

- Do the conclusions make sense?
- Can you rule out alternative explanations for the results?
- Are the results consistent with previous studies?
- Is the study convincing enough to affect your beliefs or actions?

If the answers to any of these questions are *no*, then you should demand very strong evidence before you accept the conclusions. In the words of noted scientist and writer Carl Sagan (1934–1996):

Extraordinary claims require extraordinary evidence.

EXAMPLE 11   *Car Phones and Crashes*

In a study reported in May 1997, researchers from the University of Toronto examined 26,798 calls from car phones. They concluded that phoning from a car quadruples the risk of a crash. Moreover, they claimed that the accidents were generally caused by the distraction of talking on a phone, and not by having only one hand on the steering wheel. They also found one benefit to driving with a cell phone: a call can be made immediately after an accident occurs!

Analysis: The conclusions of this study seem reasonable: most of us have seen people who appear distracted while talking on a car phone. It also appears that the researchers ruled out alternative explanations for the higher crash rate among phone users, such as having only one hand on the wheel. We are not told of any previous studies on this topic, so unless we happen to know of a similar study, we have no reason to think that this one contradicts previous work. All in all, this study seems believable, which leads us to the final question: will this report change your use of car phones, or change your opinion about other drivers using car phones?   ■

REVIEW QUESTIONS

1. Make your own summary list of the eight guidelines presented in this unit for evaluating statistical studies. For each, give your own example of how the guideline can be applied.

2. What is *peer review*? Why is research reported in peer-reviewed journals generally more reliable than research that has not been peer reviewed?

3. Describe and differentiate between *selection bias* and *participation bias*. How can a study avoid each of these biases?

4. What do we mean by the *quantities of interest* in a statistical study? Why is it so important that they be defined and measured carefully.

5. What is the *availability error*? How can it cause poorly worded survey questions to give misleading results?

6. Describe what you think Carl Sagan meant by the statement: *Extraordinary claims require extraordinary evidence*.

PROGRAMS •

Bias in Sampling. *For Problems 1–7, explain why bias might be introduced because of the way in which the sample is chosen (or state that the method is unbiased).*

1. In a quality control study of computer chips, you test every 100th chip that comes off the assembly line.

2. In a telephone survey of TV network preference, you call all people in your local phone directory whose last name starts with A.

3. In a telephone survey of TV network preference, you call a randomly selected person whose last name starts with Z in your local directory, then a randomly selected person whose last name starts with Y, and so on through the alphabet, repeating the process if necessary.

4. You mail a survey form to every student at your college asking for the students' choice of their all-time best and worst professors. Students are asked to return the survey in the campus mail.

5. A start-up pharmaceutical company conducts its own trials on 1000 subjects to determine whether its new allergy drug is better than its competitors' drugs.

6. A start-up pharmaceutical company provides funding to University scientists to conduct trials on 1000 subjects to determine whether its new allergy drug is better than its competitors' drugs.

7. You conduct a survey at a supermarket on a weekday between 10:00 A.M. and noon to determine which of two brands of beer customers prefer.

Stat-Bytes. *Politicians must make their political statements (often called sound-bytes) very short because the attention span of listeners is so short. A similar effect has happened in reporting statistical news: major statistical studies are often reduced to one or two sentences. The statistical reports in Problems 8–11 are quoted directly from* Time *magazine (January 12, 1998). Discuss what crucial information is missing and what more you would want to know before you would act on the report.*

8. "In a study, patients who donned gold bands had significantly less arthritis in their ring fingers than in corre-

sponding fingers of the opposite hand." Source: *Annals of Rhuematic Diseases.*

9. "Another one for wine. As little as one glass a month—red or white—may cut in half the risk of macular degeneration" (an eye disease). Source: *Journal of American Geriatrics.*

10. "To achieve maximum longevity, finds a study, weight should be kept at a stable and low level throughout adulthood." Source: *New England Journal of Medicine.*

11. "Women who use [an electric blanket] at the time of conception or in early pregnancy raise their odds of spontaneous abortion 75%." Source: *Epidemiology.*

12. **Preelection Poll Bias.** A week before the annual student government election, you decide to conduct a telephone poll to determine which of two candidates, Walker or Castilla, is likely to win the race for president.

    a. What is the population for this study? Suggest a method for selecting an unbiased sample for your telephone poll.

    b. Consider the following two questions:
    - Do you plan to vote for Walker or Castilla for president?
    - For whom do you plan to vote for president?

    What are the advantages and disadvantages of using each question for your phone survey?

13. **Open vs. Closed Questions.** Following carefully constructed methods for random sampling, two surveys are given, one to each of two groups of 1000 Americans. The surveys are designed to find out which issues are most important in an upcoming election. The first survey gives specific responses (often called a *closed question* since specific options are listed); it asks participants to "rank each of the following issues in order of their importance to you."

Inflation	Pollution	Other
Crime	Immigration	
Unemployment	Abortion	

The second survey simply asks the respondents to "rank the six issues that are most important to you" (often called an *open question* since no options are listed).

State the advantages and disadvantages of the two surveys in terms of sources of bias, the response rate, and the honesty of responses. Which do you think will be a more effective survey? Why?

14. **Biased Questioning About the Holocaust.** Nazi Germany killed six million Jews—more than two-thirds of Europe's prewar Jewish population—in the horror known as the Holocaust. However, some people claim that the Holocaust never occurred. A 1992 Roper poll asked the following question about the Holocaust.

"Does it seem possible or does it seem impossible to you that the Nazi extermination of the Jews never happened?"

Only 65% of those surveyed answered "impossible it never happened." About a year later, a Gallup poll asked the following rephrased question about the Holocaust.

"The term Holocaust usually refers to the killing of millions of Jews in Nazi death camps during World War II. In your opinion, did the Holocaust: definitely happen, probably happen, probably not happen, or definitely not happen?"

The results of the Gallup poll: 83% said the Holocaust definitely happened, 13% said it probably happened, 2% said it probably did not happen, 1% said it definitely did not happen, and 1% had no opinion.

a. The Roper poll suggested that a significant fraction of Americans doubt the reality of the Holocaust. Do the results of the Gallup poll support this conclusion? Explain.

b. Identify a double negative in the 1992 question. Could this double negative have affected responses to the question? Explain.

c. Another source of potential bias in the 1992 question lies in respondents' interpretation of the words *impossible* and *extermination*. Could someone who accepts the reality of the Holocaust have chosen the answer "possible" in the 1992 survey? Explain.

d. Some college newspapers have printed advertisements from groups selling literature denying the reality of the Holocaust. Is it ethical to accept such ads? Defend your opinion.

15. **Evaluating a Statistical Study.** A poll conducted by the American Management Association asked how employers monitor the activities of employees. The percentage of employers surveyed that engaged in various surveillance activities were as follows:

Keep records of employee phone calls	35%
Videotape employees	15%
Store and review e-mail	14%
Store and review computer files	13%
Tape phone conversations	10%
Tape and review voice mail	5%

A report on the survey (*Time*, June 2, 1997) had the single statement: "A new survey shows that nearly two-thirds of companies spy on their employees."

a. Based on the percentages in the table, what is the *largest* possible percentage of employers that "spy on employees"? (Note that employers may be counted in more than one category.)

b. Based on the percentages in the table, what is the *smallest* possible percentage of employers that "spy on employees"?

c. Does the statement that nearly two-thirds of companies spy on their employees follow *directly* from the percentages given in the table? Could it be correct with the given data? Explain.

d. For each of the eight guidelines given in this unit, discuss whether the guideline is relevant to this study and, if so, how it applies.

16. **Opinion Poll Bias.** Find a recent news report of a single-question opinion poll. State the exact words of the question and the results of the poll. Analyze the question and the reported results for potential biases. At the end of your analysis, state whether you believe the results, and defend your opinion.

17. **Poor Sampling.** Find an article in a newspaper or magazine of the past two weeks that attempts to describe some characteristic of a population, but which you believe involved poor sampling (e.g., a sample that was too

small, or unrepresentative of the population under study). Attach a photocopy of the article.

  a. Describe the population and the sample for the study. Why do you think the sampling was poor?

  b. Summarize the conclusions reported in the article, and discuss whether you believe the end results. Defend your opinion.

18. **Good Sampling.** Find an article in a newspaper or magazine from the last two weeks that describes a statistical study in which the sample was well-chosen. Attach a photocopy of the article.

  a. Describe the population and the sample for the study. How was the sample chosen? Why do you think this sample was a good one?

  b. Summarize the conclusions reported in the article, and discuss whether you believe the end results. Defend your opinion.

19. **Analysis of a Statistical Study.** Find a recent newspaper article or television report about a statistical study on a topic that you find interesting. As you read, apply the eight guidelines given in this unit. Write a short summary of the study, and clearly describe what aspects of the study appear to have been done well and what (if any) may have been done poorly. Describe any information that you feel is missing from the news report. At the conclusion of your analysis, state whether you believe the study's conclusions, and why.

20. **Project: Journals and Statistical Studies.** Medical journals are one of the best sources of statistical studies. Visit your library and find the area where recent journals are kept. Two medical journals with many general interest articles are *The New England Journal of Medicine* and the *Journal of the American Medical Association*. Find a statistical study in one of these journals (or any other journal) and write a summary of the study and its findings. Identify key aspects of the study, as discussed in this chapter (e.g., identify the population and the sample, comment on the sampling methods and whether the sample is well-chosen, and identify sample statistics that are presented). Conclude your report with your own personal reaction to the study. Do you believe the findings or did you find reasons to be skeptical of the results?

21. **Project: Twin Studies.** Researchers doing statistical studies in biology, psychology, and sociology are grateful for the existence of twins. They can be used to study whether certain traits are inherited from parents (nature) or acquired from their surroundings during upbringing (nurture). Identical twins are formed from the same egg in the mother and have the same genetic material. Fraternal twins are formed from two separate eggs and share roughly half of the same genetic material. Find a published report of a twin study. Discuss how identical and fraternal twins are used to form case and control groups. Apply the eight guidelines in the text to the study and comment on whether you find the conclusion of the report convincing.

22. **Project: The Effects of Using Car Phones.** Find and read the article *Is Using a Car Phone Like Driving Drunk* in the journal *Chance* (Vol. 10, No. 2, 1997). Apply the eight guidelines in the text to this study and discuss the various ways that the researchers addressed bias in this study.

23. **Project: Shuffling Horoscopes.** Find a newspaper column that gives a horoscope for each of the astrological signs. Copy the horoscope for each sign onto a separate piece of paper, *without* any information identifying the sign. Shuffle the horoscopes randomly, and ask someone to guess which one represents their own personal horoscope. Does the person choose the right one for his or her astrological sign? Repeat this experiment with at least ten people, and preferably more. (Be sure that the people you ask did not already read the same source from which you obtained the horoscopes.) Discuss your results.

UNIT 2C

....

BASIC STATISTICAL GRAPHS

Statisticians have always found innovative ways to produce striking figures and graphs, but the computer age has made advanced graphical techniques even more available. The result is that modern media are filled with imaginative graphics. These graphics can make complex ideas much easier to understand; properly executed, a picture truly is worth a thousand words! However, graphics also can be confusing or deceptive. Thus it is very important to be able to interpret the wide variety of graphics that confronts us every day.

In this unit, we examine a few of the most basic types of statistical graphs—the types you are most likely to draw yourself in preparing business or financial reports, or reports for other college classes. The easiest way to draw these graphs is with a computer; however, we'll look at creating graphs by hand to illustrate the principles behind each type of graph.

BAR GRAPHS

Suppose that you were given the following data on the number of immigrants to the United States from various regions in 1995.

Region	Europe	Asia	Mexico	West Indies	Other Americas	Africa
Immigrants	128,185	267,931	89,932	96,788	90,472	42,456

The data consist of six different *categories* (in this case, regions) and a number for each category. A good way to display these numbers and to compare them is with a **bar graph**. Each of the six regions is represented by a bar, which can be either vertical or horizontal; Figure 2.2 shows vertical bars. The bars on a bar graph generally do not touch one another, because the categories are distinct and have no particular order. That is, the choice of putting the bar for Europe first and the bar for Asia second in Figure 2.2 is completely arbitrary. Moreover, the length or height of each bar is proportional to the data value it represents. For example, there were about three times as many immigrants from Europe as from Africa, so the bar representing Europe is about three times as long as the bar representing Africa.

Note that the vertical scale in Figure 2.2 is set so that the total height of the graph represents 300,000 immigrants—slightly more than the largest number in the data set. This choice ensures that all the bars are just about as tall as possible while fitting on the graph. Note also that the graph is drawn with a height of 6 cm, so that each 1 cm represents 50,000 immigrants on the vertical scale:

$$1 \text{ cm} = \frac{300,000}{6} = 50,000 \text{ immigrants}$$

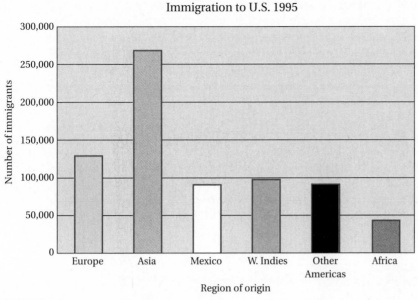

Immigration to U.S. 1995

FIGURE 2.2

This scale makes it easy to draw bars. For example, because each 1 cm represents 50,000 immigrants, the bar for Europe has a height of

$$\frac{128,185}{50,000} \approx 2.6 \text{ cm.}$$

> **TIME-OUT TO THINK:** Suppose that the number of immigrants from Russia was 22,000. If you add a bar to the graph for Russian immigrants, how tall should it be?

EXAMPLE 1   *Constructing a Bar Graph*

Make a bar graph for the following data set, which shows the *per capita* (amount per person) carbon dioxide emissions of selected countries in 1991.

Per capita emissions of CO_2 (in metric tons per person)			
Country	**Emissions**	**Country**	**Emissions**
United States	19.5	France	6.6
Canada	15.2	Germany	12.1
Japan	8.8	China	2.2
United Kingdom	10.0	Russia	12.3

Solution: We begin by setting up the axes in a way that will make the graph easy to read and interpret. The largest bar will represent the U.S. emissions of 19.5 metric tons per person, so it makes sense to set the vertical scale to a maximum of 20. We can then choose to make a graph that is 4 cm tall by letting each 1 cm represent 5 metric tons of emissions per person (because $20 \div 4 = 5$). Thus, for example, the height of the bar for Japan will be

$$\frac{8.8}{5} \approx 1.8 \text{ cm.}$$

On the horizontal scale, we need enough room for eight bars to represent the eight countries in the data set. The bars will look good if we make them each 0.5 cm in width; if we also leave 0.5 cm between each bar, then we can space the bars out along a horizontal scale that is 8 cm in length. The result is shown in Figure 2.3. Note that the very important last step in completing the graph is adding clear labels to explain what is shown in the graph.

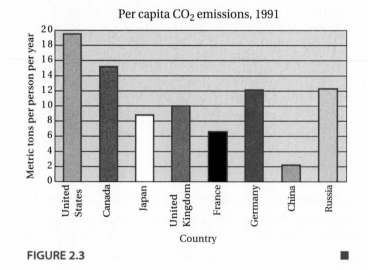

Per capita CO_2 emissions, 1991

FIGURE 2.3

TIME-OUT TO THINK: Note that China has the shortest bar in Figure 2.3. Does that mean that China emits less carbon dioxide than any of the countries shown? Why or why not? (Hint: The graph shows *per capita* emissions; China is the most populous country in the world.)

PIE CHARTS

Consider the following results from a 1997 survey of college students in the United States:

- Thirty-two percent work full-time.
- Thirty percent work part-time.
- Thirty-eight percent do not work at all.

The **pie chart** shown in Figure 2.4 offers a convenient way to represent these data. Each of the three categories gets its own slice, or wedge, in the circular pie. The entire pie represents 100% of students in the survey, and the area of each wedge is proportional to the percentage it represents. For example, the wedge corresponding to full-time working students comprises 32% of the area of the whole pie.

The only difficulty in drawing a pie chart is making the wedges the correct size. After all, it is difficult to be certain that the area of the full-time wedge in Figure 2.4 represents precisely 32%, and not 31% or 33%. The easiest way to make the wedges correctly is by measuring their *angles*.

Recall that a complete circle contains 360°. Thus we can find the angle represented by any *fraction* of a circle simply by multiplying the fraction or percentage by 360°. For example, the wedge corresponding to full-time working students is supposed to represent 32% of the full circle, so it should fill an angle of

$$32\% \times 360° = 115.2° \approx 115°.$$

You can measure this angle with the aid of a *protractor* (Figure 2.5). Similarly, the wedge representing the 30% of students that work part-time should fill an angle of

$$30\% \times 360° = 108°,$$

and the wedge corresponding to the 38% that don't work at all should have an angle of

$$38\% \times 360° = 136.8° \approx 137°.$$

EXAMPLE 2   *Making a Pie Chart*

Make a pie chart to represent the following distribution of final grades in an English class.

Grade	Number of Students	Grade	Number of Students
A	14	D	22
B	31	F	9
C	37	W (withdrawn)	7

Solution: The entire pie will represent the grades of *all* the students in the class. By adding the numbers of students with each grade, we find that the total number of students in the class was

$$14 + 31 + 37 + 22 + 9 + 7 = 120.$$

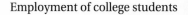
Employment of college students

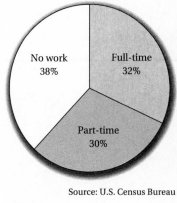

Source: U.S. Census Bureau

FIGURE 2.4

FIGURE 2.5

Thus the entire pie represents 120 students. We can find the angle needed for each wedge by multiplying the fraction of students it represents by 360°.

$$\text{A grades (14 out of 120 students)}: \frac{14}{120} \times 360° = 42°$$

$$\text{B grades (31 out of 120 students)}: \frac{31}{120} \times 360° = 93°$$

$$\text{C grades (37 out of 120 students)}: \frac{37}{120} \times 360° = 111°$$

$$\text{D grades (22 out of 120 students)}: \frac{22}{120} \times 360° = 66°$$

$$\text{F grades (9 out of 120 students)}: \frac{9}{120} \times 360° = 27°$$

$$\text{W grades (7 out of 120 students)}: \frac{7}{120} \times 360° = 21°$$

Distribution of grades

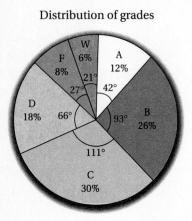

FIGURE 2.6

We begin by drawing a circle and a single radius extending from the center to the edge. Using this radius as a baseline, we use a protractor to measure an angle of 42° for the A wedge. The end of the A wedge becomes the baseline for measuring the 93° angle for the B wedge, and so on. The result is shown in Figure 2.6. ■

HISTOGRAMS

Consider the following table showing the 1995 U.S. population according to age categories.

Age Category	Population (millions)	Age Category	Population (millions)
< 5	19.6	50–54	13.6
5–9	19.2	55–59	11.1
10–14	18.9	60–64	10.0
15–19	18.1	65–69	9.9
20–24	17.9	70–74	8.8
25–29	19.0	75–79	6.7
30–34	21.9	80–84	4.5
35–39	22.2	85–89	2.3
40–44	20.2	90–94	1.0
45–49	17.4	> 95	0.2

Unlike the data categories in the bar graphs of Figures 2.2 and 2.3, in which the order of the bars was arbitrary, age categories fall into a natural numerical order. Thus when we make a bar graph of this data, our only choice is whether to show the age categories in ascending or descending order; Figure 2.7 shows the ages in ascending order. A bar graph of this type, in which the bars follow an ordered progression, is called a **histogram**.

1995 U.S. population by age

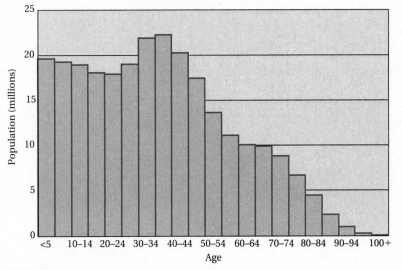

FIGURE 2.7

Mathematical Note: Different books define the terms *histogram* and *bar graph* differently, and there is no universally accepted guideline for distinguishing between them. In this book, a bar graph is any graph that uses bars, and we use histograms only for ordered, quantitative data.

Thus a histogram is a special case of a bar graph, and the process of constructing a histogram is essentially the same as constructing any other bar graph. In the case of the U.S. population data, the height of each bar is proportional to the population it represents. The total height of the histogram in Figure 2.7 corresponds to 25 million people—a nice round number that is slightly greater than the largest data value.

Each data category in a histogram is called a **bin**. For example, everyone between 35 and 39 years of age is grouped into the bin labeled 35–39. Note that everyone falls into *some* age bin; that is, there are no gaps between the bins. This is usually the case with histograms. As a result, the bars in a histogram usually touch.

EXAMPLE 3   *Constructing a Histogram*

Make a histogram to represent the following results from an exam.

Score	Number of Students	Score	Number of Students
90–100	12	50–59	7
80–89	20	40–49	1
70–79	45	30–39	2
60–69	19	< 30	0

Solution: The largest data value is 45 students, for scores between 70–79. The graph will be easy to draw if we choose 50 as the total height; then, by making the graph 5 cm tall, each 1 cm of height represents 10 students. The result is shown in Figure 2.8 on the following page.

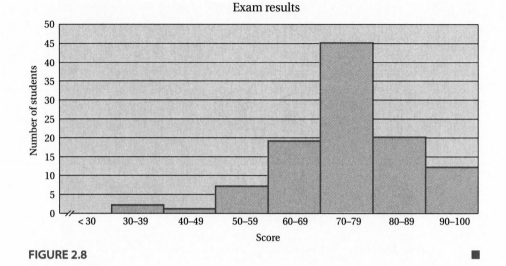

FIGURE 2.8   ■

LINE CHARTS

A **line chart** serves the same purposes as a histogram: it shows the number of people or objects that appear in a set of numerical categories. However, instead of using bars to represent data values, a line chart connects a series of dots. The height of each dot represents the corresponding data value.

Figure 2.9 shows the U.S. population data of Figure 2.7 converted into a line chart; for comparison, the histogram also is shown with dotted lines. Note that the horizontal scale is now labeled to show a continuous range of numbers, rather than a series of distinct data bins. This can give a somewhat misleading impression, unless you are careful in interpreting the line chart. For example, the dot representing those 25–29 is located at a position of 27.5 on the horizontal axis—the center of the bin for ages 25–29. Therefore, if you didn't know better, you might think that it means there are 19 million people of age 27.5, rather than 19 million people in the age range of 25–29.

EXAMPLE 4   *Constructing a Line Chart*

Make a line chart from the data in Example 3.

Solution: Because we are using the same data, the axes and labels of the line chart are the same as those for the histogram in Figure 2.8. But instead of drawing a bar for each data value, we place a point at the appropriate height. For example, 12 students scored between

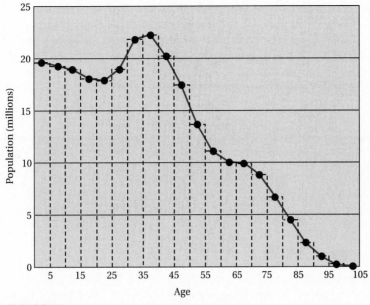

FIGURE 2.9

90 and 100, so we put a dot at a height of 12 on the vertical axis and centered between 90 and 100 on the horizontal axis. The final result is shown in Figure 2.10. For comparison, the histogram from Figure 2.8 is shown with dotted lines.

Grade distribution

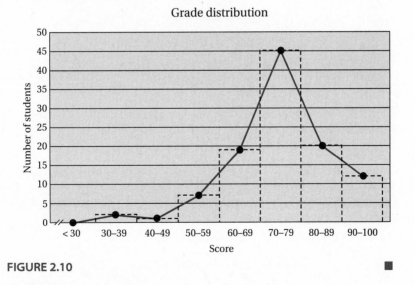

FIGURE 2.10   ■

TIME-SERIES DIAGRAMS

A particularly common use of histograms and line charts is to show how a quantity varies with time. Such a graph, in which the horizontal axis represents *time*, is called a **time-series diagram**. For example, Figure 2.11 on the following page shows relative values of

By the Way ....
Standard and Poor is a firm established in 1941 by Henry V. Poor and John Moody, two New York financial analysts. Its listing of 500 stocks is often used to track movements in the stock market.

stock, bond, and gold prices over a 12-week period. The stock prices shown are an average for the 500 stocks tracked by the Standard and Poor's 500 stock list (S.&P. 500). If you wanted to invest money that would reflect this average, you could purchase shares in a *stock index fund* that spreads its investments over the S.&P. 500 stock list. Similarly, the bond prices shown are an average for many bonds, as tracked by the Lehman Treasury Bond Index. Gold prices are from New York traders.

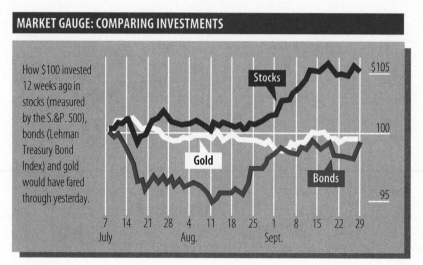

FIGURE 2.11
The New York Times, 9/30/95. Copyright © 1995 by The New York Times Company. Reprinted with permission.

EXAMPLE 5   *Reading the Investment Graph*

Study Figure 2.11. Suppose that, on July 7, you had invested $100 each in a stock index fund that tracks the S&P 500, a bond fund that follows the Lehman Index, and gold. If you sold all three funds on August 4, how much did you gain or lose?

Solution: The graph shows that the $100 in the stock fund would have been worth about $101 on August 4. The $100 bond investment would have declined in value to about $96. The gold investment would have held its initial value of $100. Thus on August 4 your complete portfolio would have been worth

$$\$101 + \$96 + 100 = \$297.$$

Thus you would have suffered a loss of $3 on your total investment of $300.   ∎

SCATTERPLOTS AND CORRELATION

We often measure *two* quantities for each object or person in a sample. For example, we might record the height and weight of every student in a class. Or a wildlife biologist might collect fish, then record their length and weight. When data are collected in this manner, the easiest way to display them is with a **scatterplot**. As an example, consider the following table showing rainfall and snowfall data for 15 cities in the United States.

City	Rainfall (in.)	Snowfall (in.)
Albany	36	65
Boston	44	41
Casper	11	80
Cleveland	35	54
Denver	15	60
El Paso	8	5
Fairbanks	10	67
Honolulu	23	0
Juneau	53	102
Miami	58	0
Nashville	48	11
Omaha	30	31
Reno	7	25
Seattle	39	13
Tucson	11	1

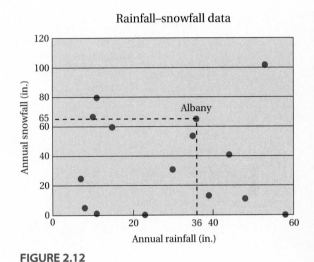

FIGURE 2.12

We can make a scatterplot for these data as follows. We draw a *horizontal axis* to represent annual rainfall. Because the greatest rainfall value is 58 inches for Miami, a scale going from 0 to 60 inches will fit all the data. We let the *vertical axis* represent annual snowfall, so a scale from 0 to 120 inches will fit all the data. (Note that 0 to 110 inches also would work, but using 0 to 120 allows us to place the tick marks every 20 inches, rather than every 10 inches.) The decision to put rainfall on the horizontal axis and snowfall on the vertical axis is arbitrary—the graph would work equally well the other way around.

We then represent each city with a single dot placed appropriately on the graph. For example, the dot for Albany is located to represent 36 inches of rainfall on the horizontal axis and 65 inches of snowfall on the vertical axis. Proceeding similarly for all the other cities, we get the scatter plot shown in Figure 2.12 above on the right.

> **TIME-OUT TO THINK:** Label each of the dots in Figure 2.12 with the first letter of the city to which it corresponds.

Scatterplots can be used to evaluate the question of whether two quantities are related. For example, does a high rainfall always imply a high snowfall? If there is a clear relationship, we say the quantities are **correlated**. We will discuss the concept of correlation in more detail in Chapter 9, but for now we can answer the question of correlation visually. The dots in Figure 2.12 look truly scattered. Some cities have high annual rainfall and low annual snowfall, while others have both high rainfall and high snowfall. Based on these data, there appears to be little if any correlation between the amount of rainfall a city receives and its annual snowfall.

EXAMPLE 6   *Life Expectancy and Infant Mortality*

The following table shows the life expectancy and the infant death rate for 16 countries (1995 data). Note that life expectancy is measured in years—the number of years a person can be expected to live. Infant death rate is measured as the number of infants that die for each 1000 infants born alive. Make a scatter plot of the data, and discuss whether life expectancy and the infant death rate are correlated.

Country	Life Expectancy (years)	Infant Death Rate (per 1000)	Country	Life Expectancy (years)	Infant Death Rate (per 1000)
Canada	79	6.1	India	60	71
Mexico	74	25	Pakistan	58	97
Brazil	62	55	Kenya	56	55
Czech	78	8.4	Australia	79	5.5
Greece	78	7.4	Egypt	61	73
Russia	63	25	Israel	78	8.5
Guatemala	65	51	S. Korea	73	8.2
Bangladesh	56	102	Peru	69	52

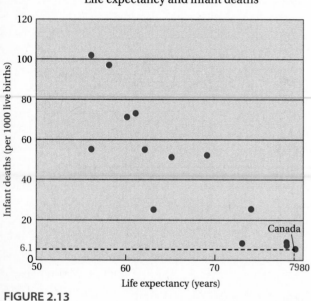

Life expectancy and infant deaths

FIGURE 2.13

Solution: We set up the graph by arbitrarily choosing to put life expectancy on the horizontal axis and the infant death rate on the vertical axis. The life expectancies range from 56 to 79 years. Thus we can run the horizontal scale from 50 to 80 years, which makes it easier to see individual data points than if it ran from, say, 0 to 80. To include all the data values for the infant death rate, we run the vertical scale from 0 to 120 deaths per 1000 live births. Then we simply plot the data points. For example, the dot for Canada is in the lower right at 79 along the horizontal axis and 6.1 along the vertical axis. The final result is shown in Figure 2.13.

This scatterplot shows a general downward trend among the data points: although there are a few exceptions, the countries with higher life expectancies tend to have a lower infant death rate. We therefore say that there is a correlation between the infant death rate and life expectancy. This correlation makes sense because both a higher life expectancy and a lower infant death rate result from better medical care. ▪

GRAPHS AND COMPUTERS

The principles behind the graphs discussed in this unit are all fairly simple. The hard part is to make the graphs carefully with paper, pencil, and ruler, plus compass and protractor in the case of pie charts. Fortunately, computers make it very easy to create graphs. For example, every graph you see in this book was generated with a computer! Many different types of software allow you to make graphs of data sets. Spreadsheets and presentation programs almost always include graphing capabilities. There also are many more specialized programs for graphing and statistics that you can use to generate graphs. Each package works a bit differently, but all follow the same basic principles.

- First, you must create a table of the data you wish to graph.
- Next, you instruct the computer program to create a graph of the type you want.
- Finally, you make adjustments to make the graph look good and add labels to explain what the graph shows.

REVIEW QUESTIONS

1. Describe the design and construction of each of the following types of graphs: bar graph, pie chart, histogram, line chart, time-series diagram, and scatterplot.

2. Describe how you can use a protractor to help you make a *pie chart*. When you are given a set of data, how do you calculate the angles to use for each wedge?

3. How is a *histogram* related to a *bar graph*? When can you use a histogram, and when must you use an ordinary bar graph?

4. What do we mean by a *bin* when making a histogram or line chart?

5. How can you make a *line chart* from a *histogram*? How can the line chart be misleading when using binned data?

6. What is a *scatterplot*? How can a scatterplot reveal a *correlation*?

PROBLEMS

1. **Bar Graph Analysis.** Consider the bar graph on immigration in Figure 2.2.

   a. Approximately how many more Asians than Africans immigrated to the United States in 1995?

   b. During 1995, 54,494 people immigrated to the U.S. from the former Soviet Union, and 45,666 people immigrated from South America. If we added bars for these two regions in Figure 2.2, how tall would each be?

2. **DWI Repeat Offenders.** A 1997 study by the National Highway Traffic Safety Administration was conducted to determine how to reduce the number of repeat offenders in DWI (driving while intoxicated) cases. In states where good records were available, the following percentages of repeat offenders were found.

Iowa: 21%	Nebraska: 26%
North Carolina: 32%	California: 34%
Louisiana: 24%	Wisconsin: 31%
Ohio: 33%	New Mexico: 47%

   a. Display these data graphically using a bar graph.

   b. The study carried the disclaimer "These states may not necessarily be representative of all 50 states in the United States." Explain why other states might differ from the states in the study and why this disclaimer was included.

3. **Income by Education.** A 1993 survey of Americans showed the following average incomes according to level of education. Make a bar graph to display this information. Comment on how this information might have been collected.

No high school diploma	$9000
High school graduate	$14,500
Bachelor's degree	$27,000
Master's degree	$35,000
Professional degree	$56,000
Doctoral degree	$49,000

4. **U.S. Age Groups.** Consider the graph in Figure 2.7.

   a. Approximately what is the population shown for those 30–34 years old?

   b. Approximately what is the population shown for those 65–69 years old?

   c. What is the most populous age group?

   d. How do you account for the "bulge" in the age groups between 30 and 50 years of age?

   e. Could you estimate the total population of the United States in 1995 from this graph? If so, how?

5. **Pie Chart for Language.** According to the U.S. Census Bureau (1993), 31.2 million people spoke a language other than English at home. The leading languages other than English and the number of speakers (in millions) were as follows:

Spanish 17.3	French 1.7	German 1.5	Italian 1.3	Chinese 1.2

   a. Because these figures do not total 31.2 million people, there must be a category for "All Other Languages." How many people are in this category?

   b. Convert the data, including the "all other" category from part (a), to percentages of the total number of people speaking languages other than English at home. If you used these percentages to make a pie chart, what angle would each wedge have?

   c. Make a pie chart that shows this language distribution.

6. **Pie Chart for Religions.** The 1992 National Survey of Religion and Politics revealed the following distribution of religious affiliations of Americans.

Evangelical Protestants: 25.9%	Other Christians: 3.3%
Mainline Protestants: 18.0%	Jews: 2.0%
Black Protestants: 7.8%	Others: 1.1%
Roman Catholics: 23.4%	Nonreligious: 18.5%

a. Calculate the angles that each category should have in a pie chart.

b. Make a pie chart that shows this religious distribution.

7. Women Workers. The following data show the percentage of women workers who had year-round, full-time jobs in periods since 1970. Note that each data value is given in a bin representing the average percentage over a 5-year period.

Years	Percentage Full-time	Years	Percentage Full-time
1970–1974	40%	1985–1989	50%
1975–1979	42%	1990–1995	52%
1980–1984	47%		

a. Make a histogram of these data.

b. Superimposed on the histogram of part (a), draw a line chart.

c. In words, describe the trend shown in your graphs. Do you think the trend will continue? Why or why not?

8. Home Run Leaders. The following table gives the most home runs hit by a single player in every season between 1986 and 1997 for major league baseball. Draw a histogram for this set of data.

Year	Home Runs	Year	Home Runs
1986	40	1992	43
1987	49	1993	46
1988	42	1994	43
1989	47	1995	50
1990	51	1996	52
1991	44	1997	58

9. Voter Turnout. The figure in the next column shows voter turnout and unemployment in the years of national elections between 1964 and 1996. Voter turnout is the percentage of eligible voters who actually voted.

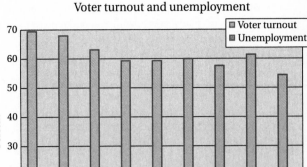

Voter turnout and unemployment

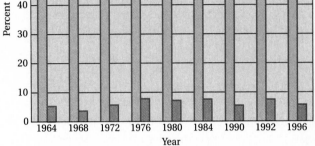

a. In what year was voter turnout the lowest?

b. In what years does unemployment appear to be the highest?

c. Does there appear to be a correlation between voter turnout and unemployment? Explain.

d. Use the data in the above graph to make a line chart (time-series) for voter turnout and unemployment separately.

10. Nobel Prize Statistics. The following chart shows the number of American Nobel Prize recipients and the value of the prize between 1986 and 1996.

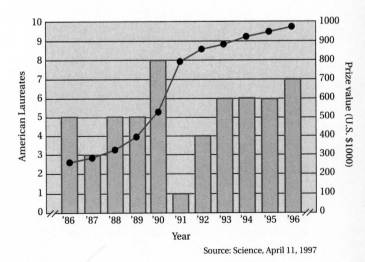

Source: Science, April 11, 1997

a. To which graph (bars or line chart) do the left and right vertical scales refer? What does 900 on the right vertical scale mean in actual dollars?

b. How many American laureates were there in 1990. What was the prize value in 1990?

c. In what year were there the fewest laureates? In what year did the prize value have the largest increase? Is there a connection between these two observations?

11. Time-Series Business Data. Find an example of a time-series diagram in the business section of a newspaper from this week. Attach a copy of the graph and cite its source. Briefly explain the quantity shown in the graph, then interpret any trends shown by the graph.

12. Scatterplot. The following table lists the heights and weights of players on a college basketball team. Make a scatterplot with height (inches) on the horizontal axis and weight (pounds) on the vertical axis. Does there appear to be a strong correlation? Explain.

Height	Weight	Height	Weight
6 ft. 2 in.	175	6 ft. 2 in.	202
6 ft. 7 in.	255	5 ft. 11 in.	170
5 ft. 9 in.	155	6 ft. 1 in.	181
6 ft. 4 in.	188	6 ft. 6 in.	215
6 ft. 1 in.	190	6 ft. 7 in.	230

Correlations. *For the pairs of quantities listed below, state whether you expect a scatterplot to show a strong correlation. Explain your reasoning.*

13. *Age* and *height* of 50 adults.

14. *Maximum annual temperature* and *population* of 50 cities of the world.

15. *Average number of children per family* and *life expectancy* in 50 countries of the world.

16. *Latitude* and *annual snowfall* for 50 cities of the world.

17. Choosing the Best Display. The following chart shows the percentage of web sites on the Internet in various languages (1997). Choose the graphing method from this unit that you think would be most effective for displaying these data. Explain your choice and draw the graph.

English: 82.3%	German: 4.0%	Japanese: 1.6%
Italian: 0.8%	Portuguese: 0.7%	Swedish: 0.6%
French: 1.5%	Spanish: 1.1%	
Dutch: 0.4%	Norwegian: 0.3%	

18. Displaying the World Population. The following data show regional populations in 1996; data are in millions of people. Choose the graphing method from this unit that you think would be most effective for displaying these data. Explain your choice and draw the graph.

North America	295
Latin America	488
Europe	507
Former USSR	293
Asia	3428
Africa	731
Oceania	29

19. Project: Height–Weight Survey in Class. Carry out a statistical study of your class to determine how well height and weight are correlated. (If you prefer, you may study height and shoe size instead.) Make a scatterplot of your data, and write a brief summary of your study and your conclusions. Be sure to describe how you collected your data, and how you decided whether there is a correlation. Do you think that your results are representative of the population of all human beings? Why or why not?

20. Project: Blood Types and Personal Traits. Some studies have found human blood type (i.e., types A, B, AB, and O) is correlated with certain diseases and conditions. Choose a human trait or condition that is easy to measure or detect (e.g., hair color, allergies, left-handedness, tall/short). Then design a statistical study to search for a link between blood type and that trait. Choose an appropriate sample, and carry out your study; you will have to ask people their blood type, and discard data from people who don't know it. Make an appropriate graph of your results, and write a report summarizing your findings. Could your findings be biased as a result of having to throw out data from people who don't know their blood type? Explain.

21. **Project: A Correlation Between Study Habits and Grades?** Describe how you would conduct a study to determine the extent to which study habits affect grades in mathematics classes. Be sure to explain what quanti-

ties you would measure, how you would measure them, and why you chose them. If possible, conduct your study with students from your current class by surveying them about the relevant study habits and relating those to their grades on a recent homework assignment or exam.

GRAPHICS IN THE MEDIA

UNIT 2D

In the previous unit we studied the most basic types of statistical graphs. Now we are ready to explore some of the fancier graphics that show up regularly in news and research reports. We begin by exploring a few of the many types of graphics, and then examine some subtleties and pitfalls that can occur if graphs are not interpreted carefully.

MULTIPLE BAR GRAPHS

A **multiple bar graph** is a simple extension of a regular bar graph: it has two or more sets of bars, usually to facilitate comparisons. Figure 2.14 shows a multiple bar graph that compares the 1970 and 1996 populations of four regions of the world. The multiple bars provide an easy comparison of both the total populations and the *change* in the populations. For example, it is clear that the current population of Africa is more than double that of North America. We also see that Africa experienced the greatest population increase.

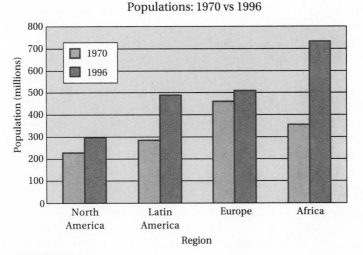

FIGURE 2.14

By the Way ············
Asia is not shown in Figure 2.14 because its bars are so large in comparison to the others: the population of Asia rose from 2.1 billion in 1970 to 3.4 billion in 1996!

EXAMPLE 1   *Grain Production and Consumption*

Net grain production is the difference (subtraction) between the amount of grain a country produces and the amount of grain its citizens consume. The multiple bar graph in Figure 2.15 shows the net grain production of the four most populous countries in the years 1990 (real data) and 2030 (projections). Interpret the graph and comment.

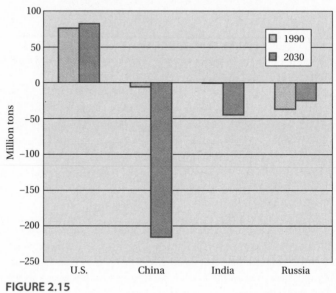

Net grain production: 1990 and 2030 (projected)

FIGURE 2.15

Solution: If the net grain production is positive, the country produced more grain than it consumed. In that case, the country can store the excess grain or export it to other countries. If the net grain production is negative, the country produced *less* grain than it consumed, and must import grain to make up the difference. If the net grain production is zero, the country is self-sufficient in grain, producing exactly as much grain as its citizens consume. Of the four countries shown, only the United States has a positive net grain production. India is self-sufficient, while Russia and China do not produce enough grain to feed their populations.

The graph also shows projected trends. The United States is expected to have more excess grain in 2030 than it did in 1990. Russia is expected to move closer to self-sufficiency by 2030. Both China and India are expected to require substantial grain imports by 2030 to feed their populations. Because China and India are the world's two most populous countries, this graph suggests that it may become increasingly difficult to feed the world's population.   ■

STACK PLOTS

The graph in Figure 2.16 is a **stack plot** that shows death rates from four diseases over the period 1900–1995. The death rates are measured as the number of people who died from a particular disease out of every 100,000 people. Each disease has its own region, or wedge,

Death rates for various diseases: 1900–1995

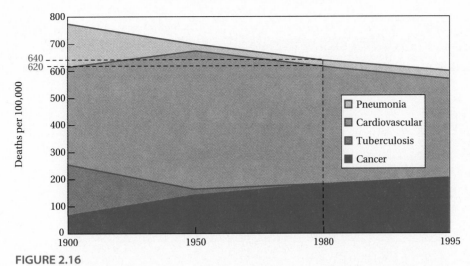

FIGURE 2.16

that can be shown either by shading or color-coding. The *thickness* of a wedge at a particular time tells you its value at that time. For example, for 1980, the wedge for pneumonia extends from about 620 to 640 on the vertical axis. Thus the thickness of this wedge for 1980 is 20, meaning that the death rate from pneumonia in 1980 was 20 deaths per 100,000 people.

Although it can be a bit tedious to determine the precise thickness of a wedge for a given year, the stack plot makes it easy to see long-term trends. For example, the dramatic decline in the thickness of the tuberculosis wedge shows that this disease was once a major killer, but that it has been nearly wiped out since about 1950. Meanwhile, we see that cancer has caused an increasing fraction of deaths over time since 1900. The top of the graph, which shows the total impact of all the wedges, shows another important trend: overall, the death rate from these four diseases has decreased substantially, from nearly 800 deaths per 100,000 in 1900 to about 600 in 1995.

EXAMPLE 2   *Interpreting a Stack Plot*

Figure 2.17 on the next page shows the changes in major spending categories of the federal budget from 1960 through 1999. The chart was created in 1995, so the figures for 1996 through 1999 are projections. Note that the *net interest* category represents interest payments on the national debt, and the *all other* category includes spending on such things as education, environmental cleanup, and scientific research. Interpret the stack plot and discuss some of the trends it reveals.

Solution: This stack plot shows only *percentages* of the total budget. Thus its total height is always 100%, and we cannot directly determine the actual amount of money spent by the government. However, it is easy to see long-term trends in how the government spends its money. For example, the proportion of the federal budget going to *national defense* has fallen substantially: For 1960, this wedge extended from about 30% to 80% on the vertical

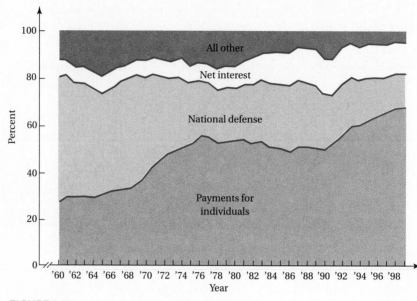

Percentage of composition of federal government outlays

FIGURE 2.17

scale, so national defense represented about 80% − 30% = 50% of the federal budget. For 1999, this wedge extends from about 67% to about 80%, so the proportion of the budget going to national defense had declined to about 80% − 67% = 13%. During the same period, the proportion of the budget going to *payments to individuals* (e.g., Social Security, Medicare, and welfare) rose dramatically, from about 27% to about 67%. The proportion spent on *net interest* more than doubled. Spending for the *all other* category had ups and downs as a proportion of the national budget, but the overall trend is clearly downward since about 1978. ■

> **TIME-OUT TO THINK:**   In a sense, Figure 2.17 represents changing priorities of the federal government. Do you think the changes in spending priorities make sense given changes in the challenges facing the United States? Why or why not? If it were up to you, how would you change the priorities?

THREE-DIMENSIONAL GRAPHICS

In all the graphs we have studied so far, data points or bars have represented only two related quantities. For example, Figure 2.7 shows population (one quantity) for various age groups (the second quantity). A different case arises when three (or more) quantities are related in some way. In that case, each data point must represent the values of all three quantities simultaneously. Such graphs require three axes, and hence are called **three-dimensional graphics**.

Figure 2.18 shows a three-dimensional graphic of data concerning the migration patterns of a bird called the *bobolink* over New York state. The vertical axis represents the

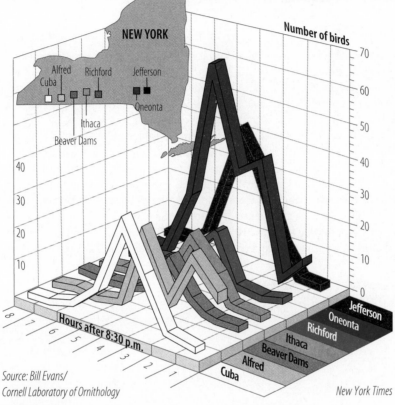

SONIC MAPPING TRACES BIRD MIGRATION

Sensors across New York State counted each occurrence of the nocturnal flight call of the bobolink to trace the fall migration on the night of Aug. 28 - 29, 1993. Computerized, the data showed the heaviest swath passing over the eastern part of the state.

Source: Bill Evans/
Cornell Laboratory of Ornithology

New York Times

FIGURE 2.18

The New York Times. 10/3/95. Copyright © 1995 by The New York Times Company. Reprinted with permission.

number of birds flying over a particular city at a particular time. The other two axes plot the time of night and the city for which the data were recorded. Note that the cities were selected to form an east-west line across New York state, representing a *sample* of all locations along that line. Thus each point on this graph represents *three* quantities: the *number* of birds, the *time* of night, and the *location* of the city.

EXAMPLE 3   *Interpreting the Three-Dimensional Graph*

Answer each of the following questions by studying Figure 2.18. At about what time did the sensors detect the most bird calls? Approximately how many birds passed over Oneonta at about 12:30 AM? Over what part of the state did most of the birds fly?

Solution: The number of birds detected in all the cities peaked between four and six hours after 8:30 PM, or between about 12:30 and 2:30 AM.

On the time axis, 12:30 AM is four hours after 8:30 PM. That time appears to align with the lower peak on the line for Oneonta. Reading the exact height of this peak is somewhat difficult, but an estimated 30–40 birds were flying over the city at that time.

Clearly, more birds flew over the two easternmost cities of Oneonta and Jefferson than cities farther west. Thus most of the birds were flying over the eastern part of the state.   ■

CONTOUR PLOTS

What if you want to show three-dimensional data without fancy three-dimensional graphics? Figure 2.19 shows one possibility, in which temperature over the United States is represented with **contour curves**. Temperature is one of the three dimensions shown on this graph, and the other two dimensions are geographic directions (north-south and east-west).

On such a **contour plot**, the quantity represented by the curves has the same value everywhere along each contour curve. The values are labeled by small numbers embedded within the curves. For example, Figure 2.19 shows that the temperature is 50° F everywhere along the contour labeled 50°, and it is 60°F everywhere along the contour labeled 60°. Between these two contours, the temperature varies between 50°F and 60°F. Note that the more closely contours are packed, the greater the change in temperature across a similar-sized geographical area. To make the graph easier to read, the regions between adjacent contours are often color-coded or shown in varying shades of gray.

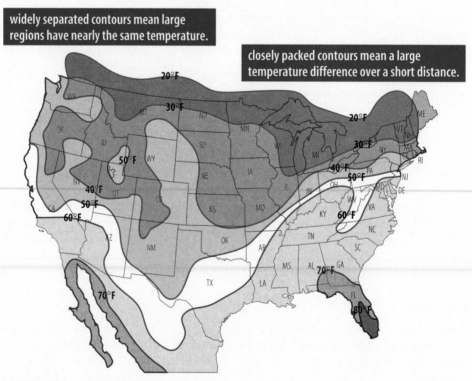

FIGURE 2.19

EXAMPLE 4   *A Contour Elevation Map*

Another common use of contour plots is showing geographical elevations. Figure 2.20 shows elevation contours around Boulder, Colorado. The elevation change is 100 feet from one contour to the next. Discuss a few of the key features shown on the map.

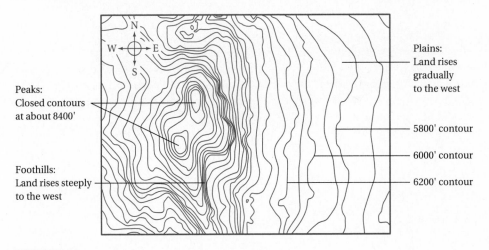

FIGURE 2.20

Solution: Note that the contours are widely spaced toward the east, indicating that elevations are fairly constant in this region. That is, the eastern part of the map shows nearly flat plains. Going westward, the contours suddenly become very close together near the center of the map. This represents the region where the mountains first rise up from the plains, and hence elevations increase dramatically within very short distances. Around the peaks to the west, the contours are nearly circular. Each successive circle is smaller and represents a higher altitude; the center of the circles represents the top of the mountain. ■

OTHER WAYS TO REPRESENT GEOGRAPHICAL DATA

Contour plots are only one of many possible ways to show data that varies over geographical regions. Another common method of representing geographical data is to use color-coding or shading to represent different data values. For example, Figure 2.21 on the following page shows how the mortality from *melanoma* (a form of skin cancer) varies on a county-by-county basis across the United States. The accompanying legend shows that the darker the shading in a county, the higher the mortality rate.

EXAMPLE 5   *Melanoma Mortality*

Discuss a few of the trends revealed in Figure 2.21. If you were researching skin cancer, which regions might warrant special study?

Solution: There appears to be a general trend of higher melanoma mortality in southern states and western states. This might be the result of people in these states spending more

By the Way ············
Malignant *melanoma* is the most dangerous type of skin cancer. It affects about 32,000 people a year in the U.S., resulting in 6,700 deaths (1996 data). Its incidence is increasing and researchers expect the number of cases to double in ten years.

Female melanoma mortality rates by county

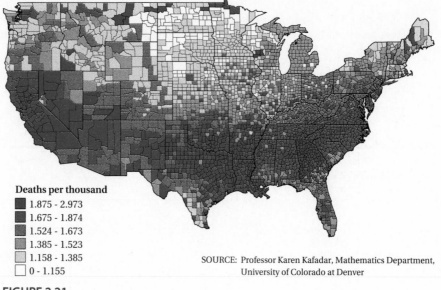

Deaths per thousand

- 1.875 - 2.973
- 1.675 - 1.874
- 1.524 - 1.673
- 1.385 - 1.523
- 1.158 - 1.385
- 0 - 1.155

SOURCE: Professor Karen Kafadar, Mathematics Department, University of Colorado at Denver

FIGURE 2.21

time outdoors, exposed to sunlight. As a researcher, you might be particularly interested in regions that deviate from general trends. For example, counties in northern Maine and Minnesota stand out with a very high melanoma mortality. You might first want to verify that the data are accurate, and not an error of some type. If they are accurate, you may want to find out why these counties have a higher melanoma mortality than surrounding counties.   ■

A FEW CAUTIONS ABOUT GRAPHICS

Most people who make displays of statistical data have good intentions and want to communicate information accurately. Nevertheless, graphics often involve subtleties that make them misleading unless you analyze them very carefully. In addition, attempts to make a graph visually appealing can introduce distortions. Before we leave the topic of graphics, let's investigate a few examples in which we must exercise special care when interpreting graphics.

Slicing the Pie Fairly

The pie chart in Figure 2.22(a) illustrates the age distribution of college students in 1996. According to the labels, the under-25 and the 25–35 age groups happen to both represent 41% of college students. However, because of the three-dimensional projection, the wedge for the 25–35 age group *appears* to be much larger than the wedge for the under-25 age group. In fact, if you measure the angles of these two wedges, you'll find that the under-25 wedge fills an angle of about 115°, while the 25–35 wedge fills an angle of more than 160°. If the pie chart is drawn flat, as in Figure 2.22(b), each 41% wedge fills an angle of 147.6°.

$$41\% \times 360° = 147.6°$$

Ages of college students

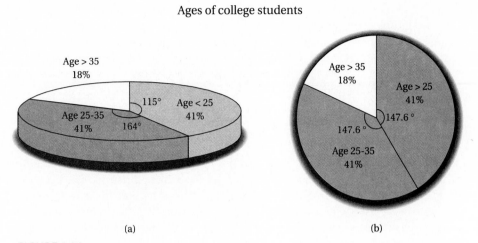

(a)   (b)

FIGURE 2.22

TIME-OUT TO THINK: As demonstrated by Figure 2.22, adding three-dimensional effects to pie charts always adds some visual distortion when you look at the pie chart on a flat page. Given this fact, why do you think pie charts are so often shown with 3-D effects? Do you think it is ever done with the aim of deliberately deceiving readers?

Watch the Scales!

Figure 2.23(a) shows how the fraction of women enrolled in colleges and universities changed between 1910 and 1990.

Percentage of all women enrolled in Higher Education

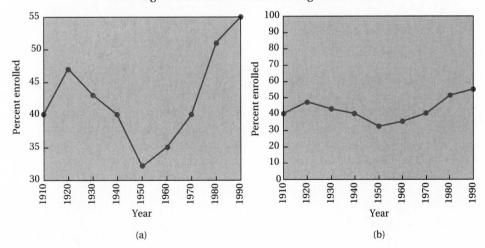

(a)   (b)

FIGURE 2.23

At first glance it appears that there was a huge increase between 1950 and 1990 in the fraction of women attending college—until you look carefully at the scale on the vertical axis.

The scale does not start at zero, as we usually expect. If we redraw the graph with the vertical axis starting from zero, and ending at 100%, the increase looks far less dramatic (Figure 2.23b). Not including the zero point on a scale is perfectly honest from a mathematical point of view, and it can make it easier to see trends in data points that are relatively close together. Nevertheless, it can be visually deceptive, and you must exercise great care in interpreting graphs in which the scales do not begin at zero.

Unadjusted Economic Data

You've probably noticed that prices tend to rise with time, an effect known as **inflation**. Fortunately, because wages also tend to rise with time, rising prices do not necessarily affect your standard of living. For example, suppose that, sometime in the past, the average price of a car was $1000 and average annual wages were $10,000. Sometime later, suppose that the average price of a car tripled to $3000, but average annual wages also tripled to $30,000. In this case we would say that the **real cost** of a car did not change.

The only way to measure changes in *real costs* over time is by first adjusting the prices of an item for the average effects of inflation. Economic data that is not adjusted for inflation can be very misleading. The black curve in Figure 2.24 shows the average price of gasoline in the United States over many decades without any adjustment for inflation. This curve gives the impression that gasoline is much more expensive today than it was a few decades ago. The colored curve shows the same data adjusted for inflation. Now we

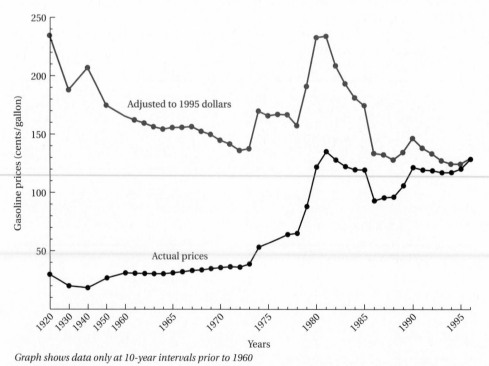

Graph shows data only at 10-year intervals prior to 1960

FIGURE 2.24

see that the *real cost* of gasoline in the mid-1990s was about as cheap as it has ever been! In general, you can tell whether economic data have been properly adjusted for inflation by looking for code words such as *adjusted to 1995 dollars*. If the data are not adjusted for inflation, they can be very difficult to interpret.

Percentage Change Graphs

Figure 2.25 is a multiple bar graph in which the three sets of bars represent the cost of public colleges, the cost of private colleges, and the consumer price index, which is a measure of the effects of inflation. However, this graph differs in a subtle, but important way from the graphs we have studied previously. Rather than showing actual data values, this graph shows *changes* in data values. In particular, each bar represents the *percentage change* in a data value from the previous year. For example, the first set of bars on the left shows that:

- The consumer price index increased by about 4% from 1986–1987.

- The average cost of a public college increased by about 6% from 1986–1987.

- The average cost of a private college increased by about 8% from 1986–1987.

By the Way ············

The *consumer price index* (CPI) represents an average of the prices of many goods purchased by typical consumers. Thus changes in the CPI can be used to estimate the level of inflation. We'll discuss the CPI further in Unit 4E.

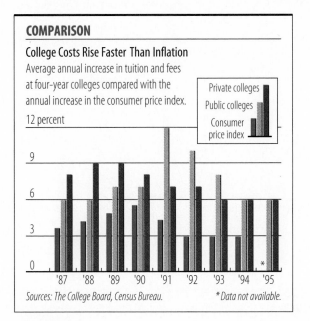

FIGURE 2.25
The New York Times, 9/29/95. Copyright © 1995 by The New York Times Company. Reprinted with permission.

This graph must be interpreted with special care. If you looked quickly at the graph, you might think that the shrinking bars for public colleges from 1991–1994 mean that costs were going down. However, the shrinking bars mean only that the annual *increases* in tuition declined. Because there still was an increase every year, the actual cost of tuition continued to rise. Moreover, because the increase in public college costs was greater than the increase due to inflation in every year (as measured by the consumer price index), the *real cost* of public colleges rose every year. Graphs that show percentage change are very common, particularly with economic data. Be sure you study them carefully, so that you are not accidentally misled.

Exponential Scales

Consider the growth in speed of the fastest computers. Around 1950, the fastest computers could perform about 10 operations (e.g., additions or multiplications) per second. Since that time, computer speed has increased *tenfold* approximately every five years, to around 100 billion operations per second today. Figure 2.26(a) shows the change in computer speed, but the dramatic rise makes it very difficult to see any detail. Figure 2.26(b) shows the same data displayed on a graph where each tick mark on the vertical scale represents a value ten times higher than the previous one. This type of scale is called an **exponential scale** because it grows by powers of 10 and powers of 10 are written with *exponents* (e.g., 10^3). In general, exponential scales are useful for displaying data that vary over a huge range of values. However, they must be interpreted carefully: if you didn't recognize the exponential scale, Figure 2.26(b) might make you think that computer speeds were rising only gradually. (The type of increase shown in these figures is called *exponential growth*, which is the topic of Chapter 7.)

By the Way ·············
Mechanical computers were invented in the 1800s, but the first electronic computer was built in 1946 at the University of Pennsylvania. Known as ENIAC, it filled a room and could do about 400 operations per minute. Today, some hand-held calculators are far more powerful.

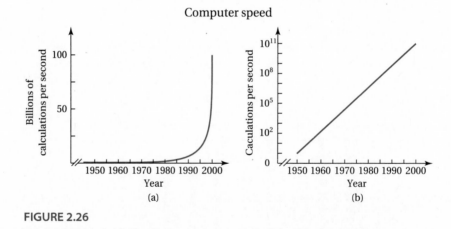

FIGURE 2.26

TIME-OUT TO THINK: Can you determine how much computer speed rose from 1950 to 1970 by looking at Figure 2.26(a)? How about by looking at Figure 2.26(b)? Why is the exponential scale so much easier to read in this case?

Pictographs

A **pictograph** is any graph that is embellished with additional art work. The art work may make the graph more appealing, but it can also distract or misinform the reader. Figure 2.27 shows the percent decline in the average price of stocks for the last six times that the stock market had a steep plunge prior to 1997. (Note that the figure states "the past *five*

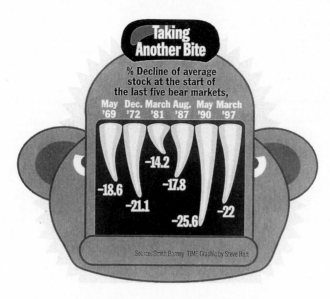

FIGURE 2.27
© 1997 Time Inc. Reprinted by permission

bear markets," which is a clear error.) The teeth of the bear represent the declines, rather like bars on a bar graph. However, these teeth are very difficult to interpret: no scale is given, and it's not clear whether it is the *length* of the curved teeth that corresponds to the amount that stock prices decreased, or something else, such as the height or even the area of the teeth. Fortunately, the actual declines are labeled. Pictographs are very common, but you'll have to study them carefully to extract the essential information and not be distracted by the cosmetic effects!

By the Way ············
When the price of stocks drops sharply for a long period, Wall Street people call it a *bear market*. When stocks are rising, it is called a *bull market*.

REVIEW QUESTIONS ●

1. What is a multiple bar graph? Give an example of a situation in which it is useful.

2. Describe the use and interpretation of a *stack plot*. What kind of information is difficult to read from a stack plot, and what kind stands out clearly?

3. Describe at least two ways of showing three-dimensional data. Give an example of each.

4. How do pie charts drawn with three-dimensional effects distort their data?

5. Why are graphs sometimes drawn with scales that do not begin at zero? How can such graphs be deceptive?

6. What is *inflation*? Why is it important in a graph that economic data be adjusted for the effects of inflation?

7. Explain how a graph that shows *percentage change* can show *descending* bars (or a descending line) even when the quantity of interest is *rising*.

8. What is an exponential scale? When is it useful to make a graph with an exponential scale? How can it be misleading?

9. What is a pictograph? Give an example of how a pictograph can enhance a graph, and also an example of how it can make a graph misleading.

PROBLEMS ●

1. **Net Grain Production.** Consider Figure 2.15.

   a. The graph shows net grain production. Based on this information, can you tell how much grain the United States produced and consumed in 1990? Why or why not?

   b. In 1990 the grain production for Brazil was 37 million tons and its grain consumption was 43 million tons. What was the net grain production for Brazil in 1990? How would this bar appear on Figure 2.15?

2. **Disease Stack Plot.** Study Figure 2.16.

   a. State whether the death rate for each of the four diseases decreased or increased individually between 1900 and 1995.

   b. When was the death rate due to cardiovascular diseases the greatest, and what was it?

   c. What was the death rate due to cancer in 1995?

   d. Based on the trends in the graph, which of these four diseases do you think will be responsible for the most deaths in 2050? Explain.

3. **Government Outlays Stack Plot.** Study Figure 2.17.

   a. Use a ruler to determine the percentage of the budget that went to net interest on the debt in 1960, 1970, 1980, 1990, and 1999 (projected).

   b. Use a ruler to determine the percentage of the budget that went to defense in 1960, 1970, 1980, 1990, and 1999 (projected).

   c. If the total budget in 1988 was $1.0 trillion, about how much was spent on each of the four categories?

   d. If the total budget in 1995 was $1.5 trillion, about how much was spent on each of the four categories?

4. **Interpreting a Stack Plot.** Use the stack plot of college degrees awarded to men and women shown at the top of the next column.

   a. Estimate the number of college degrees that were awarded to men and to women (separately) in 1930 and in 1990.

   b. Compare the number of degrees awarded to men and to women (separately) in 1980 and in 1990.

   c. During what decade did the *total* number of degrees awarded increase the most?

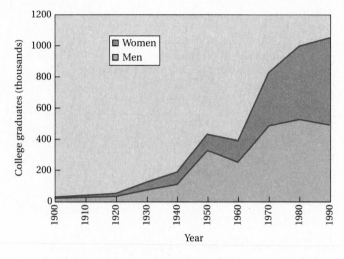

   d. Compare the *total* number of degrees awarded in 1950 to those awarded in 1990.

5. **Three-Dimensional Bird Graph.** Study Figure 2.18.

   a. What times of day correspond to the labels 1, 2, . . . , 8 along the left axis of the graph?

   b. Approximately how many birds passed over Richford at about 9:30 PM, 11:30 PM, and 1:30 AM?

   c. Approximately how many birds passed over Oneonta at about 9:30 PM, 11:30 PM, and 1:30 AM?

   d. Approximately how many birds passed over Cuba at about 9:30 PM, 11:30 PM, and 1:30 AM?

   e. Over which town did sensors detect the largest number of birds over the course of the entire night?

   f. Do you think this graph presents its data effectively? Explain.

6. **Three-Dimensional Data.** Consider the following two three-dimensional histograms on the next page. Both show the same data, but with the age groups in opposite orders.

   a. Why are two histograms needed to get a complete picture of the data?

   b. Estimate the percentage of those under 5 years in 1860 and in 1990.

   c. Estimate the percentage of those 20–44 years in 1860 and in 1990.

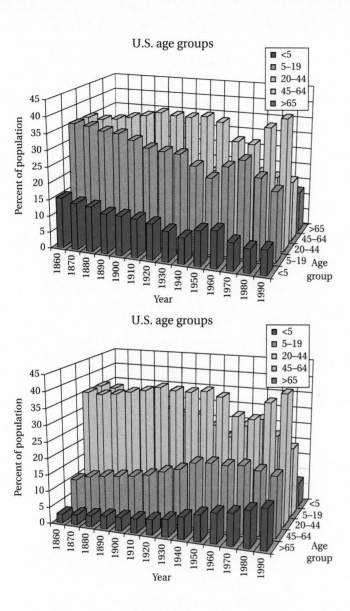

U.S. age groups

U.S. age groups

d. Estimate the percentage of those over 65 years in 1860 and in 1990.

e. Describe in general terms how the percentage of Americans in the over-65 age group has changed since 1860.

f. Describe in general terms how the percentage of Americans in the under-5 age group has changed since 1860.

g. What happened to the number of those under 5 years in 1950 and in 1960? Can you explain this change?

7. Tuition Increases. Study Figure 2.25.

   a. Did tuition in either category decrease between 1987 and 1995? Explain.

   b. In which year did average public college tuition rise the most compared to the consumer price index?

   c. In which year did average private college tuition rise the most compared to the consumer price index?

   d. In which year did average private college tuition rise the least compared to the consumer price index?

   e. Suppose that tuition at an average private college was $10,000 in 1990. Estimate the tuition in 1991.

   f. Suppose that tuition at an average public college was $4,000 in 1990. Estimate the tuition in 1991.

8. Age of Marriage. Consider the following table that shows the median age at first marriage for Americans between 1900 and 1994.

Year	Men	Women	Year	Men	Women
1900	25.9	21.9	1960	22.8	20.3
1910	25.1	21.6	1970	23.2	20.8
1920	24.6	21.2	1980	24.7	22.0
1930	24.3	21.3	1990	26.1	23.9
1940	24.3	21.5	1993	26.5	24.5
1950	22.8	20.3	1994	26.7	24.5

   a. Display the data as a multiple bar graph, with one set of bars for men and one for women.

   b. Describe any patterns or trends that you see in the data.

9. U.S. Suicide Rates. Examine the table below that gives suicide rates in the United States (measured in deaths per 100,000) for eight different age groups in 1970 and in 1991.

Age	1970	1991
5–14	0	1
15–24	9	13
25–34	14	15
35–44	17	14
45–54	20	15
55–64	21	15
65–74	20	17
75–84	20	23

a. Display these data in a multiple bar graph.

b. Describe any patterns or trends that you see in these data.

10. **Baseball Standings.** The 1994 baseball season ended on August 11 because of a players' strike. The following table shows the standings for the American League East at the end of the season. Using any of the graph types shown in Units 2C or 2D, how would you present these data as completely and as informatively as possible? Draw the graph that you think best represents these data, and explain why you chose this type of graph over other possibilities.

Team	Games Won	Games Lost
New York	70	43
Baltimore	63	49
Toronto	55	60
Boston	54	61
Detroit	53	62

11. **Distorted Pies.** The pie chart shown below displays the percentage of World War II casualties in the three branches of the military (Air Force was included in Army). The total number of casualties was 16,112,566. Comment on the usefulness of this graph. Can you tell how many casualties there were in each branch? What would you add to the graph to make it more informative?

World War II casualties

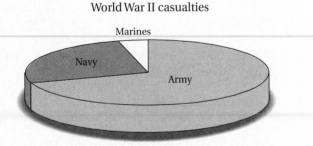

12. **Tumor Growth with an Exponential Scale.** It is not unusual for the cancer cells of a tumor to double their population on a scale of weeks or months. Suppose that the cell population in a tumor grows as shown in the table below. Make a graph of the cell population by using a uniform scale for months and an exponential scale for the cell count.

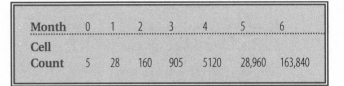

Month	0	1	2	3	4	5	6
Cell Count	5	28	160	905	5120	28,960	163,840

13. **Find a Rate of Change Graph.** Look through current newspapers and magazines to find an example of a graph that shows rates of change rather than absolute numbers. Attach a copy of the graph and cite its source.

a. Briefly explain the quantities shown in the graph. Why do you think the graph plotted rates of change rather than absolute numbers?

b. If someone mistakenly believed that the graph showed absolute numbers rather than rates of change, what incorrect conclusions might that person draw?

c. Briefly summarize the purpose of the graph.

14. **Find a Three-Dimensional Graph.** Look through current newspapers and magazines to find an example of a graph that shows three-dimensional data. (Do not use a graph in which the 3-D effects are purely cosmetic.) Attach a copy of the graph and cite its source. Briefly summarize the purpose of the graph.

15. **Find a Deceptive Graph.** Look through current newspapers and magazines to find an example of a statistical graph that either presents some subtleties or is deceptive to your eye. Attach a copy of the graph and cite its source.

a. Explain the subtlety or deception in the graph.

b. Either draw or explain how you would make the graph clearer.

16. **Find a Pictograph.** Look through current newspapers and magazines to find an example of a pictograph. Attach a copy of the graph and cite its source.

a. Are the pictorial features of the graph used to a clear advantage or do they confuse the graph?

b. Draw a simpler graph that presents the same data in a clearer way.

17. **Project: Graphing World Population Data.** Consider the data on world population growth in the table below for the years 1985 (real) and 2025 (projected). Suppose that you are asked to present and summarize these data for a conference. Choose a method to display these data as effectively as possible. Create your display, and explain why you chose this type of display.

	World	Africa	Asia	Americas	Europe	Oceania
Population 1985 (billions)	4840	560	2819	666	770	25
Population 2025 (billions)	8188	1495	4758	1035	863	36
Growth rate 1985 (%)	1.71	3.05	1.80	1.58	0.45	1.37
Growth rate 2025 (%)	0.94	1.74	0.89	0.72	0.15	0.59
Birth rate 1985 (per 1000)	26.9	45.0	27.4	23.4	14.7	19.6
Birth rate 2025 (per 1000)	17.6	24.1	17.0	15.3	13.0	15.0
Death rate 1985 (per 1000)	9.8	14.5	9.2	7.9	10.3	8.2
Death rate 2025 (per 1000)	8.2	6.7	8.1	8.2	11.5	9.1

CAUSAL CONNECTIONS

**UNIT
2E**

A major goal of many statistical studies is to determine whether one factor *causes* another. For example, does smoking cause lung cancer? Or does human use of fossil fuels threaten the environment?

Unfortunately, establishing **causality**—that one thing causes another—is extremely difficult. The search for causality usually begins with the identification of a correlation. As we saw in Unit 2C, correlations can often be revealed by displaying data on a scatter plot. Unfortunately, correlation does not necessarily imply causality. In general, a correlation may have at least three possible explanations.

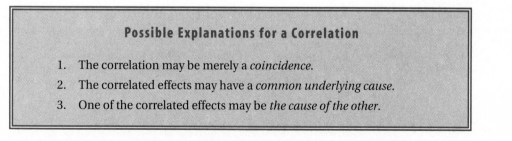

Possible Explanations for a Correlation

1. The correlation may be merely a *coincidence.*
2. The correlated effects may have a *common underlying cause.*
3. One of the correlated effects may be *the cause of the other.*

An example of *coincidence* is found in a reported correlation between the performance of the stock market (as measured by the Dow-Jones Industrial Average) and the winner of the Super Bowl. This correlation shows that when the stock market rises from November until Super Bowl Sunday in January, the winner of the Super Bowl usually is the

$\mathcal{T}$HINKING ABOUT...

Meaningless Correlations

In 1988, Oxford physician Richard Peto submitted a paper to the British medical journal *Lancet* showing that heart attack victims had a better chance of survival if they were given aspirin within a few hours after their heart attacks. The editors of *Lancet* asked Peto to break down the data into subsets, to see whether the benefits of the aspirin were different for different groups of patients. For example, was aspirin more effective for patients of a certain age or for patients with certain dietary habits?

Peto objected to the request, claiming that breaking the sample into too many subgroups would result in purely coincidental correlations. Writing about this story in the *Washington Post*, journalist Rick Weiss said, "When the editors insisted, Peto capitulated, but among other things he divided his patients by zodiac birth signs and demanded that his findings be included in the published paper. Today, like a warning sign to the statistically uninitiated, the wacky numbers are there for all to see: Aspirin is useless for Gemini and Libra heart-attack victims but is a lifesaver for people born under any other sign."

The moral of this story is that a "fishing expedition" for correlations usually produces some. That doesn't make the correlations meaningful, even though they may appear significant by standard statistical measures.

team whose city name comes later in the alphabet. This correlation successfully matched the Super Bowl result in all but 3 of the last 23 Super Bowl games through 1997. Nevertheless, it seems preposterous to believe that this is a real cause and effect relationship, and few people would bet on the Super Bowl by checking the stock market and the alphabet.

An example of *common underlying cause* is found in the correlation between infant death rates and longevity that we saw in Figure 2.13. Surely, the fact that infant death rates are low cannot itself be the cause of greater longevity. Instead, the correlation between low infant death rates and greater longevity probably has a common underlying cause: better health care.

The third possible explanation for a correlation is the most important one because discovering causality helps us understand the world around us. But it is also the most problematic: How can we establish that one event truly is the cause of another?

TYPES OF CAUSE

Before we can determine whether some event is *the* cause of another event, we must first ask whether it is *a* cause. Broadly speaking, we can divide causes into three types.

Types of Cause

- A cause is **necessary** if the effect *cannot* happen in its absence.
- A cause is **sufficient** if the effect *always* happens whenever the event occurs.
- A cause is **probabilistic** if it is neither necessary nor sufficient, but its presence increases the probability of the effect.

We can illustrate these concepts with some examples.

- Eating is *necessary* to survival. But it is not sufficient, because many other factors also are needed (such as air to breathe).

- Starvation is *sufficient* to cause death. But it is not *necessary* because there are many other ways to die.

- Smoking is a *probabilistic* cause for lung cancer because smokers have a higher probability than nonsmokers of contracting the disease. (It is not *necessary* because lung cancer can arise even in nonsmokers, nor is it *sufficient* because many smokers do not get lung cancer.)

ESTABLISHING CAUSALITY

There is no general procedure for establishing causality, and each case must be examined on its own merits. However, one useful set of guidelines gives some method to common-sense ideas about causality. The guidelines were developed by John Stuart Mill (1806–1873), and hence are known as **Mill's methods**.

The first method is called the **method of agreement**. Suppose that your stomach is frequently upset. The method of agreement suggests looking for a common factor among your cases of upset stomach. Perhaps you realize that an upset stomach always occurs on days that you drink coffee. The *agreement* is that every case of upset stomach is associated with consumption of coffee. This agreement doesn't *prove* that coffee is a causal factor, but it suggests that further investigation is warranted.

The **method of difference** looks for factors whose absence eliminates the effect. Perhaps there are some days when you drink coffee but do not get an upset stomach. What is the *difference* between your good days and your bad days? You might eventually realize that on the good days you drank decaffeinated coffee. This difference leads you to suspect that the causal factor is caffeine, rather than the coffee itself.

The **method of concomitant variation** looks for a *quantitative* relationship between the suspected causal factor and the effect. Suppose that when your caffeine consumption (the suspected causal factor) increases or decreases, the pain of your upset stomach (the effect) also increases or decreases. Then you have shown concomitant variation. The existence of concomitant variation provides strong evidence that the suspected casual factor is at least partially responsible for the observed effect.

The **method of residues** involves subtracting *known* effects in order to test what is left (the "residue") as a possible cause. The method of residues can be applied to the caffeine example only if you have prior knowledge of the precise effects of *everything* else that you eat (and if you have ruled out nonfood causes such as stress). If you discover that nothing you consume besides caffeine can be causing your upset stomach, then you'll know that caffeine must be the cause because it is the only possible cause that has not been eliminated.

By the Way ············
John Stuart Mill was a leading philosopher and economist. He studied Greek at the age of three and had read Plato by the age of ten. He was a prolific writer and an early advocate of women's suffrage (the right to vote). You can find more detail on Mill's methods in many logic textbooks.

> **TIME-OUT TO THINK:**   Suppose that you find yourself sneezing every time you go to a friend's house, and you suspect the cause may be an allergy to your friend's cat. Briefly describe how you could test your suspicion with each of the methods of agreement, difference, concomitant variation, and residues.

Going Astray

Suppose that you always put sugar in your coffee but don't add sugar to anything else that you eat. Further, suppose that your first guess about what caused your upset stomachs was *sugar*, rather than coffee. The method of agreement would seem to substantiate your incorrect guess, because sugar would always be present when you had an upset stomach. The method of difference would seem to lend further support: on days that you did not use sugar (no sugar means no coffee), your stomach would feel fine. Even the method of concomitant variation would support the mistaken belief that sugar is the causal factor: when you use more sugar (because you had more coffee) you are sicker. The lesson should be clear. If you use Mill's methods with a preconception about cause, or simply apply the methods blindly, you can be easily led astray.

A Physical Model

An alternative way to establish causality is by discovering *how* the suspected factor causes the observed effect. That is, we can seek a **physical model** to explain the cause and effect relationship. Caffeine generally is accepted as a probabilistic cause for upset stomachs because there *is* a physical model that explains the causal connection: Caffeine inhibits certain nerve signals in the human body, which can lead to secretions that irritate the stomach wall. Still, although this model shows that caffeine *can* cause an upset stomach, you cannot be certain that this model applies to *your* particular case.

CONFIDENCE IN CAUSALITY

It is very difficult, if not impossible, to prove causality beyond all doubt. Thus we are forced to think about our *level of confidence* in a causal link. Statisticians usually express levels of confidence quantitatively, by determining a margin of error that defines a 95% confidence interval (see Unit 2A). For our purposes, it is easier to work with three broad levels of confidence in causality that are commonly used in the legal system.

Legal Levels of Confidence in Causality

- **Possible Cause:** We have discovered a correlation, but cannot yet determine whether the correlation implies causality.

- **Probable Cause:** We have good reason to suspect that the correlation is causal, perhaps because one or more of Mill's methods suggests causality. Probable cause to suspect that a crime was committed is generally the legal standard for obtaining a warrant for a search, wiretap, or other method of seeking evidence.

- **Cause Beyond Reasonable Doubt:** We have found a model that is so successful in explaining the linkage between events that it seems unreasonable to doubt the causal connection. Whereas scientists seek a physical model to explain the action of the causal factor, in court the prosecution presents a theoretical model designed to show how the defendant committed the crime.

A FEW CASE STUDIES

The best way to learn about establishing causality is through case studies from real investigations. Each of the following case studies offers a slightly different twist on the issue of causality.

CASE STUDY 1   *Fluoride and Tooth Decay*

In the 1940s, researchers discovered that people in some cities averaged fewer tooth cavities than people in other cities. In other words, the number of cavities in different populations was *correlated* with the cities in which the populations lived. Further investigation of this correlation found that the common factor (using Mill's method of agreement) among the populations with fewer cavities was a high natural level of fluoride in the water supply. Scientists eventually concluded that fluoride is a causal factor for resistance to cavities. As a result, fluoride is now an ingredient in most dental products.   ■

---

*T*HINKING ABOUT …

Reasonable Doubt

The idea of using *beyond a reasonable doubt* to establish causality sounds good in principle, but can be very difficult in practice. Traditionally, judges instruct juries on how to interpret reasonable doubt. Until a 1994 Supreme Court decision, a common instruction involved the term *moral certainty*. This instruction dated from an 1850 decision that defined reasonable doubt as a mental state in which a juror "cannot say he or she feels an abiding conviction to a *moral certainty* of the truth of the charge." The author of the 1994 decision, Justice Sandra Day O'Connor, traced how the common meaning of the phrase moral certainty had changed since 1850 and concluded that the term today conveys a *lesser* degree of certainty than it did in 1850. The 1994 decision therefore warned against continued use of the phrase.

Although the Court agreed that the term *moral certainty* should no longer be used, the Justices were unable to agree on a replacement. Justice Ruth Bader Ginsburg endorsed the following set of jury instructions on reasonable doubt:

> *Proof beyond a reasonable doubt is proof that leaves you firmly convinced of the defendant's guilt. There are very few things in this world that we know with absolute certainty, and in criminal cases the law does not require proof that overcomes every possible doubt. If, based on your consideration of the evidence, you are firmly convinced that the defendant is guilty of the crime charged, you must find him guilty. If on the other hand, you think there is a real possibility that he is not guilty, you must give him the benefit of the doubt and find him not guilty.*

Justice Ginsburg stated that this instruction "surpasses others I have seen in stating the reasonable doubt standard succinctly and comprehensibly." Nevertheless, she pointed out that it still leaves some questions unanswered. For example, how firm is "firmly convinced"? What is a "real possibility that he is not guilty"? Like all questions of causality, determining guilt always involves at least some level of subjectivity.

CASE STUDY 2   *The Milwaukee Water Contamination Incident*

In 1993, thousands of people in Milwaukee mysteriously began to suffer from acute diarrhea. Researchers soon discovered that an infectious organism called *Cryptosporidium* had contaminated the water supply. That finding may be regarded as an example of the method of difference. The normal water supply didn't cause illness and the only known change in the water supply was the presence of *Cryptosporidium*. So the *Cryptosporidium* represents the difference believed to cause the illnesses.

Cryptosporidium can cause the symptoms observed during the outbreak, but it is not necessary; other organisms produce similar symptoms. Moreover, not everyone drinking the contaminated water got sick; thus the presence of *Cryptosporidium* was not sufficient to cause illness. Thus it was a *probabilistic* cause of illness during the outbreak. As a result, it was very difficult to blame *Cryptosporidium* for any *particular* case of illness, even though it seemed clearly to be responsible for many cases of illness.   ■

TIME-OUT TO THINK: Suppose that an individual died during the outbreak and an autopsy found that *Cryptosporidium* was present in the person's bloodstream. Could you then say that *Cryptosporidium* was the cause of death? Why or why not?

CASE STUDY 3   *What Caused the Challenger Explosion?*

In January 1986, the Space Shuttle Challenger exploded shortly after liftoff. Seven astronauts were killed, and the remaining space shuttle fleet was grounded for the next three years. Because the explosion was a single event, Mill's methods weren't applicable—they generally require contrasting situations. Hence a commission was established to determine the cause by developing a physical model. Their model indicated that a small hole had formed in one of the "O-rings" used to seal joints in the shuttle's solid rocket boosters. When the hole formed, a jet of hot material shot through the side of one booster. That, in turn, ignited the shuttle's large liquid fuel tank, causing the catastrophic explosion.

This physical model accounts for all known aspects of the Challenger explosion, and the O-ring failure was therefore considered to be the immediate cause of the accident. But was it the ultimate cause? Further study found a basic design flaw in the joints containing the O-rings; the design was changed after Challenger. Thus it might be argued that a faulty design caused the failure. However, some of the engineers involved in building the solid rocket boosters apparently knew about the problem and claimed that they warned of potential failure in cold weather. In that case, because the temperature was below freezing at the time of the accident, it might be argued that the failure was caused by a negligent launch decision. Another possibility was lack of communication between the engineers and the NASA team responsible for the launch decision, who may not have been aware of the engineers' concerns. In that case, blame might be assigned to a poor management structure or to the people who created the management structure. What would you say caused the Challenger accident? ■

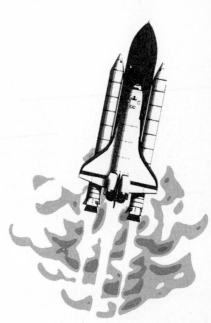

CASE STUDY 4   *The Ozone Hole*

Ozone is a gas that naturally exists high in the atmosphere, where it absorbs harmful ultraviolet radiation from the Sun. During the 1980s, researchers discovered a major reduction in the concentration of ozone each spring over Antarctica—the so-called **ozone hole** (Figure 2.28).

   A debate quickly arose over the cause of the ozone hole. Existing models of ozone chemistry suggested that human-made chemicals called **chlorofluorocarbons (CFCs)** can destroy ozone. These light gases can rise high into the atmosphere where ultraviolet light from the Sun breaks down their molecules. According to the models, the by-products of CFC breakdown can lead to ozone depletion. However, no existing model at the time could explain the enormous, localized depletion represented by the ozone hole.

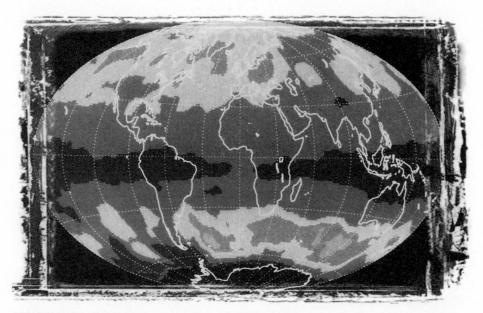

FIGURE 2.28   This NASA photograph shows the global depletion of ozone. The darker areas indicate low ozone levels, and the lighter areas indicate high ozone levels.

By the Way ....
CFCs have been used in many products including Styrofoam, air conditioners, refrigerators, and industrial solvents. Since the discovery of the ozone hole, industries have worked to replace CFCs with other chemicals. International treaties call for completely eliminating CFC production.

The first step in establishing causality involved the discovery of a *negative* correlation between the atmospheric concentration of the chemical *chlorine monoxide* and ozone: more chlorine monoxide meant less ozone in the ozone hole. Chlorine monoxide is a known by-product of CFC breakdown, but it also can come from other sources. However, the model that explains the breakdown of CFCs also predicts the presence of fluorine products. Finding these fluorine products added strong evidence that CFCs are a causal factor in the ozone hole.

However, because CFCs are distributed globally throughout the atmosphere at all times of the year, they could not be the only causal factor; the ozone hole was found only in the Antarctic spring. Eventually, scientists developed a model to explain the existence of the ozone hole by invoking another causal factor: ice crystals that form in the swirling, springtime atmosphere over Antarctica. According to this model, CFCs are trapped on these ice crystals, which thereby facilitates the chemical reactions that destroy ozone. ■

REVIEW QUESTIONS

1. What are the three possible explanations for a correlation? Give an example of each.

2. What is the purpose of *Mill's methods*? Describe each method, and explain how it is used with an example.

3. Describe three legal levels of confidence in causality. Give an example of each.

4. What do we mean by a *physical model*, and why is having a physical model important to establishing causality?

5. For each of the case studies in this unit, describe how causality was established and the implications of determining causality.

PROBLEMS

Interpreting Correlations. *Each of Problems 1–11 describes a correlation. State whether you believe that the correlation occurs because of mere coincidence, because of an underlying cause for both phenomena, or because one is the cause of the other. Explain your answers. Propose a cause when you believe one is involved.*

1. During the 1980s, the crime rate in the United States rose at the same time that the number of people in prison increased. That is, the crime rate was correlated with the number of people sent to prison.

2. The longer Jane exercises vigorously, the more weight she loses. That is, weight loss is correlated with duration of exercise.

3. Over the past three decades, the number of freeways and freeway lanes in Los Angeles has grown, and traffic con-

gestion has worsened. That is, there is a correlation between increased availability of freeways and increased traffic congestion.

4. Astronomers have discovered that, outside the Milky Way's relatively small neighborhood of the universe, all galaxies are moving away from us. Moreover, the farther the galaxy, the faster it is moving away. That is, there is a correlation between the distance to a galaxy and the speed at which it is receding.

5. In one class, grades were found to be correlated with the age of the students' mothers. That is, the older the mother, the higher was the grade.

6. In some studies, the incidence of melanoma (the most dangerous form of skin cancer) increases with latitude. That is, skin cancer is correlated with living in the generally cooler climates of more northerly latitudes.

7. When gasoline prices rise, attendance at National Parks declines. That is, there is a correlation between gasoline price and the number of visitors to National Parks.

8. Sales of ice tea in a local restaurant are positively correlated with ticket sales at the local swimming pool.

9. When a ban on deer hunting was imposed in a mountainous region, the incidence of mountain lion sightings increased.

10. As the volume of water flowing through mountain streams increases, the sale of snow shovels in nearby communities decreases.

11. A grocer notices that if she raises the price of peaches, fewer peaches are sold in a single day.

Causal Conditions. *Each of Problems 12–25 lists a cause and an effect. State whether the cause is necessary, sufficient, or probabilistic. Explain your answers.*

12. Cause: poor grades. Effect: expelled from school.

13. Cause: getting caught for speeding. Effect: paying a fine.

14. Cause: a high fat diet. Effect: coronary disease.

15. Cause: being a famous movie star. Effect: getting the lead role in a movie whose script you love.

16. Cause: temperature above freezing. Effect: snow is melting.

17. Cause: it is winter in New York. Effect: it is cold.

18. Cause: air bags in cars. Effect: people surviving head-on collisions.

19. Cause: ozone depletion. Effect: increase in cases of skin cancer.

20. Cause: sexual intercourse with an AIDS-infected person. Effect: contraction of AIDS.

21. Cause: exposure to a friend with flu. Effect: getting flu.

22. Cause: suspect at the scene of the murder. Effect: suspect found guilty of murder.

23. Cause: rising interest rates. Effect: falling stock market.

24. Cause: always have trouble being on time. Effect: being unemployed.

25. Cause: poor mathematical skills. Effect: job applications for managerial positions are routinely rejected.

26. **Identifying Causes: Headaches.** Suppose that you are trying to identify the cause of late-afternoon headaches that plague you several days each week. For each of the following tests and observations, explain which of Mill's methods you used and what you concluded.

   a. The headaches occur only on days that you go to work.

   b. If you stop drinking Coke at lunch, the headaches persist.

   c. In the summer, the headaches occur less frequently if you open the windows of your office slightly. They occur even less often if you open the windows of your office fully.

   d. Having made all these observations, what reasonable conclusion can you reach about the cause of the headaches?

27. **Physical Models.** Each of the following statements describes a generally accepted causal connection. Suggest a physical model that explains each connection.

   a. Running out of gas causes a car to stop.

   b. Being unable to breathe for half an hour causes death.

   c. Dropping a book causes it to fall.

   d. Inflating a balloon and then letting it go (without tying the end) causes it to fly about the room.

   e. Lack of rain creates a forest fire hazard.

28. **Smoking and Lung Cancer.** There is a strong correlation between tobacco smoking and incidence of lung cancer, and most physicians believe that tobacco smoking causes lung cancer. Yet, not everyone who smokes gets lung cancer. Briefly describe how smoking could cause cancer when not all smokers get cancer.

29. Other Causal Factors for Lung Cancer. Several things besides smoking have been shown to be probabilistic causal factors in lung cancer. For example, exposure to asbestos and exposure to radon gas, both of which are found in many homes, can cause lung cancer. Suppose that you meet a person who lives in a home that has a high radon level and insulation that contains asbestos. The person tells you, "I smoke, too, because I figure I'm doomed to lung cancer anyway." What would you say in response?

30. Suing the Tobacco Companies. Many of the families of smokers who died of lung cancer have sued tobacco companies, claiming that actions of the tobacco companies were the cause of death. Despite the strong evidence showing that smoking is a probabilistic causal factor in lung cancer, few (if any) of the plaintiffs have won such cases. Why do you think that winning such a case is so difficult? What kinds of evidence might lead a jury to rule in favor of the plaintiffs? Would such a case be easier to win if it were made on behalf of a large group of individuals (such as airline attendants)?

31. Older Moms Live Longer. A study reported in *Nature*, September 1997, claims that women who give birth later in life tend to live longer. Of 78 women who were at least 100 years old at the time of the study, 19% had given birth after their fortieth birthday. Of 54 women who were 73 years old at the time of the study, only 5.5% had given birth after their fortieth birthday. A researcher stated that "if your reproductive system is aging slowly enough that you can have a child in your forties, it probably bodes well for the fact that the rest of you is aging slowly too."

  a. Was this an observational study or experiment?

  b. Does the study suggest that later child rearing *causes* longer lifetimes or does later child rearing reflect an underlying cause?

  c. Comment on how persuasive you find the conclusions of the report.

32. High Voltage Power Lines. Suppose that people living near a particular high-voltage power line have a higher incidence of cancer than people living farther from the power line. Can you conclude that the high-voltage power line is the cause of the elevated cancer rate? If not, what other explanations might there be for it? What other types of research would you like to see before you conclude that high-voltage power lines cause cancer?

33. Do Guns Cause a Higher Murder Rate? Those who favor gun control often point to a correlation between the availability of handguns and murder rates to support their position that gun control would save lives. Does this correlation, by itself, indicate that handgun availability causes a higher murder rate? Suggest some other factors that might support or weaken this conclusion.

34. Seasonal Affective Disorder. Some people tend to experience mood swings and a generally depressed and lethargic state at certain times of year, a malady known to psychiatrists as *seasonal affective disorder* (SAD). Medical researchers found that in the northern states (those above a certain latitude) 100 of 100,000 individuals suffered from SAD, whereas only 6 of 100,000 in the southern states suffered from this disorder. They concluded that lack of sunlight causes the disorder. What do you think of this conclusion? What further information might be needed before you would accept it?

35. The Cricket and the Scientist. Here is a macabre parable that carries a lesson about causality. A scientist studying the behavior of crickets commands "Jump!" Seeing the cricket jump, he notes in his book that "crickets with six legs can jump." The scientist then removes a leg of the cricket and commands "Jump!" Seeing the cricket jump, he notes that "crickets with only five legs can jump." The scientist then removes a leg of the cricket and commands "Jump!" Seeing the cricket jump, he notes that "crickets with only four legs can jump." The scientist continues the experiment until the cricket has no legs. The scientist commands the legless cricket to jump. Seeing the cricket remain still, the scientist notes that "crickets with no legs cannot hear." Comment on which of Mill's methods the scientist used in this experiment and whether his conclusion is well-founded.

36. Causal Connections in the News. Search newspapers from the past week and find at least three instances in which a causal connection is claimed. For each case, briefly explain the claimed causal link. Then, write a paragraph or more explaining whether you believe a cause and effect relationship is involved.

37. Marie Curie Discovers Radium. In 1898, a young physicist named Marie Curie began experimenting with the elements uranium and thorium to learn about their curious emissions. Curie measured the amount of energy emitted by lumps of pure uranium and thorium. She then tried larger "lumps" that contained the same amounts of uranium or thorium as in her earlier experiment but was mixed with other materials. She found that the added materials had no effect on the emitted energy. She did the experiment again, this time heating the uranium and thorium and then cooling them. She did many more variations of the experiment. Each time, regardless of the conditions, she found that the emitted energy was affected only by the amounts of uranium and thorium in the lumps of material. This led her to assert that the energy emitted by a substance is proportional only to the amount of uranium and thorium it contains.

Next, Curie tested a substance called *pitchblende*, a shiny, black-brown mineral. Knowing how much uranium and thorium were present in pitchblende, she predicted (based on her earlier results) how much energy it would emit. To her surprise, her measurements showed a much higher level of energy emission than she had expected. To explain this discrepancy, Curie concluded that the pitchblende must contain at least one more, previously unknown, energy-emitting element besides uranium and thorium. She eventually found that the pitchblende contained two previously unidentified elements. She named one of them *Polonium*, after her native Poland, and the other *Radium*, after the property radioactivity (a term she also coined).

a. Explain how Curie used the method of agreement to support her assertion that the energy emitted by a substance is proportional only to the amount of uranium or thorium that it contains.

b. Explain how the method of concomitant variations also helped support her assertion that the energy emitted by a substance is proportional only to the amount of uranium or thorium that it contains.

c. Suggest a simple experiment that would have represented the method of difference that might have further supported her assertion about the energy-emitting properties of uranium and thorium.

d. Explain how Curie used the method of residue in her work with the pitchblende to propose the existence of at least one new element.

e. Do you think that Curie's initial conclusion that the pitchblende contained unidentified elements was reasonable? What other explanations could there have been for the unexpectedly high energy emissions she measured?

38. **Project: Marie Curie.** Research Curie's career. In particular, discuss obstacles that she faced and how she overcame them as a woman in science at a time when women scientists were rare.

39. **Project: Causality and the Law.** Legal proceedings involve many examples of attempts to establish causal connections. Watch a set of legal proceedings for a couple of hours (either by going to court or on TV). Analyze what you observe. Describe any causal connections sought and how evidence for the causal connections is introduced and used.

CHAPTER 2 •

SUMMARY

*R*eports of statistical research appear frequently in the news, and results from statistical research can impact our lives. This chapter provided an overview of statistical data and research, with emphasis on its interpretation. As you read this text, keep in mind the following key ideas.

- Statistical research involves careful measurements on a relatively small *sample* from which we try to infer something about a larger *population*. The research will be valid only if the sample is truly representative of the population.

- Critical errors may occur at any stage of a statistical study and can invalidate the entire study. Before accepting any statistically based conclusion, you should ask several questions to ensure that the study was carried out properly and that its results are meaningful.

- One of the best ways to present statistical data is with graphs. Many different types of graphs can be used, and the best one for a given situation depends on what you wish to show.

- Graphics are great! They convey a lot of information in a small amount of space. But some graphics involve subtleties or distortions, and must be interpreted with great care.

- One of the main goals of statistical research is to establish that one event causes another. The search for causality usually begins with the discovery of a correlation. However, correlations do not necessarily imply causality; they may be coincidental or reflect a common underlying cause. Establishing genuine causality can be very difficult.

SUGGESTED READING

A Mathematician Reads the Newspaper, J. A. Paulos (New York: Basic Books, 1995).

All the Math That's Fit to Print: Articles from the Manchester Guardian, Keith Devlin (Washington, D.C.: Mathematical Association of America, 1994).

Envisioning Information, E. Tufte (Cheshire, Connecticut: Graphics Press, 1983).

How to Lie with Statistics, D. Huff (New York: W. W. Norton, 1954).

How to Tell the Liars from the Statisticians, R. Hooke (New York: Dekker, 1983).

"Is Using a Car Phone Like Driving Drunk" in the journal *Chance*, Vol. 10, No. 2, 1997.

Misused Statistics: Straight Talk for Twisted Numbers, A. Jaffe and H. Spirer (New York: Dekker, 1987).

Seeing Through Statistics, J. M. Utts (Belmont, California: Duxbury Press, 1996).

Statistics with a Sense of Humor, F. Pyrczak (Los Angeles: Fred Pyrczak Publisher, 1989).

Tainted Truth: The Manipulation of Fact in America, C. Crossen (New York: Simon and Schuster, 1994).

The Cartoon Guide to Statistics, L. Gonick and W. Smith (New York: Harper Collins, 1993).

The Visual Display of Quantitative Information, E. Tufte (Cheshire, Connecticut: Graphics Press, 1983).

Chapter 3
PROBLEM-SOLVING TOOLS

Nearly every important personal and social issue involves mathematical or quantitative information at some level. In the first two chapters we explored ideas and examples intended to help you think critically about such issues. In this chapter, we begin to apply critical thinking to the quantitative problems associated with these issues.

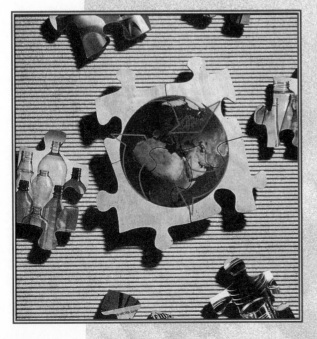

Today's world is more mathematical than yesterday's, and tomorrow's world will be more mathematical than today's. As computers increase in power, some parts of mathematics become less important while other parts become more important. While arithmetic proficiency may have been "good enough" for many in the middle of the century, anyone whose mathematical skills are limited to computation has little to offer today's society that is not done better by an inexpensive machine.

FROM *EVERYBODY COUNTS*

Nothing in life is to be feared. It is only to be understood.

MARIE CURIE

*P*roblem solving is more of an art than a science. No fixed rules are prescribed and no set of tools will always work. Problems may be practical or abstract, and may involve many different creative processes.

Consider the types of problems a plumber faces. Confronted with a leaky pipe, the plumber must determine the source and cause of the leak, and the best way to repair it. If the source is clear and the pipe accessible, the plumber might solve the problem quickly and easily. However, if the pipe is in a difficult location, the plumber may need to devise a unique method for repair. The process also may need to be revised if water still leaks after the first repair attempt.

The process of solving a *quantitative* problem also requires ingenuity, experience, and a bag of tools. Whereas a plumber may use mechanical tools like wrenches and saws, quantitative problem solving calls for mental tools such as organization, drawing pictures, adding, subtracting, solving equations, or using statistics.

The quantitative problems associated with modern issues may be as straightforward as calculating a sales tax or as complex as assessing the risks associated with global warming. Some problems require tools or experience that go beyond this book. But no matter the nature of the problems, experience with problem solving and the use of mathematical tools will help you make intelligent decisions on personal and social issues.

$\mathcal{T}$HINKING ABOUT …
Dreaded Word Problems

Many people look back with dread at the occasional attempts made to show them the practical utility of mathematics. You may still remember (or be unable to forget) "word problems" (or "story problems") like:

> *Bob is three years older than his sister Jill will be when she is twice the present age of her niece Sarah who was born when Bob's dad was 30. Bob's dad is five years older than his mom, who is now 29. How old was Bob's mom when he was born?*

(Although you might be tempted to answer "why not ask Bob's mom?" the daring among you can verify that the answer is 18.)

These problems probably seemed perverse at the time and, indeed, their relevance to daily life is hard to appreciate.

The irony in this state of affairs is striking. A common misconception is that mathematics has nothing to do with "real" issues. Surely, the *purpose* of word problems is to change that perception and to introduce some relevance into mathematics courses. But with rare exceptions, word problems have had the opposite effect. The great challenge to future teachers and textbook writers is to develop problems that will help students appreciate the deep connections between mathematics and real life, rather than inspiring dread. We hope that you will find the problems in *this* book to be more interesting and worthy of your time than those that led Gary Larson to draw the "Hell's Library" cartoon.

THE FAR SIDE By GARY LARSON

Hell's library

USING UNIT ANALYSIS

If we add five apples and three apples we get eight apples. But if we add five apples and three oranges, we get . . . five apples and three oranges! This simple illustration of the old saying that you can't compare apples and oranges points to one of the most powerful techniques for problem solving.

Numbers in real problems almost always represent a quantity of *something*, such as the number of apples in a basket, the length of a room in feet, or the distance in miles that a car travels in an hour. Terms that describe a quantity—*apples*, *feet*, or *miles*—are called its **units**. Problems that may seem difficult at first often are much easier to solve if you identify and keep track of the units.

IDENTIFYING AND WORKING WITH UNITS

Units come in two basic types. **Simple units** such as *apples*, *feet*, or *miles* involve a single concept. **Compound units** arise when simple units are divided, raised to powers, or multiplied.

Compound Units Formed by Division

Compound units formed by division usually can be identified by the key word *per*, which means *divided by*. (Recall that *percent* means "divided by 100.") When you work with such units in problems, you should always express the division of units as a fraction. For example:

- a speed of 65 miles per hour becomes $65\dfrac{\text{mi}}{\text{hr}}$ or $\dfrac{65\ \text{mi}}{1\ \text{hr}}$.

- a price of \$1.22 per pound becomes $1.22\dfrac{\$}{\text{lb}}$ or $\dfrac{\$1.22}{1\ \text{lb}}$.

- 12 inches in a foot means 12 inches *per* foot; it becomes $12\dfrac{\text{in.}}{\text{ft}}$ or $\dfrac{12\ \text{in.}}{1\ \text{ft}}$.

- traveling 100 kilometers in 2 hours means a speed of $50\dfrac{\text{km}}{\text{hr}}$.

Some units have standard abbreviations, such as "km" for kilometers, "in." for inches, "s" for seconds, or "lb" for pounds. For units with no standard abbreviations, you may make up your own abbreviations if it is helpful, but be sure that you define your abbreviations clearly.

Compound Units Formed by Powers: Area and Volume

The most common examples of units raised to powers occur with area and volume. **Area** describes the size of a surface such as a floor, a wall, or a piece of cloth. The most basic way to measure area is by counting squares of a known size. For example, if you use squares that are one centimeter on a side, the units of area will be *square centimeters*, abbreviated cm^2 (Figure 3.1).

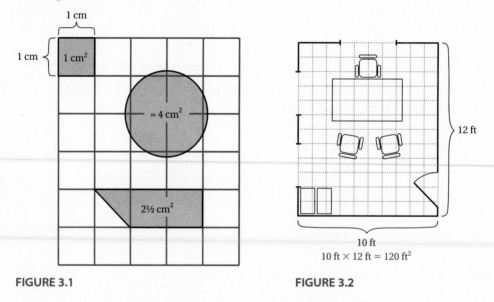

FIGURE 3.1   **FIGURE 3.2**

We can see why units of area involve a power by recalling that the area of a rectangle is the product of its length and width (Figure 3.2). For example, the area of a room that measures 10 feet by 12 feet is

$$10\ \text{ft} \times 12\ \text{ft} = (10 \times 12)\,(\text{ft} \times \text{ft}) = 120\ \text{ft}^2 = 120\ \text{square feet}.$$

Note that we work with units just as we do with numbers: *feet times feet* gives *feet*^{2}, or *feet to the second power*, showing that area involves the second power of a length unit.

Volume describes the capacity of any container or space. The most basic way to measure volume is by counting *cubes* of a known size (Figure 3.3). For example, the volume of a box measuring 6 inches by 4 inches by 10 inches is

$$4 \text{ in.} \times 6 \text{ in.} \times 10 \text{ in.}$$
$$= (4 \times 6 \times 10)\,(\text{in.} \times \text{in.} \times \text{in.})$$
$$= 240 \text{ in.}^3$$

Thus the units of this volume are *inches*^{3}, or *cubic inches*, showing that volume involves the third power of a length unit.

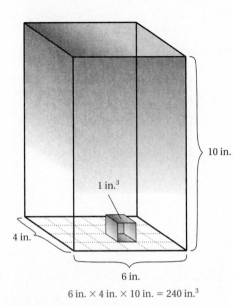

6 in. × 4 in. × 10 in. = 240 in.^{3}

FIGURE 3.3

Compound Units Formed by Multiplication

Another way to form compound units is to multiply two *different* units together. Multiplication of units occurs all the time in problems, but only occasionally do the units have a special name. In those cases, the multiplication is usually indicated by a hyphen. For example:

- Utility companies bill for electrical energy usage in units of kilowatts × hours, or kilowatt-hours.

- Auto mechanics measure the torque produced by an engine in units of feet × pounds, or foot-pounds.

- Water engineers measure the volume of water in a reservoir using units of acres × feet, or acre-feet.

EXAMPLE 1

You want to buy 30 acres of land that costs $12,000 per acre. How much will it cost?

Solution: It's clear from the question that the units of the final answer should be *dollars*. The "$12,000 per acre" means 12,000 $/acre. The cost of the land is

$$30 \text{ acres} \times 12,000\,\frac{\$}{\text{acre}} = \$360,000.$$

Note how the unit *acre(s)* "cancels" to yield the answer in units of *dollars*. The land will cost $360,000. ■

> ### Guidelines for Working with Units
>
> Before you begin any problem, think ahead and identify the units you expect for the final answer. Then operate on the units along with the numbers as you solve the problem. If the answer does not come out with the expected units, you've probably done something wrong! The following guidelines may be helpful in working with units.
>
> - You cannot add or subtract numbers unless they have the *same* units. For example, 5 apples + 3 apples = 8 apples, but the expression 5 apples + 3 oranges cannot be simplified further.
> - You *can* multiply units, divide units, or raise units to powers. Look for key words that tell you what to do. For example:
>   - *per* suggests division.
>   - *of* suggests multiplication.
>   - *square* suggests raising to the second power.
>   - *cube* suggests raising to the third power.
> - Mathematically, it doesn't matter whether a unit is singular (e.g., *foot*) or plural (e.g., *feet*); we can use the same abbreviation (e.g., *ft*) for both.
> - It's easier to work with units if you replace division with multiplication by the reciprocal. For example, instead of dividing by $60\,\dfrac{\text{s}}{\text{min}}$, multiply by $\dfrac{1\ \text{min}}{60\ \text{s}}$.

EXAMPLE 2

You need 10 square yards of cloth priced at $2 per square yard. How much will the cloth cost?

Solution: The problem asks for a cost, so we expect the answer to have units of dollars ($). The word *per* indicates division, so that $2 per square yard means $\dfrac{\$2}{1\ \text{yd}^2}$ or $2\,\dfrac{\$}{\text{yd}^2}$. Thus the cost is

$$10\ \text{yd}^2 \times 2\,\frac{\$}{\text{yd}^2} = 20\,\$ \text{ (usually written \$20)}.$$

Note how the unit yd^2 cancels to yield the answer in dollars. The 10 square yards of cloth will cost $20. ∎

EXAMPLE 3

Suppose that a car travels 25 miles every half-hour. How fast is it going?

Solution: The "how fast?" suggests that the final answer should have units of *miles per hour*. The problem can be rephrased to state that the car is covering 25 miles *per* half-hour, indicating division:

$$25\ \text{mi} \div \frac{1}{2}\text{hr} = 25\ \text{mi} \times \frac{2}{1\ \text{hr}} = 50\,\frac{\text{mi}}{\text{hr}}$$

3A   Using Unit Analysis   **133**

𝒜 BRIEF REVIEW

Reciprocals and Division

A number times its *reciprocal* is 1. For example:

3 hr and $\dfrac{1}{3\text{ hr}}$ are reciprocals because 3 hr $\times \dfrac{1}{3\text{ hr}} = 1$.

$\dfrac{2\text{ ft}}{1\text{ s}}$ and $\dfrac{1\text{ s}}{2\text{ ft}}$ are reciprocals because $\dfrac{2\text{ ft}}{1\text{ s}} \times \dfrac{1\text{ s}}{2\text{ ft}} = 1$.

$60\,\dfrac{\text{s}}{\text{min}}$ and $\dfrac{1\text{ min}}{60\text{ s}}$ are reciprocals because

$$60\,\frac{\text{s}}{\text{min}} \times \frac{1\text{ min}}{60\text{ s}} = 1.$$

In general:

- the reciprocal of the number a is $\dfrac{1}{a}$, and

- the reciprocal of $\dfrac{a}{b}$ is $\dfrac{b}{a}$ (a, b both nonzero).

The latter rule shows that you can find a number's reciprocal simply by inverting it; that is, by interchanging its numerator and denominator. (Recall that the *numerator* is the top of a fraction and the *denominator* is the bottom.)

Dividing by a number is equivalent to multiplying by the number's reciprocal; this process is sometimes referred to as *invert* (the fraction) *and multiply*. For example:

$$10 \div \frac{1}{2} = 10 \ \times \ \underbrace{\frac{2}{1}}_{\text{invert}} = 20$$

and multiply

Recall that dividing by a number is the same as multiplying by its reciprocal. Rewriting as multiplication makes it easier to see the final units. The car is traveling at a speed of 50 miles per hour. ∎

Units Can Help You Solve Problems

Consider the following problem: How much gas does a car need to go 90 miles if it gets 25 miles per gallon?

The problem asks "how much gas?" so the answer should have units of *gallons*. The only two quantities involved in the problem are the distance of 90 *miles* and the gas mileage of 25 *miles per gallon*. Thus the only way to end up with an answer in *gallons* is by dividing:

$$90\text{ mi} \div 25\,\frac{\text{mi}}{\text{gal}} = 90\text{ mi} \times \frac{1\text{ gal}}{25\text{ mi}} = 3.6\text{ gal}$$

Although that's all there is to it, we can demonstrate the power of unit analysis by imagining that we were completely lost on how to approach this problem. There are only four possible ways to combine (multiplicatively) the two given quantities of *90 miles* and *25 miles per gallon*. Let's look at the units we get with each of the four combinations shown on the next page.

Combination	*Units of Problem*	*Units of Answer*
$25\,\dfrac{\text{mi}}{\text{gal}} \div 90\text{ mi}$	$\dfrac{\text{mi}}{\text{gal}} \times \dfrac{1}{\text{mi}}$	$\dfrac{1}{\text{gal}}$
$25\,\dfrac{\text{mi}}{\text{gal}} \times 90\text{ mi}$	$\dfrac{\text{mi}}{\text{gal}} \times \text{mi}$	$\dfrac{\text{mi}^2}{\text{gal}}$
$90\text{ mi} \div 25\,\dfrac{\text{mi}}{\text{gal}}$	$\text{mi} \times \dfrac{\text{gal}}{\text{mi}}$	gal
$90\text{ mi} \times 25\,\dfrac{\text{mi}}{\text{gal}}$	$\text{mi} \times \dfrac{\text{mi}}{\text{gal}}$	$\dfrac{\text{mi}^2}{\text{gal}}$

Because we know that the answer must have units of *gallons,* the third combination must be the correct solution to the problem.

EXAMPLE 4

A jet traveling at an average speed of 830 kilometers per hour made a particular trip in 3.7 hours. How far did it travel?

Solution: The question "how far?" suggests an answer with units of *kilometers.* The problem involves two quantities: a speed of 830 *kilometers per hour* and a time of 3.7 *hours.* The only way to end up with an answer in *kilometers* is by multiplying km/hr by hr:

$$830\,\frac{\text{km}}{\text{hr}} \times 3.7\text{ hr} = 3071\text{ km}$$

The jet traveled about 3100 kilometers during its trip. ∎

EXAMPLE 5

You are a grader for a math course. An exam question reads: "Cheryl purchased 5 pounds of apples at a price of 50 cents per pound. How much did she pay for the apples?"

Suppose that a student has written the answer: "50 ÷ 5 = 10. She paid 10 cents." Write a note to the student explaining what went wrong.

Solution: Dear student: Does your answer make sense? If one pound of apples costs 50¢, how could five pounds cost only 10¢? You could have prevented your error by keeping track of the units. The number 50 has units of *cents per pound* and the number 5 has units of *pounds.* Thus your calculation of "50 ÷ 5 = 10" means

$$50\,\frac{\text{¢}}{\text{lb}} \div 5\text{ lb} = 50\,\frac{\text{¢}}{\text{lb}} \times \frac{1}{5\text{ lb}} = 10\,\frac{\text{¢}}{\text{lb}^2}.$$

This cannot be correct because the answer must have units of *cents,* not *cents per square pound.* The correct method yields units of cents as follows:

$$50\,\frac{\text{cents}}{\text{lb}} \times 5\text{ lb} = 250\text{ cents, or }\$2.50$$

The 5 pounds of apples cost $2.50. ∎

UNIT CONVERSIONS

Many problems involve converting one set of units to another. For example, you might want to convert a distance in kilometers to miles, or a measurement from quarts to cups.

The basic "trick" of unit conversions is to devise an appropriate way of multiplying by 1. For example, the following are all different ways of writing 1.

$$1 = \frac{1}{1} = \frac{8}{8} = \frac{\frac{1}{4}}{\frac{1}{4}} = \frac{\sqrt[7]{\pi}}{\sqrt[7]{\pi}} = \frac{1 \text{ kilogram}}{1 \text{ kilogram}} = \frac{12 \text{ inches}}{1 \text{ foot}}$$

This expression shows the necessity of stating units: $12 \div 1$ is *not* 1, but 12 inches $\div$ 1 foot *is* 1 because 12 inches and 1 foot are the same thing.

Conversion Factors

The statement 12 inches = 1 foot is an example of a **conversion factor**. We can write this conversion factor in three equivalent ways:

$$12 \text{ in.} = 1 \text{ ft} \quad \text{or} \quad \frac{12 \text{ in.}}{1 \text{ ft}} = 1 \quad \text{or} \quad \frac{1 \text{ ft}}{12 \text{ in.}} = 1$$

> **TIME-OUT TO THINK:** Translate the three forms of the conversion factor between feet and inches into words (e.g., 12 in./1 ft means 12 inches per foot). Do they all have the same meaning? Explain.

The key to unit conversions is to identify the needed conversion factor, and use it in the form that helps you change from one set of units to another.

EXAMPLE 6

Convert a distance of 7 feet into inches.

Solution: We begin with *feet* and want to end up with *inches*. We therefore need the conversion factor between *inches* and *feet* in a form that has *feet* in the denominator, so *feet* will cancel:

$$7 \text{ ft} \times \underbrace{\frac{12 \text{ in.}}{1 \text{ ft}}}_{1} = 84 \text{ in.}$$

Note that we really just multiplied by a form of 1 that allowed us to change the units but not the amount. Seven feet is the same as 84 inches. ∎

EXAMPLE 7

Convert a length of 102 inches to feet.

Solution: This time, we use the conversion factor in the form with *inches* in the denominator:

$$102 \text{ in.} \times \underbrace{\frac{1 \text{ ft}}{12 \text{ in.}}}_{1} = 8.5 \text{ ft}$$

Thus 102 inches is equivalent to 8.5 feet.   ∎

EXAMPLE 8   *Using a Chain of Conversions*

How many seconds are there in one day?

Solution: Most of us don't immediately know the answer to this question, but we do know that 1 day = 24 hr, 1 hr = 60 min, and 1 min = 60 s. We can answer the question by setting up a *chain* of unit conversions in which we start with *day* and end up with *seconds*.

$$1 \text{ day} \times \frac{24 \text{ hr}}{1 \text{ day}} \times \frac{60 \text{ min}}{1 \text{ hr}} \times \frac{60 \text{ s}}{1 \text{ min}} = 86{,}400 \text{ s}$$

By using the conversion factors needed to cancel the appropriate units, we are left with our answer in *seconds*. There are 86,400 seconds in one day.   ∎

> **Mathematical Note:**
> As always, conversions are just fancy ways to write the number one:
> $$\frac{24 \text{ hr}}{1 \text{ day}} = \frac{60 \text{ min}}{1 \text{ hr}}$$
> $$= \frac{60 \text{ s}}{1 \text{ min}} = 1$$

Conversions with Units Raised to Powers

Special care is required in converting units that are raised to a power. For example, suppose that we want to know the number of square feet in a square yard. We may not know the conversion factor between square yards (yd^2) and square feet (ft^2), but we know that 1 yard = 3 feet. We can find the conversion factor between square yards and square feet by writing:

$$1 \text{ yd}^2 = 1 \text{ yd} \times 1 \text{ yd} = 3 \text{ ft} \times 3 \text{ ft} = 9 \text{ ft}^2$$

That is,

$$1 \text{ yd}^2 = 9 \text{ ft}^2.$$

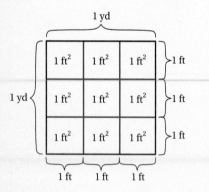

FIGURE 3.4

Figure 3.4 confirms that, indeed, 9 square feet fit into 1 square yard. As usual, we can write the conversion factor in three equivalent forms:

$$1 \text{ yd}^2 = 9 \text{ ft}^2 \quad \text{or} \quad \frac{1 \text{ yd}^2}{9 \text{ ft}^2} = 1 \quad \text{or} \quad \frac{9 \text{ ft}^2}{1 \text{ yd}^2} = 1$$

Note that we can also find this conversion factor by squaring both sides of the yards-to-feet conversion:

$$1 \text{ yd} = 3 \text{ ft} \xrightarrow{\text{square both sides}} (1 \text{ yd})^2 = (3 \text{ ft})^2, \quad \text{or} \quad 1 \text{ yd}^2 = 9 \text{ ft}^2$$

EXAMPLE 9   *The Cost of Carpet*

The area of a room in a house usually is measured in units of square feet. However, carpet usually is sold in units of square yards. Suppose that you have a room that measures 12 feet by 15 feet and you wish to install carpet that costs $15 per square yard. How much will the carpet cost?

Solution: The area of the room is

$$12 \text{ ft} \times 15 \text{ ft} = 180 \text{ ft}^2.$$

Now we need to convert 180 ft^2 to square yards. We use the conversion factor, 1 yd^2 = 9 ft^2, in the form that has ft^2 in the denominator:

$$180 \text{ ft}^2 \times \frac{1 \text{ yd}^2}{9 \text{ ft}^2} = 20 \text{ yd}^2$$

The room needs 20 square yards of carpet. Knowing that the price of the carpet is $15 per square yard, we can find its total cost.

$$20 \text{ yd}^2 \times \frac{\$15}{\text{yd}^2} = \$300$$

The cost of the carpet for the room will be $300.   ∎

EXAMPLE 10   *Cubic Unit Conversions*

Convert a volume of 12 cubic yards to cubic feet.

Solution: We can find the conversion factor between yd^3 and ft^3 by cubing both sides of the yards-to-feet conversion and simplifying.

$$(1 \text{ yd})^3 = (3 \text{ ft})^3 \qquad \text{becomes} \qquad 1 \text{ yd}^3 = (3 \text{ ft}) \times (3 \text{ ft}) \times (3 \text{ ft}) = 27 \text{ ft}^3$$

We use this conversion factor in the form that has yd^3 in the denominator.

$$12 \text{ yd}^3 \times \frac{27 \text{ ft}^3}{1 \text{ yd}^3} = 324 \text{ ft}^3$$

Thus 12 cubic yards is the same as 324 cubic feet.   ∎

TIME-OUT TO THINK: Make a picture or model to confirm that 27 cubic feet fit into 1 cubic yard. (Hint: your picture should be similar to Figure 3.4, but showing a three-dimensional box that is 1 yard on a side, rather than a two-dimensional area.)

CURRENCY CONVERSIONS

Different countries use different money, or **currency**. For international travelers, converting from one currency to another is a frequent and important problem. Fortunately, currency conversions are just another type of unit conversion. You need to know only the conversion factors.

Table 3.1 on the next page shows a typical currency exchange rate table. The *Dollars per Foreign* column gives the conversion factor in the form useful for converting from foreign currency to U.S. dollars. For example, the row for Mexico shows 0.1269 in this column, which means

$$1 \text{ peso} = \$0.1269, \quad \text{or} \quad \frac{\$0.1269}{1 \text{ peso}} = 1.$$

• Web • Watch •
Find currency exchange rates on the web.

Table 3.1 International Currency Exchange Rates (May 17, 1997)					
Country/Currency	Dollars per Foreign	Foreign per Dollars	Country/Currency	Dollars per Foreign	Foreign per Dollars
Australia/Dollar	0.7742	1.2917	Japan/Yen	0.00868	115.2
Austria/Schilling	0.08361	11.956	Malaysia/Ringit	0.4024	2.4850
Belgium/Franc	0.02853	35.09	Mexico/Peso	0.1269	7.88
Britain/Pound	1.6406	0.6095	Netherlands/Guilder	0.5234	1.911
Canada/Dollar	0.7330	1.3643	New Zealand/Dollar	0.6939	1.4411
Denmark/Krone	0.1545	6.47	Norway/Krone	0.1416	7.0627
Finland/Mark	0.1950	5.1294	Portugal/Escudo	0.005832	171.48
France/Franc	0.1747	5.723	Singapore/Dollar	0.6982	1.4322
Germany/Deutsche mark	0.5888	1.6984	S. Ireland/Punt	1.5269	0.6549
Greece/Drachma	0.003694	270.72	Spain/Peseta	0.00698	143.2
Hong Kong/Dollar	0.1292	7.7380	Sweden/Krone	0.1314	7.612
India/Rupee	0.0279	35.80	Switzerland/Franc	0.7003	1.4280
Italy/Lire	0.000598	1673.4	Thailand/Baht	0.7324	1.37

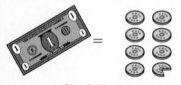

$1 = 7.88 pesos

1 peso = $0.1269

The *Foreign per Dollars* column gives the reciprocal conversion that is useful for converting U.S. dollars into foreign currency.

$$\$1 = 7.88 \text{ pesos}, \quad \text{or} \quad \frac{7.88 \text{ peso}}{\$1} = 1$$

When traveling, it can be very useful to use approximate conversions. For example, we can approximate the conversion from pesos to dollars as

$$1 \text{ peso} = 13\cent.$$

Now you can quickly realize that 10 pesos is about $1.30, 20 pesos is about $2.60, 100 pesos is about $13, and so on.

By the Way ....

You may notice that the *Dollars per Foreign* and *Foreign per Dollars* columns in Table 3.1 are not always precise reciprocals. These figures were taken directly from a newspaper currency table, and the discrepancies appear to occur because of rounding.

EXAMPLE 11   *Using the Exchange Rate Table*

a)   Which is larger, one U.S. dollar or one Irish (S. Ireland) punt?

b)   How many Hong Kong dollars are there in one U.S. dollar?

c)   How many U.S. dollars are there in one Swedish krone?

Solution:

a)   The *Dollars per Foreign* column for S. Ireland shows $1.5269 per punt. Thus it takes *more* than $1 to buy one Irish punt, so one Irish punt is *larger* than one U.S. dollar.

b)   The *Foreign per Dollars* column for Hong Kong shows that there are 7.7380 Hong Kong dollars per U.S. dollar. That is,

$$\$1 \text{ (U.S.)} = 7.7380 \text{ Hong Kong dollars.}$$

c)   The *Dollars per Foreign* column for Sweden shows 0.1314 U.S. dollars per Swedish krone, or

$$1 \text{ Swedish krone} = \$0.1314 \ (= 13.14\text{¢}).$$ ■

Beware! Currency exchange rates fluctuate daily, and sometimes even hourly when a country experiences particular economic upheaval. Be sure to use up-to-date tables when planning a trip.

EXAMPLE 12   *Comparing Prices*

While shopping in a Finnish department store, you find that Levi jeans cost 145 marks. What is the price in U.S. dollars?

Solution: We use the conversion factor from Finnish marks to U.S. dollars from the table:

$$1 \text{ mark} = \$0.1950 \quad \text{or} \quad \frac{\$0.1950}{1 \text{ mark}} \quad \text{or} \quad \frac{5.13 \text{ marks}}{\$1}$$

Because we want to convert from Finnish marks to dollars, we use the second form of the conversion factor. Starting with the price of 145 Finnish marks, we find

$$145 \text{ marks} \times \frac{\$0.1950}{\text{mark}} = \$28.28.$$

The price for the Levi jeans of 145 Finnish marks is equivalent to $28.28.   ■

EXAMPLE 13   *Conversions Between Two Foreign Currencies*

How many Japanese yen are there in 12 German marks?

Solution: If you were in Japan or Germany, you'd find tables giving direct conversions. But we can still answer this question with the data in Table 3.1 by using U.S. dollars as a stepping-stone between the two other currencies. The table shows that

$$1 \text{ German mark} = \$0.5888 \quad \text{and} \quad 115.2 \text{ Japanese yen} = \$1.$$

We now make a chain of conversions from German marks to U.S. dollars to Japanese yen:

$$12 \text{ marks} \times \frac{\$0.5888}{\text{mark}} \times \frac{115.2 \text{ yen}}{\$1} = 813.96 \text{ yen}$$

Thus 12 German marks are equivalent to 813.96 Japanese yen.   ■

By the Way ············
As this book goes to print, Germany and other countries in the European Monetary Union plan to adopt a new currency, called the *Euro*, in 1999.

REVIEW QUESTIONS

1. What is the difference between simple units and compound units? Give examples of compound units formed by division, powers, and multiplication.

2. What mathematical operations are suggested by the words *per*, *of*, *square*, and *cube*? Give an example in each case.

3. Explain how units can help you solve a problem. Give an example.

4. Explain how a unit conversion really is just a way of multiplying by 1.

5. How many square inches are in a square foot? Why? How many cubic inches in a cubic foot?

6. Explain the columns in Table 3.1.

PROBLEMS

1. **Your Problem-Solving Experience.** Write two or three paragraphs reflecting on your own experience at problem solving in prior mathematics classes. Have you enjoyed "word problems," or do you relate to the "Hell's Library" cartoon? How do you think your attitudes towards word problems were formed? What steps do you think you need to take to improve your problem-solving ability?

2. **Identifying Compound Units.** Rewrite each statement below in terms of compound units. *Example:* Detergent sells for $2 per ounce. *Solution:* Detergent sells for $\frac{\$2}{oz}$.

   a. Apples sell for 50 cents per pound.

   b. A mile is 1760 yards long.

   c. The tile installation will cost $3.50 per square foot.

   d. The river is flowing at a rate of 5000 cubic feet per second.

   e. An acre contains 43,560 square feet.

3. **Identifying Compound Units.** Rewrite each statement below in terms of compound units. *Example:* There are 3 feet in a yard. *Solution:* There are 3 ft/yd.

   a. A liter holds 1000 cubic centimeters.

   b. Atmospheric pressure is 14.7 pounds per square inch.

   c. Earth's average density is 5.5 grams per cubic centimeter.

   d. The rocket accelerated at 7 meters per second per second.

   e. The ore contains 5 ounces of gold in every cubic foot.

4. **The Key Words *Of* and *Per*.** Solve the following problems. Show units clearly in your work!

   a. Sam used one-third of a dozen eggs for his Sunday morning omelet. How many eggs did he use?

   b. The 25 scouts in Troop 11 raised $200. How much was raised (on average) per scout?

   c. Two-fifths of the 25,000 people in Midway are Republicans, and one-third of the Republicans are women. How many Republican women are there in Midway?

   d. A total of 550 apples were picked and divided evenly among 11 crates. How many apples were there per crate?

5. **U.S. Areas–Colorado.** The United States, excluding Alaska and Hawaii, is shaped very roughly like a rectangle

with a length of about 3500 miles and height of about 1000 miles. The state of Colorado is rectangular, with a length of about 400 miles and a height of about 300 miles. (Hint: $1 \text{ mi}^2 = 640$ acre.)

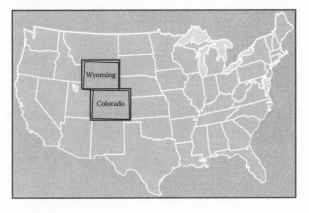

a. What is the approximate area of the United States in square miles? in square feet? in acres?

b. What is the area of Colorado in square miles? in square feet? in acres?

c. What fraction of the area of the United States does Colorado represent? State your answer as a percentage.

6. **U.S. Areas–Wyoming.** Repeat Problem 5 for Wyoming, which is rectangular with a length of about 320 miles and a height of about 300 miles.

7. **Units of Volume.** Calculate the volumes of the following objects.

a. A swimming pool that is 3 meters deep, 10 meters long, and 5 meters wide.

b. A package measuring 22 inches by 15 inches by 12 inches.

c. A skyscraper that is 1000 feet high with a 25,000 square foot base.

8. **Units of Volume.** Calculate the volumes of the following objects.

a. A box-shaped reservoir that is 200 feet long, 85 feet wide, and 12 feet deep.

b. The storage space of a van that is 3.5 feet high, 8 feet long, and 4 feet wide.

c. A human shaped roughly like a box, with a height of 5.5 feet, width of 1.25 feet, and a depth of 0.5 feet.

9. **Working with Units.** Solve each of the following problems. Show your work and use units.

a. Suppose that you buy 3 pounds of apples priced at 50 cents per pound. How much will you pay?

b. There are 100 centimeters in a meter and 1000 meters in a kilometer. How many centimeters are in a kilometer?

c. There are 3 feet in a yard, and 5280 feet in a mile. How many yards are in a mile?

d. Through a stroke of great luck, you find a 252-milligram ruby. With 200 milligrams per carat (for precious stones), what is its weight in carats?

10. **Working with Units.** Solve each of the following problems. Show your work and use units.

a. Suppose that you work 40 hours per week and are paid $6 per hour. How much money will you make in a week?

b. Suppose that you take a trip driving 1200 miles in 20 hours. What is your average speed for the trip?

c. If you sleep an average of 7.5 hours each night, how many hours do you sleep in a year?

d. An average human heart beats 60 times per minute. If an average human being lives to the age of 75, how many times does the average heart beat in a lifetime?

Unit Analysis: What Went Wrong? *Problems 11–14 state an exam question and a solution given by a student. State whether each solution is right or wrong. If it is wrong, write a note to the student explaining why the answer is wrong and how to solve it correctly. (Hint: See Example 5.)*

11. *Exam Question:* A candy store sells chocolate for $7.70 per pound. The piece you want to buy weighs 0.11 pounds. How much will it cost, to the nearest cent? (Neglect sales tax.) *Student Solution:* $0.11 \div 7.70 = 0.014$. It will cost 1.4¢.

12. *Exam Question:* You ride your bike up a steep mountain road at 5 miles per hour. How far do you go in 3 hours? *Student Solution:* $5 \div 3 = 1.7$. I rode 1.7 miles.

13. *Exam Question:* You can buy a 50-pound bag of flour for $11 or you can buy a 1-pound bag for $0.39. Compare the per pound cost for the large and small bags. *Student Solution:* The large bag price is $50 \div \$11 = \$4.50$ per pound, which is much more than the 39¢ per pound price of the small bag.

14. *Exam Question:* The average person needs 1500 calories a day. A can of Coke contains 140 calories. How many Cokes would you need to drink to fill your daily caloric needs? (Note: this diet may not meet other nutritional needs!) *Student Solution:* 1500 ÷ 140 = 210,000. You would need to drink 210,000 Cokes to meet your daily caloric needs.

15. Unit Conversions.

   a. There are 8 ounces in a cup, 4 cups in a quart, and 4 quarts in a gallon. How many ounces are in a gallon?

   b. How many minutes are in a year?

   c. A car is driving at 55 miles per hour. What is this speed in feet per hour? feet per second?

16. Unit Conversions.

   a. There are 12 inches in a foot and 5280 feet in a mile. How many inches are in a mile?

   b. How many minutes are in a century?

   c. A car is driving at 40 kilometers per hour. What is this speed in kilometers per minute? kilometers per second?

17. Conversions with Units Raised to Powers.

   a. Find a conversion factor between square inches and square feet.

   b. A football field is 100 yards long and 55 yards wide. Find its area in square yards and square feet.

   c. How many cubic feet in 5 cubic yards?

18. Conversions with Units Raised to Powers.

   a. How many square feet are in two square yards?

   b. Find a conversion factor between cubic inches and cubic feet.

   c. How many cubic inches in 3 cubic yards?

19. New Flooring. Imagine that you own a house and want to install new flooring. The rooms to be covered are the kitchen, which measures 10 ft by 12 ft; the master bedroom, which is 14 ft by 15 ft; the living room, which is 16 ft by 12 ft; the bathroom, in which the floor area to be covered is 3 ft by 5 ft; and the hallway, which is 3 ft by 8 ft. You plan to cover the kitchen with linoleum, at a price of $4 per square foot. The bathroom will be tiled, at $7 per square foot. The living room and hallway will be carpeted, at $18 per square yard. And the bedroom will be carpeted at $14.25 per square yard. What is the total price for your planned flooring?

20. Monetary Conversions. Suppose that you are traveling in Europe with the following monetary exchange rates in effect: 1 British pound = $1.60, 1 French franc = $0.21, 1 German mark = $0.72, and 1580 Italian lire = $1.00.

   a. Which is larger, 1 franc or 1 dollar?

   b. Convert 25 francs to dollars.

   c. Convert 2000 lire to dollars.

   d. Convert $20 to marks.

   e. Convert 20 francs to British pounds.

21. Monetary Conversions. Suppose that you are traveling in Europe with the following monetary exchange rates in effect: 1 British pound = $1.60, 1 French franc = $0.21, 1 German mark = $0.72, and 1580 Italian lire = $1.00.

   a. Which is smaller, 1 lira or 1 dollar?

   b. Convert 17 marks to dollars.

   c. Convert 5.25 British pounds to dollars.

   d. Convert $19.95 to lire.

   e. Convert 200 marks to francs.

22. Full of Hot Air. Assume that the average person breathes 6 times per minute (at rest), inhaling and exhaling half a liter of air each time. How much "hot air" (the air is warmed by the body) does the average person exhale each day?

23. Dog Years. Sometimes the age of dogs is described in a unit called "dog years." A commonly used conversion is that 1 real year equals 7 dog years.

   a. If your dog is 15 real years old, what is her age in dog years?

   b. People often refer to the third year in the life of a human child as the "terrible twos" stage. If dogs have a terrible twos stage in their third dog year, how old are they, in real years, during this stage?

c. Why do you think that anyone ever started using the unit of dog years? Do you think the common conversion of 1 real year to 7 dog years is reasonable? Explain.

24. Explain This One! A Goodyear tire commercial shown during the 1995 NFL playoffs began by stating that the Goodyear Aquatread tire can "channel away" 1 gallon of water per second. The announcer then goes on to state: "1 gallon per second—that's 396 gallons per mile." What's wrong with this statement? Hypothesize about how the advertising agency was able to come up with such a bizarre statement.

25. Computer-Stored Books. Computer memory is measured in units of "bytes," where one byte is enough memory to store one character (a letter in the alphabet or a number). How many typical pages of text can be stored on a 500-megabyte hard drive? (A megabyte is one million bytes.)

26. Global Energy Consumption at the U.S. Rate? (1994 data) The average American consumes fossil fuels at a rate equivalent to about 40 barrels of oil per year. The estimated global reserve of recoverable fossil fuels was the equivalent of 10 trillion barrels of oil. World population is about 6 billion. If everyone in the world consumed fossil fuels at the same rate as Americans, how long would the reserves last?

27. The Glen Canyon Dam. In April 1996, the Department of Interior released a "spike flood" from the Glen Canyon Dam on the Colorado River. The purpose of the release was to restore the river and the habitats along its banks, particularly in the Grand Canyon. The reservoir behind the dam contains about 1.2 trillion (1,200,000,000,000) cubic feet of water. The release from the dam lasted a week at a rate of 25,800 cubic feet of water per second.

a. About how much water was released during the one-week flood?

b. What fraction (percent) of the total water in the reservoir was released during the flood? Opponents of the release argued that it would compromise other uses of the water (agriculture and hydroelectric power). Proponents argued that the amount of water released was so small it would not affect other uses. Which side do you take? Defend your opinion.

28. Project: Glen Canyon Dam. Refer to Problem 27. Some people have proposed removing the Glen Canyon Dam altogether. Research the pros and cons of this proposal, and write a two to three page summary of your findings along with your personal opinion of the merit of this proposal.

29. Project: Textbook Analysis. Although research shows that most adults today have difficulty with "word problems," we might hope that the next generation will have less difficulty. Find a current textbook in mathematics that is used at the upper elementary school level (grades 4–6). Read through the "word problems" in the textbook. Write a critical analysis of the problems and conclude with an opinion as to whether the problems make mathematics meaningful.

UNIT 3B

SYSTEMS OF STANDARDIZED UNITS

Suppose that you want to describe the content of a basket of apples. You might choose units of *apples* because you eat them one at a time. However, a store probably charges for them in units of *pounds*. The pound is an example of a **standardized unit**: *Two pounds* has the same (standard) meaning at all stores. In contrast, *apples* is not a standardized unit because apples come in different sizes and shapes. In this section, we describe the two systems of standardized units in common use today.

THE U.S. CUSTOMARY SYSTEM OF MEASUREMENT

By the Way ·············
The U.S. customary system is sometimes referred to as the "English system" because of its English origins. However, England (Great Britain) has abandoned this system in favor of metric units.

The units used in the United States, such as feet, yards, miles, pounds, and quarts, make up the **U.S. customary system** (USCS). The origin of the U.S. customary units goes back to ancient systems of measurement used by Middle Eastern civilizations, including the Egyptians, Sumerians, Babylonians, and Hebrews. These systems of measurement were further developed in ancient Greece and Rome, and in Europe during the Middle Ages. Eventually, the English government standardized these ancient measures and brought them to the colonies whose revolution created the United States.

Because the USCS developed over thousands of years and over large geographical regions, it is almost hopelessly complicated. Nevertheless, we will try to make some sense of this ancient system because its units are still used commonly in the United States.

USCS Measures of Length

Units of length in the U.S. customary system come from ancient units originally based on body parts. For example, the *foot* originally referred to the length of *your* foot, or the foot of whoever was doing the measuring! The ancient Romans discovered that most adult feet are about as long as 12 thumb-widths. The Latin word for thumb-width is *uncia,* from which we get our word *inch*. For measuring longer distances, the Romans used pacing. Our word mile comes from the Latin *milia passuum*, meaning one thousand paces.

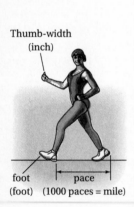

Thumb-width
(inch)

foot   pace
(foot)   (1000 paces = mile)

> **TIME-OUT TO THINK:**   Measure the number of your thumb-widths that fit along your bare foot. Is it close to 12?

Lengths based on body parts or pacing vary from person to person. Many cultures therefore based the lengths on the body parts of a particular person. For example, the English long regarded the *foot* as the length of the *King's* foot—so this definition changed every time the King changed! The first permanent standardization began when King Henry I (1100–1135) set the *yard* as the measurement from the tip of *his* nose to the tip of *his* thumb on *his* outstretched arm. Table 3.2 summarizes USCS units of length.

USCS Measures of Weight

Measures of weight in the USCS are more complex than units of length because three distinct sets of measures were used traditionally. All three sets are summarized in Table 3.3.

Table 3.2 USCS Lengths (common abbreviations in parentheses)	
1 inch (in.) (basic unit of length)	1 furlong = 40 rods (= 1/8 mile)
1 foot (ft) = 12 inches	1 mile (mi) = 1760 yards = 5280 feet
1 yard (yd) = 3 feet	1 nautical mile = 1.15 mile = 6076.1 feet
1 rod = 5.5 yards	1 land league = 3 miles
1 fathom = 6 feet	1 marine league = 3 nautical miles

Table 3.3 USCS Weights (common abbreviations in parentheses)		
1 grain = 0.0648 grams (basic unit of weight)		
Avoirdupois Measures	**Troy Measures**	**Apothecary Measures**
1 ounce (oz) = 437.5 grains	1 carat = 3.086 grains = 0.2 g	1 scruple = 20 grains
1 pound (lb) = 16 oz	1 pennyweight = 24 grains	1 dram = 3 scruples
1 ton = 2000 lb	1 troy ounce = 480 grains	1 apoth. ounce = 8 drams
1 long ton = 2240 lb	1 troy lb = 12 troy oz	1 apoth. pound = 12 ounces

By the Way
The *international nautical mile* is based on the distance along one minute of arc on the Earth's surface: 60 nautical miles correspond to 1° of arc. A related unit, the *knot*, measures speed in nautical miles per hour.

By the Way
The term *avoirdupois* comes from French words meaning *goods of weight*. The *troy* system probably was named for the town of Troyes, France. *Apothecary* is a synonym for *pharmacy*, and apothecary measures were developed by medieval pharmacists.

The only measures in common use today are *avoirdupois* weights, originally standardized by London merchants around the year 1300. Jewelers still sometimes use troy measures for weighing precious metals and stones. The *apothecary* measures are no longer in use, but you may encounter them in folk remedies and literature. Note that the basic unit of weight in all three sets is the *grain*, an ancient unit that originally was based on the weight of a typical grain of wheat.

Only one USCS weight unit is still used internationally: the **carat**, now defined to be 200 milligrams (0.2 gram). Diamonds and other jewels are commonly weighed in carats. For example, a 5-carat diamond weighs 5×200 mg = 1 gram. Note that the carat as a unit of weight is *not* the same as the **karat** used to describe the purity of gold! Pure gold is defined to be 24 karats. Thus, for example, *12-karat gold* means gold that is only 50% pure (because $12 \div 24 = 50\%$); the other 50% is a mixture of other metals.

USCS Measures of Volume

The basic USCS unit of volume is the cubic inch (in.^{3}), but other USCS volume units are more complicated because their meanings depend on whether dry or liquid materials are being measured. For example, a *dry pint* is 33.60 in.^{3}, while a *liquid pint* is only 28.88 in.^{3} (Figure 3.5). Thus a container that holds one pint of water is too small for one pint of flour! Table 3.4, shown on the next page summarizes USCS measures of volume.

Flour   Beer

Volume of dry pint   Volume of liquid pint
= 33.60 in.^{3} = 28.88 in.^{3}

FIGURE 3.5

By the Way ············
British customary volume units differ from U.S. units. For example, the British fluid ounce is slightly smaller than the U.S. fluid ounce (1.734 in.^{3} versus 1.804 in.^{3}), but the British pint is 20 fluid ounces, rather than 16 fluid ounces. Overall, that makes a *pint* bigger in Britain than in the U.S. (34.68 in.^{3} versus 28.88 in.^{3}).

Table 3.4	USCS Volumes (common abbreviations in parentheses)		
Liquid Measures		**Dry Measures**	
1 tablespoon (tbsp or T) = 3 teaspoons (tsp or t)		1 in.^{3}	= 16.387 cm^3
1 fluid ounce (fl oz) = 2 tablespoons = 1.805 in.^{3}		1 ft^3	= $(12$ in.$)^3$ = 1728 in.^{3}
1 cup (c) = 8 fluid ounces		1 yd^3	= $(3$ ft$)^3$ = 27 ft^3
1 pint (pt) = 16 fluid ounces = 28.88 in.^{3}		1 dry pint (pt)	= 33.60 in.^{3}
1 quart (qt) = 2 pints = 57.75 in.^{3}		1 dry quart (qt)	= 2 dry pints = 67.2 in.^{3}
1 gallon (gal) = 4 quarts		1 peck	= 8 dry quarts
1 barrel of petroleum = 42 gallons		1 bushel	= 4 pecks
1 barrel of liquid = 31 gallons		1 cord	= 128 ft^3

TIME-OUT TO THINK: Check the labels on several containers of food. You'll see that dry food, such as cereal, is labeled with net weight in *ounces*. Liquids, such as milk or oil, are labeled with volumes in *fluid ounces*. Which units appear on "in-between foods," such as peanut butter, spaghetti sauce, or salsa? Why?

EXAMPLE 1   *Using the Conversion Tables*

a)   Convert a distance of 2.5 miles into feet.

b)   How many square feet are in a square mile?

c)   You've lost your measuring spoons and have only a measuring cup. What should you measure when a recipe calls for 4 tablespoons of water?

Solution:

a)   Table 3.2 shows the conversion factor from miles to feet, which we can write in three equivalent forms:

$$1 \text{ mi} = 5280 \text{ ft} \quad \text{or} \quad \frac{1 \text{ mi}}{5280 \text{ ft}} = 1 \quad \text{or} \quad \frac{5280 \text{ ft}}{1 \text{ mi}} = 1$$

We use the latter form to convert from miles to feet:

$$2.5 \text{ mi} \times \frac{5280 \text{ ft}}{1 \text{ mi}} = 13{,}200 \text{ ft}$$

A distance of 2.5 miles is equivalent to 13,200 feet.

b)   Squaring both sides of this conversion factor, 1 mile = 5280 feet, we find

$$1 \text{ mi} = 5280 \text{ ft} \xrightarrow{\text{square both sides}} (1 \text{ mi})^2 = (5280 \text{ ft})^2.$$

Simplifying the squares gives $1 \text{ mi}^2 = 27{,}878{,}400 \text{ ft}^2$. A square mile is almost 28 million square feet.

c)   Table 3.4 shows there are 2 tablespoons in 1 fluid ounce. Thus 4 tablespoons is the same as 2 fluid ounces, which is the same as 1/4 cup.   ∎

EXAMPLE 2   *Carats and Karats*

Suppose that you buy a gold chain made out of 12-karat gold. If the entire chain weighs 100 carats, how many carats of pure gold do you have? How many grams of pure gold do you have?

Solution: Recall that 24-karat gold is pure, and 12-karat gold is 50% pure. Thus only half the weight of the gold chain is made up of pure gold. Because the chain weighs 100 carats, it contains 50 carats of pure gold. Remembering that 1 carat = 0.2 g, we find the weight of pure gold in the chain is

$$50 \text{ carats} \times \frac{0.2 \text{ g}}{\text{carat}} = 10 \text{ g}.$$ ∎

Beware! It's easy to confuse carats and karats. Remember that the *carat* is a measure of weight equal to 200 mg. Anything can be weighed in carats, though it is most commonly used for diamonds and other gems. The *karat* describes only the purity of gold; 24-karat gold is 100% pure.

BASIC SI (METRIC) UNITS

After studying customary units, you almost certainly will conclude that "there's got to be a better way!" Fortunately, there is. Recognizing the difficulties with customary systems of measurement, French politicians and scientists got together to invent the **metric system** in the late 1700s. Adopted by the Republic of France in 1795, the metric system essentially was a product of the French Revolution of 1789. The two basic ideas behind the creation of the metric system were:

- to create a coherent and sensible set of standardized units to replace the customary systems in use around the world; and
- to simplify conversions and calculations by basing the units on factors of 10 (decimal relationships).

Use of the metric system slowly spread internationally throughout the 1800s. In 1875, the French government convened a conference of delegates from twenty nations to consider metric standards. The conference produced the *Treaty of the Meter*, which created mechanisms for refining the metric system and encouraging its international use.

The modern version of the metric system, known as *Systeme Internationale d'Unites* (French for the International System of Units) or **SI**, was formally established in 1960. By 1975, SI had been adopted for everyday use by every nation except Myanmar, Liberia, and the United States. Even in the United States, SI units are legally established and widely used in science, manufacturing, and commerce.

The basic units of length, mass, time, and volume in the present SI are:

- the **meter** for length, abbreviated "m";
- the **kilogram** for mass, abbreviated "kg"; and
- the **second** for time, abbreviated "s."
- the **liter** for volume, abbreviated "ℓ."

By the Way ··········
Under Napoleon, France abandoned use of the metric system in 1812. It was readopted by the French in 1840.

By the Way ··········
Technically, the liter is a *derived* (rather than basic) unit because its definition is based on the meter:
$1\ell = 0.001 \text{ m}^3 = 1000 \text{ cm}^3$.

Decimal-Valued Prefixes

Multiples of metric units are formed by powers of 10, using a prefix to indicate the power. For example, *kilo* means 10^3 (1000) so a kilometer is 1000 meters; and a microgram is 0.000001 gram because *micro* means 10^{-6}, or one millionth. Some of the more common prefixes are listed in Table 3.5 on the next page.

Table 3.5 SI (Metric) Prefixes					
Small Values			**Large Values**		
Prefix	Abbrev.	Value	Prefix	Abbrev.	Value
Deci	d	10^{-1}	Deca	da	10^{1}
Centi	c	10^{-2}	Hecto	h	10^{2}
Milli	m	10^{-3}	Kilo	k	10^{3}
Micro	μ	10^{-6}	Mega	M	10^{6}
Nano	n	10^{-9}	Giga	G	10^{9}
Pico	p	10^{-12}	Tera	T	10^{12}

TIME-OUT TO THINK: Popular usage has adopted the prefix *mega* to mean "a lot." For example, people say that expensive things cost "megabucks," or that something really fun is "megafun." What do these statements mean literally? Do you think you can start a new trend by calling things, say, "gigafun"?

 ## BRIEF REVIEW

Powers of 10

Powers of 10 simply indicate how many times to multiply 10 by itself. For example:

$$10^2 = 10 \times 10 = 100$$
$$10^6 = 10 \times 10 \times 10 \times 10 \times 10 \times 10 = 1,000,000$$

Negative powers are the reciprocal of the corresponding positive powers. For example:

$$10^{-2} = \frac{1}{10^2} = \frac{1}{100} = 0.01$$

$$10^{-6} = \frac{1}{10^6} = \frac{1}{1,000,000} = 0.000001$$

Table 3.6 lists powers of 10 from 10^{-12} to 10^{12}. Note that powers of 10 follow two basic rules.

1. A positive exponent tells how many zeros follow the 1. For example, 10^0 is a 1 followed by no zeros, and 10^8 is a 1 followed by eight zeros.

2. A negative exponent tells how many places are to the right of the decimal point, including the 1. For example, $10^{-1} = 0.1$ has one place right of the decimal point; $10^{-6} = 0.000001$ has six places to the right of the decimal point.

Table 3.6   Powers of 10

Zero and Positive Powers			Negative Powers		
Power	**Value**	**Name**	**Power**	**Value**	**Name**
10^0	1	One			
10^1	10	Ten	10^{-1}	0.1	Tenth
10^2	100	Hundred	10^{-2}	0.01	Hundredth
10^3	1000	Thousand	10^{-3}	0.001	Thousandth
10^4	10,000	Ten thousand	10^{-4}	0.0001	Ten thousandth
10^5	100,000	Hundred thousand	10^{-5}	0.00001	Hundred thousandth
10^6	1,000,000	Million	10^{-6}	0.000001	Millionth
10^7	10,000,000	Ten million	10^{-7}	0.0000001	Ten millionth
10^8	100,000,000	Hundred million	10^{-8}	0.00000001	Hundred millionth
10^9	1,000,000,000	Billion	10^{-9}	0.000000001	Billionth
10^{10}	10,000,000,000	Ten billion	10^{-10}	0.0000000001	Ten billionth
10^{11}	100,000,000,000	Hundred billion	10^{-11}	0.00000000001	Hundred billionth
10^{12}	1,000,000,000,000	Trillion	10^{-12}	0.000000000001	Trillionth

Multiplying and Dividing Powers of 10

Multiplying or dividing powers of 10 simply requires adding or subtracting exponents. A few examples should clarify this process:

$$10^4 \times 10^7 = \underbrace{10,000}_{10^4} \times \underbrace{10,000,000}_{10^7} = \underbrace{100,000,000,000}_{10^{11}}$$

add exponents: $10^4 \times 10^7 = 10^{11}$

$$10^5 \times 10^{-3} = \underbrace{100,000}_{10^5} \times \underbrace{0.001}_{10^{-3}} = \underbrace{100}_{10^2}$$

add exponents: $10^5 \times 10^{-3} = 10^2$

$$\frac{10^5}{10^3} = \underbrace{100,000}_{10^5} \div \underbrace{1000}_{10^3} = \frac{100,000}{1000} = \underbrace{100}_{10^2}$$

subtract exponents: $10^5 \div 10^3 = 10^2$

$$\frac{10^3}{10^7} = \underbrace{1000}_{10^3} \div \underbrace{10,000,000}_{10^7} = \frac{1000}{10,000,000} = \underbrace{0.0001}_{10^{-4}}$$

subtract exponents: $10^3 \div 10^7 = 10^{-4}$

We can generalize these rules using n and m to represent any numbers.

- To multiply powers of 10, *add* exponents:
  $10^n \times 10^m = 10^{n+m}$

- To divide powers of 10, *subtract* exponents:
  $\dfrac{10^n}{10^m} = 10^{n-m}$

Adding and Subtracting Powers of 10

Unlike multiplication and division, there is no shortcut for adding or subtracting powers of 10. The values must be written in longhand notation. For example:

$$10^6 + 10^2 = 1,000,000 + 100 = 1,000,100$$
$$10^8 + 10^{-3} = 100,000,000 + 0.001 = 100,000,000.001$$
$$10^7 - 10^3 = 10,000,000 - 1000 = 9,999,000$$

EXAMPLE 3   *Using Metric Prefixes*

a)   Convert 2759 cm to meters.

b)   How many nanoseconds are in a microsecond?

c)   A *high density, 3.5-inch diskette* holds 1.4 megabytes of data. Suppose that your computer has a 4.0 gigabyte hard drive. How many diskettes would it take to fully back up your drive?

Solution:

a)   Table 3.6 shows that centi means 10^{-2}, so there are 10^2, or 100, centimeters in a meter. Thus the conversion factor from centimeters to meters is 1 m = 100 cm, and 2759 cm is the same as

$$2759 \text{ cm} \times \frac{1 \text{ m}}{100 \text{ cm}} = 27.59 \text{ m}.$$

b)   We can compare these quantities by dividing the longer one (microsecond) by the shorter (nanosecond):

$$\frac{1 \text{ } \mu\text{s}}{1 \text{ ns}} = \frac{10^{-6} \text{ s}}{10^{-9} \text{ s}} = 10^{-6-(-9)} = 10^{-6+9} = 10^3$$

A microsecond is 1000 (10^3) times longer than a nanosecond, so there are 1000 nanoseconds in a microsecond.

c)   We divide the 4.0 gigabytes of information that can be stored on your hard drive by the 1.4 megabytes on each diskette:

$$\frac{4.0 \text{ gigabyte}}{1.4 \text{ megabyte}} = \frac{4.0 \times 10^9 \text{ byte}}{1.4 \times 10^6 \text{ byte}} \approx 2.86 \times 10^3$$

It will take about 2,860 diskettes to fully back up your hard drive!   ∎

METRIC–USCS CONVERSIONS

Conversions between SI and USCS units are carried out like any other unit conversions: you need only to know the conversion factors. Table 3.7 lists a few basic conversions.

By the Way ············

Pounds are a unit of *weight* and kilograms are a unit of *mass*, so the given conversions between pounds and kilograms are valid only on Earth. That is, on Earth a 50-kg astronaut weighs about 110 pounds. In Earth orbit, the astronaut still has a mass of 50 kg, but has a weight of zero (weightless).

Table 3.7   USCS ⟺ SI Conversions

USCS to SI		SI to USCS	
1 inch	= 2.540 cm	1 cm	= 0.3937 inch
1 foot	= 0.3048 m	1 m	= 3.28 feet
1 yard	= 0.9144 m	1 m	= 1.094 yd
1 mile	= 1.6093 km	1 km	= 0.6214 mile
1 pound	= 0.4536 kg	1 kg	= 2.205 pound
1 fl. oz.	= 29.574 mℓ	1 mℓ	= 0.03381 fl oz
1 qt	= 0.9464 ℓ	1 ℓ	= 1.057 qt
1 gal	= 3.785 ℓ	1 ℓ	= 0.2642 gal

It can be useful to memorize these conversions, at least in rough terms, particularly if you plan to do much traveling or if you commonly work with metric units in sports or business. For example, if you recall that a kilometer is about 0.6 mile, you will know that a 10-kilometer road race is about 6 miles. If you recall that a gallon is a little less than 4 liters, you will know that a gasoline price of $1 per liter is a little less than $4 per gallon.

EXAMPLE 4

International athletic competitions generally use metric distances. Compare the length of a 100-meter race to a 100-yard race.

Solution: Table 3.7 shows that 1 m = 1.094 yd, so 100 meters is 109.4 yards. Note that 100 meters is almost 110 yards; a good "rule of thumb" to remember is that distances in meters are about 10% longer than the corresponding number of yards.   ■

EXAMPLE 5

Suppose you go to a gas station and find that the price of gasoline is stated as $0.275 (27.5 cents) per liter. What is the equivalent cost per gallon?

Solution: We simply convert from price per liter to price per gallon. Table 3.7 shows that there are 3.785 liters per gallon. Thus

$$\frac{\$0.275}{1\ell} \times \frac{3.785\ell}{1 \text{ gal}} = \frac{\$1.041}{1 \text{ gal}}.$$

The price of the gasoline is $1.041 (or 104.1 cents) per gallon.   ■

$\mathcal{T}$HINKING ABOUT …

Will the United States Go Metric?

Throughout most of the world, speed limits are posted in kilometers per hour, milk and gasoline are sold by the liter, and meat and fruit are sold by the kilogram. Will the United States ever join the rest of the world in the use of SI units?

In fact, the metric system has a long history in the United States. Thomas Jefferson and Benjamin Franklin both spent time in France and were familiar with the scientific discussions that led to the invention of the metric system. Jefferson was especially fond of the metric system reliance on *decimal* relationships. Indeed, Jefferson used decimal relationships himself in devising the U.S. system of currency based on the dollar, which was adopted in 1785.

In 1790, Jefferson (then serving as Secretary of State to President George Washington) formally proposed adoption of the metric system to Congress. Had Jefferson's proposal been accepted, the United States would have been the *first* country in the world to adopt the metric system—even ahead of the French! Alas, the Congress did not go along with Jefferson's plan.

Today, the biggest push toward SI units in the United States comes from industry. Because U.S. businesses hope to prosper through exports, they must use international standards. Already, nearly all machinery is built to metric specifications, and the labels on most containers of food and drink state metric equivalents. Although public resistance to metric units remains strong, the complete conversion of the United States to SI units is probably inevitable.

EXAMPLE 6

You are in an Italian market where eggplants cost 8000 lire per kilogram. What is the equivalent price in dollars per pound?

Solution: Table 3.7 shows that 1 kg ≈ 2.2 pounds and Table 3.1 shows a currency exchange rate (in May 1997) of $1 = 1673.4 lire. Starting with the price of 8000 lire per kilogram, we convert lire to dollars and kilograms to pounds:

$$\frac{8000 \text{ lire}}{\text{kg}} \times \frac{1 \text{ kg}}{2.2 \text{ lb}} \times \frac{\$1}{1673.4 \text{ lire}} = \frac{\$2.17}{\text{lb}}$$

The eggplants cost about $2.17 per pound.   ∎

EXAMPLE 7

How many square kilometers are in 1 square mile?

Solution: We use the square of the miles-to-kilometers conversion factor:

$$(1 \text{ mi})^2 \times (1.6093 \text{ km})^2 \longrightarrow 1 \text{ mi}^2 = 2.5898 \text{ km}^2$$

Therefore 1 square mile is 2.5898 square kilometers.   ∎

STANDARDIZED UNITS OF TEMPERATURE

Three temperature scales are commonly used today. Internationally, temperature is usually measured on the **Celsius** scale, which places the freezing point of water at 0°C and the boiling point at 100°C. In the United States, the **Fahrenheit** scale is more common; it is defined so that water freezes at 32°F and boils at 212°F.

By the Way ··············
The degree symbol (°) is not necessary when writing temperatures on the Kelvin scale.

The official SI unit of temperature is the **Kelvin**. The Kelvin scale is the same as the Celsius scale except in its zero point. A temperature of 0 K is **absolute zero**, which is the coldest possible temperature.

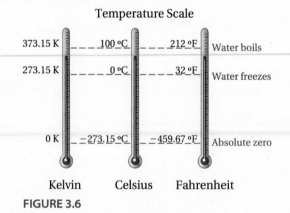

Temperature Scale

373.15 K — — — 100 °C — — — 212 °F   Water boils

273.15 K — — — 0 °C — — — 32 °F   Water freezes

0 K — −273.15 °C — −459.67 °F   Absolute zero

Kelvin   Celsius   Fahrenheit

FIGURE 3.6

As shown in Figure 3.6, any particular temperature has a Kelvin value that is numerically 273.15 larger than its Celsius value. That is,

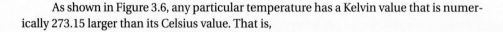

Temperature (Kelvin) = Temperature (°C) + 273.15, or

Temperature (°C) = Temperature (Kelvin) − 273.15.

We can find the conversion between Fahrenheit and Celsius by noting that the Fahrenheit scale has 180 degrees (212°F − 32°F = 180°F) between the freezing and boiling points of water, whereas the Celsius scale has only 100 degrees between these points. Each Celsius degree therefore represents a temperature change equivalent to 1.8 Fahrenheit degrees. Furthermore, the freezing point of water is 32 units larger on the Fahrenheit scale than the Celsius scale. Combining these two facts, the conversions between Fahrenheit and Celsius are as follows.

$$\text{Temperature } (°C) = \frac{\text{Temperature } (°F) - 32°F}{1.8\frac{°F}{°C}}$$

and

$$\text{Temperature } (°F) = \left(1.8\frac{°F}{°C}\right) \times \text{Temperature } (°C) + 32°F.$$

Note that we can write these formulas more simply by using symbols. Letting C represent temperatures in Celsius and F represent temperatures in Fahrenheit, the conversion formulas become:

$$C = \frac{F - 32}{1.8} = \frac{5}{9}(F - 32)$$

$$F = 1.8 \times C + 32 = \frac{9}{5}C + 32$$

> **TIME-OUT TO THINK:**   Using a temperature of 32°F, confirm that the first formula above yields the freezing point of water on the Celsius scale (0°C). Use a temperature of 100°C to confirm that the second formula yields the equivalent 212°F.

EXAMPLE 8   *Paris Weather Report*

You'll be in Paris tomorrow, and the weather forecast calls for a temperature of 30°C. What is this temperature in Kelvin and Fahrenheit? Will you need a winter jacket?

Solution: We begin by doing the temperature conversions.

$$T \text{ (Kelvin)} = T \text{ (°C)} + 273.15$$
$$= 30 + 273.15 = 303.15 \text{ K}$$
$$T \text{ (°F)} = \left(1.8\frac{°F}{°C}\right) \times 30°C + 32°F$$
$$= 54°F + 32°F = 86°F$$

With a temperature of 86°F, there'll be no need for a winter jacket!   ■

EXAMPLE 9   *Human Body Temperature*

Normal average human body temperature is 98.6°F. What is this temperature in Celsius and Kelvin?

Solution: We convert the temperature from Fahrenheit to Celsius:

$$T\,(°C) = \frac{98.6°F - 32°F}{1.8\frac{°F}{°C}}$$

$$= \frac{66.6°F}{1.8\frac{°F}{°C}}$$

$$= 66.6°F \times \left(\frac{1}{1.8}\right)\frac{°C}{°F}$$

$$= 37.0°C$$

Human body temperature is 37°C. We can also convert this to Kelvin by adding 273.15, which makes it 310.15 K. ∎

REVIEW QUESTIONS

1. What do we mean by *standardized units*? Why are they useful?

2. What is the origin of the units in the U.S. customary system?

3. Describe several ways in which U.S. customary units can be confusing.

4. Explain the difference between *carats* and *karats*. How are these units important for buying jewelry?

5. Describe several ways in which metric units are simpler than U.S. customary units.

6. For each of the conversions shown in Table 3.7, state an approximate conversion that is easier to remember (e.g., 1 inch is about 2.5 centimeters).

7. Describe the differences between the Fahrenheit, Celsius, and Kelvin temperature scales. In what situations is each scale commonly used?

PROBLEMS

1. **Everyday Metric.** Describe three ways that you use metric units in your everyday life.

2. **Grocery Metric.** Find three examples of the use of metric units in the grocery store.

3. **USCS Units.** Make the following conversions.

   a. Convert your own height in feet to inches.

   b. A basketball player might be 6.75 feet tall. Express this height in inches and yards.

   c. The Kentucky Derby horse race is 10.2 furlongs in length. What is this distance in miles? How does it compare to a road race of 6.2 miles or a marathon of 26 miles?

d. A gallon of water weighs about 128 ounces. How many pounds is that?

e. You can send a letter overseas with a weight of up to 154 grains for the lowest postal rate. How many ounces is this in avoirdupois measures?

f. Suppose that you have a 1/2 gallon milk jug. How many liquid pints can it hold? How many dry pints can it hold?

g. Agricultural products such as corn and wheat often are traded in units of 5000 bushels. How many cubic inches does this quantity represent? If 150 million bushels of wheat are traded in one day, how many cubic inches is this? How many cubic feet? Describe, in words, the size of a building that would hold 5000 bushels of wheat.

4. USCS Units. Make the following conversions.

a. Convert your own weight in pounds to ounces using avoirdupois measures. How much do you weigh in tons?

b. Deep sea trenches can reach a depth of 6000 fathoms. How deep is that in feet? in nautical miles? in marine leagues?

c. If a boat is moving at 30 knots (*nautical miles* per hour), how fast is it going in miles per hour?

d. Most soda cans contain 12 fluid ounces. How many cubic inches is that?

e. A large car's gas tank might hold 20 gallons. How many cubic inches is 20 gallons?

f. Suppose a small city produces 500,000 cubic feet of garbage per week. If all of this garbage were stacked neatly (in a nice vertical pile) on a 100 yard by 60 yard football field, how high would the pile be?

5. Metric Prefixes. For each of the following, state how much larger or smaller the first unit is than the second.

a. millimeter, meter

b. gram, kilogram

c. milliliter, deciliter

d. kilometer, micrometer

6. Metric Prefixes. For each of the following, state how much larger or smaller the first unit is than the second. (Hint: Be especially careful with units of area or volume.)

a. centimeter, millimeter

b. cubic meter, cubic centimeter

c. square millimeter, square kilometer

d. decaliter, centiliter

7. USCS–SI Conversions. Convert each measurement to the units specified.

a. 10 meters to feet.

b. 880 yards to kilometers.

c. 20 gallons to liters.

d. 5 milliliters to cubic inches.

e. 150 pounds to kilograms.

8. USCS–SI Conversions. Convert each measurement to the units specified.

a. 105 centimeters to yards.

b. 1200 square feet to square meters.

c. 100 kilometers per hour to miles per hour.

d. 5.5 grams per cm^3 to pounds per cubic foot.

e. 25 miles per hour to kilometers per hour.

9. Sensible or Ridiculous? Determine whether the following statements are sensible or patently ridiculous. Explain why. *Example 1:* I ate 2 meters of apples at lunch. *Solution:* The statement is ridiculous. A meter is a unit of length, so talking about "meters of apples" makes no sense. *Example 2:* My brother is 4 meters tall. *Solution:* The statement is ridiculous. A meter is slightly longer than a yard, so 4 meters is slightly more than 12 feet; no one is 12 feet tall.

a. I drank 2 liters of water today.

b. A professional football player weighs 300 kilograms.

c. Bill drove along the interstate at 100 kilometers per hour.

d. Fred ran 35 liters per second.

e. The world record high jump for men is 7 meters.

f. Sue ran 10,000 meters in less than an hour.

g. The book I sent you weighs 3 milligrams.

h. An infant eats 2500 grams of food each day.

10. Celsius–Fahrenheit Conversions. In each of the following convert, as appropriate, Fahrenheit into Celsius or Celsius into Fahrenheit. State answers to the nearest tenth of a degree.

a. 0°F   **b.** 200°C   **c.** 100°C

d. 10,000°C   **e.** 70°F   **f.** −273.15°C

g. 415°F   **h.** 15°C   **i.** 98.6°F

11. Celsius–Kelvin Conversions. In each of the following convert, as appropriate, Kelvin into Celsius or Celsius into Kelvin. State answers to the nearest degree.

   a. 50 K

   b. 240 K

   c. 500,000 K

   d. 10°C

   e. 100°C

   f. 320 K

12. A Pint Is a Pint? A British fluid ounce is 1.734 in.^{3} and a U.S. fluid ounce is 1.805 in.^{3}. A U.S. pint contains 16 fluid ounces, whereas a British pint contains 20 British fluid ounces. How much more do you get in a British pint than in a U.S. pint? Explain.

13. Mountains and Trenches.

   a. Mt. Whitney, the tallest mountain in California and in the continental United States, is 14,494 feet above sea level. How high is that in miles? in meters? in kilometers?

   b. The tallest mountain in the world, Mt. Everest, rises 29,028 feet above sea level. How high is that in miles? in meters? in kilometers?

   c. Mauna Kea, the highest mountain on the island of Hawaii, rises 13,796 ft above sea level. It extends an additional 18,200 ft from sea level to its base on the ocean floor. How tall is Mauna Kea from its base to its peak, in feet, miles, meters, and kilometers? Compare its total extent to the height of Mt. Everest. Would it be fair to call Mauna Kea the highest mountain in the world? Why or why not?

   d. The deepest point in the oceans is a gorge called the *challenger deep* that lies within the Marianas trench in the western Pacific ocean. It reaches a depth of 36,201 feet. How deep is that in miles, meters, and kilometers? Compare the depths of the oceans to the heights of the tallest mountains on Earth.

14. A Jules Verne Novel. A Jules Verne novel is titled *20,000 Leagues Under the Sea*. Is an ocean *depth* of 20,000 marine leagues possible? Explain. (Bonus: What did Jules Verne mean by the title?)

15. Price Conversions. Suppose that you are traveling in Europe with the following monetary exchange rates in effect: 1 British pound = $1.60, 1 French franc = $0.21, 1 German mark = $0.72, and 1580 Italian lire = $1.00.

   a. Convert a price of 4 francs per liter to dollars per gallon.

   b. Convert a price of 2000 lire per kilogram to dollars per pound.

   c. Convert a price of 2 marks per kilogram to dollars per pound.

   d. Convert a price of 0.5 British pounds per liter to dollars per gallon.

16. The Metric Mile. In track and field, the 1500-meter race is sometimes called the *metric mile*.

   a. Compare the metric mile to a USCS mile. How much longer or shorter is it, by percentage?

   b. Look up the current men's and women's world records for the mile. If you assume that the runners maintain the same pace for the metric mile, what should their times be for the metric mile?

   c. Look up the current world records for the metric mile. Do they agree with the expected times you calculated in part (b)? If not, why do you think they differ?

17. Carats and Karats. Answer the following questions.

   a. If you find a nugget that is 75% gold, what is its purity in karats?

   b. Suppose that you purchase a 14-karat gold chain that weighs 15 grams. How much gold have you purchased?

c. Find the price of gold today and use it to calculate the value of the gold chain from part (b).

d. Is it possible to have jewelry made of 30-karat gold? Why or why not? Is it possible to have a 30-carat gold nugget? Explain.

e. Diamonds are sold according to their weight in carats. How much does a 23-carat diamond weigh in grams?

f. Can diamonds be sold in units of karats? Why or why not?

18. Gold Pendant. Suppose that you have a 4-gram pendant made of gold that is 50% pure.

a. How many karats is its gold?

b. How many carats does the pendant weigh?

c. How many carats of gold does the pendant contain?

19. The Cullinan Diamond and the Star of Africa. The largest single rough diamond ever found, the Cullinan diamond, weighed 3106 carats; it was used to cut the world's largest diamond gem, the Star of Africa (530.2 carats), which is part of the British crown jewels collection. How much did the Cullinan diamond weigh in milligrams? in pounds? How much does the Star of Africa weigh in milligrams? in pounds?

20. Metric Tools. Many tools come in both USCS and SI standards. In a standard socket set, the smallest USCS subdivision is $\frac{1}{16}$ inch; the smallest SI subdivision is 0.5 millimeters. Which of these subdivisions is smaller? Explain.

21. Temperature Algebra. Starting with the conversion formula from Fahrenheit to Celsius, show the algebraic steps needed to find the conversion formula from Celsius to Fahrenheit.

22. Project: Will the United States Go Metric? Investigate the current prospects of the United States converting to the metric system. What are the obstacles? Which agencies or organizations support the change? Which agencies or organizations oppose the change? Do you believe the change would be beneficial in the long run?

• •

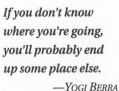

THE PROCESS OF PROBLEM SOLVING

Now that we've discussed the importance of units to problem solving and the most common systems of standardized units, we are ready to return to the problem-solving process. Perhaps the most important principle of problem solving is this: Be flexible, because no particular strategy always works!

If you don't know where you're going, you'll probably end up some place else.

—*Yogi Berra*

Nevertheless, the process of problem solving can be thought of in terms of four basic steps shown on the following page. You should think of this four-step process as a set of general guidelines that will help you stay organized for problem solving, rather than as an explicit set of rules. Note that only some of the detailed actions described in the four-step process may be needed for a particular problem.

By the Way ....

These four steps are modified from a process in *How to Solve It*, by George Polya (Princeton University Press, Princeton, N.J., 1957). First published in 1945, this book has sold more than one million copies and is available in at least 17 different languages.

A Four-Step Problem-Solving Process

Step 1.   **Understand the problem.**   The first step in problem solving is to determine where you are going. Be sure that you understand what the problem is asking.

- Think about the context of the problem (that is, how it relates to other problems in the real world) to gain insight into its purpose.
- Make a list or table of the specific information given in the problem.
- Draw a picture or diagram to help you make sense of the problem.
- Restate the problem in different ways to clarify its question.
- Make a mental or written model of the solution, into which you will fill in details as you work through the problem.

Step 2.   **Devise a strategy for solving the problem.**   Once you understand the problem, the next step is to decide how to go about solving it. This step is the most difficult, and is the one that requires creativity, organization, and experience.

- Obtain needed information that is not provided in the problem statement, using recall, estimation, or research.
- Make a list of possible strategies and hints that will help you select your overall strategy.
- Map out your strategy with a flow chart or diagram.
- If you have a mental model of the solution, think of your strategy as an argument whose conclusion is the solution; then, use your tools for constructing arguments to help formulate the strategy.

Step 3.   **Carry out your strategy, and revise it if necessary.**   After you have selected a strategy, the next step is to carry it out to find the result. In this step you are likely to use analytical and computational tools as you work through the mathematical details of the problem.

- Keep an organized, neat, and written record of your work, which will be helpful if you later need to review or study your solution.
- Double-check each step you take so that you do not risk carrying errors through to the end of your solution.
- Constantly reevaluate your strategy as you work; if you find a flaw in your strategy, return to step 2 and create a revised strategy.

> **Step 4. Look back to check, interpret, and explain your result.**   Although you may be tempted to feel like you have finished after you find a result in Step 3, this final step is the most important. After all, a result is useless if it is wrong, misinterpreted, or cannot be explained to others.
>
> - Be sure that your result makes sense; for example, be sure that it has the expected units, that its numerical value is sensible, and that it is a reasonable answer to the original problem.
>
> - Once you are sure that your result is reasonable, recheck your calculations or find an independent way of checking the result.
>
> - Identify and understand potential sources of uncertainty in your result.
>
> - Write your solution clearly and concisely, including discussion of any relevant uncertainties or assumptions.
>
> - Consider and discuss any pertinent implications of your result.

STRATEGIC HINTS FOR PROBLEM SOLVING

Using the four-step process will help you become better organized. But the only sure way to become more creative and improve your problem solving is through experience. We now offer several hints about problem solving, each illustrated with one or more examples. The hints themselves relate more to *attitudes* about problem solving than to actual techniques. Indeed, their primary purpose is to help you develop a "mind-set" that is conducive to enjoyable and successful problem solving.

Hint 1: There May Be More Than One Answer

How can society best reduce the total amount of greenhouse gases emitted into the atmosphere? We won't even attempt to answer this question, but it should make the point that no *single* best answer may be available. Indeed, many different political and economic strategies could yield similar reductions in greenhouse gas emissions.

Most people recognize that policy questions do not have unique answers, but the same is true of many mathematical problems. For example, both $x = 4$ and $x = -4$ are solutions to the equation $x^2 = 16$. Without further information and context, we have no way to determine whether both solutions are valid for a particular problem. Nonunique solutions often occur because not enough information is available to distinguish among a variety of possibilities.

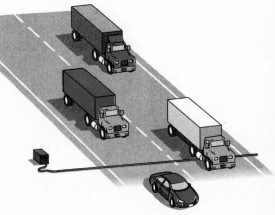

EXAMPLE 1   *Traffic Counters*

A traffic counter is a device designed to count the number of vehicles passing along a street. It usually is a thin black tube stretched across a street or highway and connected to a "brain box" at the side of the road. The device registers one "count" each time a set of wheels (that is, wheels on a single axle) rolls over the tube. A normal automobile registers two counts: one for the front wheels and one for the rear wheels. A light truck with three axles registers three counts. A large semitrailer truck might register four or five counts.

Suppose that, during a one-hour period, a particular traffic counter registers 35 counts on a residential street on which only two-axle vehicles (cars) and three-axle vehicles (light trucks) are allowed. How many cars and light trucks passed over the traffic counter?

Solution: We might begin by trial and error. For example, we can rule out the possibility of 12 light trucks, because at 3 counts per truck, 12 light trucks would yield $12 \times 3 = 36$ counts—more than the 35 registered. However, 11 light trucks would yield only $11 \times 3 = 33$ counts. Because a car registers 2 counts, 11 light trucks and 1 car would produce the 35 counts. We've found a solution, but it is not the only one. Note that 9 light trucks and 4 cars also is a solution: the 9 trucks yield $9 \times 3 = 27$ counts, and the four cars yield $4 \times 2 = 8$ counts, for a total of $27 + 8 = 35$ counts.

In fact, it turns out that six different combinations of cars and light trucks will produce a total count of 35. Yet, during a particular hour on a particular street, only one combination of cars and trucks actually passed that point. Unfortunately, with only the traffic counter information, we cannot determine which of the six solutions represents the actual traffic flow during the particular hour. ■

Hint 2: There May Be More Than One Strategy

Just as we must often admit more than one right answer, we should also expect more than one strategy for finding an answer. Mathematics and the human mind are far too rich and diverse to expect that everyone will follow the same path to a solution. As illustrated in the following example, an efficient strategy can save a lot of time and work.

EXAMPLE 2   *Jill and Jack's Race*

Jill and Jack ran a 100-meter race. When Jill crossed the finish line Jack had run only 95 meters. They decided to race again, but this time Jill started 5 meters behind the starting line. Assuming that both runners ran at the same pace as before, who won the second race?

Solution—Strategy 1: One approach to this problem is analytical, in which we analyze each race quantitatively. We were not told how fast either Jill or Jack ran, so we can choose some reasonable numbers. For example, we might assume that Jill completed the 100 meters in the first race in 20 seconds. In that case, her pace was: $100 \text{ m} \div 20 \text{ s} = 5 \text{ m/s}$, or 5 meters per second. Because Jack ran only 95 meters in the same 20 seconds, his pace was: $95 \text{ m} \div 20 \text{ s} = 4.75 \text{ m/s}$, or 4.75 meters per second.

We can analyze the second race with these numbers. Jill must run 105 meters in the second race (because she starts 5 meters behind the starting line) to Jack's 100 meters. Their times would be as follows:

$$\text{Jill:} \qquad 105 \text{ m} \div 5 \frac{\text{m}}{\text{s}} = 105 \text{ m} \times \frac{1 \text{ s}}{5 \text{ m}} = 21 \text{ s}$$

$$\text{Jack:} \qquad 100 \text{ m} \div 4.75 \frac{\text{m}}{\text{s}} = 100 \text{ m} \times \frac{1 \text{ s}}{4.75 \text{ m}} = 21.05 \text{ s}$$

Thus Jill will win the second race because she has the faster time.

Solution—Strategy 2: Although the analytical method works, we can use a much more intuitive and direct solution. We simply note that Jill runs 100 meters in the same time that Jack runs 95 meters. Therefore, in the second race, Jill will pull even with Jack 95 meters from the starting line. In the remaining 5 meters, Jill's faster speed will allow her to pull away and win. Note how this insight avoids the calculations needed in Strategy 1!   ■

Hint 3: Use Appropriate Tools

You don't need a computer to check your tab in a restaurant, and you don't need calculus to find the area of a rectangular room! For any given task there is an appropriate level of power that is needed, and it is a matter of style and efficiency to neither underestimate nor overestimate that level. You usually will have a choice of tools to use in any problem. Choosing the ones most suited to the job will make your task much easier.

EXAMPLE 3   *The Cars and the Canary*

Two cars, 120 miles apart, begin driving toward each other on a long straight highway. One car travels 20 miles per hour and the other 40 miles per hour (Figure 3.7). At the same time a canary, starting on one car, flies back and forth between the two cars as they approach each other. If the canary flies 150 miles per hour and turns around instantly at each car, how far has it flown when the cars collide?

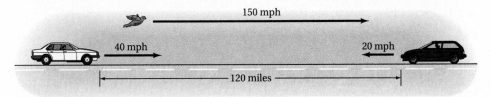

FIGURE 3.7

Solution: Because the problem asks "how far?" we might be tempted to calculate the *distance* traveled by the canary on each back and forth trip between the cars. However, these trips get shorter as the cars approach one another and we would have to add up all the individual distances. In principle, we would need to add up an *infinite* number of ever-smaller distances—a task that involves the mathematics of *calculus*.

But note what happens if we focus on *time* rather than distance. The cars will collide with each other at a speed of 60 mi/hr (because one car is traveling at 20 mi/hr and the other is traveling in the opposite direction at 40 mi/hr). Because they initially are 120 miles apart, they will collide in precisely 2 hours:

$$120 \text{ mi} \div 60 \frac{\text{mi}}{\text{hr}} = 120 \text{ mi} \times \frac{1 \text{ hr}}{60 \text{ mi}} = 2 \text{ hr}$$

We are told that the canary is flying at a speed of 150 mi/hr. Thus, in the two hours until the cars collide, the canary will fly

$$2 \text{ hr} \times 150 \frac{\text{mi}}{\text{hr}} = 300 \text{ mi.}$$

We've found the answer: the canary will fly 300 miles before the cars collide. We could have found the answer with calculus, but why bother when we were able to do it with just multiplication and division?   ■

*T*HINKING ABOUT...

Zeno's Paradox

The Greek philosopher Zeno of Elea (c. 460 B.C.) posed several paradoxes that defied solution for thousands of years. (A *paradox* is a situation or statement that seems to violate common sense or to contradict itself.) One begins with an imaginary race between the warrior Achilles and a slow-moving tortoise. The tortoise is given a head start, but our common sense says that the swift Achilles will soon overtake the tortoise and win.

Zeno suggested a different way to look at the race. Suppose that, as shown in Figure 3.8, Achilles starts from point P0 and the tortoise starts from P1. During the time it takes Achilles to reach P1, the slow-moving tortoise will move ahead a little bit to P2. While Achilles continues on to P2, the tortoise will move ahead to P3. And so on. Thus Achilles must cover an infinite set of ever-smaller distances to catch the tortoise (i.e., from P0 to P1, from P1 to P2, etc.). From this point of view, it seems that Achilles will never catch the tortoise.

This paradox puzzled philosophers and mathematicians for more than 2000 years. Its resolution depends on a key insight that became clear only with the invention of calculus in the seventeenth century: It does *not* necessarily require an infinite amount of time to cover an infinite set of distances. For example, imagine that the infinite set of distances covered by Achilles begins with 1 mile, then 1/2 mile, then 1/4 mile, and so

on. Then the total distance he covers, in miles, is

$$1 + \frac{1}{2} + \frac{1}{4} + \frac{1}{8} + \frac{1}{16} + \frac{1}{32} + \frac{1}{64} + \frac{1}{128} + \frac{1}{256} + \frac{1}{512}$$
$$+ \frac{1}{1024} + \frac{1}{2048} + \ldots$$

This sum is called an **infinite series** because it is the sum of an infinite number of terms. Note what happens when we add just the first four terms, then the first eight terms, and then the first twelve terms.

$$1 + \frac{1}{2} + \frac{1}{4} + \frac{1}{8} = 1.875$$

$$1 + \frac{1}{2} + \frac{1}{4} + \frac{1}{8} + \frac{1}{16} + \frac{1}{32} + \frac{1}{64} + \frac{1}{128} = 1.9921875$$

$$1 + \frac{1}{2} + \frac{1}{4} + \frac{1}{8} + \frac{1}{16} + \frac{1}{32} + \frac{1}{64} + \frac{1}{128} + \frac{1}{256} + \frac{1}{512}$$
$$+ \frac{1}{1024} + \frac{1}{2048} = 1.99951171875$$

Confirm these results with a calculator. If you continue to add more terms in this infinite series, you will find that the sum gets closer and closer to 2, but never exceeds it. It is possible to prove deductively that the sum of this infinite series is 2. So, if the fractions represent distances in miles run by Achilles, then he runs a total distance of only 2 miles (even though this sum involves an infinite set of distances.) Clearly, it won't take him long to run a finite distance of 2 miles, at which point he will pass the slower tortoise and win the race.

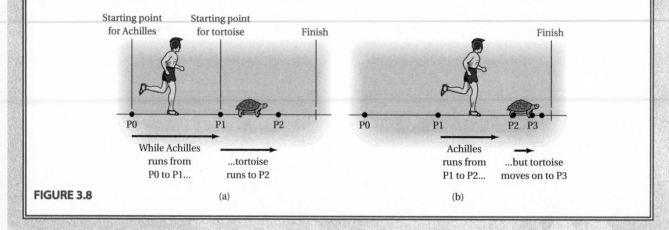

FIGURE 3.8   (a)   (b)

Hint 4: Consider Simpler, Similar Problems

Sometimes we are confronted with a problem that at first may seem daunting. Our fourth hint is to consider a simpler, but similar, problem. The insight gained from solving the easier problem may then help you understand the original problem.

EXAMPLE 4   *Coffee and Milk*

Suppose you have two cups in front of you: One holds coffee and one holds milk (Figure 3.9). You take a teaspoon of milk from the milk cup and stir it into the coffee cup. Next, you take a teaspoon of the mixture in the coffee cup and put it back into the milk cup. After the two transfers, there will be either: (1) more coffee in the milk cup than milk in the coffee cup; (2) less coffee in the milk cup than milk in the coffee cup; or (3), equal amounts of coffee in the milk cup and milk in the coffee cup. Which of these three possibilities is correct?

One teaspoon of milk is stirred into the coffee.

One teaspoon of coffee-milk is stirred into the milk.

Coffee   Milk   Coffee   Milk   Coffee   Milk

FIGURE 3.9

Solution: A cup of either milk or coffee contains something like a *trillion trillion* molecules of liquid. Clearly, it would be very difficult to visualize how such enormous numbers of molecules mix together, let alone to calculate the result. However, the essence of this problem is the *mixing* of two things. Thus one approach is to try a similar mixing problem that is much easier: mixing two piles of marbles.

Suppose that the "black pile" has ten black marbles and represents the coffee. The "white pile" has ten white marbles and represents the milk. In this simpler problem we can represent the first transfer—the teaspoon of milk into the coffee cup—by moving two white marbles to the black pile. This leaves the white pile with just eight white marbles, while the black pile now has ten black and two white marbles (Figure 3.10).

Two white marbles are moved to the black pile.

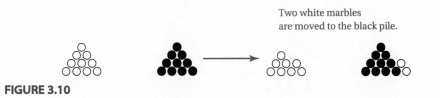

FIGURE 3.10

Representing the second transfer—the teaspoon from the coffee cup into the milk cup—involves taking two marbles from the black pile and putting them in the white pile. We can then ask a question analogous to the original question: Are there more black marbles in the white pile or white marbles in the black pile?

Because the marbles represent molecules that mix thoroughly, the two marbles for the second transfer must be drawn at random. This opens three possible cases: the two marbles in the second transfer can be either both black, both white, or one of each. However, as shown in Figure 3.11, all three cases yield the same essential result: we end up with the same number of white marbles in the black pile as black marbles in the white pile.

Two black marbles transfered

Two white marbles transfered

One of each transfered

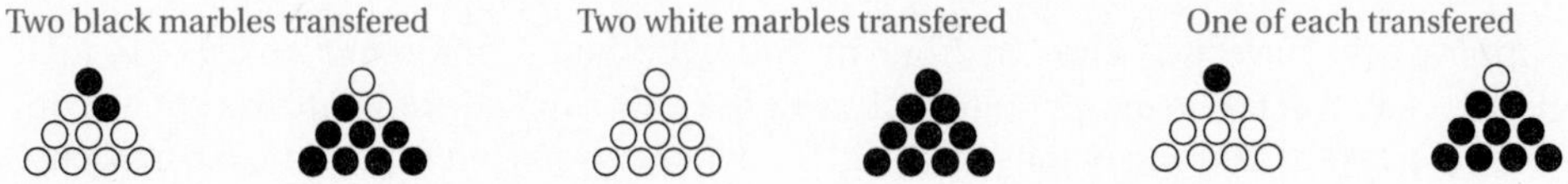

FIGURE 3.11

By analogy, we have the answer to our original question: After the two transfers, the amounts of coffee in the milk cup and milk in the coffee cup are equal.

The only remaining step is to confirm that the simpler problem is a reasonable representation of the real problem. Our choice of using two marbles to represent a teaspoon was arbitrary. If we redo this example with transfers of one, three, or any other number of marbles, we will find the same result: We always end up with the same number of black marbles in the white pile as white marbles in the black pile. Starting with ten marbles in each pile also was arbitrary; the conclusion remains the same if we start with twenty, fifty, or a trillion trillion marbles. Because molecules can be thought of as tiny marbles for the purpose of this problem, the real problem has no essential differences from the marble problem.   ■

> **TIME-OUT TO THINK:** Most people are surprised by the result of the coffee and milk problem. Are you? Now that you know the solution, can you give a simple explanation of the real problem that would satisfy surprised friends?

Hint 5: Consider Equivalent Problems with Simpler Solutions

Replacing a problem with a similar, simpler problem can reveal essential insights about a problem, as we've just seen. However, "similar" is not good enough when we need a numerical solution. In that case, a useful approach to a problem that appears difficult is to look for an *equivalent* problem. An equivalent problem will have the same numerical solution but may be easier to solve.

EXAMPLE 5   *A Coiled Wire*

Let's consider a problem of a type often encountered by plumbers, electricians, and engineers: measuring or wrapping a wire around a cylindrical pipe. Suppose that eight turns of a wire are wrapped around a pipe with a length of 20 centimeters and a circumference of 6 centimeters. What is the length of the wire?

Solution: This problem is difficult because it involves three-dimensional geometry. However, we can convert it into an equivalent problem by imagining that we cut the pipe along its length and unfold the pipe and wire into a flat rectangle (Figure 3.12). The width of the rectangle is the 6-cm circumference of the pipe, and its length is the 20-cm length of the pipe.

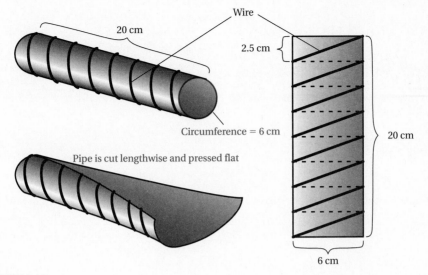

FIGURE 3.12

Now, instead of dealing with wire wrapped around a three-dimensional pipe, we have a simple rectangle with eight diagonal segments of wire. The total length of the eight segments must still be the same as in the original problem. As shown in the figure, each wire segment is the hypotenuse of a right triangle. The height of each triangle is $\frac{1}{8}$ of the total length of the rectangle, or 20 cm ÷ 8 = 2.5 cm. The base of each triangle is the 6-cm width of the rectangle. The Pythagorean theorem tells us that

$$\text{base}^2 + \text{height}^2 = \text{hypotenuse}^2 \qquad \text{or} \qquad \text{hypotenuse} = \sqrt{\text{base}^2 + \text{height}^2}.$$

Substituting the 6-cm base and 2.5-cm height yields

$$\text{hypotenuse} = \sqrt{(6\text{ cm})^2 + (2.5\text{ cm})^2} = 6.5\text{ cm}.$$

Thus the length of each wire segment is 6.5 cm, and the total length of the wire is 8 × 6.5 cm = 52 cm. Note how much easier it was to solve this equivalent problem than the original one! ◼

Hint 6: Approximations Can Be Useful

Another useful strategy is to make problems easier by using approximations. Most real problems involve approximate numbers to begin with, so an approximation often is good enough for a final answer. In other cases, an approximation will reveal the essential character of a problem, making it easier to reach an exact solution. Approximations also provide a useful check: if you come up with an "exact solution" that isn't close to the approximate one, something must have gone wrong!

EXAMPLE 6   *A Bowed Rail*

Imagine a mile-long bar of metal such as a railroad rail. Suppose that the rail is anchored on both ends (a mile apart) and that, on a hot day, its length expands by 1 foot. If the added length causes the rail to bow upward in a circular arc as shown in Figure 3.13(a) on the following page, about how high would the center of the rail rise above the ground?

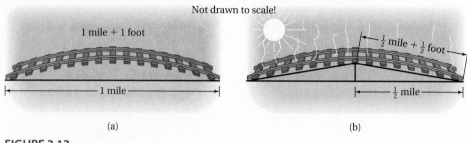

Not drawn to scale!

1 mile + 1 foot

$\frac{1}{2}$ mile + $\frac{1}{2}$ foot

1 mile

$\frac{1}{2}$ mile

(a)   (b)

FIGURE 3.13

Solution: Because the added length is short compared to the original length, we can approximate the curved rail with two straight lines (Figure 3.13b). We now have two right triangles and the Pythagorean theorem applies. The bases of the two right triangles together give the original rail length of 1 mile, so each base is $\frac{1}{2}$ mile long. The two hypotenuses together represent the expanded length of 1 mile + 1 foot, so each hypotenuse is $\frac{1}{2}$ mile + $\frac{1}{2}$ foot long. Because there are 5280 feet in a mile, $\frac{1}{2}$ mile equals 2640 feet. The height of the rail off the ground is approximately the height of the triangles.

$$\text{Height of triangle} = \sqrt{(2640.5 \text{ ft})^2 - (2640 \text{ ft})^2} = 51.4 \text{ ft}$$

Based on our approximation, the center of the rail would rise more than 50 feet off the ground! Because a triangle will stick up higher than a curve of the same base and length, the actual height would be less than what we found with the approximation; an exact solution shows that the top of the curved rail would be about 48 feet off the ground. However, real rails aren't a mile long, so we can conclude that the original question was not very realistic. ■

TIME-OUT TO THINK: Are you surprised by the answer to Example 6? How does the use of the approximation tell you that the original question contained at least some unrealistic assumptions? Which assumptions do you think were unrealistic?

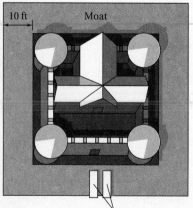

10 ft   Moat

Both planks - too short

FIGURE 3.14

Hint 7: Try Alternative Patterns of Thought

Try to avoid rigid patterns of thought that tend to suggest the same ideas and methods over and over again. Instead, you should approach every problem with an openness and freshness that allows innovative ideas to percolate. In its most wondrous form, this approach is typified by what Martin Gardner, a well-known popularizer of mathematics, calls "aha!" problems. These are problems whose best solution involves a penetrating insight that reduces the problem to its essential parts.

EXAMPLE 7   *The Moat*

A castle is surrounded by a deep moat 10 feet across (Figure 3.14). A knight on a rescue mission must cross the moat, but he has only two $9\frac{1}{2}$-foot planks (and no glue or nails). How does he do it?

Solution: If you are having difficulty with this problem, try thinking about different ways to place both planks together. The solution is shown in Figure 3.15 (see page 170). As with most "aha!" problems, the solution is undeniable once it is seen.   ■

EXAMPLE 8   *China's Population Policy*

Since 1978, the government of China has officially allowed only one child per family. The stated goal of this policy is to reduce China's population from more than 1.1 billion today to about 700 million by 2050. However, the policy has had other unintended social consequences. One of the most serious concerns is an apparent shortage of girls: instead of roughly equal numbers of boys and girls, boys considerably outnumber girls in China. The reasons for this shortage of girls are controversial, but in at least some cases, Chinese families apparently make sure that their one child is a boy with selective abortions.

One proposal for stopping such practices asks the government to replace its one-child policy with a one-*son* policy. For example, if the first child is a boy, a family has met its limit of children. However, if the first child is a girl, a family can have additional children until one is a boy. Suppose that the one-son policy were implemented. How would it affect the overall birth rate and the numbers of boys and girls?

Solution: One way to address this problem is by counting. Half the families would have a boy as their first and hence only child. Of the remaining families, half would have a boy for their second child (reaching their limit), while the other half would have a girl and go on to a third child. Half of these families would have a boy on this third try, while the other half would continue on after a third girl. And so on.

However, a moment of insight allows us to answer the two questions quickly. The issue of when a family stops having children cannot affect the natural probability that any child will be a boy or a girl: nearly equal numbers of boys and girls must be born. Moreover, *all* families would eventually have one boy under this policy. Because there will be nearly equal numbers of boys and girls overall, the average number of girls per family must also be one. Thus the average number of children per family must be two (one boy and one girl). A one-son policy would lead to nearly equal numbers of boys and girls, but would double the average number of children per family from one to two.   ■

> *By the Way* ·············
>
> Hospital birth records show that boys and girls are born nearly, but not precisely, in equal numbers: roughly 106 boys are born for every 100 girls. However, males have higher mortality rates than females, so the numbers even out in adulthood and women outnumber men in old age.

> **TIME-OUT TO THINK:**   China occupies roughly the same amount of land as the United States, but has more than four times as many people. Given these circumstances, do you think that the one-child policy is a good idea? Would a one-son policy be better? Defend your opinions.

Hint 8: Do Not Spin Your Wheels

Finally, everyone has had the experience of getting "bogged down" with a problem. When your wheels are spinning, let up on the gas! Often the best strategy in problem solving is to put a problem aside for a few hours or days. You will be amazed at what you might see (and what you overlooked) when you return to it.

REVIEW QUESTIONS

1. Describe the four basic steps of problem solving.

2. Summarize the strategic hints for problem solving given in this unit, with an example of the meaning of each one.

PROBLEMS

1. **Traffic Counters.** Refer to Example 1. Be sure to look for patterns to simplify your work.

   a. Find, by trial and error, all six solutions to the problem.

   b. Suppose that the same traffic counter registers 41 counts during another period of time. Find all possible combinations of two- and three-axle vehicles that might have passed.

   c. A similar traffic counter is placed on a major highway, where vehicles with two-, three-, and four-axles can travel (that is, vehicles that register 2, 3, or 4 counts). Suppose that, during a 5-minute period late at night, the counter registers 10 counts. Find all possible combinations of two-, three-, and four-axle vehicles that might have passed that point.

   d. Suppose that the traffic counter from part (c) registers 66 counts during a 1-minute period of heavy traffic. Find all possible combinations of two-, three-, and four-axle vehicles that might have passed.

2. **More on Jack and Jill.** Refer to Example 2.

   a. Suppose that Jill's time was 10 seconds in the first race. Following the solution of strategy 1, by how much would she win the second race?

   b. Hack and Quill race 200 meters and Hack wins by 10 meters. They race a second time, with Hack starting 10 meters behind the starting line. Who wins the second race? Explain.

3. **The Cars and Canary Revisited.** Refer to Example 3. Two cars, 150 kilometers apart, begin driving toward each other on a long, straight highway. One car travels 80 kilometers per hour and the other 100 kilometers per hour. At the same time a canary, starting on one car, flies back and forth between the two cars as they approach each other. If the canary flies 120 kilometers per hour and spends no time to turn around at each car, how far has it flown when the cars collide?

4. **Mixing Marbles.** Refer to Example 4. Consider the case in which each pile initially has 15 marbles. Suppose that on the first transfer *three* black marbles are moved to the white pile. On the second transfer, any three marbles are taken from the white pile and put into the black pile. Demonstrate, with diagrams and words, that you will always end up with as many white marbles in the black pile as black marbles in the white pile.

5. **The Coiled Wire Revisited.** Refer to Example 5. Suppose that 10 turns of a wire are wrapped around a pipe having a length of 20 centimeters and a circumference of 6 centimeters. What is the length of the wire?

6. **The Bowed Rail Revisited.** Consider a 1-kilometer-long rail, anchored on both ends. On a hot day, its length expands by one centimeter, causing the rail to bow upward. Use the approximation technique of Example 6 to determine about how high the center of the rail rises above the ground. Discuss your results.

7. **China's One-Child Policy.**

   a. To convince yourself that a one-son policy would lead to an average of two children per family, with equal numbers of boys and girls, do the following. Suppose that 100,000 families are having children according to the one-son policy. Describe the general makeup of all of the families (that is, start with the fact that 50,000 families have a boy as their first and therefore only child, and continue on). Use this process to show that the average number of children is two and that boys and girls are equal in number.

b. Suppose that, as the current generation matures, China's population of young adults has more men than women by a ratio of 118 to 100. With 400 million young adults in China, how many men will be unable to find a spouse?

Puzzle Problems. *Problems 8–22 are puzzle problems that require careful thinking and can help your problem-solving skills. Hint: look for* aha! *solutions.*

8. It takes you 30 seconds to walk from the first (ground) floor of a building to the third floor. How long will it take to walk from the first floor to the sixth floor (at the same pace, assuming that all floors have the same height)?

9. Reuben says, "Two days ago I was 20 years old. Later next year I will be 23 years old." How is this possible?

10. There are three kinds of apples all mixed up in a basket. How many apples must you draw (without looking) from the basket to be sure of getting at least two of one kind?

11. "Brothers and sisters I have none, but that man's father is my father's son." Who is "that man?"

12. "I am the brother of the blind fiddler, but brothers I have none." How can this be?

13. A woman bought a horse for $500 then sold it for $600. She bought it back for $700 then sold it again for $800. How much did she gain or lose on these transactions?

14. Three boxes of fruit are labeled *Apples*, *Oranges*, and *Apples and Oranges*. Each label is wrong. By selecting just one fruit from just one box, how can you determine the correct labeling of the boxes?

15. Each of ten large barrels is filled with golf balls that all look alike. The balls in nine of the barrels weigh one ounce and the balls in one of the barrels weigh two ounces. With only *one* weighing on a scale, how can you determine which barrel contains the heavy golf balls?

16. A woman is traveling with a wolf, a goose, and a mouse. She must cross a river in a boat that will hold only herself and one other animal. If left to themselves, the wolf will eat the goose and the goose will eat the mouse. How many crossings are required to get all four creatures across the river alive?

17. How do you measure nine minutes with a seven-minute and a four-minute hourglass?

18. A 150-foot rope is suspended at its two ends from the tops of two 100-foot flagpoles. The lowest point of the rope is 25 feet from the ground. What is the distance between the flagpoles?

19. You are considering buying 12 gold coins that look alike but have been told that one of them is a heavy counterfeit. How can you find the heavy coin in three weighings on a balance scale?

20. Suppose you have 40 blue socks and 40 brown socks in a drawer. How many socks must you take from the drawer (without looking) to be sure of getting a pair of the same color?

21. Alma, Bess, Cleo, and Dina visited Elf on Saturday. Alma visited at 8:00, Bess visited at 9:00, Cleo visited at 10:00, and Dina visited at 11:00. Some clues: (1) The times may be either AM or PM; (2) At least one woman visited Elf between Alma and Bess; (3) Alma did not visit Elf before both Cleo and Dina; (4) Cleo did not visit Elf between Bess and Dina. Who visited Elf last?

22. Three prisoners know that the jailer has three white hats and two red hats. The jailer gives one hat to each prisoner and says, "If you can deduce the color of your own hat, you will be freed." Each prisoner can see the hats of the other two prisoners, but not his own. The first prisoner says, "I cannot tell the color of my hat." The second prisoner says, "I cannot tell the color of my hat." The third prisoner is blind, but he is freed. What color hat does he have and how did he know?

Projects: Real Problems in the Real World. *Problems 23–31 describe complex problems that do not have single, straightforward solutions. For each, describe how you would apply the four-step problem-solving process described in the text (without actually carrying out the process to obtain a solution). You may choose either to enumerate the four steps or to describe your process in essay form. Either way, be sure that you list as many relevant factors as possible and discuss sources of uncertainty associated with each. Also, describe how you would work from these factors to find a solution. Conclude by describing your overall impression of whether the problem can be solved and whether any solution would likely generate controversy.*

23. You are asked to calculate the cost of installing enough bike racks on campus to solve a bicycle parking problem.

24. You want to know the number of new faculty members that would be needed, and the total cost to the university, of making sure that all classes have twenty or fewer students (in order to replace large lecture classes with smaller classes).

25. You want to know whether having a top-quality football program means more money for academics at a university.

26. You want to figure out how much taxes would have to be increased to provide public school teachers with twice their present salaries.

27. You decide that, in the interest of the environment, you will convert your home heating and hot water system to solar power. How much will this conversion cost or save over the next ten years?

28. Suppose that China and India decide to use their extensive coal reserves to supply energy to their populations at the same per capita level as in the United States. How much carbon dioxide would be added to the atmosphere?

29. Are automobile insurance companies gouging drivers? Suppose that you want to figure out whether they are justified in raising insurance rates as rapidly as they have during the past few years.

30. A large city of the American Southwest claims that it soon will be facing a severe shortage of water. Would that still be the case if people replace their current lawns with grasses or other plants that require less water?

31. Suppose that a city added new bus routes and handed out free bus passes. How many people would give up driving in favor of the bus? How much money, overall, would this cost or save the city?

Solution to puzzle given in Example 7.

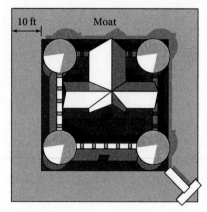

FIGURE 3.15

CHAPTER 3
SUMMARY

*I*n this chapter we focused on the problem-solving process. With the experience gained in this chapter, you will be able to apply new mathematical tools as you learn them. Keep in mind a few key ideas.

- Working with both numbers and the units of a problem is one of the best problem-solving techniques. Take advantage of unit analysis both to help you in setting up problems and in checking your answers.

- If you're going to work with units, it's important to be familiar with common standardized units. Although ancient customary units are still common in the United States, knowing metric units is crucial for international travel, commerce, and communication. Also, metric units are the units of choice in all scientific work.

- Problem solving is more of an art than a science, and requires both creativity and organization. The only sure way to improve at problem solving is by doing it!

SUGGESTED READING

Challenging Lateral Thinking Puzzles, P. Sloane, D. MacHale (New York: Sterling Publishing Co., Inc., 1993).

Great Lateral Thinking Puzzles, P. Sloane, D. MacHale (New York: Sterling Publishing Co., Inc., 1994).

How To Solve It, G. Polya (Princeton, New Jersey: Princeton University Press, 1945, 1985).

Mathematical Puzzles of Sam Loyd, S. Loyd (New York: Dover, 1959).

Mathematics: Problem Solving Through Recreational Mathematics, B. Averbach, O. Chein (San Francisco, California: W. H. Freeman, 1980).

More Mathematical Puzzles of Sam Loyd, S. Loyd (New York: Dover, 1960).

The Scientific American Book of Mathematical Puzzles and Diversions, Martin Gardner (New York: Simon and Schuster, 1959).

The Second Scientific American Book of Mathematical Puzzles and Diversions, Martin Gardner (New York: Simon and Schuster, 1961).

Chapter 4
NUMBERS IN THE REAL WORLD

*L*ife is filled with numbers that may at first seem incomprehensible: a population measured in billions of people, a national debt measured in trillions of dollars, and distances ranging from nanometers to light-years. Moreover, the numbers encountered in the real world nearly always involve uncertainties. In this chapter, we discuss the interpretation of the numbers we encounter in our daily lives. We emphasize common uses of percentages, techniques for putting large and small numbers into perspective, and methods for dealing with the uncertainty that nearly always arises in real-world problems.

And now for some temperatures around the nation: 58, 72, 85, 49, 77.
GEORGE CARLIN, COMEDIAN

The concept of number is the obvious distinction between the beast and man. Thanks to number, the cry becomes song, noise acquires rhythm, the spring is transformed into a dance, force becomes dynamic, and outlines figures.
JOSEPH DE MAISTRE, NINETEENTH-CENTURY FRENCH PHILOSOPHER

None of us really understands what's going on with all these numbers.
DAVID STOCKMAN, BUDGET DIRECTOR FOR PRESIDENT REAGAN, 1981

*M*athematics is sometimes called the language of nature. Just as words are fundamental to spoken languages, numbers are fundamental to mathematics. However, the numbers encountered in daily life often are different from the numbers you may remember from mathematics textbooks. Most school mathematics problems involve relatively small numbers that are exact, and those problems usually have exact solutions. Real problems involve numbers that may be extremely large, ambiguous, or highly uncertain. In this chapter, we investigate the concepts needed to understand numbers in the real world.

UNIT 4A

CONCEPTS OF NUMBER

We tend to take the use of numbers for granted, but numbers can have different meanings and can be used in different ways. We use numbers daily for at least three very different purposes: as *cardinal numbers* for counting, as *ordinal numbers* for ordering, and as *nominal numbers* for naming or labeling. Fortunately, the purpose of a number usually is obvious from its context, as the following examples show.

Using numbers for counting: If you board an airplane and a sign says that it has 75 seats, it means that you could *count* 75 seats on the airplane.

Using numbers for ordering: If your airplane ticket says you have seat 10, it means that the seats are arranged in some kind of order and you have the *tenth* seat in this order.

Using numbers for labeling: If a passenger boards wearing a basketball jersey with a large number 32, it simply is a label that identifies the person who wears it.

A BRIEF HISTORY OF NUMBERS

Humans first used numbers for counting. Animal bones with notches that appear to be tally marks for counting were carved as early as 35,000 B.C. in Africa and 30,000 B.C. in Europe. Going beyond simple tally marks for counting required the invention of more sophisticated **numerals**, or symbols for numbers. The first true numeral system probably was an **additive system** developed by the ancient Egyptians, prior to 3000 B.C. (Figure 4.1). In an additive system, a particular symbol always has the same value no matter where it is written, and the value of a number is obtained by adding up the values of each individual numeral.

*T*HINKING ABOUT...

Roman Numerals

Not all ancient numeral systems have disappeared. For example, **Roman numerals** remain popular in decorative applications, such as on clock faces. The Roman numeral system is additive and uses seven symbols:

$$I = 1, V = 5, X = 10, L = 50, C = 100,$$
$$D = 500, \text{ and } M = 1000.$$

Like the ancient Egyptian hieroglyphic numerals, the values of the Roman numerals always are the same. In general, we read Roman numerals by adding the values of the symbols. For example,

$$III = 1 + 1 + 1 = 3, \quad \text{and}$$
$$XXVII = 10 + 10 + 5 + 1 + 1 = 27.$$

However, when a symbol is followed immediately by a symbol of greater value, we *subtract* the smaller value from the larger. (Subtraction was not part of the original Roman numeral system; it was introduced in the sixteenth or seventeenth centuries—long after the decimal system had replaced it for most uses.) For example,

$$IV = 5 - 1 = 4, \quad \text{and}$$
$$XLIX = (50 - 10) + (10 - 1) = 49.$$

To see why mathematical operations are more difficult with additive numeral systems than with place-value systems, try adding MXLIX + CMLI. No simple rule gives you the answer MM. In contrast, it's easy to add the same numbers expressed as Hindu-Arabic numerals: 1049 + 951 = 2000.

For example, the Egyptians wrote the number 162 as follows:

$$\text{𓍢𓎆𓎆𓎆𓎆𓎆𓎆𓏤𓏤} = 100 + 6 \times 10 + 2 \times 1 = 162$$

In contrast, we use a **place-value system** in which the *placement* of a numeral affects its value. For example, each "7" in the number 777 represents a different number, depending on its *column*:

$$\underbrace{7}_{700} \quad \underbrace{7}_{70} \quad \underbrace{7}_{7}$$

The columns increase in value by factors of 10 (i.e., 1s, 10s, 100s, . . .).

The ancient Babylonians developed the earliest known place-value system by about 2000 B.C. Each column in the Babylonian system increased in value by a factor of 60 (i.e., 1s, 60s, 3600s, . . .). We still use this *base-60* system for keeping time (60 seconds = 1 minute and 60 minutes = 1 hour) and measuring angles (60 arcseconds = 1 arcminute and 60 arcminutes = 1 degree).

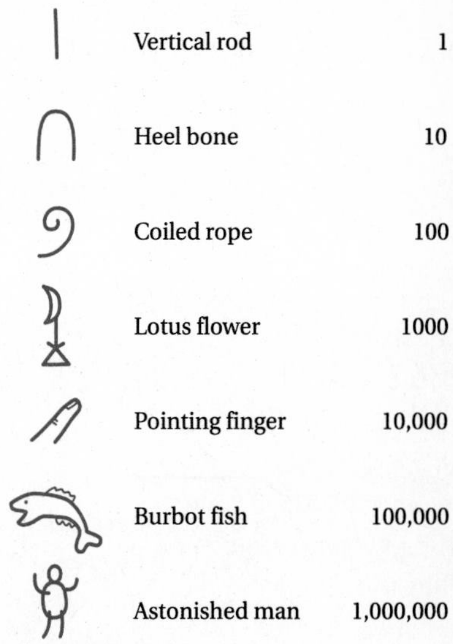

Symbol	Name	Value
|	Vertical rod	1
∩	Heel bone	10
𓏢	Coiled rope	100
	Lotus flower	1000
	Pointing finger	10,000
	Burbot fish	100,000
	Astonished man	1,000,000

FIGURE 4.1

Egyptian hieroglyphic numerals.

By the Way ············

The city of Babylon (meaning *gate of God*) was located between the Tigris and Euphrates Rivers in modern-day Iraq. It rose to prominence about 2500 B.C. and was conquered by the Persians in 539 B.C.

Mathematical **operations**—such as addition, subtraction, multiplication, and division—are much easier in place-value systems. The Babylonian system was not as easy as it might have been because it lacked a symbol for *zero*, a number necessary for identifying "empty" columns in a place-value system. Surprisingly, most ancient cultures never developed either the concept of zero or a symbol for it. The first people known to develop zero as a mathematical concept were the Mayans in Central America.

The concept of zero was introduced in the Eurasian world by Hindu mathematicians around A.D. 600. The Hindu symbols for zero and other numerals were adopted by Arab scholars around A.D. 800. The Arab world became the center of mathematical advancement for the next several centuries, as Arab mathematicians invented algebra and trigonometry. The works of the Arab scholars reached Western Europe in about A.D. 1200. By this time, the old Hindu numerals had been transformed into roughly their modern forms: the symbols 0, 1, 2, 3, 4, 5, 6, 7, 8, and 9. Because of their origin, our modern symbols are referred to as **Hindu-Arabic numerals**.

The ten Hindu-Arabic numerals are called **digits**, from the Latin *digitus*, meaning *finger*. Because there are ten different numerals and ten different place values, these numerals form a **decimal**, or **base-10**, number system.

THE MODERN NUMBER SYSTEM

By the Way ············

The word *decimal* comes from the Latin *decimus*, meaning *tenth*. The word *decimate* has the same root, and came from a Roman army practice in which every tenth man of a platoon was killed.

Over time, mathematicians expanded the concepts of numbers to include fractions, negative numbers, and more. We can build the entire number system used in modern mathematics by beginning with the counting numbers, or **natural numbers**. Because counting always begins at 1, we define the natural numbers as the set {1, 2, 3, 4, . . .}, where the symbol ". . ." means the pattern continues endlessly. We can represent the natural numbers on a *number line* with equally spaced dots beginning at 1 and continuing to the right forever.

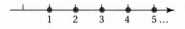

If we start from 0 instead of from 1, we define what mathematicians call the **whole numbers**: the set {0, 1, 2, 3, 4, . . .}. The number line shows that the whole numbers are the same as the natural numbers except for the addition of the number zero.

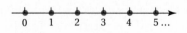

By the Way ············

On tax forms, negative numbers often are denoted by placing them in parentheses. In accounting, negative numbers are often written in red ink. Hence the expression "in the red" means being in debt, while "in the black" means having net assets.

Adding two whole numbers always results in another whole number; for example, 5 + 3 = 8. However, if we subtract a larger whole number from a smaller one, the result is a **negative number**. Mathematicians generally did not think of negative numbers as "true" numbers before about the sixteenth century, but negative numbers arise naturally in many applications. Their first use probably was in commerce, where debts or losses can be expressed as negative numbers. For example, if your business has income of $1000 and expenses of $1500, your net income is −$500; that is, you have a *loss* of $500. Other common uses of negative numbers are in temperature and elevation measurements. On the

Celsius scale, 0° is the temperature at which water freezes to ice; the temperature in your freezer must be below 0°C (say, −4°C) to keep ice cubes frozen. Elevations usually are measured with sea level as 0 elevation. Thus the elevation of a place below sea level, such as Death Valley, California, is −282 feet, or 282 feet *below* sea level.

The set of numbers that includes the whole numbers and their negatives is called the **integers**. We can write the set as $\{\ldots, -3, -2, -1, 0, 1, 2, 3, \ldots\}$, or represent it on a number line that shows the integers extending forever both to the left and right.

What happens when we multiply two integers? Suppose you have a *debt* of $100 in your checking account (a balance of −$100) and you double your debt. The new balance is $2 \times (-\$100) = -\$200$, which makes sense because your debt is twice as large. We see that multiplying a positive number and a negative number gives a negative number. It also says that if you multiply one number (such as the number 2) by a negative number (such as the number −$100), the sign of the first number and the sign of the product are opposite. This rule tells us that if we start with a negative number and multiply it by another negative number, the sign of the product must be opposite the sign of the first number; that is, the product of two negative numbers is positive. For example, $(-1) \times (-1) = 1$. To summarize:

- Multiplying or dividing two positive numbers yields another positive number: (positive) × (positive) = positive, **and** (positive) ÷ (positive) = positive.

- Multiplying or dividing a positive and negative number yields a negative number: (positive) × (negative) = negative, **and** (positive) ÷ (negative) = negative.

- Multiplying or dividing two negative numbers yields a positive result: (negative) × (negative) = positive, **and** (negative) ÷ (negative) = positive.

> **Mathematical Note:**
> Notice the similarity between the facts that (a) the product of two negative numbers is positive, and (b) the negation of the negation of a proposition is the original proposition (See Unit 1A).

Rational and Real Numbers

Dividing two integers yields a *fraction*, or **rational number** (as long as we do not divide by zero). The word rational comes from the word *ratio*, which refers to the comparison of two numbers by division. More technically, the rational numbers are the set of all numbers that can be expressed in the form

$$\frac{x}{y}, \quad \text{where } x \text{ and } y \text{ are integers and } y \neq 0.$$

Note that integers are rational numbers because any integer x can also be expressed as $\frac{x}{1}$. Ordinary fractions arise naturally whenever we divide something into pieces, and were used by merchants in ancient times—even before an accepted way of writing fractions was developed.

Not all numbers can be expressed as rational numbers. For example, the Pythagorean theorem tells us that a right triangle with two sides of length 1 has a hypotenuse of length $\sqrt{2}$ (Figure 4.2).

FIGURE 4.2

However, $\sqrt{2}$ cannot be expressed exactly in a form $\dfrac{x}{y}$, which means its decimal representation is neither terminating nor repeating. It can be written as $1.414213562\ldots$, where the dots mean the digits continue with no pattern forever. Hence, we say that $\sqrt{2}$ is an **irrational number**—a number that cannot be expressed as a ratio.

> **Mathematical Note:**   Terminating decimals, such as 0.25, represent rational numbers. Decimals that don't terminate also represent rational numbers if they have a repeating *pattern*. For example, $\frac{1}{3}$ is the never-ending decimal $0.3333\ldots$, and $\frac{1}{7}$ is the never-ending decimal $0.142857142857\ldots$ (the pattern 142857 repeats endlessly).

Pythagoras (c. 500 B.C.) inspired a group of followers who formed a "secret society" in ancient Greece that studied mathematics, music, and numbers. This Pythagorean Brotherhood (which admitted women as members) held great power for about two hundred years, and developed a mystical belief that *everything* could be understood in terms of numbers. One of the most sacred early beliefs of the Pythagoreans was that all numbers were rational. However, by applying their cherished Pythagorean theorem, they soon discovered irrational numbers like $\sqrt{2}$. Legend has it that this discovery was so devastating to the Pythagoreans that they attempted to keep it secret. Supposedly they even killed one of their members for telling others of the discovery of irrational numbers.

> **TIME-OUT TO THINK:**   The word *irrational*, literally, means "inexpressible as a ratio." Considering the story of the Pythagoreans, how do you think the word *irrational* came to mean *unreasonable* thinking or behavior?

Today, the combination of the rational and irrational numbers is called the **real numbers**. Each point on the number line has a corresponding real number, and each real number has a corresponding point on the number line. Therefore, the real numbers are the rational numbers and "everything in between." A few selected real numbers are shown on the number line below.

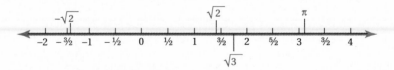

Imaginary and Complex Numbers

The operations addition, subtraction, multiplication, and division on real numbers always give another real number. But consider what happens when we take the square root of a real number. There is no problem if the number is positive. For example,

$$\sqrt{9} = 3 \qquad \text{because} \qquad 3 \times 3 = 9.$$

THINKING ABOUT...

Numerology

The mystical beliefs of the Pythagoreans were the beginnings of a long line of superstitions about numbers that we call **numerology**. For example, the Pythagoreans believed the number 7 had special meaning because there are 7 objects visible to the naked eye that move among the stars in the sky: the Sun, the Moon, Mercury, Venus, Mars, Jupiter, and Saturn. (The names of the 7 days of the week also have their roots in these 7 objects.) This bit of ancient mysticism remains with us today in the common belief that 7 is a "lucky" number. Other examples of ancient numerology include 13 being an unlucky number, and expressions such as

"the third time's the charm." Another form of numerology arises when letters are assigned numerical values. This form of numerology is especially prevalent when dealing with ancient texts, because ancient cultures often used letters to represent numbers. For example, the Caballah of Jewish mysticism interprets the Bible by looking at the numerical values of its Hebrew letters.

A more recent example of numerology concerns four great composers—Beethoven, Shubert, Bruckner, and Mahler. Each man died shortly after completing his ninth (and last) symphony. Speculating that nine symphonies must be some sort of natural limit, twentieth-century composer Arnold Schönberg suggested that nine symphonies must bring a composer "too close to the hereafter."

Dilbert ® by Scott Adams

DILBERT reprinted by permission of United Feature Syndicate, Inc.

However, no real number can be multiplied by itself to yield a negative number such as -4, so $\sqrt{-4}$ is not a real number. In fact, the square root of a negative number is *never* a real number.

To solve the problem of finding square roots of negative numbers, mathematicians invented **imaginary numbers**, or numbers that represent the square roots of negative numbers. A special number called i (for "imaginary") is defined to be the square root of -1. That is,

$$i = \sqrt{-1} \qquad \text{or, equivalently,} \qquad i^2 = \sqrt{-1} \times \sqrt{-1} = -1.$$

Using i, we can find the square root of any negative number. For example,

$$\sqrt{-4} = 2i \qquad \text{because} \qquad 2i \times 2i = 4 \times i^2 = 4 \times (-1) = -4.$$

> **TIME-OUT TO THINK:**   What happens if you try to take the square root of a negative number on your calculator? Why?

Imaginary numbers cannot be shown on a real number line because they are *not* real numbers. The **complex numbers** are the set of numbers that include all the real numbers *and* all the imaginary numbers. We will not work with complex numbers in this book, but they are very important and useful in mathematics, science, and engineering. Figure 4.3 summarizes the relationships among the various sets that make up our complete modern number system.

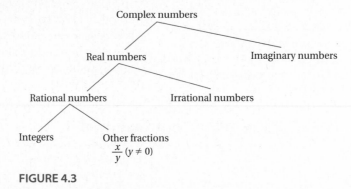

FIGURE 4.3

> **TIME-OUT TO THINK:**   Think about how the imaginary numbers are defined, and about the concept of numbers. Do you think that imaginary numbers are any less "real" than other numbers? Why or why not?

REVIEW QUESTIONS

1. Describe the three basic uses of numbers. Give an example of each.

2. What is the difference between a *number* and a *numeral*?

3. What do we mean by a *place-value system*? How does it differ from an *additive system* like that used by the ancient Egyptians?

4. Why is the number *zero* so important to place-value systems?

5. Why do we call our modern numerals *Hindu-Arabic numerals*? Why do we say that this numeral system is a *base-10* system?

6. Briefly define each of the following sets: natural numbers, whole numbers, integers, rational numbers, real numbers, and complex numbers.

PROBLEMS

• •

1. **Uses of Numbers.** For the numbers expressed in each statement, indicate whether they are being used for counting, ordering, or labeling. Explain your reasoning.

   a. "I have three boxes here for a Ms. Jones."

   b. Deli worker to crowd: "Next up is number 45!"

   c. "The price of the car is $19,200."

   d. "Go left at the next light; you'll see signs for Highway 66."

2. **Uses of Numbers.** For the numbers expressed in each statement, indicate whether they are being used for counting, ordering, or labeling. Explain your reasoning.

   a. The population of the United States now exceeds 250 million.

   b. "This is the fourth time I've told you to clean up your room!"

   c. National Football League guidelines: Any player wearing number 80 is supposed to be a tight end.

   d. Joe came in second in the 10-kilometer race.

3. **Classifying Numbers.** Consider this list of sets: natural numbers, integers, rational numbers, and real numbers. For each number given, identify from the list the *first* set that describes the type of number. Explain.

   a. 2.3   b. $-3/2$

   c. 3   d. -5

   e. 100.1   f. -6.1

   g. 5/3   h. π

4. **Classifying Numbers.** Consider this list of sets: natural numbers, integers, rational numbers, and real numbers. For each number given, identify from the list the *first* set that describes the type of number. Explain.

   a. 23.8   b. $\sqrt{3}$

   c. 15   d. -25

   e. 4/5   f. -14.1

   g. 22/7   h. $0.123123123\ldots$

5. **How Operations Affect Numbers.** For each of the following operations (i) explain whether the result is negative, positive, or zero; and (ii) explain whether the result would be a(n) integer, rational number, or real number.

   a. multiplying three negative integers

   b. dividing a positive integer by a negative rational number

   c. multiplying a positive rational number by a positive irrational number

   d. adding two irrational numbers

6. **How Operations Affect Numbers.** For each of the following operations (i) explain whether the result is negative, positive, or zero; and (ii) explain whether the result would be a(n) integer, rational number, or real number.

   a. multiplying a negative integer by a negative integer by a positive integer.

   b. subtracting a negative rational number from a positive irrational number

   c. dividing a positive integer by a negative integer and multiplying the result by a negative integer

   d. adding two positive rational numbers and dividing the result by a negative integer.

7. **Fractions to Decimals.** Convert each fraction to decimal form. If the decimal does not terminate, identify the repeating pattern of digits.

   a. $\dfrac{3}{4}$ b. $\dfrac{4}{11}$ c. $\dfrac{5}{8}$ d. $\dfrac{1}{6}$

   e. $\dfrac{4}{7}$ f. $\dfrac{11}{20}$ g. $\dfrac{37}{60}$ h. $\dfrac{23}{37}$

8. **Fractions to Decimals.** Convert each fraction to decimal form. If the decimal does not terminate, identify the repeating pattern of digits.

   a. $\dfrac{2}{5}$ b. $\dfrac{5}{6}$ c. $\dfrac{17}{51}$ d. $\dfrac{5}{11}$

   e. $\dfrac{3}{7}$ f. $\dfrac{7}{15}$ g. $\dfrac{11}{30}$ h. $\dfrac{19}{101}$

Thinking About Numbers. *Problems 9–14 describe a particular use of a number. For each problem do the following.*

  i) *Identify the first set that could always describe this number: natural numbers (positive integers), integers, rational numbers, or real numbers.*

  ii) *State whether the number is likely to be exact or approximate. Explain.*

  Example: *The balance in your checking account.*

  Solution: *(i) Rational numbers describe the account balance because it is expressed in dollars and fractions of dollars. (ii) The balance is approximate, rounded to the nearest cent.*

9. The number of students in your English class.

10. The number of people in Brazil.

11. The temperature outside your kitchen.

12. Your supply of chips in a poker game (you may not always be winning!).

13. The fraction of students at your school who are women.

14. The national debt.

A Language of Symbols. *Mathematics uses symbols to write general statements. For practice with symbols, rewrite the statements in Problems 15–18 in symbolic form, using the symbol indicated.*   Example: *Any number* n, *except 0, raised to the fifth power.*   Solution: $n^5, n \neq 0$.

15. The quotient of 10 divided by any number x except 0.

16. The product of 5 times any even number y.

17. The difference when 12 is subtracted from any number a.

18. The sum of 5 plus twice any number.

Mathematical Patterns. *A common character in mathematics is the* ellipsis (. . .), *which indicates the continuation of a pattern. In Problems 19–28, describe the pattern in words. Then either fill in the ellipsis or, if the ellipsis is at the end, write out the next three terms.*

Example: $1 + 2 + 3 + \cdots + 10$ Solution: *The pattern is adding a set of consecutive numbers. The pattern continues only from 1 through 10, so the entire statement would read:* $1 + 2 + 3 + 4 + 5 + 6 + 7 + 8 + 9 + 10$. *Note that the sum is 55.*

Example: $2, 4, 6, 8, \ldots$ Solution: *The pattern is counting by 2, or listing the even natural numbers. No stopping point is indicated, so the pattern continues forever; the next three entries following 8 would be 10, 12, 14.*

19. $1 + 2 + 3 + \cdots + 6$

20. $3, 6, 9, 12, \ldots$

21. $1 + 10 + 100 + 1000 + \cdots + 1{,}000{,}000$

22. $1 + 3 + 5 + 7 + 9 + 11 + \cdots$

23. $2^0 + 2^1 + 2^2 + 2^3 + \cdots + 2^{10}$

24. $2, 3, 5, 7, 11, 13, 17, 19, 23, 29, 31, \ldots$

25. $2, 4, 8, \ldots, 1024$

26. $1^2 + 2^2 + 3^2 + 4^2 + 5^2 + \cdots$

27. $1 + 5 + 14 + 30 + 55 + \cdots$

28. $1, 2, 3, 5, 8, 13, 21, 34, 55, 89, \ldots$ (Hint: Look at the *two* previous entries for each number. This pattern is called the *Fibonacci sequence*.)

29. **Reading and Writing Roman Numerals.** Convert each number either to or from a Roman numeral.

a. LIII	**b.** LXVI	**c.** CXXI
d. CLXXXI	**e.** MCDXLIII	**f.** XLIV
g. 4	**h.** 37	**i.** 41
j. 49	**k.** 106	**l.** 334

30. **Reading and Writing Roman Numerals.** Convert each number either to or from a Roman numeral.

a. CMLIII	**b.** CV	**c.** MMMCCXLIV
d. MMCMIII	**e.** CLXXII	**f.** CDLXXVIII
g. 972	**h.** 1995	**i.** 2001
j. 2462	**k.** 3789	**l.** 3540

31. **Roman Numerals in Use.**

    a. On the back of a U.S. one dollar bill is a picture of a pyramid with an eye. What date is written along the base of the pyramid? What is the significance of that date?

    b. Walking through Oxford University, England, you discover a building with MCCCLXXIX chiseled on the wall. What do you think it means? Would you expect to find a building with this Roman numeral in the United States? Why or why not?

32. **Project: Finding Square Roots.** Suppose you want to find $\sqrt{a}$, where a is any positive number. How can you find an answer without a calculator? Here are two approximation methods for finding $\sqrt{a}$.

Method 1 (attributed to the ancient Babylonians): Begin by choosing numbers b and c that satisfy the equation $a = b^2 + c$ (b and c can be any numbers, but it's easiest to choose natural numbers). For example, if $a = 10$, you could choose $b = 2$ and $c = 6$, since $2^2 + 6 = 10$, or you could choose $b = 3$ and $c = 1$, since $3^2 + 1 = 10$. Many other choices of b and c are possible, and some choices give better approximations than others. This method gives the approximation

$$\sqrt{a} \approx b + \frac{c}{2b}$$

Method 2 (attributed to Isaac Newton): Make an initial guess for $\sqrt{a}$, call it x, and then improve it with the new approximation

$$y = \frac{x}{2} + \frac{a}{2x}.$$

Then repeat the process by using the new approximation, y for x in this formula, and computing a new y. Each time you find a new y, it should be closer to the exact value of $\sqrt{a}$.

a. Use Method 1 to approximate $\sqrt{23}$ ($a = 23$). Try at least four different sets of values for b and c. Be sure to record your b and c values in each case, along with your result. How does each approximation compare to the exact answer you find on your calculator? Which values for b and c give the best approximation?

b. Use Method 2 to approximate $\sqrt{23}$. Record your starting guess x, and carry out the procedure, finding better approximations for y at least three times. How does this approximation compare to the ones found by Method 1 in part (a)? Does your initial guess for x matter? Explain.

c. Try both methods to approximate $\sqrt{66}$. Describe your work clearly, and compare your approximations to a precise answer from your calculator.

....

USES AND ABUSES OF PERCENTAGES

**UNIT
4 B**

One of the most common methods for communicating quantitative information is with percentages. Although a percentage is simply a fraction, percentages are often used in subtle ways. Consider the following statement that appeared in a front-page article in the *New York Times* (4/20/97):

The many percentages in this statement are all used correctly, but they may be very confusing. For example, what does the statement "the rate for eighth graders is up 44%, to 10.4%" mean? In this unit, we will learn to make sense of statements like this, and investigate some of the many ways that percentages can be used and abused.

THREE WAYS OF USING PERCENTAGES

Consider the following statements from recent news reports.

- A total of 13,000 newspaper employees, 2.6% of the newspaper work force, lost their jobs.

- Time-Warner stock rose 5.7% last week, to $41.75.

- High Definition Television sets (HDTV) have 125% more resolution than conventional TV, but will initially cost 400% more.

On close examination, each of these statements uses a percentage in a different way. The first statement uses a percentage as a *fraction*: "2.6%" is a fraction of the total work force. The second statement uses a percentage to describe a *change*: "5.7%" reflects an increase in the value of the stock. The third statement uses percentages to *compare* two quantities: the resolutions and costs of televisions.

Using Percentages as Fractions

Recall that *percent* means *divided by 100*. Thus percentages are easy to interpret or use as fractions.

EXAMPLE 1   *A Taste Test*

A taste test asks 125 people to choose between Clear Cola and Cloudy Cola. Forty-nine people prefer Clear Cola, and the rest prefer Cloudy Cola. Express the results as percentages.

Solution: Of the 125 people polled, 49 people preferred Clear Cola; that is, the fraction of the people who preferred Clear Cola is 49/125. We convert this fraction to a percent by first converting it to decimal form and then multiplying it by 100%.

$$\frac{49}{125} = 0.392 \times 100\% = 39.2\%$$

Thus 39.2% of the people sampled preferred the taste of Clear Cola. Because the total of the percentages favoring the two colas must be 100%, the percentage favoring Cloudy Cola was

$$100\% - 39.2\% = 60.8\%. \qquad \blacksquare$$

EXAMPLE 2   *Eighth Grade Smoking*

Suppose that 10.4% of eighth graders smoke. If there are 100,000 eighth graders in a particular state, how many smoke?

Solution: The "10.4 percent" represents the *fraction* of eighth graders who smoke. Thus, if there are 100,000 eighth graders, the number who smoke is 10.4% of 100,000, or

$$10.4\% \times 100,000 = 0.104 \times 100,000 = 10,400. \qquad \blacksquare$$

Using Percentages to Describe Change

Table 4.1 lists the approximate world population by decade since 1950. From 1950 to 1960, the population rose by

$$3.1 \text{ billion} - 2.6 \text{ billion} = 0.5 \text{ billion} = 500 \text{ million}.$$

We can also express this rise as a percentage. The population in 1950 was 2.6 billion, so a rise of 0.5 billion represents a percent increase of

$$\frac{0.5}{2.6} = 0.19 = 19\%.$$

(Note that we rounded this answer to two decimal places.) Thus the population from 1950 to 1960 rose by 19%. This example shows the two basic ways to describe change.

Table 4.1	
Year	**World Population (billions)**
1950	2.6
1960	3.1
1970	3.7
1980	4.5
1990	5.3
2000	6.1 (est.)

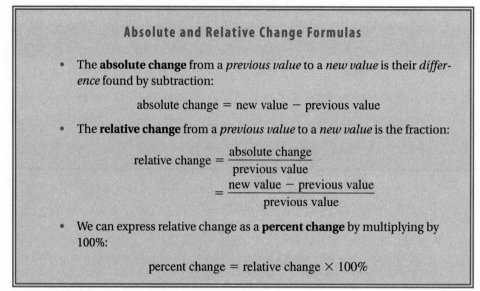

Note that the change found with these formulas is an *increase* (positive answer) if

$$\text{new value} > \text{previous value}.$$

The change is a *decrease* (negative answer) if

$$\text{new value} < \text{previous value}.$$

EXAMPLE 3   *Stock Gains*

Suppose that you invest $800 in the stock market. At the end of one year you sell the stock and receive $875 (after paying extra costs such as fees and commissions). What was the absolute change in the investment value? What was the relative change?

Solution: The *previous value* of your stock was the price you paid, or $800. The *new value* was the price you received, or $875. The absolute change was

$$\text{new value} - \text{previous value} = \$875 - \$800 = \$75.$$

Thus you gained $75 from your sale of the stock. The relative change in the value of the stock was

$$\frac{\text{absolute change}}{\text{previous value}} = \frac{\$75}{\$800} = 0.09375,$$

or $0.09375 \times 100\% = 9.375\%$.   ■

EXAMPLE 4   *Depreciation in Value*

You bought a computer three years ago for $2000. Today, it is worth only $600. Describe the absolute and relative change in the computer's value.

Solution: The *previous value* is the original price of $2000. The *new value* is the current value of $600. The absolute change in the computer's value is

$$\text{new value} - \text{previous value} = \$600 - \$2000 = -\$1400.$$

The answer has a negative sign because the new value is *less* than the previous value. **The relative change is**

$$\frac{\text{absolute change}}{\text{previous value}} = \frac{-\$1400}{\$2000} = -0.7 = -70\%.$$

The computer is now worth 70% *less* than it was three years ago. Note that we can also say that the computer's new value is 30% of its original value, because

$$\frac{\text{new value}}{\text{previous value}} = \frac{\$600}{\$2000} = 0.3 = 30\%.$$

Thus the statement that "the new value is 30% *of* the previous value" is equivalent to "the new value is 70% *less* than the previous value." ∎

Using Percentages for Comparisons

Suppose that we want to compare the price of a $50,000 Mercedes to the price of a $40,000 Cadillac. The difference between the Cadillac and Mercedes prices is

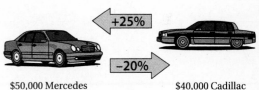

$50,000 Mercedes $40,000 Cadillac

$$\$50,000 - \$40,000 = \$10,000.$$

That is, the Mercedes costs $10,000 *more than* the Cadillac.

We can also express this difference as a percentage of the Cadillac's price:

$$\frac{\$10,000}{\$40,000} = 0.25 = 25\%.$$

Thus, in relative terms, the Mercedes costs 25% *more than* the Cadillac.

Because we are *comparing to* the Cadillac price, we say that the Cadillac price is the **reference quantity**. The Mercedes price is the **compared quantity**. We can now define the absolute and relative difference between the quantities, much as we defined absolute and relative change:

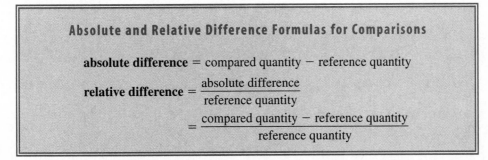

These formulas give positive answers if the compared quantity is *greater than* the reference quantity, and negative answers if the compared quantity is *smaller than* the reference quantity.

That's essentially all there is to comparisons, except for one subtlety: There's no particular reason why we chose to compare to the Cadillac price. We can just as easily compare to the Mercedes price. In that case, the Mercedes price is the reference quantity and the Cadillac price is the compared quantity, and the absolute difference is

$$\text{absolute difference} = \text{Cadillac price} - \text{Mercedes price}$$
$$= \$40{,}000 - \$50{,}000 = -\$10{,}000.$$

The negative sign tells us that the Cadillac costs $10,000 *less than* the Mercedes. The relative difference is

$$\text{relative difference} = \frac{\text{absolute difference}}{\text{Mercedes price}}$$
$$= \frac{-\$10{,}000}{\$50{,}000} = -0.2 = -20\%.$$

The negative sign indicates that the Cadillac costs 20% *less than* the Mercedes.

Note that we now have two ways to express the relative difference:

- The Mercedes costs 25% more than the Cadillac; or
- The Cadillac costs 20% less than the Mercedes.

Both statements are correct, yet they contain different percentages! Thus we see the importance of keeping careful track of the reference and compared quantities. In general, the reference quantity is preceded by the key word *than*.

EXAMPLE 5   *Comparing Incomes by State*

In 1995, the per capita income was $16,531 in Mississippi (lowest in the nation) and $30,303 in Connecticut (highest in the nation). In absolute and relative terms, how much higher is the per capita income in Connecticut than in Mississippi?

Solution: The words *than in Mississippi* tell us that the reference quantity is *income in Mississippi* and the compared quantity is *income in Connecticut*. We find the following values.

$$\text{absolute difference} = \text{Connecticut income} - \text{Mississippi income}$$
$$= \$30{,}303 - \$16{,}531 = \$13{,}772$$
$$\text{relative difference} = \frac{\text{absolute difference}}{\text{Mississippi income}} = \frac{\$13{,}772}{\$16{,}531}$$
$$= 0.8331 = 83.31\%$$

In absolute terms, per capita income in Connecticut is $13,772 higher than in Mississippi. In relative terms, per capita income in Connecticut is about 83% higher than in Mississippi.   ■

EXAMPLE 6   *Reversing the Compared and Reference Quantities*

Using the data from the previous example, how much lower in relative terms is the per capita income in Mississippi than in Connecticut?

Solution: This time we are using Connecticut income as the reference quantity. The absolute difference is −$13,772 because Mississippi income is lower than Connecticut income. The relative difference is

$$\text{relative difference} = \frac{\text{absolute difference}}{\text{Connecticut income}} = \frac{-\$13,772}{\$30,303} = -0.45 = -45\%.$$

The negative sign indicates that Mississippi's per capita income is 45% *lower* than Connecticut's.   ■

THE *OF VERSUS MORE THAN* RULE FOR PERCENTAGES

Consider again a $50,000 Mercedes and a $40,000 Cadillac. The Mercedes costs 25% *more than* the Cadillac. But note also that the Mercedes price is 1.25 *times* the Cadillac price, or 125% *of* the Cadillac price:

$$\frac{\text{Mercedes price}}{\text{Cadillac price}} = \frac{\$50,000}{\$40,000} = 1.25 = 125\%$$

This example illustrates an important fact about percentages, which we will call the *of versus more than* rule:

If the compared quantity is *P% more than* the reference quantity, then it is $(100 + P)\%$ *of* the reference quantity.

This rule also applies to change problems if we identify the *new value* as the compared quantity and the *previous value* as the reference quantity.

EXAMPLE 7   *Applying the* Of Versus More Than *Rule*

Suppose that Carol earns 50% *more than* William. How many times larger is her income than his?

Solution: Using the *of versus more than* rule, Carol's income is   (100 + 50)%,   or 150% of William's income. Because   150% = 1.5,   Carol's income is 1.5 *times* William's income.   ■

EXAMPLE 8   *Two Hundred Percent More*

Suppose that the retail cost of a diamond is 200% more than its wholesale cost. If the wholesale cost is $2000, what is the retail cost?

Solution: A retail cost of 200% *more than* the wholesale cost means that the retail cost is $(200 + 100)\% = 300\%$ *of* the wholesale cost. That is,

$$\text{retail cost} = 300\% \times \text{wholesale cost} = 3 \times \text{wholesale cost.}$$

Substituting $2000 for the wholesale cost, we find

$$\text{retail cost} = 3 \times \$2000 = \$6000.$$ ■

> **TIME-OUT TO THINK:**   Example 8 shows that it's possible for a retail price to be 200% *more than* a wholesale price. Is it possible for a wholesale price to be 200% *less* than the retail price? Why or why not?

EXAMPLE 9   *Sale!*

A store is having a "25% off" sale. How does a sale price compare to the original price?

Solution: In this case, the "25% off" means that a sale price is 25% *less* than the original price. If 25% is taken from the original price, then 75% of the original price remains. Thus a sale price is 75% of the original price. For example, an item with an original price of $100 will have a sale price of $75. Note that we can find the same answer with the *of versus more than* rule by observing that the sale price is −25% *more* than the original price, which means that it is

$$(100 + [-25])\% = 75\%$$

of the original price.   ■

> **TIME-OUT TO THINK:**   One store advertises "$\frac{1}{3}$ off everything!" Another store advertises "sale prices just $\frac{1}{3}$ of original prices!" Which store is having the bigger sale? Explain.

SOLVING PERCENTAGE PROBLEMS

Consider the Cadillac and Mercedes problem once more. Suppose that we know the Mercedes costs 25% more than the Cadillac, but are given only the Cadillac price of $40,000 (the reference quantity). How do we calculate the Mercedes price?

Using the *of versus more than rule*, the "25% more than" means that the Mercedes price is $(100 + 25)\% = 125\%$ *of* the Cadillac price:

$$\text{Mercedes price} = 125\% \times \text{Cadillac price}$$

Substituting the $40,000 Cadillac price, we find

$$\text{Mercedes price} = 1.25 \times \$40,000 = \$50,000.$$

Now, suppose instead that we were given the Mercedes price of $50,000 (the compared quantity) and asked to calculate the Cadillac price. Again, we have:

$$\text{Mercedes price} = 125\% \times \text{Cadillac price}$$

This time, we solve for the Cadillac price by dividing both sides by 125%:

$$\text{Cadillac price} = \frac{\text{Mercedes price}}{125\%} = \frac{\$50,000}{125\%} = \frac{\$50,000}{1.25} = \$40,000$$

We can now generalize our results.

> If the compared quantity is *P% more* than the reference quantity, the following equation holds:
>
> compared quantity $= (100 + P)\% \times$ reference quantity

Note that if we seek the compared quantity and are given the reference quantity, we simply substitute the numbers and *multiply*. We call this the *forward solution* to the percentage problem. If, instead, we seek the reference quantity and are given the compared quantity, we must solve the equation by *dividing* both sides by $(100 + P)\%$:

> $$\text{reference quantity} = \frac{\text{compared quantity}}{(100 + P)\%}$$

We call this the *backward solution* to the percentage problem. The forward and backward solutions also apply to change problems if we identify the *new value* as the compared quantity and the *previous value* as the reference quantity.

EXAMPLE 10   *HDTV*

Consider the statement:

> *High definition television (HDTV) sets cost 400% more than conventional TV sets.*

Suppose that a particular conventional TV set costs $250. How much would a comparable HDTV set cost?

Solution: The words *more than conventional TV* tell us that the reference quantity is the conventional TV price and the compared quantity is the HDTV price. We are given that the HDTV price is 400% *more than* the conventional TV price, so the following equation holds:

$$\text{HDTV price} = (100 + 400)\% \times (\text{conventional TV price})$$
$$= 500\% \times (\text{conventional TV price})$$

We are given the conventional TV price of $250, so we substitute the numbers to find that

$$\text{HDTV price} = 500\% \times \$250 = 5 \times \$250 = \$1250.$$

The HDTV would cost $1250, or 5 times as much as the conventional TV.   ∎

EXAMPLE 11   *Tax Calculations*

a) Suppose that the local sales tax rate is 5% and you purchase a shirt with a pretax price of $17. What is the total price of the shirt?

b) Suppose that the local sales tax rate is 7% and the total price of a compact disk is $19.40. What is the pretax price?

Solution: For both parts, we can consider the pretax price as the *previous value* and the aftertax price as the *new value*. The tax rate is *P*%. Thus the following equation holds:

$$\text{aftertax price} = (100 + P)\% \times \text{pretax price}$$

a) The tax rate of 5% means we set $P = 5$. Substituting the pretax price of $17, we solve for the aftertax price of the shirt:

$$\text{aftertax price} = (100 + 5)\% \times \$17 = 1.05 \times \$17 = \$17.85$$

The total price of the shirt with tax is $17.85.

b) The tax rate of 7% means we set $P = 7$ and the equation becomes

$$\text{aftertax price} = 107\% \times \text{pretax price}.$$

This time we are given the aftertax price of $19.40, so we solve for the pretax price by dividing both sides by 107%.

$$\text{pretax price} = \frac{\text{aftertax price}}{(107)\%} = \frac{\$19.40}{107\%} = \frac{\$19.40}{1.07} = \$18.13$$

The pretax price of the compact disk was $18.13. ■

EXAMPLE 12   *Percentages in the News*

Consider the statement that *a total of 13,000 newspaper employees, 2.6% of the newspaper work force, lost their jobs.* How large was the total work force before the layoffs? How large is it now?

Solution: The statement tells us that 2.6% of the old work force was 13,000 employees. Mathematically, this statement becomes

$$2.6\% \times \text{old work force size} = 13,000.$$

This is a slight variation on our previous percentage problems. To solve for the old work force size, we divide both sides by 2.6% to find

$$\frac{2.6\% \times \text{old work force size}}{2.6\%} = \frac{13,000}{2.6\%}, \quad \text{or}$$

$$\text{old work force size} = \frac{13,000}{0.026} = 500,000.$$

The old work force had 500,000 employees. The new work force has 13,000 fewer, or

$$500,000 - 13,000 = 487,000 \text{ employees.} ■$$

ABUSES OF PERCENTAGES

We've seen that it's important to be aware of several subtleties when working with percentages. Unfortunately, some people are less careful than others, which can lead to percentages being used incorrectly. Let's look at just a few of the many ways that percentages can be abused.

More Is Less: Shifts in the Previous Value

Imagine that you are forced to accept a temporary 10% pay cut. A few weeks later, you receive a 10% pay raise. Is your pay back where it started?

Suppose that your original weekly pay was $200. A 10% pay cut means that the new value of your pay is 10% *less than*, or $(100 - 10)\% = 90\%$ *of* its previous value:

$$\text{new value} = 90\% \times \text{previous value} = 0.9 \times \$200 = \$180$$

When you get the 10% raise, the new value of your pay is 10% more than, or $(100 + 10)\% = 110\%$ *of,* its previous value. However, the previous value now is $180, so your new pay is

$$\text{new pay} = 110\% \times \text{previous value} = 1.1 \times \$180 = \$198.$$

Thus the 10% pay cut followed by the 10% pay raise leaves you short of where you began! This result arises because the *previous value* shifted during the problem: it was $200 in the first calculation and $180 in the second.

> **TIME-OUT TO THINK:** Suppose your original weekly pay had been $300. Would you still lose money if you get the 10% pay cut followed by the 10% pay raise? Would you lose money for *any* weekly pay? What if you first get the 10% pay raise and then get the 10% pay cut?

EXAMPLE 13   *Shifting Investment Value*

A stockbroker is confronted by angry investors and offers the following defense: "I admit that the value of your investments under my management fell by 60% during my first year on the job. This year, however, their value has increased by 75%! So stop complaining." Evaluate the stockbroker's defense.

Solution: The stockbroker's defense is clever because it states that your percentage gain in the second year was larger than your percentage loss in the first year. However, the 75% second-year gain leaves you well short of recovering your 60% first-year loss.

Imagine that you began with an investment of $1000. During the first year, your investment lost 60% of its value, or $600, leaving you with $400. During the second year, you gained 75% of this $400, or

$$0.75 \times \$400 = \$300,$$

bringing your investment value to $400 + $300 = $700. This is still well short of your original investment of $1000. ∎

EXAMPLE 14   *Tax Cuts*

A politician promises, "If elected, I will cut your taxes by 20% for each of the first three years of my term, for a total cut of 60%." Evaluate the promise.

Solution: The politician neglected the effects of the shifting previous value. A cut of 20% in each of three years will *not* make an overall cut of 60%. To see what really happens, suppose that you currently pay $1000 in taxes. The following table shows how your taxes change over the three years.

Year	Tax paid in previous year	20% of previous year tax	New tax this year
1	$1000	$200	$800
2	$800	$160	$640
3	$640	$128	$512

Over three years, your taxes decline by

$$\$1000 - \$512 = \$488.$$

But $488 is only 48.8% of $1000, so your tax bill declines by 48.8% of its original value, not by 60%! Three consecutive 20% cuts add up to a total cut of only 48.8%. ■

Percentages of Percentages

Misunderstanding often arises when a changing quantity is *itself* a percentage. Suppose that your bank increases the interest rate on your savings account from 3% to 4%. It's tempting to say that the interest rate increased by 1%, but that statement is ambiguous at best. The interest rate increased by 1 *percentage point*, but the relative change in the interest rate is

$$\text{relative change} = \frac{\text{new value} - \text{previous value}}{\text{previous value}} = \frac{4\% - 3\%}{3\%} = 0.33 = 33\%.$$

Thus you can say that the bank raised your interest rate by 33%, even though the actual rate increased only by one percentage point from 3% to 4%!

EXAMPLE 15   *Newspaper Readership Declines*

According to a *Time* magazine story: *the percentage of adults reading daily newspapers fell from 78% in 1970 to 64% in 1995.* Describe the change in newspaper readership.

Solution: The absolute change in readership is

$$\text{new value} - \text{previous value} = 64\% - 78\% = -14\%.$$

The negative sign indicates a *decrease* in readership. To avoid confusion with the relative change, we can say that newspaper readership decreased by 14 *percentage points*. The relative change is

$$\frac{\text{new value} - \text{previous value}}{\text{previous value}} = \frac{64\% - 78\%}{78\%} = -0.18 = -18\%.$$

Thus readership declined by 18%, from an old value of 78% to a new value of 64%. Note how carefully this statement must be phrased! ■

EXAMPLE 16   *Up 44%, to 10.4%*

Consider the following statement from the quote at the beginning of this unit: *the [tobacco use] rate for eighth graders is up 44 percent, to 10.4 percent.* What was the previous rate of tobacco use for eighth graders?

Solution: The "10.4 percent" means that 10.4% of all eighth graders now smoke. The "44 percent" expresses the change from the previous rate of tobacco use to the new rate. It tells us that the new rate is 44% *more than* the previous rate, which means it is $(100 + 44)\% = 144\%$ *of* the previous rate. Thus the following equation holds:

$$\text{new rate} = 144\% \times \text{previous rate} = 1.44 \times \text{previous rate}$$

We solve for the previous rate by dividing both sides by 1.44, and substitute 10.4% for the new rate.

$$\text{old rate} = \text{previous rate} = \frac{\text{new rate}}{1.44} = \frac{10.4\%}{1.44} = 7.2\%$$

Thus the previous rate of tobacco use among eighth graders was 7.2%. Note that, in absolute terms, the new rate is $10.4 - 7.2 = 3.2$ percentage points higher than the old rate.   ∎

Telling Only Part of the Story

Relative change by itself can lead to misleading interpretations. Be wary of relative numbers when you don't have at least some of the absolute numbers for comparison.

EXAMPLE 17   *Does Marijuana Use Lead to Harder Drugs?*

A 1994 survey conducted by the National Center on Addiction and Substance Abuse at Columbia University asked children whether they had smoked marijuana and whether they had tried cocaine. Among the marijuana users, 17% said they had tried cocaine. Among nonmarijuana users, only 0.2% had tried cocaine. The survey report concluded that "children who smoke marijuana are 85 times more likely to use cocaine than those who don't." Is the conclusion accurate? Does it tell the whole story?

Solution: If we divide the 17% of marijuana users who had tried cocaine by the 0.2% of nonmarijuana users who had tried cocaine, we find $\frac{17\%}{0.2\%} = 85$. Thus it is true that the children who smoke marijuana were 85 times more likely to use cocaine than those who didn't. By itself, this fact might seem like very strong evidence that marijuana use leads to cocaine use. However, it is also true that $100\% - 17\% = 83\%$ of the marijuana users had *not* tried cocaine. With this additional piece of the story, the question of whether marijuana use leads to cocaine use is much more debatable.   ∎

REVIEW QUESTIONS

1. Describe the three basic uses of percentages. Give a sample statement that uses percentages in each of the three ways.

2. What is the difference between *absolute* and *relative* change? Explain the usage of the absolute and relative change formulas.

3. What do we mean by absolute and relative *differences*? Explain the usage of the absolute and relative difference formulas.

4. When given a statement of comparison between two quantities, how can you tell which one is the reference quantity and which one is the compared quantity?

5. What is the *of versus more than* rule? Give an example of its use.

6. Suppose we are given that a compared quantity is some percentage greater than a reference quantity. What equa-

tion must hold? How do we solve it if we are given the reference quantity? How do we solve it if we are given the compared quantity?

7. Suppose that the stock market declines by 20% one week, then gains 20% the next week. Has it returned to its original value? Explain why or why not.

8. Consider the following statement: From 1997 to 1998, the number of students passing the exam improved by 50%, from 20% to 30%. Does this statement make any sense? Explain.

PROBLEMS

1. Percentages as Fractions. Express the following fractional quantities as percentages.

 a. 25 women in a room of 113 people.

 b. 345 blooming tulips in a field of 398 tulips.

 c. 1234 people who voted for the losing candidate out of 3009 voters.

2. Percentages as Fractions. Express the following fractional quantities as percentages.

 a. 38 men at a convention of 236 people.

 b. 23 purple jelly beans in a jar of 123 jelly beans.

 c. 2345 people who voted for the winning candidate out of 4023 voters.

3. Working with Percentages as Fractions. In the following statements, you are told that an amount is some percentage of a total. In each case, find the total.

 a. 23 is 3% of the total.

 b. 100 is 15% of the total.

 c. 150 is 89% of the total.

4. Working with Percentages as Fractions. In the following statements, you are told that an amount is some percentage of a total. In each case, find the total.

 a. 23 is 97% of the total.

 b. 100 is 45% of the total.

 c. 150 is 15% of the total.

5. Percentages for Change. Express each change in terms of absolute change, relative change, and percent change. Don't forget that a decrease is indicated by a negative change.

 a. a change from 50 to 55

 b. a change from 0.1 to 0.5

 c. a change from 1000 to 1001

6. Percentages for Change. Express each change in terms of absolute change, relative change, and percent change. Don't forget that a decrease is indicated by a negative change.

 a. a change from 23 to 20

 b. a change from 50,000 to 45,000

 c. a change from 0.0001 to 0.001

7. Change Everywhere. Translate each (factual) statement into a statement about percent change.

 a. The U.S. population has changed from 220 million people in 1974 to 265 million people in 1997.

 b. The number of deaths from heart disease decreased from 250,000 in 1970 to 140,000 in 1993.

 c. Defense spending by the U.S. government decreased from $266 billion in 1995 to $254 billion in 1997.

8. Change Everywhere. Translate each (factual) statement into a statement about percent change.

 a. Consumer debt in the United States increased from $160 billion in 1974 to $850 billion in 1995.

b. The percentage of bachelor's degrees awarded to women increased from 44% in 1972 to 54% in 1992. The percentage of doctoral degrees awarded to women changed from 18% to 38%.

c. The percentage of single-parent households in the United States increased from 12% (of all households) in 1970 to 30% (of all households) in 1995.

9. Fifty Years of World Population Growth. Using the data in Table 4.1, describe the relative change in world population between the following years:

a. 1950 and 1970   **b.** 1950 and 2000

c. 1970 and 2000   **d.** 1990 and 2000

10. Percentages for Comparison. Answer the comparison question that accompanies each given fact. Express the answer in both absolute and percentage terms.

a. Given that 1 meter = 39.37 inches and 1 yard = 36 inches, how much larger is 1 meter than 1 yard?

b. Given that 1 quart = 0.94 liters, how much smaller is a quart than a liter?

c. Colorado has an area of 104,000 square miles and Norway has an area of 125,000 square miles. How much larger is Norway than Colorado?

d. The life expectancy in Canada is 79.1 years (highest in the world for 1997). The life expectancy in the United States is 76.0 years. How much higher is the life expectancy in Canada than in the United States?

11. Percentages for Comparison. Answer the comparison question that accompanies each given fact. Express the answer in both absolute and percentage terms.

a. Given that $1 U.S. = 0.6 English pounds, how much larger is 1 pound than $1?

b. Given that 1 mile = 1.6 kilometers, how much smaller is a kilometer than a mile?

c. The population of California is 31.6 million and the population of New York is 18.1 million. How much smaller is New York than California?

12. Using the *Of Versus More Than* Rule.

a. Betsy earns 30% more than Kevin. How many times larger is her income than his?

b. James earns 2.3 times as much as Lisa. How much more is James' income than Lisa's, by percentage?

c. The wholesale cost of a car is 50% less than its retail cost. How many times greater is the retail cost than the wholesale cost?

d. A store is having a 15% off sale. How does a sale price compare to the original price?

13. Using the *Of Versus More Than* Rule.

a. Brian earns 300% more than Wilson. How many times larger is Brian's income than Wilson's?

b. Kathy earns 0.8 times as much as Martha. How much less is Kathy's income than Martha's, by percentage?

c. The retail cost of a TV is 40% more than its wholesale cost. How many times greater is the retail cost than the wholesale cost?

d. An item costs $32.40 during a 30% off sale. What was the item's original price?

14. Gold Pendant. A gold pendant costs $600 at The Gold Store and $700 at The Pendant Store.

a. How much less is the price at The Gold Store than at The Pendant Store, as a percentage?

b. How much more is the price at The Pendant Store than at The Gold Store, as a percentage?

15. Percent Larger and Smaller. Is it possible for Quantity A to be 25% larger than Quantity B and Quantity B to be 25% smaller than Quantity A? Explain.

16. Comparing populations. The populations of China, India, and the United States in 1996 were 1.20 billion, 0.95 billion, and 0.27 billion, respectively. Complete each statement. Comment on the ambiguity that may arise in the resulting statement.

a. China has ____% more people than India.

b. China is a factor of ____ times more populous than the United States.

c. The United States has ____% fewer people than India.

d. India has ____ times more people than the United States.

17. Sales Tax. Suppose that your local sales tax is 8.3% on nonfood purchases. You purchase a TV set and a videodisk player for $495 and $429 (pretax prices), respectively. What is your total bill, tax included?

18. Decimals, Fractions, and Percentages. Suppose that you buy a $250 CD player at a "$\frac{1}{3}$ off" sale. The sales tax is 5.5%. How much do you actually pay for the CD player?

19. **Comparison Shopping.** In Niwot, you pay $5 in tax for a lawn mower, which is 5% of the purchase price. In Longmont, you pay $5 in tax for the same lawn mower, but it is 7% of the purchase price. In which town do you get a better price for the lawn mower and why?

20. **Pay Cut.** Suppose that you are earning a salary of $1000 per week.

   a. Your company experiences a slowdown in earnings, and asks all workers to take a 20% pay cut. What is your salary after the cut?

   b. Six months later, the company recovers and it offers you a 20% increase on your current salary. How much will you be earning after the increase?

   c. Explain why your salary after the increase is not back to its original level.

   d. What percentage pay raise would you need to restore your original salary after a 20% cut?

21. **Profitable Company?** In defense of apparent losses, the president of a company states: "I realize that last year our profits dropped by 25%. Fortunately, we increased our profits by 45% this year over last year. So I'd say we are doing pretty well!" Evaluate the president's analysis.

22. **Changing Test Scores.** A high school reports that its students' SAT scores were down by 20% for one year. The next year, however, they rose by 30%. As a result, the high school principal announced at the PTA meeting, "Overall, test scores have improved by 10% over the past two years."

   a. Is the principal's announcement correct? Describe what the principal has done.

   b. How much have test scores actually changed over the two-year period?

23. **Shifting Percentages.** Suppose that you pay a flat $10 monthly service charge on your checking account. In March, that fee is 1% of your balance. In April, the (same) fee is 4% of your balance.

   a. What was your balance in March?

   b. What was your balance in April?

   c. By what percent did your balance change from March to April?

24. **Analyzing Percentage Statements.** Consider each statement and decide if it is true or false. Give your reasoning.

   a. If 60% of the class consists of women and 10% of the women have blond hair, then blond-haired women comprise $60\% \times 10\% = 6\%$ of the class.

   b. If your weight increases by 10% this year and 10% next year, the increase over two years is 20%.

   c. If your grade point average decreases by 5% this semester and increases by 5% next semester, it will be unchanged after two semesters.

25. **Analyzing Percentage Statements.** Consider each statement and decide if it is true or false. Give your reasoning.

   a. If 40% of the class consists of men and 20% of the men are bald, then bald men comprise $40\% \times 20\% = 80\%$ of the class.

   b. If 70% of the hotels in town have a restaurant and 20% have a swimming pool, then 90% of the hotels in town have a restaurant or a pool.

   c. If 40% of the class drives to school and 10% takes a bus, then 50% of the class neither drives nor takes a bus.

26. **Drug Test Accuracy.** Suppose that 1000 people are given a drug test that is 98% accurate and that 50 of the people actually are drug users.

   a. If the test is 98% accurate, we can expect approximately 98% of the 50 drug users to test positive. How many of the 50 drug users can we expect to test positive?

   b. Similarly, we can expect 98% of the 950 nonusers to test negative. How many of the nonusers can we expect to test positive?

   c. Using your answers from parts (a) and (b), how many of the 1000 people can we expect to test positive? Of these, how many are users?

   d. What percentage of the positive tests are, in fact, false positives (nonusers who test positive because of the inherent inaccuracy of the test)? Discuss.

27. **Adding Percentages.** A political candidate pitches this argument to his campaign staff: "In this district 35% of the voters are Union members, and 40% of the voters in the district are Democrats. We can capture virtually all the votes in each of these two groups. That will give us 75% of the votes in the district and an easy win." Suppose the percentages of Union members and Democrats are correct. Explain how the conclusion could be false and how the candidate could, in fact, lose the election in this district. (Hint: It may help to give a numerical example.)

28. **Combining Percentages.** On a certain Boy Scout outing, 70% of the scouts carried a compass, 75% carried a knife, 80% carried a map, and 85% carried a flashlight. What percentage of scouts, at a maximum, carried all four items? Explain. (Extra credit: What percentage of scouts, at a *minimum*, carried all four items?)

29. **Opening Quote.** Refer to the quote at the beginning of this unit. Use the given information to answer each question. Explain your answers clearly.

    a. What percentage of twelfth graders smoked in 1991?

    b. What percentage of tenth graders smoked in 1991?

30. **Ambiguous News.** The average annual precipitation on Mt. Washington, New Hampshire, is 90 inches. During one particularly wet year, more rain and snow fell than usual. The news carried both of the following statements.
    - The precipitation this year is 200% of normal.
    - The precipitation this year is 200% above normal.

    What do each of these statements imply about the precipitation during this year? Do the two statements have the same meaning? Explain.

31. **Percentages in the News.** Each given statement is a quote from a news article. Answer the question that follows the quote. If the question cannot be answered unambiguously, explain why.

    a. "Almost 75% of the $100 billion in U.S. trade with Mexico is delivered by truck and most of that cargo travels through Texas."   *Question:* What is the value of U.S. trade with Mexico (in dollars) that is delivered by truck?

    b. "In 1994 Zaire's per capita GNP was $125 (70% lower than it was in 1958)."   *Question:* What was Zaire's (now the Republic of the Congo) per capita GNP in 1958? (GNP stands for gross national product.)

    c. "In the first six months of 1996, the number of deaths [from AIDS] fell 12% to 22,000 [compared to the same period a year earlier]."   *Question:* How many AIDS deaths were there during the first six months of 1995?

    d. "Between 1994 and 1995, the number of Americans studying in Britain increased by 16% to 19,410."   *Question:* How many Americans studied in Britain in 1994?

    e. "Personal bankruptcies rose by 6% to 832,000 by mid-1995 from the prior-year period."   *Question:* How many bankruptcies were there in 1994?

32. **Percentages in the News.** Each given statement is a quote from a news article. Answer the question that follows the quote. If the question cannot be answered unambiguously, explain why.

    a. "Overall Americans gambled away more than $40 billion in 1995—up from $10.4 billion in 1982."   *Question:* By what percentage have gambling outlays increased between 1982 and 1995?

    b. "Bagel sales have reached $3 billion and are growing 20% annually according to Lehman Brothers."   *Question:* How much will be spent on bagels next year?

    c. In 1996, 317,000 Americans will be told they have prostate cancer . . . a 30% increase over last year's new prostate cancer cases.   *Question:* How many prostate cancer cases were there last year?

    d. "The newcomers increased their share of the $30 billion U.S. pork industry from 7% to 17% between 1988 and 1994."   *Question:* What was the newcomer's share (in billions of dollars) in 1994?

    e. "Consumer debt has soared 39% in the last five years and now exceeds $1 trillion."   *Question:* What (approximately) was consumer debt five years ago?

33. **Find Your Own Percentages in the News.** Find at least three examples in which a percentage is used in a news report in a way that might be confusing or ambiguous. Describe each case, and explain the source of potential confusion.

UNIT
4C

PUTTING NUMBERS IN PERSPECTIVE

At the polling booth, we choose among candidates who talk about spending and taxation in units of billions of dollars. The size of the global economy is measured in trillions of dollars. Some of the most important policy issues of our time involve the impact of six billion people on the environment, and the impact of the billions of tons of carbon dioxide that humans release into the atmosphere each year. On a more personal level, we choose among computers with gigabytes of memory and processing times measured in nanoseconds. Here, we explore techniques for putting large or small numbers in perspective.

WRITING LARGE AND SMALL NUMBERS

Beware! Above the hundred millions, the same words refer to different numbers in different countries. For example, the word billion means 1,000,000,000 (10^9) in the United States, but it means 1,000,000,000,000 (10^{12}) in Great Britain. The British call 1,000,000,000 (10^9) a *thousand million*. This book uses the common names of numbers in the United States.

Before we begin putting extreme numbers in perspective, it's useful to think about how we express them. Consider the numbers in the following statements.

- World population is approximately 6,000,000,000 people.
- The U.S. federal debt is about $5,000,000,000,000.
- The nucleus of a hydrogen atom has a diameter of about 0.000000000000001 meters.

The many zeros make these numbers very difficult to read. A much better way to express the numbers is with powers of 10, as the following examples show.

- World population is approximately 6×10^9 people.
- The US federal debt is about $\$5 \times 10^{12}$.
- The nucleus of a hydrogen atom has a diameter of about 10^{-15} meters.

It's now much easier to get a sense of the size of each number. The number 6×10^9 shows that the number involves *billions* (10^9), and 5×10^{12} involves trillions (10^{12}). This format, in which a number *between* 1 and 10 is multiplied by a power of 10, is called **scientific notation**.

Another advantage of scientific notation is that it makes it easy to *approximate* answers without a calculator. For example, we can quickly approximate the answer to 5795×326 by rounding 5795 to 6000 and 326 to 300. Then, writing the rounded numbers in scientific notation, we can easily see that

A Brief Review: The symbol ≈ means *approximately equal to*.

$$5795 \times 326 \approx (6 \times 10^3) \times (3 \times 10^2) = 18 \times 10^5 = 1,800,000.$$

Since the exact answer is 1,889,170, this approximate answer is a good estimate and gives us a good sense of the size of the answer. This technique can be especially useful for checking answers.

The only drawback to scientific notation is that it makes extremely large or small numbers deceptively easy to write. In particular, numbers that hardly *look* any different on paper can have vastly different meanings. For example, 10^{26} doesn't look much different from, say, 10^{20}, but it is *a million times* larger because 10^{26} is 10^6 times larger than 10^{20}. Or consider the number 10^{80}, which doesn't look so incredibly large on paper—yet it is a number larger than the total number of atoms in our entire universe!

A BRIEF REVIEW

Working with Scientific Notation

Working with numbers in scientific notation is very easy because it involves powers of 10. Let's review a few simple rules.

Converting a Number to Scientific Notation

Converting a number into scientific notation involves a simple, two-step process.

Step 1. Move the decimal point to come after the *first* nonzero digit.

Step 2. The number of places the decimal point moves tells you the power of 10; the power is *positive* if the decimal point moves to the left and *negative* if it moves to the right.

Examples:

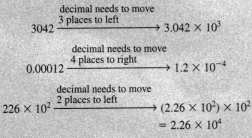

$$3042 \xrightarrow[\text{3 places to left}]{\text{decimal needs to move}} 3.042 \times 10^3$$

$$0.00012 \xrightarrow[\text{4 places to right}]{\text{decimal needs to move}} 1.2 \times 10^{-4}$$

$$226 \times 10^2 \xrightarrow[\text{2 places to left}]{\text{decimal needs to move}} (2.26 \times 10^2) \times 10^2$$
$$= 2.26 \times 10^4$$

Converting a Number from Scientific Notation

A number written in scientific notation can be quickly converted to ordinary notation by the reverse process.

Step 1. The power of 10 indicates how many places to move the decimal point; move it to the *right* if the power of 10 is positive and to the *left* if it is negative.

Step 2. If moving the decimal point creates any open places, fill them with zeros.

Examples:

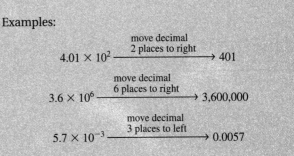

$$4.01 \times 10^2 \xrightarrow[\text{2 places to right}]{\text{move decimal}} 401$$

$$3.6 \times 10^6 \xrightarrow[\text{6 places to right}]{\text{move decimal}} 3,600,000$$

$$5.7 \times 10^{-3} \xrightarrow[\text{3 places to left}]{\text{move decimal}} 0.0057$$

Multiplying or Dividing Numbers in Scientific Notation

Multiplying or dividing numbers in scientific notation simply requires operating on the powers of 10 and the other parts of the number separately. Examples:

$$(6 \times 10^2) \times (4 \times 10^5) = (6 \times 4) \times (10^2 \times 10^5) = 24 \times 10^7$$
$$= (2.4 \times 10^1) \times 10^7 = 2.4 \times 10^8$$

$$\frac{4.2 \times 10^{-2}}{8.4 \times 10^{-5}} = \frac{4.2}{8.4} \times \frac{10^{-2}}{10^{-5}} = 0.5 \times 10^{-2-(-5)} = 0.5 \times 10^3$$
$$= (5 \times 10^{-1}) \times 10^3 = 5 \times 10^2$$

Note that, in both these examples, we first found an answer in which the number multiplied by a power of 10 was *not* between 1 and 10. We therefore followed the process for converting the final answer into scientific notation.

Addition and Subtraction with Scientific Notation

In general, we must write numbers in ordinary notation before adding or subtracting. Examples:

$$(3 \times 10^6) + (5 \times 10^2) = 3,000,000 + 500 = 3,000,500$$
$$= 3.0005 \times 10^6$$

$$(4.6 \times 10^9) - (5 \times 10^8)$$
$$= 4,600,000,000 - 500,000,000$$
$$= 4,100,000,000 = 4.1 \times 10^9$$

When both numbers have the *same* power of 10, we can factor out the power of 10 first. Examples:

$$(7 \times 10^{10}) + (4 \times 10^{10}) = (7 + 4) \times 10^{10} = 11 \times 10^{10}$$
$$= 1.1 \times 10^{11}$$

$$(2.3 \times 10^{-22}) - (1.6 \times 10^{-22}) = (2.3 - 1.6) \times 10^{-22}$$
$$= 0.7 \times 10^{-22}$$
$$= 7.0 \times 10^{-23}$$

EXAMPLE 1 *Checking Your Work with an Approximation*

You and a friend are working on a problem that involves finding the product 9,768,221 × 0.0047. Your friend quickly presses calculator buttons, and tells you that the answer is 459.1. Without using your calculator, is this answer reasonable?

Solution: You can approximate 9,768,221 as about 10 million, or 10^7, and 0.0047 as about 0.005, or 5×10^{-3}. Thus the answer should be approximately

$$10^7 \times (5 \times 10^{-3}) = 5 \times 10^{7-3} = 5 \times 10^4 = 50,000.$$

Clearly, your friend's answer of 459.1 is far too small, and the calculation should be tried again. Note that this checking technique cannot tell you whether an answer is exactly correct, but it can confirm that you are close to the correct answer, and it can uncover mistakes. ∎

Calvin and Hobbes   by Bill Watterson

PERSPECTIVE THROUGH ESTIMATION

Making estimates is one of the best techniques for putting numbers into perspective. For example, how high is 1000 feet? For most people, the quantity 1000 feet has little meaning by itself. However, if we estimate that a story is about 10 feet from floor to ceiling, then we can put 1000 feet in perspective by thinking of it as the height of a 100-story building.

A very important consideration in making estimates is how close we want the estimate to be to the exact value. For example, the 102 stories of the Empire State Building in New York City stand 1250 feet tall. This is 250 feet, or 25%, taller than our estimate of 1000 feet for a 100-story building. Nevertheless, our estimate gives us a reasonable "sense" of the quantity 1000 feet, even if it is off by 25%.

Suppose that you want to tell a distant friend about the size of the town or city in which you live. Clearly, a town with a population in the *ten thousands* will have a very different character than a city with a population in the *millions*. Thus you can convey a great deal about your hometown if you simply say that its population is in the *ten thousands*, without worrying about whether the exact population is closer to 20,000 or 70,000. Estimates such as this, in which we care only about the number of zeros, are called **order of magnitude** estimates. Technically, an *order of magnitude* is a power of 10.

For example,

- 100 is one order of magnitude larger than 10;
- 10,000 (10^4) is two orders of magnitude larger than 100 (10^2); and
- 10^{23} is five orders of magnitude larger than 10^{18}.

However, we often use the term *order* when referring to any broad range. For example, the population of the United States as of the beginning of 1998 is within 3 or 4 million of 270 million. However, we can give a good sense of the U.S. population by saying that it is on the *order* of 300 million. The term order implies that 300 million is in the right general neighborhood, even if it may be too high or too low by a few tens of millions.

EXAMPLE 2   *US Ice Cream Spending*

Estimate the total amount of money spent on ice cream in the United States each year.

Solution: We know that the U.S. population is about 270 million, so we can find the total annual spending on ice cream if we can estimate how much an average person spends on ice cream per year, which is called the *annual per capita spending.*

$$\text{total annual spending} = (\text{U.S. population}) \times (\text{annual per capita spending})$$

We can break the problem down further by noting that the annual per capita spending depends on the price of an average serving of ice cream and the number of servings that an average person consumes in a year.

$$\text{annual spending per capita} = (\text{price per serving}) \times (\text{annual number of servings})$$

Without a detailed survey of ice cream prices and ice cream consumption, we cannot make a very accurate estimate. However, we can make an *order of magnitude* estimate. For example, we know that the price for a serving of ice cream from the grocery store may be under \$1, while the price from a specialty store may be several dollars. Thus it is reasonable to say that the average price of a serving of ice cream is on the *order* of \$2.

Similarly, we know that some people have ice cream every day, while others have it rarely or never. But it seems reasonable to say that the average person has on the *order* of 50 servings of ice cream per year (about once a week). Using these estimates, we find

$$\text{annual spending per capita} \approx \frac{\$2}{\text{serving}} \times \frac{50 \text{ servings}}{\text{person}} = \frac{\$100}{\text{person}}.$$

Returning to our original equation, we find the total annual spending on ice cream in the United States.

$$\text{total annual spending} \approx 2.7 \times 10^8 \text{ persons} \times \frac{\$100}{\text{person}} = \$2.7 \times 10^{10}$$

Our answer is \$27 billion, but because of the nature of our assumptions, it is better stated as follows: Annual spending on ice cream in the United States is *on the order of tens of billions* of dollars. ∎

> **TIME-OUT TO THINK:**   Suppose that you work for a company that distributes ice cream to stores throughout the United States. Market research has told you that a $10 million advertising campaign could increase your market share by 5% (i.e., you'd gain 5% of the total market for ice cream in the United States). Based on the estimate in Example 2, is it worth spending $10 million to get a 5% increase in market share? Explain.

Using Your
Calculator   Scientific Notation

Most calculators have a special key for entering scientific notation. The key, usually labeled either *exp* (short for "exponent") or *EE* (short for "enter exponent"), is used to enter a power of 10. For example, you enter the number 3.5×10^6 with the key sequence

   3.5 (EE) 6.

Think of the *EE* (or *exp*) as "times 10 to the power entered next." Note that you must be especially careful in entering "pure" powers of 10. For example, to enter the number 10^5 you would press

   1 (EE) 5.

Beware of two common mistakes: First, some people try to enter 10^5 by pressing   10 (EE) 5,   but this key sequence actually means 10×10^5, or 10^6. Second, some people try to enter a number like 7×10^6 by pressing 7 (×) 10 (EE) 6,   which actually means $7 \times 10 \times 10^6$,   or 7×10^7.   The correct way to enter 7×10^6 is   7 (EE) 6.

PERSPECTIVE THROUGH SCALING

Another technique for giving meaning to numbers that might otherwise seem incomprehensible is through the process of *scaling*. You probably are familiar with ideas of scaling from reading maps and working with scale models. Scales usually are expressed in one of three ways.

- *Verbally:* A scale can be described in words such as "one centimeter represents one kilometer." Sometimes, this description is written simply as 1 cm = 1 km.

- *Graphically:* A scale can be shown with a "miniruler" marked to show the represented distances.

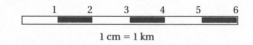

1 cm = 1 km

- *With a Scale Factor:* Because there are 100,000 centimeters in a kilometer, a scale where 1 centimeter represents 1 kilometer means that scaled lengths are $\frac{1}{100,000}$ of actual distances. We say that the scale is $\frac{1}{100,000}$, or 1 to 100,000, where the number 100,000 is the **scale factor**. That is,

$$\text{scaled distance} = \frac{\text{actual distance}}{\text{scale factor}}.$$

EXAMPLE 3   *Scale Factor*

Suppose that you have a map with a scale expressed as "one inch on the map represents one mile." What is the scale factor for this map?

Solution: The map scale can be written as

$$1 \text{ inch} = 1 \text{ mile}.$$

One way to find the scale factor is to ask, how much bigger is an actual distance of 1 mile than a scaled distance of 1 inch? We find the answer by dividing and converting units:

$$\frac{\text{actual distance}}{\text{scaled distance}} = \frac{1 \text{ mi}}{1 \text{ in.}} = \frac{1 \text{ mi} \times 5280 \frac{\text{ft}}{\text{mi}} \times 12 \frac{\text{in.}}{\text{ft}}}{1 \text{ in.}} = \frac{6.34 \times 10^4 \text{ in.}}{1 \text{ in.}} = 63{,}400$$

That is, actual distances are 63,400 times larger than the scaled distances shown on the map. The scale of the map is 1 to 63,400. We say that 63,400 is the scale factor for the map. ∎

Using Scales

Scales have many uses besides maps. Architects and engineers often build scale models as a way of visualizing what a finished product will look like. History often is represented on *time lines*, on which each unit of distance along the time line represents a certain number of years. Time also can be scaled against other times: for example, many TV stations use *time-lapse photography* to show short videos of an entire day's weather. The station might compile satellite photographs showing weather systems at one-hour intervals throughout the day. If these photographs are put together into a four-second video clip, a full day's weather can be presented in just four seconds. In this case, each second of the video represents six hours of real time.

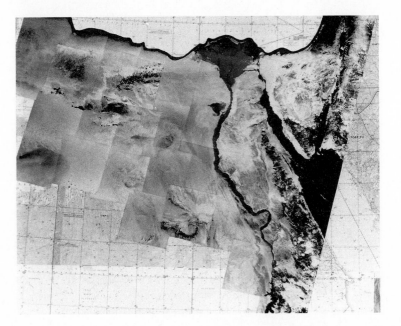

Beware!   Many people mistakenly use the term light-year to describe *time*, but it actually is a unit of *distance*. You can check whether it is being used properly by substituting "10 trillion km" whenever you hear the term *light-year*.

Scaling is particularly useful for gaining perspective on extremely large or small sizes. In describing distances in our universe, we often use a unit called the **light-year**, which is the *distance* that light can travel in one year. The speed of light is 300,000 km/s (3×10^5 km/s). We can calculate how far light travels in one year by multiplying this speed by the number of seconds in one year:

$$1 \text{ light-year} = \text{speed of light} \times 1 \text{ yr}$$

$$= 3 \times 10^5 \frac{\text{km}}{\text{s}} \times 1 \text{ yr} \times \frac{365 \text{ day}}{1 \text{ yr}} \times \frac{24 \text{ hr}}{1 \text{ day}} \times \frac{60 \text{ min}}{1 \text{ hr}} \times \frac{60 \text{ s}}{1 \text{ min}}$$

$$= 9.5 \times 10^{12} \text{ km}$$

Thus *one light-year* is just an easy way of saying *9.5 trillion kilometers* or *almost 10 trillion kilometers.*

EXAMPLE 4   *Earth and Sun*

The distance from the Earth to the Sun is about 150 million km. The diameter of the Sun is about 1.4 million km and the diameter of the Earth is about 12,760 km. Put these numbers in perspective by imagining a scale model of the solar system with a 1 to 10 billion scale.

Solution: We are given that the scale factor for our model should be 10 billion (10^{10}), so we find scaled lengths by dividing the actual lengths by 10^{10}. For the Earth-Sun distance, we find

$$\text{scaled distance} = \frac{\text{actual distance}}{10^{10}}$$

$$= \frac{1.5 \times 10^8 \text{ km}}{10^{10}}$$

$$= 1.5 \times 10^{-2} \text{ km} \times 10^3 \frac{\text{m}}{\text{km}} = 15 \text{ m}.$$

For the Sun and Earth diameters, we find the following values.

$$\text{scaled Sun diameter} = \frac{\text{actual Sun diameter}}{10^{10}}$$

$$= \frac{1.4 \times 10^6 \text{ km}}{10^{10}}$$

$$= 14 \times 10^{-4} \text{ km} \times 10^5 \frac{\text{cm}}{\text{km}} = 14 \text{ cm}$$

$$\text{scaled Earth diameter} = \frac{\text{actual Earth diameter}}{10^{10}}$$

$$= \frac{1.276 \times 10^4 \text{ km}}{10^{10}}$$

$$= 1.276 \times 10^{-6} \text{ km} \times 10^6 \frac{\text{mm}}{\text{km}} = 1.276 \text{ mm}$$

The model Sun, at 14 cm in diameter, is roughly the size of a grapefruit. The model Earth is about the size of the ball point in a pen, and the distance between them is 15 meters. At this scale, it is easy to visualize relative sizes and distances.   ■

By the Way ············
The Earth's distance from the Sun varies between a minimum of 147.1 million km each January and 152.1 million km each July. The Earth's *average* distance from the Sun, about 149.6 million km, is called an *astronomical unit*, or 1 AU.

> **TIME-OUT TO THINK:** Find a grapefruit or similar-sized ball and the ball point from a pen. Set them 15 meters apart to represent the Earth and Sun on a 1 to 10 billion scale. How does a scale model like this one make it easier to understand our solar system? Discuss.

EXAMPLE 5   *Distances to the Stars*

The distance from the Earth to the nearest stars besides the Sun (the three stars of the *Alpha Centauri* system) is about 4.3 light-years. On the 1 to 10 billion scale of Example 4, how far away would these stars be located from the Earth?

Solution: Because a light-year is about 9.5×10^{12} km, 4.3 light-years is equivalent to

$$4.3 \text{ light-years} \times \frac{9.5 \times 10^{12} \text{ km}}{1 \text{ light-year}} = 4.1 \times 10^{13} \text{ km}.$$

We calculate the scaled distance to the nearest stars by dividing this actual distance by the scale factor of 10 billion:

$$\text{scaled distance} = \frac{\text{actual distance}}{10^{10}} = \frac{4.1 \times 10^{13} \text{ km}}{10^{10}} = 4.1 \times 10^3 \text{ km} = 4100 \text{ km}$$

The distance to even the nearest stars on this scale is more than 4000 kilometers, or approximately the same as the east-west distance across the entire United States!   ■

> **TIME-OUT TO THINK:** Suppose that an Earth-like planet is orbiting a nearby star. Based on the results of Examples 4 and 5, discuss the challenge of trying to detect such a planet.

EXAMPLE 6   *Time Line*

Human civilization, at least since the time of ancient Egypt, is on the order of 5000 years old. The age of the Earth is on the order of 5 billion years. Suppose we make a time line the length of a football field, or about 100 meters, to represent the age of the Earth. If we put the birth of the Earth at the start of the time line, how far from the line's end would human civilization begin?

Solution: First, we compare the 5000-year history of human civilization to the 5 billion-year age of the Earth:

$$\frac{5000 \text{ yr}}{5 \text{ billion yr}} = \frac{5 \times 10^3 \text{ yr}}{5 \times 10^9 \text{ yr}} = 10^{-6}$$

That is, 5000 years is about 10^{-6}, or one *millionth*, of the age of the Earth. One millionth of a 100-meter time line is

$$10^{-6} \times 100 \text{ m} = 10^{-6} \times 10^2 \text{ m} = 10^{-4} \text{ m},$$

or 0.1 mm. On a time line where the Earth's history is stretched along the length of a 100-meter football field, human civilization would show up only in the final 0.1 mm!   ■

By the Way ··········
According to modern science, the Earth and solar system formed from the collapse of a large cloud of interstellar gas. Meteorites represent some of the first rocks that formed as the cloud collapsed. Radioactive dating shows the oldest meteorites to be 4.6 billion years old, so scientists conclude that the Earth formed 4.6 billion years ago.

PUTTING IT ALL TOGETHER: CASE STUDIES

Estimation and scaling are both useful techniques for putting numbers in perspective. However, there may be many different ways to put any particular number in perspective. Thus putting a number into perspective is more of an art than a science. The only general guideline to remember is that you must make a large or small number comprehensible by relating it to numbers with which you already are familiar. A few more case studies will help illustrate some of the ways that this can be accomplished.

CASE STUDY *What Is a Billion Dollars?*

The ledgers of large corporations, the impacts of tax policies, and even the assets of a few individuals involve money measured in *billions* of dollars.

One way to put $1 billion in perspective is to ask a question like: How many people can you employ with $1 billion per year? Let's take $25,000 as a reasonable estimate of a typical starting salary for a college graduate. Further, let's assume that it costs a business an additional $25,000 per year to have a $25,000 employee on the payroll (costs for office space, computer services, health insurance, and other benefits). At $50,000 per employee, $1 billion would allow a business to hire

$$\frac{\$1 \text{ billion}}{\$50,000 \text{ per employee}} = \frac{\$10^9}{\$5 \times 10^4 / \text{ employee}} = 2 \times 10^4 \text{ employees.}$$

Thus, $1 billion per year could support a work force of 20,000 employees at a cost of $50,000 per employee.

A billion here, a billion there; soon you're talking real money.

—Senator Everett Dirksen

Another way to put $1 billion in perspective also points out how different numbers can be, even when they sound similar (like million, billion, and trillion). Suppose, that you become a sports star and earn a salary of $1 million per year. How long would it take you to earn a billion dollars?

$$\frac{\$1 \text{ billion}}{\$1 \text{ million} / \text{ yr}} = \frac{\$10^9}{\$10^6 / \text{ yr}} = 10^3 \text{ yr} = 1000 \text{ yr}$$

Even at a salary of $1 million per year, earning a billion dollars would take a thousand years! ∎

CASE STUDY *The National Debt*

The national debt of the United States is about $5 trillion. Let's try to put this huge number in perspective.

Suppose that the government decided to pay off the national debt with a one-time tax on everyone in the United States. To keep the numbers simple, let's call the US population about 250 million. Then the tax on each person would be

$$\frac{\$5 \times 10^{12}}{2.5 \times 10^8 \text{ persons}} = \frac{\$2 \times 10^4}{\text{person}} = \frac{\$20,000}{\text{person}}.$$

Every U.S. citizen would have to pay about $20,000 to the federal government to retire the national debt. A family of four would owe $80,000!

Next, imagine the debt as a stack of $1 bills laying flat on top of each other. If you make a stack of ten $1 bills, you'll find that it is about 2 mm thick. Thus each $1 bill is about 0.2 mm thick, and a pile of 5 trillion bills would have a thickness of

$$5 \times 10^{12} \text{ bills} \times \frac{0.2 \text{ mm}}{\text{bill}} = 1 \times 10^{12} \text{ mm}.$$

It's more meaningful to convert this answer to kilometers:

$$10^{12} \text{ mm} \times \frac{1 \text{ m}}{10^3 \text{ mm}} \times \frac{1 \text{ km}}{10^3 \text{ m}} = 10^6 \text{ km}$$

Thus a stack of $1 bills equaling the national debt would rise to a height of about 1 million kilometers, or more than twice the 380,000-kilometer distance from the Earth to the Moon!

Suppose that you could drive your car in space along this stack of money. How long would the drive take? At a typical highway cruising speed of about 100 km/hr (62 mi/hr), a drive of 1 million kilometers would take

$$\text{time} = \frac{\text{distance}}{\text{speed}} = \frac{10^6 \text{ km}}{100 \text{ km/hr}} = 10^4 \text{ hr} \times \frac{1 \text{ day}}{24 \text{ hr}} \times \frac{1 \text{ month}}{30 \text{ day}} \approx 14 \text{ months}.$$

It would take more than a year to "visit" the national debt as a stack of $1 bills.

Tour the National Debt in $1 bills! (Fun for over a year!)

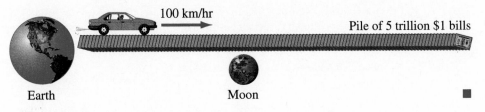

Earth   Moon

CASE STUDY   *The Scale of the Atom*

Everything in the world is made from *atoms*, which consist of a nucleus (made from *protons* and *neutrons*) surrounded by a "cloud" of *electrons*. A typical atom has a diameter of about 10^{-10} m (as defined by its electron cloud), while its nucleus is about 10^{-15} m in diameter. Can we put these numbers in perspective?

Let's begin with the atom itself. Its diameter of 10^{-10} m, or one *ten billionth* of a meter, means that we could fit 10 billion atoms in a line along a meter stick. Or, because a centimeter is $\frac{1}{100}$ of a meter, we could fit

$$\frac{10 \text{ billion}}{100} = \frac{10^{10}}{10^2} = 10^8 = 100 \text{ million}$$

atoms in a line 1 cm long. An inch is about 2.5 cm, so we could fit 250 million atoms in a 1-inch line. In other words, if we could shrink people down to the size of atoms, the entire population of the United States could form a line only about an inch long!

By the Way ············

The size of an atom is determined by the size of its electron cloud, but most of the mass of an atom is contained in its nucleus. Because the nucleus is so tiny compared to the atom itself, we have the surprising fact that atoms consist mostly of empty space!

Next, let's consider the scale of the atom and its nucleus.

$$\frac{\text{diameter of atom}}{\text{diameter of nucleus}} = \frac{10^{-10}\,\text{m}}{10^{-15}\,\text{m}} = 10^{-10-(-15)} = 10^5$$

That is, the atom itself is about 10^5, or 100,000, times larger than its nucleus. Thus if we made a scale model of an atom in which the nucleus were 1 cm in diameter, the electrons would be found at distances of about 100,000 cm, or 1 km! ∎

CASE STUDY   *Until the Sun Dies*

We can expect that living things will continue to flourish on Earth until the Sun dies, which astronomers estimate will be in about 5 billion years. How long is 5 billion years?

If we take a human lifetime to be about 100 years, the Sun's remaining lifetime is 50 million (5×10^7) times longer since

$$\frac{5 \times 10^9\,\text{yr}}{100\,\text{yr}} = 5 \times 10^7.$$

We can put this comparison in perspective by dividing a human lifetime by 50 million:

$$\frac{100\,\text{yr}}{5 \times 10^7} = 2 \times 10^{-6}\,\text{yr} \times \frac{365\,\text{days}}{1\,\text{yr}} \times \frac{24\,\text{hr}}{1\,\text{day}} \times \frac{60\,\text{min}}{1\,\text{hr}} \approx 1\,\text{min}$$

Thus a human lifetime in comparison to the life expectancy of the Sun is roughly the same as *one minute*—about 60 heartbeats.

How about human creations? The Egyptian pyramids have often been described as "eternal." But they are slowly eroding due to wind, rain, air pollution, and the impact of tourists; all traces of them will have vanished within a few million years. If we call the lifetime of the pyramids about 5 million years, then

$$\frac{\text{Sun's remaining lifetime}}{\text{lifetime of pyramids}} \approx \frac{5 \times 10^9\,\text{yr}}{5 \times 10^6\,\text{yr}} = 10^3 = 1000.$$

By the Way ············

A 200-km wide crater near Mexico's Yucatán Peninsula is thought to be from the asteroid that wiped out the dinosaurs. The asteroid's impact sent debris raining across North America, and hot embers of "fallout" ignited wildfires around the world. Huge quantities of dust were blasted high into the atmosphere, blocking sunlight and cooling the surface for several years. During this period, about 70% of all living species went extinct, including all dinosaurs.

A few million years may seem like a long time, but the Sun's remaining lifetime is a thousand times longer!

On a more somber note, we can gain perspective on 5 billion years by considering evolutionary time scales. During the past century, we have acquired sufficient technology and power to destroy human life totally, if we so choose. However, even if we make that unfortunate choice, some species (including many insects) are likely to survive. Would another intelligent species ever emerge on the Earth? There is no way to know, but we can look to the past for guidance. It's been 65 million years since the dinosaurs were wiped out in what many scientists believe was a fairly sudden cataclysm caused by the impact of an asteroid. Imagine that another intelligent species could evolve some 65 million years after a human extinction. If they also destroyed themselves, another species could evolve 65 million years after that, and so on. How many more chances would the Earth have to produce intelligent species? The answer is

$$\frac{5\,\text{billion years}}{65\,\text{million years}} = \frac{5 \times 10^9\,\text{years}}{6.5 \times 10^7\,\text{years}} = 77.$$

Even at 65 million years per shot, the Earth would have *nearly 80* more chances to evolve an intelligent species in 5 billion years. Perhaps one of those species will not destroy itself, and descendants of the Earth might move on to other star systems by the time the Sun finally dies. Perhaps it will be us.

I met a traveller from an antique land
Who said: Two vast and trunkless legs of stone
Stand in the desert . . . Near them, on the sand,
Half sunk, a shattered visage lies, whose frown,
And wrinkled lip, and sneer of cold command,
Tell that its sculptor well those passions read
Which yet survive, stamped on these lifeless things,

The hand that mocked them, and the heart that fed:
And on the pedestal these words appear:
'My name is Ozymandias, king of kings:
Look on my works, ye Mighty, and despair!'
Nothing beside remains. Round the decay
Of that colossal wreck, boundless and bare,
The lone and level sands stretch far away.
—*FROM* OZYMANDIAS, *BY* PERCY BYSSHE SHELLEY

REVIEW QUESTIONS

1. When and why is it useful to put numbers in scientific notation? Give some examples.

2. Explain how you can use scientific notation to check answers by making approximations.

3. What is an *order of magnitude* estimate? Explain why such an estimate can be useful even though it may be as much as 10 times too large or too small.

4. What is a scale factor? How do we find a scaled size from an actual size?

5. Describe some uses of scaling besides scales on maps.

6. Suppose that the Sun were the size of a grapefruit. How big and how far away would the Earth be on this scale?

7. What is a *light-year*? How can you tell whether the term light-year is being used properly as a distance or incorrectly as a time?

8. Describe several ways of putting $1 billion in perspective.

9. Describe ways of putting each of the following in perspective: the national debt; the size of atoms; a time of 5 billion years.

PROBLEMS

1. **Large Numbers in the News.** Search today's newspaper for as many instances of numbers larger than 100,000 that you can find. Briefly explain the context within which each large number is used.

2. **Perspective in the News.** Find an example in the recent news in which the reporter uses a technique similar to one of those in this Unit to put a number in perspective. Describe the example. Do you think the technique was effective? Can you think of a better way to put the number in perspective? Explain.

3. **Reading Powers of 10.** Convert each number from scientific notation and write its name. *Example:* 2×10^3. *Solution:* $2 \times 10^3 = 2000 =$ two thousand.

   a. 5×10^6 **b.** 7×10^9 **c.** -2×10^{-2}
   d. 8×10^{11} **e.** 1×10^{-7} **f.** 9×10^{-4}

4. **Reading Powers of 10.** Convert each number from scientific notation and write its name.   *Example:* 2×10^3. *Solution:* $2 \times 10^3 = 2000 =$ two thousand.

   a. 4.6×10^9 **b.** 3.95×10^{-6} **c.** -6.02×10^{-10}
   d. 2.3×10^8 **e.** -3.33×10^{-3} **f.** 8.122×10^8

5. **Writing Powers of 10.** Write each number in scientific notation.

   a. 600   **b.** 0.9   **c.** 50,000
   d. 0.003   **e.** 0.0005   **f.** 70,000,000,000

6. **Writing Powers of 10.** Write each number in scientific notation.

   a. 0.00000002   **b.** 7 million   **c.** 9 billionths

   d. 4 trillion   **e.** 80 billion   **f.** 600 million

7. **Converting to Scientific Notation.** Convert each number to scientific notation.

   a. 1,000,000   **b.** 150,000   **c.** 45×10^{-1}

   d. 18 hundredths   **e.** 540×10^6 **f.** 530×10^{23}

8. **Converting to Scientific Notation.** Convert each number to scientific notation.

   a. 645   **b.** 0.92   **c.** 500.098

   d. 0.002×10^6 **e.** 250 million   **f.** $-23,800$

9. **Converting from Scientific Notation.** Convert each number from scientific notation.

   a. 2.2×10^{-4} **b.** 2×10^{-1} **c.** 9.828×10^7

   d. 6.667×10^1 **e.** 3.5×10^4 **f.** 1.501×10^{-10}

10. **Converting from Scientific Notation.** Convert each number from scientific notation.

   a. 7.0×10^3 **b.** 1.5×10^9 **c.** 3.0906×10^3

   d. 2.3×10^6 **e.** 2.2×10^{-4} **f.** 4.44×10^{-1}

11. **Large and Small Numbers Everywhere.** Rewrite each statement with a number in scientific notation.

   a. The U.S. national debt in 1997 was about 5.3 trillion dollars.

   b. Corporate profits in the United States in 1996 were 632 billion dollars.

   c. Consumer debt (loans and credit cards) in the United States in 1994 was 800 billion dollars.

   d. The hard drive on my computer has a capacity of 5.4 gigabytes. (Hint: Recall that the prefix "giga" means 1 billion).

12. **Large and Small Numbers Everywhere.** Rewrite each statement with a number in scientific notation.

   a. In 1995, cardiovascular diseases (heart diseases and strokes) caused 739,860 deaths in the United States.

   b. The area of the Earth's surface is 509,600,000 square kilometers.

   c. In 1994, Americans used 2.9 million gigawatt-hours of electricity; express this quantity in watt-hours. (Hint: Recall that the prefix "giga" means 1 billion).

   d. In 1995, the United States contributed $1,173,000,000 to international organizations.

13. **Approximation with Powers of 10.**

   a. Suppose that you add $10^{26} + 10^7$. What, approximately, is the answer? Explain.

   b. Suppose that you subtract $10^{81} - 10^{62}$. What, approximately, is the answer? Explain.

Practice with Scientific Notation. *Do the operations in Problems 14 and 15 without a calculator and show your work clearly. Be sure to express the final answers in scientific notation. You may round your answers by writing only two digits (as in 3.2×10^5).*

14. **a.** $(3 \times 10^4) \times (8 \times 10^5)$ **b.** $(6.3 \times 10^2) + (1.5 \times 10^1)$

   c. $(9 \times 10^3) \times (5 \times 10^{-7})$ **d.** $(4.4 \times 10^{99}) \div (2.0 \times 10^{11})$

   e. $(8 \times 10^{12}) \div (4 \times 10^9)$ **f.** $(7.5 \times 10^{21}) \div (1.5 \times 10^{13})$

   g. $(3.2 \times 10^{22}) \div (1.6 \times 10^{-14})$

   h. $(6 \times 10^{10}) - (5 \times 10^9)$

15. **a.** $(9 \times 10^8) + (5 \times 10^9)$ **b.** $(8.1 \times 10^{30}) \times (9 \times 10^{15})$

   c. $(2.5 \times 10^{-4}) \times (3 \times 10^{-4})$

   d. $(8.1 \times 10^{14}) \div (3 \times 10^{-7})$

   e. $(6.6 \times 10^{-2}) \div (4.4 \times 10^{-3})$

   f. $(2.1 \times 10^4) - (1.5 \times 10^5)$

   g. $(2.4 \times 10^2) \div (8.1 \times 10^8)$

   h. $(9.8 \times 10^2) \times (2.0 \times 10^{-5})$

Approximation with Scientific Notation. *In Problems 16 and 17, (i) make an estimate of the answer without a calculator, showing your method of estimation; (ii) do the exact calculation (with a calculator if necessary); (iii) compare the approximation to the exact answer and say how well your approximation technique worked.*

16. **a.** $20,000 \times 100$ **b.** $9,642 \div 31$

   c. $-12.5 \times 11,890$ **d.** 250 million $\times 40$

   e. 7.453×291 **f.** $6,570,999 \div 32.7$

17. **a.** 5.6 billion $\div 200$ **b.** 4 trillion $\div$ 260 million

   c. $9,000 \times 54,986$ **d.** 3 billion $\div 25,000$

   e. $5,987 \times 341$ **f.** $43 \div 765$

They Don't Look Very Different! *In Problems 18 and 19, compare each pair of numbers. By what factor is one number larger than the other? Example: 10^6, 10^4. Solution: 10^6 is 10^2, or 100, times larger than 10^4.*

18. a. 10^{35}, 10^{26} **b.** 10^{17}, 10^{27}

    c. 1 billion, 1 million   **d.** 7 trillion, 7 thousand

    e. 2×10^{-6}, 2×10^{-9} **f.** 6.1×10^{27}, 6.1×10^{29}

19. a. 250 million, 5 billion   **b.** 9.3×10^{2}, 3.1×10^{-2}

    c. 10^{-8}, 2×10^{-13} **d.** 3.5×10^{-2}, 7×10^{-8}

    e. 1 thousand, 1 thousandth   **f.** 10^{12}, 10^{-9}

20. Orders of Magnitude. In each statement given, compare the two quantities by order of magnitude. *Example:* A small car (weight about 1500 to 2000 pounds) is about one order of magnitude heavier than a typical person (weight about 150 pounds).

    a. The population of China (1.1 billion) and the population of Japan (125 million)

    b. The population of the United States (260 million) and the population of Iceland (250,000)

    c. The area of Canada (10 million square kilometers) and the area of Lebanon (10,500 square kilometers)

    d. The mass of the Sun (2×10^{30} kg) and the mass of the Earth (6×10^{24} kg)

    e. The diameter of a cell (10^{-6} m) and the diameter of a proton (10^{-15} m)

21. Estimation Practice. Explain each estimate clearly.

    a. Estimate total annual spending on magazines by students at *your* school.

    b. Estimate the total amount of money spent each year by Americans in going to the movies.

22. Estimation Practice. Explain each estimate clearly.

    a. Estimate the total number of words in this textbook.

    b. Estimate the total amount you spend on food in a year.

    c. Estimate the amount of gasoline an average adult uses per year.

23. Comparisons Through Estimation. Make estimates as needed to answer clearly each question.

    a. Which is bigger, the height of a 10-story apartment building or the length of a football field? By how much?

    b. Could a person walk across the United States (New York to California) in a year? If not, about how long would it take?

    c. Which is more, the number of miles Americans fly each year or the number of miles Americans drive each year?

    d. Which holds more people, a football stadium or 10 movie theaters?

24. Topographic Map Scale. Assume you are using a topographic map with a scale of 1 to 24,000.

    a. If you measure the distance between two locations on a map as 7.5 cm, how far apart are they actually?

    b. If you knew that two towns on the map were 5 km apart, how far would they be separated on this map?

    c. If the smallest reasonably measurable distance on the map is 2 mm, what actual distance does this represent?

25. Scale Factors. What is the scale factor described in each statement?

    a. 1 cm on the map represents 1 km on the ground.

    b. 2 inches on the map represents 0.5 mile on the ground.

    c. 5 cm (map) = 100 km (actual).

    d. 1 foot (map) = 100 meters (actual).

26. The New York Marathon. A marathon is about 42 kilometers (about 26 miles). In typical years, about 20,000 runners complete the New York marathon.

    a. Put the total amount of running in perspective by calculating the total distance run by all the runners combined, and comparing your answer to the circumference of the Earth (about 40,000 km).

b. Suppose that the marathon were run as a relay with one runner going at a time. Estimate how long the relay would last.

27. Atoms in a Sugar Cube. A sugar cube is about 1 cm on a side, so we could fit 100 million atoms along its length, width, and depth. How many atoms could we fit in the *volume* of a sugar cube? Compare your answer to the total number of stars in the universe, which is on the order of 10^{22}.

28. Scale Model Solar System. In Example 4, we used a 1 to 10 billion scale to represent the Sun and the Earth. The following table gives size and distance data for all the planets. Calculate the scaled sizes and distances for each planet in a 1 to 10 billion scale model solar system. Give your results in table form. Then write one or two paragraphs that describe your findings in words and give perspective to the size of our solar system.

Planet	Diameter	Average Distance from Sun
Mercury	4,880 km	57.9 million km
Venus	12,100 km	108.2 million km
Earth	12,760 km	149.6 million km
Mars	6,790 km	227.9 million km
Jupiter	143,000 km	778.3 million km
Saturn	120,000 km	1,427 million km
Uranus	52,000 km	2,870 million km
Neptune	48,400 km	4,497 million km
Pluto	2,260 km	5,900 million km

29. Universal Time Line. Suppose that the universe is 15 billion years old. Imagine that you make a time line 100 meters long. What distance along the time line represents 1 billion years? Written human history extends back only about 10,000 years. How far would that be on the time line?

30. Visiting the National Debt. In the case study on the national debt, we found that it would make a stack of $1 bills about 1 million kilometers tall. Suppose that you could fly through space along this stack of money in a 747 airplane. Assuming a speed of 1000 km/hr (620 mi/hr), how long would the flight take? Write one or two paragraphs that use your results to put the national debt in perspective.

31. Printing Money. Suppose that the government decides to pay off the national debt simply by printing more money and decides to print it in $1 bills. Assume that the printing starts when the national debt is $5 trillion. If $1 bills were printed at a rate of 5 trillion bills per year, what would this rate be in bills per second? How many bills would have to be printed each second if $100 bills, instead of $1 bills, were printed? Comment briefly on the economic repercussions if the government were to pay off the debt by printing money.

32. Counting Your Gigabuck of Cash. Suppose that you were given $1 billion, in $1 bills. How long (in years) would you need to count your fortune? Explain any assumptions you make.

33. Paving with Dollar Bills. Measure the length and width of a $1 bill and use your result to find its area.

a. Suppose that you began laying $1 bills to cover the ground. If you had the 1995 debt of $5 trillion in $1 bills, how much total area could you cover?

b. The total land area of the United States is about 10 million square kilometers. Could you cover the United States with the national debt in $1 bills?

34. Counting the Stars. Our solar system is just one of more than 100 billion star systems that make up the *Milky Way galaxy*. Calculate how long it would take to count to 100 billion at a rate of one count per second (with no breaks and no stopping). Give your answer in years. Then write one or two paragraphs that use your result to put the number of stars in the Milky Way in perspective.

35. Zipper Money. Suppose that you invented and patented a useful, inexpensive product such as the zipper. Imagine that, on average, every person in the United States buys about 10 items using your product each year. Further imagine that you earn a royalty of 1¢ on each item. Estimate how much money will you earn each year from your invention?

(The first zipper, called the Hookless Fastener, was invented in 1893 by Whitcomb L. Judson. It was used on boots and shoes and consisted of two thin metal chains that could be locked together with a metal slider. In 1910, Judson developed the C-Curity Fastener, for trousers and skirts. B. F. Goodrich bought Judson's company in 1923 and used the zipper for its rubber galoshes.)

36. How Old Are You? Calculate your age in hours, in minutes, and in seconds. Use your results to put the numbers 100,000, 1 million, and 1 billion into perspective. (E.g., how does your age compare to 1 million hours? 1 million minutes? 1 million seconds?)

37. Water Use. In the United States, about 340 billion gallons of fresh water are used *per day*, of which approximately 10% is used for public tap water; 11% is used by industry; 38% is used to cool electric power plants; and 41% is used for agricultural irrigation.

  a. What is the total per capita use of fresh water in the United States per day?

  b. What is the per capita use of tap water?

  c. Use these results and any others you wish to put the overall use of fresh water in the United States in perspective. Your answer should be expressed clearly in two or three paragraphs.

38. The Amazing Amazon. The February 1995 issue of *National Geographic* contains the following statement: "Dropping less than two inches per mile after emerging from the Andes, the Amazon drains a sixth of the world's runoff into the ocean. One day's discharge at its mouth— 4.5 trillion gallons—could supply all U.S. households for five months."

  a. Calculate the daily discharge of water from the Amazon in units of gallons per second and cubic feet per second. (1 cubic foot = 7.5 gallons.)

  b. The river with the next greatest daily discharge is the Mississippi River, with an outflow rate of 1,600,000 cubic feet per second. Compare the discharge rate of the two rivers with percentages.

  c. Do you think the image of U.S. households used for perspective is effective? Explain.

  d. Find another way to put the discharge of the Amazon River into perspective, and describe it in one or two paragraphs.

39. Automobile Travel.

  a. Estimate the total number of miles that you traveled in a car (yours or someone else's) during the past year.

  b. Estimate the cost of gasoline for all your traveling by car during the past year.

  c. Estimate the total cost of operating an automobile for all your traveling by car in the past year. Be sure to include all costs (e.g., maintenance, insurance, repairs).

  d. Estimate the total amount of money spent on gasoline each year in the United States. Compare this amount to the federal debt.

  e. For every liter of gasoline burned by an automobile, approximately 2.8 kilograms of carbon dioxide are emitted into the atmosphere. Estimate the total amount of carbon dioxide added to the atmosphere by your automobile travel over the past year.

  f. Estimate the total amount of carbon dioxide emitted into the atmosphere each year by all the cars in the United States.

40. Ethanol Fuel. Because the supply of oil in the world is limited, alternative fuels for automobiles are being sought. One alternative fuel is ethanol, which can be produced from corn.

  a. Assuming typical driving mileage and fuel efficiency, approximately 6 hectares (1 hectare = 10,000 square meters) of corn-producing land are needed to make enough ethanol for one car for one year. About how much land, in hectares, must be devoted to corn production in order to supply ethanol for all automobiles in the United States?

  b. About 2 million square kilometers of land currently are devoted to agriculture in the United States. Compare this area to the land needed for ethanol production under the assumptions of part (a). Based on your results, do you think that ethanol is a feasible replacement fuel for gasoline? Explain.

41. Sampling Problems. Sampling techniques can be used to estimate physical quantities. To estimate a large quantity, you might measure a representative small sample and find the total quantity by "scaling up." To estimate a small quantity, you might measure several of the small quantities together and "scale down." Carry out the necessary measurements and use sampling techniques to answer the following questions on page 216. Explain your method in each case. Create another problem whose solution requires a sampling technique.

Example: How thick is a sheet of a paper.　*Solution:* One way to estimate the thickness of a sheet of paper is to measure the thickness of a ream (500 sheets) of paper. A particular ream was 7.5 centimeters thick. Thus a sheet of paper from this ream was　7.5 cm ÷ 500 = 0.015 centimeters　thick, or 0.15 millimeters.

a. *Paper Weight.* How much does a sheet of paper weigh?

b. *Coin Counting.* How thick is a penny? a nickel? a dime? a quarter? Would you rather have your height stacked in pennies, nickels, dimes, or quarters? Explain.

c. *Grains of Sand.* How much does a grain of sand weigh? How many grains of sand are in a typical playground sandbox?

d. *Star Counting.* How many stars are visible in the sky on the clearest, darkest nights? How could astronomers estimate the total number of stars in the universe?

42. Project: Your Own Perspective. Choose a large or small number that you've heard quoted recently; it can be an amount of money, a population, a size, or anything else you wish. Using estimates or scaling as needed, write one or two paragraphs that put the number in perspective.

UNIT 4D

DEALING WITH UNCERTAINTY

CASE STUDY　　*Deficit Projections*

Like a business, the federal government either takes in more money than it spends (thereby running a surplus) or it spends more money than it takes in (thereby running a *deficit*). In the 1991 *Budget of the United States Government*, the administration of President Bush projected that the government would run a $6.0 billion *surplus* in fiscal year 1993. Because the number was stated to the nearest tenth of a billion dollars, it seemed to imply that the actual 1993 figure would be within $0.1 billion, or $100 million, of $6 billion. By the time the 1992 version of the *Budget of the United States Government* was printed, the projection for 1993 had been revised to a *deficit* of $201.5 billion. In the end, the 1993 budget deficit turned out to be $254.7 billion. The point should be clear: The original number in 1991 was offered with confidence that it was within $0.1 billion of the correct value when, in fact, it was off by more than $250 billion. Indeed, not even the *sign* of the number was right, as the supposed surplus turned out to be a huge deficit! ∎

> **• Web • Watch •**
> You can find a wealth of information about the federal budget on the web.

As the case study on deficit projections illustrates, projected numbers often bear little resemblance to the ultimate reality. Why, then, are the projected numbers reported in such a way that they imply great certainty?

Alas, there is no satisfactory answer to this question. Moreover, this problem extends far beyond the government and budgets. The sad reality is that many of the numbers bandied about in the news are far more uncertain than anyone admits. As a result, one of the most important aspects of quantitative reasoning is learning how to interpret and deal with uncertainty in a world where it is often not acknowledged.

ACCURACY AND PRECISION

The goal of any measurement or estimate is to obtain an **approximate value** that is as close as possible to the **true value**. The closer that the approximate value lies to the true value, the greater is the **accuracy** of the estimate. When the government projected a budget surplus of $6.0 billion that turned out to be a deficit of $254.7 billion, the accuracy of the projection was extremely poor. In contrast, if a company projects sales of $7.3 billion and actual sales turn out to be $7.32 billion, we would say that projection was quite accurate.

The **precision** of a number describes how *precisely* it is reported. The numbers $6.0 billion and $7.3 billion both have the *same precision* because both are *precise* to the nearest $0.1 billion, even though one turned out to be much more accurate than the other. The number $7.32 billion has greater precision than $7.3 billion because it is precise to the nearest $0.01 billion.

EXAMPLE 1   *Accuracy and Precision in Your Weight*

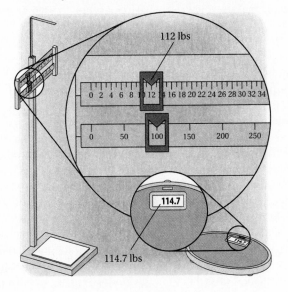

112 lbs

114.7

114.7 lbs

The scale at the doctor's office can be read only to the nearest pound, and says that you weigh 112 pounds. The scale at the gym is digital and says that you weigh 114.7 pounds. Suppose that your actual weight is 111.69 pounds. Which scale is more *precise*? Which is more *accurate*?

Solution: The scale at the gym is more precise because it reported your weight to the nearest tenth of a pound, whereas the doctor's scale reported your weight only to the nearest pound. However, the scale at the doctor's office is more accurate because its value is closer to your true weight. ∎

TIME-OUT TO THINK: We needed to know your *true* weight in order to determine which scale was more accurate. But the only way to determine your weight is by getting on a scale. Given these facts, can we really be sure that one scale is more accurate than another? Explain.

SIGNIFICANT DIGITS

Suppose that a candy bar costs exactly 75¢. Because this price is exact, it doesn't matter whether we write the price as 75¢, 75.0¢, or 75.00000¢. Now suppose you measure your weight to be 132 pounds on a scale that can be read only to the nearest pound. Reporting that you weigh 132.0 pounds would be misleading because it would incorrectly imply that you know your weight to the nearest *tenth* of a pound. When we are dealing with measured or estimated numbers, "132 pounds" and "132.0 pounds" do *not* have the same meaning!

One way to ensure that we don't accidentally imply more precision than we can justify is by keeping track of **significant digits**: the digits in a number that reflect what is actually known. The number "132 pounds" has three significant digits and implies a weight known to the nearest pound; "132.0 pounds" has four significant digits and implies a weight known to the nearest tenth of a pound.

Note that zeros that serve only as place holders are *not* significant digits. For example, "6 mm," "0.006 m," or "0.000006 km" all mean 6 millimeters and hence have only one significant digit. The zeros that arise when we express the measurement in meters or kilometers are merely place holders for the decimal point and are not significant.

The only difficulty in counting significant digits arises when we cannot be sure whether zeros are place holders or truly significant. For example, suppose that your professor states that there are 200 students in your class. Without further information, you have no way to know whether this number is precise to the nearest hundred, the nearest ten, or exact. Given this ambiguity, the general rule is to assume that the zeros are place holders unless told otherwise. Thus we should assume that "200 students" has only one significant digit and therefore is precise only to the nearest hundred students.

We can avoid this kind of ambiguity by writing numbers in scientific notation. In that case, zeros appear only when they are significant. For example, an enrollment stated as 2×10^2 implies precision to the nearest hundred students, 2.0×10^2 implies precision to the nearest ten students, and 2.00×10^2 implies exactly 200 students.

These examples also show the connection between significant digits and precision. The location of the last significant digit in a number tells us the precision of the number. The last significant digit of 200 is the "2" in the *hundreds* place, so the number is precise to the nearest *hundred*. The last significant digit of 2.0×10^{-2} is the "0" in the *thousandths* place, so the number is precise to the nearest *thousandth*.

SUMMARY: WHEN ARE DIGITS SIGNIFICANT?

Nonzero digits	always significant
Zeros *between* nonzero digits (as in 4002 or 3.06)	always significant
Zeros that follow a nonzero digit *and* lie to the right of the decimal point (as in 4.20 or 30.00)	always significant
Zeros to the *left* of the first nonzero digit (as in 0.006 or 0.00052)	never significant
Zeros to the *right* of the last nonzero digit but before the decimal point (as in 40,000 or 210)	not significant unless stated otherwise

EXAMPLE 2   *Counting Significant Digits*

For each of the following measurements or estimates, state the number of significant digits and the implied precision.
a)   a volume of 6.05 liters
b)   a time of 11.90 seconds
c)   a length of 0.0000067 meters
d)   a population reported as 240,000
e)   a population reported as 2.40×10^5

Solution:
a)   The number 6.05 liters has three significant digits. It is precise to the nearest 0.01 liter because the last significant digit is in the hundredths place.
b)   The number 11.90 seconds has four significant digits. It is precise to the nearest 0.01 second because the last significant digit is in the hundredths place. The final zero is significant because, had it not actually been measured, the number should have been reported as 11.9 seconds.
c)   The number   0.0000067 meters $= 6.7 \times 10^{-6}$ meters   has two significant digits. It is precise to the nearest 0.0000001 meter because the last significant digit is in the ten millionths place. The leading zeros are not significant because they are merely place holders.
d)   Following the rules for counting significant digits, we assume that zeros in the population of 240,000 are not significant because we were not told otherwise. Thus the number has two significant figures and is precise to the nearest 10,000 people.
e)   The number 2.40×10^5 also is 240,000. However, the use of scientific notation tells us that it has three significant digits and therefore is precise to the nearest thousand. ∎

BRIEF REVIEW

Rounding

The basic process of rounding numbers takes just two steps.

1. Decide which decimal column (e.g., tens, ones, tenths, or hundredths) is the smallest that should be kept.
2. Look at the number in the next column to the *right* (e.g., if rounding to tenths, look at hundredths). If the value in the next column is *less than 5*, round *down*; if it is *5 or greater*, round *up*.

For example, the number 382.2593 is given to the nearest ten thousandth. It can be rounded in the following ways:

382.2593 rounded to the nearest thousandth is 382.259.

382.2593 rounded to the nearest hundredth is 382.26.

382.2593 rounded to the nearest tenth is 382.3.

382.2593 rounded to the nearest one is 382.

382.2593 rounded to the nearest ten is 380.

382.2593 rounded to the nearest hundred is 400.

(Note that some statisticians use a more complex rounding rule if the value in the next column is exactly 5: They round up if the last digit being kept is odd and down if it is even. We won't worry about this subtlety in this book, but you may encounter it if you take a course in statistics.)

EXAMPLE 3   *Rounding with Significant Digits*

For each of the following operations, give your answer with the specified number of significant digits.

a)   7.7 mm $\times$ 9.92 mm; give your answer with two significant digits

b)   240,000 $\times$ 72,106; give your answer with four significant digits

Solution:

a)   7.7 mm $\times$ 9.92 mm = 76.384 mm^2. However, because we are asked to give the answer with just two significant digits, we round the answer to 76 mm^2.

b)   240,000 $\times$ 72,106 = 1.730544 $\times$ 10^{10}. However, because we are asked to give the answer with four significant digits, we round to   1.731 $\times$ 10^{10}.   ∎

COMBINING APPROXIMATE NUMBERS

Let's now look at what needs to be considered when performing operations with approximate numbers.

Adding or Subtracting Approximate Numbers

Suppose that you live in a city with a population of about 300,000. After months of long-distance phone calls, your best friend moves to your city to share an apartment with you. What is the population of the city now?

You might be tempted to add your friend to the city's population, so that the new population would be 300,001. However, the number 300,000 has only one significant digit, implying that the population is known only to the *nearest 100,000*. The number 300,001 has six significant digits, which implies that you know the population *exactly*. Clearly, your friend's move cannot change the fact that the population is known only to the nearest 100,000. Thus the correct way to add your friend to the population is

$$300,000 + 1 \approx 300,000.$$

Indeed, it would be *wrong* in this case to say that   300,000 + 1 = 300,001:   An answer of 300,001 would imply that your friend's arrival suddenly gave you the ability to know the population exactly when you knew it only to the nearest 100,000 before.

As this example illustrates, adding and subtracting approximate numbers requires special care to prevent inadvertently stating an answer with more certainty than it deserves. Fortunately, the following simple rule works well in most cases.

> **Rounding Rule for Addition or Subtraction**
>
> An answer obtained by adding or subtracting approximate numbers should be rounded to the same precision as the *least precise* number in the problem.

EXAMPLE 4   *Adding and Subtracting Approximate Numbers*

a) A book written in 1960 states that the oldest Mayan ruins are 2000 years old. How old are they now?

b) You are driving to a party at a friend's house. The instructions say to continue 1.7 miles past the town library. At a gas station, the attendant tells you that you are about 7 miles from the library. How far do you have to go?

c) A swimmer in a 100-meter race is timed by a hand-held timer in 58.7 seconds. A few weeks later, an electronic timing system clocks her in 57.34 seconds. How much has she improved?

Solution:

a) The book says that the ruins are 2000 years old, but a book written in 1960 is about 40 years old itself. Thus we might be tempted to add 40 years to 2000 years to get 2040 years for the age of the ruins. However, 2000 years is the least precise of the two numbers: It is precise only to the nearest 1000 years because its only significant digit is in the thousands place. The number 40 years is precise to the nearest 10 years. Thus the answer also should be precise only to the nearest 1000 years:

> **Beware!** Round only in the final step of a problem, not during intermediate steps. Otherwise you may compound the uncertainties further.

$$\underbrace{2000 \text{ yr}}_{\text{precise to nearest 1000}} + \underbrace{40 \text{ yr}}_{\text{precise to nearest 10}} = \underbrace{2040 \text{ yr}}_{\text{must round to nearest 1000}} \approx \underbrace{2000 \text{ yr}}_{\text{correct final answer}}$$

Despite the 40-year age of the book, the ruins still are 2000 years old to the given precision.

b) The instructions to continue 1.7 miles past the town library imply precision to the nearest 0.1 mile. The gas station attendant's statement of 7 miles to the library implies precision to the nearest mile. Thus the answer should be precise only to the nearest mile. The distance you have to go is about

$$\underbrace{7 \text{ mi}}_{\text{precise to nearest 1}} + \underbrace{1.7 \text{ mi}}_{\text{precise to nearest 0.1}} = \underbrace{8.7 \text{ mi}}_{\text{must round to nearest 1}} \approx \underbrace{9 \text{ mi}}_{\text{correct final answer}}.$$

You can expect a total distance of about 9 miles to the party. (Note that, while it is possible that you have precisely 8.7 miles to go, this is only one of a range of distances consistent with 9 miles.)

c) The first time of 58.7 seconds is precise to the nearest 0.1 second, while the next time of 57.34 seconds is precise to the nearest 0.01 second. Thus, when we subtract the later time to determine the swimmer's improvement, the answer should be precise only to the nearest 0.1 second.

$$\underbrace{58.7 \text{ s}}_{\text{precise to nearest 0.1}} - \underbrace{57.34 \text{ s}}_{\text{precise to nearest 0.01}} = \underbrace{1.36 \text{ s}}_{\text{must round to nearest 0.1}} \approx \underbrace{1.4 \text{ s}}_{\text{correct final answer}}$$

She has improved her time by about 1.4 seconds.   ∎

TIME-OUT TO THINK: In Example 4, part (c), suppose that the swimmer's first time was measured as 58.70 seconds. In that case, what was her improvement?

Multiplying or Dividing Approximate Numbers

Slightly different considerations apply when we multiply or divide approximate numbers. Suppose that a survey finds that, each month, Americans spend an average of $120 per person on entertainment. How much is spent on monthly entertainment in a city of 300,000 people?

If we multiply the two numbers together, the monthly entertainment total comes out as

$$300{,}000 \text{ persons} \times \frac{\$120}{\text{person}} = \$36{,}000{,}000.$$

However, because we know the city's population only to the nearest 100,000, the actual population might be anywhere in the range of about 250,000 to 350,000. Thus our calculation of the monthly entertainment bill could be off by as much as the total spending of 50,000 people, which is about

$$50{,}000 \text{ persons} \times \frac{\$120}{\text{person}} = \$6{,}000{,}000.$$

Clearly, we should not claim to have found an answer of $36 million if we know that it could be off by as much as $6 million!

One way to deal with this uncertainty is to report our answer as 36 ± 6 million, indicating that the actual answer could be anywhere between $30 million and $42 million. This type of detailed accounting is required in careful scientific work. However, in most practical applications we can apply a much simpler rule.

Rounding Rule for Multiplication or Division

An answer obtained by multiplying or dividing approximate numbers should be rounded to the same number of significant digits as the number in the problem with the *fewest significant digits*.

In this case, we round our answer to the one significant digit that appears in the population of 300,000, so the monthly entertainment spending in the city is

$$\underbrace{300{,}000 \text{ persons}}_{\text{1 significant digit}} \times \underbrace{\$120/\text{person}}_{\text{2 significant digits}} = \underbrace{\$36 \text{ million}}_{\substack{\text{must round to} \\ \text{1 significant digit}}} \approx \underbrace{\$40 \text{ million}}_{\text{correct final answer}}.$$

EXAMPLE 5   *Multiplying and Dividing Approximate Numbers*

a) Suppose that you sleep an average of about 8 hours per night. How much time do you spend sleeping in a year?

b) The government in a town of 82,000 people plans to spend $41.5 million this year. Assuming all this money must come from taxes, how much must the city collect from each resident?

c) Suppose that you measure the side length of a square room to be 3.3 meters. What is the area of the room?

Solution:

a) The number of nights in a year is 365, which has three significant digits. But the 8 hours per night of sleeping has only one significant digit. Therefore the answer should have only one significant digit:

$$\underbrace{8 \text{ hr/night}}_{1 \text{ significant digit}} \times \underbrace{365 \text{ nights}}_{3 \text{ significant digits}} = \underbrace{2920 \text{ hr}}_{\substack{\text{must round to} \\ 1 \text{ significant digit}}} \approx \underbrace{3000 \text{ hr}}_{\text{correct final answer}}$$

You sleep about 3000 hours per year.

b) The average tax that must be collected per resident is the $41.5 million divided by the population of 82,000. But we must round the answer to the two significant digits of the population, because it is the number with the fewest significant digits in this problem:

$$\underbrace{\$41,500,000}_{3 \text{ significant digits}} \div \underbrace{82,000 \text{ persons}}_{2 \text{ significant digits}} = \underbrace{\$506.10}_{\substack{\text{must round to} \\ 2 \text{ significant digits}}} \approx \underbrace{\$510}_{\text{correct final answer}}$$

The average resident must pay about $510 in taxes to balance this town's budget.

c) The area of the square room is simply its side length squared. Because squaring is a form of multiplication, the final answer should have the same two significant digits as the measured side length:

$$(3.3 \text{ m})^2 = 10.89 \text{ m}^2 \approx 11 \text{ m}^2$$

Thus the area of the room is about 11 square meters.   ■

CASE STUDIES IN UNCERTAINTY

We've seen that numbers often are reported with far more certainty than they deserve, particularly when they involve projections into the future. The inherent uncertainty in numbers underlies many of the most significant controversies of our time. The following case studies may help to illuminate the consequences of uncertainty and how we can recognize and deal with it.

CASE STUDY   *The Census*

The Constitution of the United States mandates a census of the population every ten years. Upon completing the 1990 census, the U.S. Census Bureau reported a population of 248,709,873. Was this a reasonable number to report?

Not a chance. The population was stated with nine significant digits—to the nearest person—implying an *exact* count! Even if the Census Bureau were somehow able to count every person in the United States instantaneously, their count would be off within a matter of minutes: about eight births and four deaths occur every minute in the United States.

The 1990 census primarily relied on a survey that was supposed to be delivered to and returned by every household in the United States. But the count was affected by

By the Way ············
There are an average of 7.7 births, 4.3 deaths, 4.4 marriages, and 2.3 divorces per minute in the United States (1997 data).

numerous sources of error. For example, some people were counted twice, such as college students counted by their parents and counted again at their school residence. Others were not counted at all because they moved during the survey period or they discarded their census forms. Errors also occurred because survey forms were filled out incorrectly or because responses were recorded incorrectly by Census Bureau employees. These types of errors are called **random errors** because their impact cannot be predicted.

The census also was subject to **systematic errors**—errors that arise because of problems in the *system* being used to make the count. For example, the homeless were certainly undercounted because it is very difficult to deliver surveys to them.

The Census Bureau eventually used statistical techniques to estimate the severity of the errors. The statistical results suggested that the census had missed about 10 million people and counted 6 million people twice. Overall, then, it undercounted the population by about 4 million people. Thus the initially reported population of 248,709,873 was misleading: It implied that the population was known with far more precision than could be justified. ■

> • Web • Watch •
> Check population facts at the Census Bureau web site.

> **TIME-OUT TO THINK:**   The Constitution calls for an "actual enumeration" of the population every ten years (Article 1, Section 2, Subsection 2). Some people claim that this wording precludes the Census Bureau from using statistical techniques that do not actually count each individual person. Others claim that it demands the use of statistical techniques because they make the final count more accurate. Which side do you take? Defend your choice.

CASE STUDY   *Inflation and the Consumer Price Index*

Suppose that you earn $30,000 this year. How much will you need to earn to maintain the *same* standard of living next year? The answer to this question is tied to the rate of *inflation,* or increase in the cost of living (see discussion of inflation in Unit 2D).

For decades, the U.S. Bureau of Labor Statistics has tracked the cost of living with the **Consumer Price Index (CPI)**. The CPI is calculated monthly by averaging prices of goods, services, and housing collected from more than 60,000 sources. The annual CPI tabulated in Table 4.2 is shown relative to a value of 100 that represents average prices during the years 1982–1984.

By the Way ············
The government measures two consumer price indices: *CPI-U* is based on products thought to reflect the purchasing habits of *all* urban consumers, whereas *CPI-W* is based on the purchasing habits of only wage earners. Table 4.2 shows the CPI-U.

Table 4.2	Consumer Price Index (1982–1984 = 100)		
Year	CPI	Year	CPI
1984	103.9	1991	136.2
1985	107.6	1992	140.3
1986	109.6	1993	144.5
1987	113.6	1994	148.2
1988	118.3	1995	152.4
1989	124.0	1996	156.9
1990	130.7	1997	160.5

We can use these data to calculate the change in the CPI in any period. For example, from 1995 to 1996 the CPI rose by

$$\text{change in CPI 1995 to 1996} = \frac{\text{CPI}_{1996} - \text{CPI}_{1995}}{\text{CPI}_{1995}} \times 100\% = \frac{156.9 - 152.4}{152.4} \approx 3.0\%.$$

If this change truly reflects the underlying rate of inflation, you would have needed to earn 3.0% more money in 1996 than in 1995 to maintain the same standard of living.

However, many economists argue that the CPI overstates inflation. One concern is that the CPI compares monthly changes in prices of the same items at the same stores. Thus when an item's price rises at one store, it tends to increase the CPI. However, consumers may not experience an increase in their cost of living if they can find the same item for a lower price at a different store, or if they can substitute a similar but lower-priced item. Another possible bias occurs because the CPI tracks prices of items purchased by typical consumers. Thus today's CPI includes items such as VCRs, cable television, and computers that were not available just a couple of decades ago. The CPI does not factor in any improvement in the standard of living that these items represent.

A 1996 commission (the Boskin Commission) concluded that the CPI overstates the actual rate of inflation by between 0.8 to 1.6 percentage points per year. In that case, the actual inflation rate from 1995 to 1996 was between 1.4% to 2.2%, rather than the 3.0% suggested by the rise in the CPI.

These small percentages may not sound very important, but they have a tremendous impact on national finances. For example:

- Annual increases in individual Social Security benefits are supposed to match the rate of inflation so that recipients can maintain a constant standard of living. But benefits actually have *risen* in value if the CPI really does overstate inflation (Figure 4.4).

> **• Web • Watch •**
> Learn more about the Consumer Price Index at the Bureau of Labor Statistics web site.

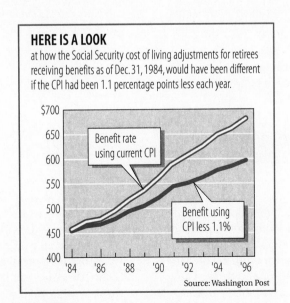

HERE IS A LOOK
at how the Social Security cost of living adjustments for retirees receiving benefits as of Dec. 31, 1984, would have been different if the CPI had been 1.1 percentage points less each year.

Benefit rate using current CPI

Benefit using CPI less 1.1%

Source: Washington Post

FIGURE 4.4

- The levels of income at which different income tax rates take effect (*tax brackets*) are supposed to rise with the rate of inflation. Otherwise, people with no change in their living standards would gradually move to higher tax brackets. But if inflation is overstated, tax rates have effectively come down.

Because Social Security is a large fraction of the federal budget, and income taxes are the federal government's primary source of revenue, these effects add up to huge amounts of money. The Boskin Commission recommended that the government adjust Social Security benefits and tax rates by an annual amount that is 1.1 percentage points lower than the change in the CPI. Incredibly, this adjustment could save the federal government over *$1 trillion* between 1998 and 2008!   ■

> **TIME-OUT TO THINK:**   Discuss some of the political ramifications of the Boskin Commission recommendations. For example, how would they affect individual Social Security recipients? How would they affect tax rates? If you were a member of Congress, would you vote to accept the Boskin Commission recommendations? Why or why not?

REVIEW QUESTIONS

1. What is the difference between *accuracy* and *precision*? Give an example in which a number is quite accurate but not very precise. Give an example in which a number is very precise but not very accurate.

2. What are *significant digits*? How can you tell when zeros are significant?

3. When is 100,000 + 1 ≠ 100,001? Explain.

4. If you add or subtract approximate numbers, how should you write your answer so that it does not imply more precision than it deserves?

5. If you multiply or divide approximate numbers, how should you write your answer so that it does not imply more knowledge than you actually had?

6. What is the difference between a *random error* and a *systematic error*? Give an example of each type of error and its effects in the 1990 census.

7. What is the *consumer price index* (CPI)? How is it related to *inflation*? Why is it controversial?

PROBLEMS

1. **Numbers in the News.** Look through the newspapers from the past week for articles that mention numbers in any context. Find at least two numbers from each of the following sections: national/international news, local news, sports, and business. In each case, describe the number and its context, and discuss any uncertainty that you think is associated with the number.

2. **Failures to Report Uncertainty in the News.** Search through the past week's newspapers. Determine at least three instances in which reported numbers clearly were stated with more precision or accuracy than justified and without any indication of the uncertainty in the numbers. Describe each instance and, for each, briefly explain what might have been a more realistic way to report the numbers.

3. **Hometown Population.** According to local groups (such as the city council or chamber of commerce), what is the current population of your hometown? Discuss the uncertainty in this value.

4. **Sources of Error.** Briefly identify and discuss possible sources of error in each given claim.

   a. The life expectancy in Japan is 78.7 years.

   b. Twenty-three and four-tenths percent of Americans are Roman Catholic.

   c. A violent crime occurs in the United States once every 16 seconds.

   d. The population of the United States has been increasing at a rate of approximately 5 million people per year.

5. **Sources of Error.** Briefly identify and discuss possible sources of error in each given claim.

   a. The price of memory for microcomputers is decreasing at a rate of approximately $10 per megabyte per year.

   b. Six hundred twenty-three students in my high school class graduated.

   c. Two-thirds of the town voted for the Republican mayoral candidate.

   d. The literacy rate in Mongolia is 90%.

6. **Rounding Practice.** Round each number to the nearest whole number.

   a. 2.4   **b.** 14.500   **c.** 779.49

   d. 4.1999   **e.** 13.5001   **f.** 234.5

   g. 1999.5   **h.** 88.71   **i.** −13.998

7. **Rounding Practice.** Round each number to the nearest thousandth, tenth, ten, and hundred.

   a. 2365.98521   **b.** 322354.09005   **c.** 6000

   d. 34/3   **e.** 578.555   **f.** 0.45232768

   g. −12.1   **h.** −850.7654   **i.** −10,995.6239

8. **Real-World Rounding.** Find at least three different examples of numbers that are rounded in news stories. Explain how and why each number is rounded. (Hint: You might consider baseball batting averages, times in athletic events, or the stock market.)

9. **Accuracy and Precision.** Suppose that you purchase a standard weight at a store that actually weighs 1.500 kilograms. You use the standard weight to check the calibration of two different scales. Using the first scale, you are able to measure to the nearest hundredth of a kilogram; it shows 1.51 kg for the standard weight. With the second scale, you can measure to the nearest ten thousandth of a kilogram, and it shows 1.5340 kg for the standard weight. Which scale gives the more accurate measurement? Which gives the more precise measurement? Explain.

10. **Counting Significant Digits.** State the number of significant digits and the implied precision in each given number.

    a. 96.2 km/hr   **b.** 100.020 seconds

    c. 0.00098 mm   **d.** 0.0002020 meters

11. **Counting Significant Digits.** State the number of significant digits and the implied precision in each given number.

    a. 401 people   **b.** 200.0 liters

    c. 1.00098 mm   **d.** 0.000202 meters

12. **Ambiguity in Zeros.** State the number of significant digits in the number 300,000 if the number is correct to

    a. within a single unit.

    b. the nearest 10.

    c. the nearest tenth.

    d. the nearest hundredth.

13. **Ambiguity in Zeros.** State the number of significant digits in the number 300,000 if the number is correct to

    a. the nearest 100.

    b. the nearest 1000.

    c. the nearest 10,000.

    d. the nearest 100,000.

14. **Removing Ambiguity with Scientific Notation.** Write 300,000 in scientific notation with

    a. one significant digit.

    b. two significant digits.

    c. five significant digits.

    d. eight significant digits.

15. Removing Ambiguity with Scientific Notation. Write 500,000 in scientific notation with

  a. one significant digit.

  b. two significant digits.

  c. five significant digits.

  d. eight significant digits.

16. Confidence and Implied Precision. Consider each given statement and state whether you believe the implied precision in each number. Give your reasons.

  a. The U.S. Census Bureau reports that the population of a city is 1,452,332.

  b. A real estate agent, trying to convince you to list your home with him, tells you that he has sold 500 homes.

  c. A stockbroker tells you that she has run some calculations and that by investing with her you will earn $1100 more over the next year than if you put the money in the bank.

  d. In 1998, the president states that by 2005 the government budget will run a surplus of $23.2 billion.

  e. A group of population experts states that world population will level off at 11 billion people in 2050.

17. Confidence and Implied Precision. Consider each given statement and state whether you believe the implied precision in each number. Give your reasons.

  a. My little brother is 4 feet, 2.9051 inches tall.

  b. In the 1996 presidential election, Bill Clinton received 7,756,824 votes more than Bob Dole.

  c. The average amount of money spent on textbooks by students is $120 \pm 50 per semester.

  d. There are 260 million people in the United States.

  e. The owners of your favorite, but last-place, football team state that their rebuilding program will bring a Super Bowl victory in 5 years.

18. Rounding Rule for Addition and Subtraction. Assume that each given number is measured to the indicated precision. Carry out the following operations, using the rounding rule for addition and subtraction.

  a. $65.7832 + 7.112 + 51,009$

  b. $8.409 + 1.227 + 13.19$

  c. $48.49 + 4.237 + 12.1$

  d. $(5 \times 10^3) + (2 \times 10^2)$

  e. $(4.326 \times 10^{-6}) + (9.36478 \times 10^{-9})$

  f. $(8.599 \times 10^9) + (7.62 \times 10^7)$

  g. $(6.5 \times 10^2) - (4.2 \times 10^2)$

19. Rounding Rule for Addition and Subtraction. Assume that each given number is measured to the indicated precision. Carry out the following operations, using the rounding rule for addition and subtraction.

  a. Add the distances 36 centimeters and 8.22 centimeters.

  b. Add the weights 260 kilograms and 17 kilograms.

  c. Subtract 1.09 liters from 140 liters.

  d. Subtract 1 hour, 22 minutes, 15 seconds from 2 hours, 37 minutes.

  e. Add the distances 4.093×10^{10} kilometers and 6.1×10^8 kilometers.

  f. Subtract 3.5 kilograms from 72 kilograms.

20. Rounding Rule for Multiplication and Division. Assume that each given number is stated with the proper number of significant digits. Carry out the indicated operations, using the rounding rule for multiplication and division.

  a. $(1.3 \times 10^{21}) \times (4.1 \times 10^{-12})$

  b. $(2.871 \times 10^{35}) \times (3 \times 10^{-33}) \times (5.78 \times 10^7)$

  c. $(3.43 \times 10^{-7}) \times (5.661 \times 10^{-5})$

  d. $(3 \text{ million}) \div (1.56 \text{ thousand})$

  e. $(4.448921 \times 10^{13}) \times (1 \times 10^1)$

21. Rounding Rule for Multiplication and Division. Assume that each given number is stated with the proper number of significant digits. Carry out the indicated operations, using the rounding rule for multiplication and division.

  a. Multiply the lengths 105 meters and 26 meters.

  b. Divide the distance 110 kilometers by the time 55 minutes.

  c. Multiply the weights 9.7 kilograms and 165 kilograms.

  d. Divide the weight 5 grams by the volume 1.3 cubic centimeters.

22. Scale Calibration. Suppose that after measuring your weight to be 52.3 kilograms, you check the calibration of the scale. To your surprise, you find that the scale reads 2.4 kilograms even when nothing is on it! How would this fact have affected your measured weight? Is this an example of a systematic error or a random error? Explain.

23. **Minimizing Random Errors.** Suppose that 25 people, including yourself, are asked to measure the length of a room to the nearest tenth of a millimeter. Assume that everyone uses the same well-calibrated measuring device, such as a tape measure.

    a. All 25 measurements are not likely to be exactly the same; thus the measurements will contain some sources of error. Are these errors systematic or random? Explain.

    b. If you want to minimize the effect of random errors in determining the length of the room, which is the better choice: to report your own, personal measurement as the length of the room, or to report the average of all 25 measurements? Explain.

    c. Describe any possible sources of systematic errors in the measurement of the room length.

    d. Can the process of averaging all 25 measurements help reduce any systematic errors? Why or why not?

24. **Should You Get Off the Elevator?** Suppose that you step onto an elevator that already has nine people in it. You look at the safety panel and see a sign that says "Maximum Capacity 1300 Pounds." Should you leave the elevator? Explain. If all nine people are members of a girls' high school gymnastics team, how will that affect your decision? What if the nine people are linemen on your school's football team?

25. **Race Timing.** Modern timing devices can easily measure time to the nearest thousandth of a second, yet international sports federations recognize world records (in events such as running or swimming) only to the nearest hundredth of a second. Why do you think that this is so?

26. **Inflation with the Consumer Price Index.**

    a. Use the CPI data in Table 4.2 to calculate the rate of inflation (as a percentage) from 1996 to 1997.

    b. Suppose that you earned $25,000 in 1996. How much would you have to have earned in 1997 to retain the same purchasing power?

    c. Suppose that an item cost $75.00 in 1996 and $78.50 in 1997. How did the price increase for this item compare to the overall rate of inflation?

27. **Overflowing Trash.** Answer the following questions and be sure to discuss the uncertainty in each of your estimates.

    a. Estimate the amount of trash you produce in an average week, in units of both volume and weight. You could begin with a family estimate.

    b. Estimate the total amount of trash that individuals produce each week in the United States. How much trash is produced each year?

    c. Put the total amount of trash from individuals in perspective. (Hint: You may use any method you choose for this perspective. For example, try estimating how many football stadiums the trash would fill.)

    d. We have not included trash from business and industry in this estimate. How do you think this trash affects the total? Explain.

28. **Recycling Newspapers.** Newsprint is made from trees, so recycling it means that fewer trees must be cut down to produce newspapers. In the following problems, be sure to include an uncertainty with each of your estimates. Explain your work.

    a. Estimate the weight of one of your local newspapers and the total daily circulation of this newspaper. Use these numbers to estimate the total weight of this newspaper printed each day.

    b. Estimate the percentage of newspapers in your area that are recycled.

    c. For every 50 kg of newspaper that is recycled, rather than made from virgin wood, approximately one tree is saved. Approximately how many trees are saved each year from the recycling of your local newspaper?

29. **Car Cost Per Mile.** The Saturn car company advertises that its 1995 Saturn SW2 costs $0.24 per mile to drive. This figure is based on adding the purchase price of the car to the cost of gasoline, maintenance, insurance, and registration, and then subtracting the estimated resale value of the car at the end of five years. To get the units of dollars per mile, the company divides the cost by 60,000 miles; this figure is based on the assumption that the owner will drive the car 60,000 miles in five years, then sell or trade it in for a new car.

   a. Based on the assumptions stated, what is the total cost of owning a Saturn SW2 for five years?

   b. Analyze each of the assumptions made. Are they realistic? Overall, is the advertisement reasonable? Explain.

30. **Project: Social Security.** A major economic policy question in the United States concerns whether the Social Security system is sustainable. Learn about how Social Security works, and why some people believe the system is in danger while others believe it is sustainable for decades to come. Write a one- to two-page essay about the uncertainties involved in projecting the future of Social Security. Conclude your essay by giving your own opinion of what policies should be enacted concerning Social Security.

31. **Project: Global Warming.** One of the most pressing issues of our time surrounds the question of whether governments should mandate reductions in carbon emissions to help prevent global warming. Learn about the current evidence regarding how much the Earth has warmed over the past century, and how much human activity contributes to that warming. Based on what you learn, write a one- to two-page essay about the uncertainty involved in understanding how human activity is contributing to global warming. Conclude your essay by giving your own opinion as to whether, in light of the current uncertainties, we should mandate reductions in carbon emissions.

32. **Project: Species Extinction.** Another important environmental issue involves the question of how many species are being driven to extinction by human activity. This turns out to be a very difficult number to estimate, in part because we don't even know how many different species of plant and animal inhabit the Earth! Research the methods used to estimate the species extinction rate, and how these estimated rates compare to the rates expected in the absence of human activity. Summarize your findings in a one- to two-page essay. Also include your opinion as to what, if anything, should be done to reduce the rate of species extinction.

CHAPTER 4
SUMMARY

$\mathcal{W}$e began this chapter by building the number system from the natural numbers through the real numbers. We then explored the ways numbers are really used in the modern world and the uncertainty that accompanies most numbers. Here are a few key ideas to keep in mind.

- Our modern system of expressing numbers developed over thousands of years and with contributions from many different cultures around the world.

- Statements that involve percentages must be studied with great care because percentages are so frequently used in misleading or incorrect ways.

- Large and small numbers will be meaningless unless you find a way to put them in perspective. You can put a number in perspective by using techniques such as estimation or scaling to compare the number to other numbers with which you are already familiar.

- The numbers we encounter in the real world often have far more uncertainty than is generally acknowledged. If we want to understand such numbers, we must first examine their uncertainties.

SUGGESTED READING

Innumeracy, J. A. Paulos (New York: Hill and Wang, 1988).
A Mathematician Reads the Newspaper, J. A. Paulos (New York: Basic Books, 1995).
Number: The Language of Science, T. Dantzig (New York: Macmillan, 1930).
Pi in the Sky: Counting, Thinking, and Being, J. D. Barrow (Boston: Little Brown and Company, 1992).
200% of Nothing, A. K. Dewdney (New York: John Wiley and Sons, 1993).
Billions and Billions, C. Sagan (New York: Random House, 1997).

Chapter 5

FINANCIAL MANAGEMENT

M anaging your personal finances is a complex task in the
modern world. If you are like most Americans, you
already have a bank account and at least one credit card.
You may also have student loans, a home mortgage, or various investment
plans. In this chapter we discuss key issues in personal financial management, including savings, loans, taxes, and investments.

Remember that time is money.
BENJAMIN FRANKLIN

A fool and his money are soon parted.
ENGLISH PROVERB

*P*ersonal financial management used to be easy: everything was paid in cash, and banks were the only reliable place for savings. Today, we are confronted with numerous choices. We can invest in banks, bonds, stocks, mutual funds, and much more. We can borrow money in many different ways, such as student loans for college, home mortgages to buy a home, or credit cards. We may have thousands of choices for retirement savings.

The most remarkable aspect of personal finance is that the same amount of money today can have different values in the future. If you spend $1000 today, it's gone. If you take the same $1000 and invest well, it may become $10,000 in a few years. If you use the $1000 as part of a down payment on a house, you may find your living situation improved and also find a decreased income tax bill. If you get duped into a crackpot investment scheme, you may find that your $1000 simply disappears—just as if you were robbed. Indeed, two people who earn identical amounts of money may have very different standards of living in the future, depending on how well they manage their respective finances.

Your future financial success depends critically on understanding the basic principles involved in savings, loans, taxes, and investments, which are the topics of this chapter. Investing the time needed to understand these topics will *pay*, literally, in terms of your future financial well-being.

UNIT
5A

THE POWER OF COMPOUND INTEREST

CASE STUDY   *New College Wants the Queen to Pay Up*

In 1996, administrators at New College of Oxford discovered paperwork showing that King Edward IV of England had borrowed the equivalent of $384 from the College on July 18, 1461. The King soon paid back $160, but never repaid the remaining $224. A College official wrote to the Queen of England asking for repayment of the 535-year-old debt—

with interest. Assuming an interest rate of 4% per year, the official calculated that the college was owed about $290 billion. However, he indicated a willingness to settle for a lower interest rate of 2% per year, in which case the college was owed only $8.9 million. This, he said, would be just about enough to pay for a modernization project at the College. ■

This case study illustrates what is sometimes called the "power of compound interest": the remarkable way that money grows when interest continues to accumulate year after year. In the New College case, there is no clear record of a promise to repay the debt with interest and, even if there were, the Queen might not feel obliged to pay a debt that was forgotten for more than 500 years. But anyone can take advantage of compound interest simply by opening a savings account. With diligence and patience, the results may be truly astonishing.

SIMPLE VERSUS COMPOUND INTEREST

Imagine that you deposit $1000 in Honest John's Money Holding Service, which promises to pay 5% interest each year. At the end of the first year, Honest John's sends you a check for

$$5\% \times \$1000 = 0.05 \times \$1000 = \$50.$$

You also get $50 at the end of the second and third years. Over the three years, you've received a total of

$$3 \times \$50 = \$150$$

in interest. Thus your original $1000 has grown in value to $1150. Honest John's method of payment represents **simple interest**, in which interest is paid only on your original investment, or **principal**.

Now, suppose that you had placed the $1000 in a bank account that pays the same 5% interest once a year, but deposits it directly into your account. At the end of the first year, the bank deposits the $50 interest into your account, raising your balance to $1050. At the end of the second year, the bank again pays you 5% interest. This time, however, the 5% interest is paid on the balance of $1050:

$$5\% \times \$1050 = 0.05 \times \$1050 = \$52.50$$

Adding the $52.50 interest payment to your account gives you a new balance of

$$\$1050 + \$52.50 = \$1102.50.$$

At the end of the third year, the interest on this balance of $1102.50 is

$$5\% \times \$1102.50 = 0.05 \times \$1102.50 = \$55.13,$$

making your new balance

$$\$1102.50 + \$55.13 = \$1157.63.$$

Despite identical interest rates, you end up with $7.63 more by using the bank instead of Honest John's. The difference comes about because the bank pays you *interest on the interest* as well as on the original principal. This type of interest payment is called **compound interest**.

CALCULATING COMPOUND INTEREST

Consider again the New College case study. We can calculate the amount owed to the College by pretending that the $224 borrowed by King Edward was deposited into an interest-bearing account. Let's assume, as did College officials in their calculation, that the interest rate was 4% per year. The calculations for the first three years are shown in the following table.

After n years	Interest	Balance
1 year	4% × $224 = $8.96	$224 + $8.96 = $232.96
2 years	4% × $232.96 = $9.32	$232.96 + $9.32 = $242.28
3 years	4% × $242.28 = $9.69	$242.28 + $9.69 = $251.97

All we need to do is continue the calculations until we reach the 535 years since the loan was taken!

Fortunately, there's a much easier way. Note that the balance increases by 4% = 0.04 per year, so we call this rate the **annual percentage rate**, or **APR**. Because the balance increases by 4% when interest is paid at the end of the first year, the new balance is 1.04 times the original balance of $224:

$$1.04 \times \$224 = \$232.96$$

We can find the balance at the end of year 2 by multiplying by 1.04 again:

$$1.04 \times 1.04 \times \$224 = (1.04)^2 \times \$224 = \$242.28$$

In fact, we can find the balance after n years by multiplying the original balance of $224 by 1.04 n times; that is, we multiply by 1.04 raised to the nth power. For example, the balance after $n = 10$ years is

$$(1.04)^{10} \times \$224 = \$331.57.$$

We can generalize this result as the **compound interest formula**.

By the Way ············
Financial planners often call the principal, P, the *present value* (PV) of the money and the accumulated amount, A, the *future value* (FV).

Compound Interest Formula for Interest Paid Once a Year

$$A = P \times (1 + APR)^Y, \quad \text{where} \begin{cases} P = \text{starting principal} \\ APR = \text{annual percentage rate (as a decimal)} \\ Y = \text{number of years} \\ A = \text{accumulated balance after } Y \text{ years} \end{cases}$$

In the New College case, the original principal is P = \$224 and APR = 4% = 0.04. Note that we *must* write 4% as the decimal 0.04. The number of years is Y = 535. Thus the accumulated balance is

$$A = P \times (1 + APR)^Y = \$224 \times (1 + 0.04)^{535}$$
$$= \$224 \times (1.04)^{535}$$
$$= \$224 \times 1{,}296{,}691{,}084$$
$$\approx \$2.9 \times 10^{11}$$

or \$290 billion!

Using Your
___ *Calculator*

Most calculators have a key for raising to powers labeled (y^x) or $(\triangle)$ or similar. For example, calculate 1.04^{535} by pressing 1.04 (y^x) 535 $(=)$ or 1.04 $(\triangle)$ 535.

EXAMPLE 1   *New College Debt at 2%*

Calculate the amount due to New College if the interest rate is 2% using (a) simple interest and (b) compound interest.

Solution: (a) At a rate of 2%, the simple interest due each year would be

$$2\% \times \$224 = 0.02 \times \$224 = \$4.48.$$

Over 535 years, the total interest due would be

$$535 \times \$4.48 = \$2396.80.$$

Adding this total interest to the original loan principal of \$224, the payoff amount would be

$$\$224 + \$2396.80 = \$2620.80.$$

b)   With compound interest, the accumulated balance (amount due) is

$$A = P \times (1 + APR)^Y = \$224 \times (1 + 0.02)^{535}$$
$$= \$224 \times (1.02)^{535}$$
$$= \$224 \times 39911$$
$$\approx \$8.94 \times 10^6$$

or about \$8.94 million.   ■

TIME-OUT TO THINK: Compare the debt owed to New College using compound interest at 4% with the amount owed using compound interest at 2%. Are you surprised that cutting the interest rate in half can make such an enormous difference? Discuss how this difference illustrates the "power of compound interest."

EXAMPLE 2   *Mattress Investments*

Your grandfather put $100 under his mattress 50 years ago. If he had instead invested it in a bank account paying 3.5% interest compounded yearly (roughly the average U.S. rate of inflation during that period), how much would it be worth now?

Solution: The original principal is $P = \$100$. The annual percentage rate is $APR = 3.5\% = 0.035$. The number of years is $Y = 50$. We use the compound interest formula to find the accumulated amount:

$$A = P \times (1 + APR)^Y = \$100 \times (1 + 0.035)^{50}$$
$$= \$100 \times (1.035)^{50}$$
$$= \$558.49$$

Invested at a rate of 3.5%, the $100 would be worth well over $500 today. Unfortunately, the $100 was put under a mattress, so it still has a face value of only $100. Because of inflation, the buying value of the $100 is much less than it was 50 years ago. ■

Compound Interest Paid More Than Once A Year

Suppose that you deposit $1000 into a bank that pays interest at an annual percentage rate of $APR = 8\%$. If the interest is paid all at once at the end of a year, you'll receive interest of

$$8\% \times \$1000 = 0.08 \times \$1000 = \$80,$$

making your year-end balance $1080.

Now, assume instead that the bank pays the interest *quarterly*, or four times a year (that is, once every three months). The quarterly interest rate is one-fourth of the annual interest rate, or

$$\text{quarterly percentage rate} = \frac{\text{annual percentage rate}}{4} = \frac{8\%}{4} = 2\%.$$

The following table shows how quarterly compounding affects the $1000 starting balance during the first year.

After *n* quarters	Interest Paid	New Balance
First quarter (3 months)	2% × $1000 = $20	$1000 + $20 = $1020
Second quarter (6 months)	2% × $1020 = $20.40	$1020 + $20.40 = $1040.40
Third quarter (9 months)	2% × $1040.40 = $20.81	$1040.40 + $20.81 = $1061.21
Fourth quarter (1 full year)	2% × $1061.21 = $21.22	$1061.21 + $21.22 = $1082.43

Note that the year-end balance with quarterly compounding ($1082.43) is *greater* than the year-end balance with interest paid all at once ($1080).

Is there an easier way to get these results? Yes. The quarterly interest rate of 2% means the balance increases by a factor of 1.02 after each quarter. Thus after *n* quarters, the balance is the original balance multiplied by $(1.02)^n$. For example, after $n = 4$ quarters, the balance is

$$(1.02)^4 \times \$1000 = \$1082.43.$$

Note that this is the same result found by working one step at a time in the previous table.

If you compare these quarterly calculations to the earlier calculations for compound interest paid at the end of each year, you'll see that the only difference is that we are using the *quarterly* interest rate and the number of *quarters*, rather than the annual interest rate and number of years. Because the quarterly interest rate is $\dfrac{APR}{4}$ and the number of quarters is four times the number of years, the compound interest formula for this case becomes:

$$\text{quarterly compounding: } A = P\left(1 + \frac{APR}{4}\right)^{4 \times Y}$$

Generalizing, we can write the compound interest formula in a form that works no matter how many times interest is paid per year.

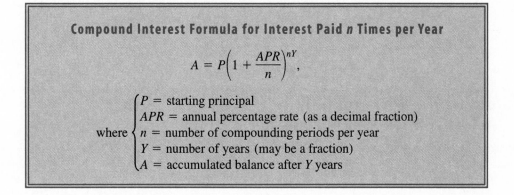

Compound Interest Formula for Interest Paid n Times per Year

$$A = P\left(1 + \frac{APR}{n}\right)^{nY},$$

where $\begin{cases} P = \text{starting principal} \\ APR = \text{annual percentage rate (as a decimal fraction)} \\ n = \text{number of compounding periods per year} \\ Y = \text{number of years (may be a fraction)} \\ A = \text{accumulated balance after } Y \text{ years} \end{cases}$

TIME-OUT TO THINK: Verify that if you put $n = 1$ into the above formula, which means that interest is paid only one time each year, you get our earlier formula for compound interest paid once a year.

EXAMPLE 3   *Monthly Compounding at 6%*

You deposit $5000 in a bank account that pays an *APR* of 6% and compounds interest monthly. How much money will you have after five years? Compare to the amount you'd have if interest were paid only once each year.

Solution: The principal is $P = \$5000$ and the annual percentage rate is $APR = 0.06$. Monthly compounding means that interest is paid $n = 12$ times a year, and we are considering a period of $Y = 5$ years. Substituting into the compound interest formula gives

$$A = P \times \left(1 + \frac{APR}{n}\right)^{nY} = \$5000 \times \left(1 + \frac{0.06}{12}\right)^{12 \times 5}$$
$$= \$5000 \times (1.005)^{60}$$
$$= \$6744.25.$$

For interest paid only once each year, we find the balance after five years by using the formula for compound interest paid once a year.

$$A = P \times (1 + APR)^Y = \$5000 \times (1 + 0.06)^5$$
$$= \$5000 \times (1.06)^5$$
$$= \$6691.13$$

The difference is

$$\$6744.25 - \$6691.13 = \$53.12.$$

Thus you would have \$53.12 more with monthly compounding than with annual compounding after five years. ∎

Using Your Calculator
The Compound Interest Formula
(For Interest Paid More Than Once Per Year)

You can do compound interest calculations on any calculator that includes a $\boxed{y^x}$ or similar key for raising to powers. Here's a five-step procedure that will work on most calculators, along with an example in which $P = \$1000$, $APR = 8\% = 0.08$, $Y = 10$ years, and $n = 12$ (monthly compounding). Your calculator may have keys that enable you to take shortcuts not shown here; some business calculators even have built-in functions for calculating compound interest in a single step. *Note: It is very important that you do not round any answers until completing all the calculations.*

		in general	example	display
Starting Formula:		$A = P \times \left(1 + \dfrac{APR}{n}\right)^{nY}$	$\$1000 \times \left(1 + \dfrac{0.08}{12}\right)^{12 \times 10}$	——
Step 1.	Multiply factors in exponent.	$n \; \boxed{\times} \; Y \; \boxed{=}$	$12 \; \boxed{\times} \; 10 \; \boxed{=}$	120.
Step 2.	Store product in memory (or write down).	$\boxed{\text{Store}}$	$\boxed{\text{Store}}$	120.
Step 3.	Add terms 1 and $\dfrac{APR}{n}$.	$1 \; \boxed{+} \; APR \; \boxed{\div} \; n \; \boxed{=}$	$1 \; \boxed{+} \; 0.08 \; \boxed{\div} \; 12 \; \boxed{=}$	1.0066666667
Step 4.	Raise result to power in memory.	$\boxed{y^x} \; \boxed{\text{Recall}} \; \boxed{=}$	$\boxed{y^x} \; \boxed{\text{Recall}} \; \boxed{=}$	2.219640235
Step 5.	Multiply result by P.	$\boxed{\times} \; P \; \boxed{=}$	$\boxed{\times} \; \$1000 \; \boxed{=}$	2219.640235

With the calculation complete, you can round to the nearest cent to write the answer as \$2219.64. Finally, because it's easy to push the wrong buttons by accident, you should always check the calculation (preferably check it twice).

Annual Percentage Yield (*APY*)

We've seen that money grows by *more* than the *APR* when it is compounded more than once a year. For example, we found that on a $1000 principal with 8% *APR* compounded quarterly, the year-end balance is $1082.43. This represents a percentage increase in the balance of

$$\text{percent increase in balance} = \frac{\text{(year-end balance)} - \text{(starting balance)}}{\text{starting balance}} \times 100\%$$

$$= \frac{\$1082.43 - \$1000}{\$1000} \times 100\%$$

$$= \frac{\$82.43}{\$1000} \times 100\%$$

$$= 8.243\%.$$

This actual percentage increase in balance, which is *more* than the *APR*, is called the **annual percentage yield (*APY*)** of the bank account.

By the Way ············

Banks are required by law to state the *APY* on interest-paying accounts. The *APY* is sometimes called the *effective yield*, or simply the *yield*.

EXAMPLE 4   *More Compounding Means a Higher Yield*

You deposit $1000 into an account with *APR* = 8%. Calculate your balance after one year and the annual percentage yield with *monthly* compounding and *daily* compounding. Can you form a general conclusion about compounding and annual yield?

Solution: We have $P = \$1000$, $APR = 8\% = 0.08$, and $Y = 1$ year. For monthly compounding, we set $n = 12$. Substituting these values into the compound interest formula gives

$$A = P \times \left(1 + \frac{APR}{n}\right)^{nY} = \$1000 \times \left(1 + \frac{0.08}{12}\right)^{12 \times 1}$$

$$= \$1000 \times (1.006666667)^{12}$$

$$= \$1083.00.$$

The annual yield in this case is

$$APY = \%\text{ increase in balance} = \frac{\$1083.00 - \$1000}{\$1000} \times 100\% = 8.300\%.$$

With monthly compounding, your balance grows to $1083 in one year and your annual percentage yield is 8.300%.

Daily compounding means that interest is paid $n = 365$ times per year. The compound interest formula now gives

$$A = P \times \left(1 + \frac{APR}{n}\right)^{nY} = \$1000 \times \left(1 + \frac{0.08}{365}\right)^{365 \times 1}$$

$$= \$1000 \times (1.000219178)^{365}$$

$$= \$1083.28.$$

Your annual yield in this case is

$$APY = \frac{\$1083.28 - \$1000}{\$1000} \times 100\% = 8.328\%.$$

By the Way ············

Most banks divide the *APR* by 360, rather than 365, when calculating the interest rate and *APY* for daily compounding.

With daily compounding, your balance grows to $1083.28 in one year and your annual percentage yield is 8.328%.

Note that, with a single annual interest payment, the annual yield (APY) is simply $APY = APR = 8\%$. With quarterly compounding, $APY = 8.243\%$; with monthly compounding $APY = 8.300\%$; and with daily compounding $APY = 8.328\%$. In general, the more frequently interest is compounded for a given APR, the higher the annual yield. ■

Continuous Compounding

Suppose that interest were compounded more often than daily; say every second, or every trillionth of a second. We know that this more frequent compounding should lead to a higher annual yield, but how much higher?

For a given APR, it's straightforward to calculate the APY for any number of compounding periods. Table 5.1 lists some annual yields at an 8% APR, and Figure 5.1 shows a graph of the results.

Table 5.1	Annual Yield for $APR = 8\%$ with Various Numbers of Compounding Periods (n)			
n	**APY**	**n**	**APY**	
1	8%	1000	8.3283601%	
4	8.2432%	10,000	8.3286721%	
12	8.29995%	1,000,000	8.3287064%	
365	8.32776%	10,000,000	8.3287067%	
500	8.32801%	1,000,000,000	8.3287068%	

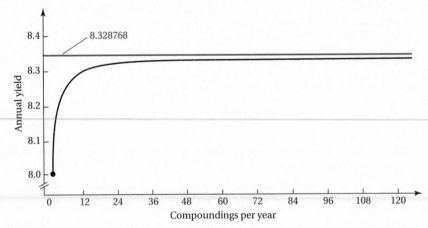

FIGURE 5.1

The annual yield for $APR = 8\%$ depends on the number of times interest is compounded per year.

Note that the annual yield does *not* grow indefinitely. Instead, it approaches a *limit* that must be very close to the 8.3287068% found for $n = 1$ billion. In other words, even if we could compound *infinitely* many times per year, the annual yield would not go much above 8.3287%. Compounding infinitely many times per year is called **continuous compounding**, and represents the best possible compounding for a particular *APR*. With continuous compounding, the compound interest formula takes the following form.

Compound Interest Formula for Continuous Compounding

$$A = P \times e^{(APR \times Y)},$$

where $\begin{cases} P = \text{starting principal} \\ APR = \text{annual percentage rate (as a decimal fraction)} \\ Y = \text{number of years} \\ A = \text{accumulated balance after } Y \text{ years} \end{cases}$

The number e is a special irrational number with a value of $e \approx 2.71828$. You can compute e to a power with the e^x key on your calculator.

By the Way

Named after the great mathematician Leonhard Euler (1707–1783), the number e is one of the fundamental constants of mathematics.

TIME-OUT TO THINK: Look for the e^x key on your calculator. Use it to enter e^1, and thereby verify that $e \approx 2.71828$.

EXAMPLE 5   *Continuous Compounding at 5.5%*

You deposit $100 in an account with an *APR* of 5.5% and continuous compounding. How much will you have after 10 years?

Solution: We have $P = \$100$, $APR = 5.5\% = 0.055$, and $Y = 10$ years of continuous compounding.

$$A = P \times e^{(APR \times Y)} = \$100 \times e^{(0.055 \times 10)}$$
$$= \$100 \times e^{0.55}$$
$$= \$173.33.$$

Your balance will be $173.33 after 10 years.   ∎

Using Your **Calculator**

This calculation is easiest on most calculators if you first multiply $0.055 \times 10 = 0.55$, then use the e^x key to get the result.

PLANNING AHEAD WITH COMPOUND INTEREST

Suppose that you have a new baby and want to make sure that you'll have $100,000 for your baby's college education. Assuming your baby will start college in 18 years, how much money should you deposit now?

If we know the interest rate, this problem is simply a "backwards" compound interest problem. We start with the amount A that you want to end up with after 18 years, and then calculate the starting principal P that will get you to that amount.

By the Way

The process of finding the principal (present value) that must be deposited today to yield some particular future amount is called *discounting* by financial planners.

EXAMPLE 6   *College Fund at 9%*

Assume that you are counting on receiving interest at a rate of $APR = 9\%$, compounded just once each year. How much must you deposit now to realize $100,000 after 18 years for your baby's college fund?

Solution: We know the interest rate ($APR = 0.09$), the number of years of compounding ($Y = 18$), and the amount desired after 18 years ($A = \$100,000$). Using the compound interest formula for interest paid once a year

$$A = P \times (1 + APR)^Y,$$

we can calculate the starting principal P by dividing both sides by $(1 + APR)^Y$:

$$\frac{A}{(1 + APR)^Y} = \frac{P \times (1 + APR)^Y}{(1 + APR)^Y} \quad \Rightarrow \quad P = \frac{A}{(1 + APR)^Y}$$

Now we substitute the values we know.

$$P = \frac{\$100,000}{(1 + 0.09)^{18}} = \frac{\$100,000}{(1.09)^{18}} = \$21,199.37$$

Depositing about $21,200 now will yield the desired $100,000 in 18 years—assuming that you get the 9% *APR* and make no withdrawals or additional deposits.   ■

TIME-OUT TO THINK: In reality, it is extremely difficult to find an investment in which you can count on a steady 9% interest rate for 18 years. Nevertheless, financial planners often make such assumptions when exploring investment options. Explain why such calculations can be useful, despite the fact that you can't be sure of a steady interest rate.

EXAMPLE 7   *College Fund at 10%, Compounded Monthly*

Repeat Example 6, but this time assuming an interest rate of $APR = 10\%$ with monthly compounding.

Solution: This time we must solve for P in the compound interest formula for interest paid more than once a year.

$$A = P \times \underbrace{\left(1 + \frac{APR}{n}\right)^{nY}}_{\substack{\text{divide both sides} \\ \text{by this term}}} \quad \Rightarrow \quad P = \frac{A}{\left(1 + \frac{APR}{n}\right)^{nY}}$$

Again, we know the interest rate ($APR = 0.10$), the number of years of compounding ($Y = 18$), and the balance after 18 years ($A = \$100,000$). With monthly compounding, we have $n = 12$. Now we substitute these values.

$$P = \frac{\$100,000}{\left(1 + \frac{0.10}{12}\right)^{12 \times 18}} = \frac{\$100,000}{(1.008333333)^{216}} = \$16,653.64$$

Depositing about $16,650 now will yield the desired $100,000 in 18 years—assuming that you get the 10% *APR*, compounded monthly.   ■

REVIEW QUESTIONS

1. What is the difference between *simple interest* and *compound interest*? Why do you end up with more money with compound interest?

2. In your own words, explain how New College could claim that a debt of $224 from 535 years ago is worth $290 billion today.

3. In the New College case study, we saw that a 2% interest rate makes the debt worth $8.9 million and a 4% interest rate makes it worth $290 billion—which is more than 30,000 times greater than $8.9 million. Explain how merely doubling the interest rate from 2% to 4% can change the value by a factor of more than 30,000.

4. What is the compound interest formula? This unit gave three versions of the compound interest formula: for interest paid once a year, for interest paid *n* times per year, and for continuous compounding. Describe a situation in which each of the three formulas should be used.

5. What is an *annual percentage rate*? What is an *annual percentage yield*? Under what conditions is the annual percentage rate the same as the annual percentage yield? Under what conditions are they different?

6. For a given *APR*, how does the number of compounding periods affect the annual yield (*APY*)?

7. Give an example in which you might want to solve the compound interest formula to find the principal *P* that must be invested now to yield a particular amount *A* in the future.

PROBLEMS

Simple Vs. Compound Interest. *In Problems 1 and 2, make a table that shows the performance of your account and your friend's account for five years. The table should show the amount of interest earned each year and the balance in each account. Compare the balance in the accounts in percentage terms after five years.*

1. Suppose that you invest $500 in a way that earns simple interest at an annual rate of 5% per year. At the same time your friend invests $500 in a savings account with annual compounding at a rate of 5% per year. Assuming that neither of you withdraws any money, compare how much you each will have in your accounts after five years.

2. Suppose that you invest $1000 in an account that earns simple interest at an annual rate of 6% per year. At the same time your friend invests $1000 in a savings account with annual compounding at a rate of 6% per year.

Assuming that neither of you withdraws any money, compare how much you each will have in your accounts after five years.

3. **Calculating Compound Interest.** Use the compound interest formula to determine the accumulated balance in each given case. Assume that interest is compounded annually.

   a. $2000 is invested at an *APR* of 3% for 10 years.

   b. $10,000 is invested at an *APR* of 5% for 20 years.

   c. $30,000 is invested at an *APR* of 7% for 25 years.

4. **Calculating Compound Interest.** Use the compound interest formula to determine the accumulated balance in each given case. Assume that interest is compounded annually.

   a. $3000 is invested at an *APR* of 4% for 12 years.

   b. $10,000 is invested at an *APR* of 6% for 25 years.

   c. $40,000 is invested at an *APR* of 8.5% for 30 years.

5. **Compounding More Than Once a Year.** Use the compound interest formula for compounding more than once a year to determine the accumulated balance in each investment account described.

   a. You deposit $1000 at an *APR* of 5.5% with monthly compounding for 10 years.

   b. You deposit $2000 at an *APR* of 3% with daily compounding for 5 years.

   c. You deposit $5000 at an *APR* of 7.3% with quarterly compounding for 20 years.

   d. You deposit $10,000 at an *APR* of 6.2% with monthly compounding for 5 years.

6. **Compounding More Than Once a Year.** Use the compound interest formula for compounding more than once a year to determine the accumulated balance in each investment account described.

   a. You deposit $1000 at an *APR* of 7% with monthly compounding for 15 years.

   b. You deposit $3000 at an *APR* of 5% with daily compounding for 10 years.

   c. You deposit $5000 at an *APR* of 6.2% with quarterly compounding for 30 years.

   d. You deposit $15,000 at an *APR* of 7.8% with monthly compounding for 15 years.

Compound Interest and Annual Yield. *Use the compound interest formula to answer the questions in Problems 7–9.*

7. Suppose that you deposit $500 in a bank that offers an *APR* of 6.5% compounded daily. What is your balance after one year? What is the annual yield for this account?

8. Suppose that you deposit $500 in a bank that offers an *APR* of 4.5% compounded monthly. What is your balance after one year? What is the annual yield for this account?

9. Suppose that you deposit $800 in a bank that offers an *APR* of 7.25% compounded quarterly. What is your balance after one year? What is the annual yield for this account?

10. **Annual Yield Formula.** In the text and the previous problems, the *APY* was found by calculating the percent change in the balance over one year. The following formula also gives the *APY*.

$$APY = \left(1 + \frac{APR}{n}\right)^n - 1,$$

where *n* is the number of compoundings per year. Verify that this formula gives the same results as you found in Problems 7–9.

11. **Annual Yield.** Use the formula from Problem 10 to find the annual percentage yield (*APY*) for each given case.

   a. An account with quarterly compounding at an *APR* of 6.6%.

   b. An account with monthly compounding at an *APR* of 6.6%.

   c. An account with daily compounding at an *APR* of 6.6%.

12. **Annual Yield.** Use the formula from Problem 10 to find the annual percentage yield (*APY*) for each given case.

   a. An account has quarterly compounding at an *APR* of 8%.

   b. An account has monthly compounding at an *APR* of 8%.

   c. An account has daily compounding at an *APR* of 8%.

13. **Rates of Compounding.** Compare the accumulated balance in two accounts that both start with an initial deposit of $1000. Both accounts have an *APR* of 5.5%, but one account compounds interest annually while the other account compounds interest daily. Make a table that shows the interest earned each year and the accumulated balance in both accounts for the first ten years. Compare the balance in the accounts, in percentage terms, after ten years.

14. **Understanding Annual Percentage Yield (*APY*).**

   a. Explain why *APR* and *APY* are the same with annual compounding.

   b. Explain why *APR* and *APY* are different with daily compounding.

   c. Does *APY* depend on the starting principal *P*? Explain.

   d. Does *APY* depend on the number of compoundings *n* during a year? Explain.

   e. Does *APY* depend on the number of years *Y* that the account is held? Explain.

15. Continuous Compounding. Use the compound interest formula for continuous compounding to determine the accumulated balance in each given account after 5 years and after 20 years. Also find the *APY* for each account.

  a. You deposit $1000 in an account with continuous compounding and an *APR* of 4%.

  b. You deposit $2000 in an account with continuous compounding and an *APR* of 5%.

  c. You deposit $10,000 in an account with continuous compounding and an *APR* of 6%.

16. Continuous Compounding. Use the compound interest formula for continuous compounding to determine the accumulated balance in each given account after 5 years and after 20 years.

  a. You deposit $3000 in an account with continuous compounding and an *APR* of 4%.

  b. You deposit $5000 in an account with continuous compounding and an *APR* of 6.5%.

  c. You deposit $500 in an account with continuous compounding and an *APR* of 7%.

17. APY for Continuous Compounding. Find the annual percentage yield (*APY*) for accounts with continuous compounding and the given *APR*.

  a. *APR* = 4%

  b. *APR* = 5%

  c. *APR* = 6%

18. APY Formula for Continuous Compounding. The following formula allows you to determine the annual percentage yield (*APY*) with continuous compounding.

$$APY = e^{APR} - 1$$

Verify that the *APY*s that you computed in Problem 17 are consistent with this formula.

19. Continuous Compounding. The following questions allow you to explore continuous compounding further.

  a. Make a table similar to Table 5.1 for an *APR* of 12% in which you display the *APY* for n = 1, 4, 12, 365, 500, 1000.

  b. Use the formula for continuous compounding to find the *APY* at an *APR* of 12%.

  c. Show the results of parts (a) and (b) on a graph similar to that of Figure 5.1.

  d. Compare, in words, the *APY* with continuous compounding to the *APY* with other types of compounding.

  e. Suppose that you deposit $500 in an account with an *APR* of 12%. With continuous compounding, how much money will you have at the end of one year? at the end of five years?

20. Comparing Investment Plans. Bernard deposits $1600 in a savings account that compounds interest annually at an *APR* of 4%. Carla deposits $1400 in a savings account that compounds interest daily at an *APR* of 5%. Who will have the higher accumulated balance after 5 years and after 20 years?

21. Comparing Investment Plans. Brian invests $1600 in an account with annual compounding and an *APR* of 4.5%. Celeste invests $1400 in an account with continuous compounding and an *APR* of 5.5%. Determine who has the highest accumulated balance after 5 years and after 20 years.

22. Death and the Maven (A True Story). In December, 1995, 101-year-old Anne Scheiber died and left $22 million to Yeshiva University. This fortune was accumulated through shrewd and patient investment of a $5000 nest egg over the course of 50 years. Suppose that she had simply deposited the $5000 in a high-interest bank account in which interest was compounded annually. By trial and error with various interest rates, estimate the interest rate she would have needed to turn the $5000 into $22 million in 50 years.

23. Planning Ahead with Compounding. Suppose you start saving today for a $10,000 house down payment that you plan to make in 10 years. Assume that you make no deposits into the account after your initial deposit. In order to reach your $10,000 goal, how much will you need to deposit in each account described?

   a. An account with annual compounding and an *APR* of 9%.

   b. An account with quarterly compounding and an *APR* of 9%.

   c. An account with monthly compounding and an *APR* of 9%.

   d. An account with continuous compounding and an *APR* of 9%.

24. College Fund. Suppose that you want to have a $100,000 college fund in 18 years. How much will you have to deposit now under each of the following scenarios?

   a. An *APR* of 6%, compounded daily.

   b. An *APR* of 7.5%, compounded continuously.

   c. An *APR* of 11%, compounded monthly.

25. Retirement Planning. You have two choices for an investment plan that you want to accumulate $75,000 for your retirement in 35 years. Plan A is an account with annual compounding and an *APR* of 5%. Plan B is an account with continuous compounding and an *APR* of 4.5%. Which plan requires the smallest starting principal?

26. Loans Vs. Investments. Consider the following two situations. *Situation (a):* You borrow $1000 at an annual rate of 8%, but do not make any payments (as in the New College example). *Situation (b):* You invest $1000 in a savings account at an annual rate of 8%. Describe in words why the same mathematics applies to the balance of the loan in *Situation (a)* and the balance in the savings account in *Situation (b)*.

27. Bank Advertisement. Find two advertisements from banks that refer to compound interest rates. Explain the terms in each advertisement. Which bank has the better deal? Explain.

28. Your Bank Account. Find the current interest rate (*APR*) for your personal savings account. (Choose just one account if you have more than one, or pick a rate from a nearby bank if you don't have an account.)

   a. How often is interest compounded on your account?

   b. Calculate the annual yield on your account by using the compound interest formula. What is the annual yield for your account as stated by the bank? Does your calculation agree with the bank's claim?

   c. Suppose that you receive a gift of $10,000 and place it in your account. If you leave the money there, and the interest rate never changes, how much will you have in 10 years?

   d. Pick a compounding period that is different from yours. For example, if your bank compounds daily, you might pick monthly compounding. Calculate how much you would have after 10 years from the $10,000 in part (c) with this different compounding period. Briefly discuss your results.

   e. Suppose that you find another account that offers interest at an *APR* that is 2% higher than yours, with the same compounding period as your account. Calculate how much you would have after 10 years from the $10,000 in part (c) with this higher *APR*. Briefly discuss your results.

29. Working with the Compound Interest Formula. Use a calculator and possibly some trial and error to answer the following questions.

   a. How long will it take your money to triple at an *APR* of 8% compounded annually?

   b. How long will it take your money to grow by 50% at an *APR* of 7% compounded annually?

   c. If you deposit $1000 in an account that pays an *APR* of 7% compounded annually, how long will it take for your balance to reach $100,000?

   d. Suppose that you give your niece a 10-year certificate of deposit that pays an *APR* of 9% compounded annually. If you want the value of the account to be $10,000 at the end of the 10 years, how much should you deposit now?

SAVINGS PLANS (ANNUITIES)

Suppose that you want to save for retirement, your child's college expenses, or any other reason. You could deposit a "lump sum" of money today, and let it grow through the power of compound interest. But what if you don't have a large lump sum to start such an account?

For most people, a more realistic way to save is by depositing smaller amounts on a regular basis. For example, you might put $50 a month into savings. Such long-term **savings plans** probably are the best way to prepare for your financial future. Savings plans are so popular that many have special names—and some even get special tax treatment (see Unit 5D). Popular types of savings plans include Individual Retirement Accounts (IRAs), 401(k) plans, and employee pension plans.

We can understand savings plans with a simple example. Suppose that you deposit $100 into your savings plan at the end of each month. To make the numbers simple, let's suppose that your plan pays interest at an annual percentage rate of $APR = 12\%$, which is equivalent to 1% per month.

- You begin with $0 in the account; at the end of Month 1 you make the first deposit of $100.

- At the end of Month 2, you receive the monthly interest of 1% on the $100 already in the account.

$$1\% \times \$100 = \$1$$

In addition, you deposit another $100. Thus your balance at the end of Month 2 is

$$\underbrace{\$100}_{\text{prior balance}} + \underbrace{\$1.00}_{\text{interest}} + \underbrace{\$100}_{\text{new deposit}} = \$201.00.$$

- At the end of Month 3, you receive 1% interest on the $201 already in the account.

$$1\% \times \$201 = \$2.01$$

In addition, you deposit another $100. Thus your balance at the end of Month 3 is

$$\underbrace{\$201.00}_{\text{prior balance}} + \underbrace{\$2.01}_{\text{interest}} + \underbrace{\$100}_{\text{new deposit}} = \$303.01.$$

We can continue these calculations with the aid of the following table. The end of each row shows the new balance, which is the sum of the prior balance, the interest, and the new deposit.

By the Way ....
Financial planners call any series of equal, regular payments an *annuity*. Thus savings plans are a type of annuity, as are loans that you pay with equal monthly payments.

End of...	Prior Balance	Interest on Prior Balance	New Deposit	New Balance
Month 1	$0	$0	$100	$100.00
Month 2	$100	1% × $100 = $1	$100	$201.00
Month 3	$201	1% × $201 = $2.01	$100	$303.01
Month 4	$303.01	1% × $303.01 = $3.03	$100	$406.04
Month 5	$406.04	1% × $406.04 = $4.06	$100	$510.10
Month 6	$510.10	1% × $510.10 = $5.10	$100	$615.20

In principle, we could extend this table indefinitely—but it would take a lot of work! Fortunately, there's a much easier way: the **savings plan formula**.

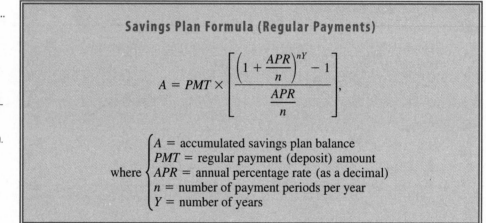

By the Way ··········
This formula is often called the *ordinary annuity formula*. The accumulated amount (*A*) is called the *future value* of the annuity. Most computer spreadsheet programs have built-in functions for using this formula.

Savings Plan Formula (Regular Payments)

$$A = PMT \times \left[\frac{\left(1 + \frac{APR}{n}\right)^{nY} - 1}{\frac{APR}{n}} \right],$$

where $\begin{cases} A = \text{accumulated savings plan balance} \\ PMT = \text{regular payment (deposit) amount} \\ APR = \text{annual percentage rate (as a decimal)} \\ n = \text{number of payment periods per year} \\ Y = \text{number of years} \end{cases}$

This formula assumes that the payment and compounding periods are the same. For example, if payments are made monthly, interest also is calculated and paid monthly. (If the compounding period is different from the payment period, the term APR/n should be replaced by the effective yield for each payment period.)

EXAMPLE 1   *Using the Savings Plan Formula*

Use the savings plan formula to calculate the balance after 6 months for an *APR* of 12% and monthly payments of $100.

Solution: We have monthly payments of $PMT = \$100$, annual interest rate of $APR = 0.12$, $n = 12$ because the payments are made monthly, and $Y = \frac{1}{2}$ because 6 months is half a year. Putting these values into the savings plan formula, we find the accumulated balance.

$$A = PMT \times \left[\frac{\left(1 + \dfrac{APR}{n}\right)^{nY} - 1}{\dfrac{APR}{n}}\right] = \$100 \times \left[\frac{\left(1 + \dfrac{0.12}{12}\right)^{(12 \times 1/2)} - 1}{\dfrac{0.12}{12}}\right]$$

$$= \$100 \times \left[\frac{(1.01)^6 - 1}{0.01}\right]$$

$$= \$615.20$$

Note that this answer agrees with what we found in our earlier table. ■

Using Your **Calculator**   **The Savings Plan Formula**

There are many ways to do the savings plan formula on your calculator; some business calculators have built-in functions that allow you to make savings plan calculations in a single step. However, the following procedure will work on most standard calculators. The example uses numbers from Example 2: $n = 12$ payments (monthly payments), $PMT = \$100$, $APR = 8\% = 0.08$, and $Y = 35$ years. *It is very important that you do not round any numbers until the end of the calculation.*

		in general	example	display
Starting Formula:		$A = PMT \times \left[\dfrac{\left(1 + \dfrac{APR}{n}\right)^{nY} - 1}{\dfrac{APR}{n}}\right]$	$\$100 \times \left[\dfrac{\left(1 + \dfrac{0.08}{12}\right)^{(12 \times 35)} - 1}{\dfrac{0.08}{12}}\right]$	——
Step 1.	Multiply factors in exponent.	$n \times Y =$	$12 \times 35 =$	420.
Step 2.	Store product in memory (or write down).	Store	Store	420.
Step 3.	Add terms 1 and $\dfrac{APR}{n}$.	$1 + APR \div n =$	$1 + 0.08 \div 12 =$	1.0066666667
Step 4.	Raise result to power in memory.	y^x Recall $=$	y^x Recall $=$	16.2925499
Step 5.	Subtract 1 from result.	$- 1 =$	$- 1 =$	15.2925499
Step 6.	Store result in memory (or write down).	Store	Store	15.2925499
Step 7.	Compute denominator, then take its reciprocal	$APR \div n = 1/x$	$0.08 \div 12 = 1/x$	150
Step 8.	Multiply by result in memory and payment.	$\times$ Recall $\times PMT =$	$\times$ Recall $\times 100 =$	229,388.249

With the calculation complete, you can round to the nearest cent to write the answer as $229,388.25. Be sure to check the calculation.

EXAMPLE 2   *Retirement Plan*

At age 30, Michelle starts an IRA to save for her retirement. She deposits $100 into the account at the end of each month. If she can count on an *APR* of 8%, how much will she have when she retires 35 years later (at age 65)? Compare the IRA's value to the total amount of her deposits over this time period.

Solution: Michelle's IRA will accumulate deposits and interest for $Y = 35$ years. Her monthly payment (deposit) amount is $PMT = \$100$, the interest rate is $APR = 0.08$, and $n = 12$ because the deposits are made monthly. We put these values into the savings plan formula.

$$A = PMT \times \left[\frac{\left(1 + \frac{APR}{n}\right)^{nY} - 1}{\frac{APR}{n}} \right] = \$100 \times \left[\frac{\left(1 + \frac{0.08}{12}\right)^{(12 \times 35)} - 1}{\frac{0.08}{12}} \right]$$

$$= \$100 \times \left[\frac{(1.006666667)^{420} - 1}{0.006666667} \right]$$

$$= \$229{,}388.25$$

To compute Michelle's total deposits, note that 35 years is the same as $35 \times 12 = 420$ months. Thus, with deposits of $100 per month, Michelle's total contribution over 35 years is

$$420 \text{ months} \times \frac{\$100}{\text{month}} = \$42{,}000.$$

However, thanks to compounding, her IRA will have a balance of almost $230,000—*more than five times* the amount of her contributions!   ■

PLANNING AHEAD WITH SAVINGS PLANS

Most people start savings plans with particular goals in mind, such as saving enough for retirement, for a college fund, or to buy a new car in a couple of years. If we know the financial goal (which is the amount to be accumulated), we can rearrange the savings formula to determine how much money should be deposited on a regular basis. We do this by dividing both sides of the savings plan formula by the quantity in brackets:

$$A \div \frac{\left(1 + \frac{APR}{n}\right)^{nY} - 1}{\frac{APR}{n}} = PMT \times \frac{\left(1 + \frac{APR}{n}\right)^{nY} - 1}{\frac{APR}{n}} \div \frac{\left(1 + \frac{APR}{n}\right)^{nY} - 1}{\frac{APR}{n}}$$

Remembering that dividing is the same as multiplying by the reciprocal, this equation becomes

$$A \times \frac{\frac{APR}{n}}{\left(1 + \frac{APR}{n}\right)^{nY} - 1} = PMT \times \frac{\left(1 + \frac{APR}{n}\right)^{nY} - 1}{\frac{APR}{n}} \times \frac{\frac{APR}{n}}{\left(1 + \frac{APR}{n}\right)^{nY} - 1}.$$

$$= PMT \times 1$$

Switching the two sides and simplifying, we have the savings plan formula solved for *PMT*.

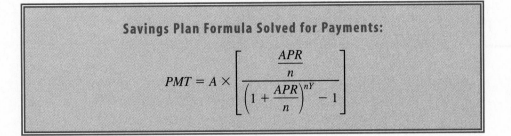

Savings Plan Formula Solved for Payments:

$$PMT = A \times \left[\frac{\dfrac{APR}{n}}{\left(1 + \dfrac{APR}{n}\right)^{nY} - 1} \right]$$

EXAMPLE 3   *College Savings Plan at 10%*

Suppose that you want to build a $100,000 college fund in 18 years by making regular, end-of-month deposits. Assuming an *APR* of 10%, how much should you deposit monthly? How much of the final value comes from actual deposits, and how much from compounded interest?

Solution: The amount to be accumulated is $A = \$100,000$, the time period is $Y = 18$ years, the interest rate is $APR = 0.10$, and the number of deposits (payments) per year is $n = 12$. We use the savings plan formula solved for the payments.

$$PMT = A \times \left[\frac{\dfrac{APR}{n}}{\left(1 + \dfrac{APR}{n}\right)^{nY} - 1} \right] = \$100,000 \times \left[\frac{\dfrac{0.10}{12}}{\left(1 + \dfrac{0.10}{12}\right)^{(12 \times 18)} - 1} \right]$$

$$= \$100,000 \times \left[\frac{0.008333333}{(1.008333333)^{216} - 1} \right]$$

$$= \$166.51$$

Assuming an interest rate of 10%, monthly payments of $166.51 will provide the needed $100,000 after 18 years. The total amount of your contributions over the 18 years would be

$$18 \text{ yr} \times 12 \frac{\text{mo}}{\text{yr}} \times \$166.51 \frac{1}{\text{mo}} = \$35,966.16,$$

or about $36,000. The rest of the $100,000—about $64,000—comes from compounded interest.   ■

EXAMPLE 4   *A Comfortable Retirement*

Suppose that you are 25 years old now, and would like to retire at age 50. Furthermore, you would like to have a retirement fund from which you can draw an income of $50,000 per year—forever! How can you do it? Assume an *APR* of 9% for all parts of this problem.

Using Your
*__ **Calculator***

On most calculators, it is easiest to calculate the denominator first, then take its reciprocal and multiply by the other terms.

Solution: You can achieve your goal by building a retirement fund that is large enough to earn $50,000 per year *from interest alone*. In that case, you can withdraw the interest for your living expenses while leaving the principal untouched (for your heirs!). The principal will then continue to earn the same $50,000 interest year after year (assuming no change in interest rates).

How large a balance do you need to earn $50,000 from interest? We can make a good estimate by assuming that simple interest applies during the pay-out period after retirement. Then we are looking for the balance that gives simple interest of $50,000 at an *APR* of 9%. That is, we are looking for the needed balance in the equation:

$$\$50{,}000 = 9\% \times \text{(needed balance)}$$

Dividing both sides by 9%, we get

$$\text{needed balance} = \frac{\$50{,}000}{9\%} = \frac{\$50{,}000}{0.09} = \$555{,}556.$$

Now we can calculate the monthly payments needed to achieve a balance of $A = \$556{,}000$. Let's assume that you'll make monthly deposits into your retirement fund, so $n = 12$. You want to retire in 25 years, so $Y = 25$. The interest rate is $APR = 0.09$. Using the Savings Plan Formula Solved for Payments, we find that the needed payments are

$$PMT = \$556{,}000 \times \left[\frac{\dfrac{0.09}{12}}{\left[\left(1 + \dfrac{0.09}{12}\right)^{(12 \times 25)} - 1 \right]} \right]$$

$$= \$556{,}000 \times \left[\frac{0.0075}{(1.0075)^{300} - 1} \right] = \$495.93.$$

You'll achieve your goal for retirement at age 50 if you deposit about $500 per month into your retirement plan—*and* if you can count on an *APR* of 9%. Although $500 per month may seem like a lot, it can be easier than it sounds thanks to special tax treatment for retirement plans (discussed in Unit 5D). ∎

By the Way ············
The lump sum deposit that would give you the same end result as regular payments into a savings plan is called the *present value* of the savings plan. For example, you can build the $100,000 college fund in 18 years either with $166 monthly payments or a lump sum deposit of $16,650. Thus $16,650 is the present value of the $166/month savings plan.

LUMP SUM OR SAVINGS PLAN?

In Example 7 of Unit 5A, we found that you could build a $100,000 college fund in 18 years with an initial, lump sum deposit of about $16,650 and an *APR* of 10%. In Example 3 above, we found that you could build the same fund with regular monthly deposits of about $166, also at *APR* = 10%. These examples illustrate that there are two basic ways to save:

1. By depositing a lump sum now and letting it grow.

2. By making regular payments into a savings plan.

Your best choice in any particular situation will depend on your financial circumstances. In the college fund case, most people don't have $16,650 to put into a savings account—particularly when facing all the expenses associated with a new baby—and probably would choose the option of making regular payments of $166 per month.

However, if you are lucky enough to get a sudden windfall of at least $16,650, such as an inheritance, you might decide to use it as a lump sum for the college fund.

TIME-OUT TO THINK: Suppose that you want a $100,000 college fund, and you are counting on an *APR* of 10%. Further suppose that you *can* afford the needed $166 monthly payments, but you also have enough savings to deposit a lump sum of $16,650. Which would you choose: the lump sum or the savings plan? Why?

---

*T*HINKING ABOUT…

Compound Interest: Ideals Versus Reality

We use the compound interest and savings plan formulas primarily to calculate how money will grow in the future. The formulas give straightforward answers but, as with all formulas, the reliability of the answers depends on the validity of the assumptions built into the formulas.

The most basic assumption in the compound interest and savings formulas is that of a steady interest rate. In reality, however, interest rates can vary substantially over time. For example, many bank accounts were paying an *APR* of more than 10% in the late 1970s. By the mid-1990s, 4% was considered a high interest rate. The savings plan formula also assumes equal, regular payments. But most people vary their monthly payments over time. If you get a raise, you might increase your monthly deposit; if you get laid off, you might reduce or suspend your monthly deposits, or even withdraw some of the accumulated money from your savings plan.

A few hidden assumptions may be even more important in interpreting results from the formulas. For example, the value of money tends to decline over time because of inflation. Saving $100,000 for your baby's college may sound like enough now, but may be far too little if the country goes through a bout of high inflation. And, of course, the formulas mean nothing at all if your money is lost by the bank or company holding it. In the United States, most bank accounts are federally insured (up to $100,000) so that you are protected even if the bank fails. But most other types of investments are *not* insured, including money market funds, stocks, and bonds. In such cases, there's always a possibility that you'll end up with little or nothing—no matter what a mathematical formula tells you.

REVIEW QUESTIONS

1. What is a *savings plan*? Why might you want to create one?

2. Explain the savings plan formula. What is the easiest way to work with it on *your* calculator?

3. In Example 2, we found that a person with a particular savings plan could end up with more than five times as much money as she contributed to the plan. How is this possible?

4. Give an example in which you might use the savings plan formula solved for the payments. Explain how you use this formula.

5. Is it possible to have a retirement fund from which you could draw $50,000 per year forever? How? What would you have to do to create such a fund?

6. Suppose that you expect to need a particular amount of money some time in the future. Under what circumstances would you plan to meet that need with a lump sum deposit now? Under what circumstances would you use a savings plan?

PROBLEMS

Investment Plans. *Answer the questions in Problems 1–4 with the help of the savings plan formula.*

1. To save money for retirement, you set up an IRA (individual retirement account) with an *APR* of 8% at age 25. At the end of each month you deposit $50 in the account. How much will the IRA contain when you retire at age 65? Compare that amount to the total amount of deposits made over the time period.

2. A friend creates an IRA with an *APR* of 8.25%. She starts the IRA at age 25 and deposits $50 per month. How much will her IRA contain when she retires at age 65? Compare that amount to the total amount of deposits made over the time period.

3. You put $200 per month in an investment plan that pays an *APR* of 7%. How much money will you have after 18 years? Compare that amount to the total amount of deposits made over the time period.

4. You put $200 per month in an investment plan that pays an *APR* of 7.5%. How much money will you have after 18 years?

Who Comes Out Ahead? *In Problems 5–8, compare the balance in the two investment plan accounts after ten years. Assume that the payment and compounding periods are the same. Who deposits more money in each case? Who comes out ahead in each case? Comment on any lessons about savings plans that you find in the results.*

5. Yolanda deposits $100 per month in an account with an *APR* of 5%, while Zach deposits $1200 at the end of each year in an account with an *APR* of 5%.

6. Polly deposits $50 per month in an account with an *APR* of 6%, while Quint deposits $40 per month in an account with an *APR* of 6.5%.

7. Juan deposits $200 per month in an account with an *APR* of 6%, while Maria deposits $2500 at the end of each year in an account with an *APR* of 6.5%.

8. George deposits $40 per month in an account with an *APR* of 7%, while Harvey deposits $150 per quarter in an account with an *APR* of 7.5%.

Investment Planning. *Answer the questions in Problems 9–13 with the help of the savings plan formula.*

9. You intend to create a college fund for your baby. If you can get an *APR* of 7.5% and want the fund to have $150,000 in it after 18 years, how much should you deposit monthly?

10. At age 35 you start saving for retirement. If your investment plan pays an *APR* of 9% and you want to have $2 million when you retire in 30 years, how much should you deposit monthly?

11. You want to purchase a new car in three years, and expect the car to cost $10,000. Your bank offers a plan with a guaranteed interest rate of *APR* = 5.5%, if you make regular monthly deposits. How much should you deposit each month to end up with $10,000 in three years?

12. You want to save $50,000 so you can go to graduate school in ten years. How much will you need to deposit per month to reach your goal if you use an account with an *APR* of 6%?

13. Suppose that you can afford to put $100 per month in an investment plan that pays an *APR* of 7%. Approximately how many years will it take for the value of your investment to reach $50,000? $1 million? Find the answers by using trial and error with various values of *Y* in the savings plan formula.

Comparing Investment Plans. *Suppose that you want to accumulate $50,000 for your child's college fund within the next 20 years. Explain fully whether the investment plans in Problems 14–16 will allow you to reach your goal.*

14. You decide that the most you can afford to deposit is $50 per month. Will an account with an *APR* of 7% allow you to reach your goal?

15. You deposit $75 per month into an account with an *APR* of 7%. Will you reach your goal?

16. You deposit $75 per month into an account with an *APR* of 10%. Will you reach your goal?

17. Comfortable Retirement. Suppose that you are 30 years old now, and would like to retire at age 60. Furthermore, you would like to have a retirement fund from which you can draw an income of $50,000 per year—forever! How can you do it? Assume an *APR* of 8% both as you pay into the retirement fund and when you collect from it later.

18. Very Comfortable Retirement. Suppose that you are 25 years old now, and would like to retire at age 65. Furthermore, you would like to have a retirement fund from which you can draw an income of $200,000 per year—forever! How can you do it? Assume an *APR* of 8% both as you pay into the retirement fund and when you collect from it later.

19. Regular Deposits Vs. Lump Sum. Suppose that you want to accumulate $60,000 for a college fund to be available in 20 years.

   a. How much would you have to deposit as a *single lump sum* into an account that earns interest at an *APR* of 4.5% compounded monthly in order to reach this goal?

   b. How much would you have to deposit monthly into an account that earns interest at an *APR* of 4.5% (compounded monthly) in order to reach this goal?

   c. For the monthly payment plan of part (b) how much money will you have deposited over the 20-year period?

   d. Compare the total amount of money that you will deposit under these two plans and explain the outcome in your own words.

20. Lump Sum *Plus* Regular Deposits. It is possible to create a savings plan by making a lump sum deposit and then adding regular payments. The balance in such an account is the sum of what you would earn from the lump sum deposit alone plus what you would earn from the regular payments alone.

   a. How much will you save after 10 years in an account with an *APR* of 5% compounded monthly that you open with a $2000 initial deposit, followed by regular monthly payments of $50?

   b. How much will you save after 20 years in an account with an *APR* of 5% compounded monthly that you open with a $1000 initial deposit, followed by regular monthly payments of $100?

21. Variable Rates. Imagine that you have a savings plan with an *APR* of 6%. During the first two years you make monthly payments of $100. Thanks to a pay raise, you then increase your monthly payments to $125.

   a. Use the savings plan formula to determine the balance in the account after the first two years.

   b. Use the result of part (a) as an initial lump sum deposit into a second account with an *APR* of 6% and payments of $125. How much will you have saved after five years?

22. Personal Savings Plan. Describe something for which you would like to save money right now. How much will you need to save? How long do you have to save it? Based on these needs, calculate how much you should deposit each month in a savings plan to meet your goal. For the interest rate, use the highest rate currently available at local banks.

UNIT 5C

LOAN PAYMENTS, CREDIT CARDS, AND MORTGAGES

Do you have a credit card? Do you have a loan for your car? Do you have student loans? Do you own a house? Chances are that you owe money for at least one of these purposes. If so, you not only have to pay back the money you borrowed, but you also have to pay interest on the money that you owe.

Suppose that you borrow $1200 at an annual interest rate of *APR* = 12%, which is the same as 1% per month. At the end of the first month, you'd owe interest in the amount of

$$1\% \times \$1200 = \$12.$$

If you paid *only* this $12 in interest, but no more, you'd still owe $1200. That is, the total amount of the loan, called the **loan principal**, would still be $1200. In that case, you'd owe the same $12 in interest the next month. In fact, if you paid *only* the interest from one month to the next, you'd have to pay $12 per month forever—and never pay off the loan!

If you hope to make progress in paying off the loan, you need to pay part of the principal as well as interest. For example, suppose that you pay $200 *toward your loan principal* each month, with the goal of paying the loan off in six months. You'll also have to pay interest each month. Thus, at the end of the first month, you'd pay the $200 toward principal *plus* $12 for the 1% interest that you owe, making a total payment of $212. Because you've paid $200 toward principal, your new loan balance would be

$$\$1200 - \$200 = \$1000.$$

At the end of the second month, you'd again owe 1% interest—but only on the $1000 that you still owe. Thus your interest payment would be

$$1\% \times \$1000 = \$10.$$

Your total payment, including the $200 you pay toward the loan principal, would be $210. We can continue these calculations with the table below.

End of . . .	Prior Loan Balance	Interest on Prior Loan Balance	Payment Toward Principal	Total Payment	New Loan Balance
Month 1	$1200	1% × $1200 = $12	$200	$212	$1000
Month 2	$1000	1% × $1000 = $10	$200	$210	$800
Month 3	$800	1% × $800 = $8	$200	$208	$600
Month 4	$600	1% × $600 = $6	$200	$206	$400
Month 5	$400	1% × $400 = $4	$200	$204	$200
Month 6	$200	1% × $200 = $2	$200	$202	$0

Note that your total payment decreases from month to month because of the declining amount of interest that you owe. There's nothing inherently wrong with this method of paying off the loan, but most people prefer to pay the *same* total amount each month because it makes it easier to plan a budget. A loan that you pay off with equal regular payments is called an **installment loan** (or an *amortized loan*). If you had an installment loan, how much would you need to pay monthly to pay off your $1200 loan in six months?

It's probably clear from the table above that the monthly amount must be somewhere between $200 and $212, but the exact amount is not obvious. Fortunately, we can find the required monthly payments with the **loan payment formula**.

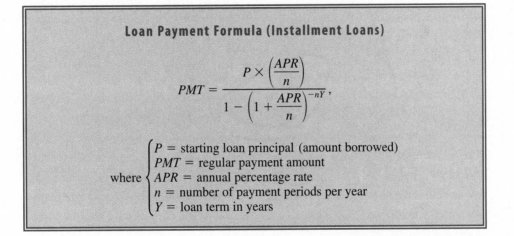

Loan Payment Formula (Installment Loans)

$$PMT = \frac{P \times \left(\dfrac{APR}{n}\right)}{1 - \left(1 + \dfrac{APR}{n}\right)^{-nY}},$$

where $\begin{cases} P = \text{starting loan principal (amount borrowed)} \\ PMT = \text{regular payment amount} \\ APR = \text{annual percentage rate} \\ n = \text{number of payment periods per year} \\ Y = \text{loan term in years} \end{cases}$

In our current example, the starting loan principal is $P = \$1200$, the interest rate is $APR = 12\%$, the number of payments per year is $n = 12$ (monthly payments), and the loan term is $Y = \frac{1}{2}$ (6 months). The loan payment formula gives

$$PMT = \frac{P \times \left(\dfrac{APR}{n}\right)}{1 - \left(1 + \dfrac{APR}{n}\right)^{-nY}} = \frac{\$1200 \times \left(\dfrac{0.12}{12}\right)}{1 - \left(1 + \dfrac{0.12}{12}\right)^{-(12 \times 1/2)}}$$

$$= \frac{\$1200 \times (0.01)}{1 - (1 + 0.01)^{-6}}$$

$$= \frac{\$12}{1 - 0.942045235}$$

$$= \$207.06.$$

The monthly payments would be $207.06 which, as we expected, is between $200 and $212.

EXAMPLE 1   *Student Loan*

Suppose that you have a student loan of $7500 with an interest rate of *APR* = 9% and a loan term of 10 years. What are your monthly payments? How much will you pay over the lifetime of the loan? What is the total interest you will pay on the loan?

Solution: The starting loan principal is $P = \$7500$, the interest rate is *APR* = 0.09, the loan term is $Y = 10$ years, and $n = 12$ for monthly payments. We put these numbers in the loan payment formula.

$$PMT = \frac{P \times \left(\dfrac{APR}{n}\right)}{1 - \left(1 + \dfrac{APR}{n}\right)^{-nY}} = \frac{\$7500 \times \left(\dfrac{0.09}{12}\right)}{1 - \left(1 + \dfrac{0.09}{12}\right)^{-(12 \times 10)}}$$

$$= \frac{\$7500 \times (0.0075)}{1 - (1.0075)^{-120}}$$

$$= \frac{\$56.25}{1 - 0.407937305}$$

$$= \$95.01$$

Your monthly payments will be $95.01 on this student loan. Over the 10-year lifetime of the loan you will pay a total of

$$10 \text{ yr} \times 12 \frac{\text{months}}{\text{yr}} \times \frac{\$95.01}{\text{month}} = \$11{,}401.20.$$

Of this amount, $7500 pays off the principal, so the total interest you will pay is

$$\$11{,}401 - \$7500 = \$3901.$$ ■

Choices of Rate and Term

You'll often have several choices of interest rate and loan term when seeking a loan. For example, a bank might give you a choice between a 3-year loan at 8%, a 4-year loan at 9%, and a 5-year loan at 10% when you want to buy a car. Clearly, you'll pay less total interest by taking the shorter-term, lower-rate loan. However, this loan will have the highest monthly payments. There is no set rule by which you can decide which loan is best for you. You'll have to evaluate your own personal situation and budget in making a decision.

EXAMPLE 2   *Choice of Auto Loans*

You need a loan of $6000 to buy a used car. The bank gives you a choice between a 3-year loan at 8%, a 4-year loan at 9%, and a 5-year loan at 10%. Calculate your monthly payments and total interest over the loan term with each option.

Solution: Let's first calculate the 3-year loan at 8%. The starting loan principal is $P = \$6000$, the interest rate is *APR* = 0.08, the loan term is $Y = 3$ years, and $n = 12$ monthly payments. Your monthly payments would be

$$PMT = \frac{P \times \left(\frac{APR}{n}\right)}{1 - \left(1 + \frac{APR}{n}\right)^{-nY}} = \frac{\$6000 \times \left(\frac{0.08}{12}\right)}{1 - \left(1 + \frac{0.08}{12}\right)^{-(12 \times 3)}}$$

$$= \frac{\$40}{1 - (1.006666667)^{-36}}$$

$$= \$188.02.$$

Over the 3-year period, or 36 months, your total payments would be

$$36 \text{ mo} \times \frac{\$188.02}{\text{mo}} = \$6768.72.$$

Thus the total amount of interest you'd pay on your $6000 loan is $768.72.

Using Your
Calculator   **The Loan Payment Formula**

As with other formulas in this chapter, there are many ways to do loan calculations on your calculator, and business calculators may have built-in functions for these calculations. Here is a procedure that will work on most calculators. The example uses $P = \$1200$, $APR = 12\%$, $n = 12$ (monthly payments), and $Y = \frac{1}{2}$ year (6 months). *It is important that you not round any numbers until the last step.*

		in general	example	display
Starting Formula:		$PMT = \left[\dfrac{P \times \left(\frac{APR}{n}\right)}{1 - \left(1 + \frac{APR}{n}\right)^{-nY}}\right]$	$\dfrac{\$1200 \times \left(\frac{0.12}{12}\right)}{1 - \left(1 + \frac{0.12}{12}\right)^{-(12 \times 1/2)}}$	———
Step 1.	Multiply factors in exponent.	n (+/−)* (×) Y (=)	12 (+/−) (×) 1 (÷) 2 (=)	−6.
Step 2.	Store product in memory (or write down).	(Store)	(Store)	−6.
Step 3.	Add denominator terms 1 and $\frac{APR}{n}$.	1 (+) APR (÷) n (=)	1 (+) 0.12 (÷) 12 (=)	1.01
Step 4.	Raise result to power in memory.	(y^x) (Recall) (=)	(y^x) (Recall) (=)	0.942045235
Step 5.	Subtract result from 1 by making result negative and adding 1.	(+/−) (+) 1 (=)	(+/−) (+) 1 (=)	0.057954764
Step 6.	Denominator is now complete; take its reciprocal	(1/x)	(1/x)	17.25483667
Step 7.	Multiply result by factors P and $\frac{APR}{n}$.	(×) P (×) APR (÷) n (=)	(×) 1200 (×) 0.12 (÷) 12 (=)	207.0580401

With the calculation complete, you can round to the nearest cent to write the answer as $207.06. Be sure to check the calculation.

*The (+/−) key is used on scientific calculators to change the sign of a number.

For the 4-year loan at 9%, we repeat the calculations with $APR = 0.09$ and $Y = 4$ years to find the monthly payment.

$$PMT = \frac{P \times \left(\dfrac{APR}{n}\right)}{1 - \left(1 + \dfrac{APR}{n}\right)^{-nY}} = \frac{\$6000 \times \left(\dfrac{0.09}{12}\right)}{1 - \left(1 + \dfrac{0.09}{12}\right)^{-(12 \times 4)}} = \$149.31$$

Your total payments over 4 years, or 48 months, would be

$$48 \text{ mo} \times \frac{\$149.31}{\text{mo}} = \$7166.88.$$

Thus you'd pay total interest of $1166.88 on your $6000 loan.

Finally, we calculate the monthly payments for the 5-year loan at 10% by setting $APR = 0.1$ and $Y = 5$ years.

$$PMT = \frac{P \times \left(\dfrac{APR}{n}\right)}{1 - \left(1 + \dfrac{APR}{n}\right)^{-nY}} = \frac{\$6000 \times \left(\dfrac{0.1}{12}\right)}{1 - \left(1 + \dfrac{0.1}{12}\right)^{-(12 \times 5)}} = \$127.48$$

Your total payments over 5 years, or 60 months, would be

$$60 \text{ mo} \times \frac{\$127.48}{\text{mo}} = \$7648.80.$$

The total interest payments would be $1648.80. As we anticipated, the monthly payment is least and the total interest is greatest on the loan with the longest term. ■

TIME-OUT TO THINK: Consider your own current financial situation. If you needed a $6000 car loan, which option from Example 2 would you choose? Why?

CREDIT CARDS

Credit card loans differ from other loans in that you are not required to pay off your balance in any set period of time. Instead, you are required to make only a "minimum monthly payment" that depends on your balance and the interest rate. However, most credit cards have high interest rates compared to other types of loans, so it is to your advantage to pay off credit card balances as quickly as possible. Once you decide to pay off a balance in a certain period of time, you can use the loan payment formula to calculate the payments you'd need to make.

EXAMPLE 3   *Credit Card Debt*

You have a credit card balance of $2300 with an annual interest rate of 21%. You decide that you'd like to pay off your balance over one year. How much will you need to pay each month, and how much total interest will you pay? Assume that you make no further purchases with your credit card.

Solution: Your starting loan principal is $P = \$2300$, the interest rate is $APR = 0.21$, and you will make $n = 12$ payments per year. You plan to pay off your balance in $Y = 1$ year. The loan payment formula gives

$$PMT = \frac{P \times \left(\dfrac{APR}{n}\right)}{1 - \left(1 + \dfrac{APR}{n}\right)^{-nY}} = \frac{\$2300 \times \left(\dfrac{0.21}{12}\right)}{1 - \left(1 + \dfrac{0.21}{12}\right)^{-(12 \times 1)}} = \$214.16.$$

You'll need to make payments of \$214.16 to get your balance paid off in one year—assuming you don't charge anything additional to your credit card!

During the 12 months, you'll pay a total of

$$12 \text{ mo} \times \frac{\$214.16}{\text{mo}} = \$2569.92.$$

Your loan principal was \$2300, so the remaining \$269.92 goes to interest. ∎

TIME-OUT TO THINK: Continuing Example 3, suppose that you can get a personal loan at a bank at an interest rate of 10% for one year. Should you take this loan and use it to pay off your \$2300 credit card debt? Why or why not?

---

*T*HINKING ABOUT…

Avoiding Credit Card Trouble

Most adults have one or more credit cards, and for good reason. Used properly, credit cards offer many conveniences: an easy method of payment, clear monthly statements that list everything you have charged to the card, a form of ID that may allow you to rent a car, and many more. But credit card trouble can compound quickly, and many people get into financial trouble as a result. A few simple guidelines can help you to avoid credit card trouble.

- Use only one credit card. People who accumulate balances on several cards often lose track of their overall debt.
- If possible, pay off your balance in full each month. Then there's no chance of getting into a financial hole.
- If you plan to pay off your balance in full each month, be sure that your credit card offers a "grace period" so that you will not have to pay any interest at all.

- Know your credit card rate, and know how it compares to other available rates. Different credit cards may charge hugely different rates. If you're carrying a balance, and you can get a lower rate somewhere else, then do it!
- When choosing a credit card, watch out for "teaser rates." These are low interest rates offered for short periods, such as 6 months, that revert to very high rates after 6 months.
- Never use your credit card for a cash advance unless it is an emergency. Nearly all credit cards charge substantial fees for cash advances, in addition to charging high interest rates on the advance. Instead, get cash directly from your own bank account by cashing a check or using an ATM card.
- If you own a home, consider replacing a common credit card with a home equity credit line. You'll generally get a lower interest rate, and the interest will be tax deductible.
- If you find yourself in a deepening financial hole, consult a financial advisor right away. The longer you wait, the worse off you'll be in the long run.

Credit Card Danger

The high interest rates on credit cards can mean particular problems for people who are unable to pay their bills. Like compound interest in reverse, failure to pay at least the interest can put a person into an ever-deepening financial hole.

EXAMPLE 4   *A Deepening Hole*

Paul has gotten into credit card trouble. He has a balance of $9500, and just lost his job. His credit card company charges interest, compounded daily, at an annual rate of $APR = 21\%$.

Suppose that the credit card company allows him to suspend his payments until he finds a new job—but continues to charge interest. If it takes him a year to find a new job, how much will he owe at that time?

Solution: First, we need to decide what type of problem we are dealing with. Because Paul is not making any payments during the year, it is not a loan payment problem. Instead, it is a compound interest problem, in which Paul's balance of $9500 is growing at an annual rate of 21%, compounded daily. Thus we use the compound interest formula, with a starting balance of $P = \$9500$, $APR = 0.21$, $Y = 1$ year, and $n = 365$ (for daily compounding). At the end of the year, his loan balance will be

$$A = P \times \left(1 + \frac{APR}{n}\right)^{nY} = \$9500 \times \left(1 + \frac{0.21}{365}\right)^{365 \times 1} = \$11{,}719.23.$$

During his year of unemployment, interest alone will make Paul's credit card balance grow from $9500 to over $11,700, an increase of more than $2200! Clearly, this increase will only make it more difficult for Paul to get back on his financial feet. ∎

MORTGAGES

By the Way ··············

The idea of a mortgage contract originated in early British real estate law. The curious word *mortgage* comes from Latin and old French. It literally means *dead pledge*.

One of the most popular types of installment loan is designed specifically to help you buy a home—a home **mortgage**. Mortgage interest rates generally are lower than interest rates on other types of loan because your home itself serves as a payment guarantee. If you fail to make your payments, the bank can take possession of your home and sell it to recover the amount loaned to you.

Securing a home loan usually consists of two basic parts. First, the bank will require you to make a **down payment**, typically 10% to 20% of the purchase price. Then the bank pays the rest of the purchase price to the seller of the home, while you choose some type of home mortgage with which you will repay the bank.

Fixed-Rate Mortgages

The simplest type of home loan is a **fixed-rate** mortgage, in which you are guaranteed an interest rate that will not change over the life of the loan. The most common fixed-rate loans have terms of either 15 or 30 years, with lower interest rates on the shorter-term loans. We can calculate payments on fixed-rate mortgages with the loan payment formula.

EXAMPLE 5   *Fixed-Rate Home Mortgage*

Suppose you need a loan of $100,000 to buy your new home. The bank offers a choice of a 30-year loan at an *APR* of 8%, or a 15-year loan at 7.5%. Compare your monthly payments by the two options.

Solution: The starting loan principal is $P = \$100,000$ and $n = 12$. For the 30-year loan, we set $APR = 0.08$ and $Y = 30$. The monthly payments are

$$PMT = \frac{P \times \left(\dfrac{APR}{n}\right)}{1 - \left(1 + \dfrac{APR}{n}\right)^{-nY}} = \frac{\$100,000 \times \left(\dfrac{0.08}{12}\right)}{1 - \left(1 + \dfrac{0.08}{12}\right)^{-(12\times30)}} = \$733.76.$$

For the 15-year loan, we set $APR = 0.075$ and $Y = 15$. The monthly payments are

$$PMT = \frac{P \times \left(\dfrac{APR}{n}\right)}{1 - \left(1 + \dfrac{APR}{n}\right)^{-nY}} = \frac{\$100,000 \times \left(\dfrac{0.075}{12}\right)}{1 - \left(1 + \dfrac{0.075}{12}\right)^{-(12\times15)}} = \$927.01.$$

Your payments of $927.01 on the 15-year loan would be almost $200 higher than the payments of $733.76 on the 30-year loan. But let's compare the total amount paid with the two options:

$$\text{30-year loan: } 30 \text{ yr} \times \frac{12 \text{ mo}}{\text{yr}} \times \frac{\$733.76}{\text{mo}} \approx \$264,150$$

$$\text{15-year loan: } 15 \text{ yr} \times \frac{12 \text{ mo}}{\text{yr}} \times \frac{\$927.01}{\text{mo}} \approx \$166,860$$

You'd end up paying a total of almost $100,000 more with the longer-term loan! Your choice is simple, if difficult: the 15-year loan saves you nearly $100,000 in the long run, but it's a good plan only if you can afford the additional $200 per month that it will cost you for the next 15 years. ■

EXAMPLE 6   *An Alternative Strategy*

An alternative strategy for the mortgage in Example 5 is to take the 30-year loan at 8%, but to try to pay it off in 15 years by making larger payments than required. How much would you have to pay each month? Discuss the pros and cons of this strategy.

Solution: This strategy essentially means taking an 8% loan and paying it off in 15 years. Thus we set $APR = 0.08$ and $Y = 15$, and calculate the needed monthly payments:

$$PMT = \frac{P \times \left(\dfrac{APR}{n}\right)}{1 - \left(1 + \dfrac{APR}{n}\right)^{-nY}} = \frac{\$100,000 \times \left(\dfrac{0.08}{12}\right)}{1 - \left(1 + \dfrac{0.08}{12}\right)^{-(12\times15)}} = \$955.65$$

You could pay off your 30-year loan in just 15 years by making payments of $955.65 per month.

Note that this payment is about $30 per month *more* than the payment of $927.01 required with the 15-year loan because of the higher interest rate—8% versus 7.5%. Clearly, if you're going to pay off the loan in 15 years, you should take the lower-interest 15-year loan. However, this alternative strategy has one advantage: with a 30-year loan, you are required to make payments of only $733.76; the rest of your payment is the extra amount that allows you to pay off your loan in less than 30 years. If at some point you cannot afford this extra amount, you can go back to paying only the required $733.76.   ■

> **TIME-OUT TO THINK:**   Compare the strategies in Examples 5 and 6. Which would you recommend to someone with a new job and very little savings? Which would you recommend to someone who has a stable, long-term job and a lot of savings? Explain.

Principal and Interest Payments

We calculated that, on a $100,000 loan at a 30-year fixed rate of 8%, the total payments over the life of the loan would be about $264,000—more than two and a half times the original loan amount! In fact, the portion of the payments going to interest and principal varies as the loan is paid off. Early payments, when most of the loan balance is unpaid, go mostly to interest. Toward the end of the loan period, when the remaining loan balance is relatively small, most of the payments go toward principal.

Figure 5.2 shows the overall pattern of principal and interest for the 30-year, $100,000 loan at 8%. Note that more than 90% of your payment amounts go to interest in the early payments. By the end of the loan term, more than 90% of the payment amounts go to principal.

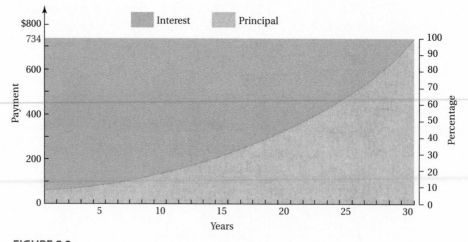

FIGURE 5.2
Principal and interest payments on a 30-year, $100,000 loan at 8%.

It is, of course, possible to calculate exactly how much of each payment goes to principal and how much goes to interest over the life of the loan. However, the formulas are more complex than others in this chapter, and it is generally easier to do these calculations with the aid of a computer spreadsheet. In addition, most banks will provide you with a table of principal and interest payments over the life of a loan (called an *amortization schedule*) if you ask.

Adjustable-Rate Mortgages

A fixed-rate mortgage is advantageous for you because your monthly payments never change. However, it poses a risk to the lender. Imagine that you take out a fixed, 30-year loan of $100,000 from *Great Bank* at an 8% interest rate. Initially, the loan may seem like a good deal for *Great Bank*. But suppose that, two years later, prevailing interest rates have risen to 10%. If *Great Bank* still had the $100,000 that it lent to you, it could lend it out to someone else at this higher 10% rate. Instead, it's stuck with the 8% rate that you are paying. In effect, *Great Bank* loses potential future income if prevailing rates rise substantially and you have a fixed-rate loan.

Lenders can lessen the risk of rising interest rates by charging higher rates for longer-term loans. That is why rates generally are higher for 30-year loans than for 15-year loans. But an even lower risk strategy for the lender is an **adjustable-rate mortgage (ARM)**, in which the interest rate you pay changes whenever prevailing rates change. Because of their reduced long-term risk, lenders generally offer ARMs with much lower initial interest rates than fixed-rate loans. For example, a bank offering an 8% rate on a *fixed* 30-year loan might offer an ARM that begins at 6%. Most ARMs guarantee their starting interest rate for the first six months or one year, but interest rates can change in subsequent years to reflect prevailing rates. In addition, most ARMs also include a *rate cap* that cannot be exceeded. For example, if your ARM begins at an interest rate of 6%, you may be promised that your interest rate can never go higher than a rate cap of 12%. Making a decision between taking a fixed-rate loan or an ARM can be one of the most important financial decisions of your life.

Beware!   Under normal circumstances, your rate on an ARM will not rise unless prevailing interest rates rise. However, some lenders offer low "teaser" rates for the first few months of an ARM; that is, rates that are certain to rise as soon as the teaser period is over. While these teasers may sound attractive at first, the longer-term policies of the ARM are far more important.

EXAMPLE 7   *Rate Approximations for ARMs*

Suppose that you have a choice between a 30-year fixed-rate loan at 8% and an ARM with a first-year rate of 5%. By ignoring changes in principal, estimate your monthly savings during the first year on a $100,000 loan. Suppose that the rate rises to 11% by the fourth year. How will your payments be affected?

Solution: Because mortgage payments are mostly interest in the early years of a loan, we can make approximations by pretending that the principal remains unchanged. The ARM will save you 3% of the interest amount during the first year (5% instead of 8%). On a $100,000 loan, this translates to approximately

$$3\% \times \$100,\!000 = \$3000.$$

Dividing this savings of $3000 over 12 months gives

$$\$3000 \div 12 \text{ mo} = \$250/\text{mo}.$$

Thus you'd save about $250 per month during the first year.

By the fourth year, when rates reach 11%, the situation is reversed. The rate on the ARM is now 3 percentage points above the rate of the fixed-rate loan. Instead of saving $250 per month, you'd be paying $250 per month *more* on your ARM than on the 8% fixed-rate loan. Moreover, if interest rates remain high, you'd continue these high payments for many years to follow. While ARMs reduce risk for the lender, they add risk for the borrower. ■

Additional Mortgage Considerations

Because a home mortgage is likely to be the largest loan you ever take, it's particularly important to shop around for the best rates and terms. Besides just the monthly payments, your decision on a loan may be affected by a number of other factors. Some of the most important include:

- How much does the lender charge in fees, or **closing costs**, for giving you the loan? High closing costs may offset the advantages of a slightly lower rate.

- You may have the option to deliberately pay higher up-front costs, called **points** (each point is 1% of your loan amount), in order to get a lower interest rate. You'll need to decide whether the additional costs are worth the long-term benefit of the lower rate.

- Are there any **prepayment penalties** if you decide to pay off your loan early? If interest rates drop, you might want to **refinance** your loan; that is, take out a new loan at a lower interest rate and use this loan to pay off your original one at the higher rate. Prepayment penalties may make refinancing costly.

- Is your lender reliable? The lender effectively owns your house until you pay off the mortgage, so you want a lender with a solid reputation for honesty and integrity.

As with any major financial decision, be sure to learn all you can before making a decision on a mortgage. And always keep in mind what is sometimes called the first rule of finance:

If it sounds too good to be true, it probably is!

EXAMPLE 8   *Closing Costs*

Great Bank is offering you a $100,000, 30-year, 8% fixed-rate loan with closing costs of $500. *Big Bank*, is offering you a lower rate of 7.9%, but with closing costs of $1000. Evaluate the two options.

Solution: We already calculated the payments on the 8% loan to be $733.76 (Example 5). At the lower 7.9% rate, the payments would be

$$PMT = \frac{P \times \left(\dfrac{APR}{n}\right)}{1 - \left(1 + \dfrac{APR}{n}\right)^{-nY}} = \frac{\$100{,}000 \times \left(\dfrac{0.079}{12}\right)}{1 - \left(1 + \dfrac{0.079}{12}\right)^{-(12 \times 30)}} = \$726.80.$$

Thus you'll save about $7 per month with the lower interest rate, but it will cost you an additional $500 in closing costs ($1000 versus $500). Note that

$$\$500 \div \$7/\text{mo} = 71 \text{ months (almost 6 years)}.$$

Thus it will take you about 6 years to save the extra $500 you paid up front. Unless you are sure that you will be staying in your house (and keeping the same loan) for much more than 6 years, you probably should go with the lower closing costs, even though your monthly payments will be slightly higher. ■

By the Way ....
The average mortgage in the United States is paid off after 7 years, usually because the home is sold.

EXAMPLE 9   *Points Decision*

Continuing Example 8, you've decided to go with *Great Bank*'s 8% loan. Just as you are about to sign, you learn that there's another option at *Great Bank*: If you're willing to pay 1 *point*, the bank will give you the same loan at a rate of 7.5%. What should you do?

Solution: At 7.5% for 30 years, the payments on the $100,000 loan are

$$PMT = \frac{P \times \left(\dfrac{APR}{n}\right)}{1 - \left(1 + \dfrac{APR}{n}\right)^{-nY}} = \frac{\$100{,}000 \times \left(\dfrac{0.075}{12}\right)}{1 - \left(1 + \dfrac{0.075}{12}\right)^{-(12 \times 30)}} = \$699.21.$$

This payment is smaller than the payments at 8% by

$$\$733.76 - \$699.21 = \$34.55.$$

One point means a fee of 1% on your $100,000 loan. This translates to a fee of

$$1\% \times \$100{,}000 = \$1000.$$

Thus you must decide whether it is worth an extra $1000 up front for a monthly savings of just under $35. Let's calculate how long it will take to make up the added up-front costs:

$$\$1000 \div \$34.55/\text{mo} = 30 \text{ months} = 2.5 \text{ years}$$

If you think it's likely that you will sell or refinance within a couple of years, you should not pay the point. However, if you think it's likely that you'll be keeping this loan for a long time, it might well be worth paying the point. If you pay the loan for the full 30 years, you'll save a total of

$$360 \text{ mo} \times \$34.55/\text{mo} = \$12{,}438$$

in interest over the life of the loan—far more than the extra $1000 you will pay for the lower rate today.   ∎

By the Way ···········
Most financial analysts say that you should pay additional up-front costs only if it will be made up in savings on your monthly payments within two to three years.

REVIEW QUESTIONS

1. Suppose you pay only the interest on a loan. Will the loan ever be paid off? Why not?

2. What is an *installment loan*? Do you currently have any installment loans?

3. Explain the meaning and use of the *loan payment formula*.

4. Suppose that you need a loan of $10,000, and are offered a choice of a 3-year loan at 7% interest or a 5-year loan at 8% interest. Discuss the pros and cons of each choice.

5. How do credit card loans differ from ordinary installment loans? Why are credit card loans particularly dangerous?

6. What is a *mortgage*? What is a *down payment* on a mortgage?

7. Explain the difference between a *fixed-rate* mortgage and an *adjustable-rate* mortgage. What are the pros and cons of each for the borrower? for the lender?

8. Suppose that you hope to pay off a mortgage in 15 years. Under what circumstances might it still be advantageous to take a 30-year loan? What is the disadvantage of this strategy?

9. Explain, in general terms, how the portions of installment loan payments going to principal and interest change over the life of the loan.

10. Explain how *closing costs* and *points* can affect a decision in which you have a choice of loans with different interest rates.

11. Why might you want to *refinance* a loan? How might your decision be affected by *prepayment penalties*?

PROBLEMS

1. **Loan Terminology.** Suppose you apply for a $40,000 loan for a house purchase. You will pay $310 per month for 20 years at an *APR* of 7%.

   a. What is the principal of the loan?

   b. What is the annual percentage (interest) rate?

   c. What are the monthly payments?

   d. What is the term of the loan?

   e. How many payments will you make?

   f. How much money will you actually pay the lender over the course of the loan?

   g. Of the total amount you pay, how much of it (in dollars and in percentage terms) will go toward interest?

2. **Loan Terminology.** Suppose you apply for a $7500 student loan. You will pay $95 per month for 10 years at an *APR* of 9%.

   a. What is the principal of the loan?

   b. What is the annual percentage (interest) rate?

   c. What are the monthly payments?

   d. What is the term of the loan?

   e. How many payments will you make?

   f. How much money will you actually pay the lender over the course of the loan?

   g. Of the total amount you pay, how much of it (in dollars and in percentage terms) will go toward interest?

3. **Loan Payments.** Calculate the monthly payments on each loan described.

   a. A student loan of $25,000 at a fixed *APR* of 10% for 20 years.

   b. A home mortgage of $150,000 with a fixed *APR* of 9.5% for 30 years.

   c. A home mortgage of $150,000 with a fixed *APR* of 8.75% for 15 years.

4. **Loan Payments.** Calculate the monthly payments on each loan described.

   a. A student loan of $12,000 at a fixed *APR* of 8% for 15 years.

   b. A home mortgage of $100,000 with a fixed *APR* of 9.5% for 30 years.

   d. A $3000 credit card bill to be paid off in 2 years at an *APR* of 16%.

Monthly Payments. *In Problems 5–8, determine (i) your monthly payments; (ii) your total payments over the term of the loan; and (iii) how much you will pay in interest over the loan term, both in dollars and as a percentage of your total payments.*

5. You borrow $5000 over a period of 3 years at an *APR* of 12%.

6. You borrow $10,000 over a period of 5 years at an *APR* of 10%.

7. You borrow $50,000 over a period of 15 years at an *APR* of 8%.

8. You borrow $100,000 over a period of 30 years at an *APR* of 7%.

9. **Accelerated Student Loan Payment.** Suppose that you have a student loan of $25,000 with an *APR* of 9% for 20 years.

   a. What are your required monthly payments?

   b. Suppose that you would like to pay the loan off in 10 years instead of 20. What monthly payments will you need to make?

c. Compare the total amounts you'll pay over the loan term if you pay the loan off in 20 years versus 10 years.

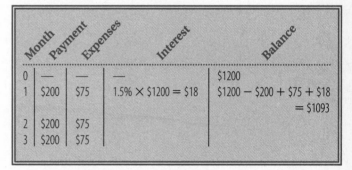

10. Accelerated Student Loan Payment. Suppose that you have a student loan of $60,000 with an *APR* of 8% for 25 years.

a. What are your required monthly payments?

b. Suppose that you would like to pay the loan off in 15 years instead of 25. What monthly payments will you need to make?

c. Compare the total amounts you'll pay over the loan term if you pay the loan off in 25 years versus 15 years.

11. Student Loan Consolidation. Suppose that you have the following three student loans: $10,000 with an *APR* of 8% for 15 years; $15,000 with an *APR* of 8.5% for 20 years; $12,500 with an *APR* of 9% for 10 years.

a. Calculate the monthly payments for each loan individually.

b. Calculate the total you'll make in payments during the life of all three loans.

c. Suppose that, just as you are about to begin payments on these loans after graduation, a company offers to consolidate your three loans into a single loan with an *APR* of 8.5% and a loan term of 20 years. What will your monthly payments be in that case? What will your total payments be over the 20 years? Discuss the pros and cons of accepting this loan consolidation.

12. Credit Card Debt. Imagine that you have a balance of $2500 on your credit card that you decide to pay off. Calculate your monthly payments and total payments under the following conditions. In all cases assume that you charge no additional expenses to the card.

a. The *APR* for the credit card is 18% and you want to pay off the balance in one year.

b. The *APR* for the credit card is 20% and you want to pay off the balance in two years.

c. The *APR* for the credit card is 21% and you want to pay off the balance in three years.

13. Credit Card Payments. Assume that you have a balance of $1200 on a credit card that carries an *APR* of 18%, or 1.5% per month. You start making monthly payments of $200, but at the same time you charge an additional $75 per month (on average) to the credit card. Assume that interest for a given month is based on the balance for the previous month. The following table shows how you can calculate your monthly balances.

Month	Payment	Expenses	Interest	Balance
0	—	—	—	$1200
1	$200	$75	1.5% × $1200 = $18	$1200 − $200 + $75 + $18 = $1093
2	$200	$75		
3	$200	$75		

a. Complete and extend the table to show your balances at the end of each of the first six months.

b. Continue to extend the table until the debt is paid off. How long does it take to pay off the credit card debt?

14. Deeper Trouble. Repeat the table of Problem 13, but this time assume that you make monthly payments of only $125. Extend the table as long as necessary until your debt is paid off. How long does it take to pay off your debt?

15. Credit Card Woes. The following table shows the expenses and payments on a credit card for eight months, starting with an initial balance of $300. Fill out the last two columns that show the interest and balance on the account. Assume that the *APR* for the credit card is 18%, and that interest for a given month is based on the balance for the previous month. After eight months, what is the balance on the credit card? Comment on the effect of interest and initial balance, in light of the fact that for seven of the eight months, expenses never exceeded payments.

Month	Payment	Expenses	Interest	Balance
0	—	—	—	$300
1	$300	$175		
2	$150	$150		
3	$400	$350		
4	$500	$450		
5	0	$100		
6	$100	$100		
7	$200	$150		
8	$100	$80		

16. Teaser Rate. You have a total credit card debt of $4000. You receive an offer to transfer this debt to a new card with an introductory *APR* of 6% for the first 6 months. After that, the rate becomes 24%.

  a. What is the monthly *interest* payment on $4000 during the first 6 months. (Assume you pay nothing toward principal and don't charge any further debts.)

  b. What is the monthly *interest* payment on $4000 *after* the first 6 months.

17. Choosing an Auto Loan. Suppose that you can afford monthly car payments of $220 and need to borrow $10,000 to buy the car you want. The bank offers three choices of car loans with the following rates and terms: 7% for a 3-year loan; 7.5% for a 4-year loan; or 8% for a 5-year loan. Which loan best meets your needs? Explain your reasoning.

18. Mortgage Strategies. Suppose you apply for a $75,000 house mortgage. You have two loan options from which to choose. One is a 15-year fixed-rate loan at 7% and the other is a 30-year fixed-rate loan at 8%.

  a. What will your monthly payments be for each loan?

  b. If you estimate that you can afford $600 per month for loan payments, which loan should you choose?

  c. If you estimate that you can afford $800 per month for loan payments, which loan should you choose?

  d. Discuss the advantages and disadvantages of choosing the 30-year loan and then making payments as if it were a 15-year loan.

19. Mortgage Options. Suppose that you need a loan of $120,000 to buy a home. Your bank offers a choice of a 15-year fixed-rate loan at an *APR* of 6.75%, a 20-year fixed-rate loan at an *APR* of 7.0%, or a 30-year fixed-rate loan at an *APR* of 7.15%. Calculate both your monthly payments and your total payments over the loan term for each scenario. Discuss the pros and cons of each choice.

20. ARM Rate Approximations. Suppose that you have a choice between an ARM with a first-year rate of 5% and a 30-year fixed-rate with an *APR* of 7%.

  a. By ignoring changes in principal, estimate your monthly savings during the first year on a $140,000 loan.

  b. Suppose that the rate on the ARM rises to 8.5% at the start of the third year. Approximately how much extra will you then be paying compared to if you had taken the fixed-rate loan?

Closing Costs and Points. *You need a loan of $80,000 to buy a home. In Problems 21–23 you are offered two choices. Calculate your monthly payments in each case, and discuss how you would decide between the two choices.*

21. *Choice 1:* 30-year fixed-rate at 8% with closing costs of $1200 and no points. *Choice 2:* 30-year fixed-rate at 7.5% with closing costs of $1200 and 1.5 points.

22. Choice 1: 30-year fixed-rate at 8.5% with no closing costs and no points. Choice 2: 30-year fixed-rate at 7.5% with closing costs of $1200 and 2 points.

23. Choice 1: 30-year fixed-rate at 7.25% with closing costs of $1200 and 1 point. Choice 2: 15-year fixed-rate at 6.5% with closing costs of $1200 and 3 points.

24. Car-Title Lenders. Some "car-title lenders" offer quick cash loans in exchange for allowing them to hold the title to your car as collateral (you lose your car if you fail to pay off the loan). In many states, these lenders operate under pawnbroker laws that allow them to charge fees as a percentage of unpaid balance. Suppose that you need $2000 in cash, and a car-title company offers you a loan at an interest rate of 2% *per month plus monthly fees* of 20% of the unpaid balance.

a. How much will you owe in interest and fees on your $2000 loan at the end of the first month?

b. Suppose that you pay only the interest and fees each month. How much will you pay over the course of a full year?

c. Suppose instead that you obtain a loan from a bank with a term of 3 years and an *APR* of 10%. What are your monthly payments with this loan? Compare these payments to the payments due the car-title lender.

25. How Much House Can You Afford? Suppose that you can afford monthly payments of $500. If current mortgage rates are 9% for a 30-year fixed-rate loan, what loan principal can you afford? If you are required to make a 20% down payment and have the cash on hand for it, what price home can you afford? (Hint: You will need to solve the loan payment formula for *P*.)

26. Project: Choosing a Mortgage. Imagine that you work for an accounting firm and a client has told you that he is buying a house and needs a loan of $120,000. His monthly income is $4000 and he is single with no children. He has $14,000 in savings that can be used for a down payment. Find the current rates available from local banks for both fixed-rate mortgages and adjustable-rate mortgages (ARMs). Analyze the offerings and summarize orally or in writing the best options for your client, along with the pros and cons of each option.

· ·

INCOME TAXES

There's an old saying that the only sure things in life are death and taxes. Although it's rather cynical, it contains at least a grain of truth. There's no legal way to avoid taxes in the modern world.

However, there are many different types of taxes, and different people may pay taxes at vastly different rates. Some of the more common forms of tax include:

- Sales tax, in which you pay some percentage of an item's cost in tax.

- Gasoline tax, usually added as a particular amount of tax per gallon.

- Property tax, in which you pay some percentage of the value of your property in tax each year.

- Income tax, in which you pay some percentage of your earnings each year in tax.

- Social security and Medicare taxes, in which you and your employer each pay some percentage of your wages in tax.

In the United States, only state and local governments levy sales and property taxes. Gasoline taxes are levied primarily by the federal and state governments, although some localities also collect them. Income taxes are also levied by both the federal and state governments. For the federal government, income tax is the largest source of revenue. Federal income tax also is the largest tax burden faced by most people. In this unit, we focus exclusively on income taxes.

By the Way ·············

An income tax was first levied in the United States in 1862 (during the Civil War) but was abandoned a few years later. The Sixteenth Amendment to the Constitution, ratified in 1913, gave the federal government full authority to levy an income tax.

INCOME TAX BASICS

It's quite possible that *no one* fully understands federal income taxes. The complete tax code consists of thousands of pages of detailed regulations. Many of the regulations are subject to interpretation, and disputes about their meaning often must be settled in court. Congress frequently tinkers with tax laws, and occasionally undertakes major reforms. For example, tax law was greatly simplified by Congress in 1986. Unfortunately, politicians were unable to resist making modifications to the simplified tax code, so it gradually became more complex once again.

Nevertheless, a few basic ideas underlie most of income tax law. Taxes are collected by the **Internal Revenue Service (IRS)**, which is part of the United States Department of Treasury. We file federal taxes by completing a tax form, such as Form 1040, 1040A, or 1040EZ. In all cases the basic procedure for calculating taxes is as shown in Figure 5.3.

• Web • Watch •
You can learn more about income tax rules and obtain tax forms and filing instructions directly from the IRS Web site.

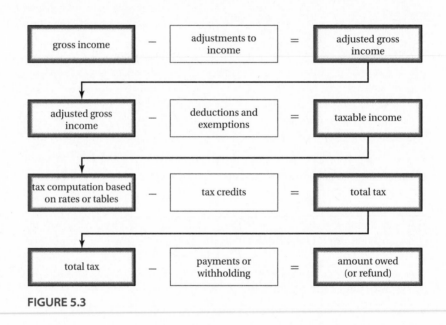

FIGURE 5.3

Gross income means all your income including wages, tips, profits from a business, interest or dividends from investments, and more. Some gross income is not subject to taxation (at least not in the year it is received), and is deducted to yield the **adjusted gross income**, on which further computations are based. For most people, the most important adjustment to gross income comes from contributions to tax-deferred savings plans, which we'll discuss shortly.

Everyone is entitled to certain **deductions** and **exemptions**—amounts that are subtracted from your adjusted gross income before your tax is calculated. *Exemptions* are a fixed amount per person ($2650 in 1997) and you can take an exemption for yourself and each of your **dependents** (for example, children that you support) on your taxes. For *deductions*, you are allowed to deduct either:

- the **standard deduction**—the amount of which depends on whether you are single, married, or a head of household. If you are married, it also depends on whether you and your spouse file separately or jointly. (For example, in 1997, the standard deduction was $4150 for single people and $6900 for married couples filing jointly.)
- **itemized deductions**—the sum total of deductible expenditures, such as interest on home mortgages and contributions to charity.

By the Way ············
The amounts of exemptions and standard deductions rise each year by amounts that are supposed to reflect the effects of inflation.

Once you have subtracted your deductions and exemptions from your adjusted gross income, you are left with your **taxable income**. A tax table or tax rate computation allows you to calculate how much tax you owe. However, you may not have to pay this full amount if you are entitled to any **tax credits**. Tax credits are deducted directly from your tax bill, and hence are more valuable than tax deductions: A $1 tax credit reduces your tax bill by $1, but a $1 tax deduction only reduces your tax bill by $1 times your tax rate.

Most people are required either to have their employer *withhold* a portion of their wages (which the employer sends to the IRS) or to pay *estimated taxes* over the course of the year. Thus, at the end of the year, you may have already paid some or all of your taxes—and you may even be owed a refund.

TIME-OUT TO THINK: Tax terms can be confusing! Before you go on, check your understanding by imagining that you are explaining the term *income* to a friend. How would you explain the difference between *gross income, adjusted gross income,* and *taxable income*?

EXAMPLE 1   *Taxable Income*

In 1997, Karen was single and earned wages of $24,200. She also received $750 in interest from a savings account. She contributed $1200 to a tax-deferred retirement plan. Her only tax deductible expenditures were $900 in charitable contributions. What is Karen's taxable income for 1997? If her tax rate is 15%, what is her 1997 tax bill?

For a single person in 1997, the personal exemption is $2650 and the standard deduction is $4150.

Solution: Karen's two sources of income, wages and interest, were both part of her gross income:

$$\text{Gross income} = \text{sum of all income} = \$24{,}200 + \$750 = \$24{,}950$$

Her $1200 contribution to a tax-deferred retirement plan counts as an adjustment to her gross income, so her adjusted gross income (AGI) is

$$\text{AGI} = \text{gross income} - \text{adjustments} = \$24{,}950 - \$1200 = \$23{,}750.$$

She is entitled to deduct her personal exemption of $2650 and the larger of the $4150 standard deduction or her itemized deductions. Because her only itemized deduction is the $900 in charitable contributions, she is better off with the standard deduction. Thus her taxable income is

$$\text{Taxable income} = \text{AGI} - \text{exemptions} - \text{deductions}$$
$$= \$23{,}750 - \$2650 - \$4150 = \$16{,}950.$$

At a tax rate of 15%, she owes

$$15\% \times \$16{,}950 = 0.15 \times \$16{,}950 = \$2542.50$$

in federal income tax.   ∎

EXAMPLE 2   *Tax Credits Vs. Tax Deductions.*

Suppose that your taxable income is being taxed at a rate of 15%. How much will you save in taxes if you find an additional $500 tax *credit* (such as the $500 tax credit per child allowed to most families)? How much will you save if you find an additional $500 tax *deduction* (such as a contribution to charity)?

Solution: The entire $500 tax credit is deducted from your actual tax bill, and therefore saves you a full $500. In contrast, the $500 tax deduction reduces your *taxable income*, not your total tax bill, by $500. Thus, if you are taxed at a rate of 15%, *at best* this $500 deduction can save you only

$$15\% \times \$500 = \$75.$$

Moreover, you will save this $75 only if you are itemizing deductions. If your total itemized deductions are less than the standard deduction, like Karen in Example 1, you will still be better off with the standard deduction. Thus the additional $500 deduction will save you nothing at all.   ∎

TAX RATES

The United States has a **progressive income tax**, meaning that people with higher taxable incomes pay a higher tax *rate*. The system works by assigning different **marginal tax rates** to income in different ranges (or *margins*). In 1997, the marginal tax rates were 15%, 28%, 31%, 36%, and 39.6%. The ranges to which these rates apply depend on your **filing status**; most people fall into one of four filing status categories:

- *Single*. This status applies if you are unmarried, divorced, or legally separated.
- *Married filing jointly*. This status applies if you are married and you and your spouse choose to file a single tax return.
- *Married filing separately*. This status applies if you are married and you and your spouse choose to file two separate tax returns.
- *Head of household*. This status applies if you are unmarried but are paying more than half the cost of supporting a dependent child or dependent parent.

Table 5.2 shows the income levels to which each of the marginal tax rates applied in 1997. For example, a single person earning a *taxable income* of $34,650 would have paid tax at a 15% rate on the first $24,650, or

$$15\% \times \$24{,}650 = \$3698.$$

Beware! Gross income, adjusted gross income, and taxable income are very different. Taxes are collected only on *taxable* income. Thus two people with the same gross income will not pay the same overall tax rate if they have different adjustments, exemptions, and deductions.

The remaining $10,000 of income would be subject to the higher marginal rate of 28% because it was above the *margin* of $24,650. This marginal tax would have been

$$28\% \times \$10,000 = \$2800,$$

making the person's total tax

$$\$3698 + \$2800 = \$6498.$$

Table 5.2   1997 Marginal Tax Rates

Filing Status	15% tax rate	28% tax rate	31% tax rate	36% tax rate	39.6% tax rate
Single	$0 — $24,650	$24,651 — $59,750	$59,751 — $124,650	$124,651 — $271,050	over $271,050
Head of household	$0 — $33,050	$33,051 — $85,350	$85,351 — $138,200	$138,201 — $271,050	over $271,050
Married filing jointly	$0 — $41,200	$41,201 — $99,600	$99,601 — $151,750	$151,751 — $271,050	over $271,050
Married filing separately	$0 — $20,600	$20,601 — $49,800	$49,801 — $75,875	$75,876 — $135,525	over $135,525

EXAMPLE 3   *Marginal Tax Computations*

Calculate the tax owed by each of the following people using the 1997 rates.
a)   Diedre is single; her *taxable income* was $67,000.
b)   Robert is a head of household taking care of two dependent children; his *taxable income* also was $67,000.

Solution:

a) We calculate Diedre's taxes using the *single* rates in Table 5.2. She owes 15% on the first $24,650 of her taxable income, 28% on her taxable income between $24,651 and $59,750, and 31% on her taxable income above $56,551:

$$\underbrace{(15\% \times \$24,650)}_{\substack{\text{15\% marginal rate} \\ \text{on first \$24,650} \\ \text{of taxable income}}} + \underbrace{(28\% \times [\$59,750 - \$24,650])}_{\substack{\text{28\% marginal rate on taxable income} \\ \text{between \$24,651 and \$59,750}}} + \underbrace{(31\% \times [\$67,000 - \$59,750])}_{\substack{\text{31\% marginal rate on her taxable} \\ \text{income above \$59,750}}}$$

$$= \$3697.50 + \$9828 + \$2247.50$$
$$= \$15,773$$

Her tax is $15,773.

b)   We calculate Robert's taxes using the *head of household* rates:

$$\underbrace{(15\% \times \$33,050)}_{\substack{\text{15\% marginal rate} \\ \text{on first \$33,050} \\ \text{of taxable income}}} + \underbrace{(28\% \times [\$67,000 - \$33,050])}_{\substack{\text{28\% marginal rate on his taxable} \\ \text{income above \$33,050.}}} = \$4957.50 + \$9506 = \$14,464$$

His tax is $14,464.   ■

By the Way ··············
The IRS saves you the trouble of calculating your own taxes by providing tax tables that tell you what you owe. However, understanding marginal rates is crucial for understanding how additional income or deductions will affect your finances.

> **TIME-OUT TO THINK:**   Note that Robert and Diedre had the same *taxable income* in **Example 3**. Does this also mean that they had the same *gross income*? Why or why not?

EXAMPLE 4   *The Marriage Penalty*

In 1997, Jessica and Frank earned *adjusted gross incomes* of $42,000 and $34,000, respectively. Calculate their combined tax if Jessica and Frank are *engaged* to be married. Calculate their tax if they already are married and file jointly. Assume they take the standard deductions.

Note: In 1997, the standard deduction was $4150 for a single person and $6900 for a married couple filing jointly. Both Jessica and Frank were entitled to a $2650 personal exemption, regardless of their filing status.

Solution: Tax-wise, being *engaged* means *still single*. Thus Frank and Jessica would file taxes separately as single individuals. Each would have a taxable income found by subtracting the $2650 personal exemption and $4150 standard deduction from their respective adjusted gross incomes:

$$\text{Jessica's taxable income} = \$42,000 - \$2650 - \$4150 = \$35,200$$
$$\text{Frank's taxable income} = \$34,000 - \$2650 - \$4150 = \$27,200$$

We now calculate their taxes using the *single* marginal rates in Table 5.2.

$$\text{Jessica: } (15\% \times \$24,650) + (28\% \times [\$35,200 - \$24,650]) = \$6651.50$$
$$\text{Frank: } (15\% \times \$24,650) + (28\% \times [\$27,200 - \$24,650]) = \$4411.50$$

If they are engaged, their combined taxes are

$$\$6651.50 + \$4411.50 = \$11,063.$$

If Jessica and Frank are married, we compute tax on their combined *adjusted gross income* of

$$\$42,000 + \$34,000 = \$76,000.$$

We get their taxable income by subtracting two personal exemptions of $2650 each and the standard deduction of $6900 for a married couple filing jointly:

$$\text{Combined taxable income} = \$76,000 - (2 \times \$2650) - \$6900 = \$63,800$$

Finally, we calculate their joint taxes using the marginal rates for *married filing jointly* in Table 5.2:

$$\text{Joint tax: } (15\% \times \$41,200) + (28\% \times [\$63,800 - \$41,200]) = \$6180 + \$6328 = \$12,508$$

Note that their combined tax bill is greater if they are married than if they are still single by

$$\$12,508 - \$11,063 = \$1,445.$$

This "extra" tax paid by married couples is sometimes referred to as the **marriage penalty**. ∎

• Web • Watch •
As this book goes to press, Congress is considering legislation to eliminate the marriage penalty. Look for updates on the marriage penalty on the book web site.

> **TIME-OUT TO THINK:**   Do you think the marriage penalty makes sense as part of the tax code? If so, why? If not, what do you think would make a better policy?

Social Security and Medicare Taxes

In addition to taxes computed with the marginal rates, some income is subject to Social Security and Medicare taxes, which are collected under the obscure name of **FICA** (Federal Insurance Contribution Act) taxes. Taxes collected under FICA are used to pay Social Security and Medicare benefits, primarily to people who are retired.

FICA applies only to income from wages (including tips) and self-employment; it does not apply to income from such things as interest, dividends, or profits from sales of stock. In 1997, the FICA tax rates for individuals who are *not* self-employed were:

- 7.65% on the first $65,000 of income from wages
- 1.45% on any income from wages in excess of $65,000

In addition, the individual's employer is required to pay matching amounts of FICA taxes.

Individuals who are *self-employed* must pay both the employee and employer shares of FICA; thus the rates for self-employed individuals are double the rates paid by individuals who are not self-employed:

- 15.3% on the first $65,000 of income
- 2.9% on any income in excess of $65,000.

Note that FICA is calculated on *all* wages, tips, and self-employment income; that is, you may not subtract any adjustments, exemptions, or deductions when calculating FICA taxes.

By the Way ············
The portion of FICA going to Social Security is called OASDI (Old Age, Survivors, and Disability Insurance). The portion going to Medicare is called HI (Hospital Insurance).

By the Way ············
In 1995, 17% of the U.S. population received Social Security benefits, including 92% of people over age 65. The total cost of the benefits was $340 billion, making Social Security the single largest expense of the U.S. government.

EXAMPLE 5   *FICA Taxes*

In 1997, Jude earned $18,000 from wages and tips from her job waiting tables. Calculate her FICA taxes, and her total tax bill including marginal taxes. What is her overall tax rate on her $18,000 income? Assume she is single and takes the standard deduction.

Solution: Jude's entire income of $18,000 is subject to the 7.65% FICA tax:

FICA tax: $7.65\% \times \$18,000 = \$1377$

To find her federal tax, we first subtract her $2650 personal exemption and $4150 standard deduction to get her taxable income:

taxable income $= \$18,000 - \$2650 - \$4150 = \$11,200$

This income is taxed at the 15% marginal rate.

federal tax $= 15\% \times \$11,200 = \$1680$

Her FICA tax and federal tax total

$$\$1377 + \$1680 = \$3057,$$

which is

$$\frac{\$3057}{\$18,000} = 0.170 = 17.0\%$$

of her gross income. Thus her overall tax rate is 17.0%.   ∎

EXAMPLE 6   *Self-Employment Tax*

In 1997, Kevin earned $61,000 as a self-employed computer consultant. Calculate his FICA taxes and total tax bill including marginal taxes. What is his overall tax rate? Note that a special rule allows one-half of the self-employment tax to be taken as an *adjustment* to gross income. Assume he is single and takes the standard deduction, and has no other adjustments to gross income.

Solution: Kevin's gross income of $61,000 is all subject to the 15.3% self-employed FICA tax.

$$\text{self-employed FICA tax: } 15.3\% \times \$61,000 = \$9333$$

Half of this $9333, or $4667 can be taken as an adjustment to his gross income.

$$\text{adjusted gross income} = \$61,000 - \$4667 = \$56,333$$

We get his taxable income by subtracting the $2650 personal exemption and $4150 standard deduction from his adjusted gross income:

$$\text{taxable income} = \$56,333 - \$2650 - \$4150 = \$49,533$$

Using the *single* marginal rates, we find his federal tax is

$$(15\% \times \$24,650) + (28\% \times [\$49,533 - \$24,650]) = \$10,665.$$

His FICA and federal taxes add to

$$\$9333 + \$10,665 = \$19,998,$$

which is

$$\frac{\$19,998}{\$61,000} = 0.3278 = 32.78\%$$

of his gross income. Thus his overall tax rate is 32.78%.   ∎

Special Tax Treatment for Capital Gains

Not all income is created equal, at least not in the eyes of the tax collector! If you buy an item (such as property or a stock) at one price and later sell it at a higher price, the profit is called a **capital gain**. Most other types of income fall into the category of **ordinary income**, including wages, interest, dividends, and rent from real estate that you own.

Moreover, capital gains are divided into two subcategories as of 1998:

- **short-term capital gains** are profits on items sold within 18 months of their purchase.

By the Way ·············
The rationale behind a lower tax on capital gains is that it encourages investment in new businesses and products that involve risk on the part of the investor. Beginning in 2001, the government plans even lower rates for capital gains on items held 5 years or more.

- **long-term capital gains** are profits on items held for more than 18 months before being sold.

Long-term capital gains receive special tax treatment. As of 1997, they are taxed at a maximum marginal rate of 20%—even if Table 5.2 shows that you are in a higher marginal tax bracket. (If your marginal rate would otherwise be 15%, your capital gains are taxed at a rate of 10%.)

EXAMPLE 7   *Capital Gains Income*

In 1997, Serena was single and her gross income consisted solely of $360,000 in long-term capital gains. She is eligible to take $10,000 in itemized deductions. How much tax did she owe? What is her overall tax rate? *Note:* because of special rules that take effect for people with high incomes, she cannot claim any personal exemption.

Solution: Because her income is not from wages, she owes no FICA tax. She has no adjustments to her gross income, so subtracting her $10,000 in itemized deductions from her gross income of $360,000 makes her taxable income $350,000. Her marginal tax is 15% on the first $24,650 of this income, and the rest of her income is taxed at the maximum 20% rate for long-term capital gains—not at the higher rates that would apply to ordinary income. Thus her tax is

$$(15\% \times \$24{,}650) + (20\% \times [\$350{,}000 - \$24{,}650]) = \$68{,}768.$$

Her overall tax rate is

$$\frac{\$68{,}768}{\$360{,}000} = 0.1910 = 19.10\%$$

of her gross income.   ∎

TIME-OUT TO THINK: Note that Serena in Example 7 has an income that is more than five times higher than Kevin in Example 6. Compare their tax payments and overall tax rates. Who pays more tax? Who pays a higher tax *rate*? Do you think these facts have any important social implications? Explain.

TAX-DEFERRED INCOME

The tax code tries to encourage long-term savings by allowing you to *defer* income taxes on contributions to certain types of savings plans, called **tax-deferred savings plans**. In other words, any money that you deposit into such savings plans is not taxed now; instead, it will be taxed only at some time in the future when you withdraw the money for use.

Tax-deferred savings plans go by a variety of names, such as *Individual Retirement Accounts* (IRAs), *Qualified Retirement Plans* (QRPs), *401(k) Plans*, and more. All are subject to strict rules; for example, you generally are not allowed to withdraw money from any of these plans until you reach age $59\frac{1}{2}$. Anyone can set up a tax-deferred savings plan, and *you should*, regardless of your current age. Why? Because they offer two key advantages in saving for your long-term future.

By the Way ....
As of early 1998, Congress is considering a reduction in the time distinction between short- and long-term capital gains from 18 months to 12 months.

By the Way ....
For people with incomes above certain *threshold* amounts, a set of *phaseout rules* limits the amount that can be deducted both for personal exemptions and itemized deductions.

By the Way ....
Some investments are *tax-exempt*—you never have to pay tax on the earnings. Some government bonds offer tax-exempt interest; however, the interest rates on these bonds usually are lower than on ordinary bonds.

First, contributions to tax-deferred savings plans count as *adjustments* to your present gross income, and thus are not part of your taxable income. As a result, the contributions cost you less than contributions to savings plans without special tax treatment. For example, suppose that you earn enough money to put you in the 28% marginal tax bracket. If you deposit $100 in an ordinary savings account, your tax bill is unchanged and your spendable income goes down by $100. However, if you deposit $100 in a tax-deferred savings account, you do not have to pay tax on that $100. With your 28% marginal rate, you therefore save

$$28\% \times \$100 = \$28$$

in taxes. In other words, you end up with $100 in the tax-deferred savings account, but your *spendable income* goes down by only

$$\$100 - \$28 = \$72.$$

The second advantage of tax-deferred savings plans is that their interest *also* is tax-deferred. With an ordinary savings plan, you have to pay taxes on the interest earned each year, which effectively comes out of the interest you earn. With a tax-deferred savings plan, *all* of the compounded interest accumulates from one year to the next. Over many years, this tax savings makes the value of tax-deferred savings accounts rise much more quickly than ordinary savings accounts (Figure 5.4).

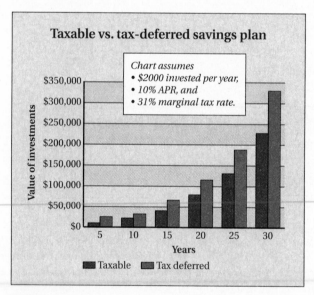

FIGURE 5.4

EXAMPLE 8   *Tax-Deferred Savings Plan*

Suppose that you are single and have a taxable income of $75,000, and make monthly payments of $500 to a tax-deferred savings plan. How do the tax-deferred contributions affect your monthly take-home pay?

Solution: Table 5.2 shows that your marginal tax rate is 31%. Thus each $500 contribution to a tax-deferred savings plan will reduce your tax bill by

$$31\% \times \$500 = \$155.$$

In other words, $500 will go into your tax-deferred savings account each month, but your monthly paychecks will go down by only

$$\$500 - \$155 = \$345.$$

Note that the special tax treatment makes it significantly easier for you to afford the monthly contributions needed to build your retirement fund. ∎

By the Way ·············
This example assumes that the tax withheld from your monthly paychecks is correctly adjusted for the effects of your tax-deferred contributions.

EXAMPLE 9   *Comparative Contributions*

Sara and Jim each make tax-deferred contributions of $100 per month to their pension plans. Sara, who has a relatively high income, is in a tax bracket with a 36% marginal tax rate. Jim earns much less and is in the 15% marginal tax rate bracket. How much is each person's take-home (aftertax) pay reduced by the $100 pension contribution?

Solution: Both Sara and Jim see their gross pay drop by $100. However, because Sara is in the 36% marginal bracket, her $100 contributions save her

$$36\% \times \$100 = \$36$$

in taxes. Therefore, her contributions reduce her take-home paychecks by

$$\$100 - \$36 = \$64.$$

Because Jim is in the 15% marginal bracket, his contributions save him only

$$15\% \times \$100 = \$15.$$

Thus his take-home pay checks are reduced by

$$\$100 - \$15 = \$85.$$

Note that the effective cost to Jim for adding $100 to his retirement plan is $85, while the effective cost to Sara is only $64. This illustrates the fact that deductions are more valuable to people in higher marginal tax brackets. ∎

THE MORTGAGE INTEREST TAX DEDUCTION

The most valuable tax deduction for most people is the **mortgage interest tax deduction**. If you itemize your deductions (rather than taking the standard deduction), all *interest* that you pay on a mortgage is tax deductible. Because mortgage payments typically are mostly interest, at least during the early years of a 15- or 30-year loan, this deduction can save you a lot in taxes.

For example, suppose that you take out a mortgage of $100,000 to buy a home. At an *APR* of 8%, you will pay *approximately*

$$8\% \times \$100,000 = \$8,000$$

By the Way ·············
About 65% of American families own their homes (1997 data).

in interest during the early years when your principal remains close to $100,000. This $8000 of interest is tax deductible; if you are single, it is nearly $4000 larger than the standard deduction by itself—and you may be able to claim other itemized deductions, such as charitable contributions and state taxes. The value of this deduction could certainly influence a decision of whether to rent or to buy a home. Indeed, the purpose of the mortgage interest tax deduction is to encourage home ownership.

EXAMPLE 10   *Mortgage Tax Savings*

Suppose that you are single and have a taxable income of $50,000 per year. If you take out a 30-year mortgage of $100,000 at an *APR* of 8%, approximately how much will you save in taxes? What is the effective cost of your monthly payments? Assume you have no other itemized deductions, and use the 1997 value of the standard deduction ($4150).

Solution: At an *APR* of 8%, you will pay *approximately*

$$8\% \times \$100,000 = \$8,000$$

By the Way ....
The money you use for a down payment could otherwise have been earning interest somewhere, a fact that is not considered in this analysis.

in interest during the early years when your principal remains close to $100,000. This $8000 of interest is tax deductible; if you are single, it is $3850 larger than the $4150 standard deduction.

Your taxable income of $50,000 puts you in the 28% marginal bracket. Thus the additional $3850 deduction saves you

$$28\% \times \$3850 = \$1078$$

in taxes over the year—which is a monthly savings of nearly $90!

In Example 5 of Unit 2C, we calculated that the monthly payments on this loan would be about $734. Because of the roughly $90 in tax savings, these payments will reduce the rest of your spendable income by only about $644. ∎

TIME-OUT TO THINK: Suppose that you are trying to decide whether to rent or buy a home. You found a house that you can afford with the loan in Example 10. You've also found an apartment you like that rents for $700 per month. Which option will cost you less money? Why?

TAX REALITIES

Our brief overview of the tax system might make taxes seem fairly straightforward, but the devil is in the details of the thousands of pages of tax code. Here are just a few examples of the complexities that can arise:

- People at high incomes are not permitted to take full advantage of the exemptions allowed to most taxpayers.

- Some itemized deductions are allowed only in the amount by which they exceed some percentage of your income; for example, only medical expenses that exceed 7% of your adjusted gross income are deductible.

- Some people with many deductions must calculate their taxes with the *alternative minimum tax*, rather than the normal marginal rates.

The hardest thing in the world to understand is the income tax.

—ALBERT EINSTEIN

- If you receive royalties on an invention, whether you must pay the self-employed FICA tax depends on whether you created the invention as a hobby or as a job.

The complexity of our tax system leads many people to call for *tax simplification*. Some argue that the current system of progressive tax rates should be replaced with a single, *flat rate* for everyone. Others have suggested replacing federal income tax by a federal sales tax. However, proposals to change the tax code invariably lead to huge political battles. Because the government needs a certain amount of revenue to meet its expense obligations, changes in the tax code generally mean that some people would end up with larger tax bills while others would get a tax cut. Those who will have to pay more are certain to object loudly.

Finally, it's worth remembering that even the simplest tax system can be complex because of the difficulty in defining the term *income*. For example, should gifts you receive be counted as income? How about benefits such as the cost of medical insurance provided by your employer? Should you have to count money that you spend on uniforms for your job as income? Should a self-employed person be able to count a trip to Hawaii as a business expense? If you look carefully at the issue of what should count as income and what should not, you may be quite surprised at the difficulties involved.

REVIEW QUESTIONS

1. Explain the basic process of calculating income taxes, as shown in Figure 5.3.

2. Explain the difference between *gross income, adjusted gross income*, and *taxable income*.

3. What is the difference between a *tax deduction* and a *tax credit*? Why is a tax credit more valuable?

4. What do we mean by a *progressive income tax*? Explain the use of *marginal tax rates* in calculating taxes.

5. What are FICA taxes? Are FICA taxes an example of a progressive tax?

6. What is the difference between a *capital gain* and ordinary income? How are capital gains treated differently by the tax code?

7. Explain how you can benefit from a *tax-deferred savings plan*. Explain why the benefits are greater for people with higher marginal tax rates.

8. What is the *mortgage interest tax deduction*? Explain how this deduction can make it cheaper for you to own a house with $800 monthly payments than to rent an $800 per month apartment.

PROBLEMS

Tax Calculations. *In Problems 1–4, compute the individual's (or couple's) gross income, adjusted gross income, and taxable income. Assume 1997 values for the personal exemption ($2650) and standard deductions ($4150 single, $6900 married filing jointly, $3450 married filing separately, and $6100 head of household).*

1. Suzanne is single and earned wages of $33,200. She received $350 in interest from a savings account. She contributed $500 to a tax-deferred retirement plan. She had $450 in itemized deductions from charitable contributions.

2. Malcolm is single and earned wages of $23,700. He had $4300 in itemized deductions from interest on a house mortgage.

3. Wanda is married, but she and her husband filed separately. Her gross salary was $35,400 and she earned $500 in interest. She had $1500 in itemized deductions and claimed three exemptions for herself and two children.

4. Emily and Bret are married and filed jointly. Their combined wages were $75,300. They also earned a net of $2000 from a rental property they own, and received $1650 in interest. They claimed four exemptions for themselves and two children. They contributed $3240 to their tax-deferred retirement plans, and their itemized deductions totaled $9610.

Tax Credits and Tax Deductions. *In Problems 5–10, state how much each individual or couple saves in taxes with the tax credit or tax deduction specified.*

5. Karen and Tremaine are in the 28% marginal tax bracket and claim the standard deduction. How much will their tax bill be reduced if they qualify for a $500 tax credit?

6. Vanessa is in the 36% marginal tax bracket and itemizes her deductions. How much will her tax bill be reduced if she qualifies for a $500 tax credit?

7. Lisa is in the 15% marginal tax bracket and claims the standard deduction. How much will her tax bill be reduced if she makes a $1000 contribution to charity?

8. Howard is in the 15% marginal tax bracket and itemizes his deductions. How much will his tax bill be reduced if he makes a $1000 contribution to charity?

9. Sebastian is in the 28% marginal tax bracket and itemizes his deductions. How much will his tax bill be reduced if he makes a $1000 contribution to charity?

10. Samantha is in the 39.6% marginal tax bracket and itemizes her deductions. How much will her tax bill be reduced if she makes a $1000 contribution to charity?

Marginal Tax Calculations. *Use the 1997 marginal tax rates in Table 5.2 to compute the taxes owed in Problems 11–18.*

11. Gene is single and had a taxable income of $35,400.

12. Sarah and Marco are married filing jointly with a taxable income of $87,500.

13. Bobbi is married filing separately with a taxable income of $77,300.

14. Abraham is single with a taxable income of $23,800.

15. Paul is a head of household with a dependent child and a taxable income of $89,300. He also is entitled to a $500 tax credit.

16. Pat is a head of household with a dependent child and a taxable income of $57,000. He also is entitled to a $500 tax credit.

17. Winona and Jim are married filing jointly with a taxable income of $105,500. They also are entitled to a $1000 tax credit.

18. Chris is married filing separately with a taxable income of $127,300.

19. **Marriage Penalty.** Joan and Peter plan to get married. Assume they know their adjusted gross incomes will be $44,500 and $33,400, respectively. Compute their total taxes under the following assumptions, using the 1997 tax rates in Table 5.2. Assume they claim the standard deductions.

   a. They get married on December 31 of this year and file jointly as a married couple.

   b. They get married on January 1 of next year and file individually as single persons.

20. **Marriage Penalty.** Daniel and Paula will be married near the end of the year. Assume they know their adjusted gross incomes will be $32,500 and $29,400, respectively. Compute their total taxes under the following assumptions, using the 1997 tax rates in Table 5.2. Assume they claim the standard deductions.

   a. They get married on December 31 of this year and file jointly as a married couple.

   b. They get married on January 1 of next year and file individually as single persons.

FICA Taxes. *In Problems 21–26, compute (a) how much each individual owes in FICA taxes and in federal incomes taxes and (b) what percent of the person's gross income is paid to federal taxes (FICA and income tax combined). Use the 1997 marginal tax rates in Table 5.2. Assume all individuals are single and take the 1997 standard deduction of $4150. (Hint: For self-employed individuals, remember that half of any self-employment tax can be taken as an adjustment to gross income.)*

21. Lars earned $28,000 from wages as a computer programmer and made $2500 in tax-deferred contributions to a retirement fund.

22. Juliette earned $34,500 in salary, $750 in interest, and made $3000 in tax-deferred contributions to a retirement fund.

23. Jack earned $44,800 in salary, $1250 in interest, and made $2000 in tax-deferred contributions to a retirement fund.

24. Henry earned $102,400 in salary, $4450 in interest, and made $9500 in tax-deferred contributions to a retirement fund.

25. Brittany is self-employed and earned $48,200 from her business. She had no other income and made no contributions to retirement plans.

26. Larae is self-employed and earned $68,200 from her business. She also earned $800 in interest and made $2500 in tax-deferred contributions to a retirement fund.

Capital Gains vs. Ordinary Income. *In Problems 27–29, (a) Calculate the total tax owed by each of the two people, including both FICA and income taxes and (b) compare the percentage of gross income that the two people pay to federal taxes (FICA and income tax combined). Use the 1997 marginal tax rates in Table 5.2, along with the maximum 20% rate for long-term capital gains. Assume all individuals are single and take the 1997 standard deduction of $4150. As a simplifying assumption, you may ignore the personal exemption in all cases (so that you will not have to deal with the phaseout rules).*

27. Pierre earned $120,000 in wages. Katharine earned $120,000, all as long-term capital gains.

28. Luis is self-employed and earned $55,000 from his business. Gina earned $55,000, all as long-term capital gains.

29. Fred is self-employed and earned $275,000 from his business. Tamara earned $275,000, all as long-term capital gains.

Tax-Deferred Savings Plans. *In Problems 30–33, calculate the effect on annual take-home pay of the tax-deferred contributions described. Use the 1997 tax rates in Table 5.2.*

30. Suppose that you are single and have a taxable income of $18,000. How will your take-home pay be affected if you make monthly contributions of $400 to a tax-deferred savings plan?

31. Suppose that you are single and have a taxable income of $45,000. How will your take-home pay be affected if you make monthly contributions of $600 to a tax-deferred savings plan?

32. Suppose that you are single and have a taxable income of $90,000. How will your take-home pay be affected if you make monthly contributions of $800 to a tax-deferred savings plan?

33. Suppose that you are single and have a taxable income of $150,000. How will your take-home pay be affected if you make monthly contributions of $800 to a tax-deferred savings plan?

34. **Comparing Tax-Deferred Contributions.** Adrienne and Alphonse each make tax-deferred contributions of $200 per month to their pension plans. Adrienne is in a tax bracket with a marginal tax rate of 28%. Alphonse's marginal tax rate is 15%. By how much do the $200 pension contributions reduce each person's take-home pay? Discuss.

35. **Mortgage Tax Savings.** Suppose that you are single and have a taxable income of $55,000 per year. If you take out a 30-year mortgage of $120,000 at an *APR* of 8%, approximately how much will you save in taxes? Assume you have no other itemized deductions, and use the 1997 value of the standard deduction ($4150).

36. **Mortgage Tax Savings.** Suppose that you are single and have a taxable income of $105,000 per year. If you take out a 30-year mortgage of $120,000 at an *APR* of 8%, approximately how much will you save in taxes? Assume you have no other itemized deductions, and use the 1997 value of the standard deduction ($4150).

37. **Home Equity Loan.** Suppose that you are in the 28% marginal tax bracket. You have $4000 in debt on a credit card at an *APR* of 18%.

    a. How much do you owe in *monthly* interest on the $4000 debt?

    b. Suppose that you own your home, and take out a *home equity loan* with an *APR* of 10% to pay off your credit card debt. With this new loan, how much do you owe in *monthly* interest on the $4000 debt?

    c. Interest paid on home equity loans is tax deductible. How much will you save in taxes (per month) because of the interest owed on the home equity loan? What is the effective cost of the monthly interest on this loan? Discuss the advantages of converting your credit card debt to a home equity loan.

38. Home Equity Vs. Car Loan. Suppose that you are in the 28% marginal tax bracket, and you need a loan of $10,000 to buy a new car. You have a choice between an auto loan at an interest rate of 8%, or a home equity loan at an interest rate of 10%. Interest on the home equity loan is tax deductible, but interest on the auto loan is not deductible.

    a. Which loan should you take? Why?

    b. Would your decision be different if you were in the 15% marginal tax bracket? Explain.

39. Project: Tax Fairness Issues. Tax policies invariably raise questions of fairness. For example, is it fair to tax married couples at higher rates than single individuals? Is it fair to tax capital gains at a different rate than ordinary income? Are tax deductions fair, considering that they yield greater tax savings for people in higher tax brackets? Discuss the issue of fairness on one of these questions, or on some other tax question. If possible, hold a class debate on the issue. Alternatively, write a short essay stating and defending your opinion on the issue.

40. Project: *Your* Taxes. Estimate your gross income for this year and any itemized deductions you expect, then calculate your adjusted gross income and taxable income using the 1997 rates (or rates for the current year, if you can find them).

    a. Based on your estimates, how much tax will you owe this year?

    b. How much (if any) tax is being withheld from your paychecks each month?

    c. Should you expect a tax refund next year? Explain.

    d. Suppose you begin making a $100 monthly contribution to a tax-deferred retirement plan. How will it affect your take-home pay?

    e. Suppose you make a $1000 contribution to charity. By how much, if at all, will this contribution reduce your tax bill?

UNIT 5E

INVESTMENTS

So far in this chapter, we've looked at the principles behind the growth of investments. We've seen that the power of compound interest, along with the tax advantages of *tax-deferred savings plans*, makes it a good idea for working people to save and invest. The hard part comes in choosing *where* to invest.

    Most investments fall into one of three basic categories.

- **Stocks** (or *equities*) give you a share of ownership in a company whose stock you hold.

- **Bonds** (or *debt*) make you the lender to a government or corporation that makes its loan payments to you.

- **Cash** investments include money you deposit into bank accounts, certificates of deposit (CD), and U.S. Treasury bills.

Beware! There are many other investment types besides the basic three, such as rental properties, precious metals, commodities futures, and stock options. These investments are higher risk than the basic three, and you should not invest in them without spending plenty of time understanding how they work.

    You can make investments yourself by purchasing individual stocks or bonds, or investing cash directly in banks or U.S. Treasury bills. (However, you may have to make your investment through a *broker* who will charge a commission.) Or, you can invest through a **mutual fund**, in which you hand over your money to a professional investor who makes the investments for you. There are *stock mutual funds* that invest primarily in stocks, *bond mutual funds* that invest primarily in bonds, *money market funds* that invest only in cash, and *diversified funds* that invest in a mixture of stocks, bonds, and cash. (Mutual funds generally charge a fee for managing your investments.)

INVESTMENT CONSIDERATIONS: LIQUIDITY, RISK, AND RETURN

No matter what type of investment you make, you should evaluate the investment in terms of three general considerations.

- How difficult is it to take out your money? An investment from which you can withdraw money easily, such as an ordinary bank account, is said to be **liquid**; the **liquidity** of an investment like real estate is much lower because real estate can be difficult to sell.

- Is your investment principal at **risk**? The safest investments are federally insured bank accounts and U.S. Treasury Bills; there's virtually no risk of losing the money you've invested. Stocks and bonds are much riskier because they can drop in value.

- How much can you expect your investment to grow? The percentage increase in an investment is called its **return**. For example, a bank account with an annual percentage yield (*APY*) of 5% means a 5% annual return. In general, riskier investments offer the prospects of higher returns—along with the possibility of losing your principal.

By the Way ⋯⋯⋯⋯

The U.S. Treasury issues *Bills, Notes,* and *Bonds. Treasury Bills* are essentially cash investments that are highly liquid and very safe. *Treasury Notes* are essentially bonds with 2- to 10-year terms, and *Treasury Bonds* have 20- to 30-year terms.

EXAMPLE 1   *Liquidity, Risk, and Return*

Evaluate each of the following investments in terms of liquidity, risk, and return:
a) A U.S. Treasury bond with a 30-year term and a 7% interest rate.
b) Shares of stock in a large, established company with more than $50 billion in annual revenue.
c) Shares of stock in a new, small company that is testing a potential cancer drug, but currently has no revenues because no products are for sale.
d) A 5-year bond with a 15% interest rate issued by a company that is on the verge of bankruptcy.

Solution:
a) When you buy a U.S. Treasury bond, you are lending money to the U.S. government. The 30-year term means that the bond is not liquid at all—it will be 30 years before you can get your principal out, unless you sell the bond to someone else before-hand. The U.S. Treasury bond is a very safe investment because it is backed by "the full faith and credit" of the United States. The annual return is fixed at 7%.
b) Large company stocks are fairly liquid because they are easy to sell to others through *stockbrokers*. All stocks are inherently risky, but stock in a large established company *usually* is among the lowest risk stock because it's rare that such a company goes out of business or suddenly plummets in value. If the company does well, you can expect a decent return on your investment. Of course, just as it is rare for a large company stock to plunge in value, a dramatic rise also is unlikely.
c) The small company stock may be more difficult to sell than a large company stock, making it somewhat less liquid. This stock also is extremely risky: if the cancer drug proves ineffective, the company and its stock may be worthless. On the other hand,

if the drug succeeds, the company may flourish, providing a huge return on your investment.

d)   The 15% annual interest rate represents a high return compared to U.S. Treasury bonds, but the company's poor financial health also makes it very risky: If the company goes bankrupt and does not recover, you'll never get your money back. This 5-year bond also has poor liquidity because the company is not obligated to pay you back for 5 years and the bond's risky nature may make it difficult to sell in the meantime.   ■

Total and Annual Return

Suppose that you invest $1000 and its value grows to $1500 in five years. Your **total return** is the percent change in value of your investment over the five-year period:

$$\text{total return} = \frac{\text{new value} - \text{previous value}}{\text{previous value}} = \frac{\$1500 - \$1000}{\$1000} = 0.5 = 50\%$$

In other words, your investment grew in value by 50% over five years.

Although total returns are easy to calculate, it's much easier to compare investments by looking at their **annual returns**, which describe the percentage change *per year* in investment values. Once we know the total return, we calculate the annual return by finding the annual percentage rate (*APR*) that *would have* given the same total return in the same amount of time (assuming annual compounding). For example, we can use the compound interest formula (for compounding once a year) to confirm that an *APR* of 8.5% will make an investment of $P = \$1000$ grow to about $1500 in $Y = 5$ years:

$$A = P(1 + APR)^Y = \$1000 \, (1 + 0.085)^5 = \$1503.66$$

Using Your **Calculator**

Raising a number to the $1/Y$ power is the same as taking the Yth root. The key for taking a root will be labeled something like $\sqrt[x]{\ }$ or $x^{1/y}$. For example, calculate $\sqrt[4]{2.8}$ by pressing 2.8 $\sqrt[x]{\ }$ 4 $=$.

Thus an investment with a total return of 50% in five years represents an *annual return* of about 8.5%. The easiest way to find the annual return from the total return is with a simple formula.

Annual Return Formula (from Total Return)

$$\text{annual return} = (\text{total return} + 1)^{1/Y} - 1,$$

where Y is the number of years the investment was held.

Let's use the formula to confirm our example above. The total return was $50\% = 0.5$ over $Y = 5$ years, so the formula gives

$$\text{annual return} = (0.5 + 1)^{1/5} - 1 = \sqrt[5]{1.5} - 1 = 0.0845 = 8.45\%.$$

Note that this is quite close to the 8.5% that we estimated above.

EXAMPLE 2   *Annual Return*

Suppose that, over a period of four years, a $3000 investment grows in value to $8400. What are the total and annual returns over the four-year period?

Solution: The total return over the four years is

$$\text{total return} = \frac{\text{new value} - \text{previous value}}{\text{previous value}} = \frac{\$8400 - \$3000}{\$3000} = 1.8 = 180\%.$$

The annual return is

$$\begin{aligned}
\text{annual return} &= (\text{total return} + 1)^{1/Y} - 1 \\
&= (1.8 + 1)^{1/4} - 1 \\
&= \sqrt[4]{2.8} - 1 \\
&= 0.294 = 29.4\%.
\end{aligned}$$

In other words, you would have gotten the same 180% total return in four years by **investing** at a fixed *APR* of 29.4%, compounded once each year. ■

EXAMPLE 3   *Investment Loss*

Suppose that you purchased shares in New Corp. for $2000. Three years later, **you sold** them for $1100. What were your total return and annual returns on this investment?

Solution: The total return over the three years was

$$\text{total return} = \frac{\text{new value} - \text{previous value}}{\text{previous value}} = \frac{\$1100 - \$2000}{\$2000} = -0.45 = -45\%.$$

The total return is negative because you *lost* money on this investment. Your **annual return** over the three-year period was

$$\begin{aligned}
\text{annual return} &= (\text{total return} + 1)^{1/Y} - 1 \\
&= (-0.45 + 1)^{1/3} - 1 \\
&= \sqrt[3]{0.55} - 1 \\
&= -0.18 = -18\%.
\end{aligned}$$

That is, your investment *lost* an average of 18% per year in value. ■

How Should *You* Invest?

Your financial *portfolio* is essentially a listing of all the investments that you hold. **If you** could be sure of future returns, putting together a financial portfolio would **be easy.** Unfortunately, the future is unpredictable.

Nevertheless, historical trends can offer at least some guidance. Table 5.3 **on the following** page shows the average annual returns for several different types of **investments** over a 70-year period from 1926–1995. By looking at both the *average annual return* **and** *worst year* columns, you can see that the highest average returns came from **investments** with the highest risk.

Because of the trade-off between risk and return, most financial advisors **recommend** a **diversified portfolio**; that is, a portfolio with a mixture of low-risk and **high-risk** investments. The particular mix that is appropriate for you depends on **many factors,** including your age, your job security, and your family status. As a result, it is very important that you develop your investment portfolio with care, and preferably with plenty of advice from financial professionals.

Table 5.3 Returns on Different Investment Categories, 1926–1995			
Investment Type	Average Annual Return*	Best Year	Worst Year
Small company stocks	12.5%	142.9% (1933)	−58.0% (1937)
Large company stocks	10.5%	54.0% (1933)	−43.3% (1931)
Long-term corporate bonds	5.7%	42.6% (1982)	−8.1% (1969)
Cash (U.S. Treasury Bills)	3.7%	14.7% (1981)	−0.02% (1938)

*Includes both increases in price and any dividends or interest.
Source: K. M. Morris and A. M. Siegel, *The Wall Street Journal Guide to Understanding Personal Finance*, New York: Lightbulb Press, 1997, p. 121.

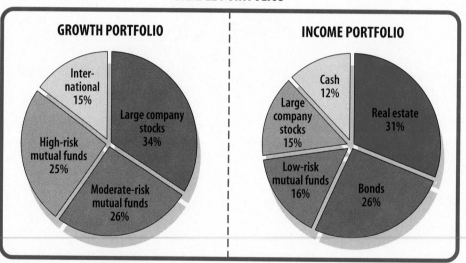

SAMPLE PORTFOLIOS

GROWTH PORTFOLIO

- International 15%
- Large company stocks 34%
- High-risk mutual funds 25%
- Moderate-risk mutual funds 26%

INCOME PORTFOLIO

- Cash 12%
- Large company stocks 15%
- Real estate 31%
- Low-risk mutual funds 16%
- Bonds 26%

EXAMPLE 4   *Historical Returns*

Suppose that your great-grandmother invested $1000 each in small company stocks, large company stocks, long-term corporate bonds, and U.S. Treasury bills at the beginning of 1926. Assuming her investments grew at the long-term average annual returns in Table 5.3, approximately how much would each investment be worth at the beginning of 2001?

Solution: We can find the value of each of the four investments by using the compound interest formula, setting the *APR* to the average annual return. In all four cases, the starting principle is $P = \$1000$ and $Y = 75$ years from the beginning of 1926 to the beginning of 2001.

Investment Type	Annual Return	Investment Value: $A = P(1 + APR)^Y$
Small company stocks	12.5% = 0.125	$A = \$1000 \times (1 + 0.125)^{75} = \$6,861,800$
Large company stocks	10.5% = 0.105	$A = \$1000 \times (1 + 0.105)^{75} = \$1,787,200$
Corporate bonds	5.7% = 0.057	$A = \$1000 \times (1 + 0.057)^{75} = \$63,900$
Treasury bills	3.7% = 0.037	$A = \$1000 \times (1 + 0.037)^{75} = \$15,300$

Note the enormous difference that the different choices would have made. The worst of the four $1000 investments would have grown to about $15,000, while the best would be worth close to $7 million! ■

EXAMPLE 5   *Bad Year*

Suppose that you invest $1000 in each of the four categories shown in Table 5.3. If all four categories matched their *worst* year performances, how much would your investments be worth a year later?

Solution: We find the new value of each investment by subtracting the amount of the loss.

Investment Type	Worst Year Loss	Value of $1000 Investment One Year After Worst Year Loss
Small company stocks	−58%	$1000 − $580 = $420
Large company stocks	−43.3%	$1000 − $433 = $567
Corporate bonds	−8.1%	$1000 − $81 = $919
Treasury bills	−0.02%	$1000 − $0.20 = $999.80

Note that, if performance is poor, the higher risk investments lose much more value. ■

THE FINANCIAL PAGES

Most newspapers carry daily financial sections, often with articles reporting on emerging trends and the recommendations of professional investors. Many newspapers also carry data tables showing the performance of various investments. Increasingly, this same information and more can be found on-line through the Internet. Let's look briefly at what you must know to understand commonly published data about stocks, bonds, and mutual funds.

Stocks

A **corporation** is a legal entity set up for the purpose of conducting business. The owners of a corporation hold stock **shares** that reflect their percentage of ownership. For example, suppose a corporation has issued a total of 1 million shares, and you own 10,000 of these shares. In that case, you own $\frac{10,000 \text{ shares}}{1,000,000 \text{ shares}} = 0.01 = 1\%$ of the corporation.

Some corporations are **privately held**, meaning that their shares are owned by only a select group of people. Others are **publicly traded**, meaning that their shares can be traded (bought and sold) by anyone through a public stock exchange. In the United States, the major public stock exchanges are the *New York Stock Exchange* (NYSE), the *American Stock Exchange* (AMEX), and the *Nasdaq Stock Market* (NASDAQ, short for *National Association of Securities Dealers Automated Quotations*).

Stock shares have no intrinsic value; instead, their value is determined by whether shareholders want to keep or sell their stock and on how much other investors are willing to pay to buy the stock. The purpose of a stock market is to make it easy for shares to be traded. Thus the **market price** of a stock reflects the current level at which at least some shareholders are willing to sell and other investors are willing to buy.

In general, there are two ways to make money on stocks.

- You can make money if you sell a stock for more than you paid for it, in which case you have a *capital gain* on the sale of the stock.

- You can make money while you own the stock if the corporation distributes part or all of its profits to stockholders as **dividends**. Each share of stock is paid the same dividend, so the amount of money you receive depends on the number of shares you own. Not all companies distribute profits as dividends; some reinvest all profits within the corporation.

Of course, you also can *lose* money on stocks if you sell shares for *less* than you paid for them, or if the company goes out of business—in which case the shares become worthless.

Daily stock tables provide a wealth of information about stocks, summarized in Figure 5.5. Nevertheless, it pays to get even more information if you are buying stocks. For example, you can learn a lot by studying a company's annual report. Many companies have web sites with information for investors. You can also get independent research reports from many investment services (usually for a fee) or by working with a stockbroker (to whom you pay commissions when you buy or sell stock).

Frank and Ernest

© 1997 Thaves / Reprinted with permission. Newspaper distribution by NEA, Inc.

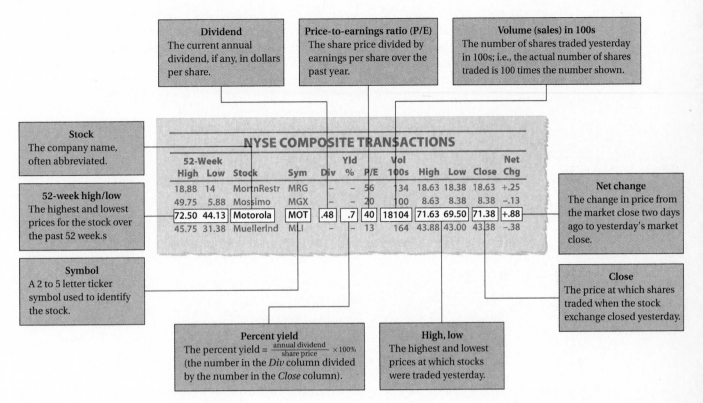

Dividend
The current annual dividend, if any, in dollars per share.

Price-to-earnings ratio (P/E)
The share price divided by earnings per share over the past year.

Volume (sales) in 100s
The number of shares traded yesterday in 100s; i.e., the actual number of shares traded is 100 times the number shown.

Stock
The company name, often abbreviated.

52-week high/low
The highest and lowest prices for the stock over the past 52 week.s

Symbol
A 2 to 5 letter ticker symbol used to identify the stock.

Net change
The change in price from the market close two days ago to yesterday's market close.

Close
The price at which shares traded when the stock exchange closed yesterday.

NYSE COMPOSITE TRANSACTIONS

52-Week High	Low	Stock	Sym	Div	Yld %	P/E	Vol 100s	High	Low	Close	Net Chg
18.88	14	MortnRestr	MRG	–	–	56	134	18.63	18.38	18.63	+.25
49.75	5.88	Mossimo	MGX	–	–	20	100	8.63	8.38	8.38	–.13
72.50	44.13	Motorola	MOT	.48	.7	40	18104	71.63	69.50	71.38	+.88
45.75	31.38	MuellerInd	MLI	–	–	13	164	43.88	43.00	43.38	–.38

Percent yield
The percent yield = $\frac{\text{annual dividend}}{\text{share price}} \times 100\%$ (the number in the *Div* column divided by the number in the *Close* column).

High, low
The highest and lowest prices at which stocks were traded yesterday.

FIGURE 5.5

EXAMPLE 6   *Motorola Stock*

Suppose that Figure 5.5 comes from today's paper.

a)   What is the ticker symbol for the Motorola Corporation?

b)   What was the range of selling prices for Motorola shares yesterday? How do these prices compare to prices over the past year?

c)   What was the closing price of Motorola shares yesterday and two days ago?

d)   How many shares of Motorola were traded yesterday?

e)   Suppose that you own 100 shares of Motorola. What total dividend payment should you expect this year?

f)   Compare what you can expect to earn from dividends alone to a bank account offering a 3% interest rate.

g)   Over the past year, how much profit (earnings) has Motorola made per share? How do the current share prices compare to the profits?

Solution:

a)   We first find Motorola in the table, then find its ticker symbol MOT in the *symbol* column.

b) The *high* and *low* columns show us that, yesterday, Motorola stock traded in the range from $69.50 to $71.63 per share. This is toward the high end of the stock price over the past year, which the *52-week high* and *low* columns tell us has been as low as $44.13 and as high as $72.50 per share.

c) The *close* column shows that Motorola closed at $71.38 per share. The *change* column shows that the share price rose $0.88 from the previous day. Thus its closing price two days ago was $0.88 lower than yesterday's closing price, or

$$\$71.38 - \$0.88 = \$70.50$$

per share.

d) The *volume* column shows that 18,104 *hundreds* of shares traded yesterday, which means the actual number of shares traded was

$$18,104 \times 100 = 1,810,400.$$

e) The *dividend* column shows that Motorola is currently paying dividends at an annual rate of $0.48 per share. If you own 100 shares, your total dividend payment this year will be

$$100 \text{ shares} \times \$0.48/\text{share} = \$48.$$

Note that dividends are usually paid quarterly, and a company may change its dividend rate at any time.

f) The *percent yield* column shows that dividends alone give you an annual return of 0.7%—much lower than the 3% interest rate offered by the bank. However, if the shares appreciate in value, the gain may make Motorola stock a better investment than the bank account.

g) The *PE* column shows that Motorola's price-to-earnings ratio is 40:

$$\text{PE ratio} = \frac{\text{share price}}{\text{earnings per share}} = 40$$

If we multiply both sides by earnings per share and divide both sides by 40 we get

$$\text{earnings per share} = \frac{\text{share price}}{40} = \frac{\$71.38}{40} = \$1.78.$$

(Note that we used the closing price as the share price.) Motorola earned a profit of $1.78 per share of its stock over the past year. The PE ratio tells us that the share price is 40 times higher than the actual earnings per share; in other words, it would take 40 years worth of current profits to equal the current share price. ■

EXAMPLE 7   *Total Return on a Stock*

Suppose that you bought Motorola stock one year ago for $66 per share and sell at the closing price shown in Figure 5.5. Ignoring any commissions, what is the total return on your investment? What is the return if you include a broker's commission of $1 per share on the sale?

Solution: The total return on a stock is the return including *both* price appreciation and dividends. Given your purchase price of $66 and the closing price of $71.38, each share gained $5.38 in value; this is your *capital gain* on each share sold. In addition, each share earned a dividend of $0.48. Thus your total gain per share in absolute terms is

$$\text{total gain} = \text{capital gain} + \text{dividend} = \$5.38 + \$0.48 = \$5.86.$$

In relative terms, your return is

$$\text{percent return} = \frac{\text{gain in value}}{\text{previous value}} = \frac{\$5.86}{\$66} = 0.089 = 8.9\%.$$

If you include the effects of a $1 per share sales commission, your gain is reduced to $4.86 per share, or

$$\text{percent return} = \frac{\text{gain in value}}{\text{previous value}} = \frac{\$4.86}{\$66} = 0.074 = 7.4\%.$$

Note that the commission can take a sizable chunk out of your return—in this case, it lowers your return by 1.5 percentage points. Thus it pays to shop around for the best commissions available at a particular level of service that you desire. ■

By the Way ⋯⋯⋯⋯
In general, a *discount broker* has low commissions but offers little advice on investments. A *full-service broker* offers advice based on in-depth research, but charges much higher commissions.

Bonds

Whereas buying a stock makes you a part-owner of a corporation, buying a bond makes you a *lender*. That is, you lend your money to the corporation or government from which you buy the bond, and then get paid back with interest.

Bonds may be issued for a variety of purposes. For example, a corporation may issue bonds as a way of borrowing money to build a new manufacturing plant. The U.S. Treasury Department issues bonds to borrow money to cover the federal deficit. State or local governments might issue bonds as a way of borrowing money to build new schools.

Most bonds are issued with three main characteristics:

- The **face value** (or *par value*) of the bond is the price you must pay to buy it at the time it is issued. This is the amount of money that you are loaning the issuer.

- The **coupon rate** of the bond is the *simple interest* rate that the issuer promises to pay. For example, a coupon rate of 8% on a bond with a face value of $1000 means that the issuer will pay interest of

$$8\% \times \$1000 = \$80$$

each year to the bond holder.

- The **maturity date** of the bond is the date on which the issuer promises to repay the face value of the bond.

Bonds would be simple if that were the end of the story. However, bonds can also be bought and sold *after* they are issued, in what is called the **secondary bond market**. For example, suppose that you own a bond with a $1000 face value and a coupon rate of 8%.

By the Way ⋯⋯⋯⋯
A company that needs cash can raise it either by issuing new shares of stock or by issuing bonds. Issuing new shares of stock reduces the ownership fraction represented by each share, and hence can depress the value of the shares. Companies usually prefer bond issues, which only obligate them to repay the bond with interest.

Further, suppose that equivalent new bonds (e.g., same level of risk and same time to maturity) are being issued with coupon rates of 9%. In that case, no one would pay $1000 for your bond because the new bonds offer a higher interest rate. However, you may be able to sell your bond at a **discount**; that is, for *less* than its face value. In contrast, suppose that new bonds are being issued with coupon rates of 7%. In that case, buyers will prefer your 8% bond to new bonds, and therefore may pay a **premium** for your bond—a price *greater* than its face value.

Because bonds may not be traded at their face values, the only way to compare returns on bonds is by comparing their **current yields**.

Current Yield of a Bond

$$\text{current yield} = \frac{\text{annual interest payment}}{\text{market price of bond}}$$

For example, suppose that a bond with a face value of $1000 and a coupon rate of 8% can be bought today for only $800. The bond issuer will still pay the promised $80 per year of interest, but you need invest only $800 to buy the bond. Therefore the current yield on the bond is

$$\text{current yield} = \frac{\text{annual interest payment}}{\text{market price of bond}} = \frac{\$80}{\$800} = 0.1 = 10\%.$$

This calculation shows that a bond selling at a discount from its face value has a current yield that is higher than its coupon rate. The reverse is also true: A bond selling at a premium over its face value has a current yield that is lower than its coupon rate. Thus the rule of thumb that

bond prices and yields move in opposite directions.

As with all investments, bond yields must be evaluated along with liquidity and risk. In general,

- bonds with longer times to maturity carry higher interest rates to compensate for the reduced liquidity, and
- bonds with higher risk carry higher interest rates to compensate for the possibility that the principal may never be repaid.

Many newspapers publish daily tables of the current prices and current yields of various bonds; a sample bond table is shown in Figure 5.6. Note that "prices" of bonds are quoted in **points**, which means percentage of face value. Most bonds have a face value of $1000 so, for example, a bond that closes at 102 points is selling for

$$102\% \times \$1000 = \$1020.$$

By the Way ·············
U.S. Treasury Notes and Bonds are the highest quality bonds. All other bonds are graded in terms of risk by independent rating services; bonds with a AAA rating have the lowest risk and bonds with a D rating have the highest risk.

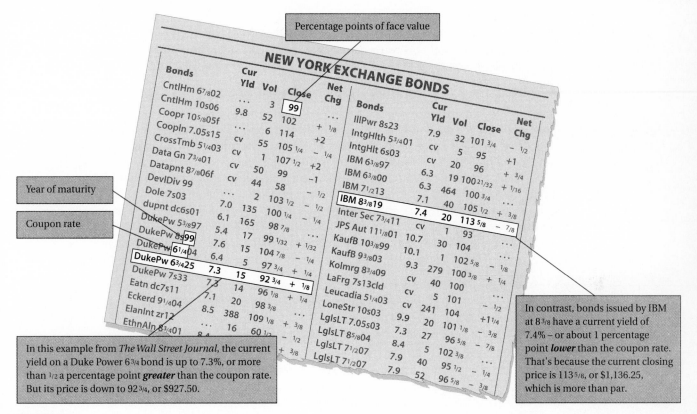

FIGURE 5.6
Source: K.M. Morris and A.M. Siegel, *The Wall Street Journal Guide to Personal Finance* (New York: Lightbulb Press, 1997).

EXAMPLE 8   *Current Yield*

Suppose that the closing price of a U.S. Treasury Bond (face value of $1000) is quoted as 105.97 points with a current yield of 6.7%. If you buy this bond, how much will you receive in interest each year until it matures?

Solution: A $1000 face value bond that closes at 105.97 points is selling for

$$105.97\% \times \$1000 = \$1059.70.$$

We now know both the market price of the bond and its current yield, so the current yield formula becomes:

$$\text{current yield} = \frac{\text{annual interest payment}}{\text{market price of bond}} \quad \Rightarrow \quad 6.7\% = \frac{\text{annual interest payment}}{\$1059.70}$$

Multiplying both sides by $1059.70, we find

$$\text{annual interest payment} = 6.7\% \times \$1059.70 = 0.067 \times \$1059.70 = \$71.00.$$

You can buy this bond for $1059.70 and will receive annual interest payments of $71. ∎

Mutual Funds

By the Way ············

When you buy shares in a mutual fund, you are essentially buying the expertise of the fund manager. A mutual fund is rated according to how the returns obtained by the manager compare to other mutual funds and market averages.

Before you buy a new television for a few hundred dollars, you'd probably do a fair amount of research to make sure that you're getting a good buy. You should be even more diligent when buying stocks or bonds whose performance may determine your entire financial future!

Unfortunately, researching individual stocks and bonds takes a lot of work. If you don't have the time or inclination to do this research for yourself, you can let a professional manager make the decisions for you by purchasing shares in a mutual fund. In a mutual fund, your money is pooled with the money of other investors. The *fund manager* invests this pool of money, often making many trades to get the largest possible returns.

Today, there are thousands of mutual funds, each with its own goals and strategies. Some specialize in high-risk investments and others in low-risk investments. Some invest only in particular industries, such as biotechnology or telecommunications. Some use "green" criteria, avoiding investments in companies with poor environmental records.

Some invest only in U.S. stocks while others invest in stocks from around the world. In general, you should look at four criteria when buying a mutual fund:

- Does the fund have an appropriate balance of risk and return for your needs?

- Does the fund manager have a track record of producing better than average returns?

- How much will it cost to own this fund? Some funds charge a commission, or **load**, when you buy or sell shares; virtually all charge an **annual fee**, usually a percentage of your investment's value. Look at the combination of loads and fees; if a fund has high costs, are the costs justified?

- Does the fund meet any other special criteria that you seek, such as investments in a particular type of company or "green" investments?

Newspapers also publish tables of mutual fund performance (Figure 5.7). Even if your investment strategy is long-term, it's a good idea to check periodically the performance of any mutual funds you own, just to make sure that the fund manager is doing a good job with your money.

EXAMPLE 9   *Mutual Fund Growth*

Suppose that Figure 5.7 represents a table from today's paper. Suppose that you had invested $500 in the Calvert Social Investment Equity Fund (SocInvEq) three years ago, and reinvested all dividends and gains. What is your investment worth now?

Solution: The annual return for the past three years is 17.0%. We find the current value of your investment by using this annual return as the *APR* in the compound interest formula (for compounding once a year).

$$A = P(1 + APR)^Y = \$500 \times (1 + 0.17)^3 = \$800.80.$$

In three years, your investment has risen in value from $500 to $800. Note that this return calculation does not include the costs of loads or fees.   ∎

Rating
A system for comparing fund performance with 1 as the worst and 5 as the best. The first number is a performance compared to a broad group of similar funds (e.g., all stock funds) and the second number is performance compared only to funds of the same type.

NAV
The net asset value of the fund's shares; that is, the amount that each fund share is currently worth.

Weekly % Return
The total return for the week, including capital gains from sales and any dividends.

Fund Family
A group of funds from the same company.

Fund Name
The name of an individual mutual fund.

Type
An abbreviation describing the type of investments the fund manages. *The New York Times* categorizes funds into about 50 different types and includes an index each Sunday.

YTD % Return
The total return for the year-to-date (i.e., since Jan.1).

1-year % Return
The total return for the past one year period.

3-year % Return
The annual return over the past 3 years, calculated assuming that any dividends and gains are reinvested into that fund.

MUTUAL FUND QUOTATIONS

Fund Family Fund Name	Type	Rating	NAV	Wkly. % Ret.		YTD % Ret.		1-Yr. % Ret.		3-Yr. % Ret.	
TaxFLnc	SL	3/3	12.78	–	0.6	+	3.1	+	9.3	+	7.9
Calvert (800) 368-2748											
CAMunInt m	SI	3/3	10.42	–	0.2	+	2.2	+	6.8	+	5.7
CapAccmC m	SG	NA	24.48	–	1.2	+	9.8	+	11.5		NA
Income m	CL	2/3	16.74	–	0.2	+	2.9	+	9.0	+	8.0
IntMuni m	MI	4/4	10.55	–	0.4	+	2.3	+	7.2	+	6.6
SocInvBd m	CI	2/2	16.33	–	0.3	+	3.2	+	8.5	+	7.2
SocInvEqA m	MV	2/1	25.47	–	1.7	+	13.6	+	27.2	+	17.0
SocInvMgA m	DH	2/2	32.60	–	0.7	+	9.4	+	19.1	+	14.5
StrGrowA m	DH	1/1	15.31	–	1.0	–	18.8	–	19.0	+	5.9
TaxFLong m	ML	4/4	16.77	–	0.5	+	2.5	+	7.8	+	7.3
TaxFLtdA f	MS	5/5	10.68	–	0.1	+	1.9	+	4.1	+	4.2
TaxFVT f	SL	4/4	16.14	–	0.3	+	2.3	+	7.5	+	6.8
WdVaIntlA m	FS	3/NA	21.51	+	0.4	+	12.7	+	19.8	+	11.3
Capstone (800) 262-6631											
GovInc b	UB	3/1	24.45	+	0.1	+	2.1	+	4.3	+	4.2

FIGURE 5.7

Indexes: The Dow and All That

If your investment returns 10% this year, is that good? It depends on how you look at it. If your return is greater than the rate of inflation, then the real value of your money has grown. On the other hand, your 10% return might not look so good if it came from a stock fund in a year when the average stock fund returned 15%.

One way you can evaluate your returns is by comparing it to an **index** that describes the average return in some category of investments. The best known index is the **Dow Jones Industrial Average** (DJIA), which reflects the average stock prices of 30 of the largest and most stable companies listed on the New York Stock Exchange (see Figure 5.8 on the next page).

TIME-OUT TO THINK: Table 5.3 showed that stocks have historically outperformed other investments over the long term. However, Figure 5.8 shows that there have been some periods of many years during which stocks have gained little or even lost value. What do *you* think will happen to the stock market over the next 5 years? the next 10 years? the next 50 years? Why?

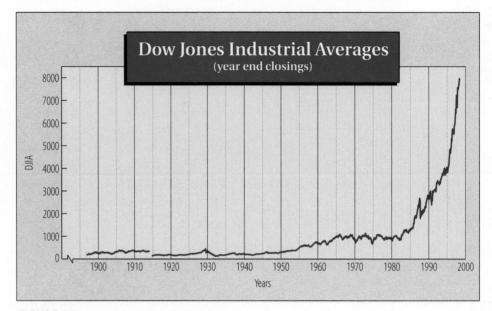

FIGURE 5.8
Historical values of the Dow Jones Industrial Average.

Other indexes frequently cited in financial reports include:

- The **Standard and Poor's 500** (S&P 500), which reflects average prices of 500 *large company* stocks.

- The **Russell 2000**, which reflects average prices of 2000 *small company* stocks.

- The **Wilshire 5000**, which reflects average prices of *all* stocks traded on the New York and American stock exchanges, plus some actively traded Nasdaq stocks.

- The **Nasdaq Composite**, which reflects average prices of 100 large company stocks listed on the Nasdaq exchange.

- The **Lehman Brothers T-Bond Index**, which tracks the performance of U.S. Treasury Bonds.

- The **Federal Funds Index**, which tracks the interest rate charged on money that banks lend each other for short terms.

However you decide to invest, make sure that you understand your investments and track their performance over time. After all, your financial future depends on it.

REVIEW QUESTIONS

1. Explain the basic differences between investments in *stocks*, *bonds*, and *cash*.

2. What is a *mutual fund*? Why might you want to invest in one?

3. What do we mean by the *liquidity* of an investment? Give an example of an investment that is very liquid, and one that is not very liquid.

4. How are risk and return for investments usually related? Give an example of an investment with high risk and high return, and one of an investment with low risk and low return.

5. Explain the meaning and use of the *annual return formula*.

6. Why do most financial advisors recommend that you keep a *diversified portfolio*?

7. What is *stock*? How can you determine your percentage ownership in a company in which you hold stock?

8. Describe the two basic ways to make money from stocks. How can you determine your total return on a stock investment?

9. Explain the meaning of the columns in a newspaper stock table.

10. What is a *bond*? Define the *face value, coupon rate*, and *maturity date* of a bond.

11. What does it mean to buy a bond at a *premium*? at a *discount*? How can you calculate the *current yield* of a bond?

12. Explain the meaning of the columns in a newspaper mutual fund table.

13. What is an *index* for investments? Why is it useful to keep track of indexes?

PROBLEMS

1. **Liquidity, Risk, and Return.** Evaluate each investment plan in terms of liquidity, risk, and return.

   a. A U.S. Treasury bond with a 30-year term and a 6% interest rate.

   b. Shares of stock in General Motors—the largest corporation in the United States.

   c. Shares of speculative stock in a new computer software company that has yet to produce revenues.

   d. A 5-year corporate bond with a 15% interest rate issued by a company that is on the verge of bankruptcy.

2. **Liquidity, Risk, and Return.** Evaluate each investment plan in terms of liquidity, risk, and return.

   a. A U.S. Treasury bond with a 10-year term and a 5% interest rate.

   b. Shares of stock in IBM—one of the largest corporations in the United States.

   c. Shares of speculative stock in a new drug company that has begun trials on an experimental cancer drug.

   d. A 10-year corporate bond with a 12% interest rate issued by a company that is on the verge of bankruptcy.

3. **Annual Returns.** Use the annual return formula to compute the annual return in each situation.

   a. Five years after buying 100 shares of XYZ stock for $55 per share, you sell the stock for $10,300.

   b. You pay $8000 for a municipal bond. When it matures after 20 years, you receive $12,500.

   c. Ten years after purchasing shares in a mutual fund for $5500, you sell them for $11,300.

   d. Three years after paying $4500 for shares in a start-up company, you sell the shares for $2500 (at a loss).

4. **Annual Returns.** Use the annual return formula to compute the annual return in each situation.

   a. Five years after paying $5000 for shares in a new company, you sell the shares for $3000 (at a loss).

   b. Three years after buying 200 shares of XYZ stock for $25 per share, you sell the stock for $8500.

c. Eight years after purchasing shares in a mutual fund for $7500, you sell them for $12,600.

d. You pay $10,000 for a municipal bond. When it matures after 10 years you receive $16,500.

5. Historical Returns. Suppose that your great-uncle invested $500 each in small company stocks, large company stocks, long-term corporate bonds, and U.S. Treasury bills in 1930. Assuming his investments grew at the long-term average annual returns in Table 5.3, approximately how much would each investment be worth in 2000?

6. Best and Worst Years. Suppose that you invest $2000 in each of small company stocks, large company stocks, long-term corporate bonds, and US Treasury bills. Using the returns shown in Table 5.3, how much would your investments be worth a year later if it had been the best of years? How much would your investments be worth a year later if it had been the worst of years?

Reading Stock Tables. *In Problems 7–11, answer the following questions for the indicated stocks.*

a. What is the ticker symbol for the company?

b. What was the range of selling prices for the company's shares yesterday? How do these prices compare to prices over the past year?

c. What was the closing price of the company's shares yesterday and two days ago?

d. How many shares were traded yesterday?

e. Suppose that you own 100 shares of the company. What total dividend payment should you expect this year?

f. Over the past year, how much profit (earnings) has the company made per share? How do the current share prices compare to the profits?

7. Mossimo Corporation, using the data in Figure 5.5 of the text (and assuming Figure 5.5 represents today's paper).

8. Muellerland Corporation, using the data in Figure 5.5 of the text (and assuming Figure 5.5 represents today's paper).

9. Motorola Corporation, using today's actual stock tables.

10. General Motors Corporation, using today's actual stock tables.

11. IBM Corporation, using today's actual stock tables.

12. Total Return on Stock. Suppose that you bought Motorola stock one year ago for $54 per share and sell at the closing price shown in Figure 5.5. Ignoring any commissions, what is the total return on your investment? What is the return if you include a broker's commission of $1 per share on the sale?

13. Total Return on Stock. Suppose that you bought Motorola stock one year ago for $72 per share and sell at the closing price shown in Figure 5.5. Ignoring any commissions, what is the total return on your investment? What is the return if you include a broker's commission of $1 per share on the sale?

14. Total Return on Stock. Suppose that you bought Mossimo stock one year ago for $5.80 per share and sell at the closing price shown in Figure 5.5. You also pay a commission of $0.25 per share on your sale. What is the total return on your investment?

15. Total Return on Stock. Suppose that you bought Mossimo stock one year ago for $46.00 per share and sell at the closing price shown in Figure 5.5. You also pay a commission of $0.25 per share on your sale. What is the total return on your investment?

16. Bond Yields. Calculate the yield on each bond.

a. A $1000 Treasury bond with a coupon rate of 6% that has a market value of $950.

b. A $1000 Treasury bond with a coupon rate of 7% that has a market value of $1050.

c. A $1000 Treasury bond with a coupon rate of 8% that has a market value of $900.

17. Bond Yields. Calculate the yield on each bond.

a. A $10,000 Treasury bond with a coupon rate of 6% that has a market value of $9500.

b. A $10,000 Treasury bond with a coupon rate of 7% that has a market value of $10,500.

c. A $10,000 Treasury bond with a coupon rate of 8% that has a market value of $9000.

18. Bond Interest. Calculate the annual interest that you will receive on each bond.

a. A $1000 Treasury bond with a current yield of 8.5% that is quoted at 105 points.

b. A $1000 Treasury bond with a current yield of 6.5% that is quoted at 98 points.

c. A $1000 Treasury bond with a current yield of 7.0% that is quoted at 102.5 points.

19. **Bond Interest.** Calculate the annual interest that you will receive on each bond.

    a. A $10,000 Treasury bond with a current yield of 8.5% that is quoted at 105 points.

    b. A $10,000 Treasury bond with a current yield of 6.5% that is quoted at 98 points.

    c. A $10,000 Treasury bond with a current yield of 7.0% that is quoted at 102.5 points.

20. **Mutual Fund Growth.** Suppose that Figure 5.7 represents a table from today's paper. Suppose also that you had invested $500 in the Calvert Strategic Growth Fund (StrGrowA) three years ago, and reinvested all dividends and gains. What is your investment worth now?

21. **Mutual Fund Growth.** Suppose that Figure 5.7 represents a table from today's paper. Suppose also that you had invested $500 in the Calvert Income Fund (Income) three years ago, and reinvested all dividends and gains. What is your investment worth now?

22. **Project: Investment Advisor.** Recently, your long-lost aunt passed away. In her will, she bequeathed to you her entire estate, worth $500,000. However, the will also specifies that you cannot spend any of the money for 10 years. Decide how to invest this money in order to maximize the return on it in ten years. Explain your investment decisions and discuss any risks involved in your investment plan.

23. **Project: Investment Picking.** Do a bit of research on stocks, bonds, and mutual funds. Choose three stocks, three bonds, and three mutual funds that you think would make good investments. Imagine that you invest $100 in each of these nine investments. Track the value of your investment over the next five weeks, then imagine that you sell them. What is your return over the five-week period? Compare your return to the returns of other students in your class for this project.

SUMMARY

*M*anaging your personal finances is one of the most important tasks you will face throughout your life. Although modern finance can be astonishingly complex, the topics covered in this chapter should provide you with the background you'll need to understand *most* financial management issues. Keep in mind a few key ideas.

- You can make most of the calculations required to manage your personal finances with the formulas given in this chapter and a standard calculator. Even if you leave such calculations to financial professionals, be sure you understand how they work so that you can control your own financial destiny.

- It is important to understand the effects of compounding in both investments and loans. Because of the effects of compounding, seemingly minor financial decisions that you make today may have large financial implications for your future.

- Tax laws and investment choices can change at almost any time, and involve subtleties that go beyond what we have discussed here. You can keep abreast of changes and learn more by making it a habit to read the financial pages in a newspaper.

SUGGESTED READING

Because financial strategies change often depending on the state of the economy and current law, the best sources for learning more about financial issues are newspapers, periodicals, and very recent books. Many investment companies also offer information, and a lot of financial information is available on the web. The following are among the most easily accessible of the many sources of financial information.

The Wall Street Journal (daily newspaper; also offers books on financial topics)
The New York Times (daily business section)
Money Magazine
Business Week Magazine

Chapter 6
MODELING OUR WORLD

*I*t may be impossible to predict the future, but we needn't go forward blindly. Analyzing trends can help us make educated guesses about the future, and perhaps even give us some control over our fates. Mathematics is the one tool that allows us to model trends with enough precision to give insight into how the world works and how it may change in the future. Just as we used equations to help understand financial options in Chapter 5, we can use mathematical ideas to model many other issues. In this chapter, we discuss the basic principles involved in using mathematics to model our world.

Nothing endures but change.
HERACLITUS, C. 500 B.C.

Nothing in the world lasts save eternal change.
HONORAT DE BUEIL, MARQUIS DE RACAN, C. 1600

There is nothing in this world constant,
but inconstancy.
JONATHAN SWIFT, C. 1707

Man's yesterday may ne'er be like his morrow;
Nought may endure but Mutability.
PERCY BYSSHE SHELLEY, 1816

The only thing that one knows about human
nature is that it changes.
OSCAR WILDE, 1895

A real office complex may not look exactly like the scale model used to design it, but the model helps the architects create the design. A road map doesn't look at all like real highways, but it serves as a model of a road system and can be extremely useful when you travel. The purpose of a **mathematical model** is similar to that of an architectural model or a road map: to represent something real and to help us understand it.

All of us are impacted by mathematical models. For example, models are used to ensure that a new bridge won't collapse. Models are used to study complex questions such as how a tax cut will affect future government revenues or why some years have more hurricanes than others. Even if you never need to create a mathematical model yourself, understanding the principles behind mathematical modeling is crucial to understanding current issues.

Mathematical models take many different forms: Some models are visual representations such as pictures, graphs, or charts. Others are sets of equations that must be solved, or mathematical simulations that are run on a computer. Indeed, in Chapter 5 we made extensive use of equations to model how money can grow or how loan balances decrease. In this chapter, we will generalize many of the ideas we used to study finance so that we can use mathematics to model many other issues. Amazingly, we will find that just a few basic principles allow us to create mathematical models for a great range of topics.

UNIT 6A

FUNCTIONS: THE BUILDING BLOCKS OF MATHEMATICAL MODELS

It is not difficult to find situations in which two or more quantities are related in some way. For example, ice cream sales are related to the time of year, and the demand for a product is related to its price. Mathematicians have developed a tool for describing relationships between quantities. This tool is called a **function**. Functions express mathematical relationships between two (or more) quantities, and they are the building blocks of mathematical models.

Suppose that we want a simple mathematical model of the temperature changes on a particular day, based on the data in Table 6.1. The first step is to recognize that *two* quantities are involved in this model: *time* and *temperature*. Our goal is to express the relationship between the time and the temperature in the form of a function. In other words, we want a function that describes how the temperature measurements are *related* to the time of day.

Table 6.1 Temperature Data for One Day			
Time	**Temperature**	**Time**	**Temperature**
6:00 A.M.	50°F	1:00 P.M.	73°F
7:00 A.M.	52°F	2:00 P.M.	73°F
8:00 A.M.	55°F	3:00 P.M.	70°F
9:00 A.M.	58°F	4:00 P.M.	68°F
10:00 A.M.	61°F	5:00 P.M.	65°F
11:00 A.M.	65°F	6:00 P.M.	61°F
12:00 noon	70°F		

The quantities related by a function are called **variables** because they change or *vary*. In this function, the temperature is called the **dependent variable** because it *depends on* the time of day. The time is called the **independent variable** because time passes independently of the temperature. We say that the temperature varies *with respect to* time and express the function with the notation

(*time, temperature*).

By convention, we always write the independent variable first.

> A *function* is a relationship describing how a *dependent variable* changes *with respect to* an *independent variable*. We express the function with the notation
>
> (*independent variable, dependent variable*).

Mathematical Note: The notation (x, y) is called an *ordered pair* because the order of the two variables is significant.

Many functions involve changes with respect to time, such as the growth of a child with respect to time or changes in the consumer price index with respect to time. But functions don't have to involve time, as in a function that describes how adjustable-rate mortgage payments change with respect to the current interest rate.

TIME-OUT TO THINK: From your everyday experiences, identify several pairs of variables that appear to be related and might be described by a function.

EXAMPLE 1   *Language and Notation of Functions*

For each situation, express the function involved in words and use the notation (*independent variable, dependent variable*).

a)   You are riding in a hot-air balloon. As the balloon rises, the surrounding atmospheric pressure decreases (which causes your ears to pop).

b)   You're on a barge headed south down the Mississippi River. You notice that the width of the river changes as you travel southward with the current.

By the Way ············

The Mississippi River runs 2340 miles (3800 km) from Lake Itasca, Minnesota, to the Gulf of Mexico. The Mississippi River *system*, which includes the Red Rock River in Montana and the Missouri River, has a length of 3700 miles (6000 km).

Solution:

a)   There is a relationship between your *altitude* and the surrounding atmospheric *pressure*: The pressure *depends on* your altitude, so we say that the pressure changes *with respect to* altitude. *Pressure* is the dependent variable and *altitude* is the independent variable, so we express this function as

(*altitude, pressure*).

b)   There is a relationship between the *river width* and your location, which we can describe as your *distance from the source* of the river. The river width *depends on* your distance from the river's source, so we say that the river width changes *with respect to* the distance from the source. *River width* is the dependent variable and *distance from the source* is the independent variable. We express the function as

(*distance from source, river width*).   ■

TIME-OUT TO THINK: Does dependence imply causality? That is, do changes in the independent variable *cause* changes in the dependent variable? Give a few examples to make your case.

REPRESENTING FUNCTIONS

There are four basic ways to represent a function.

1.   We can represent a function by using a *data table*, such as Table 6.1. A table provides detailed information but can become unwieldy with great quantities of data.

2.   We can represent a function using *words*. In the (*time, temperature*) function of Table 6.1, we could say, "The temperature increased from 50°F at 6 A.M. to a high of 73°F between 1 and 2 P.M., then steadily decreased to 61°F at 6 P.M."

3.   We can draw a *picture*, or **graph**, of a function. A graph is visual and easy to interpret, and it consolidates a great deal of information.

4.   We can write a compact mathematical representation of a function in the form of an **equation**.

The first two ways to represent a function (tables and words) are straightforward and probably quite familiar. Here, we discuss how functions are represented by pictures. We will discuss equations in Unit 6B.

BRIEF REVIEW

The Coordinate Plane

The most common way to draw a graph is to use a **coordinate plane**, which is made by drawing two perpendicular number lines. Each of the number lines is called an **axis** (plural, **axes**). Normally, numbers increase to the right on the horizontal axis and upward on the vertical axis. The intersection point of the two axes, where both number lines show the number zero, is called the **origin**.

Points in the coordinate plane are described by two **coordinates** (called an *ordered pair*) that give the horizontal and vertical distances between the point and the origin. We express the location of a particular point by writing its coordinates in parentheses in the form

(*horizontal coordinate, vertical coordinate*).

When working with functions, we always use the horizontal axis for the independent variable and the vertical axis for the dependent variable.

Figure 6.1(a) shows a sample coordinate plane with several points identified by their coordinates. Note that the origin is the point (0, 0). In Figure 6.1(b), note that the axes divide the coordinate plane into four **quadrants**, numbered counterclockwise starting from the upper right.

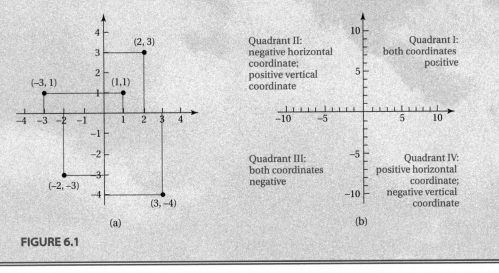

(a)   (b)

FIGURE 6.1

Domain and Range

Mathematically, each axis on a graph extends to infinity in both directions. However, most functions are meaningful only over a small region of the coordinate plane. For example, negative values of time do not make sense in the (*time, temperature*) function. In fact, the only times of interest in this function are those during which the measurements were made; that is, from 6 A.M. to 6 P.M. These times that make sense and are of interest make up the **domain** of the function.

Similarly, the only temperatures of interest for this function are those that occurred between 6:00 A.M. and 6:00 P.M. The lowest temperature recorded in this period was 50°F and the highest was 73°F. The temperatures between 50°F and 73°F make up the **range** of the (*time, temperature*) function.

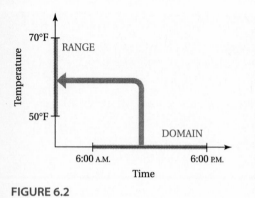

FIGURE 6.2

More generally, we can use the following definitions (Figure 6.2).

> The **domain** of a function is the set of values that both make sense and are of interest for the *independent variable*.
>
> The **range** of a function consists of the values of the dependent variable that correspond to the values in the domain.

It is straightforward to plot the individual data points for the (*time, temperature*) function (Figure 6.3a). We use the horizontal axis for the independent variable *time* and the vertical axis for the dependent variable *temperature*. We can make the graph easier to read by zooming in on the region covered by the domain on the horizontal axis and the range on the vertical axis (Figure 6.3b). Note that in both graphs the horizontal axis is labeled as *hours after 6 A.M.*, which makes the numbers somewhat more readable than if the actual time of day were labeled.

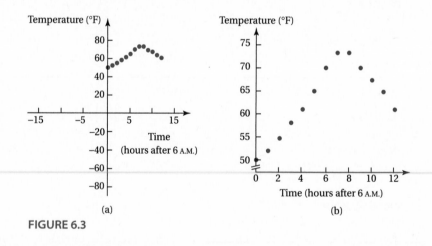

(a)   (b)

FIGURE 6.3

Completing the Model

So far, we have plotted only the thirteen data points from Table 6.1 for the (*time, temperature*) function. However, we know that the temperature *really* changes continuously, from each moment to the next, throughout the day. If we want the temperature graph to give us a *model* that describes the temperature more realistically, we should fill in the gaps between the data points. Because we don't expect any sudden spikes or dips in temperature during the day, it seems reasonable to connect the data points with a smooth curve (Figure 6.4).

The graph is now a model that we can use to "predict" the temperature at *any* time of day. For example, the model predicts that the temperature was about 67°F at 11:30 A.M. (5½ hours after 6 A.M.). Of course, because the temperature was not measured at 11:30 A.M., the model prediction may not be exact.

In most instances, mathematical models are far more complex than our (*time, temperature*) function. Mathematical models often consist of many separate functions, along with data and assumptions that help define the variables, domains, and ranges involved in the functions. Nevertheless, our simple example illustrates an important lesson: *A model's predictions can be only as good as the data and the assumptions from which the model is built.*

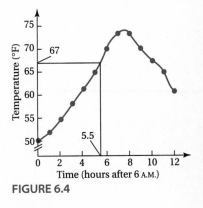

FIGURE 6.4

SUMMARY: CREATING AND USING GRAPHS OF FUNCTIONS

Step 1: Identify the independent and dependent variables in the function.

Step 2: Identify the domain (the independent variable values) and the range (the dependent variable values) of the function. Use this information to choose the scale and labels on the axes. Zoom in on the region of interest to make the graph easier to read.

Step 3: Make a graph by using any available data. If appropriate, fill in the gaps between data points.

Step 4: Before accepting any predictions of the model represented by the function, be sure to evaluate the assumptions built into the model.

EXAMPLE 2   *Pressure-Altitude Function*

Imagine riding in a hot-air balloon as it rises upward through the atmosphere. How does the atmospheric pressure change as your altitude increases? A common unit for atmospheric pressure is *inches of mercury*, or the height of a column of mercury in a pressure gauge. The following table gives some typical pressure values that you might observe as you rise. (A general rule is that the pressure drops by half with any 20,000-foot increase in altitude.)

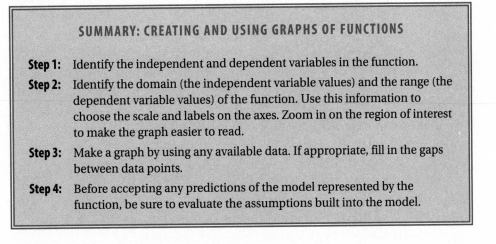

Altitude (ft)	0	5000	10,000	20,000	30,000
Pressure (in. mercury)	30	25	22	15	10

Use these data points to graph the (*altitude, pressure*) function. Then use the graph to predict the atmospheric pressure at 15,000 feet, and discuss the validity of the prediction.

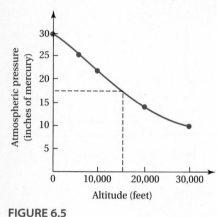

FIGURE 6.5

Solution: In Example 1 we identified *altitude* as the independent variable and *pressure* as the dependent variable. The *domain* is the set of values of interest for *altitude*. Using the *altitude* values in the table, the domain extends from 0 feet (sea level) to 30,000 feet. The *range* is the set of values taken by the pressure within this domain, and therefore extends from 10 to 30 inches of mercury.

We now plot the five data points. Between any two data points, we can reasonably assume that pressure decreases smoothly with increasing altitude. That is, the surrounding pressure doesn't suddenly jump or change as the balloon rises. Furthermore, because the Earth's atmosphere doesn't end abruptly, the pressure must decrease more gradually at higher altitudes. Thus we make a graph showing a smooth, increasingly gradual decrease in pressure (Figure 6.5).

According to this model, the atmospheric pressure at 15,000 feet is about 18 inches of mercury. Because we've sketched the function only roughly, we should not expect this prediction to be exact. Moreover, because we used data points based on average weather conditions, we should expect further deviations from the prediction if the weather is not average. ∎

EXAMPLE 3   *Hours of Daylight*

The number of hours of daylight varies with the seasons. Use the following data for 40°N latitude (e.g., San Francisco, Denver, Washington, D.C.) to model the change in the number of daylight hours with time.

By the Way ·············

The seasons arise because of the tilt of the Earth's axis. The northern and southern hemispheres alternately get more and less direct sunlight as the Earth orbits the Sun, so the seasons are opposite in the two hemispheres.

- The number of hours of daylight is greatest on the *summer solstice* (about June 21), when it is about 14 hours for latitude 40°N.

- The number of hours of daylight is smallest on the *winter solstice* (about December 21), when it is about 10 hours for latitude 40°N.

- Halfway between these extremes, on the spring and fall *equinoxes* (about March 21 and September 21, respectively), there are 12 hours of daylight at all latitudes.

Based on the model, at what times of year does the number of daylight hours change most gradually? most quickly? Do you think that predictions made by this model for the year 2050 will be accurate?

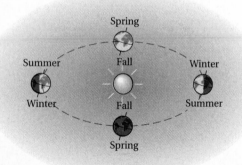

Solution: We are looking for a function between days of the year and the number of hours of daylight. *Time* is the independent variable because time marches on regard-less of other events. *Hours of daylight*, or *daylight* for short, is the dependent variable because it depends on the time of year. We can express this function as (*time, daylight*).

The times of interest are *all* days of the year. Thus the *domain* is the 365 days of a single year. The *range* extends from 10 hours to 14 hours of daylight. We know from experience that the number of hours of daylight changes smoothly with the seasons, so we can connect the four given data points for each year with a smooth curve (Figure 6.6a). Note that the same pattern will repeat from one year to the next, making this an example of a *periodic function*. The periodic nature is clear if we extend the graph for several years (Figure 6.6b).

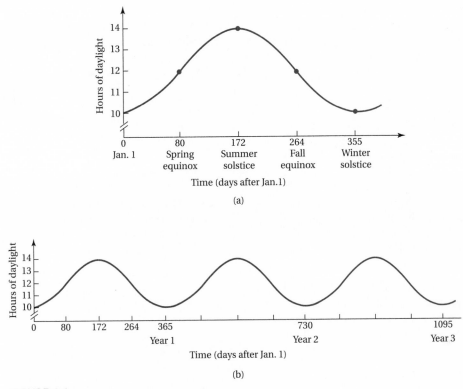

(a)

(b)

FIGURE 6.6

The shape of the curve shows that the number of hours of daylight varies slowly around the times of the solstices, and rapidly around the times of the equinoxes. Because this function is based only on the Earth's rotational and orbital characteristics, we can expect it to be a very accurate model of how the number of hours of daylight varies with the seasons. Thus we should expect it to be quite accurate for the year 2050, and well beyond.   ∎

TIME-OUT TO THINK: What is the current date? Based on the function shown in Figure 6.6, should the number of hours of daylight be changing rapidly or slowly at this time of year? Try to notice the change from one day to the next, and confirm that it matches the prediction of our model.

REVIEW QUESTIONS

1. What is a *mathematical model*? Give several examples in which mathematical models are useful.

2. What is a *function*? How do we decide which variable is the *independent variable* and which is the *dependent variable* in a function?

3. What are the four basic ways to represent a function?

4. Define *domain* and *range*, and explain how we determine them for a particular function.

5. Explain the statement: *A model's predictions can be only as good as the data and the assumptions from which the model is built.*

6. In your own words, describe the models discussed in Examples 2 and 3.

PROBLEMS

1. **Everyday Models.** Describe three different models (mathematical or otherwise) that you use or encounter frequently in everyday life. What is the underlying "reality" that these models represent? What simplifications are made in constructing these models?

2. **Functions in the News.** In today's newspaper, identify at least three different quantities (variables) that change. For each quantity, state what other quantity it changes with respect to and write a paragraph or two that explains the function relating the two quantities. At least one of your examples should involve a function in which the independent variable is *not* time.

Related Quantities. *For each part in Problems 3 and 4, write a short statement that expresses a possible function between the quantities. Example: (age, shoe size). Solution: As a person's age increases, shoe size also increases up to a point. After that point, show size remains relatively constant.*

3. **a.** (*weight of a bag of apples, price of the bag*)

   b. (*time, price of movies*), where *time* represents years between 1960 and 2000.

   c. (*price of a product, demand for a product*)

   d. (*distance from Earth, strength of gravity*)

4. **a.** (*slope of hill, speed of skateboard*)

   b. (*rate of pedaling, speed of bicycle*)

   c. (*tax rate, revenue collected*)

   d. (*number of cars on road, air quality*)

5. **Points in the Coordinate Plane.** Draw a set of axes in the coordinate plane. Plot and label the following points: $(0, 1)$, $(-2, 0)$, $(1, 5)$, $(-3, 4)$, $(5, -2)$, $(-6, -3)$.

6. **Points in the Coordinate Plane.** Draw a set of axes in the coordinate plane. Plot and label the following points: $(0, -1)$, $(2, -1)$, $(6, 5)$, $(3, -4)$, $(-5, -2)$, $(-6, 2)$.

Functions from Graphs. *Each graph in Problems 7 and 8 represents a function. In each case, do the following.*

   a. *Identify the independent and dependent variables.*

   b. *Make a copy of the graph and highlight the numbers that make up the domain and range.*

   c. *Describe the function in words.*

7.

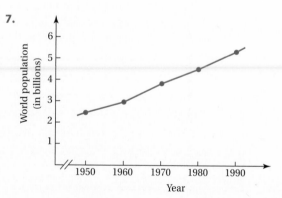

8.

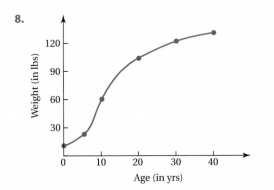

Functions from Data Tables. *Each data table in Problems 9–11 represents a function. In each case, do the following.*

   a. *Identify the independent and dependent variables.*

   b. *Describe the domain and range.*

   c. *Plot the data in a coordinate plane. Be sure to label all axes clearly and choose units appropriately. To highlight details of the function "zoom in" as appropriate.*

9.

Date	Average High Temperature	Date	Average High Temperature
Jan 1	42°F	Jul 1	85°F
Feb 1	38°F	Aug 1	83°F
Mar 1	48°F	Sep 1	80°F
Apr 1	58°F	Oct 1	69°F
May 1	69°F	Nov 1	55°F
Jun 1	76°F	Dec 1	48°F

10.

Year	Tobacco produced (billion of lbs)	Year	Tobacco produced (billion of lbs)
1975	2.2	1986	1.2
1980	1.8	1987	1.2
1982	2.0	1988	1.4
1984	1.7	1989	1.4
1985	1.5	1990	1.6

11.

Year	Projected U.S. Population (millions)	Year	Projected U.S. Population (millions)
1995	258	2015	289
2000	268	2020	294
2005*	276	2025	298
2010*	283	2030	302

* Projected data after year 2000

12. (*Time, Temperature*) Function. Study Figure 6.4.

   a. Use the graph to estimate the temperature at 9:30 A.M., 10:30 A.M., 12:30 P.M., and 1:30 P.M.

   b. Use the graph to estimate the times at which the temperature is 56°F, 60°F, and 66°F.

13. (*Altitude, Pressure*) Function. Study Figure 6.5.

   a. Use the graph to estimate the pressure at altitudes of 8000, 17,000, and 25,000 feet.

   b. Use the graph to estimate the altitude at which the pressure is 27, 20, and 12 inches of mercury.

   c. Estimating beyond the boundaries of the graph, at what altitude do you think the atmospheric pressure reaches 5 inches of mercury? Is there an altitude at which the pressure is exactly zero? Explain your reasoning.

   d. Suppose that the graph shown was based on measured data on a particular day at a particular location. Should you use the same function for other locations and other days? Why or why not?

14. Daylight Function. Study Figure 6.6, which applies to 40°N latitude.

   a. Use the graph to estimate the number of hours of daylight on April 1 (the 91st day of the year) and October 31 (the 304th day of the year).

   b. Use the graph to estimate on what days there are 13 hours of daylight.

   c. Use the graph to estimate on what days there are 10.5 hours of daylight.

   d. The graph in Figure 6.6 is valid at 40°N latitude. How do you think the graph would be different at 20°N latitude, 60°N latitude, and 40°S latitude? Why?

Rough Sketches of Functions.   *For each function in Problems 15–27, use your intuition or additional research, if necessary, to do the following.*

  a. *Describe an appropriate domain for graphing the function.*

  b. *Describe the range.*

  c. *Make a rough sketch of the function and explain the assumptions that go into your graph.*

  d. *Briefly discuss the validity of your graph as a model of the true function between the variables.*

15. (*altitude, temperature*) when climbing a mountain.

16. (*day of year, high temperature*) for the town in which you are living over a two-year period.

17. (*blood alcohol level, reflex time*) for a single person.

18. (*number of pages in a book, time to read the book*) for a single person.

19. (*mortgage interest rate, number of homes sold*) per year in a particular town.

20. (*time of day, traffic flow*) at a busy intersection over a two-day period.

21. (*price of gasoline, number of tourists in Yellowstone*).

22. (*time, world record in the 100-meter dash*) over the last 30 years.

23. (*minutes after lighting, length of candle*).

24. (*time, population of China*) where time is measured in years after 1900.

25. (*time of day, elevation of tide*) at a particular seaside port over two days.

26. (*angle of cannon, horizontal distance traveled by cannonball*).

27. (*weight of car, average gas mileage*).

28. **Project: Daylight Hours.** Make a graph, like Figure 6.6, of the function between the number of hours of daylight and the date for the town in which you live. To make the graph, you will need to know the number of hours of daylight on the summer and winter solstices and on the equinoxes. You can find this information in a variety of sources, such as almanacs. An easy method to calculate the number of hours of daylight is by comparing the times of sunrise and sunset on any particular day. Most local newspapers publish the time of sunrise and sunset each day on their weather page, and most libraries carry old newspapers either in print or on microfilm.

LINEAR GRAPHS AND MODELING

Functions can be represented by tables, words, graphs, or equations. Having discussed the first three in Unit 6A, we now turn our attention to representing functions by equations. Although equations are more abstract than pictures, they are easier to manipulate mathematically. Thus they give us greater power when creating and analyzing mathematical models.

Some functions can be represented by very simple equations, while others require equations of astonishing complexity. Fortunately, we can understand the basic principles involved in mathematical modeling by focusing on the simplest functions: those with straight-line graphs, called **linear** functions.

RATE OF CHANGE

Imagine that a steady rain is falling and we measure the depth of rain accumulating in a rain gauge. The rain subsides after six hours, and we want to describe how the rain depth varied with time. In this situation, *time* is the independent variable and *rain depth* is the dependent variable. Suppose that, based on our measurements with the rain gauge, we find the (*time, rain depth*) function shown in Figure 6.7. Note that its graph is a straight line, which we call a **linear graph**.

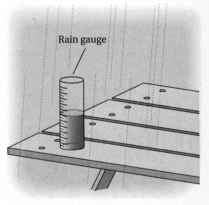

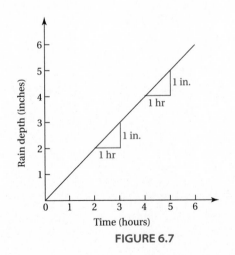

FIGURE 6.7

The graph shows that during every one-hour interval of the storm, the rain depth increased one inch. We say that the **rate of change** of the rain depth with respect to time was 1 inch per hour, or 1 in./hr. Note that this rate of change was constant throughout the storm: no matter which hour interval we choose to study, the rain depth increased by 1 inch. This demonstrates a key fact about rate of change and graphs:

A function with a constant rate of change has a linear graph.

Figure 6.8 shows linear graphs for three other steady rain storms. From these graphs we observe the following:

- In Figure 6.8(a), the rain depth increased 0.5 inch with each passing hour. Thus the *rate of change* of the rain depth with respect to time was a constant 0.5 in./hr.

- In Figure 6.8(b), the rain depth increased 1.5 inches with each passing hour, so the *rate of change* of the rain depth with respect to time was a constant 1.5 in./hr.

- In Figure 6.8(c), the rain depth increased 2 inches with each passing hour, so the *rate of change* of the rain depth with respect to time was a constant 2 in./hr.

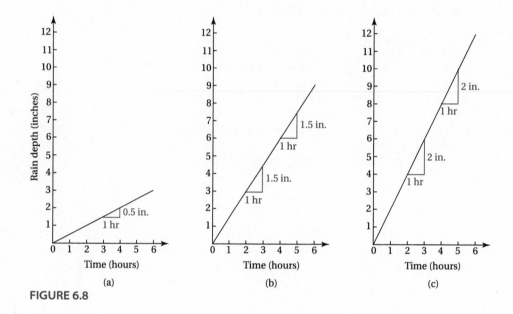

FIGURE 6.8

Comparing the three graphs in Figure 6.8 leads to another crucial observation:

The greater the rate of change, the steeper is the graph.

Moreover, note that in each case the rate of change is equal to the amount that the graph *rises* vertically for any distance that it *runs* horizontally, which we recognize as the **slope** of the linear graph.

$$\text{slope of a linear graph} = \frac{\text{vertical } \textit{rise}}{\text{horizontal } \textit{run}}$$

For the (*time, rain depth*) functions shown in Figures 6.7 and 6.8, we identify the horizontal run as the change in the independent variable *time* and the vertical rise as the corresponding change in the dependent variable *rain depth*. We can now generalize to the case of any linear function.

The rate of change of any *linear* function is constant and is the *slope* of its graph; it can be calculated as

$$\text{rate of change} = \text{slope} = \frac{\text{change in } \textit{dependent variable} \text{ from } P_1 \text{ to } P_2}{\text{change in } \textit{independent variable} \text{ from } P_1 \text{ to } P_2},$$

where P_1 and P_2 are *any* two distinct points on the straight-line graph (Figure 6.9).

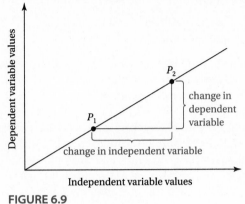

FIGURE 6.9

EXAMPLE 1   *A Price-Demand Function*

Suppose that a small store sells fresh pineapples. Based on data in which pineapple prices varied between \$2 and \$7, the store owners created a linear model of the (*price, demand*) function for the pineapples (Figure 6.10). This function describes how the *demand* (the number of pineapples that are sold per day) varies with respect to the *price*. For example, the point (\$2, 80 pineapples) means that, at a price of \$2 per pineapple, 80 pineapples can be sold on an average day. What is the rate of change for this function? Discuss the validity of this model.

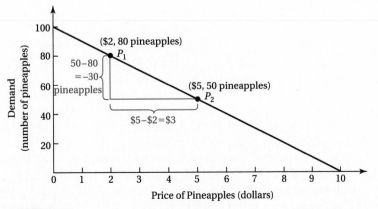

FIGURE 6.10

Solution: The rate of change of the (*price, demand*) function is the slope of its linear graph. We can calculate the slope using *any* two points on the graph. Let's choose P_1 as ($2, 80 pineapples) and P_2 as ($5, 50 pineapples). The change in *price* from P_1 to P_2 is

$$\text{change in price} = \$5 - \$2 = \$3.$$

The change in *demand* from P_1 to P_2 is

$$\text{change in demand} = 50 \text{ pineapples} - 80 \text{ pineapples} = -30 \text{ pineapples}.$$

The change in demand is negative because demand *decreases* from P_1 to P_2. The rate of change is the slope of the graph.

$$\text{rate of change} = \frac{\text{change in demand}}{\text{change in price}} = \frac{-30 \text{ pineapples}}{\$3} = \frac{-10 \text{ pineapples}}{\$1}.$$

Thus the rate of change of the (*price, demand*) function is -10 pineapples per dollar. The negative rate of change indicates that demand *decreases* as price increases. That is, for every dollar that the price increases, the number of pineapples sold decreases by 10.

This model seems reasonable within the domain for which the store owners gathered data: between prices of $2 and $7. Outside of this price interval, the model's predictions probably are not valid. For example, the model predicts that the store could sell one pineapple per day at a price of $9.90, but could *never* sell a pineapple at a price of $10. Even less believable is the model's prediction that the store could "sell" only 100 pineapples if they were free! As with many models, this price-demand model is useful only in a limited domain. ∎

EXAMPLE 2   *Drawing a Linear Model*

Imagine that you hike on a 3-mile trail starting at an elevation of 8000 feet. Along the way, the trail gains elevation at an average rate of 650 feet per mile. Draw a graph that shows a linear model of this function, using 650 feet per mile as the rate of change. Does this model seem realistic?

Solution: Because your elevation depends on the distance you've walked, *distance* is the independent variable and *elevation* is the dependent variable. We know one data point: (0 mi, 8000 ft) represents the 8000-foot elevation at the start of the trail. We also know that the elevation changes with respect to the distance at a rate of 650 feet per mile. Thus a second point on the graph is (1 mi, 8650 ft). We can draw a linear graph by connecting these two points and extending the line over the domain from 0 to 3 miles, which represents the length of the trail. As we expect, the rate of change is the slope of the graph, 650 ft/mi (Figure 6.11).

This model assumes that elevation increases at a constant rate along the entire 3-mile trail. While an elevation change of 650 feet per mile seems reasonable as an average, it's difficult to believe this rate of change never varies along the entire trail length. The model's predictions therefore are likely to be reasonable *estimates* of your elevation at different points along the trail, but the real (*distance, elevation*) function for the trail is probably more complex than this linear model suggests.

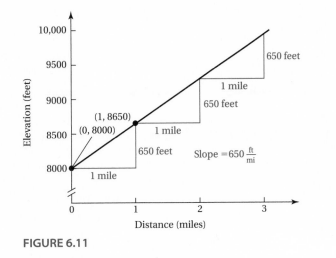

FIGURE 6.11

The Rate of Change Rule

Consider again the (*time, rain depth*) function shown in Figure 6.7. Suppose that we want to know how much the rain depth changes in a 4-hour period. Because the rate of change for this function is 1 in./hr, the total change after four hours is

$$\text{change in rain depth} = \underbrace{1\,\frac{\text{in.}}{\text{hr}}}_{\text{rate of change}} \times \underbrace{4\,\text{hr}}_{\text{elapsed time}} = 4\,\text{in.}$$

Notice how the units work out. Note also that the *elapsed time* is the change in the independent variable and the change in rain depth is the change in the dependent variable. Thus for all linear functions the following is true.

Rate of Change Rule for Linear Functions

change in dependent variable = (rate of change) × (change in independent variable)

EXAMPLE 3   *Rate of Change Rule for Price-Demand Function*

Using the linear model of the (*price, demand*) function in Figure 6.10, predict the change in demand for pineapples (the number sold) if the price increases by $3.

Solution: The independent variable is the *price* of the pineapples, and the dependent variable is the *demand* for pineapples. In Example 1, we found the rate of change of demand with respect to price is −10 pineapples per dollar. Using the rate of change rule, the change in demand for a price increase of $3 is

$$\text{change in } demand = \text{rate of change} \times \text{change in } price$$

$$= -10\,\frac{\text{pineapples}}{\$} \times \$3$$

$$= -30\,\text{pineapples.}$$

This model predicts that a $3 price increase will lead to 30 fewer pineapples being sold per day.

GENERAL FORM OF A LINEAR EQUATION

Suppose that your job is to oversee an automated assembly line that manufactures computer chips. You arrive at work one day to find a stock of 25 chips that were produced during the night. If chips are produced at the constant rate of 4 chips per hour, how large is the stock of chips at any particular time during your shift?

Answering this question requires finding a function that describes how the number of chips depends on the time of day. We begin by identifying *time*, measured in hours, as the independent variable and *number of chips* as the dependent variable. We seek a linear function (*time, number of chips*). At the start of your shift, *time* = 0 and your initial stock is 25 chips. Because the stock grows by 4 chips every hour, the *rate of change* of this function is 4 chips per hour.

We construct a graph by starting at the initial point (0 hr, 25 chips), which represents the number of chips at the start of your shift, and draw a straight line with a slope of 4 chips per hour (Figure 6.12).

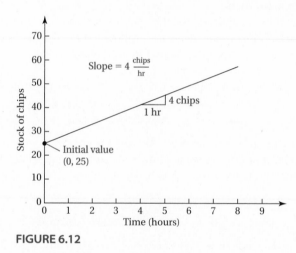

FIGURE 6.12

We are now ready to write an equation describing the function. The stock of chips at any particular time during your shift is

number of chips = initial number of chips + change in number of chips.

Now we apply the rate of change rule to find the change in the number of chips:

$$\underbrace{\text{change in } \textit{number of chips}}_{\text{change in dependent variable}} = \underbrace{4\,\frac{\text{chips}}{\text{hr}}}_{\text{rate of change}} \times \underbrace{\text{elapsed } \textit{time}}_{\text{change in independent variable}}$$

Using this result and 25 for the initial number of chips, the equation for the number of chips becomes

$$\textit{number of chips} = 25 \text{ chips} + \left(4\,\frac{\text{chips}}{\text{hr}} \times \text{elapsed } \textit{time}\right).$$

So far, we have used full names for variables (such as *time* and *number of chips*) and have been careful to include units in our equations. However, much of the power of the mathematical language lies in its compactness. So we will now follow common practice and use single letters for variables. It is helpful to choose letters that remind you of the actual variable (for example, *t* for time and *d* for distance). The only drawback to this procedure is that we must be careful to remember the units that go with each variable, since they will no longer be shown explicitly.

We may now write the computer chip equation more compactly by letting *t* stand for *time* and *N* for the *number of chips*. The equation becomes

$$N = 25 + 4t,$$

where we must remember that 25 represents a number of chips and 4 represents a rate of change with units of chips/hr. This equation provides a compact **formula** that gives the number of chips *N* at any time *t*. For example, after *t* = 3.5 hours, the stock of chips is

$$N = 25 + 4 \times 3.5 = 39.$$

At the end of your eight-hour day, *t* = 8, and the number of chips is

$$N = 25 + 4 \times 8 = 57.$$

Mathematical Note:
If you choose to use full names for variables, you will not be alone: this is the recommended practice when writing computer programs.

Mathematical Note:
Recall that, when no other operation is indicated, we assume multiplication. For example, $4t$ means $4 \times t$.

TIME-OUT TO THINK: Use the equation for chip production to find the number of chips produced after four hours. Does the result agree with the answer you read from the graph in Figure 6.12?

To generalize from this example to any linear function, note that

- The number of chips *N* is the **dependent variable**;
- The time *t* is the **independent variable**;
- The initial stock of 25 chips represents the **initial value** of the dependent variable when *t* = 0; and
- The term $4\,\dfrac{\text{chips}}{\text{hr}}$ is the **rate of change** of *N* with respect to *t*, so $4\,\dfrac{\text{chips}}{\text{hr}} \times t$ is the *change* in *N*.

With these observations, we can write the general linear equation.

General Linear Equation

dependent variable = initial value + (rate of change × *independent variable*)

EXAMPLE 4   *Rain Depth Equation*

Using the function shown in Figure 6.7, write a general equation that describes the rain depth at any time after the storm begins. Use the equation to find the rain depth three hours after the storm began.

Solution: For the function (*time, rain depth*) in Figure 6.7, the rate of change in rain depth with respect to time is 1 in./hr and the initial value of the rain depth is 0 inches when the storm begins. Thus the general equation for this function is

$$\underbrace{rain\ depth}_{\text{dependent variable}} = \underbrace{0\ in.}_{\text{initial value}} + \underbrace{1\frac{in.}{hr}}_{\text{rate of change}} \times \underbrace{time}_{\text{independent variable}}.$$

We can write this equation more compactly by letting r represent the dependent variable *rain depth* (in inches) and t represent the independent variable *time* (in hours).

$$r = 0 + 1 \times t, \qquad \text{or} \qquad r = t.$$

If we now set $t = 3$ hours in this equation, we find that the rain depth three hours after the storm began was

$$r = t = 3 \text{ (inches)},$$

where we must remember that r carries units of inches. Thus 3 hours after the storm began, the rain depth was 3 inches. ∎

EXAMPLE 5   *Heavy Loads*

Suppose that you drive a freight truck and that one of your standard routes has a long uphill section. Over the years you notice that, with no freight on board, you can drive up the hill at a maximum speed of 45 miles per hour (mph). If the truck is loaded with 5 tons of cargo, you can drive up the hill at 30 miles per hour. Assume that a linear model can be used to relate *load* to *maximum speed*. Find the equation that gives the maximum truck speed as it varies with the weight of the cargo. Use this function to predict the maximum truck speed with a 10-ton cargo and discuss the validity of this prediction.

Solution: We seek a linear function (*weight, maximum speed*), where the cargo *weight* (in tons) is the independent variable and the *maximum speed* (in mph) is the dependent variable. The observations give us two points: P_1 is (0 tons, 45 mph) and P_2 is (5 tons, 30 mph). We can use these two points to find the rate of change of the function.

$$\text{rate of change} = \frac{\text{change in } \textit{dependent variable} \text{ from } P_1 \text{ to } P_2}{\text{change in } \textit{independent variable} \text{ from } P_1 \text{ to } P_2}$$

$$= \frac{\text{change in } \textit{maximum speed}}{\text{change in } \textit{weight}}$$

$$= \frac{(30 - 45)\ \text{mph}}{(5 - 0)\ \text{tons}} = \frac{-15\ \text{mph}}{5\ \text{tons}} = -3\frac{\text{mph}}{\text{ton}}$$

That is, for each ton of freight, the maximum speed *decreases* by 3 mph. The initial value for the maximum speed is the speed when the cargo weight is zero, which is 45 mph. Thus the general linear equation for this function becomes

$$\underbrace{maximum\ speed}_{\text{dependent variable}} = \underbrace{45\ \text{mph}}_{\text{initial value}} + \underbrace{-3\,\frac{\text{mph}}{\text{ton}}}_{\text{rate of change}} \times \underbrace{weight}_{\text{independent variable}}$$

> **Mathematical Note:**
> We can abbreviate miles per hour either as mi/hr or as mph. In this example, in which the rate of change has units of *miles per hour per ton*, it is less confusing to use the abbreviation mph.

Figure 6.13 shows th graph of this function. Note that we can write this equation more compactly by letting w represent the weight in tons and s represent the maximum speed in mph:

$$s = 45 - 3w$$

If we substitute $w = 10$ tons into this equation, the equation predicts a maximum truck speed of

$$s = 45 - 3 \times 10 = 45 - 30 = 15\ \text{(mph)},$$

where we must remember that s carries units of miles per hour. Thus the model predicts that the maximum speed of the truck would be 15 mph with a load of 10 tons. Although this prediction seems reasonable, the load of 10 tons is heavier than any load in your observations. Thus we should be skeptical of the conclusion; for example, a load of 10 tons might damage your truck.

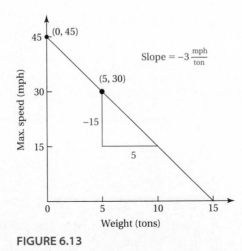

FIGURE 6.13

The Equation of a Line (Using the Variables x and y)

The general linear equation in this text may be familiar to you in a more abstract form. Traditionally, mathematicians use the letter x to represent the independent variable and y to represent the dependent variable. Then the general linear equation takes the form

$$y = \text{initial value of } y + (\text{rate of change} \times x).$$

It is customary to let b stand for the initial value of y (also called the **y-intercept**) and m stand for the rate of change (or slope), which allows us to write **the equation of a line** in an even more compact form:

$$y = mx + b$$

For example, consider the equation $y = 4x - 4$. Its graph has a slope of 4, meaning that the line *rises* 4 units vertically for each unit that it *runs* horizontally, and it has a y-intercept of -4, meaning that the line crosses the y-axis (the vertical axis) at the value -4 (Figure 6.14a).

Figure 6.14(b) shows the effects of keeping the same y-intercept but changing the slope. A positive slope ($m > 0$) means the line rises to the right. A negative slope ($m < 0$) means the line falls to the right. A zero slope ($m = 0$) means a horizontal line.

Figure 6.14(c) shows the effect of changing the y-intercept for a set of lines that have the same slope. All the lines rise at the same rate, but cut the y-*axis* at different points.

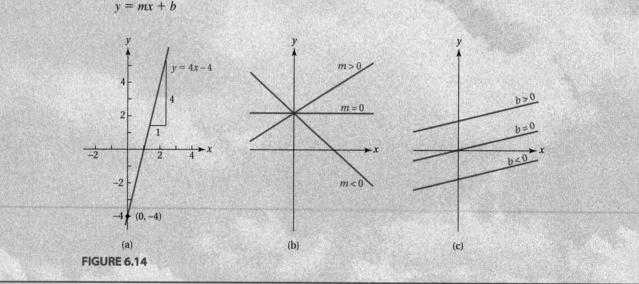

(a)   (b)   (c)

FIGURE 6.14

SOLVING LINEAR EQUATIONS

Consider once more the (*price, demand*) function in Figure 6.10. We already found that the rate of change for this function is -10 pineapples per dollar. The initial value for this function is 100 pineapples, because that is the demand that this model predicts when the price is 0. Thus the equation for this function is

$$\underbrace{demand}_{\text{dependent variable}} = \underbrace{100 \text{ pineapples}}_{\text{initial value}} + \underbrace{-10\frac{\text{pineapple}}{\$}}_{\text{rate of change}} \times \underbrace{price}_{\text{independent variable}}.$$

We can write the equation more compactly by letting p stand for *price* and d stand for *demand*, while remembering that p carries units of dollars and d carries units of pineapples:

$$d = 100 - 10p$$

Suppose that we want the demand to be 80 pineapples each day. In that case, we must *solve* the (*price, demand*) equation to find the price p in terms of demand d, so that we can find the price that gives a demand of 80 pineapples. We begin by subtracting 100 (pineapples) from both sides.

$$d - 100 = 100 - 10p - 100 = -10p$$

Next we divide both sides by -10 (pineapples per dollar).

$$\frac{d - 100}{-10} = \frac{-10p}{-10} = p$$

We now switch the left and right sides, giving us an expression for the price in terms of the demand.

$$p = \frac{d - 100}{-10}$$

We can now substitute a demand of $d = 80$ pineapples to find the needed price.

$$p = \frac{80 - 100}{-10} = \frac{-20}{-10} = \$2$$

Notice that even if you don't carry units through the calculation, your final answer must be given with correct units. According to this model, the price should be set at $2 if the store owners want to sell 80 pineapples each day.

In general, we can always solve a linear equation for the independent variable in terms of the dependent variable by applying just two simple rules from algebra. These rules are probably familiar (in fact, we've used them a few times in previous units).

Mathematical Note: It is easier to work with equations without showing units, but keeping the units at all times offers an important advantage: it provides a way to check your work by making sure that your final answer has the units you expect. Keep this advantage in mind when you decide whether to show units when you work with equations.

Two Basic Rules of Algebra

- We can always add or subtract the same quantity from both sides of an equation.
- We can always multiply or divide both sides of an equation by a (nonzero) quantity.

$\mathcal{T}$HINKING ABOUT . . .

Algebra's Baghdad Connection

After the fall of ancient Rome in the fifth century, European civilization entered the period known as the *Dark Ages*. However, it was not a dark time in the Middle East, where a new center of intellectual achievement arose in the city of Baghdad (in modern-day Iraq). Jews, Christians, and Muslims in Baghdad worked together in scholarly pursuits during this period.

One of the greatest scholars was a Muslim named Muhammad ibn Musa al-Khwarizmi (A.D. 780–850). Al-Khwarizmi wrote several books on astronomy and mathematics including one titled *Hisab al-jabr wal-muqabala*, which translates roughly as "the science of equations." This book preserved and extended the work of the Greek mathematician Diophantus (A.D. 210–290), and thereby laid the foundations of algebra. In fact, the word *algebra* comes directly from the Arabic words *al-jabr* in the book's title.

In another of his works, Al-Khwarizmi described the numeral system developed by Hindu mathematicians. Although he did not claim credit for the Hindu work, later writers often attributed such credit to him; that is why modern numerals are known as *Hindu-Arabic* rather than solely as Hindu. Some later authors even attributed the numerals to Al-Khwarizmi personally and, in a sloppy writing of his name, the use of Hindu numerals became known as *algorismi*, which later became the English words *algorism* (arithmetic) and *algorithm*.

EXAMPLE 6   *When Will Chips Be Available?*

Consider again the computer chip assembly line. Find a general equation that gives the *time* needed to produce a particular number of chips, and use it to determine when you will have a stock of 43 chips.

Solution: We start with our previous equation where N represents the number of chips and t represents time.

$$N = 25 + 4t$$

We want to solve this equation for t. First, we subtract 25 (chips) from both sides.

$$N - 25 = (25 + 4t) - 25 = 4t$$

Next, we divide both sides by 4 (chips/hr).

$$\frac{N - 25}{4} = \frac{4t}{4} = t \quad \text{or} \quad t = \frac{N - 25}{4}$$

To find out when you'll have 43 chips, we set $N = 43$ and remember that t carries units of hours.

$$t = \frac{43 - 25}{4} = \frac{18}{4} = 4.5 \text{ hours}$$

You will need 4.5 hours to have a total of 43 chips.   ■

CREATING LINEAR EQUATIONS

Suppose that we have two data points and want to find a linear function that fits them. We can find the equation for this linear function by using the two data points to determine the rate of change and the initial value for the function. Here are three steps that always lead to the equation for a straight line.

Creating Linear Equations

Step 1: Using the two given points, P_1 and P_2, find the slope or rate of change with the rule

$$\text{rate of change} = \frac{\text{change in } \textit{dependent variable} \text{ from } P_1 \text{ to } P_2}{\text{change in } \textit{independent variable} \text{ from } P_1 \text{ to } P_2}.$$

Step 2: Substitute the values of the variables from *either* data point and the rate of change from Step 1 into the general linear equation, and use this equation to solve for the initial value. For example, if you use the values at P_1, the equation takes the following form.

$$\begin{matrix} \text{value of } \textit{dependent} \\ \textit{variable} \text{ at } P_1 \end{matrix} = \text{initial value} + \text{rate of change} \times \begin{pmatrix} \text{value of } \textit{independent} \\ \textit{variable} \text{ at } P_1 \end{pmatrix}$$

Step 3: Substitute the initial value found in Step 2 into the general linear equation

$$\textit{dependent variable} = \text{initial value} + (\text{rate of change} \times \textit{independent variable}).$$

EXAMPLE 7   *Crude Oil Use Since 1850*

Until about 1850, humans used so little crude oil that we can call the amount zero—at least in comparison to the amount used since that time. By 1960, humans had used a total of 600 billion cubic meters of oil. Create a linear model that describes world oil use since 1850. Discuss the validity of the model.

Solution: We want to describe the function (*time, total oil used*), where *time* is the independent variable and *total oil used* is the dependent variable. Let t denote time in years since A.D. 0 and T denote total oil used measured in units of billions of cubic meters. The domain for this function is at all times t greater than 1850.

We are given two data points: P_1 is (1850, 0), indicating that no oil was used by 1850; P_2 is (1960, 600), which represents the 600 billion cubic meters used by 1960. Drawing a straight line through the two data points gives us a linear graph (see Figure 6.15a on the following page). We find the equation for this linear function by following the three-step procedure.

Step 1.   The rate of change (in units of billions of cubic meters per year) is

$$\text{rate of change} = \frac{\text{change in } T \text{ from } P_1 \text{ to } P_2}{\text{change in } t \text{ from } P_1 \text{ to } P_2} = \frac{600 - 0}{1960 - 1850} = \frac{600}{110} = 5.45.$$

Step 2. Substituting this rate of change and the data values from P_1 (we could also have used P_2) into the general linear equation, we find

$$\underbrace{0}_{\substack{\text{value of } \textit{dependent} \\ \textit{variable} \text{ at } P_1}} = \text{initial value} + \underbrace{5.45}_{\text{rate of change}} \times \underbrace{1850}_{\substack{\text{value of } \textit{independent} \\ \textit{variable} \text{ at } P_1}}.$$

Subtracting the product 5.45×1850 from both sides gives us the initial value:

$$\text{initial value} = 0 - 5.45 \times 1850 = -10{,}083 \text{ (billions of cubic meters)}$$

Step 3. We substitute this initial value into the general linear equation to get

$$T = -10{,}083 + 5.45t.$$

To be sure of the units of the problem, we can rewrite this equation as

$$\textit{total oil used} = -10{,}083 \text{ billion cubic meters} + 5.45 \, \frac{\text{billion cubic meters}}{\text{yr}} \times \textit{time}.$$

Note that this equation gives negative values for the total amount of oil used prior to 1850. Negative results are clearly nonsense, but they should not concern us because years before 1850 are not in the domain of this problem.

Does the function seem reasonable *within* its domain? This linear model suggests that total oil use has risen at a constant rate, which could happen only if humans have used the same amount of oil year after year since 1850. In fact, the rate of oil consumption has increased over time due to increasing population and increasing per capita use of oil. The true (*time, total oil used*) function probably looks more like the curve shown in Figure 6.15(b). Nevertheless, the linear model provides a useful first approximation to the actual function for oil use.

By the Way ·············

Annual world oil consumption in the late 1990s was more than triple the annual oil consumption in 1960.

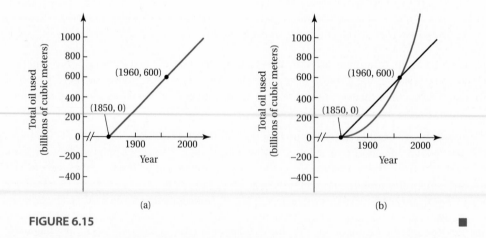

(a) (b)

FIGURE 6.15

MODELS WITH TWO LINEAR FUNCTIONS

All the models described so far consisted of only one function. But many mathematical models involve more than one function; in fact, they may involve hundreds or even thousands of functions. Let's consider an example of a model built from two linear functions.

EXAMPLE 8   *A Profit Model*

Suppose that you've written a book that you want to publish and sell yourself. The cost of publishing it is $1000 for setting up the printing equipment plus $5 per book printed. You will sell the books for $10 each. Find a function that describes the profit from the book. How many books must you sell to break even?

Solution: The cost of publishing depends on the number of books printed, so we can describe the cost with a function (*number of books, publishing cost*). The initial value of this function is the $1000 set-up cost, and the rate of change is a constant $5 per book because the cost increases by $5 for each book printed. Thus this function is linear and has the equation

$$\underbrace{publishing\ cost}_{\text{dependent variable}} = \underbrace{\$1000}_{\text{initial value}} + \underbrace{5\frac{\$}{book}}_{\text{rate of change}} \times \underbrace{number\ of\ books.}_{\text{independent variable}}$$

We can write this equation more compactly by letting n represent the number of books and C represent the publishing cost:

$$C = 1000 + 5n$$

The revenue depends on the number of books sold, so we can describe revenue with a function (*number of books, revenue*). The initial value of the revenue is zero if no books are sold and the constant rate of change is the $10 per book earned from each sale. Thus this function takes the form

$$revenue = \frac{\$10}{book} \times number\ of\ books.$$

If we let R be the revenue and again let n be the number of books, this equation takes the compact form

$$R = 10n.$$

The *profit* from the book is the difference between revenue and cost. That is, if we let P stand for profit, then $P = R - C$. We already have equations for R and C, so we substitute them to find the profit equation.

$$P = R - C = (10n) - (1000 + 5n) = 10n - 1000 - 5n = 5n - 1000,$$

or

$$P = 5n - 1000$$

As always, it's a good idea to put the units back in this equation just to make sure we remember what it represents:

$$profit = 5\frac{\$}{book} \times number\ of\ books - \$1000$$

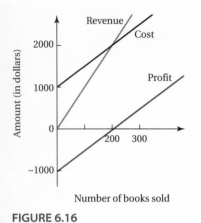

FIGURE 6.16

The three functions for cost, revenue, and profit are graphed in Figure 6.16. Note that the profit function is negative if few books are sold and positive if many books are sold. To find the *break-even point,* or the number of books that yields a profit of exactly $0, we must solve the profit function for the number of books. We first add 1000 to each side to find

$$P + 1000 = 5n.$$

Now we divide both sides by 5 to find

$$n = \frac{P + 1000}{5}.$$

We substitute a profit of $P = 0$ to find the number of books n that represent the break-even point.

$$n = \frac{P + 1000}{5}$$

$$= \frac{0 + 1000}{5}$$

$$= 200$$

According to this model, 200 books must be sold to break even.   ■

TIME-OUT TO THINK: The model in Example 8 assumes that you sell *all* the books that you print. How will your profits be affected if you are unable to sell some of the books? Explain.

REVIEW QUESTIONS

1. What does it mean to say that a function is *linear*?

2. Define *rate of change* of a function. Explain why a linear function must have a constant rate of change.

3. Suppose that we graph a linear function. How do you find the *slope* of the graph? How is the slope related to the rate of change of the function?

4. What is the *rate of change rule*? Give an example of how it is used.

5. Describe the general form of a linear equation. What is the *initial value*?

6. What two basic rules of algebra are needed to solve linear equations? Give an example of the use of these rules.

7. Describe the process of creating a linear equation from two data points. How are such models useful?

PROBLEMS •

1. **Rates of Change.** Each graph below shows a linear function (*time, rain depth*) for a particular rain storm. Find the slope of the graph and determine the rate at which rain was falling during the storm. Be sure to include units with your answers.

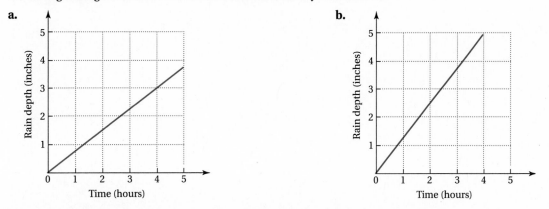

2. **Rates of Change.** Each graph below shows a linear function (*time, population*) for a particular village. Find the slope of the graph and determine the rate at which the population was growing. Be sure to include units with your answers.

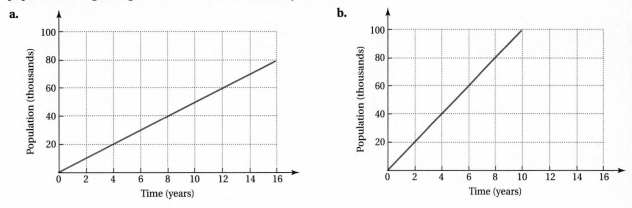

Analyzing Linear Graphs. *For each graph shown in Problems 3 and 4, do the following.*

i) *Explain the function that is represented by the graph.*
ii) *Find the slope of the graph and express it as a rate of change (be sure to include units).*
iii) *Explain whether you think a linear model is realistic for the particular function.*

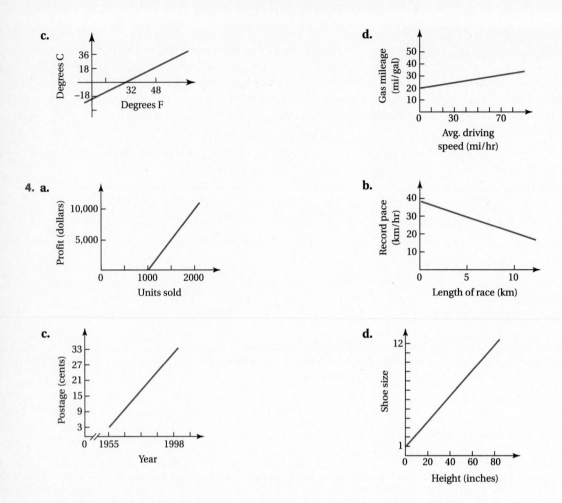

c.

d.

4. a.

b.

c.

d.

Rate of Change Rule. *Each situation in Problems 5–13 describes a rate of change that you may assume to be constant. Write a statement that describes how one variable varies with respect to another, give the rate of change numerically (with units), and use the rate of change rule to answer any questions.*

Example: *Every week your fingernails grow 5 millimeters. How much will your fingernails grow in 2.5 weeks?*

Solution: *The* length of your fingernails *varies with respect to* time, *with a rate of change of 5 mm/week. Your fingernails will grow 5 mm/wk × 2.5 wk = 12.5 mm in 2.5 weeks.*

5. The water depth in a reservoir decreases at a rate of 0.25 inches per hour due to evaporation. How much does the water depth change in 6.5 hours? in 12.5 hours?

6. You drive along the highway at a constant speed of 50 miles per hour. How far do you travel in 2.5 hours? in 5.7 hours?

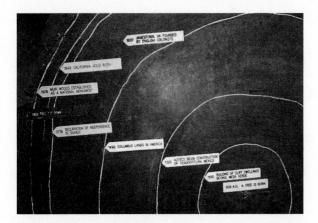

7. A tree increases in diameter by 0.2 inches per year by adding annual rings. How much does the diameter of the tree increase in 4.5 years? in 20.5 years?

8. A gas station owner finds that for every penny increase in the price of gasoline, she sells 150 fewer gallons of gas per week. How much more or less gas will she sell if she raises the price by 8 cents per gallon? If she decreases the price by 2 cents per gallon?

9. A new candle burns at a rate of 2 centimeters per hour. How much does the length of the candle change in 3.8 hours? in 4.2 hours?

10. The population of BoomTown is increasing at a rate of 200 people per year. How much does the population increase in 10.4 years? in 19.4 years?

11. Snow accumulates during a 12-hour storm at a constant rate of 4 inches per hour. How much snow accumulates in the first 5.5 hours? in the first 7.8 hours?

12. According to one formula, your maximum heart rate (in beats per minute) is 220 minus your age (in years). How much does your maximum heart rate change from age 35 to age 50? What is your maximum heart rate at age 65?

13. The boiling point of water (212°F at sea level) decreases by about 2°F for every 1000 feet increase in elevation. What is the boiling point of water at an elevation of 6000 feet? at an elevation of 12,000 feet?

Computer Chip Production. *In Problems 14 and 15, suppose that you work on an assembly line for computer chips. Based on the given information, answer the following.*

    a. *What is the initial value and rate of change of the function* (time, chips produced)? *Include units with your answers.*

    b. *Draw a graph of the linear function in part (a).*

    c. *Find the equation of this function. Write the equation both with units and in a more compact form with single letters for the variables; be sure to define your variables clearly.*

14. When you arrive at work there are 55 chips in stock. While you work, chips are produced at a rate of 6 chips per hour.

15. When you arrive at work there are 120 chips in stock. While you work, chips are produced at a rate of 14 chips per hour.

Linear Graphs. *Problems 16–21 describe a situation that can be modeled by a linear function. For each problem, do the following.*

    a. *Identify the independent and dependent variables involved in the function.*

    b. *Draw an accurate linear graph of the function.*

    c. *Use your graph to answer the additional question in the problem.*

    d. *Discuss whether a linear model is reasonable for the situation described.*

16. Your distance varies with time as you drive along a highway at 40 miles per hour for 4 hours. *Additional Question:* How far do you travel in 2.6 hours?

17. The diameter of a tree is growing by 0.2 inches with each passing year. When you started observing the tree, its diameter was 4 inches. *Additional Question:* Estimate the time at which the tree started growing.

18. The cost of publishing a poster is $2000 for setting up the printing equipment, and then increases by $3 per poster printed. *Additional Question:* What is the total cost to produce 2000 posters?

19. The amount of sugar in a fermenting batch of beer decreases with time at a rate of 0.1 gram per day, starting from an initial amount of 5 grams. *Additional Question:* When is the sugar gone?

20. The cost of a particular private school begins with an initiation fee of $2000 plus annual tuition of $10,000. *Additional Question:* How much will it cost to attend this school for 6 years?

21. A maximum speed of a semitrailer truck up a steep hill varies with the weight of its cargo. With no cargo it can maintain a maximum speed of 50 miles per hour. With 20 tons of cargo, its maximum speed drops to 40 miles per hour. *Additional Question:* At what load does the model predict a maximum speed of 0 miles per hour?

Linear Equations. *Problems 22–29 describe a situation that can be modeled by a linear function. For each problem, do the following.*

    a. *Identify the independent and dependent variables involved in the function.*

    b. *Write a linear equation to represent the function.*

    c. *Use your equation to answer the additional question in the problem.*

    d. *Discuss whether a linear model is reasonable for the situation described.*

22. The price of a particular model car is $12,000 today, and rises with time at a constant rate of $1200 per year. *Additional Question:* How much will a new car cost in 2.5 years?

23. In January 1998, a megabyte of memory cost $25. The price of a megabyte of computer memory is decreasing with time at a rate of $5 per year. *Additional Question:* What price is predicted for July 2001?

24. The world record time in the 100-meter butterfly (swimming) was 53.0 seconds in 1988. Assume that the record falls at a constant average rate of 0.05 second per year. *Additional Question:* What does the model predict for the record in 2000?

25. A new candle is 20 centimeters long. Once lit its length shortens at a rate of 2 cm per hour. *Additional Question:* How long does the candle last?

26. A snowplow has a maximum speed of 30 miles per hour on a dry highway. Its maximum speed decreases by 0.5 mile per hour for every inch of snow on the highway. *Additional Question:* According to this model, at what snow depth will the plow be unable to move?

27. The cost of renting a car is a flat $40 plus an additional 10 cents per mile that you drive. *Additional Question:* How far can you drive for $90?

28. You can rent time on computers at the local copy center for a $5 set-up charge and an additional $3 for every five minutes. *Additional Question:* How much time can you rent for $15?

29. Beginning in 1980, the population of BoomTown began increasing at a rate of 200 people per year. The 1980 population was 2000 people. *Additional Question:* What is your projection for the population in the year 2010?

Graphing General Linear Equations. *For each of the functions in Problems 30 and 31, find the slope of the graph and the y-intercept; then sketch the graph for values of* x *between* -5 *and 5.*

30. a. $y = 2x + 6$ **b.** $y = -3x + 3$

    c. $y = -5x - 5$ **d.** $y = 4x + 1$

31. a. $y = 3x - 6$ **b.** $y = -2x + 5$

    c. $y = -x + 4$ **d.** $y = 2x + 4$

Solving Linear Equations. *In Problems 32–34, show how you solve the given equation to answer the questions about each function.*

32. The depth of water in a swimming pool as it is being filled is given by the function

$$d = 2.5 \frac{\text{in.}}{\text{hr}} \times t,$$

where *d* is the water depth in inches and *t* is the time in hours. When is the water 15 inches deep? When is the water 75 inches deep?

33. The population *p* of Pleasantville is given by the function

$$p = 550 \text{ people} + 150 \frac{\text{people}}{\text{yr}} \times t,$$

where *p* is the population and *t* is the time in years since 1990. When will the population reach 1000?

34. The profit for a fund-raiser is given by the function

$$P = 8 \frac{\$}{\text{ticket}} \times n - \$200,$$

where *P* is the profit and *n* is the number of tickets sold. How many tickets must be sold for the organizers to make $400?

Creating Linear Equations. *For Problems 35 and 36, follow the procedure given in the text to determine the linear equation of the line that passes through each pair of data points. Then draw a graph of the linear function. Be sure to confirm that the line passes through the data points.*

35. a. $(2, 7)$ and $(4, 13)$ **b.** $(-1, 3)$ and $(0, 7)$
    c. $(14, 2)$ and $(-4, -20)$ **d.** $(17, 9)$ and $(4, 8)$

36. a. $(3, 1)$ and $(1, 7)$ **b.** $(1, 1)$ and $(6, 6)$
    c. $(7, -11)$ and $(-2, -7)$ **d.** $(1, 5)$ and $(2, 10)$

Creating and Solving Linear Equations. *Problems 37–42 ask you to create and use linear equations.*

37. Linear Growth. Suppose that you were 20 inches long at birth and 4 feet tall on your tenth birthday. Create a linear equation for the function (*age, height*) based on these two data points. Use your equation to predict your height at ages 2, 6, 20, and 50. Comment on the validity of this linear model.

38. Lease Vs. Purchase. Suppose that you can purchase a motorcycle for $4500 or lease it for a down payment of $100 and a monthly payment of $120. Find an equation that describes how the cost of the lease depends on time. How long can you lease the motorcycle before you've paid more than its purchase price?

39. Fund-raising Strategy. A YMCA fund-raiser offers raffle tickets for $5 each. The prize for the raffle is a $350 television set, and must be purchased with proceeds from the ticket sales. Find an equation that gives the profit/loss for the raffle as it varies with the number of tickets sold. How many tickets must be sold before the raffle begins to make a profit?

40. Fund-raising Strategy. The Psychology Club plans to pay a visitor $75 to speak at a fund-raiser. Tickets will be sold for $2 apiece. Find an equation that gives the profit/loss for the event as it varies with the number of tickets sold. How many people must attend the event for it to break even?

41. Depreciation of Equipment. A $1000 washing machine in a laundromat is depreciated for tax purposes at a rate of $50 per year. Find an equation for the depreciated value of the washing machine as it varies in time. When does the depreciated value reach $0?

42. Sales Tax. In the town of Paradise Valley a 3% local sales tax and a 2% state sales tax are charged on all retail sales. Let p be the pretax amount of a purchase in dollars. Let T be the aftertax amount of the purchase. Find a linear equation that describes how T varies with p. What is the total price of an item that costs $7.50 before taxes?

43. Pricing Strategies. The owner of a hubcap store devises the following pricing strategy to promote sales. He lets h be the number of hubcaps a customer buys and p be the corresponding price per hubcap. He then sets the prices as follows.

    If $1 \leq h < 10$, then $p = \$30$.
    If $10 \leq h < 20$, then $p = \$25$.
    If $h \geq 20$, then $p = \$20$.

Make a graph of the function (h, p). Based on your graph, does it make sense for a customer to buy exactly nine hubcaps? Explain. Also explain why this pricing strategy might promote sales.

44. Book Publishing. Suppose that the cost of publishing a book is $2000 for setting up the printing equipment plus $15 per book. You will sell the books for $20 per copy.

    a. Find the function (n, C) that gives the cost C of printing n books.

    b. Find the function (n, R) that gives the revenue R that results from selling n books.

    c. Use your results from parts (a) and (b) to find a function (n, P) that gives the profit P that results from printing and selling n books. Graph the profit function.

    d. What is the break-even point for the profit function? What is the profit (or loss) when 100 books are sold? when 500 books are sold?

45. Salesperson Strategies. Imagine that you sell greeting cards and have a choice of (A) earning a salary of $800 per month plus a 10% commission on all sales, or (B) earning no base salary and a 20% commission on all sales.

    a. Let s represent your monthly sales (in dollars) and E represent your monthly earnings. Find the function (s, E) that describes your monthly earnings under option (A). Graph the function. Does it give the earnings that you expect if $s = 0$? Explain.

b. Find the function (s, E) that describes your monthly earnings for a given amount of sales under option (B). Graph the function on the same set of axes used in part (a). Does it give the earnings that you expect if $s = 0$? Explain.

c. From your graphs, determine which option is preferable if your monthly sales are $2000. Which option is preferable if your monthly sales are $4000? Where is the "trade-off" point at which the two options give you the same earnings? Explain.

46. Project: Wildlife Management. A common technique for estimating populations of birds or fish is to tag and release individual animals in two different outings. This procedure is called *catch and release*. If the wildlife remain in the sampling area and are randomly caught, a fraction of the animals tagged during the first outing are likely to be caught again during the second outing. Based on the number caught and the fraction caught twice, the total number of animals in the area can be estimated.

a. Consider a case in which 200 fish are tagged and released during the first outing. During a second outing in the same area, 200 fish are again caught and released, of which one-half are already tagged. Estimate N, the *total* number of fish in the entire sampling area. Explain your reasoning.

b. Consider a case in which 200 fish are tagged and released during the first outing. During a second outing in the same area, 200 fish are again caught and released, of which one-fourth are already tagged. Estimate N, the *total* number of fish in the entire sampling area. Explain your reasoning.

c. Generalize your results from parts (a) and (b) by letting p be the fraction of tagged fish that are caught during the second outing. Find a formula for the function (p, N) between the total number of fish, N, and the fraction tagged during the second outing, p.

d. Graph the function obtained in part (c). Is it linear? What is the domain? Explain.

e. Suppose that 15% of the fish in the second sample are tagged. Use the formula from part (c) to estimate the total number of fish in the sampling area. Confirm your result on your graph.

f. Locate a real study in which catch and release methods were used. Report on the specific details of the study and how closely it followed the theory outlined in this problem.

UNIT 6C

FORMULAS AS MODELS

Although linear functions are very important and useful, many mathematical models involve functions that are **nonlinear**; that is, any function whose rate of change is *not* constant and whose graph is *not* a straight line. Most of the nonlinear functions you will encounter in this book and in daily life will be *formulas*—recipes that provide a mathematical description of something real.

We've already used many formulas in this book, including the financial formulas in Chapter 5. In this unit, we look in more detail at some of the methods that we can use to work with and analyze formulas.

LINEAR VS. NONLINEAR

We saw many examples of linear functions in Unit 6B. Now let's graph a nonlinear function. The function that gives the *area* of a circle, *A*, in terms of its *radius, r*, is nonlinear. This area function is $A = \pi r^2$. We can graph this function by choosing a few values of *r* and calculating the corresponding values of *A*. This gives us a table of ordered pairs as follows:

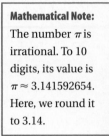

Mathematical Note:
The number π is irrational. To 10 digits, its value is $\pi \approx 3.141592654$. Here, we round it to 3.14.

$r =$	0	1	2	3	4
$A =$	0	3.14	12.56	28.26	50.24

We get a good picture of the area function by connecting just these five points (Figure 6.17). The fact that the graph is not a straight line confirms that it is a nonlinear function.

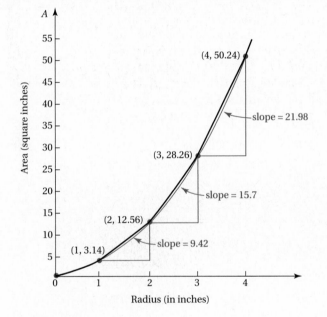

FIGURE 6.17

Note that the nonlinear graph does not have a constant slope. Instead, the slope gets gradually steeper as we look farther up the curve (larger values of *r*). We can find the average slope between any pair of points on the curve by dividing the vertical rise by the horizontal run. For example, if we choose the point (1, 3.14) as P_1 and the point (2, 12.56) as P_2, the average slope between them is

$$\text{slope} = \frac{\text{change in } \textit{dependent variable} \text{ from } P_1 \text{ to } P_2}{\text{change in } \textit{independent variable} \text{ from } P_1 \text{ to } P_2} = \frac{12.56 - 3.14}{2 - 1} = 9.42.$$

Similar calculations for other pairs of points give the results shown in Figure 6.17. The fact that we find different slopes between different points along the curve also means the rate of change of the function varies along the curve. A varying rate of change is what distinguishes nonlinear functions from linear functions.

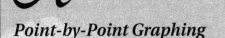

Point-by-Point Graphing

There are two basic approaches to graphing any function.

- We can make the graph using **analytical methods** by analyzing the properties of the function. For example, graphing a linear function by determining its slope and y-intercept is an analytical method.

- We can make a graph **point-by-point** by creating a table of points that satisfy the function and then plotting and connecting the points.

It requires a lot of practice to use analytical methods to graph nonlinear functions, though these methods usually are the fastest way to graph an equation. Today, you can purchase graphing calculators or specialized computer software that can graph equations easily. But if you don't have such technological aids and are out of practice with analytical methods, you can always resort to point-by-point graphing. The only "trick" is to use appropriate care in making your table of points.

Example A: Graph the function $y = 5 - x^2$.

Solution: We begin by making a table of points using integer values of x.

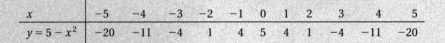

x	-5	-4	-3	-2	-1	0	1	2	3	4	5
$y = 5 - x^2$	-20	-11	-4	1	4	5	4	1	-4	-11	-20

These points are plotted in Figure 6.18(a).

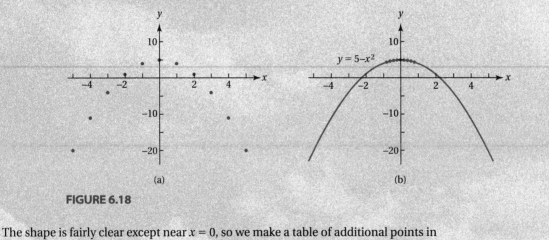

(a)   (b)

FIGURE 6.18

The shape is fairly clear except near $x = 0$, so we make a table of additional points in this region.

x	−0.8	−0.6	−0.4	−0.2	−0.1	0	0.1	0.2	0.4	0.6	0.8
$y = 5 - x^2$	4.36	4.64	4.84	4.96	4.99	5	4.99	4.96	4.84	4.64	4.36

With these additional points, it is easy to trace the shape of the graph (Figure 6.18b).

Example B: Graph the function $y = 1/x$.

Solution: We start by making a table using integer values of x.

x	−5	−4	−3	−2	−1	0	1	2	3	4	5
$y = 1/x$	−0.2	−0.25	−0.33	−0.5	−1	???	1	0.5	0.33	0.25	0.2

Even before drawing a graph, we know we will have a problem around $x = 0$ because $y = 1/x$ is undefined when $x = 0$. Thus we include additional points near $x = 0$:

x	−0.75	−0.5	−0.25	−0.1	−0.01	−0.001	0.001	0.01	0.1	0.25	0.5	0.75
$y = 1/x$	−1.33	−2	−4	−10	−100	−1000	1000	100	10	4	2	1.33

The graph is shown in Figure 6.19. Note that the graph has no value at $x = 0$, the y-values become large and positive as we approach $x = 0$ from the right, and the y-values become large and negative as we approach $x = 0$ from the left.

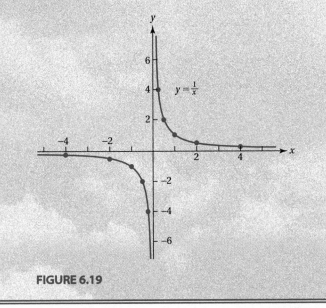

FIGURE 6.19

VARIABLES AND CONSTANTS

The only significant pitfall in reading a formula lies in distinguishing between symbols that represent *variables* and symbols that represent **constants**, or quantities with fixed numerical values. For example, consider the equation of a line written in the form

$$y = mx + b.$$

If you didn't know any better, you might think that this equation contained four variables because it has four different letters. However, only x and y are variables in this equation. The symbols m and b are constants: m is the constant slope (rate of change) and b is the fixed y-intercept (initial value). These two values never change for a particular linear function.

TIME-OUT TO THINK: What symbols represent variables in the area function for a circle, $A = \pi r^2$? What symbol represents a constant?

As another example, consider the formula relating distance, velocity and time:

$$d = v \times t,$$

where d represents *distance*, v represents *velocity*, and t represents the elapsed *time*. This formula involves three symbols, all of which may be variables depending on the context. However, in most problems involving distance, velocity, and time, one of the symbols will represent a constant with a fixed value. For example, if you travel at a constant velocity of 60 mi/hr, your distance traveled (in miles) depends on the elapsed time (in hours) according to

$$d = 60 \times t.$$

Note that this is a linear function of the form (t, d), in which t is the independent variable and d is the dependent variable. It tells us, for example, that you would travel a distance of $d = 120$ miles if you travel for $t = 2$ hours at a constant speed of 60 mi/hr.

EXAMPLE 1   *Holding Distance Constant*

Suppose you are planning a 500-mile trip. Describe how the time required for the trip depends on your velocity. Discuss any assumptions in this model.

Solution: The trip distance is 500 miles, so the distance d will be constant in this problem. Thus we are looking for a function (v, t) that describes how the dependent variable t (time) depends on the independent variable v (velocity). We therefore solve the distance formula for t by dividing both sides of the formula by v:

$$\frac{d}{v} = \frac{v \times t}{v} \quad \Rightarrow \quad \frac{d}{v} = t \quad \text{or} \quad t = \frac{d}{v}$$

Because d is held constant in this problem, we replace it with its fixed numerical value, $d = 500$ miles. The (v, t) function becomes

$$t = \frac{500 \text{ miles}}{v}.$$

By plotting just a few points, we can create a graph of this function (Figure 6.20). The velocity appears in the denominator, so a faster velocity means a shorter time for the trip. For example, at a speed of 50 mi/hr, the trip will take

$$t = \frac{500 \text{ mi}}{50 \text{ mi/hr}} = 10 \text{ hr}.$$

At a speed of 60 mi/hr, the trip will take

$$t = \frac{500 \text{ mi}}{60 \text{ mi/hr}} = 8.33 \text{ hr}.$$

> **Mathematical Note:**
> When two variables are related in such a way that an increase in one variable produces an *increase* in the other, the variables are said to be *directly proportional*. When an increase in one variable produces a *decrease* in the other, the variables are *inversely proportional*. In the (v, t) function, v and t are inversely proportional.

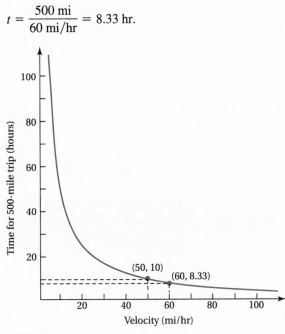

FIGURE 6.20

Several assumptions are built into this model of how your trip time depends on your velocity. For example, the formula allows only a single value of velocity for any *particular* trip. Thus the formula will be valid only if we use your *average* velocity for the trip. ■

TIME-OUT TO THINK: Note the similarity between the graph in Figure 6.20 and the graph of the function $y = 1/x$ in Figure 6.19. Explain why the two graphs are similar. Why doesn't Figure 6.20 show negative values for the velocity?

EXAMPLE 2 *Light Waves*

Light travels through space as a *wave*, somewhat like waves on water or along a string. As shown in Figure 6.21(a), a light wave is characterized by

- a **frequency** that represents the number of waves passing a fixed position each second, and
- a **wavelength** that represents the distance between two adjacent crests of the wave.

The light that our eyes can see is only the tip of the iceberg when it comes to all forms of light (Figure 6.21b).

By the Way ···········
Strangely, light can behave as a *particle* as well as a wave. Particles of light are called *photons*, and each photon is characterized by a wavelength and frequency. The complete spectrum of light, from radio waves to gamma rays, is called the *electromagnetic spectrum*.

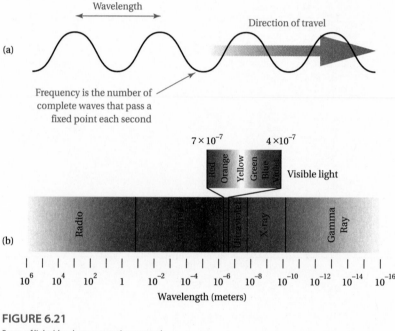

FIGURE 6.21
Forms of light (the electromagnetic spectrum).

A simple formula relates the wavelength, the frequency, and the speed of a light wave:

$$\text{wavelength} \times \text{frequency} = \text{speed of light} = 3 \times 10^8 \text{ m/s}$$

Many books represent the wavelength with the Greek letter λ (lambda) and the frequency with the Greek letter ν (nu). The speed of light, which is the same for all forms of light, is represented by the letter c. Thus the formula becomes

$$\lambda \times \nu = c.$$

Identify the variables and constants in this formula, then find a function that describes how the frequency of a light wave depends on its wavelength.

By the Way ···········
The wavelength-frequency-speed formula is easy to understand if you think of a passing train. The wavelength corresponds to the length of the train cars, the frequency is the number of cars that pass you each minute. So the speed of the train is the length of the cars times the number of cars that pass you each minute.

Solution: The speed of light, $c = 3 \times 10^8$ m/s, is a constant. The variables are the wavelength λ and the frequency ν. Because we want to find out how the frequency of a light wave depends on its wavelength, we are looking for a function (λ, ν) in which λ is the independent variable and ν is the dependent variable. We find this function by dividing both sides of the given formula by λ:

$$\frac{\lambda \times \nu}{\lambda} = \frac{c}{\lambda} \quad \Rightarrow \quad \nu = \frac{c}{\lambda}$$

Because the wavelength λ appears in the denominator, the frequency gets *larger* when the wavelength gets smaller, and vice versa. Figure 6.22 shows a graph of the function; it is *nonlinear* because the graph is not a straight line.

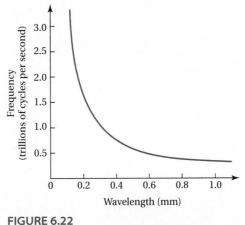

FIGURE 6.22

We can use this function to find the frequency of any light wave for which we know the wavelength. For example, Figure 6.21(b) shows that yellow light is in the middle of the visible spectrum and hence has a wavelength of about $\lambda = 5.5 \times 10^{-7}$ meter. Therefore the frequency of yellow light is

$$\nu = \frac{c}{\lambda}$$

$$= \frac{3 \times 10^8 \text{ m/s}}{5.5 \times 10^{-7} \text{ m}}$$

$$= \frac{5.5 \times 10^{14}}{\text{s}}.$$

Note that the units of the frequency are *per second*, meaning that 5.5×10^{14}, or 550 trillion, waves pass any point each second for yellow light.

By the Way ·············
Although the units of frequency are simply *per second*, they are often called *cycles per second* or *hertz*. Radio station "call numbers" are frequencies; for example, a radio station at "97 Mhz" broadcasts radio waves with a frequency of 97 megahertz, or 97 million waves *per second*.

Mathematical Note:

The square root is defined to be a positive number. For example, $\sqrt{4} = 2$. However, an equation such as $x^2 = 4$ has two solutions, $x = \sqrt{a}$ and $x = -\sqrt{a}$. So we must be careful when taking the square root of both sides of an equation. None of the examples in this book require dealing with this case.

FORMULAS WITH POWERS OR ROOTS

So far, every function we've solved has required only adding, subtracting, multiplying, or dividing both sides of an equation by the same quantity. However, if a function involves powers or roots, we may need two other rules of algebra.

Algebraic Rules for Equations with Powers or Roots

- We can raise both sides of an equation to the same power. For example, if it is true that $x = y$, it is also true that $x^2 = y^2$.
- We can take the same root of both sides of an equation. For example, if it is true that $x = y$, it is also true that $\sqrt{x} = \sqrt{y}$ (provided $\sqrt{x}$ and $\sqrt{y}$ are both defined).

EXAMPLE 3   *Soda Can Radius*

The formula for the volume of a *right circular cylinder* (Figure 6.23) is

$$V = \pi r^2 h,$$

where V is the cylinder's *volume*, r is its *radius*, and h is its *height*.

    A standard (12-ounce) soda can has a volume of 355 milliliters, or 355 cm³, and a height of 12.4 cm. What is its radius? Discuss any potential problems in modeling the soda can as a right circular cylinder.

Solution: We are given constant values for both the volume and height of the can, and π is also a constant. The only unknown variable is the radius, so we need to solve the volume formula for the radius. The first step is to isolate the term r^2 by dividing both sides by πh.

$$\frac{V}{\pi h} = \frac{\pi r^2 h}{\pi h} \quad \Rightarrow \quad r^2 = \frac{V}{\pi h}$$

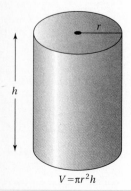

$V = \pi r^2 h$

FIGURE 6.23

Now we can solve for r by taking the square root of both sides.

$$\sqrt{r^2} = \sqrt{\frac{V}{\pi h}} \quad \Rightarrow \quad r = \sqrt{\frac{V}{\pi h}}$$

Finally, we substitute the volume $V = 355$ cm³ and height $h = 12.4$ cm.

$$r = \sqrt{\frac{355 \text{ cm}^3}{\pi \times 12.4 \text{ cm}}} = \sqrt{9.1 \text{ cm}^2} = 3.0 \text{ cm}$$

    By modeling the soda can as a right circular cylinder, we find that its radius is 3.0 cm. However, a soda can is not a perfect right circular cylinder: It has an indented base and top, and therefore the volume formula for a right circular cylinder will not be exact for a real soda can. ■

EXAMPLE 4   *Total Return on a Mutual Fund*

In Unit 5E, we used a formula that allows us to calculate the annual return on an investment from the total return:

$$\text{annual return} = (\text{total return} + 1)^{1/Y} - 1,$$

where Y is the number of years the investment was held. Suppose that a mutual fund advertises an annual return over the last five years of 11.6%, assuming all dividends were reinvested and not counting fees. What is the total return for the five-year period? If you had invested $1000 five years ago, how much would you have now? Discuss any assumptions.

Solution: First, let's write the formula more compactly by letting A stand for the annual return and T stand for the total return. It now takes the form

$$A = (T + 1)^{1/Y} - 1.$$

We are given constant values for annual return ($A = 11.6\% = 0.116$) and the number of years ($Y = 5$), so the only unknown variable is the total return T. Thus we need to solve the formula for T. First, we add 1 to both sides.

$$A + 1 = (T + 1)^{1/Y} - 1 + 1 \quad \Rightarrow \quad A + 1 = (T + 1)^{1/Y}$$

Next we raise both sides to the Y power and simplify.

$$(A + 1)^Y = \left[(T + 1)^{1/Y}\right]^Y \quad \Rightarrow \quad (A + 1)^Y = (T + 1)^{1/Y \times Y}$$
$$\Rightarrow \quad (A + 1)^Y = (T + 1)$$

Finally, we subtract 1 from both sides.

$$(A + 1)^Y - 1 = (T + 1) - 1 \quad \Rightarrow \quad (A + 1)^Y - 1 = T$$

or $$T = (A + 1)^Y - 1$$

We now find the total return by substituting the given values.

$$T = (A + 1)^Y - 1 = (0.116 + 1)^5 - 1 = 1.73 - 1 = 0.73.$$

> **A Brief Review:**
> When we raise a power to a power, we multiply the exponents. For example,
> $$(2^3)^4 = 2^{3 \times 4} = 2^{12}.$$
> In general,
> $$(a^x)^y = a^{x \times y}.$$

The total return for the five years is 73%. If you had invested $1000 five years ago, this formula predicts that you'd have a balance

$$\$1000 + (73\% \times \$1000) = \$1000 + \$730 = \$1730.$$

This number will be accurate only if you reinvested all dividends and paid the fees separately, rather than having them withdrawn from your account. It also assumes that no taxes or other withdrawals were made. ∎

FORMULAS WITH EXPONENTS OR LOGS

Whenever we want to solve a formula for a variable that appears as an *exponent*, we must make use of *logarithms*. You've probably encountered logarithms before, but let's briefly review their meaning.

Mathematical Note:
More generally, $\log_b x$ means *the power to which b must be raised to obtain x*, or *b to what power equals x?* That is, if $\log_b x = y$, it must also be true that $b^y = x$. In this book, we use only *common* logarithms with base $b = 10$.

A **logarithm**, or **log** for short, is a way of expressing a power. For example,

$$\log_{10} x \text{ means } \textit{the power to which 10 must be raised to obtain x.}$$

You may find it easier to remember the meaning of a logarithm with a less technical definition:

$$\log_{10} x \text{ means } \textit{10 to what power equals x?}$$

The following examples illustrate the meaning of a logarithm.

$$\log_{10} 1000 = 3 \text{ because } 10^3 = 1000$$
$$\log_{10} 10{,}000{,}000 = 7 \text{ because } 10^7 = 10{,}000{,}000$$
$$\log_{10} 0.1 = -1 \text{ because } 10^{-1} = 0.1$$
$$\log_{10} 30 = 1.477 \text{ because } 10^{1.477} = 30$$

In this book, we'll make use of three important rules when working with logarithms.

Three Rules for Logarithms

1. Taking the logarithm of a power of 10 gives the power. That is,

$$\log_{10} 10^x = x.$$

2. Raising 10 to a power that is the logarithm of a number gives back the number. That is,

$$10^{\log_{10} x} = x \qquad (\text{for } x > 0).$$

3. We can "bring down" an exponent within a logarithm by applying the *power rule* for logarithms:

$$\log_{10} a^x = x \times \log_{10} a \qquad (\text{for } a > 0)$$

TIME-OUT TO THINK: Use the definition of a logarithm to explain why each of the three rules *must* be true. For example, explain why $\log_{10} 10^5$ must be 5 by remembering that $\log_{10} 10^5$ means *10 to what power equals 10^5?* Similarly, explain why $10^{\log_{10} 100}$ must be 100 and why $\log_{10} 10^4$ must be $4 \times \log_{10} 10$.

Two more rules of algebra help us work with formulas that contain variables in exponents or logarithms.

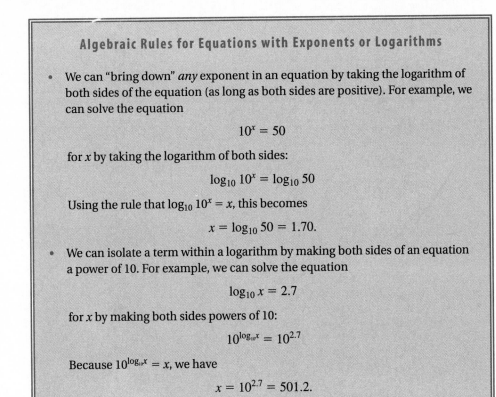

Algebraic Rules for Equations with Exponents or Logarithms

- We can "bring down" *any* exponent in an equation by taking the logarithm of both sides of the equation (as long as both sides are positive). For example, we can solve the equation

$$10^x = 50$$

for x by taking the logarithm of both sides:

$$\log_{10} 10^x = \log_{10} 50$$

Using the rule that $\log_{10} 10^x = x$, this becomes

$$x = \log_{10} 50 = 1.70.$$

- We can isolate a term within a logarithm by making both sides of an equation a power of 10. For example, we can solve the equation

$$\log_{10} x = 2.7$$

for x by making both sides powers of 10:

$$10^{\log_{10} x} = 10^{2.7}$$

Because $10^{\log_{10} x} = x$, we have

$$x = 10^{2.7} = 501.2.$$

Using Your
_____
Calculator   **Common Logs and 10^x**

You can find the common logarithm ($\log_{10}$) of any number with the ⬭log key on your calculator. The exact key sequence will vary with different calculators. For example, on most calculators you can find $\log_{10} 50$ by pressing either

  50 ⬭log ⬭=   or   ⬭log 50 ⬭=.

Check to see which sequence works on your calculator; you should get the answer $\log_{10} 50 = 1.69897$. (Note: Most calculators also have a key for taking *natural logarithms*, or $\log_e$, usually labeled ⬭ln. However, some calculators use different labels, so be sure that you know which key to use when you want $\log_{10}$.)

On most calculators, you can find any power of 10 by using the ⬭10^x key. For example, depending on your calculator, you can find $10^{2.7}$ by pressing either

  2.7 ⬭10^x ⬭=   or   ⬭10^x 2.7 ⬭=.

You should find the answer $10^{2.7} = 501.187$.

If your calculator does not have a ⬭10^x key, you can still do powers of 10 by using the ⬭y^x key. For example, you can also find $10^{2.7}$ by pressing the sequence

  10 ⬭y^x 2.7 ⬭=.

You should confirm that this gives the same answer, $10^{2.7} = 501.187$.

EXAMPLE 5   *Tumor Growth*

A cell biologist observes a small tumor in its early stages of growth. She determines that the number of cells in the tumor grows with time according to the following formula.

$$N = 3000 \times 10^{(0.12 \times t)},$$

where t is the *time in weeks* after the first observation and N is the *number of cells* in the tumor. How many cells did the tumor contain at the time of the first observation? After how many weeks will the tumor contain 500,000 cells?

Solution: The formula is defined such that $t = 0$ at the time of the biologist's first observation. Thus we substitute $t = 0$ into the formula to find the number of cells in the tumor at the time of first observation.

$$N = 3000 \times 10^{(0.12 \times 0)} = 3000 \times 10^0 = 3000 \times 1 = 3000 \text{ cells}$$

The tumor contained 3000 cells at the time of the first observation. To determine when the tumor will contain $N = 500,000$ cells, we must solve the formula for t. Because t is in the exponent, we should anticipate using logarithms. First, however, we must isolate the term containing t by dividing both sides by 3000.

$$N = 3000 \times 10^{0.12t} \quad \Rightarrow \quad \frac{N}{3000} = 10^{0.12t}$$

Next we need to "bring down" the variable t from the exponent, so we take $\log_{10}$ of both sides.

$$\log_{10}\left(\frac{N}{3000}\right) = \log_{10}\left(10^{0.12t}\right)$$

Using the rule that $\log_{10} 10^x = x$, this becomes

$$\log_{10}\left(\frac{N}{3000}\right) = 0.12t.$$

Finally, we isolate t by dividing both sides by 0.12.

$$\frac{\log_{10}\left(\dfrac{N}{3000}\right)}{0.12} = \frac{0.12t}{0.12} \quad \Rightarrow \quad t = \frac{\log_{10}\left(\dfrac{N}{3000}\right)}{0.12}$$

We now have a function that allows us to find the time at which the number of cells reaches *any* value of N. To find the time at which it reaches 500,000 cells, we substitute $N = 500,000$.

$$t = \frac{\log_{10}\left(\dfrac{500,000}{3000}\right)}{0.12} = \frac{2.22}{0.12} = 18.5$$

Remembering that t carries units of *weeks*, we see that the tumor will reach a size of 500,000 cells after about $18\frac{1}{2}$ weeks.   ■

EXAMPLE 6   *How Long to Get $100,000?*

In Unit 5A, we used the compound interest formula for interest paid once at the end of each year:

$$A = P \times (1 + APR)^Y, \quad \text{where} \begin{cases} P = \text{starting principal} \\ APR = \text{annual percentage rate (as a decimal fraction)} \\ Y = \text{number of years} \\ A = \text{accumulated balance after } Y \text{ years} \end{cases}$$

Suppose that you invest a principal of $10,000 with interest paid at an annual percentage rate of 7%. How long will it take for your money to grow in value to $100,000?

Solution: Because the question asks "how long?", we must solve for the number of years Y in the compound interest formula. First, we divide both sides by the starting principal P.

$$\frac{A}{P} = \frac{P \times (1 + APR)^Y}{P} \quad \Rightarrow \quad (1 + APR)^Y = \frac{A}{P}$$

We need to "bring down" Y from the exponent, so we take the logarithm of both sides and apply the *power rule* for logarithms ($\log_{10} a^x = x \times \log_{10} a$).

$$\log_{10} (1 + APR)^Y = \log_{10} \frac{A}{P} \quad \Rightarrow \quad Y \times \log_{10} (1 + APR) = \log_{10} \frac{A}{P}$$

Finally, we isolate Y by dividing both sides by $\log_{10} (1 + APR)$.

$$\frac{Y \times \log_{10} (1 + APR)}{\log_{10} (1 + APR)} = \frac{\log_{10} \dfrac{A}{P}}{\log_{10} (1 + APR)} \quad \Rightarrow \quad Y = \frac{\log_{10} \dfrac{A}{P}}{\log_{10} (1 + APR)}$$

We are given that the starting principal is $P = \$10,000$, the annual percentage rate is $APR = 7\% = 0.07$, and the accumulated balance (the balance you want to end up with) is $A = \$100,000$. Substituting these values, we find

$$Y = \frac{\log_{10} \dfrac{\$100,000}{\$10,000}}{\log_{10} (1 + 0.07)} = \frac{\log_{10} 10}{\log_{10} 1.07} = \frac{1}{0.029383777} = 34.0.$$

At a 7% interest rate, your $10,000 will grow to $100,000 in 34 years.   ∎

REVIEW QUESTIONS •

1. What is the difference between a linear function and a nonlinear function?

2. How can we estimate the slope between two points on a graph of a nonlinear function? Based on slope, how can we distinguish between linear and nonlinear functions? What does this tell us about the rate of change of a non-linear function?

3. What is the difference between a *variable* and a *constant*? How can we determine which is which if both involve symbols?

4. Describe the two algebraic rules given in the text for solving equations with powers or roots. Give an example using each rule.

5. What does the statement $\log_{10} x$ mean?

6. Describe each of the three rules for logarithms given in the text. Give an example using each rule.

7. Describe the two algebraic rules given in the text for solving equations with exponents or logarithms. Give an example using each rule.

PROBLEMS

1. Formulas You Use. Identify and describe at least three formulas that you have used in the past in your work, your personal life, or in another course.

Point-by-Point Graphing. *Make a table of points as needed and graph each of the functions in Problems 2 and 3.*

2. a. $y = x^2 + 1$
   b. $y = x^2 - 1$
   c. $y = -x^2 + 3$
   d. $y = 0.1x^5$
   e. $y = 4/(x + 1)$
   f. $y = 2^x$

3. a. $y = -x^2$
   b. $y = x^2 + 4$
   c. $y = 3 - x^3$
   d. $y = 3x^3 - 2x^2 + 1$
   e. $y = 10^{0.1x}$
   f. $y = 3/x$

4. Holding Time Constant. Suppose that you drive for 10 hours at a steady speed.

   a. Describe in words how the distance you travel depends on your speed.

   b. Write a function that gives the distance traveled in 10 hours for any speed.

   c. Draw a graph of the function from part (b).

5. Holding Distance Constant. Suppose that you take a 400-mile trip and drive at a steady speed.

   a. Describe in words how the time for the trip depends on your speed.

   b. Write a function that gives the time for the 400-mile trip at any speed.

   c. Draw a graph of the function from part (b).

6. Frequency from Wavelength. Use the formula that relates the wavelength and frequency of a light wave (see Example 2) to answer the following questions. (Recall that 1 nanometer = 10^{-9} meter.)

   a. What is the frequency of red light, which has a wavelength of about 700 nanometers?

   b. What is the frequency of an X-ray with a wavelength of 0.1 nanometer?

   c. What is the frequency of a gamma ray with a wavelength of 0.0005 nanometers?

7. Wavelength from Frequency. Use the formula that relates the wavelength and frequency of a light wave (see Example 2) to answer the following questions. (Hint: 1 megahertz = 1 million waves per second.)

   a. Write the function that describes how wavelength depends on frequency.

   b. What is the wavelength of radio waves from an FM radio station that broadcasts at 106.5 megahertz?

   c. What is the wavelength of radio waves from an AM radio station that broadcasts at 700 kilohertz?

8. Solving Circles. A circle with a radius of r has an area given by the formula $A = \pi r^2$, where π is the constant 3.141592. . . . Answer the following questions (with correct units).

   a. What is the area of a circle whose radius is 3 m?

   b. What is the area of a circle whose radius is 4.7 cm?

   c. What is the radius of a circle whose area is 30 m²?

   d. What is the radius of a circle whose area is 204 m²?

9. Tin Cans. Tin cans or soda cans are cylinders whose volume is given by $V = \pi r^2 h$. Answer the following questions about cylinders (with correct units).

   a. What is the volume of a cylinder whose radius is 2 cm and whose height is 12 m?

   b. What is the radius of a cylinder whose volume is 355 m^3 and whose height is 2 m?

   c. What is the radius of a cylinder whose volume is 28 cm^3 and whose height is 12 cm?

   d. What is the height of a cylinder whose volume is 355 m^3 and whose radius is 25 m?

   e. What is the height of a cylinder whose volume is 225 cm^3 and whose radius is 8.6 mm?

10. Spheres. A sphere with a radius of r has a volume given by the formula $V = (4/3)\pi r^3$. Answer the following questions (with correct units).

   a. What is the volume of a sphere whose radius is 3 m?

   b. What is the volume of a sphere whose radius is 4.7 cm?

   c. What is the radius of a sphere whose volume is 30 m^3?

   d. What is the radius of a sphere whose volume is 204 m^3?

11. Pricing Merchandise. If the price of an item is set too high, few items will be sold, and the total sales revenue will be low. Conversely, if the price of the item is set too low, many items will be sold, but the total sales revenue will be low because of the low price. At some intermediate price, we can expect sales revenue to be a maximum. Suppose that a particular deli has determined that the revenue from sales of their lunch special varies with price according to the following formula.

$$R = -5p^2 + 40\,p,$$

where R is the revenue and p is the price, both measured in dollars. (See the graph below.)

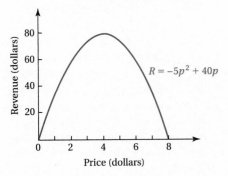
Price (dollars)

   a. What is the revenue when $p = 0$? Explain why this answer makes sense.

   b. What is the revenue when $p = 8$? Use your answer to explain why the maximum revenue must occur for a price less than $8.

   c. What is the revenue when $p = 3$? when $p = 5$?

   d. Experiment with several other values of the price. Can you find the price that gives the highest possible revenue?

Mutual Fund Return. *In Problems 12 and 13, use the formula that gives the annual return on an investment in terms of the total return (see Example 4).*

12. A mutual fund advertises an annual return over the last 10 years of 9.8%. Assume that all dividends were reinvested and ignore broker's fees. What is the total return for the 10-year period?

13. A mutual fund advertises an annual return over the last 5 years of 12.2%. Assume that all dividends were reinvested and ignore broker's fees. What is the total return for the 5-year period?

14. Tumor Growth. Suppose that the number of cells in a tumor is growing according to the formula $N = 35{,}000 \times 10^{0.09 \times t}$, where t is the *time in days* since the tumor was first observed and N is the *number of cells* in the tumor.

   a. How many cells did the tumor contain when it was first observed?

   b. How long after the first observation will the tumor reach a size of 1 million cells?

15. Tumor Growth. Suppose the number of cells in a tumor grows according to the formula $N = 2300 \times 10^{0.21 \times t}$, where t is the *time in days* since the tumor was first observed and N is the *number of cells* in the tumor.

   a. How many cells did the tumor contain when it was first observed?

   b. How long after the first observation will the tumor reach a size of 1 million cells?

Solving the Compound Interest Formula. *In Problems 16–21, use the appropriate version of the formula for the number of compounding periods per year (see Unit 5A).*

16. If you deposit $1000 in an account that pays an *APR* of 5% compounded once a year, how long will it take for an initial deposit of $1000 to grow to $1500?

17. If you deposit $2000 in an account that pays an *APR* of 5% compounded daily, how long will it take for an initial deposit of $2000 to increase to $2500?

18. If you deposit $1000 in an account that pays an *APR* of 7% compounded daily, how long will it take for your balance to reach $100,000?

19. How long will it take your money to double at an *APR* of 8% compounded daily?

20. How long will it take your money to triple at an *APR* of 8% compounded quarterly?

21. How long will it take your money to grow by 50% at an *APR* of 7% compounded daily?

22. Investment Planning. Suppose that you can afford to put $100 per month in a savings plan that pays an *APR* of 7%. When will the value of your investment reach $50,000? $1 million? Use the savings plan formula given in Unit 5B.

23. Accelerated Loan Payment. Suppose that you have a student loan of $25,000 with a fixed rate of 9% for 20 years. Use the loan payments formula (Unit 5C) to answer the following questions.

   a. Calculate your required monthly payments. Also calculate the total amount of the payments you'll make over the life of the loan.

   b. Suppose you decide to increase your total payment to $350 per month. How long will it take to pay off the loan? Compare the total amount of the payments you'll make over the life of the loan in this case with your answer from part (a).

24. Accelerated Mortgage. Suppose that you have a home mortgage of $120,000 with a fixed rate of 7.5% for 30 years. Use the loan payments formula (Unit 5C) to answer the following questions.

   a. Calculate your required monthly payments. Also calculate the total amount of the payments you'll make over the life of the loan.

   b. Suppose you decide to pay an additional $75 per month toward the principal (i.e., your required payment + $75). How long will it take to pay off the loan? Compare the total amount of the payments you'll make over the life of the loan in this case with your answer from part (a).

   c. Suppose you decide to pay an additional $150 per month toward the principal (i.e., your required payment + $150). How long will it take to pay off the loan?

Compare the total amount of the payments you'll make over the life of the loan in this case with your answers from parts (a) and (b). Discuss the results.

25. Biweekly Mortgage Payments. Some financial analysts recommend making biweekly (every two weeks), rather than monthly, mortgage payments. Let's examine this strategy. Suppose that you have a home mortgage of $100,000 at a fixed *APR* of 9% for 30 years.

   a. What are your required monthly payments? How much, in total, do you pay in a year?

   b. Suppose that you make a payment equal to half your monthly payment every two weeks, or 26 times per year. What are your biweekly payments? How much, in total, do you pay in a year?

   c. Using the biweekly payments ($n = 26$ payments per year), how long will it take to pay off the loan? Compare your total payments over the life of the loan for the biweekly payment strategy to the strategy of simply making the required payments for 30 years.

   d. In two or three paragraphs, discuss the pros and cons of the biweekly payment strategy.

26. Free Fall. The distance d that an object falls in a gravitational field t seconds after it is dropped from rest is

$$d = \frac{1}{2}gt^2,$$

where $g = 9.8 \text{ m/s}^2$ is the acceleration due to gravity. If the object is released from the top of a building 100 meters high, how long does it take to hit the ground? (Note: this formula ignores the effects of air resistance.)

27. Probability. The probability p of rolling a die n times and getting a 1 each time is $p = 6^{-n}$. What are the chances of rolling three 1s in a row? of rolling seven 5s in a row?

28. DNA. The number of possible three-letter "words" that can be formed from an alphabet with m letters (with repeated letters allowed) is $w = m^3$ words. DNA, the material that makes up genes, uses four chemical bases called adenine (A), guanine (G), thymine (T), and cytosine (C) to encode genetic information. These four bases are grouped into three-base "words" and each word codes for a specific amino acid. For example, the "word" CAG codes for the amino acid glutamine, and GAC codes for aspartic acid. How many three-letter words can be formed from this four-base genetic alphabet? As only 20 amino acids are used in the human body, are this many words needed? Explain why biologists say that the genetic code contains redundancy.

29.   Postage Inflation. The cost of postage has risen over the past several decades. A recent analysis claimed that the cost of postage for a first-class letter in any year can be roughly estimated by the formula $p = 3 \times 2^{t/10}$, where t is the number of years since 1955 and p is the postage in cents. According to this formula, what should you expect for the current cost of a first-class stamp? What is the actual cost of a first-class stamp at present? Comment on the validity of the formula.

30.   Project: Distribution of Wealth. Economists use *Lorenz functions* to describe the distribution of wealth in a culture or society. Imagine that the entire population under consideration is ranked according to wealth. Then the value, w, of the Lorenz function for a fraction, p, is the fraction of the total wealth that is owned by the poorest pth of the population. For example, if $w = 1/3$ when $p = 1/2$, one-third of the wealth is owned by one-half of the population. Zero percent of the population owns zero percent of the wealth and 100 percent of the population owns 100 percent of the wealth, so all Lorenz functions must have $w = 0$ when $p = 0$, and $w = 1$ when $p = 1$. Look up data concerning the distribution of wealth for the United States. Based on these data, draw a sketch of the Lorenz function for the United States. For additional research, do the same for several other countries. Can you draw any conclusions about how the distribution of wealth affects the overall economic health of a nation?

SUMMARY

*I*n this chapter we discussed a few basic principles of mathematical modeling and focused, in particular, on linear models and their graphs. We also showed how representing variables as symbols makes mathematical manipulations easier. Key lessons that you should remember from this chapter include the following.

- Change is ever present. Mathematically, we deal with change through functions, which describe how one variable changes with respect to another. Functions can be represented by (1) words, (2) a data table, (3) a graph, or (4) an equation.

- A linear function has a straight-line graph with a constant slope and a constant rate of change. Although linear functions are special, they illustrate many basic mathematical principles. They also are exceedingly useful because many real phenomena can be closely approximated by linear models.

- Nonlinear functions have a varying rate of change. Their graphs do not have the same slope between all pairs of points along the curve.

- We can work with functions and their formulas by knowing just a few algebraic rules, such as those for powers, roots, exponents, and logarithms.

- Functions (linear or nonlinear) can be used to construct models of real situations. Functions are commonly represented by formulas that involve constants and variables. A model based on a function is only as good as the data and assumptions on which it is built.

Chapter 7

EXPONENTIAL GROWTH AND DECAY

During the 1990s, the population of the Earth increased by more than 80 million people per year. Human consumption of vital resources also is increasing every year, but supplies of those same resources are decreasing. Radioactive wastes from nuclear weapons and power plants will remain toxic for thousands of years. Acid rain and chemicals from factories pollute our rivers and lakes. Although these facts may seem alarming, the response should not be panic but a quest for understanding and solutions. In this chapter we investigate the mathematical laws of exponential growth and decay that underlie population growth, resource depletion, growth of investments, radioactive decay, and much more.

Everyone born before 1950 has witnessed a doubling of world population, the first generation ever to do so.
LESTER R. BROWN AND HAL KANE, *FULL HOUSE*

The greatest shortcoming of the human race is [its] inability to understand the exponential relation.
ALBERT A. BARTLETT, PROFESSOR OF PHYSICS, UNIVERSITY OF COLORADO

Faith in technology as the ultimate solution to all problems can thus divert our attention from the most fundamental problem—the problem of growth in a finite system—and prevent us from taking effective action to solve it.
DONELLA A. MEADOWS, *BEYOND THE LIMITS*

*I*magine two communities, Straightown and Powertown, with initial populations of 10,000 people (Figure 7.1). Straightown grows at a constant rate of 500 people per year, so its population reaches 10,500 after one year, 11,000 after two years, 11,500 after three years, and so on. Powertown grows at a rate of 5 *percent* per year. Because 5% of 10,000 is 500, Powertown's population also reaches 10,500 after one year. In the second year, its population increases by 5% of 10,500 (or 525) to 11,025. In the third year, its population increases by 5% of 11,025 (or 551) people. If it continues to grow by 5% per year, Powertown's population will double to 20,000 in 14 years, then double again to 40,000 in the next 14 years, and double again to 80,000 after another 14 years. This pattern of repeated **doublings**, in which each doubling occurs in the same amount of time, makes Powertown's population rise ever more steeply and far outpace the growth of Straightown.

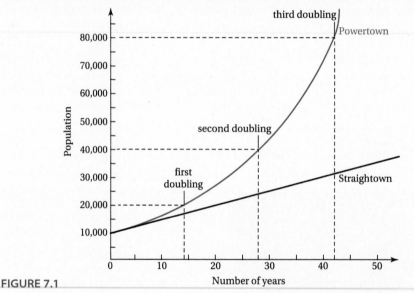

FIGURE 7.1

Straightown and Powertown illustrate two fundamentally different types of growth. Straightown grows by the same *absolute* amount—500 people per year—from one year to the next. This constant rate of change is the basic characteristic of *linear* growth, discussed in Unit 6B. In contrast, Powertown's population grows by the same *percentage* each year, which is a characteristic of **exponential growth**.

Two Basic Growth Patterns

Linear growth occurs when a quantity grows by the same *absolute* amount in each unit of time.

Exponential growth occurs when a quantity grows by the same *relative* amount—that is, by the same *percentage*—in each unit of time.

EXPONENTIAL ASTONISHMENT

UNIT 7 A

The tendency of exponentially growing quantities to "take off," as shown in Figure 7.1, leads to what we call *exponential astonishment*: the intuition-defying reality of exponential growth. The following three parables illustrate the surprising nature of exponential growth.

FROM HERO TO HEADLESS IN 64 EASY STEPS

According to legend, chess was invented in ancient times by a man who presented the game to his king. The king was so enchanted by the new game that he said to the inventor, "Name your reward."

"If you please, your Majesty, put one grain of wheat on the first square of my chessboard," said the inventor. "Then, place two grains on the second square, four grains on the third square, eight grains on the fourth square, and so on." The king gladly agreed, thinking that this man was a fool for asking only for a few grains of wheat when he could have had gold or jewels.

Let's see how it adds up for all 64 squares on a chessboard. Each square is supposed to get twice as many grains as the previous square, as shown in Table 7.1. Note that each number is a power of 2, so we are dealing with repeated doublings and hence with exponential growth.

Table 7.1	Placing Grains of Wheat on a Chessboard	
Square	**Grains on This Square**	**Total Grains Thus Far**
1	$1 = 2^0$	$1 = 2^1 - 1$
2	$2 = 2^1$	$1 + 2 = 3 = 2^2 - 1$
3	$4 = 2^2$	$3 + 4 = 7 = 2^3 - 1$
4	$8 = 2^3$	$7 + 8 = 15 = 2^4 - 1$
5	$16 = 2^4$	$15 + 16 = 31 = 2^5 - 1$
6	$32 = 2^5$	$31 + 32 = 63 = 2^6 - 1$
7	$64 = 2^6$	$63 + 64 = 127 = 2^7 - 1$
⋮	⋮	⋮
64	2^{63}	$2^{64} - 1$

Table 7.1 also shows the total number of grains on the board as we go along. The grand total for all 64 squares is $2^{64} - 1$ grains. How much wheat is this? With a calculator, you can confirm that

$$2^{64} = 1.8 \times 10^{19},$$

or about 18 *billion billion*. Aside from the fact that it would be difficult to fit so many grains on a chessboard, and that they would weigh 1 trillion tons, this number probably is larger than the total number of grains of wheat harvested in all human history. The king never finished paying the inventor, and instead had him beheaded.

THE MAGIC PENNY

One lucky day you meet a leprechaun who promises to give you fantastic wealth and hands you a penny before disappearing. You head home excited about meeting a leprechaun and place the penny under your pillow. The next morning, to your surprise, you find two pennies under your pillow. The following morning you find four pennies and the fourth morning eight pennies. Apparently, the leprechaun gave you a *magic* penny: Each night while you sleep, each magic penny turns into *two* magic pennies. Because doubling is involved, we can describe the growth of your wealth using powers of 2, as shown in Table 7.2.

Table 7.2	Wealth from the Magic Penny	
Day	**Amount under pillow**	**is the same as**
0	$0.01	0.01×2^0
1	$0.02	0.01×2^1
2	$0.04	0.01×2^2
3	$0.08	0.01×2^3
4	$0.16	0.01×2^4

We can generalize from this table to write a simple formula for your wealth. Note that "day 0" is the day you met the leprechaun. Your wealth t days after meeting the leprechaun will be

$$\text{wealth after } t \text{ days} = \$0.01 \times 2^t.$$

> **TIME-OUT TO THINK:** Confirm that the above formula gives the values for your wealth shown in Table 7.2.

When will you have fantastic wealth? After 9 days, you'll have

$$\$0.01 \times 2^9 = \$5.12,$$

barely enough to buy lunch. But after just one month, or 30 days, you'll have

$$\$0.01 \times 2^{30} = \$10,737,418.24.$$

You'll be a millionaire in a month, and you'll need a much larger pillow!

In fact, if your magic pennies keep doubling, by the end of just 50 days you'd have

$$\$0.01 \times 2^{50} \approx \$11.2 \text{ trillion},$$

more than enough to pay off the national debt of the United States.

By the Way ....
The Bacteria in a Bottle parable was developed by Professor Albert A. Bartlett, University of Colorado.

BACTERIA IN A BOTTLE

Imagine a type of bacteria that divide every minute, as long as they have enough nutrients. Suppose that you place a single bacterium in a nutrient-filled bottle at 11:00 A.M. It grows and divides into two bacteria at 11:01. These two bacteria each grow and divide into 4 bacteria at 11:02. After the next division, at 11:03, there will be $2^3 = 8$ bacteria, and $2^4 = 16$ bacteria at 11:04. In general,

$$\text{number of bacteria at } n \text{ minutes after } 11\!:\!00 = 2^n.$$

Now, suppose that the bottle is full at 12:00. Because this is 60 minutes after 11:00, the bottle will contain 2^{60} bacteria. The population growth over the hour is summarized in Table 7.3. We can learn more about the nature of exponential growth by asking a few questions regarding the fate of this bacterial colony.

- Question 1: *The colony grew from a single bacterium at 11:00 to fill the bottle at 12:00. When was the bottle half full of bacteria?*

Many people first guess 11:30—half of the time it took to fill the bottle. However, we are dealing with repeated *doublings* of the bacteria population. That is, the number of bacteria *doubled* during the last minute, from 2^{59} at 11:59 to 2^{60} at 12:00. Thus the bottle was half full at 11:59, just one minute before the colony's demise! Similarly, as shown in Table 7.3, the bottle was one-quarter full at 11:58, one-eighth full at 11:57, and so on. Figure 7.2 shows a graph of the same data.

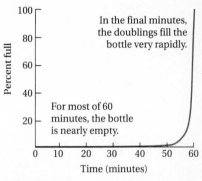

FIGURE 7.2

Table 7.3	Bacteria in a Bottle (dividing every minute)		
Time	**Minutes Since Start of Colony**	**Number of Bacteria**	**Fraction of Bottle Filled**
11:00	1	$1 = 2^0$	$1/2^{60}$
11:01	2	$2 = 2^1$	$2/2^{60} = 1/2^{59}$
11:02	3	$4 = 2^2$	$4/2^{60} = 1/2^{58}$
⋮	⋮	⋮	⋮
11:56	56	2^{56}	$2^{56}/2^{60} = 1/16$
11:57	57	2^{57}	$2^{57}/2^{60} = 1/8$
11:58	58	2^{58}	$2^{58}/2^{60} = 1/4$
11:59	59	2^{59}	$2^{59}/2^{60} = 1/2$
12:00	60	2^{60}	$2^{60}/2^{60} = 1 \text{ (Full)}$

- Question 2: *Suppose that you were one of the bacteria, and at 11:56 you began warning that the bottle would be full in just four minutes. Would your warnings be believed?*

The data in Table 7.3 make it clear that your warnings were correct. However, imagine other bacteria looking around the bottle at 11:56. They would see that the bottle is only 1/16 full; that is, the amount of unused space is *15 times* that of the used space. It might be very difficult to get your fellow bacteria to believe that they'll fill 15 times as much space in the next four minutes as they did in the first 56 minutes of the colony's history.

- Question 3: *Suppose that, thanks to a last minute technological triumph, the bacteria discover three unused, nutrient-filled bottles in the laboratory (making a total of four bottles, including the original). Further, imagine that they quickly build little bacterial spaceships and distribute their population evenly among the four bottles at 12:00. How much time will this buy for their colony?*

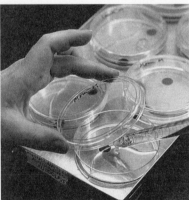

Given that it took one hour to fill one bottle, you might first guess that it would take four hours to fill four bottles. However, the bacterial population doubles each minute. If one bottle is full at 12:00, two bottles fill by 12:01, and four bottles fill by 12:02. The discovery of three new bottles buys the colony only two additional minutes of growth!

- Question 4: *Is there any hope that further discoveries of new bottles will allow the colony to continue its exponential growth?*

Let's do some calculations. Suppose that, somehow, the bacteria population continues to double each minute, thereby reaching 2^{120} bacteria after two hours (120 minutes). The smallest bacteria measure approximately 10^{-7} m (0.1 micrometer) across. If we assume that the bacteria are roughly cube-shaped, the volume of each bacterium is about

$$(10^{-7} \text{ m})^3 = 10^{-21} \text{ m}^3.$$

Therefore, the colony of 2^{120} bacteria would occupy a volume of

$$2^{120} \text{ bacteria} \times \left(10^{-21} \frac{\text{m}^3}{\text{bacteria}}\right) = 1.3 \times 10^{15} \text{ m}^3.$$

With this volume, the bacteria would cover the entire surface of the Earth in a layer more than 2 meters deep!

In fact, if the doublings continued for just $5\frac{1}{2}$ hours, the volume of bacteria would exceed the volume of the entire observable universe. Needless to say, this cannot happen. The exponential growth of the colony cannot possibly continue for long, no matter what technological advances might be imagined.

Facts do not cease to exist because they are ignored.

—ALDOUS HUXLEY

DOUBLING LESSONS

The three parables reveal at least two key lessons about the repeated doublings that characterize exponential growth. First, if you look back at Table 7.1, you'll notice that the number of grains on each square is nearly equal to the total number of grains on all previous squares combined. For example, the 64 grains on the seventh square are one more than the total of 63 grains on the first six squares combined.

Second, all three parables show quantities growing to impossible proportions. We cannot possibly fit all the wheat harvested in world history on a chessboard; $5 trillion worth of pennies is far more than the number of pennies that exist in the world; and there isn't enough food for a colony of bacteria that outgrows the Earth.

> ### SUMMARY: BASIC FACTS ABOUT EXPONENTIAL GROWTH
>
> - Exponential growth is characterized by repeated doublings. With each doubling, the amount of increase is approximately equal to the *sum* of all preceding doublings.
> - Exponential growth cannot continue indefinitely. After only a relatively small number of doublings, exponentially growing quantities reach impossible proportions.

REVIEW QUESTIONS

1. Describe the basic differences between *linear growth* and *exponential growth*.

2. In the chessboard legend, would it really be possible to double the number of grains on each successive square? Why or why not?

3. If you kept the magic penny until 52 days had passed, how much money would you have? Why?

4. In your own words, explain the answers to each of the four questions asked in the parable of the bacteria in a bottle.

5. Explain the meaning of each of the two basic facts about exponential growth given at the end of this Unit.

PROBLEMS

1. **Growth and Decay in the News.** Recall that *linear growth* occurs when a quantity grows by the same *absolute* amount in each unit of time, and *exponential growth* occurs when a quantity grows by the same *relative* (or percentage) amount in each unit of time.

   a. Identify three news stories (radio, TV, newspapers, or magazines) that describe a quantity that is increasing or decreasing exponentially. Describe the growth or decay process.

   b. Identify three news stories that describe a quantity that is increasing or decreasing linearly. Describe the growth or decay process.

2. **Linear Versus Exponential.** Do the following statements describe linear or exponential relationships? Why?

   a. The population of Danbury is increasing at a rate of 505 people per year.

   b. The price of food in Brazil is increasing at a rate of 30% per month.

   c. The birth rate in Hungary is declining at a rate of 2.5% per year.

   d. You can depreciate the value of these washing machines by $150 per year.

3. **Linear Versus Exponential.** Do the following statements describe linear or exponential relationships? Why?

   a. The population of Winesburg is increasing at a rate of 230 people per year.

   b. The price of computer memory is decreasing at a rate of 12% per year.

   c. The birth rate in Austria is declining at a rate of 1.5% per year.

   d. You can depreciate the value of office equipment by $200 per year.

4. **Approximation.** Continue the following table for values of n up to 20. Based on your results, is it true that, for large values of n, the quantities 2^n and $2^n - 1$ are nearly equal? Explain.

n	2^n	$2^n - 1$
0	$2^0 = 1$	$1 - 1 = 0$
1	$2^1 = 2$	$2 - 1 = 1$
⋮	⋮	⋮

5. **Chessboard Wheat.** According to the parable described in the text, how many grains of wheat should be placed on square 20 of the chessboard? How many total grains would there be on the board at that point?

6. **Chessboard Wheat.** According to the parable described in the text, how many grains of wheat should be placed on square 30 of the chessboard? How many total grains would there be on the board at that point?

7. **The Weight of All That Grain.** A "grain" is an ancient measure of weight, based on the weight of a typical grain of wheat; 1 pound is defined as 7000 grains.

   a. What is the weight, in tons, of the 18 billion billion grains in the chessboard parable? (Hint: 1 ton = 2,000 pounds.)

   b. According to the U.S. Department of Agriculture, the current world harvest of all grain (wheat, rice, and corn) is somewhat less than 2 billion tons per year. Use percentages to compare this amount to the grain in the chessboard parable.

8. **Magic Money.** Imagine that you receive a magic penny as described in the text.

   a. How much money would you have after 15 days?

   b. Remember that all your money is in pennies. Suppose that you stacked the pennies after 15 days. How high would the stack rise? (Hint: Find a few pennies and a ruler.)

   c. How many days would elapse before you have your first billion dollars?

   d. Suppose that you could keep making a single stack of the pennies. After how many days would the stack be long enough to reach the nearest star (beyond the Sun), which is about 4.3 light-years (4.0×10^{13} km) away?

9. **Bacteria in a Bottle.** Consider the bacteria parable described in the text.

   a. How many bacteria are in the bottle at 11:55? What fraction of the bottle is full at that time?

   b. How many bacteria are in the bottle at 11:05? What fraction of the bottle is full at that time?

10. **Bacteria in a Bottle.** Consider the bacteria parable described in the text.

    a. How many bacteria are in the bottle at 11:50? What fraction of the bottle is full at that time?

    b. How many bacteria are in the bottle at 11:10? What fraction of the bottle is full at that time?

11. **A Layer of Bacteria.** In the parable of the bacteria, we calculated that after two hours of growth, there would be 2^{120} bacteria that occupy 1.3×10^{15} m^3. Given that the surface area of the Earth is about 5.1×10^{14} m^2, how deep would the bacteria be if they were spread in a uniform layer all over the surface of the Earth? Show your work.

12. **The Volume of Bacteria.** If the bacteria in the parable continue to double their population every minute, when will their volume exceed *the total volume of the observable universe* (roughly 10^{79} m^3)? Show your work.

13. **Population Doubling.** Assume that human population in the year 2000 is about 6 billion people. Suppose that this population were to increase exponentially with a doubling time of 50 years.

    a. Continue the following table showing the population at 50-year intervals under this scenario, going out to the year 3000. Use scientific notation, as shown.

Year	Population
2000	6×10^9
2050	$12 \times 10^9 = 1.2 \times 10^{10}$
2100	2.4×10^{10}
⋮	⋮

    b. The total surface area of the Earth is about 5.1×10^{14} m^2. Assuming that people could occupy all this area (in reality, most of it is ocean), approximately when would people be so crowded that every person would have only 1 m^2 of space?

    c. Suppose that, when we take into account the area needed to grow food and to find other resources, each person actually requires about 10^4 m^2 of area to survive. About when would we reach that limit?

DOUBLING TIME AND HALF-LIFE

We've seen that exponential growth is characterized by repeated doublings in which each doubling takes the same period of time. We can also say that exponential growth occurs when a quantity grows by the same percentage from one period of time to the next. In a similar way, *exponential decay* is characterized by repeated *halvings* of a quantity and occurs when a quantity *decreases* by the same percentage in each unit of time. In this Unit, we'll see how the doubling time is related to the percentage growth rate and how the halving time is related to the percentage decay rate.

DOUBLING TIME

The time required for an exponentially growing quantity to double is called the **doubling time**. For example, the doubling time of the magic penny in Unit 7A was one day because your wealth doubled each day. The doubling time for the bacteria in the bottle in Unit 7A was one minute because the bacterial population doubled every minute.

   Once we know the doubling time of a quantity, we can quickly calculate how much the quantity grows in *any* period of time. Consider a population that begins with a size of 10,000 and grows with a doubling time of 10 years.

- After 10 years, or one doubling time, the population will have increased by a factor of 2, to a new population of $2 \times 10{,}000 = 20{,}000$.

- After 20 years, or two doubling times, the population will have increased by a factor of $2^2 = 4$, to a new population of $4 \times 10{,}000 = 40{,}000$.

- After 30 years, or three doubling times, the population will have increased by a factor of $2^3 = 8$, to a new population of $8 \times 10{,}000 = 80{,}000$.

   We can get the same answers with a simple formula. If we let t be the *actual* amount of time that has passed and T_{double} be the doubling time, then after a time of t years, the population will increase by a factor of

$$2^{t/T_{\text{double}}}.$$

For example, this formula says that after $t = 30$ years with a doubling time of $T_{\text{double}} = 10$ years, the population will increase by a factor of

$$2^{30\text{ yr}/10\text{ yr}} = 2^3 = 8,$$

as we found above. We can now generalize for any exponentially growing quantity.

Calculating Exponential Growth from the Doubling Time

After a time t, an exponentially growing quantity with a doubling time of T_{double} will increase in size by a factor of

$$2^{t/T_{\text{double}}}.$$

The new value of the quantity is related to its initial value at $t = 0$ by

$$\text{new value} = \text{initial value} \times 2^{t/T_{\text{double}}}.$$

EXAMPLE 1   *Doubling with Compound Interest*

Compound interest represents a form of exponential growth because an interest-bearing account grows by the same percentage each year. Suppose that you have a bank account that has a doubling time of 13 years. By what factor will your balance increase in 50 years?

Solution: The doubling time is $T_{\text{double}} = 13$ years. Over a period of $t = 50$ years, your balance will increase by a factor of

$$2^{50 \text{ yr}/13 \text{ yr}} = 2^{3.85} = 14.4.$$

For example, if you start with $1, you'll have $14.40 after 50 years; if you start with $10,000, you'll have $144,000 in 50 years. ∎

EXAMPLE 2   *World Population Growth with a 40-Year Doubling Time*

World population doubled from 2.6 billion in 1950 to 5.2 billion in 1990. Suppose that world population continued to grow with a doubling time of 40 years. What would the population be in 2020? in 2070? in 2190?

Solution: We are given a doubling time of $T_{\text{double}} = 40$ years. If we let $t = 0$ represent 1990, the year 2020 is $t = 30$. Using the 1990 population of 5.2 billion as the initial value, we find:

$$\text{population in 2020} = 5.2 \text{ billion} \times 2^{30 \text{ yr}/40 \text{ yr}} = 5.2 \text{ billion} \times 2^{0.75} = 8.7 \text{ billion}$$

For 2070, or $t = 80$ years after 1990, we find:

$$\text{population in 2070} = 5.2 \text{ billion} \times 2^{80 \text{ yr}/40 \text{ yr}} = 5.2 \text{ billion} \times 2^{2} = 20.8 \text{ billion}$$

Finally, for 2190, or $t = 200$ years after 1990 we find:

$$\text{population in 2190} = 5.2 \text{ billion} \times 2^{200 \text{ yr}/40 \text{ yr}} = 5.2 \text{ billion} \times 2^{5} = 166.4 \text{ billion} \quad ∎$$

TIME-OUT TO THINK: Do you think that it's really possible for human population to reach 166 billion in 200 years? Why or why not?

THE APPROXIMATE DOUBLING TIME FORMULA

We've seen how to calculate exponential growth when we know the doubling time. However, in most cases where exponential growth is involved, we are given only the percentage rate of growth in some period of time. For example, we may be told that a population grew by 2% last year, or that the Consumer Price Index rose by 0.3% in the past month. Therefore we need a way to find the doubling time once we are given the percentage growth rate.

Suppose that an ecological study of a prairie dog community shows that, at the beginning of the year, it contains 100 prairie dogs. The researchers soon determine that the prairie dog population is increasing at a rate of 10% per month. In that case, the new value of the population at the end of each month will be 110% of its previous value from the beginning of the month (see Unit 4B).

The following table tracks the growth of the population.

Month	Population at Beginning of Month	End-of-Month Population = 110% × Beginning-of-Month Population
1	100	110
2	110	121
3	121	133
4	133	146
5	146	161
6	161	177
7	177	195
8	195	214
9	214	236
10	236	259
11	259	285
12	285	314
13	314	345
14	345	380
15	380	418

The table shows that the population roughly doubles every seven months: from 100 to almost 200 by the end of month 7, and to almost 400 by the end of month 14. Note that this doubling time of about seven months happens to be related to the 10% growth rate per month as follows:

$$\text{doubling time} \approx \frac{70}{\text{percent growth rate}} = \frac{70}{10/\text{month}} = 7 \text{ months}$$

We can generalize to an approximate formula for the doubling time that applies to an exponentially growing quantity with a relatively small growth rate (less than about 15%).

The Approximate Doubling Time Formula (the Rule of 70)

If an exponentially growing quantity has a **percentage growth rate** of *P*%, its doubling time is *approximately*

$$T_{\text{double}} \approx \frac{70}{P}.$$

Note: *This approximation works best for very small growth rates, and breaks down for growth rates over about 15%.*

This formula is often called the **Rule of 70**.

Mathematical Note: Derivations of both the exact and approximate doubling time formulas are given in Unit 7C.

Although this formula is only an approximation, it is useful because it is so easy to remember and allows us to get a "feel" for exponentially growing quantities. We will discuss the exact doubling time formula later in Unit 7B.

EXAMPLE 3   *Population Doubling Time*

World population reached 5.8 billion in 1997, and was growing at a rate of about 1.4% per year. What is the approximate doubling time at this growth rate? If this growth rate continues, what will the world population be in 2020?

Solution: First, we note that the approximate doubling time formula is useful in this case because the growth rate is much less than 15%. The percentage growth rate of 1.4% per year means we set $P = 1.4/\text{yr}$ in the approximate doubling time formula:

$$T_{\text{double}} \approx \frac{70}{1.4/\text{yr}} = 50 \text{ yr}$$

The population would double in about 50 years. We calculate the population in 2020 from the doubling time by using the formula

$$\text{new value} = \text{initial value} \times 2^{t/T_{\text{double}}}.$$

If we let $t = 0$ represent 1997, then the *initial value* is the 1997 population of 5.8 billion. Since 2020 is $t = 23$ years after 1997, the new value of the population in 2020 would be

$$5.8 \text{ billion} \times 2^{23 \text{ yr}/50 \text{ yr}} = 5.8 \text{ billion} \times 2^{0.46} = 8.0 \text{ billion}.$$

With a 1.4% growth rate, the world population will reach about 8 billion in 2020. ■

EXAMPLE 4   *Solving the Doubling Time Formula*

Use the approximate doubling time formula to calculate the percentage rate of growth in world population from 1950 to 1990.

Solution: In Example 2 we were told that the world population doubled from 1950 to 1990, giving a doubling time of $T_{\text{double}} = 40$ years. We can solve the approximate doubling time formula for P by multiplying both sides by $\dfrac{P}{T_{\text{double}}}$:

$$T_{\text{double}} \approx \frac{70}{P} \quad \Rightarrow \quad \frac{P}{T_{\text{double}}} \times T_{\text{double}} \approx \frac{P}{T_{\text{double}}} \times \frac{70}{P} \quad \Rightarrow \quad P \approx \frac{70}{T_{\text{double}}}$$

Now we substitute $T_{\text{double}} = 40$ years to find:

$$P \approx \frac{70}{40 \text{ yr}} = 1.75/\text{yr}$$

The result that $P \approx 1.75$ per year means the growth rate of world population was about $P\% = 1.75\%$ per year between 1950 and 1990. Because this growth rate is relatively small, our use of the approximate formula was justified. ■

> **TIME-OUT TO THINK:** Note that the world population growth rate of 1.4% in the mid-1990s is lower than the 1.75% that prevailed before 1990. Do you think the growth rate will continue to drop? Do you think the actual population will begin to drop? Explain.

EXPONENTIAL DECAY AND HALF-LIFE

Exponential *decay* occurs whenever the size of a quantity is *decreasing* by the same percentage each year (or any other unit of time). The best-known example of exponential decay involves radioactive materials such as uranium or plutonium, but exponentially decaying quantities actually are quite common. For example, if inflation is making prices rise by 3% per year, then the *value* of a $1 bill is falling, or exponentially decaying, by 3% per year. Occasionally, nations even go through periods of *deflation*, where prices fall with time.

If a quantity is exponentially *decaying*, then instead of periodically doubling, its value is periodically halved. The fixed amount of time that it takes a quantity to halve is called its **half-life**. For example, radioactive Plutonium-239 (Pu-239) has a half-life of about 24,000 years. If one pound of Pu-239 is deposited at a nuclear waste site:

- After 24,000 years, or one half-life, only $\frac{1}{2}$ pound of the Pu-239 will remain. (The rest will have decayed into other substances.)
- After 48,000 years, or two half-lives, only $\frac{1}{4}$ pound of the Pu-239 will remain.
- After 72,000 years, or three half-lives, only $\frac{1}{8}$ pound of the Pu-239 will remain.

We can get the same answers by saying that, after a time of t years, the amount of radioactive Pu-239 is

$$\left(\frac{1}{2}\right)^{t/T_{\text{half}}}$$

of its initial value, where t is the *actual* amount of time that has passed and T_{half} is the half-life. For example, this formula says that after $t = 72,000$ years with a half-life of $T_{\text{half}} = 24,000$ years, the amount of Pu-239 is

$$\left(\frac{1}{2}\right)^{72,000 \text{ yr}/24,000 \text{ yr}} = \left(\frac{1}{2}\right)^{3} = \frac{1}{8}$$

of its initial value. This is the same answer we found above.

We can now generalize for any exponentially decaying quantity.

By the Way ....

The description "plutonium-239" means the *element* plutonium in a form (or *isotope*) with *atomic weight* 239. Atomic weight is the total number of protons and neutrons in the nucleus. A nucleus of plutonium has 94 protons. Thus Pu-239 has $239 - 94 = 145$ neutrons. Plutonium is produced as a by-product of nuclear reactions.

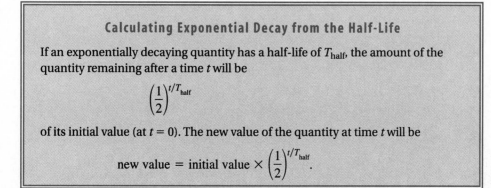

Calculating Exponential Decay from the Half-Life

If an exponentially decaying quantity has a half-life of T_{half}, the amount of the quantity remaining after a time t will be

$$\left(\frac{1}{2}\right)^{t/T_{\text{half}}}$$

of its initial value (at $t = 0$). The new value of the quantity at time t will be

$$\text{new value} = \text{initial value} \times \left(\frac{1}{2}\right)^{t/T_{\text{half}}}.$$

EXAMPLE 5   *Radioactive Decay*

If the half-life of a radioactive substance is 40 years, what fraction of it remains after 200 years?

Solution: For a half-life of $T_{half} = 40$ years, the fraction of the initial amount remaining after $t = 200$ years is

$$\left(\frac{1}{2}\right)^{200\ yr/40\ yr} = \left(\frac{1}{2}\right)^{5} = \frac{1}{32}.$$

For example, if the radioactive material initially weighed 32 grams, only one gram remains after 200 years. ∎

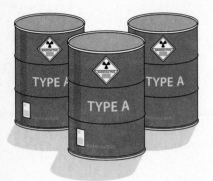

EXAMPLE 6   *Plutonium After 100,000 Years*

Suppose that 100 pounds of Pu-239 is deposited at a nuclear waste site. How much of it will still be radioactive in 100,000 years?

Solution: The half-life of Pu-239 is $T_{half} = 24,000$ years. The new value is the amount of Pu-239 remaining after $t = 100,000$ years, and the initial value is the original 100 pounds deposited at the waste site:

$$new\ value = 100\ lb \times \left(\frac{1}{2}\right)^{100,000\ yr/24,000\ yr} = 100\ lb \times \left(\frac{1}{2}\right)^{4.17} = 5.6\ lb$$

About 5.6 pounds of the original 100 pounds of Pu-239 will still be radioactive in 100,000 years. ∎

> **TIME-OUT TO THINK:** Plutonium-239 is a major ingredient in nuclear weapons. Based on the 24,000 year half-life of Pu-239, explain why the safe disposal of waste from dismantled nuclear weapons poses a challenge.

THE APPROXIMATE HALF-LIFE FORMULA

The approximate doubling time formula (the rule of 70) found earlier works equally well for exponential decay if we replace the doubling time with the half-life and the percentage growth rate with the percentage decay rate.

> ### The Approximate Half-Life Formula
>
> If an exponentially decaying quantity has a **percentage decay rate** of $P\%$, its half-life is *approximately*
>
> $$T_{half} \approx \frac{70}{P}.$$
>
> Note: *This approximation works best for very small decay rates, and breaks down for decay rates over about 15%.*

EXAMPLE 7   *Devaluation of Currency*

Suppose that inflation is causing the value of the (Russian) ruble to fall at a rate of 12% per year (relative to the dollar). How long does it take for the ruble to lose half its value?

Solution: The 12% decay rate means we set $P = 12/\text{yr}$:

$$T_{\text{half}} \approx \frac{70}{12/\text{yr}} = 5.8 \text{ yr}$$

The half-life is about 6 years, meaning that the ruble loses half its value (against the dollar) in 6 years. ∎

EXACT FORMULAS FOR DOUBLING TIME AND HALF-LIFE

The approximate formulas for the doubling time and half-life are useful because they are easy to remember. For more exact work or for cases of larger rates where the approximate formulas break down, we need exact formulas. Here are the formulas (derived in Unit 7C). Note that these formulas involve logarithms (discussed in Unit 6C).

The Exact Doubling Time and Half-Life Formulas

The doubling time, T_{double}, of an exponentially growing quantity is

$$T_{\text{double}} = \frac{\log_{10} 2}{\log_{10}(1 + r)},$$

where r is the *fractional growth rate*; r is related to the *percentage* growth rate, $P\%$, by $r = \dfrac{P}{100}$.

The half-life, T_{half}, of an exponentially decaying quantity is

$$T_{\text{half}} = -\frac{\log_{10} 2}{\log_{10}(1 - r)},$$

where r is the *fractional decay rate*; r is related to the *percentage* decay rate, $P\%$, by $r = \dfrac{P}{100}$. Because $\log_{10}(1 - r)$ is negative, this formula will give a half-life that is positive.

Note: *The units of the doubling time or half-life are determined by the units of r.*

Using Your **Calculator**

Your calculator should have a button for computing common (base-10) logarithms. It is usually labeled *log*, but be careful because some calculators use this label for natural (base-*e*) logarithms. With the right button, you will find $\log_{10} 2 = 0.301\ldots$.

Mathematical Note: An important subtlety in these formulas is related to compounding in financial formulas. From Unit 5A, we know that the growth of money depends not only on the annual percentage rate (*APR*), but also on the compounding period. The *actual* amount that your money grows in a year is the annual percentage *yield* (*APY*). In the doubling time and half-life formulas, *r* is analogous to the annual yield (*APY*), *not* to the *APR*. That is, *r* is the *actual* fraction by which a quantity grows (or decays) in a given period of time. In this book, you may assume that you are always given actual growth rates.

EXAMPLE 8   *Compound Interest*

Suppose that an investment has an annual percentage yield (*APY*) of 7.72%. What is the doubling time for the value of this investment? Compare to the approximate value found with the rule of 70.

Solution: The growth rate of 7.72% per year means $P = 7.72/\text{yr}$, or equivalently that the fractional growth rate is $r = 0.0772/\text{yr}$. The doubling time for the investment is

$$T_{\text{double}} = \frac{\log_{10} 2}{\log_{10} (1 + 0.072)} = \frac{0.301030}{\log_{10} (1.0772)} = 9.32 \text{ yr.}$$

The approximate doubling time found with the rule of 70 is

$$T_{\text{double}} \approx \frac{70}{7.72/\text{yr}} = 9.07 \text{ yr.}$$

The actual doubling time is slightly longer than the approximate value found with the rule of 70. If we were doing exact financial work, this difference would be very important.   ∎

EXAMPLE 9   *Ruble Revisited*

Suppose that the ruble is falling in value against the dollar at 12% per year. How long does it take the ruble to lose half its value? Compare to the approximate answer found in Example 7.

Solution: The percentage decay rate is 12%/yr, so the fractional decay rate is $r = 0.12/\text{yr}$. The half-life is

$$T_{\text{half}} = -\frac{\log_{10} 2}{\log_{10} (1 - 0.12)} \approx 5.42 \text{ yr.}$$

The ruble loses half its value against the dollar in 5.42 years, or slightly less than the 5.8 years obtained with the approximate formula.   ∎

REVIEW QUESTIONS ••

1. What is a *doubling time*? Suppose that a population has a doubling time of 25 years. How much will it grow in 25 years? in 50 years? in 75 years? in 100 years?

2. Explain how, if you know the doubling time, you can calculate the value of an exponentially growing quantity at any time *t*.

3. What is the approximate doubling time formula? Under what conditions does it apply?

4. What is a *half-life*? Suppose that a radioactive substance has a half-life of 1000 years. How much will be left after 1000 years? after 2000 years? after 3000 years? after 4000 years?

5. Explain how, if you know the half-life, you can calculate the value of an exponentially decaying quantity at any time *t*.

6. What is the approximate half-life formula? Under what conditions does it apply?

7. Describe the meaning of each variable in the exact doubling time and half-life formulas. Give an example in which each formula is used.

PROBLEMS

1. **Change and Doubling Time.**

   a. If the doubling time of a population of bacteria is 3 hours, by what factor does the population increase in 24 hours? in 1 week?

   b. If the doubling time of a bank account balance is 10 years, by what factor does it grow in 30 years? in 50 years?

   c. If the doubling time of a city's population is 22 years, how long does it take for the population to quadruple?

2. **Change and Doubling Time.**

   a. If prices are increasing with a doubling time of 4 weeks, by what factor do prices increase in a year?

   b. If the doubling time of a groundhog community is 4 months, by what factor does the groundhog population increase in 12 months? in 20 months?

   c. If the doubling time of a mutual fund is 5 years, how long does it take for the fund to increase by a factor of 8?

3. **Growth from Doubling Time.**

   a. Suppose you deposit $500 in a bank account that has a doubling time of 15 years. What will your balance be after 20 years? after 30 years?

   b. Suppose the population of a town doubles every 8 years. What will the population be 12 years after the time when its population was 15,600? 24 years after the population was 15,600?

   c. The number of cells in a particular tumor is known to double every 1.5 months. If the tumor begins as a single cell, how many cells will there be after 3 years? after 4 years?

4. **Growth from Doubling Time.**

   a. Suppose you deposit $1000 in a bank account that has a doubling time of 12 years. What will your balance be after 20 years? after 30 years?

   b. Suppose the population of a town doubles every 30 years. What will the population be 12 years after the time when its population was 15,600? 40 years after the population was 15,600?

   c. The number of cells in a particular tumor is known to double every 6 months. If the tumor begins with a single cell, how many cells will there be after 4 years? after 6 years?

5. **World Population Growth.** The world's population doubled from 2.6 billion in 1950 to 5.2 billion in 1990. If this doubling time of 40 years remains constant, what will the population be in 2010? in 2060? in 2100?

6. **Doubling Time Practice.** Answer each question using the approximate doubling time formula.

   a. Suppose that the consumer price index of a country is increasing at a rate of 7% per year. What is its doubling time? By what factor will prices increase in 3 years?

   b. If a certificate of deposit increases its value by 5.5% per year, what is its doubling time? If you deposit $100 today, how much will you have after 5 years? after 50 years?

   c. Suppose that a population is growing at a rate of 2.0% per year. What is its doubling time? By what factor will the population increase in 75 years?

7. **Doubling Time Practice.** Answer each question using the approximate doubling time formula.

   a. If prices are rising at a rate of 0.7% per month, what is their doubling time? By what factor will prices increase in 1 year? in 8 years?

   b. Suppose that the world population is rising at a rate of 1.6% per year. What is its doubling time? If the world population is 6 billion now, what will it be in 10 years? in 100 years? in 1000 years? Can such growth continue for 1000 years? Why or why not?

   c. Suppose the use of oil is increasing at a steady rate of 1.9% per year. What is the doubling time of oil use? By what factor will oil use increase in a decade?

8. **The Spread of Rabbits.** Make a table of the population of a rabbit community with an initial population of 100 that increases by 7% per month; that is, show the population for the first 15 months after observations begin. From the table, what is the doubling time of the population? Does it agree with the approximate doubling time formula (the rule of 70)?

9. **Change and Half-Life.**

   a. If the half-life of a radioactive sample is 35 years, by what factor does the amount of the radioactive substance decrease after 70 years? after 140 years?

   b. If the half-life of the population of an endangered species is 10 years, by what factor does its population decrease in 40 years? in 70 years?

   c. If the half-life of a drug in the bloodstream is 12 hours, by what factor does the concentration of the drug decrease in 24 hours? in 36 hours?

10. **Change and Half-Life.**

    a. If the half-life of a radioactive sample is 250 years, by what factor does the amount of the radioactive substance decrease after 500 years? after 1500 years?

    b. If the half-life of the population of an endangered species is 15 years, by what factor does its population decrease in 45 years? in 90 years?

    c. If the half-life of a drug in the bloodstream is 20 hours, by what factor does the concentration of the drug decrease in 40 hours? in 100 hours?

11. **Decay from Half-Life.**

    a. If the half-life of a radioactive sample is 1000 years, how much of a 200-gram sample of a radioactive substance remains after 5550 years? after 10,200 years?

    b. If the half-life of the population of a species of an endangered tree is 15 years, how many trees remain after 35 years in a forest that originally had 2500 trees? after 90 years in a forest that originally had 2500 trees?

    c. If the half-life of a drug in the bloodstream is 20 hours, how much drug is left in the bloodstream 30 hours after a 100-milligram dose? 70 hours after a 100-milligram dose?

12. **Decay from Half-Life.**

    a. If the half-life of a radioactive sample is 2000 years, how much of a 100-kilogram sample of a radioactive substance remains after 9000 years? after 18,000 years?

    b. If the half-life of the population of an endangered species of mouse is 12 years, how many mice remain after 25 years in a community that originally had 3000 mice? after 85 years in a community that originally had 3000 mice?

    c. If the half-life of a drug in the bloodstream is 10 hours, how much drug is left in the bloodstream 25 hours after a 100-milligram dose? 60 hours after a 100-milligram dose?

13. **Half-Life Practice.** Answer each question using the approximate half-life formula.

    a. Suppose that the consumer price index of a country is decreasing at a rate of 2% per year. What is its half-life? By what factor will prices decrease in 5 years?

    b. If a population decreases by 3.3% per year, what is its half-life? By what factor will the population decrease after 5 years? after 50 years?

    c. If the concentration of a drug in the bloodstream decreases by 10% per hour, what is its half-life? By what factor does the concentration decrease in 10 hours? in 24 hours?

    d. Suppose that predation is causing a population of rabbits to decline at a rate of 6% per year. What is the half-life for the rabbit population? Approximately when will it fall to 50% of its present size?

    e. Suppose that you have a radioactive substance that decays at a rate of 0.0005% per year. What is its half-life? If you start with 100 kilograms, how much is left after 10^5 years?

14. **Half-Life Practice.** Answer each question using the approximate half-life formula.

    a. Suppose that a low birth rate is causing the population of a country to decline at a rate of 0.3% per year. What is the half-life of the population? When will the population fall to half its current size?

    b. If the value of the dollar is falling against the value of the yen at a rate of 6% per year, how long will it take before the dollar loses half its value against the yen?

    c. Suppose that deforestation is causing the area of a particular forest to decline at a rate of 15% per year. What is the half-life of the forest? When will half the forest be gone?

    d. A clean-up project is causing the concentration of a particular pollutant in the water supply to decline at a rate of 8% per week. What is the half-life of the concentration of the pollutant? When will it fall to 1% of its original concentration?

    e. In each of the past several years, the production of a particular gold mine has declined by about 10%. When will the mine be producing only half as much gold ore as it is at present?

15. **Using the Formulas.** For each situation described, give the doubling time or half-life. In each case state whether your answer is exact, a good approximation, or a poor approximation.

   a. A quantity doubles every 4 days.

   b. A quantity increases four-fold every 4 days.

   c. A quantity is halved every 4 days.

   d. A quantity increases by 5% per year.

   e. A quantity decreases by 5% per year.

   f. A quantity declines by 5% per hour.

16. **Using the Formulas.** Answer each question using the doubling time and half-life formulas. In each case state whether your answer is exact, a good approximation, or a poor approximation.

   a. If a quantity falls by 3% per week, what is its half-life?

   b. If a quantity has a doubling time of 20 years, what is its growth rate?

   c. If a quantity has a half-life of 200 years, what is its decay rate?

   d. If a quantity rises by 25% per year, what is its doubling time?

   e. If a quantity drops by 18% per hour, what is its half-life?

17. **Working with Growth Rates.** State whether each statement is true or false. Explain your reasoning.

   a. If the inflation rate is 10% per year, average items that cost $1000 last year now cost $1100.

   b. If a country's population a year ago was 100 million people and it is growing at a rate of 3% per year, the population is now 103 million people.

   c. If hyperinflation is driving up prices at a rate of 100% per month, items that cost $100 two months ago now cost $400.

   d. If a bank account increases its value at a rate of 1% per month, an initial deposit of $100 will increase to $112 in a year.

18. **Exact Formula for the Doubling Time.** Rework each part of Problems 6 and 7 by using the exact formula for the doubling time given in the text. In each case, compare the answers obtained from the exact and approximate formulas.

19. **Exact Formula for the Half-Life.** Rework each part of Problems 13 and 14 by using the exact formula for the half-life given in the text. In each case, compare the answers obtained from the exact and approximate formulas.

20. **Investments.** Suppose that a bank account increases in value by 9% per year. If you invest $100 today, how much will be in the account when your great-great-grandchildren inherit it in 100 years?

21. **Investment Doubling Times.** Use the approximate and the exact formulas for the doubling time to answer each question. Compare the results and comment on the accuracy of the approximate formula.

   a. Suppose that an investment has an annual percentage yield of $APY = 8.50\%$. What is the doubling time for the value of this investment?

   b. Suppose that an investment has an annual percentage yield of $APY = 4.65\%$. What is the doubling time for the value of this investment?

22. **Termites!** Unbeknownst to you, a family of 100 termites invades your house. Suppose that the termite population grows at a rate of 15% per week. How many termites would be in your house after a year?

23. **Group Activity/Project: Simulating an Epidemic with Dice.** The following group activity demonstrates the exponential spread of a disease transmitted through personal contact. A group of 30–50 people is ideal. Each person in the group has a die, a tally sheet, and a personal ID number (which could be determined by rolling dice). The activity consists of five 3-minute stages. During each stage, everyone mingles and interacts. Each time a pair of people interact they each record the other's ID number on the tally sheet and then roll their dice: If the total on the dice is 5 or less, they have been "exposed" and the ID number of the other person is circled; otherwise, the ID number is left uncircled. The action breaks at the end of each stage and then resumes.

   At the end of five stages, one ID number is selected to represent an infected person. Everyone looks at his or her interactions from the first stage, and then anyone who interacted with the infected person *and* was exposed (total of 5 or less on the dice) becomes infected. The total number of infected people at the end of Stage 1 is recorded. The number of new infected people is recorded after the interactions of Stages 2 through 5. Once infected a person remains infected.

a. Having carried out this activity, graph the six data points (initial number of infected people is one, and the five stages are the other data points).

b. Does the graph look like exponential growth? Explain.

c. Repeat the activity with different conditions for expo-

sure (e.g., the total on the dice is 6 or 4). How does this change affect the outcome?

d. Can you incorporate the effect of public education programs into the activity? Can you incorporate the effect of a cure into the activity? Explain.

UNIT 7C

EXPONENTIAL MODELING

In Unit 6A, we introduced the notation (*independent variable, dependent variable*) for a *function* between two quantities, or variables. The actual function expresses *how* the value of the dependent variable depends on the value of the independent variable. In Unit 6B, we investigated equations that describe *linear* functions. Now that we have discussed the basic principles of exponential growth, we are ready to investigate exponential functions and to see how they can be used in models of exponential growth or decay.

EXPONENTIAL GROWTH AND DECAY LAWS

Consider again the case of Powertown, which begins with a population of 10,000 and grows at a rate of 5% per year (see Figure 7.1). We say that the population has an *initial value* of 10,000. The percentage growth rate is $P\% = 5\%$ per year, so the fractional growth rate is $r = \dfrac{P}{100} = 0.05$ per year. The population at the end of the first year is 5% *more than* the initial value of the population, or $105\% = 1.05$ *times* the initial population. Thus we can express the population at the end of one year as:

$$\text{population after 1 year} = 10,000 \times 1.05$$

The population again grows by 5% in the second year, so we can express the population at the end of the second year by again multiplying by 1.05:

$$\begin{aligned}
\text{population after 2 years} &= \text{population after 1 year} \times 1.05 \\
&= (10,000 \times 1.05) \times 1.05 \\
&= 10,000 \times 1.05^2
\end{aligned}$$

With the 5% per year increase continuing in the third year, we express the population at the end of the third year by multiplying once more by 1.05:

$$\begin{aligned}
\text{population after 3 years} &= \text{population after 2 years} \times 1.05 \\
&= (10,000 \times 1.05^2) \times 1.05 \\
&= 10,000 \times 1.05^3
\end{aligned}$$

The pattern may now be clear. If we let t represent the time in years, we find that

$$\text{population after } t \text{ years} = \text{initial population} \times 1.05^t.$$

For example, after $t = 25$ years, Powertown's population is

$$\text{population after 25 years} = 10{,}000 \times 1.05^{25} = 33{,}860.$$

We can now generalize to any exponentially growing quantity.

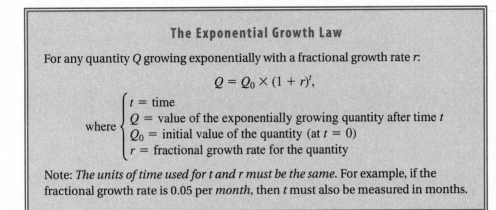

The Exponential Growth Law

For any quantity Q growing exponentially with a fractional growth rate r:

$$Q = Q_0 \times (1 + r)^t,$$

where $\begin{cases} t = \text{time} \\ Q = \text{value of the exponentially growing quantity after time } t \\ Q_0 = \text{initial value of the quantity (at } t = 0) \\ r = \text{fractional growth rate for the quantity} \end{cases}$

Note: *The units of time used for t and r must be the same.* For example, if the fractional growth rate is 0.05 per *month*, then t must also be measured in months.

Mathematical Note: As discussed in Unit 7B, r must be the *actual* fraction by which a quantity grows (or decays) in a given period of time. That is, r is analogous to the annual yield *APY* in financial formulas, rather than the *APR*.

You may notice that this exponential growth equation is identical to the compound interest formula (Unit 5A) if we identify Q as the accumulated balance A, Q_0 as the starting principal P, r as the interest rate per compounding period $\frac{APR}{n}$, and t as the number of compounding periods nY. Thus you have already worked with this equation as it applies to finance, and we now are simply generalizing to other examples of exponential growth.

A very similar argument can be used to find a general equation for exponential decay. For example, if a quantity decays by $P\% = 10\%$ per month, then the fractional decay rate is $r = \frac{P}{100} = 0.1$ per month. Each month the value of the quantity falls to $1 - r = 0.9$ of its value from the previous month. After $t = 1$ month, the value of the quantity is $0.9 \times Q_0$ (where Q_0 is the initial value of the quantity at $t = 0$). After $t = 2$ months, the value of the quantity is $(0.9)^2 \times Q_0$. We can generalize this pattern to any exponentially decaying quantity.

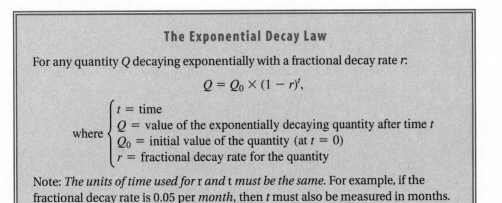

The Exponential Decay Law

For any quantity Q decaying exponentially with a fractional decay rate r:

$$Q = Q_0 \times (1 - r)^t,$$

where $\begin{cases} t = \text{time} \\ Q = \text{value of the exponentially decaying quantity after time } t \\ Q_0 = \text{initial value of the quantity (at } t = 0) \\ r = \text{fractional decay rate for the quantity} \end{cases}$

Note: *The units of time used for r and t must be the same.* For example, if the fractional decay rate is 0.05 per *month*, then t must also be measured in months.

GRAPHING AN EXPONENTIAL FUNCTION

We can graph exponential functions using point-by-point techniques (see Unit 6C). For exponential growth, the easiest way is to plot points representing several doubling times. For example, we can start at the point $(0, Q_0)$ that represents the initial value at $t = 0$. We know that the value of Q reaches $2Q_0$ (double its initial value) after the doubling time T_{double}, $4Q_0$ after the two doubling times $(2T_{double})$, $8Q_0$ after the three doubling times $(3T_{double})$, and so on. We simply fit a steeply rising curve between these points (Figure 7.3a).

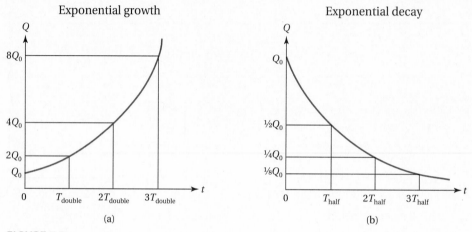

FIGURE 7.3

For exponential decay, we instead plot points representing several half-lives. Again we start at the initial value point $(0, Q_0)$. The value of Q falls to $\frac{1}{2}Q_0$ (half its initial value) after the half-life T_{half}, to $\frac{1}{4}Q_0$ after two half-lives $(2T_{half})$, and so on. Thus we fit a sharply falling curve between these points (Figure 7.3b). Note that this curve gets closer and closer to the horizontal axis, but never reaches it, because Q never quite reaches zero.

EXAMPLE 1   *U.S. Population Growth*

The 1990 census found a U.S. population of about 250 million, and the population is growing at a rate of about 0.7% per year. If this growth rate is maintained, what will the U.S. population be in 2050?

Solution: The quantity Q is the U.S. population. We are given an initial value $Q_0 = 250$ million in 1990, and we are looking for the value of Q in 2050, which is $t = 60$ years after 1990. The growth rate is 0.7% per year, so

$$r = \frac{P}{100} = \frac{0.7/\text{yr}}{100} = 0.007/\text{yr}.$$

Thus the exponential growth equation for this situation is:

$$Q = Q_0 \times (1 + r)^t = 250 \text{ million} \times (1 + 0.007)^{60}$$
$$= 250 \text{ million} \times (1.007)^{60} = 380 \text{ million}$$

At its present growth rate of 0.7% per year, the U.S. population will swell to 380 million by 2050.   ■

EXAMPLE 2   *Sensitivity to Growth Rate*

The growth rate of the U.S. population has varied substantially during the past century. It is sensitive to immigration rates as well as to birth and death rates. To see the effects of growth rate changes, project the population in 2050 using growth rates that are just 0.5 percentage points lower and higher than the 0.7% used in Example 1. Discuss the results.

Solution: A growth rate 0.5 percentage points lower than 0.7% is 0.2%, or $r = 0.002$. At this growth rate, the 2050 population would be

$$Q = 250 \text{ million} \times (1.002)^{60} = 282 \text{ million}.$$

A growth rate 0.5 percentage points higher than 0.7% is 1.2%, or $r = 0.012$, which gives a 2050 population of

$$Q = 250 \text{ million} \times (1.012)^{60} = 511 \text{ million}.$$

Within just a one percent range, from 0.2% to 1.2%, the projected population for 2050 varies by nearly 230 million people (from 282 million to 511 million). This variation alone is nearly as much as the entire current population of the United States. Clearly, population projections are very sensitive to changes in the growth rate. The population growth with the three growth rates is shown in Figure 7.4.

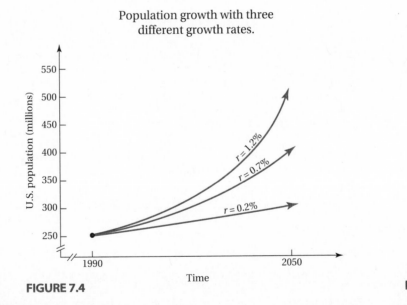

Population growth with three different growth rates.

FIGURE 7.4   ■

EXAMPLE 3   *Declining Population*

China's one-child policy was implemented in 1978 with a goal of reducing China's population to 700 million by 2050. China's 2000 population is about 1.2 billion. Suppose that China's population declines at a rate of 0.5% per year. Will this rate of decline be sufficient to meet the original goal?

Solution: The population is *declining* at a rate of $P\% = 0.5\%$ per year, so the decay rate is:

$$r = \frac{P}{100} = \frac{0.5/\text{yr}}{100} = 0.005/\text{yr}$$

If we use the year 2000 as $t = 0$, the initial value of the population is $Q_0 = 1.2$ billion. The year 2050 is 50 years after 2000, so we find the population in 2050 by setting $t = 50$ years in the exponential decay equation.

$$\begin{aligned}
Q &= Q_0 \times (1 - r)^t \\
&= 1.2 \text{ billion} \times (1 - 0.005)^{50} \\
&= 1.2 \text{ billion} \times (0.995)^{50} = 0.93 \text{ billion} = 930 \text{ million}
\end{aligned}$$

A decline of 0.5% per year will reduce China's population to 930 million by 2050, well short of the original goal of 700 million. ∎

SELECTED APPLICATIONS

In this unit, we have seen several examples using exponential functions to model population growth and decline. We also have used exponential equations in Chapter 5 for compound interest calculations. But exponential models apply to many other growth and decay processes. Before looking at a few more selected applications, recall that in Unit 7B we introduced the following equations:

For exponential growth: new value $=$ initial value $\times 2^{t/T_{\text{double}}}$

For exponential decay: new value $=$ initial value $\times \left(\dfrac{1}{2}\right)^{t/T_{\text{half}}}$

Letting Q represent the new value and Q_0 represent the initial value, we can rewrite these equations as follows:

For exponential growth: $Q = Q_0 \times 2^{t/T_{\text{double}}}$

For exponential decay: $Q = Q_0 \times \left(\dfrac{1}{2}\right)^{t/T_{\text{half}}}$

Mathematically, these statements are equivalent to the exponential growth and decay laws introduced in this unit. Thus we actually have *two* alternative ways to express the exponential growth law and *two* alternative ways to express the exponential decay law. The following table summarizes the circumstances in which each form is easier to use.

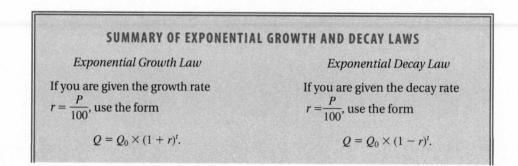

SUMMARY OF EXPONENTIAL GROWTH AND DECAY LAWS

Exponential Growth Law	*Exponential Decay Law*
If you are given the growth rate $r = \dfrac{P}{100}$, use the form	If you are given the decay rate $r = \dfrac{P}{100}$, use the form
$Q = Q_0 \times (1 + r)^t$.	$Q = Q_0 \times (1 - r)^t$.

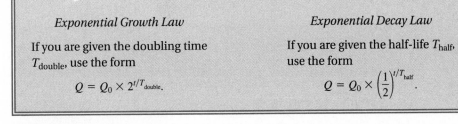

Exponential Growth Law

If you are given the doubling time T_{double}, use the form

$$Q = Q_0 \times 2^{t/T_{\text{double}}}.$$

Exponential Decay Law

If you are given the half-life T_{half}, use the form

$$Q = Q_0 \times \left(\frac{1}{2}\right)^{t/T_{\text{half}}}.$$

Inflation-Adjusted Dollars

Because prices tend to change with time, price comparisons from one time to another are meaningful only if the prices are adjusted for the effects of *inflation* (see discussion of the consumer price index in Unit 4D). We can adjust prices for inflation with the general exponential equation by setting the following:

- r = rate of inflation
- Q_0 = initial price of item(s)
- Q = adjusted price of item(s) at a time t later

EXAMPLE 4   *Converting from 1990 to 1995 Dollars*

Between 1990 and 1995, the average rate of inflation was about 3% per year (as measured by the consumer price index). If a cart of groceries cost $100 in 1990, what did it cost in 1995?

Solution: For the 3% inflation rate, we set $r = 0.03$ per year and $Q_0 = \$100$ for the price in 1990. We are looking for the price Q in 1995, or $t = 5$ years later:

$$Q = Q_0 \times (1 + r)^t = \$100 \times (1 + 0.03)^5$$
$$= \$100 \times (1.03)^5 = \$116$$

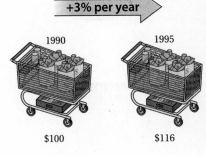

+3% per year

1990   1995

$100 $116

The groceries costing $100 in 1990 would cost $116 in 1995. In the terms used by economists, we would say that $116 in "1995 dollars" is equivalent to $100 in "1990 dollars." ∎

EXAMPLE 5   *Real Changes in Cost*

Suppose that the average price for a gallon of gasoline was $1.20 in 1990 and $1.35 in 1995. With an inflation rate of 3% per year, compare the inflation-adjusted prices of the gasoline.

Solution: We begin by calculating how the 1990 price of $1.20 per gallon adjusts to 1995 dollars. We set the inflation rate to $r = 0.03$ per year, the 1990 price to $Q_0 = \$100$, and $t = 5$ years from 1990 to 1995:

$$Q = Q_0 \times (1 + r)^t$$
$$= \$1.20 \times (1 + 0.03)^5 = \$1.39$$

That is, the price of $1.20 in 1990 is equivalent to $1.39 in "1995 dollars." Because the *actual* price in 1995 was $1.35, the real price of gasoline *fell* by $0.04 over the 5-year period. (Note that the $0.04 real decline also is measured in 1995 dollars.) ∎

EXAMPLE 6   *Hyperinflation in Brazil*

During the 1970s and 1980s, Brazil underwent tremendous *hyperinflation*—periods of extraordinarily large inflation in prices. During the worst periods, prices rose as rapidly as 80% per month. At this rate, how much would prices have risen in one year? in one day?

Solution: We can compare prices at two different times by letting Q_0 represent a price at time $t = 0$. For an inflation rate that makes prices rise by 80% per month, we set $r = 0.8$ per month. The price after 1 year, or $t = 12$ months, is

$$Q = Q_0 \times (1 + 0.8)^{12} = Q_0 \times 1157.$$

The new price after a year is 1157 times the original price. That is, prices increase more than a thousand-fold over a year! Note that we set t in units of *months* because the inflation rate was given *per month*.

If we assume a month with 30 days, we can find the price change in one day by setting $t = 1/30$ month:

$$Q = Q_0 \times (1 + 0.8)^{1/30} = Q_0 \times 1.02$$

Thus prices at the end of any day are 1.02 *times*, or 102% *of*, prices at the end of the previous day. In other words, prices rise by 2% in a single day! To keep pace with this rapid inflation, Brazilian merchants had to mark up prices every day; in some cases, they even closed stores for a couple of hours and marked up prices at midday. ∎

$\mathcal{By\ the\ Way}$ ⋯⋯⋯⋯

Overall, Brazil's currency dropped in value by a factor of more than 1 trillion in less than two decades. To avoid having to print bills with large numbers of zeros, Brazil *devalued* its currency several times by factors of 1000. By the mid-1990s, Brazil's hyperinflation had stopped.

> **TIME-OUT TO THINK:**   Discuss some of the social upheavals likely under hyperinflation. Are any of the world's countries currently experiencing this phenomenon? If so, what are the consequences?

EXAMPLE 7   *Monthly and Annual Inflation Rates*

The U.S. government reports the rate of inflation both monthly and annually (often in terms of changes in the consumer price index). Suppose that the July rate of inflation is reported as 0.8% per month. What *annual* rate of inflation does this imply?

Solution: In this case, we have an inflation rate of $r = 0.008$ per month. As in Example 6, we let Q_0 represent a price at time $t = 0$. Thus the new price after a year, or $t = 12$ months, is

$$Q = Q_0 \times (1 + 0.008)^{12} = Q_0 \times 1.100.$$

That is, the new price Q after one year is 1.1 times the old price Q_0, which means that prices rose by 10%. Thus the *annual* inflation rate is 10%, based on the July increase of 0.8%. ∎

Beware! It is tempting to say that a monthly growth rate of 0.8% implies a yearly growth rate of

$12 \times 0.8\% = 9.6\%$.

Because of the effects of compounding, this is *not* true, as this example illustrates.

Environment and Resources

Among the most important applications of exponential models are those involving the environment and resource depletion. Global concentrations of many pollutants in the water and atmosphere have increased exponentially, as has consumption of nonrenewable resources such as oil and natural gas. Two basic factors can be responsible for such exponential growth.

1. The *per capita* demand for a resource, or creation of waste products (pollution), often increases exponentially. For example, per capita energy consumption in the United States increased exponentially during most of the twentieth century.

2. An exponentially increasing population can lead to an exponentially increasing demand for resources (or creation of waste products) even if per capita demand remains constant.

In most cases, the growth or decay rate is determined by a combination of both factors.

EXAMPLE 8   *World Oil Production*

In 1950, world oil production was 518 million tons. From then until the early 1970s production increased at a rate of 7% per year. Had growth continued at this rate, how much oil would have been produced in 1993? Comment on the fact that world oil production in 1993 was *actually* about 3000 million tons.

Solution: We use 1950 as $t = 0$, with its initial value of oil production $Q_0 = 518$ million tons; 1993 was $t = 43$ years after 1950. Using a 7% growth rate, or $r = 0.07$ per year, oil production in 1993 would have been:

$$Q = Q_0 \times (1 + r)^t = 518 \text{ million tons} \times (1 + 0.07)^{43}$$
$$= 9500 \text{ million tons}$$

This model prediction of 9500 million tons overestimated the actual figure of 3000 million tons by a factor of more than 3. Apparently, the growth rate in oil production fell substantially after the mid-1970s. Total use of fossil fuel has continued to rise, however, as production of coal and natural gas now provides a larger share of energy than in the past (Figure 7.5). ■

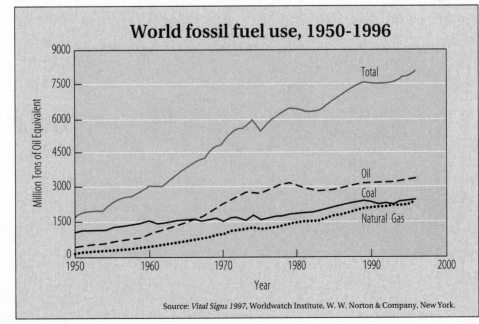

FIGURE 7.5

EXAMPLE 9   *Carbon Emissions from China and the United States*

The burning of fossil fuels releases greenhouse gases into the atmosphere. In 1995, the United States emitted about 1400 million tons of carbon into the atmosphere, nearly one-fourth of the world total. China was the second largest contributor to greenhouse emissions, with about 850 million tons. However, emissions from China were rising during the 1990s at a rate of about 4% per year, while U.S. emissions were rising at about 1.3% per year. Using these growth rates, project greenhouse gas emissions from the United States and China in 2020. Graph the projected trends in emissions for both countries.

Solution: We use 1995 as $t = 0$, so 2020 is $t = 25$ years. For the United States, we set the 1995 emissions as $Q_0 = 1400$ million tons of carbon, and $r = 0.013$ per year to find:

$$\text{U.S. carbon emissions in 2020} = 1400 \text{ million tons} \times (1 + 0.013)^{25}$$
$$= 1900 \text{ million tons}$$

For China, we set the 1995 emissions as $Q_0 = 850$ million tons of carbon, and $r = 0.04$ per year to find:

$$\text{China carbon emissions in 2020} = 850 \text{ million tons} \times (1 + 0.04)^{25}$$
$$= 2300 \text{ million tons}$$

If current trends continue, China will become the largest emitter of greenhouse gases before 2020. We can graph the trends by using equations that represent the emissions in each country over time. These come directly from the general exponential equations used for the 2020 calculations:

$$\text{U.S. carbon emissions } (t \text{ years after 1995}) = 1400 \text{ million tons} \times (1.013)^{t}$$
$$\text{China carbon emissions } (t \text{ years after 1995}) = 850 \text{ million tons} \times (1.04)^{t}$$

Figure 7.6 shows the graphs of these equations. The graphs show that Chinese carbon emissions will surpass U.S. carbon emissions in 2014.

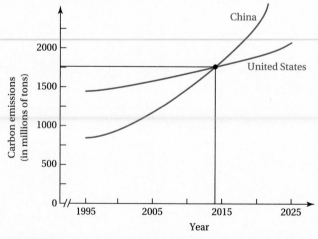

FIGURE 7.6

Radioactive Dating

Many naturally occurring substances are **radioactive**—meaning that the nuclei of their atoms tend to break apart over time. When a nucleus undergoes **radioactive decay**, the substance is transformed into something else. For example, uranium ultimately decays into lead (with several intermediate steps); we therefore say that lead is the **decay product** of uranium.

We can take advantage of the exponential nature of radioactive decay to measure the ages of rocks, bones, pottery, or other solid objects that contain radioactive elements. The process is called **radioactive dating**. We can date an object only if we know:

- both the current and original amounts of the radioactive substance in the object; and

- the half-life of the radioactive substance.

If we know both of these things, we can find the age of the object from the exponential decay law. In this case, because we are given the half-life, it is easier to use the exponential decay law in the form:

$$Q = Q_0 \times \left(\frac{1}{2}\right)^{t/T_{\text{half}}}$$

If we identify the Q as the current amount of the radioactive substance and the initial value Q_0 as the original amount, then we have

$$\text{current amount} = \text{original amount} \times \left(\frac{1}{2}\right)^{t/T_{\text{half}}}.$$

We must solve this equation to find the time t since the object was formed. We begin by dividing both sides by the *original amount*, and then switching sides:

$$\left(\frac{1}{2}\right)^{t/T_{\text{half}}} = \frac{\text{current amount}}{\text{original amount}}$$

Now we can "bring down" the exponent by taking the log of both sides (see Unit 6C):

$$\log_{10}\left(\frac{1}{2}\right)^{t/T_{\text{half}}} = \log_{10}\left(\frac{\text{current amount}}{\text{original amount}}\right)$$

Using the rule that $\log_{10} a^x = x \times \log_{10} a$, this equation becomes

$$\frac{t}{T_{\text{half}}} \times \log_{10}\left(\frac{1}{2}\right) = \log_{10}\left(\frac{\text{current amount}}{\text{original amount}}\right).$$

Finally, we isolate t by multiplying both sides by T_{half} and dividing both sides by $\log_{10}\left(\frac{1}{2}\right)$ to find the following equation.

$$t = T_{\text{half}} \times \frac{\log_{10}\left(\dfrac{\text{current amount}}{\text{original amount}}\right)}{\log_{10}\left(\dfrac{1}{2}\right)}$$

By the Way

Most atomic nuclei are *stable*, meaning they do not spontaneously fall apart. By definition, radioactive substances have unstable nuclei that *do* fall apart spontaneously. The term *radioactive* came about for historical reasons and has little to do with radio waves.

By the Way

American scientist Willard Libby won the Nobel Prize in Chemistry in 1960 for inventing the method of radioactive dating.

The Radioactive Dating Equation

If we know both the current and original amounts of a radioactive substance in an object, then the age t (time since formation) of the object is

$$t = T_{half} \times \frac{\log_{10}\left(\dfrac{\text{current amount}}{\text{original amount}}\right)}{\log_{10}\left(\dfrac{1}{2}\right)},$$

where T_{half} is the half-life of the radioactive substance.

EXAMPLE 10   *Uranium Decay*

Uranium-238 has a half-life of 4.5 billion years, and ultimately decays into lead. Suppose that you find a rock containing a mixture of uranium-238 and lead. The distribution of lead atoms among the uranium atoms appears random, suggesting that the rock originally contained *only* uranium and that the lead was formed by uranium decay. By measuring the relative quantities of lead and uranium and taking into account their different atomic weights, you determine that 65% of the original uranium-238 remains and the other 35% decayed into lead. When did the rock form?

Solution: We are given the ratio of the current to original amounts of uranium:

$$\frac{\text{current amount}}{\text{original amount}} = 65\% = 0.65$$

We use the half-life of T_{half} = 4.5 billion years, and calculate the age t of the rock:

$$\begin{aligned}
t &= T_{half} \times \frac{\log_{10}\left(\dfrac{\text{current amount}}{\text{original amount}}\right)}{\log_{10}\left(\dfrac{1}{2}\right)} \\[2em]
&= 4.5 \text{ billion yr} \times \frac{\log_{10}(0.65)}{\log_{10}\left(\dfrac{1}{2}\right)} \\[2em]
&= 4.5 \text{ billion yr} \times 0.62 \\[0.5em]
&= 2.8 \text{ billion yr}
\end{aligned}$$

The rock formed (i.e., solidified from a prior molten state) about 2.8 billion years ago.   ∎

EXAMPLE 11   *The Allende Meteorite*

The famous *Allende meteorite* lit up the skies of Mexico as it fell to Earth on February 8, 1969. Scientists melted and chemically analyzed small pieces of the meteorite and found traces of both radioactive potassium-40 and its decay product, argon-40. Because argon-40 is a gas, any argon trapped in the rock must have been released *after* the rock formed;

as a gas, it could not have solidified at the time the rock originally formed. The scientists determined that 8.5% of the potassium-40 originally present in the rock remains today. The half-life of potassium-40 is 1.3 billion years. How old is the rock that makes up the Allende meteorite?

Solution: In this case, the ratio of the current to original amounts of potassium-40 is 8.5%, or 0.085. We can now use the radioactive dating equation with the potassium-40 half-life of $T_{half} = 1.3$ billion years to calculate the age t of the meteorite:

$$t = T_{half} \times \frac{\log_{10}\left(\frac{\text{current amount}}{\text{original amount}}\right)}{\log_{10}\left(\frac{1}{2}\right)}$$

$$= 1.3 \text{ billion yr} \times \frac{\log_{10}(0.085)}{\log_{10}\left(\frac{1}{2}\right)} = 4.6 \text{ billion yr}$$

The Allende meteorite solidified about 4.6 billion years ago. ∎

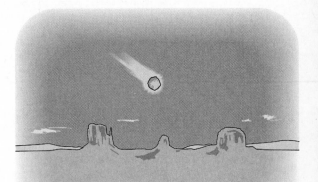

By the Way ············

Many meteorites date to 4.6 billion years ago, but none have ever been found that are older. Scientists therefore conclude that the solar system itself must have formed 4.6 billion years ago, and hence that the Allende meteorite is a remnant from the formation of the solar system.

EXAMPLE 12   *Radiocarbon Dating*

Ordinary carbon, called carbon-12, is stable and does not decay. However, carbon-14 is a relatively rare form (isotope) of carbon that decays radioactively with a half-life of about 5,700 years. Carbon-14 is constantly produced in the Earth's atmosphere by sunlight. When a living organism breathes or eats, it ingests carbon-14 along with ordinary carbon-12. After the organism dies, no more carbon-14 is ingested, so the age of its remains can be calculated by determining how much carbon-14 has decayed. Suppose that you find a human bone at an archaeological site. You analyze the bone and discover that it contains only one-tenth of the carbon-14 that it contained when the person died (various methods may be used to determine how much carbon-14 the bone originally contained). How long ago did the person die?

Solution: The fact that only one-tenth of the original carbon-14 remains means that

$$\frac{\text{current amount}}{\text{original amount}} = 0.1.$$

Using the radioactive dating equation with the half-life of $T_{half} = 5700$ years gives

$$t = T_{half} \times \frac{\log_{10}\left(\frac{\text{current amount}}{\text{original amount}}\right)}{\log_{10}\left(\frac{1}{2}\right)} = 5700 \text{ yr} \times \frac{\log_{10}(0.1)}{\log_{10}\left(\frac{1}{2}\right)} = 19{,}000 \text{ yr.}$$

The bone comes from a person who died about 19,000 years ago. ∎

> **TIME-OUT TO THINK:** Would carbon-14 be useful for dating a fossil 100 million years old? Why or why not? (Hint: Consider its half-life in comparison to 100 million years.)

Physiological Processes

Many physiological processes are exponential. For example, a cancer tumor grows exponentially, at least in its early stages, and most drugs in the bloodstream break down exponentially.

EXAMPLE 13 *Drug Metabolism*

Consider an antibiotic that has a half-life in the bloodstream of 12 hours. A 10-milligram injection of the antibiotic is given at 1:00 P.M. How much antibiotic remains in the blood at 9:00 P.M.? Draw a graph that shows the amount of antibiotic remaining as the drug is metabolized by the body.

Solution: Although this problem involves the decay of a drug in the bloodstream, rather than radioactive decay, the principles are the same. Thus we can find the current amount at any time t after the injection from the equation

$$\text{current amount} = \text{original amount} \times \left(\frac{1}{2}\right)^{t/T_{\text{half}}}.$$

In this case, 9:00 P.M. is $t = 8$ hours after the injection, the original amount is 10 mg, and the half-life is given as is $T_{\text{half}} = 12$ hours. We find that

$$\text{current amount} = 10 \text{ mg} \times \left(\frac{1}{2}\right)^{8 \text{ hr}/12 \text{ hr}} = 10 \text{ mg} \times \left(\frac{1}{2}\right)^{2/3} = 6.3 \text{ mg}.$$

Eight hours after the injection, 6.3 mg of the antibiotic remains in the bloodstream.

More generally, at any time t after the injection, the amount of antibiotic remaining will be

$$\text{current amount} = 10 \text{ mg} \times \left(\frac{1}{2}\right)^{t/12 \text{ hr}}.$$

This equation is graphed in Figure 7.7, where a is the current amount.

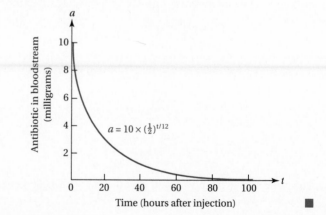

FIGURE 7.7

REVIEW QUESTIONS

1. Describe the meaning of all the variables in the exponential growth equation. Explain how the equation is used.

2. How do you graph an exponential equation? Give examples both for growth and decay.

3. Explain how you can use an exponential equation to adjust prices for the effects of inflation.

4. What is *hyperinflation*? Why is it a serious problem?

5. Why does human use of resources and creation of waste products tend to increase exponentially?

6. Explain the process of *radioactive dating*. How does radioactive dating help us know the age of the Earth? How does it help archaeologists date past human settlements?

7. Give an example of a physiological process that is exponential.

PROBLEMS

1. **Linear Versus Exponential Population Growth.**

   a. In 1990, the population of Linear City was 100,000, and it has been growing at a constant rate of 10,000 people per year ever since. Make a table that shows the population of Linear City each year from 1990 to 2005, and make a graph from the data.

   b. In 1990, the population of Exponential City was 100,000, and it has bee 10% per year ever since. Make a table that shows the population of Exponential City each year from 1990 to 2005, and make a graph from the data.

   c. In words, contrast the growth of Linear City and Exponential City. Do you think that either of these fictional cities represents the way a city would really grow? Explain.

2. **Linear Versus Exponential Monetary Growth.**

   a. Suppose that you decide to save money by putting $100 under your mattress each year. Make a table that shows how much money you have saved in each of the first 10 years and make a graph of how your money has grown in time. How much will you have at the end of 10 years?

   b. Suppose that you place the $100 in a bank account that increases its value by 8% per year (i.e., *APY* = 8%). Make a table that shows how much money you have saved in each of the first 10 years and make a graph of how your money has grown in time. How much will you have at the end of 10 years?

   c. In words, contrast the results of the linear growth of your savings under the mattress and the exponential growth involved in the savings account. How much of your own money (not counting investment interest) did you contribute in each case?

Exponential Growth and Decay Laws. *For Problems 3–12, do the following:*

   a. *Create an exponential function of the form*
   $$Q = Q_0 \times (1 + r)^t \quad \text{(for growth)}$$
   or $\quad Q = Q_0 \times (1 - r)^t \quad$ *(for decay)*
   to model the situation described. Be sure to clearly identify each of the variables in your function.

   b. *Create a table showing the value of the quantity Q for the first 10 units of time (either years, months, weeks, or hours) of growth or decay.*

   c. *Make a graph of the exponential function.*

3. The population of a town with a 1998 population of 85,000 grows at a rate of 2.4% per year.

4. In 1999, you deposit $10,000 in a mutual fund that then increases in value at a rate of 7.25% per year.

5. The homicide rate in a city that had 800 homicides in 1995 is increasing at a rate of 3% per year.

6. A privately owned forest that had 1 million acres of old-growth forest is being clear cut at a rate of 7% per year.

7. A town with a population of 10,000 is losing residents at a rate of 0.3% per month because of a poor economy.

8. The average price of a home in a town was $125,000 in 1995, but home prices are rising by 7% per year.

9. A peso is currently worth 25¢, but its value is falling against the value of the dollar at a rate of 10% per week.

10. A particular drug breaks down in the human body at a rate of 15% per hour. The initial concentration of the drug in the bloodstream is 8 milligrams per liter.

11. Your starting salary at a new job is $2000 per month, and you get annual raises of 5% per year.

12. You hid 100,000 rubles in a mattress at the end of 1991, when they had a value of $10,000. However, the value of the ruble against the dollar then fell 50% per year.

13. **Metropolitan Population Growth.** A small city had a population of 85,000 in 1990. Concerned about rapid growth, the residents passed a growth control ordinance limiting population growth to 3% each year. If the population grows at this 3% annual rate, what will the population be in 2010? in 2100? Discuss the implications of your results to the region in one or two paragraphs.

14. **Adjusting for Inflation.**

   a. Suppose that the average price for a gallon of gasoline was $1.15 in 1990 and $1.40 in 1998. With an inflation rate of 3% per year, compare the inflation-adjusted prices of the gasoline.

   b. Suppose that the price of a movie was $5.50 in 1990 and $7.75 in 1998. With an inflation rate of 3% per year, compare the inflation-adjusted prices of movies.

15. **Effects of Inflation.** Assume that between 1990 and 1998, the average rate of inflation was about 3% per year (as measured by the consumer price index). Also assume that the increase in prices is due to inflation only.

   a. If a cart of groceries cost $100 in 1990, what would they cost in 1998?

   b. If tuition at a particular college was $1440 in 1990, what would it be in 1998?

   c. If a concert ticket cost $35 in 1990, how much would it cost in 1998?

16. **Leaving Your Descendants Dollars.** Suppose that you invest $1 in a bank account that increases its value annually by 7% (i.e., *APY* = 7%). Amazingly, the bank is able to hold its interest rate constant for a very long time. In your will, you leave this bank account to your great-great-great-great-great-great-great-great-great-great-great-great-great-great-great grandchildren, who ought to be your present age in about 400 years.

   a. How much money will be in the account for them 400 years from now?

   b. Unfortunately, inflation depreciates the value of the dollar by a constant rate of 5% per year. In 400 years, how much money or "buying power" will the $1 have?

17. **Annual Inflation Rates from Monthly Rates.** Answer the following questions about monthly and annual inflation rates. Remember that the annual rate is never exactly 12 times the monthly rate.

   a. If prices increase at a monthly rate of 0.6%, how much do they increase in a year?

   b. If prices of gold decrease at a monthly rate of 0.8%, how much do they decrease in a year?

18. **Hyperinflation.**

   a. During the war in Bosnia, inflation sometimes drove prices up at a rate of 100% per month. At this rate, how much would prices have risen in 1 year? in 1 day?

   b. Suppose that a country experiences inflation at a rate of 50% per month. How much would prices rise in 1 year? in 10 years?

19. **Extinction by Poaching.** Suppose that poaching reduces the population of an endangered animal by 10% per year. Further, suppose that when the population of this animal

falls below 30, its extinction is inevitable (owing to the lack of reproductive options without severe in-breeding). If the current population of the animal is 1000, when will it face extinction? Comment on the validity of the exponential model.

20. Discovering Radioactive Waste. A toxic radioactive substance with a density of 2 milligrams per square centimeter is detected in the ventilating ducts of a nuclear processing building that was used 45 years ago. If the half-life of the substance is 20 years, what was the density of the substance when it was deposited 45 years ago?

21. Pesticide Decay and Your Cat. The concentration of a particular pesticide that is applied to lawns declines exponentially with a half-life of one week. Suppose that the pesticide is applied with an initial concentration of 10 grams per square meter of lawn.

   a. When does the concentration reach 1 gram per square meter?

   b. Suppose that your cat eats about 5 square centimeters of grass each day. On the day the pesticide is applied, how much pesticide, in grams, would the cat consume?

   c. Suppose that the pesticide is toxic to cats if consumed in amounts greater than 100 milligrams per day. For how many days should you keep your cat in the house after the pesticide is applied to ensure the cat's safety?

22. World Oil Production. World oil production was 518 million tons in 1950. Between 1950 and 1972 production increased at a rate of 7% per year. World oil production in 1993 was approximately 3000 million tons.

   a. What was the world oil production in 1972?

   b. Using the result of part (a), had growth in production continued at a rate of 6% between 1972 and 1993, how much oil would have been produced in 1993? Compare this result to the actual figure given above.

   c. Using the result of part (a), had growth in production continued at a rate of 3% between 1972 and 1993, how much oil would have been produced in 1993? Compare this result to the actual figure given above.

   d. Estimate the annual growth rate in world oil production between 1972 and 1993 needed to give a production of 3000 million tons in 1993.

23. Radioactive Uranium Dating. Uranium-238 has a half-life of 4.5 billion years.

   a. You find a rock containing a mixture of uranium-238 and lead. You determine that 85% of the original uranium-238 remains; the other 15% decayed into lead. How old is the rock?

   b. Analysis of another rock shows that it contains 55% of its original uranium-238; the other 45% decayed into lead. How old is the rock?

24. Radioactive Dating with C-14. The half-life of carbon-14 is about 5,700 years.

   a. You find a piece of cloth painted with organic dyes. By analyzing the dye in the cloth, you find that only 77% of the carbon-14 originally in the dye remains. When was the cloth painted?

   b. A well-preserved piece of wood found at an archaeological site has 6.2% of the carbon-14 that it must have had when it was alive. Estimate when the wood was cut.

   c. Is carbon-14 useful for establishing the age of the Earth? Why or why not?

25. Valium Metabolism. The drug Valium is eliminated from the bloodstream exponentially with a half-life of 36 hours. Suppose that a patient receives an initial dose of 20 milligrams of Valium at midnight.

   a. How much Valium is in the patient's blood at noon the next day?

   b. Estimate when the Valium concentration will reach 10% of its initial level?

26. Aspirin Metabolism. Assume that for the average individual, aspirin has a half-life of 12 hours in the bloodstream. At 12:00 noon, you take a 200-milligram dose of aspirin.

   a. How much aspirin will be in the blood at 6:00 P.M. later the same day? at midnight? at 12:00 noon the next day?

   b. Estimate when the amount of aspirin will decay to 5% of its original amount.

27. Project: Periodic Drug Doses. It is common to take a drug (such as aspirin or an antibiotic) repeatedly at a fixed-time interval. Suppose that an antibiotic has a half-life of 12 hours and a 100-milligram dose is taken every 8 hours.

   a. Write an exponential function that represents the decay of the antibiotic from the moment of the first dose to just *prior* to the next dose (i.e., 8 hours after the first dose). How much antibiotic is in the bloodstream just *prior* to this next dose? How much antibiotic is in the bloodstream just *after* this next dose?

   b. Following a procedure similar to that in part (a), calculate the amounts of antibiotic in the bloodstream just prior to and just after the dose at 16 hours, 24 hours, and 32 hours.

   c. Make a graph of the amount of antibiotic in the bloodstream for the first 32 hours after the first dose of the drug. What do you predict will happen to the amount of drug if the doses every 8 hours continue for several days or weeks? Explain.

   d. Consult a pharmacist (or read the fine print on the information sheet enclosed in many medicines) to find the half-lives of a common drug. Create a model for the metabolism of that drug using the above procedure.

28. Project: Increasing Atmospheric Carbon Dioxide. Between 1860 and 1990, carbon dioxide (CO_2) concentration in the atmosphere rose from roughly 290 parts per million to 350 parts per million. Assume that this growth can be modeled with an exponential function of the form $Q = Q_0 \times (1 + r)^t$.

   a. By experimenting with various values of the fractional growth rates r, find an exponential function that fits the given data for 1860 and 1990.

   b. Use your exponential model to predict when the CO_2 concentration will be double its 1990 level.

   c. In 1997, an international treaty was proposed with the aim of slowing the growth in CO_2 concentration. Research the current status of this treaty, and whether any progress has been made in slowing the growth in CO_2 concentration.

29. Project: Radon in the Home. One of the leading causes of lung cancer in the United States is radon gas that accumulates in well-sealed homes. Radon-222 is a gas created as a natural decay product of uranium-238, which is present in the ground in much of the United States. Outside, even in areas with the highest uranium content, the concentration of the radon gas released into the air generally is too low to pose a health risk. In well-sealed homes, however, radon gas can leach into the house through the foundation, and then remain trapped indoors. As a result, the concentration of radon gas can rise to levels that pose a health risk. The U.S. Environmental Protection Agency (EPA) measures the concentration of radon gas in units of picocuries per cubic meter. (The curie is a unit used in measuring radioactivity that corresponds to the decay of 37 billion atoms per second; a picocurie is 10^{-12} curie.) According to the EPA, concentrations of radon gas above 4 picocuries per cubic meter are unsafe and steps should be taken to mitigate against the radon buildup.

   a. The half-life of radon-222 is 3.8 days. Suppose that a sealed box has a radon gas concentration of 100 picocuries per cubic meter. If no gas is allowed in or out of the box, when will the concentration fall below the safe level of 4 picocuries per cubic meter?

   b. The easiest way to mitigate radon buildup is to open doors and windows so that ventilation allows the radon to escape. Unfortunately, this solution can be very energy inefficient (if the house is being heated or air conditioned); it also may not help in basements that have little ventilation. Suggest several other methods of mitigating radon buildup. You might contact local builders, environmental consultants, or the EPA to find out about effective means of radon mitigation.

c. Investigate whether radon gas is a problem in your community. If so, determine whether most homes and apartments are adequately mitigating radon buildup. Research the most cost-effective methods of mitigation in your community.

d. To determine the concentration of radon gas in a home you need to analyze three factors: the rate of inflow of radon gas to the home (through the foundation), the rate of outflow through ventilation, and the rate at which the radon decays. Discuss how a mathematical model could be created to account for all three factors. Why would such a model be useful?

REAL POPULATION GROWTH

Perhaps the most important application of exponential modeling is in population growth. For most of human history, from the earliest humans more than 2 million years ago to the start of agriculture less than 10,000 years ago, human population probably never exceeded about 10 million. The start of agriculture brought about the development of cities and more rapid population growth. By A.D. 1, human population had reached 250 million. The increase continued slowly, to about half a billion by 1650.

The beginning of the industrial age greatly increased the ability to grow food and use natural resources. Improvements in the medical and health sciences lowered death rates dramatically. As a result, human population began to grow exponentially, although the rate of growth has varied. Figure 7.8 shows estimated human population over the past 2000 years.

Let's put current world population growth in perspective. Every three years the world adds as many people as live in the entire United States. Each *month* the population increases by the equivalent of the population of Switzerland (or New York

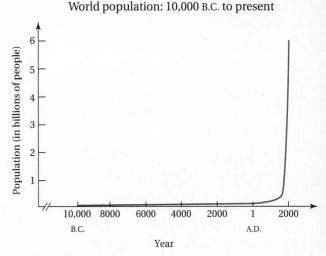

FIGURE 7.8

City). In 1992, 300,000 people died of starvation in Somalia; worldwide, that amount was replaced by new births in just 29 *hours*. During the hour or so it takes to read this chapter, world population will increase by about 10,000 people.

Although projections of future population growth have large uncertainties, most projections based on current trends suggest that the world population will reach 8 billion people by 2020—more than *four times* the population just a century earlier and nearly double the population at the time that most of today's college students were born. Most of that growth (9 out of 10 people) will take place in the developing regions of the world. More than 1 billion people will be added to Asia alone, where India is expected to overtake China as the most populous nation. Significant growth also is expected in Latin America (where Mexico City is expected to overtake Tokyo as the world's most populous city) and Africa (where Ethiopia, Egypt, and Nigeria will double their populations). Even in the United States, where population is growing at a slower rate, the population is expected to increase by *100 million* people within the next 50 years.

EXAMPLE 1   *Changing Growth Rate*

Given a world population of 6 billion in the year 2000, calculate the average annual rate of growth in world population since 1650. (Assume exponential growth.)

Solution: We can use the exponential equation to solve for the growth rate r. We set $Q_0 = 0.5$ billion for the population in 1650, $Q = 6$ billion for the approximate population in 2000, and $t = 350$ years from 1650 to 2000. With these substitutions, the exponential equation becomes

$$Q = Q_0 \times (1 + r)^t \quad \Rightarrow \quad 6 \text{ billion} = 0.5 \text{ billion} \times (1 + r)^{350}.$$

The first step in solving for r is dividing both sides by 0.5 billion:

$$(1 + r)^{350} = \frac{6 \text{ billion}}{0.5 \text{ billion}} = 12$$

Using Your
___ ***Calculator***

Next, we take the 350th root of both sides:

$$\sqrt[350]{(1 + r)^{350}} = \sqrt[350]{12} \quad \Rightarrow \quad 1 + r = \sqrt[350]{12} = 1.0071$$

You can take the 350th root by using the root key on your calculator, usually labeled something like $\sqrt[x]{y}$ or $y^{1/x}$.

Finally, we subtract 1 from both sides to get r:

$$r = 1.0071 - 1 = 0.0071 = 0.71\%$$

The average growth rate in world population since 1650 has been 0.71% per year—considerably lower than the current growth rate of about 1.4% per year.   ∎

WHAT CAUSES THE GROWTH RATE?

The rate of world population growth is simply the difference between the birth rate and the death rate. For example, suppose that there are an average of 8.5 births per 100 people and 6.5 deaths per 100 people per year. Then the population growth rate is

$$\text{growth rate} = \text{birth rate} - \text{death rate} = \frac{8.5}{100} - \frac{6.5}{100} = \frac{2}{100} = 0.02 = 2\%.$$

Most people are surprised to learn that birth rates have dropped rapidly throughout the world during the past 50 years—the same period that has seen the largest population growth in history. Indeed, worldwide birth rates have never been lower than they are today. However, death rates have fallen even faster in recent years, particularly among children, resulting in the continued exponential growth of the world population.

EXAMPLE 2   *Changes in Birth and Death Rates*

In 1950, the world birth rate was 3.7 births per 100 people and the death rate was 2.0 deaths per 100 people. By 1975, the birth rate had fallen to 2.8 births per 100 people and the death rate to 1.1 deaths per 100 people. Contrast the growth rates in 1950 and 1975.

Solution: In 1950, the growth rate was

$$\text{growth rate} = \text{birth rate} - \text{death rate} = \frac{3.7}{100} - \frac{2.0}{100} = \frac{1.7}{100} = 1.7\%.$$

In 1975, the growth rate was

$$\text{growth rate} = \text{birth rate} - \text{death rate} = \frac{2.8}{100} - \frac{1.1}{100} = \frac{1.7}{100} = 1.7\%.$$

Despite a dramatic fall in birth rates, the growth rate remained unchanged because death rates fell equally dramatically. ■

CARRYING CAPACITY AND LOGISTIC GROWTH

Exponential growth cannot continue indefinitely; indeed, at its current rate, the human population cannot continue to grow much longer. As a result, most projections assume that human population will level off at somewhere between about 8 and 15 billion during the next century. How do models account for the *end* of exponential growth?

A model can limit exponential growth by taking into account the **carrying capacity** of the Earth—the number of people that the Earth can support for long periods of time. For example, suppose that the carrying capacity of the Earth is 10 billion people. Then a reasonable assumption might be that the rate of population growth will slow as the carrying capacity is approached. One way to account for this assumption is to modify the rate of growth as follows:

$$\text{growth rate} = r \times \left(1 - \frac{\text{population}}{\text{carrying capacity}}\right)$$

This formula yields a growth rate very close to the exponential growth rate r when the population is small. However, the growth rate decreases as the population approaches the carrying capacity. Note that, when the population *equals* the carrying capacity, this formula yields a growth rate of zero. A model based on the assumption that the rate of growth decreases smoothly, and becomes zero when the carrying capacity is reached, is called a **logistic growth model**. Figure 7.9 on the following page contrasts a logistic growth model with an exponential growth model.

By the Way ············
The growth rate for a single country (as opposed to the world) consists not only of the birth and death rates, but the immigration rates (in and out of the country) as well. In the United States, immigration comprises about half the overall growth rate.

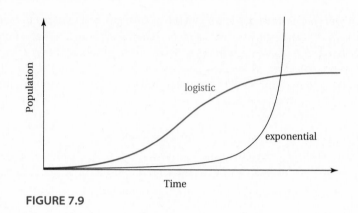

FIGURE 7.9

EXAMPLE 3   *Logistic Growth Rates*

The world population growth rate in the 1960s was about 2.1% per year, when world population was about 3.3 billion. Assume that world population is following a logistic growth model with a carrying capacity of 12 billion. What do these data predict for the exponential growth rate r? What does this model predict for the growth rate when the population is 6 billion?

Solution: We are told that, in creating the logistic model, we should use a growth rate of 2.1% = 0.021 for a population of 3.3 billion and a carrying capacity of 12 billion. With these values, the logistic growth rate formula reads:

$$0.021 = r \times \left(1 - \frac{3.3 \text{ billion}}{12 \text{ billion}}\right) = r \times 0.725.$$

We can now solve for r by dividing both sides by 0.725:

$$r = \frac{0.021}{0.725} = 0.029.$$

We can use this value of r in the logistic model to predict the growh rate for a population of 6 billion and a capacity of 12 billion:

$$\text{growth rate} = 0.029 \times \left(1 - \frac{6 \text{ billion}}{12 \text{ billion}}\right) = 0.0145.$$

This logistic model predicts a growth rate of 0.0145, or 1.45%, for a population of 6 billion—about the same as the actual growth rate (1.4%) for this population in the late 1900s. Thus, the logistic model is reasonable for a carrying capacity of 12 billion, at least through the late 1990s. If population growth continues to follow this logistic model, the growth rate will continue to decline in the twenty-first century, with world population leveling off at about 12 billion people.   ■

OVERSHOOT AND COLLAPSE

The logistic model is based on the assumption that the growth rate *automatically* adjusts as the population approaches the carrying capacity. However, because of the astonishing rate of exponential growth, real populations often increase *beyond* the carrying capacity in a relatively short period of time. This phenomenon is called **overshoot**.

If a population overshoots the carrying capacity of its environment, a decrease in the population is inevitable. If the overshoot is substantial, the decrease can be rapid and severe—a phenomenon known as **collapse**. Figure 7.10 contrasts a logistic growth model with overshoot and collapse.

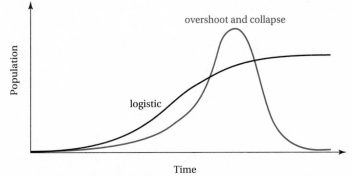

FIGURE 7.10

> **TIME-OUT TO THINK:** The concept of carrying capacity can be applied to any localized environment. Consider the decline of past civilizations such as the ancient Greeks, Romans, Mayans, Anasazi, and others. Do overshoot and collapse models describe the fall of any of these or other civilizations? Explain.

𝒯HINKING ABOUT…

People in a Bottle?

If human population continued growing at its current rate, humans would be forced to stand elbow-to-elbow in only about 650 years. Thus, like bacteria in a bottle (Unit 7A), the exponential growth of human population *will* stop; the only questions are *when* and *how*.

First consider the question of *when*. The largest estimates of carrying capacity are in the range of 40 billion, which we would reach within about 140 years at current growth rates. Thus, regardless of other assumptions, population growth must stop within about the next 140 years, quite soon on the scale of human history.

Turning to *how*, there are only two basic ways to slow the growth of a population:

- a decrease in the birth rate, or
- an increase in the death rate.

Most people are choosing the first option, as birth rates now are at historic lows. Indeed, population actually is *decreasing* in a few European nations. Nevertheless, worldwide birth rates still are much higher than death rates, and exponential growth continues.

If a decrease in birth rates doesn't halt the exponential growth, an increase in the death rate will. If population significantly overshoots the carrying capacity by the time this process begins, the increase in the death rate will be dramatic—probably on a scale never before seen. This forecast is not a threat, a warning, or a prophecy of doom. It is simply a law of nature: *Exponential growth always stops*. As human beings, we can choose to halt our population growth through intelligent and careful decisions. Or, we can do nothing, leaving ourselves at the mercy of natural forces over which we can have no more control than we do over hurricanes, tornadoes, earthquakes, or the explosion of a distant star.

WHAT IS THE CARRYING CAPACITY?

Because exponential growth cannot continue indefinitely, future world population must follow either a logistic model or some type of overshoot and collapse model, at least in broad terms. A logistic model is clearly preferable, as overshoot and collapse might well mean the collapse of our civilization.

The most fundamental question about population growth therefore concerns if and when the actual population will exceed the carrying capacity. If the carrying capacity is well above current population, then we have plenty of time to figure out a way to settle into logistic growth and long-term population stability. On the other hand, collapse may be an imminent threat if current population is near the carrying capacity, or has already exceeded it.

Unfortunately, any estimate of carrying capacity is subject to great uncertainty, for at least four important reasons:

- The carrying capacity depends on assumptions about the amount of resources consumed by the average person. For example, one factor in the carrying capacity is the availability of energy. But the same total amount of energy can support more people if we assume a lower per capita energy requirement.

- The carrying capacity also depends on assumptions about the environmental impacts of the average person. Assuming that the average person has a substantial impact on the environment leads to a lower carrying capacity than assuming a minimal impact.

- The carrying capacity can change over time, depending on both human technology and the environment. For example, many people base estimates of the carrying capacity on the availability of fresh water; however, if new sources of energy (e.g., fusion) are developed, nearly unlimited amounts of fresh water may be obtained through the desalinization of seawater. Conversely, climate change induced by human activity might alter the environment and reduce our ability to grow food, thereby lowering the carrying capacity.

- Even if we could account for the many individual factors in the carrying capacity (such as food production, energy, and pollution), the Earth is such a complex system that precisely predicting the carrying capacity may well be impossible. For example, no one can predict whether or how much the loss of rain forest species affects the carrying capacity.

The history of attempts to guess the carrying capacity of the Earth is full of missed predictions. Among the most famous was that made by English economist Thomas Malthus (1766–1834). In a 1798 paper entitled *An Essay on the Principle of Population As It Affects the Future Improvement of Society*, Malthus argued that mass deaths through starvation and disease would soon hit Europe and America. He based his argument on the fact that the populations of Great Britain, France, and America were growing rapidly at that time, and he didn't believe that food production could keep up. Malthus's prediction did not come true, primarily because advances in technology *did* allow food production to keep pace with population growth.

TIME-OUT TO THINK: Malthus wrote that the greatest problem facing humanity is "that the power of population is indefinitely greater than the power in the Earth to produce subsistence for man." Although his immediate predictions didn't come true, do you think his overall point is valid? Defend your opinion.

Finally, it is important to keep in mind that real patterns of population growth usually are much more complex than the patterns predicted by any particular model. For example, Figure 7.11 shows historical data for the population of Egypt. Imagine that, somewhere back in ancient Egypt, scientists had sought to predict the future population of the region. Do you think that anyone could have accurately predicted what transpired over a period of even a hundred years, let alone several thousand years?

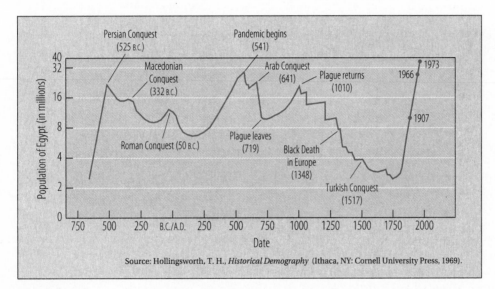

Source: Hollingsworth, T. H., *Historical Demography* (Ithaca, NY: Cornell University Press, 1969).

FIGURE 7.11

This example offers an important lesson about mathematical models. They are useful for gaining insight into the processes being modeled. However, mathematical models can be used to *predict* future changes only when the processes are relatively simple. For example, it is easy to use mathematical modeling to predict the course of a spaceship because the law of gravity is relatively simple. But the growth of human population is such a complex phenomenon that we have little hope of ever being able to predict it reliably.

REVIEW QUESTIONS

1. How much did human population increase from 10,000 years ago until 2,000 years ago? How much did it increase from 2,000 years ago to the year 1650? How much did it increase from 1650 to today?

2. How is the growth rate determined by birth and death rates? How do today's birth and death rates compare to those in the past?

3. What is a *logistic growth model*? Why would it be good if human population growth followed a logistic model in the future?

4. What is *overshoot and collapse*? Under what conditions does it occur?

5. What do we mean by *carrying capacity*? Why is it so difficult to determine the carrying capacity of the Earth?

6. Who was Thomas Malthus, and why do his arguments from 200 years ago still generate controversy today?

PROBLEMS

1. **Population Growth in Your Lifetime.** The world population in 1996 was about 5.8 billion. Assume that the population currently is growing, and will continue to grow, by about 1.4% per year. What is the doubling time? What will be the world population when you are 50 years old? 80 years old? 100 years old? Discuss the challenges or benefits of population growth that might be expected over your lifetime.

2. **Slower World Population Growth.** Suppose that the growth of world population slows from its current rate of nearly 1.4% to 1% per year. Given that the year 2000 population will be about 6 billion, what will the world population be when you are 50 years old? 80 years old? 100 years old? Do you think this slower rate of growth would eliminate problems due to overpopulation? Explain.

3. **Changing Population Growth Rate.**

   a. Find the average growth rate of the world population between 1900 (population 1.5 billion) and 2000 (population 6 billion). Compare this result to the average growth rate between 1850 and 1950 which was about 0.92%.

   b. Find the average growth rate of the world population between 1950 (population 2.5 billion) and 2000 (population 6 billion). Compare this result to the average growth rate between 1900 and 2000 from part (a).

   c. Find the average growth rate of the world population between 1970 (population 3.7 billion) and 2000 (population 6 billion). Compare this result to the average growth rate between 1950 and 2000 from part (b).

   d. Based on your answers to parts (a)–(c), discuss any conclusions you can draw about changes in the world population growth rate during the twentieth century.

Birth and Death Rates. *The following table gives the birth and death rates for four countries in three different years.*

Country	Birth Rate (per 1,000)			Death Rate (per 1,000)		
	1975	1985	1995	1975	1985	1995
Czech Republic	19.6	14.5	9.3	11.5	11.8	11.4
Israel	28.2	23.5	21.0	7.1	6.6	6.3
Sweden	12.8	11.8	11.7	10.8	11.3	11.0
United States	14.0	15.7	15.1	8.9	8.7	8.8

For the country given in Problems 4–7, do the following:

a. *Describe the general trend in the country's birth rate between 1975 and 1995.*

b. *Describe the general trend in the country's death rate between 1975 and 1995.*

c. *State the country's net growth rate due to births and deaths in 1975, 1985, and 1995 (i.e., ignore the effects of immigration.)*

d. *Based on your answers to parts (a) through (c), predict how the country's population will change over the next 20 years. Do you think your prediction is reliable? Explain.*

4. The Czech Republic   **5.** Israel

6. Sweden   **7.** The United States

8. Net Growth Rates. Suppose that in a particular country the birth rate is 3.2 births per 1,000 people, the death rate is 2.3 deaths per 1,000 people, the immigration rate into the country is 2.9 people per 1,000, and the emigration rate out of the country is 1.8 people per 1,000.

a. What is the net (overall) growth rate of the country per 1,000 people? What is the net growth rate as a percentage?

b. If the population of the country was 35 million in 1990 and the above growth rates remained constant, what will the population be in the year 2010?

9. Logistic Growth Rates. Suppose a country has an exponential growth rate of $r = 3\%$ and a carrying capacity of 50 million. If the population growth follows a logistic growth model, calculate the actual growth rate when the population is 10 million, 30 million, and 45 million.

10. Logistic Growth in the United States. In 1997, the United States had a growth rate of about 0.7% and a population of about 270 million. Suppose that the carrying capacity of the United States is 350 million.

a. Suppose that the growth rate remains at 0.7% per year. What will the population be in the year 2017? in 2047? Under this assumption, will the United States exceed its carrying capacity by 2047? Explain.

b. Suppose instead that United States population follows a logistic growth model. What will the growth rate be when the population reaches 300 million? 340 million? Under this assumption, will the United States ever exceed its carrying capacity? Explain.

11. Logistic Carrying Capacity. Suppose the base exponential growth rate of population is $r = 4\%$ per year and the actual growth rate is 3% per year when the population reaches 100,000 people. Assuming the population follows a logistic growth model, estimate the carrying capacity.

12. Growth Control Mediation. A city with a 1990 population of 100,000 has a growth control policy that limits the increase in residents to 2% per year. Naturally, this policy causes a great deal of dispute. On one side, some people argue that *growth* costs the city its small-town charm and clean environment. On the other side, some people argue that *growth control* costs the city jobs and drives up housing prices. Finding their work limited by the policy, developers suggest a compromise of raising the allowed growth rate to 5% per year. Contrast the population of this city in 2000, 2010, and 2050 for 2% annual growth and 5% annual growth. If you were asked to mediate the dispute between growth control advocates and opponents, explain the strategy you would use.

13. Project: Carrying Capacity Estimates. Do some research to obtain several different opinions concerning the Earth's human population carrying capacity. Based on your research, draw some conclusions about whether overpopulation presents an immediate threat. Write a short essay detailing the results of your research and clearly explaining your conclusions.

14. Project: U.S. Population Growth. Research population growth in the United States to determine the relative proportions of the growth resulting from birth rates and from immigration. Then research both the problems and benefits of the growing U.S. population. Form your own opinions about whether the United States has a population problem. Write an essay covering the results of your research and stating and defending your opinions. If you believe that the United States has a population problem, discuss some of the implications for society, and suggest policies that might be implemented to alleviate the situation. If you believe that the United States doesn't have a population problem, state how large a population the country could sustain, and explain your reasoning.

15. Project: Overshoot and Collapse. Do research as needed to find a case where a population of an animal species has followed a model of overshoot and collapse. Write a short essay describing the change in the population, and explaining theories as to why the population followed the overshoot and collapse model.

CHAPTER 7

SUMMARY

*E*xponential growth or decay occurs whenever a quantity changes by a fixed percentage in each unit of time. Because so many important issues involve exponential growth, it is one of the most important mathematical topics. Key ideas to remember from this chapter include the following.

- Exponential growth and decay are characterized by a constant doubling time (for exponential growth) or half-life (for exponential decay).

- Given either a growth (or decay) rate or a doubling time (or half-life), we can construct general growth (or decay) laws to model a variety of exponential processes.

- Exponentially growing quantities quickly reach incredible proportions, at which point something will force them to stop growing. Exponential growth *cannot* continue indefinitely.

- Human population has grown exponentially for the past several centuries. The survival of our civilization depends on how this exponential growth eventually stops, and on whether the population will exceed the carrying capacity of the Earth before the growth stops.

SUGGESTED READING

How Many People Can the Earth Support, J. E. Cohen (New York: W. W. Norton, 1995).

Vital Signs 1998, compiled by the Worldwatch Institute (New York: W. W. Norton, 1998)(updated annually).

Beyond the Limits: Confronting Global Collapse, Envisioning a Sustainable Future, D. H. Meadows, D. L. Meadows, J. Randers (Vermont: Chelsea Green Publishing Company, 1992).

The Population Explosion, P. R. Ehrlich and A. H. Ehrlich (New York: Simon and Schuster, 1990).

"Forgotten Fundamentals of the Energy Crisis," *American Journal of Physics*, A. A. Bartlett (September 1978).

PROBABILITY: LIVING WITH THE ODDS

P robability is involved in virtually every decision that we make. Sometimes the role of probability is clear, as in deciding whether to plan a picnic based on the probability of rain. In other cases, probability guides decisions on a deeper level. For example, you might choose a particular college because you believe that it is most likely to meet your personal needs. In this chapter we will see just how practical and powerful probability can be in our everyday lives.

The principal means for ascertaining truth—induction and analogy—are based on probabilities; so that the entire system of human knowledge is connected with the theory of probability.
PIERRE SIMON, MARQUIS DE LAPLACE (1819)

Anybody can win unless there happens to be a second entry.
GEORGE ADE (1866–1944)

*You can take it as understood
That your luck changes only if it's good.*
OGDEN NASH, *ROULETTE US BE GAY*

*P*robability plays a role in virtually every decision we make. Unfortunately, it is also a fact that most people, including many mathematicians, have a poor intuition for probability. This may explain why history is full of people who have gambled and lost—individuals who lost their fortunes at casino games and lotteries, financial "experts" who gambled on stock prices or interest rates and lost their fortunes, or national leaders who gambled on a new social policy and found unintended consequences.

Probability is particularly significant in the insurance business. Companies sell policies to protect people from events that have a low probability of happening but are bound to happen to *someone*. And everyone has personal decisions to make about insurance. Is it worthwhile to spend $10 to insure your life on a 500-mile airplane flight? Is it wise to spend $200 a year to include flood insurance on your homeowner's policy?

You will find that probability is full of surprises. For example, the average American spends several hundred dollars per year on lottery entries, yet the number of people who win the grand prize is roughly the same as the number of people struck by lightning! In this chapter, we look systematically at methods of calculating probabilities, and see how those methods can be applied to daily decision making.

UNIT 8A

PRINCIPLES OF COUNTING

Probability basically involves calculating the chance of getting some *particular* outcome from all *possible* outcomes. Therefore we begin our discussion of probability by studying methods of counting possible outcomes.

SELECTIONS FROM TWO OR MORE COLLECTIONS

How many outfits can you form from three pairs of shorts and four shirts? One way to approach this problem is to enumerate the possibilities in an **array**. We make four rows to correspond to the four choices of shirts, and three columns for the three choices of shorts. Each of the $3 \times 4 = 12$ squares in the array represents a different outfit, as shown below.

	Shorts #1	Shorts #2	Shorts #3
Shirt #1	Outfit #1	Outfit #2	Outfit #3
Shirt #2	Outfit #4	Outfit #5	Outfit #6
Shirt #3	Outfit #7	Outfit #8	Outfit #9
Shirt #4	Outfit #10	Outfit #11	Outfit #12

Next, imagine that a restaurant menu offers a choice of soup or salad for the first course and a choice of a vegetarian, fish, or meat dish for the main course. We could make an array as before, but instead let's enumerate the possible meals with a **tree** showing that each of the two first-course choices can be combined with any of the three second-course choices (Figure 8.1). Note that the total number of possible meals is $2 \times 3 = 6$.

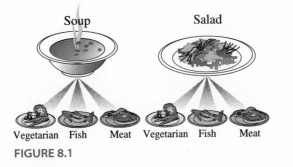

FIGURE 8.1

TIME-OUT TO THINK: Explain how we could have used an array, instead of a tree, to find the number of possible meals. Could we have used a tree to find the number of outfits from the four shirts and three shorts? How?

Both the outfit and restaurant examples involved arrangements from two groups of items. We can generalize to any two groups of items with the **multiplication principle**.

The Multiplication Principle

If you choose one item from a group of M items and another item from a group of N items, the total number of two-item choices is $M \times N$.

EXAMPLE 1   *Applying the Multiplication Principle*

a) Suppose that you manage an apartment complex and label the units with a letter of the alphabet and a single digit. How many units can you label?

b) A dancing class has eight boys and nine girls. How many boy-girl pairs are possible?

c) A restaurant offers 7 appetizers and 12 main courses. How many different two-course meals are possible?

Solution:

a) The labels each have one of the 26 letters of the alphabet (A–Z) and one of the 10 digits (0–9). Thus the total number of possible labels is

$$26 \times 10 = 260.$$

b)   With eight boys and nine girls,

$$8 \times 9 = 72$$

different boy-girl combinations are possible.

c)   Choosing from one of 7 appetizers and one of 12 main courses, the number of possible two-course meals is

$$7 \times 12 = 84.$$ ■

EXTENDING THE MULTIPLICATION PRINCIPLE (MORE THAN TWO GROUPS OF ITEMS)

How many meals would be possible if a restaurant menu offered four desserts in addition to two first courses and three main courses? We can use a tree to answer the question by adding a third row showing how each of the four desserts can be chosen with the six meals we found earlier (Figure 8.2).

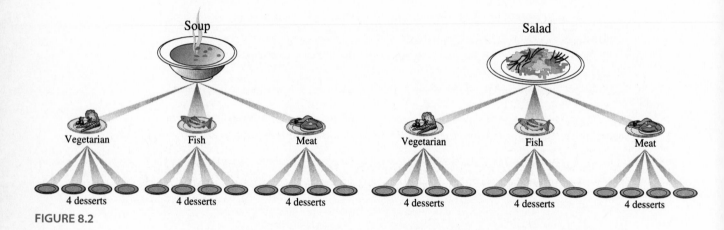

FIGURE 8.2

An easier way to solve the problem is by extending the multiplication principle. For *each* of the $2 \times 3 = 6$ ways to choose the first course and main course (found in Figure 8.1), there are now 4 ways to choose the dessert. Thus there are

$$2 \times 3 \times 4 = 24$$

different meals. In fact, the multiplication principle may be extended to any number of groups of items.

EXAMPLE 2   *Extending the Multiplication Principle*

a)   A college offers 12 natural science classes, 15 social science classes, 10 English classes, and 8 fine arts classes. Suppose that, to meet your core requirements, you must take one class of each type. How many possible ways are there to meet the core requirements?

b) A computer catalog offers processors from five different companies, each offering a choice of three different speeds; both color and monochrome (black-and-white) monitors from four different companies; six different sizes of hard drive; and a choice of seven different printers. How many different systems can you create?

Solution:

a) We find the total number of possible ways of meeting the requirements by multiplying the number of choices from each subject area. Thus the number of ways to meet the requirement is

$$12 \times 15 \times 10 \times 8 = 14{,}400.$$

b) We begin by splitting the computer problem into smaller pieces. For the processor, you can choose from five companies in three different speeds, making a total of $5 \times 3 = 15$ choices. For the monitor, you have two choices (color or monochrome) from each of four companies making a total of $4 \times 2 = 8$ choices. With the 15 choices of a processor, 8 choices of a monitor, 6 choices of a hard drive, and 7 choices of a printer, the total number of systems you can create is

$$15 \times 8 \times 6 \times 7 = 5040.$$ ■

ARRANGEMENTS WITH REPETITION

Imagine that license plates in your county display seven numerals. How many different license plates are possible? There are 10 possibilities (the numerals 0–9) for the first numeral and the same 10 possibilities for the second numeral. So there are $10 \times 10 = 100$ possible choices for the first two numerals. With 10 possibilities for the third numeral, there are $10 \times 10 \times 10 = 1000$ possible choices for the first three numerals. With 10 choices for each of the seven numerals, the total number of choices is

$$10 \times 10 \times 10 \times 10 \times 10 \times 10 \times 10 = 10^7.$$

That is, this system allows for 10 million different license plates.

The license plate problem involves selecting items from only a *single* group, namely the numerals 0–9. However, the items in the group may be chosen over and over again. We can generalize the license plate problem to state a rule for counting the number of **arrangements with repetition** of items.

> ### Arrangements with Repetition
>
> If we make r selections (for example, the seven digits of the license plates) from a group of n items (for example, the numerals 0–9 for the license plates), then n^r different arrangements are possible.

EXAMPLE 3   *Arrangements with Repetition*

a) Suppose that your state uses letters of the alphabet *and* numerals, in any order, to make seven-symbol license plates. How many plates are possible?

b) How many six-character passwords can be made by combining lowercase letters, uppercase letters, numerals, and the characters @, $, and &?

Solution:

a) With 26 letters in the alphabet and 10 numerals, 36 choices are possible for each of the seven symbols. Thus we are selecting $r = 7$ symbols from $n = 36$ choices, so the total number of possible license plates is

$$36^7 = 78,364,164,096$$

or about 78.4 billion! Note that this number is far greater than the total number of people on Earth.

b) Because there are 26 letters in the alphabet, there are $2 \times 26 = 52$ possibilities for both uppercase and lowercase letters. Adding the 10 numeral choices and 3 symbols @, $, and & to the 52 letters, we have 65 characters to use in the six-character passwords. That is, we are choosing $r = 6$ characters from $n = 65$ choices, which gives a total of

$$65^6 = 75,418,890,625$$

or 75.4 billion possible passwords. Again, note the tremendous number of possibilities for passwords with only six characters.   ■

> **TIME-OUT TO THINK:**   Suppose that the computer passwords consist of 7 characters, rather than 6. How many passwords are possible in this case?

PERMUTATIONS

Suppose that you coach a team of four swimmers. How many different ways can you put together a four-person relay team? Repetition is *not* allowed in this case: a swimmer can swim only one leg of the relay. You can choose any of the four swimmers for the first leg. Once you've chosen the first swimmer, you'll have three swimmers left to choose from for the second leg. You'll then have two swimmers left to choose from for the third leg. After the first three legs are filled, you'll have only one swimmer left for the last leg. Thus the total number of possible arrangements for the relay is

$$4 \times 3 \times 2 \times 1 = 24.$$

We can enumerate the 24 different relay orders with a tree diagram in which we label the four swimmers A, B, C, and D (Figure 8.3). On the first line we place the four possible swimmers for the first leg. Each of these choices leaves three remaining choices for the second leg, which we show as possible branches from the first line to the second line. From the second to third lines there are two possible choices remaining, and only one choice remaining for the fourth line.

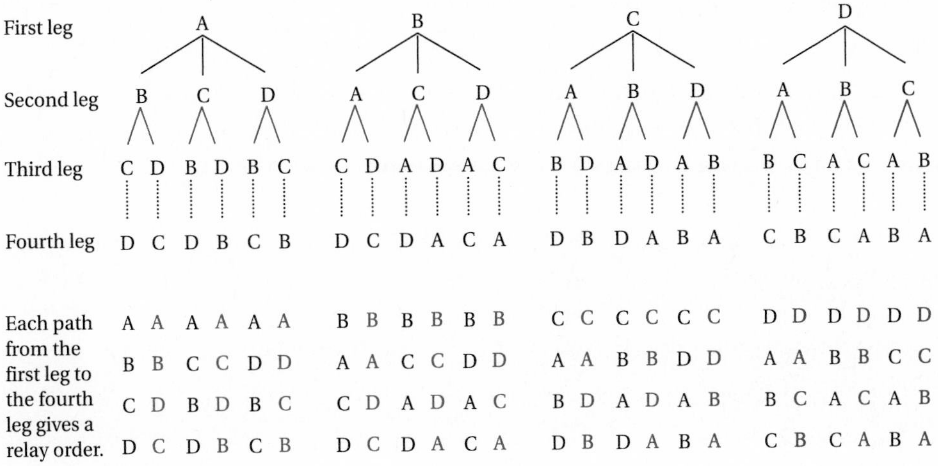

First leg		A				B				C				D		
Second leg	B	C	D	A	C	D	A	B	D	A	B	C
Third leg	C D	B D	B C	C D	A D	A C	B D	A D	A B	B C	A C	A B
Fourth leg	D C	D B	C B	D C	D A	C A	D B	D A	B A	C B	C A	B A

	A	A	A	A	A	A	B	B	B	B	B	B	C	C	C	C	C	C	D	D	D	D	D	D
Each path from the first leg to the fourth leg gives a relay order.	B	B	C	C	D	D	A	A	C	C	D	D	A	A	B	B	D	D	A	A	B	B	C	C
	C	D	B	D	B	C	C	D	A	D	A	C	B	D	A	D	A	B	B	C	A	C	A	B
	D	C	D	B	C	B	D	C	D	A	C	A	D	B	D	A	B	A	C	B	C	A	B	A

FIGURE 8.3

Each of the 24 different relay orders is called a **permutation** of the four letters A, B, C, and D. Mathematically, we are dealing with **permutations** whenever

- all items are selected from the same group,
- no item may be used more than once, and
- *the order of arrangement matters* (for example, ABCD is a different relay team than DCBA).

By the Way

The word *permute* (root of *permutation*) comes from a Latin term meaning *to change throughout*.

EXAMPLE 4   *Class Schedules*

A high-school principal needs to schedule six different classes in six periods: algebra, English, history, Spanish, science, and gym classes. How many different arrangements of classes are possible?

Solution: This is a case of permutations in which the "items" are the six classes. Each class is scheduled only once, and the order of the classes matters because a schedule that begins with Spanish is different from one that begins with gym. The principal may choose any of the six classes for the first period. That will leave five choices for the second period. After the first two periods are filled, there are four choices left for the third period, then three choices for the fourth period, two choices for the fifth period, and only one choice for the sixth period. Altogether, there are

$$6 \times 5 \times 4 \times 3 \times 2 \times 1 = 720$$

different arrangements for the classes.

FACTORIAL NOTATION

Products of the form $6 \times 5 \times 4 \times 3 \times 2 \times 1$ come up so frequently in counting problems that they have a special name. Whenever a positive integer n is multiplied by all the preceding positive integers, the result is called *n factorial* and is denoted by $n!$ (the exclamation mark is read as "factorial").

For example:

$$1! = 1$$
$$2! = 2 \times 1 = 2$$
$$3! = 3 \times 2 \times 1 = 6$$
$$4! = 4 \times 3 \times 2 \times 1 = 24$$
$$5! = 5 \times 4 \times 3 \times 2 \times 1 = 120$$

In general,

$$n! = n \times (n - 1) \times (n - 2) \times \cdots \times 2 \times 1.$$

Mathematical Note:
By convention, $0!$ is defined to be 1. That is,
$$0! = 1.$$

Note that $n!$ grows very rapidly with n. In fact, factorial growth is even more explosive than exponential growth, which we can see by comparing graphs of the equations $y = n!$ and $y = 2^n$ (Figure 8.4).

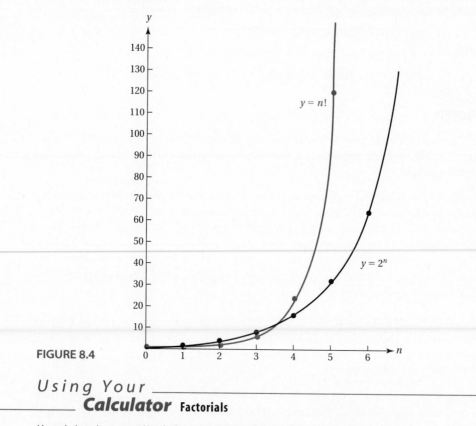

FIGURE 8.4

Using Your *Calculator* Factorials

Most calculators have a special key for factorials, usually labeled $n!$. To calculate a factorial (such as 5!) press

5 $(n!)$ $(=)$

EXAMPLE 5   *Factorial Growth*

Make a table in which you compute values of $n!$ for $n = 6$ through $n = 12$.

Solution: We make the table by using the definition of a factorial for each value of n.

n	$n!$	
6	$6 \times 5 \times 4 \times 3 \times 2 \times 1 =$	720
7	$7 \times 6 \times 5 \times 4 \times 3 \times 2 \times 1 =$	5040
8	$8 \times 7 \times 6 \times 5 \times 4 \times 3 \times 2 \times 1 =$	40,320
9	$9 \times 8 \times 7 \times 6 \times 5 \times 4 \times 3 \times 2 \times 1 =$	362,880
10	$10 \times 9 \times 8 \times 7 \times 6 \times 5 \times 4 \times 3 \times 2 \times 1 =$	3,628,800
11	$11 \times 10 \times 9 \times 8 \times 7 \times 6 \times 5 \times 4 \times 3 \times 2 \times 1 =$	39,916,800
12	$12 \times 11 \times 10 \times 9 \times 8 \times 7 \times 6 \times 5 \times 4 \times 3 \times 2 \times 1 =$	479,001,600

EXAMPLE 6   *Working with Factorials*

Calculate each of the following *without* using the factorial key on your calculator.

a) $\dfrac{6!}{4!}$ b) $\dfrac{25!}{22!}$ c) $\dfrac{300!}{299!}$

Solution:

a) We can write out the entire calculation, but it is easier if we recognize that it can be simplified as follows.

$$\frac{6!}{4!} = \frac{6 \times 5 \times 4 \times 3 \times 2 \times 1}{4 \times 3 \times 2 \times 1} = \frac{6 \times 5 \times 4!}{4!} = 6 \times 5 = 30$$

b) It would take a while to write out 25!, but we can simplify the calculation by recognizing that $25! = 25 \times 24 \times 23 \times 22!$. Therefore,

$$\frac{25!}{22!} = \frac{25 \times 24 \times 23 \times 22!}{22!} = 25 \times 24 \times 23 = 13,800.$$

c) Again, we can make this calculation very simple by recognizing that $300! = 300 \times 299!$. Therefore,

$$\frac{300!}{299!} = \frac{300 \times 299!}{299!} = 300.$$

TIME-OUT TO THINK: Confirm the calculations in parts (a) and (b) of Example 6 by using the factorial and division keys on your calculator. Now try part (c): What happens when you try to do 300! on your calculator? Why? Explain why it is easy to do part (c) without a calculator, but difficult to do it with a calculator.

THE PERMUTATIONS FORMULA

Suppose that a swimming coach has 10 swimmers on her team, from which she must select a four-person relay team. This time, she can choose any of 10 swimmers for the first leg of the relay. Then she'll have 9 swimmers from which to choose for the second leg, 8 choices for the third leg, and 7 choices for the fourth leg. The total number of relay possibilities is

$$10 \times 9 \times 8 \times 7 = 5040.$$

Each of the 5040 relays represents a different *permutation* because each relay swimmer is selected from the same group of 10 swimmers, no swimmer can swim more than once, and the order of the swimmers in the relay is important. Thus we say that the number of *permutations of 10 swimmers selected 4 at a time* is 5040.

We can find a general formula for permutations by writing the product $10 \times 9 \times 8 \times 7$ in a slightly more complex way:

$$10 \times 9 \times 8 \times 7 = \frac{10 \times 9 \times 8 \times 7 \times 6 \times 5 \times 4 \times 3 \times 2 \times 1}{6 \times 5 \times 4 \times 3 \times 2 \times 1}$$

$$= \frac{10!}{6!} = \frac{10!}{(10-4)!}$$

Note that:

- The number 10 in the numerator is the number of swimmers from which the coach may choose the four relay participants, and

- The number $10 - 4 = 6$ in the denominator is the number of swimmers who do *not* swim.

Now we introduce a special notation: we read $_{10}P_4$ as *the number of permutations of 10 swimmers selected 4 at a time*. Using this compact notation, we have

$$_{10}P_4 = \frac{10!}{(10-4)!} = \frac{10 \times 9 \times 8 \times 7 \times 6!}{6!} = 5040.$$

Generalizing, we find a simple formula for permutations.

The Permutations Formula

If we make *r* selections from a group of *n* items, the number of possible permutations is

$$_nP_r = \frac{n!}{(n-r)!},$$

where $_nP_r$ is read as "the number of permutations of *n* items taken *r* at a time."

Using Your _____
_____ *Calculator* **Permutations**

Some calculators have a special key for calculating permutations, usually labeled $_nP_r$. For example, to find the number of permutations of $n = 10$ objects taken $r = 4$ at a time, press

$$10 \; \boxed{_nP_r} \; 4 \; \boxed{=}$$

You should get the result 5040. If your calculator does not have a permutations key, you can still use the permutations formula.

_____

EXAMPLE 7   *Leadership Election*

A city has 12 candidates running in an election for three leadership positions. The top vote-getter will become the Mayor, the second biggest vote-getter will become the Deputy Mayor, and the third biggest vote-getter will become the Treasurer. How many outcomes are possible for the three leadership positions?

Solution: The voters are choosing $r = 3$ leaders from a group of $n = 12$ candidates. The order in which the leaders are chosen matters because each of the three leadership positions is different. Thus we are looking for the number of permutations of the 12 candidates taken 3 at a time:

$$_{12}P_3 = \frac{12!}{(12-3)!} = \frac{12!}{9!} = \frac{12 \times 11 \times 10 \times 9!}{9!} = 12 \times 11 \times 10 = 1320$$

There are 1320 possible outcomes for the three leadership positions.   ■

EXAMPLE 8   *Batting Orders*

How many ways can the manager of a baseball team form a (9-player) batting order from a roster of 15 players?

Solution: The manager forms a batting order by selecting $r = 9$ players from the roster of $n = 15$ players. Thus the number of possible batting orders is the number of permutations of 15 players taken 9 at a time:

$$_{15}P_9 = \frac{15!}{(15-9)!} = \frac{15!}{6!} = 1{,}816{,}214{,}400$$

Nearly 2 *billion* batting orders are possible for a baseball team with a roster of 15 players!   ■

COMBINATIONS

Suppose that the five members of a city council—let's call them Zeke, Yolanda, Wendy, Vern and Ursula—decide that a three-person committee is needed to study the impact of a new shopping center. How many committees could be formed from the five members of the council?

Note that this problem is *not* a permutation problem because we are interested only in the makeup of the committee, *not* the order in which the names are listed. Let's use the letters U, V, W, Y, and Z to represent Ursula, Vern, Wendy, Yolanda, and Zeke, respectively. If we were dealing with permutations, we would consider an arrangement such as ZWU to be different from WZU. In this case, however, both arrangements are equivalent because they both represent the committee with members Zeke, Wendy, and Ursula. Problems of this type are called **combinations** problems. In general, a counting problem involves combinations if

- all items are selected from the same group,
- no item may be used more than once, and
- the *order of arrangement does not matter* (for example, ZWU is the same committee as WZU).

We could answer the question by listing all possible three-person committees. Let's go in alphabetical order and begin by listing all the committees that include Ursula:

$$UVW \qquad UVY \qquad UVZ \qquad UWY \qquad UWZ \qquad UYZ$$

Next, we list the committees that include Vern but not Ursula:

$$VWY \qquad VWZ \qquad VYZ$$

This leaves only one committee that includes neither Ursula nor Vern:

$$WYZ$$

We've found a total of 10 different three-person committees that can be made from a five-person council.

This method of listing all the possible committees works, but it is a bit tedious. Fortunately, we can get the same answer with an alternative strategy. First, we find the number of *permutations* of the $n = 5$ council members selected $r = 3$ at a time for a committee:

$$_5P_3 = \frac{5!}{(5-3)!} = \frac{5!}{2!} = 60$$

That is, there are 60 possible ways of making three-person arrangements from the five council members. However, because we counted permutations, we have many duplicate committees. Once we choose three people for a committee, there are $3! = 3 \times 2 \times 1 = 6$ ways to arrange them in different orders. For example, the committee consisting of Zeke, Yolanda, and Wendy has six different permutations:

$$ZYW, ZWY, YZW, YWZ, WZY, \text{ and } WYZ$$

Thus the permutations formula *overcounts* the actual number of committees by a factor of $3! = 6$. To correct for this overcounting, we must divide the number of permutations by $3!$. Therefore the number of committees is

$$\frac{_5P_3}{3!} = \frac{60}{6} = 10.$$

This is the same result we obtained by listing the three-person committees, but we can generalize the equation above to find a formula for counting the number of combinations. Note that the permutation part of the equation is $_nP_r$, where $n = 5$ and $r = 3$. We then divided this permutation result by $r! = 3!$ to correct for the overcounting. Thus the general formula for the number of combinations of n items taken r at a time is the number of permutations, $_nP_r$ divided by $r!$.

The Combinations Formula

If we make r selections from a group of n items, the number of possible *combinations*, in which order does not matter, is

$$_nC_r = \frac{_nP_r}{r!} = \frac{n!}{(n-r)! \times r!},$$

where $_nC_r$ is read as "the number of combinations of n items taken r at a time."

Using Your _____
_____ ***Calculator*** **Combinations**

Some calculators also have a key for combinations, usually labeled $_nC_r$. For example, to find the number of combinations of $n = 10$ objects taken $r = 4$ at a time, press

10 $\boxed{_nC_r}$ 4 $\boxed{=}$

You should get the result 210. If your calculator does not have a combinations key, you can still use the combinations formula.

EXAMPLE 9   *Ice Cream Combinations*

Suppose that you are selecting three different flavors of ice cream in a shop that has 12 flavors. How many flavor combinations are possible?

Solution: We are looking for the number of combinations of the $n = 12$ flavors that can be selected $r = 3$ at a time. Using the combinations formula, we find that there are

$$_{12}C_3 = \frac{12!}{(12-3)! \times 3!} = \frac{12!}{9! \times 3!} = \frac{12 \times 11 \times 10 \times 9!}{9! \times 3 \times 2 \times 1} = \frac{1320}{6} = 220$$

three-flavor combinations of the 12 flavors.   ∎

EXAMPLE 10   *Poker Hands*

How many different five-card poker hands can be dealt from a standard deck of 52 cards?

Solution: The order of arrangement does *not* matter in a card hand, so we have a combinations problem. The number of combinations of $n = 52$ cards drawn $r = 5$ at a time is

$$_{52}C_5 = \frac{52!}{(52-5)! \times 5!} = \frac{52!}{47! \times 5!} = 2,598,960$$

The 52 cards may be combined into more than 2.5 *million* five-card hands.   ∎

Table 8.1	Summary of Counting Techniques			
	Selections from Two Collections	**Arrangements with Repetition**	**Permutations**	**Combinations**
Description	One choice from a group of M items and one choice from a group of N items	r items selected from a group of n items, where the same item may be selected over and over again. Order of arrangement matters.	r items selected from a group of n items, where any item may be chosen only once. Order of arrangement matters.	r items selected from a group of n items, where any item may be chosen only once. Order of arrangement does *not* matter.
Number of possibilities	$M \times N$	n^r	$_nP_r = \dfrac{n!}{(n-r)!}$	$_nC_r = \dfrac{n!}{(n-r)! \times r!}$

REVIEW QUESTIONS

1. Suppose that you have to choose from two different groups of items or objects. Explain how you can use an *array* to enumerate all the choices. Also explain how you can use a *tree* to enumerate all the choices.

2. What is the *multiplication principle*. Explain *why* it works, and give an example of its use.

3. What are *arrangements with repetition*? Give an example of something in which the n^r formula gives the number of possible arrangements.

4. What do we mean by *permutations*? Give an example in which we would want to know the number of permutations of something.

5. What is the definition of a *factorial*? How do you calculate factorials on your calculator?

6. Explain the meaning of each term in the *permutations formula*. Give an example of its use.

7. What do we mean by *combinations*? Give an example in which we would want to know the number of combinations of something.

8. Explain the meaning of each term in the *combinations formula*. Give an example of its use.

9. Study Table 8.1. In your own words, summarize the circumstances under which each formula should be used.

PROBLEMS

Choosing from Two Groups. *Solve Problems 1–4 by (a) making an array; (b) drawing a tree; and (c) using the multiplication principle. Be sure that you get the same answer by all three methods.*

1. How many different choices of cars do you have if a particular model comes in eight different colors and three different styles (sedan, station wagon, or hatchback)?

ACE AUTOS

2. Assume that your car radio has five preset buttons, each of which can be tuned to an AM or FM station. How many different stations can you preset?

3. The local paint store offers wallpaper in eight colors, each of which comes in four different patterns. How many different styles of wallpaper are available?

4. You have a choice of six TV sets that can be matched with any of five VCR players. How many different TV/VCR systems can you assemble?

Choosing from More Than Two Groups. *Use the multiplication principle to answer Problems 5–8.*

5. A ski sale features eight different kinds of skis, on each of which six different bindings can be mounted. In addition, seven different brands of boots are on sale. How many different ski/boot packages could be bought on sale?

6. Of the nine members of a university's governing board, three members are up for reelection. The first member up for reelection is running against only one other candidate, the second member is running against two other candidates, and the third member is running against three other candidates. After the election, how many different governing boards are possible?

7. Suppose that you need to take five courses next semester, one each in humanities, sociology, science, math, and music. You have a choice of four humanities courses, three sociology courses, five science courses, two math courses, and three music courses. Assuming that scheduling conflicts can be avoided, how many different five-course schedules are possible?

8. The car model you are considering comes with or without air conditioning, with or without a sun roof, with or without a CD player, and in eight stock colors. How many different versions of the car are available?

9. Arrangements with Repetition. Answer each question about arrangements of items in which the same item can be chosen over and over.

  a. How many different four-digit house addresses can be formed with the numerals 0–9?

  b. How many different 10-note "tunes" can be created from the notes C, D, E, F, G, A, and B?

  c. Supposedly, if a monkey types randomly on a typewriter long enough, it eventually will type the works of Shakespeare. How many different five-letter words could a monkey type from the 26 alphabet letters?

  d. Imagine a lock that consists of three dials, each of which can be set at the letters A, B, C, D, E, F, G, and H. How many different three-dial combinations are possible?

10. Arrangements with Repetition. Answer each question about arrangements of items with repetition. In each case, be sure that you understand why the problem involves arrangements with repetition.

  a. How many different four-digit telephone extensions can be formed with the numerals 1–9?

  b. How many different three-letter "words" can be formed with the four genetic code letters A, C, G, T?

  c. How many different six-letter words can be formed from the 26 letters of the alphabet?

  d. Imagine a lock that consists of a dial that can be set to any of the numbers 1–30. How many different three-number combinations are possible?

11. Factorial Practice. Find the value of the following quantities *without* using the factorial key on your calculator (you may use the multiplication key). Show your work.

  a. $7!$ **b.** $\dfrac{7!}{3!}$ **c.** $\dfrac{7!}{4!\,3!}$

  d. $\dfrac{12!}{5!}$ **e.** $\dfrac{12!}{8!(12-8)!}$ **f.** $\dfrac{21!}{20!}$

12. Factorial Practice. Find the value of the following quantities *without* using the factorial key on your calculator (you may use the multiplication key). Show your work.

  a. $6!$ **b.** $\dfrac{6!}{3!}$ **c.** $\dfrac{6!}{4!\,3!}$

  d. $\dfrac{11!}{5!}$ **e.** $\dfrac{11!}{4!\,(11-4)!}$ **f.** $\dfrac{16!}{15!}$

13. Permutations. Answer each question about permutations. In each case, also explain why the order of arrangement must be considered in the problem.

  a. Only 5 of 12 photographs will fit on a shelf above your desk. How many different ways can you arrange the photographs if the order makes a difference to you?

  b. How many different five-letter passwords can be formed from the letters A, B, C, D, E, F if no repetition of letters is allowed?

  c. Twelve acts at a comedy club all perform on one evening. How many different ways can their appearances be scheduled?

d. You deal ten cards from a standard deck of cards in order, face-up on a table. How many different sequences of cards can you deal?

14. **Permutations.** Answer each question about permutations. In each case, also explain why the order of arrangement must be considered in the problem.

   a. Only 6 of your 13 cookbooks will fit on a small shelf above your stove. How many different ways can you arrange the books if the order of the books makes a difference to you?

   b. How many different four-letter passwords can be formed from the five vowels in the alphabet if no repetition of letters is allowed?

   c. Ten finalists in a talent show must give their final performance. How many different ways can their appearances be scheduled?

   d. Ten finalists in a talent show must give their final performance. Five contestants will perform on the first night of the show. How many different ways can the schedule for the first night be made?

15. **Combinations.** Answer each question about combinations. In each case, also explain why the order of arrangement is *not* important in the problem.

   a. How many different five-person subcommittees can be formed from a group of ten people?

   b. How many different four-card hands can be dealt from a deck that has only 40 cards?

   c. You own ten compact discs, but plan to take only six with you on a road trip. How many different sets of six discs can you take?

   d. The coach of the track team must choose a squad of eight people to travel to the next meet. How many different squads can be chosen from the twenty members of the track team?

16. **Combinations.** Answer each question about combinations. In each case, also explain why the order of arrangement is *not* important in the problem.

   a. How many different four-person subcommittees can be formed from a group of twelve people?

   b. How many different five-card hands can be dealt from a deck that has only face cards and aces (16 cards)?

   c. You own twelve compact discs, but plan to take only five with you on a road trip. How many different sets of five discs can you take?

   d. The coach of the debate club must choose a team of 6 people to travel to the next meet. How many different teams can be chosen from the 12 members of the club?

17. **Permutation and Combination Practice.** Rewrite each permutation or combination expression in terms of factorials and then find its numerical value. In each case, make up a question that is answered by the expression. For example, $_5C_3$ could answer the question "How many three-person committees can be formed from a group of five people?" Explain why order of arrangement is or is not important in your question.

   a. $_9C_3$ **b.** $_9P_3$ **c.** $_4P_2$ **d.** $_4C_2$ **e.** $_{12}C_8$ **f.** $_{12}P_8$

18. **Lessons in the Ice Cream Shop.** Josh and John's Ice Cream Shop offers 12 different flavors of ice cream and 6 different toppings. Answer the following questions by using the appropriate counting technique (selections from more than one collection, arrangements with repetitions, permutations, or combinations). Explain why you chose the particular counting technique.

   a. How many different sundaes can you create using one ice cream flavor and one topping?

   b. How many different triple cones can you create from the 12 flavors if the same flavor may be used more than once? Assume that you specify which flavor goes on the bottom, middle, and top.

   c. Using the 12 flavors, how many different triple cones can you create with 3 *different* flavors if you specify which flavor goes on the bottom, middle, and top?

   d. Using the 12 flavors, how many different triple cones can you create with 3 *different* flavors if you don't care about the order of the flavors on the cone?

   e. Repeat parts (a)–(d), but assume that Josh and John's carries 15 flavors, rather than 12 flavors.

19. **Manager's Job.** On a baseball team with 25 players, a manager must choose a batting order having 9 players. How many different nine-player batting orders are possible? Suppose that the manager tried a different batting order every day (with no breaks for rest days or the off-season). How long (in years) would it take to try every possible batting order?

20. **License Plates.**

   a. Suppose that license plates are made with three letters followed by three numbers. How many different license plates are possible?

b. How many license plates are possible if they just consist of any six characters (either letters or numbers)?

c. How many license plates are possible if they consist of six letters or numbers, none of which can be used twice on a plate?

21. Making a Toast. If eight people make a complete toast (each pair of people touches glasses), how many different chimes of glass should you hear? Are you counting permutations or combinations? Why?

22. Choosing Finalists. Ten students are deadlocked after the regular competition of the South Dakota Super Spelling Bee. The winners will be determined by a drawing. How many different ways can the judges award the top three places?

23. House Numbers. The houses on the west side of the 800 block of Sierra Drive can have any odd three-digit house number that begins with 8. How many different house numbers are possible?

24. Seating Arrangements. Five women and four men must be seated on a platform at a graduation ceremony. How many different seating arrangements are possible? How many different seating arrangements are possible if the men and women must sit in alternate seats?

25. Passwords of Symmetric Letters. Each of the 11 letters A, H, I, M, O, T, U, V, W, X, and Y appears the same when it is flipped right to left (or looked at in a mirror). They are called the *symmetric letters*. How many six-letter computer passwords can be formed using only the symmetric letters of the alphabet? Assume that the same letter can be used more than once.

26. Senate Committee. How many different ways are there to choose the 18 members of the Senate Foreign Relations Committee from the 100 Senate members?

27. Telephone Numbers. A seven-digit phone number in the United States consists of a three-digit *exchange* followed by a four-digit number.

a. The first digit of the exchange is not allowed to be 0 or 1. How many different seven-digit phone numbers can be formed? Can a city of 2 million people be served by a single area code? Explain.

b. How many exchanges are needed to serve a city of 80,000 people? Explain.

28. Pizza Hype. Luigi's Pizza Parlor advertises 84 different three-topping pizzas. How many individual toppings does Luigi actually use? Ramona's Pizzeria advertises 45 different two-topping pizzas. How many toppings does Ramona actually use? (Hint: In these problems, you are given the total number of combinations, and you must find the number of toppings that are used.)

29. ZIP Codes. The U.S. Postal Service uses both five-digit and nine-digit ZIP codes.

a. How many five-digit ZIP codes are available to the U.S. Postal Service?

b. For a U.S. population of 270 million people, what is the average number of people per five-digit ZIP code, if all possible ZIP codes are used? Explain.

c. How many nine-digit ZIP codes are available to the U.S. Postal Service? Could everyone in the United States have their own personal nine-digit zip code? Explain.

30. More Card Hands. You are given a standard deck of 52 cards. How many ways can the following be dealt?

a. five hearts

b. three jacks

c. two kings

d. three 10s and two aces in one five-card hand

31. Tournaments. At the state basketball tournament 16 teams are vying for the championship. Teams are matched up in pairs for the 8 games of the first round. The winners will advance and the losers will be eliminated.

a. How many different matchups are possible in the first round?

b. The second round of the tournament has only eight teams remain to be matched in pairs for four games. How many different matchups are possible?

c. Consider the third round of the tournament when only four teams remain to be matched in pairs for two games. How many different matchups are possible?

32. Forming a Soccer Team. Assume that a soccer team consists of three front-line positions, four midfield positions, three defensive positions, and a goalie. On the entire team, a coach has five players that can play on the front line, seven that play at midfield, five that play at the defensive positions, and two goalies. How many different lineups can be formed? (Hint: Work this problem in stages. First, determine how many ways there are to choose players for each of the four positions.)

33. Shuffling Cards. Each of the 52 cards in a deck is distinct (different face value or suit).

a. How many arrangements of the 52 cards in a deck are possible?

b. Suppose that, through the magic of modern medicine, you become immortal. Wondering what to do with your time, you decide to arrange a deck of cards in as many different ways as possible. You get pretty good at it, so that you can try one arrangement every minute. Further, you take miracle drugs that alleviate your need for sleep and food. If you continue to rearrange the cards every minute, could you try all of the possible arrangements before the Sun dies in about 5 billion years? Explain.

UNIT 8B

FUNDAMENTALS OF PROBABILITY

Now that we have developed techniques for counting, we are ready to investigate how we calculate probabilities. Mathematically, a probability can be assigned to any individual **event** or group of events. If an event is certain to occur, it has a probability of 1; for example, the probability of the Sun rising tomorrow is essentially 1, or 100%. If an event is impossible, its probability is 0; for example, the probability that you'll land on the Moon the next time you jump off a chair is 0. The probability of an event is indicated by a fraction. For example, we might say that the probability of rain tomorrow is 0.4, or 40%. Note that probabilities must always fall in the range of 0 to 1.

There are three basic ways to find the probability of an event or group of events:

By the Way ·············
The Latin words *a priori* mean "before the fact." An *a priori* probability does not depend on any observations or experiments, and thus is based on our prior ideas about possible outcomes.

- We can use a theoretical *model* based on assumptions about the events in question. For example, when we say that the probability of heads on a coin toss is $\frac{1}{2}$, we are assuming that the coin is equally likely to land heads or tails. A probability based on assumptions or models is called an **a priori** probability.

- We can base the probability on the results of observations or experiments. For example, if we observe that it rains an average of 100 days per year, we might say that the probability of rain on any particular day is $\frac{100}{365}$. Probabilities based on observations or experiments are called **empirical** probabilities.

- We can make a **subjective** estimate of the probability through personal judgment or intuition. For example, you could make a subjective estimate of the probability that going to college today will make you more successful tomorrow.

There's not much we can say mathematically about subjective probabilities because their validity depends on personal experience and judgment. So let's turn our attention to *a priori* and empirical probabilities.

A PRIORI TECHNIQUES

The probability of a coin landing heads is $\frac{1}{2}$ only if it is a **fair** coin: one that is equally likely to land heads or tails. If a coin is dented, dirty, or weighted on one side, it may land heads more often than tails or vice versa. In general, *a priori* techniques for calculating probabilities are valid only in fair situations. However, the possible outcomes may not always be equally likely. For example, if you spin an arrow on a wheel that is $\frac{3}{4}$ blue and $\frac{1}{4}$ black, the *a priori* probability of the arrow landing on blue is $\frac{3}{4}$ (Figure 8.5). The easiest *a priori* probabilities to calculate are those in which all the possible outcomes *are* equally likely.

FIGURE 8.5

Calculating *A Priori* Probabilities for Equally Likely Outcomes

Step 1:   Count the total number of possible outcomes of an event.

Step 2:   Count the number of outcomes that represent **success**—that is, the number of outcomes that represent the desired result.

Step 3:   Determine the probability of success by dividing the number of successes by the total number of possible outcomes:

$$\text{probability of success} = \frac{\text{number of outcomes that represent success}}{\text{total number of possible outcomes}}$$

As a simple example, consider a coin toss in which we define *success* to mean that the coin lands heads. For Step 1, we find that there are two possible outcomes: heads and tails. For Step 2, we recognize that only one of these outcomes, heads, is a success. For Step 3, we divide the one success (heads) by the two possible outcomes (heads or tails) to find a probability of $\frac{1}{2}$.

EXAMPLE 1   *A Priori Die Probability*

Use the three-step process to find the probability of getting a 4 on a single roll of a six-sided die.

Solution:

Step 1.   The six faces of the die are the six possible outcomes of rolling a die.

Step 2.   Because we are looking for the probability of getting a 4, we define a roll of 4 to represent *success*. The number 4 shows on only one face of the die, so only one outcome represents success.

Step 3.   The probability of success is

$$\frac{\text{number of ways to roll a 4}}{\text{total number of ways the die can land}} = \frac{1}{6}.$$

This *a priori* probability assumes the die is fair. If the die is loaded (weighted on any particular side), the actual probability of rolling a 4 may not be $\frac{1}{6}$. ∎

EXAMPLE 2   *(Not) Winning the Lottery*

Suppose that a lottery involves drawing 6 balls at random from a drum containing 52 numbered balls (numbered 1–52). If you pick 6 numbers, what are the chances that you'll choose the 6 winning numbers?

Solution:

Step 1.   To find the total number of possible outcomes, we need to know all the ways that 6 balls can be drawn. Assuming that the order in which the balls are drawn does not matter, this is a *combinations* problem: $r = 6$ balls are selected from a total of $n = 52$ balls. The number of combinations is

$$_{52}C_6 = \frac{52!}{(52 - 6)! \times 6!} = \frac{52 \times 51 \times 50 \times 49 \times 48 \times 47 \times 46!}{46! \times 6!} = 20{,}358{,}520.$$

Step 2.   Success means choosing the 6 winning numbers. There is only one set of numbers that matches the numbers on the 6 winning balls.

Step 3.   The probability of success is

$$\frac{\text{number of ways to match the 6 winning numbers}}{\text{total number of possible outcomes from choosing 6 balls}} = \frac{1}{20{,}358{,}520}.$$

That is, your chance of winning is less than 1 in 20 million. ∎

EXAMPLE 3   *Spade Flushes*

A standard deck of 52 cards contains 13 cards each of four different *suits*: spades, hearts, clubs, and diamonds. In five-card poker, a *flush* is any set of five cards of the same suit. How many different spade flushes are there? What are the chances of being dealt a spade flush?

Solution:

Step 1.   We calculated the total number of possible five-card hands in Example 10 of Unit 8A: it is 2,598,960.

Step 2.   Success in this case means getting a spade flush (5 cards that are all spades), so we must count the number of ways we can get a spade flush. Because we are drawing 5 cards from the deck, and 13 of the cards in the deck are spades, we are looking for the number of *combinations* of $r = 5$ cards that can be made from the $n = 13$ spades in the deck:

$$_{13}C_5 = \frac{13!}{(13 - 5)! \times 5!} = \frac{13 \times 12 \times 11 \times 10 \times 9 \times 8!}{8! \times 5!} = 1287$$

Step 3. The chance of being dealt a spade flush (or a flush of any suit for that matter) is

$$\frac{\text{number of possible spade flushes}}{\text{number of possible hands}} = \frac{1287}{2,598,960} = 0.0005$$

or about 1 chance in 2000.

PROBABILITY DISTRIBUTIONS

Suppose that you toss two coins simultaneously. What are the probabilities of the possible outcomes?

Because *each* coin can land in two possible ways (heads or tails), the multiplication principle says that *two* coins can land in $2 \times 2 = 4$ different ways. We can make a simple table with a row for each of the four ways the coins can fall.

Coin 1	Coin 2	Outcome	Probability
H	H	HH	$\frac{1}{4}$
H	T	HT	$\frac{1}{4}$
T	H	TH	$\frac{1}{4}$
T	T	TT	$\frac{1}{4}$

If we don't care which coin is which, there are three different outcomes to this experiment: (1) 2 heads, (2) 2 tails, and (3) 1 head and 1 tail. The probability of getting two heads is $\frac{1}{4}$ or 25%; the probability of two tails also is $\frac{1}{4}$ or 25%; and the probability of one head and one tail is $\frac{2}{4} = \frac{1}{2}$ or 50%. We can summarize these results in a table called a **probability distribution**.

Note: *the sum of the probabilities in a probability distribution must always be 1 (or 100%).*

Probability Distribution for Two Tossed Coins

Result	Probability
2 heads	$\frac{1}{4}$
1 head, 1 tail	$\frac{1}{2}$
2 tails	$\frac{1}{4}$
Total	**1**

EXAMPLE 4   *Three Coins*

Make a probability distribution for the number of heads and tails that occur when three coins are tossed simultaneously.

Solution: By the multiplication principle, the total number of outcomes from three tossed coins is $2 \times 2 \times 2 = 8$. We can show all eight outcomes by making a table with a column for each of the three coins. The table needs eight rows for the eight possible outcomes; we can ensure that each row is different by using the pattern shown: for Coin 1 we write heads in four rows followed by tails in four rows; for Coin 2 we alternate two rows of heads with two rows of tails; for Coin 3 we alternate rows of heads and tails.

Coin 1	Coin 2	Coin 3	Outcome	Probability
H	H	H	HHH	$\frac{1}{8}$
H	H	T	HHT	$\frac{1}{8}$
H	T	H	HTH	$\frac{1}{8}$
H	T	T	HTT	$\frac{1}{8}$
T	H	H	THH	$\frac{1}{8}$
T	H	T	THT	$\frac{1}{8}$
T	T	H	TTH	$\frac{1}{8}$
T	T	T	TTT	$\frac{1}{8}$

If we care only about the number of heads and tails, we can consolidate our results in a probability distribution. Rows 2, 3, and 5 all represent two heads and one tail, so this result has a probability of $\frac{3}{8}$ (3 successes divided by 8 outcomes). Similarly, rows 4, 6, and 7 represent one head and two tails, so this result also has a probability of $\frac{3}{8}$. The overall probability distribution is shown to the right. Note that the sum of all the probabilities is 1, as it must be.

Probability Distribution for Three Tossed Coins

Result	Probability
3 heads	$\frac{1}{8}$
2 heads, 1 tail	$\frac{3}{8}$
1 head, 2 tails	$\frac{3}{8}$
3 tails	$\frac{1}{8}$
Total	**1**

EXAMPLE 5   *Two Dice*

Make a probability distribution for the sum of the dice when two dice are rolled. What is the most probable sum?

Solution: Because there are six ways for a single die to land, there are $6 \times 6 = 36$ ways for two dice to land. We can enumerate all 36 possibilities by making an array in which we list the numbers of one die along the rows and the other die's numbers along the columns. In each cell, we show the sum of the numbers on the two dice.

	1	**2**	**3**	**4**	**5**	**6**
1	$1+1=2$	$1+2=3$	$1+3=4$	$1+4=5$	$1+5=6$	$1+6=7$
2	$2+1=3$	$2+2=4$	$2+3=5$	$2+4=6$	$2+5=7$	$2+6=8$
3	$3+1=4$	$3+2=5$	$3+3=6$	$3+4=7$	$3+5=8$	$3+6=9$
4	$4+1=5$	$4+2=6$	$4+3=7$	$4+4=8$	$4+5=9$	$4+6=10$
5	$5+1=6$	$5+2=7$	$5+3=8$	$5+4=9$	$5+5=10$	$5+6=11$
6	$6+1=7$	$6+2=8$	$6+3=9$	$6+4=10$	$6+5=11$	$6+6=12$

By the Way ············

Dice have been used since antiquity and probably originated in Asia. Dice marked with dots have been recovered from Egyptian tombs, and Greek and Roman literature contains many references to dice playing.

The possible outcomes from the roll of two dice are sums between 2 and 12. We can find the probability of each sum by dividing the number of times that sum occurs by the 36 possible ways for the dice to land. For example, there are five ways for the dice to show a sum of 8, so the probability of getting a sum of 8 from the roll of two dice is

$$\frac{\text{number of ways to get 8}}{\text{number of ways two dice can land}} = \frac{5}{36}.$$

In a similar way, we can find the probability of all other results and summarize them in the following probability distribution.

Result (Sum)	2	3	4	5	6	7	8	9	10	11	12	**Total**
Probability	$\frac{1}{36}$	$\frac{2}{36}$	$\frac{3}{36}$	$\frac{4}{36}$	$\frac{5}{36}$	$\frac{6}{36}$	$\frac{5}{36}$	$\frac{4}{36}$	$\frac{3}{36}$	$\frac{2}{36}$	$\frac{1}{36}$	**1**

The most likely result is a 7, which has a probability of $\frac{6}{36}$, or $\frac{1}{6}$.   ∎

EMPIRICAL TECHNIQUES

Suppose that we want to know the probability that a particular river will rise ten feet above its flood level in some region. There is no *a priori* way to calculate this probability, so we turn to empirical techniques. Imagine that we study historical and geological records, and discover that the river has risen above flood level nine times in the past 1000 years. We might therefore conclude that the probability of such a flood in any year is about 9 in 1000, or 0.009. Because this probability is close to 0.01, or 1%, such an event is sometimes called a "100-year flood" to indicate that a flood of this severity happens *on average* about once every 100 years.

> **TIME-OUT TO THINK:**   Is it possible for a region to experience a 100-year flood in two consecutive years? Explain.

Empirical techniques are useful whenever there is no *a priori* way to find a probability. In addition, empirical techniques are useful for checking the validity of an *a priori* probability. For example, if you suspect that a particular coin is unfair, you might perform an *experiment* in which you toss the coin 100 times. If the coin lands heads 75 times and tails only 25 times, you'd have good reason to believe that the coin is unfair and that the *a priori* probability of 50% does not apply to this coin.

EXAMPLE 6   *500-Year Flood*

Suppose geological records indicate that a river has crested above flood level just four times in the past 2000 years. What is the empirical probability that the river will crest above flood level next year?

Solution: Based on the empirical data, the probability of the river cresting above sea level in any single year is

$$\frac{\text{number of years with flood}}{\text{total number of years}} = \frac{4}{2000} = \frac{1}{500}.$$

Because a flood of this magnitude occurs an average of once every 500 years, it could be called a "500-year flood." ∎

EXAMPLE 7   *Empirical Coin Testing*

A single round of an experiment in which two coins are tossed has three possible outcomes: 0 heads, 1 head, or 2 heads. Suppose that you repeat the two-coin toss 100 times and that the following table presents your results. Compare the empirical probabilities from the experiment to the *a priori* probabilities for two tossed coins. Are the coins fair?

Result	Number of Occurrences
0 heads	19
1 head	54
2 heads	27
Total	**100**

Solution: We find the empirical probabilities by dividing the number of occurrences of each result by the total number of times the experiment of tossing two coins was performed. For example, the empirical probability of zero heads in this data is

$$\frac{\text{number of occurrences of 0 heads}}{\text{total number of two–coin tosses}} = \frac{19}{100} = 0.19.$$

Similarly, the empirical probability for one head in this data is 0.54, and for two heads is 0.27.

Earlier, we found that the *a priori* probabilities for zero heads, one head, and two heads are 0.25, 0.50, and 0.25, respectively. The empirical probabilities do not match the *a priori* probabilities exactly, but we should not expect a perfect match. Because the empirical and *a priori* probabilities are relatively close and the coins were tossed only 100 times, we have no reason to suspect that either coin is unfair. ∎

TIME-OUT TO THINK: Suppose that you tossed the coins in Example 7 a *million* times, and still found the same empirical probabilities. Would you now conclude that one or both coins is unfair? Explain.

PROBABILITY OF AN EVENT *NOT* OCCURRING

Suppose that we are interested in a particular event—say, the occurrence of a 4 on the roll of a die. Let's represent this event by the letter A. We can denote the probability of event A by $P(A)$. Because the probability of rolling a 4 is $\frac{1}{6}$, we write

$$P(A) = \frac{1}{6}.$$

The probability that anything *except A* occurs—in this case, rolling any number but 4 on the die—is $\frac{5}{6}$ because there are five other possible rolls: 1, 2, 3, 5, or 6. Therefore we say that

$$P(\text{not } A) = \frac{5}{6}.$$

We can generalize this observation to any event A.

Probability of an Event *not* Occurring.

If the probability of an event A is $P(A)$, the probability that anything *except A* occurs is

$$P(\text{not } A) = 1 - P(A).$$

EXAMPLE 8   *Not an 8*

In Example 5, we found that the probability of rolling two dice and getting a sum of 8 is $\frac{5}{36}$. What is the probability of getting a sum of anything but 8?

Solution: Let's call event A getting a sum of 8 on two dice. Then $P(A) = \frac{5}{36}$. Therefore the probability of getting anything but a sum of 8 is

$$P(\text{not } A) = 1 - \frac{5}{36} = \frac{31}{36}.$$

∎

STATING THE ODDS

You've probably noticed that there are several ways to express the idea of *probability*. For example, saying that the *chance* of getting heads on a coin toss is $\frac{1}{2}$ is equivalent to saying that the *probability* is $\frac{1}{2}$. Another common term for probability is *odds*. However, odds are usually expressed as a ratio rather than as a fraction. For example, we might say that the odds of a coin landing heads are 1 in 2 (or 1 to 2).

A more common way of stating odds is by comparing the probability that a particular event happens to the probability that it does not happen. That is, the odds for an event *A* are:

$$\text{Odds}(A) = \frac{\text{probability of } A}{\text{probability of not } A} = \frac{P(A)}{P(\text{not } A)}$$

Beware! As this discussion shows, the term *odds* is used in several different ways. Be sure to examine the context of its use so you will interpret the odds correctly.

For example, the probability of a coin landing heads is $\frac{1}{2}$ and the probability of a coin *not* landing heads also is $\frac{1}{2}$. Thus by this definition of odds:

$$\text{Odds}(\text{heads}) = \frac{P(\text{heads})}{P(\text{not heads})} = \frac{\frac{1}{2}}{\frac{1}{2}} = \frac{1}{1} = 1$$

This result is usually phrased as *the odds of a coin landing heads are 1 to 1, or even.*

In gambling, the term *odds* usually expresses how much a bet will pay if you win. For example, suppose that the odds on a particular horse at a horse race are 3 to 1. Then for each $1 you bet on this horse, you will gain $3 if the horse wins. (If you place a $1 bet and the horse wins, you will receive $4: your $1 back, plus a gain of $3.) If you place a $2 bet and the horse wins, you will gain $3 \times \$2 = \$6$.

EXAMPLE 9   *Odds in Probability*

a)   What are the odds of getting two heads in two coin tosses?

b)   Suppose that, for some event, the probability of success is $\frac{1}{16}$. What are the odds *against* this event?

c)   Suppose that the odds *against* winning a particular game are 8 to 5. What is the *probability* of losing? What is the *probability* of winning?

Solution:

a)   As we found earlier, the probability of getting two heads in two coin tosses is $\frac{1}{4}$; that is, $P(2 \text{ heads}) = \frac{1}{4}$. The probability of not getting two heads in two coin tosses therefore is $P(\text{not 2 heads}) = \frac{3}{4}$. Thus the odds of getting two heads are:

$$\text{Odds}(2 \text{ heads}) = \frac{P(2 \text{ heads})}{P(\text{not 2 heads})} = \frac{\frac{1}{4}}{\frac{3}{4}} = \frac{1}{3}$$

The odds of getting two heads in two coin tosses are 1 *to* 3. We could also express the odds by turning them around: the odds *against* getting two heads in two coin tosses are 3 to 1.

b)   If the probability of success is $\frac{1}{16}$, the probability of failure (*not* success) is $\frac{15}{16}$.

$$\text{Odds (success)} = \frac{P(\text{success})}{P(\text{failure})} = \frac{\frac{1}{16}}{\frac{15}{16}} = \frac{1}{15}$$

The odds of success are 1 to 15, so the odds *against* success are 15 to 1.

c)   The 8 to 5 odds against winning mean that, on average, you will lose 8 times for every 5 times you win. Thus you will lose an average of 8 out of every $8 + 5 = 13$ times you play this strategy. Therefore your probability of losing is $\frac{8}{13}$, and your probability of winning is $1 - \frac{8}{13} = \frac{5}{13}$.   ■

EXAMPLE 10   *Odds in Gambling*

Suppose that the odds on a particular horse in a horse race are given as 7 to 2. If you bet $10 and win, how much will you gain?

Solution: The odds of 7 to 2 on the horse mean that, for each $2 you bet on this horse, you gain $7 if you win. You made the equivalent of five $2 bets (because $10 = 5 \times \$2$) so you gain $5 \times \$7 = \$35$.   ■

REVIEW QUESTIONS

1. Briefly describe the differences between *a priori, empirical*, and *subjective* techniques for finding probabilities.

2. What do we mean by a *success* when calculating an *a priori* probability? Explain the basic steps in calculating the probability of success.

3. What is a *probability distribution*? Explain how to make a table of all possible outcomes, and how to convert such a table into a probability distribution.

4. What does it mean when we talk about a "100-year flood?"

5. How is the probability of an event *not* occurring related to the probability that it *does* occur? Why?

6. Explain the common usage of the term *odds*. Suppose you roll a single die. What are the odds *against* rolling a 6?

PROBLEMS

1. **Probability in Your Life.** If you look deeply, you will find that probability is involved in *every* decision you make. For parts (a)–(f) trace your decision, identifying and briefly discussing the places where probability entered into the decision.

a. Why did you decide to enroll in college?

b. If you exercise regularly, explain why; if you don't, explain why not.

c. If you are like most drivers, you probably exceed the posted speed limit occasionally. Explain how you decide when to drive at or below the speed limit and when to exceed it.

d. Explain your choice for president in the last election. If you did not vote, explain why not.

e. Most grocery stores offer a choice between paper and plastic bags; in addition, you can always bring your own reusable bags for groceries. Explain your choice.

f. Automobiles have a significant impact on the Earth's environment. For example, they require many raw materials for construction, they use some of the world's limited remaining supply of oil to operate, and they emit pollutants into the air. Nevertheless, most people drive cars, or at least travel in cars driven by others. Explain why *you* drive a car (or travel in cars driven by others).

2. **Certainty and Impossibility.** We often hear "nothing is impossible," or "anything is possible if you try hard enough." Do such sayings mean that if you try hard enough, you can live without food for 10 years? Can you jump out of an airplane and fall up? Briefly discuss the concepts of *impossible* and *certain*. How are they related to probability?

3. **Subjective Probabilities.** State your own subjective probabilities for each given event and justify with a brief explanation.

a. The probability that your school's football team will rank in the top 10 at the end of the next football season.

b. The probability that, 20 years from now, you will feel that your collegiate experience was of great value to your life and career.

c. The probability that you will be married in 2010.

d. The probability that a Democrat will be elected president in the next presidential election.

e. The probability that terrorists will obtain and use a nuclear weapon within the next 10 years.

f. The probability that your state legislature's latest "get tough on crime" legislation will significantly reduce the crime rate.

g. The probability that the United States will suffer a major catastrophe (e.g., a world war, anarchy, or environmental devastation) during the next 75 years.

A Priori **Card Probabilities.** *A deck of 52 cards has four suits (13 cards each of diamonds, hearts, clubs, and spades). Each suit has cards numbered 2–10, jack, queen, king, and ace. For Problems 4–10, follow the three-step process in this unit to find the* a priori *probabilities when you draw one card from the deck.*

4. What is the probability that it will be a king?

5. What is the probability that it will be a 2?

6. What is the probability that it will be a card *other than* a king?

7. What is the probability that it will be a red card (a heart or a diamond)?

8. What is the probability of drawing a spade?

9. What is the probability of drawing a card other than a spade?

10. What is the probability of drawing a card with a number on it (i.e., *not* an ace, jack, queen, or king)?

11. **Bag of Marbles.** Suppose that you have a bag of colored marbles containing 10 white marbles, 5 blue marbles, and 2 red marbles. You reach into the bag and draw out 1 marble. Follow the three-step process to find the *a priori* probability that the marble is

a. red   b. blue

c. white   d. not white.

12. **Bag of Candy.** A large bag contains 10 chocolates, 15 mints, and 20 gumdrops. You reach into the bag and draw out 1 candy. Follow the three-step process to find the *a priori* probability that the candy is

a. a chocolate   b. a mint

c. a gumdrop   d. not a mint.

13. **Probability of Failure.** Give the probability for each situation.

a. Flipping two fair coins and not getting two heads.

b. Rolling a fair die and not getting an odd number.

c. Drawing one card from a standard deck of cards and not getting a jack, queen, or king.

d. Rolling two dice and not getting a double number (for example, double ones or double twos).

14. **Probability of Failure.** Give the probability for each situation.

a. Flipping two fair coins and not getting a head and a tail.

b. Rolling a fair die and not getting an even number.

c. Drawing one card from a standard deck of cards and not getting a card with two through ten on it.

d. Rolling two dice and not getting a total of seven.

15. Four-Coin Probability Distribution.

a. Make a table showing all the possible outcomes from tossing four coins at once.

b. Consolidate the table in part (a) to show the probability distribution for 4 H (four heads), 3 H 1 T (three heads and 1 tail), 2 H 2 T, 1 H 3 T, and 4 T.

c. What is the probability of getting three heads and one tail when you toss four coins at once?

d. What is the probability of getting *anything but* all tails?

16. Two-Dice Probabilities. Suppose two fair dice are rolled as in Example 5.

a. What is the probability that the sum of the two dice is 5?

b. What is the probability that the sum of the two dice is *anything but* a 5?

c. What is the probability of rolling a 6 on one die and a 1 on the other die?

d. What is the probability that one of the two dice is a 5?

e. Suppose that someone offers you 10 to 1 odds in betting that you won't roll two dice that have a total of 7. That is, if you don't roll a total of 7 you lose $1, but if you do roll a total of 7 you win $10. In the long run, would you expect to win or lose at this game? Why?

17. Two-Dice Probabilities. Suppose two fair dice are rolled as in Example 5.

a. What is the probability that the sum of the two dice is 4?

b. What is the probability that the sum of the two dice is *anything but* a 4?

c. What is the probability of rolling a 5 on one die and a 2 on the other die?

d. What is the probability that one of the two dice is a 3?

e. Suppose that someone offers you 10 to 1 odds in betting that you won't roll two dice that have a total of 6. That is, if you don't roll a total of 6 you lose $1, but if you do roll a total of 6 you win $10. In the long run, would you expect to win or lose at this game? Why?

18. Coin Expectations. Suppose that you toss a fair coin 1000 times. Should you *expect* it to come up heads exactly 500 times? Why or why not? Explain without making any calculations.

19. Coin Experiment. Take three coins and toss them simultaneously. Record your results as either zero heads (0 H), one head (1 H), two heads (2 H), or three heads (3 H). Repeat this experiment 40 times. Make a table of your results, and state your empirical probabilities for each of the four possible results: $P(0\ H)$, $P(1\ H)$, $P(2\ H)$, and $P(3\ H)$. Compare your empirical probabilities to the *a priori* probabilities obtained in Example 4. Explain any differences. If your experiment produced the *a priori* probabilities exactly, should you be surprised?

20. Fair Coins? Imagine that you do an experiment in which you toss three coins 1000 times and that the following table shows the results.

Result	Number of occurrences
0 heads	260
1 head	495
2 heads	245
3 heads	0
Total	1000

a. Compute the empirical probability for each outcome.

b. Compare the empirical probabilities from the experiment to the *a priori* probabilities for three tossed coins. Are the coins unfair?

c. Someone suggests that the reason for your strange result is that one of the three coins has tails on both sides. Do you believe him or her? Why or why not?

21. Computing the Odds. Using the definition of odds given in the text, find the odds of each event described.

a. Rolling a fair die and getting a 1 or a 2.

b. Flipping two fair coins and getting two tails.

c. Drawing a single card from a standard deck and getting a spade.

d. Drawing a single card from a deck and getting a face card.

22. Computing the Odds. Using the definition of odds given in the text, find the odds of each event described.

   a. Rolling a fair die and getting a 5 or a 6.

   b. Flipping two fair coins and getting a head and a tail.

   c. Drawing a single card from a standard deck and getting a heart.

   d. Drawing a single card from a deck and getting a jack or a queen.

23. Gambling Odds. Use the definition of odds that is used in betting.

   a. Suppose that you are given odds of 3 to 4 on a bet. If you bet $20 and win, how much do you gain?

   b. Suppose that you are given odds of 5 to 4 on a bet. If you bet $20 and win, how much do you gain?

24. Counting Cards. In blackjack, the object of the game is to receive cards whose value totals as close to 21 as possible, without exceeding it. The probability that you are dealt, say, an ace depends on which cards have already been dealt.

   a. Suppose that you receive the first card dealt from a deck of 52 cards. What is the probability that it is an ace?

   b. Suppose that you can see 10 of the cards that have already been dealt (either because they are showing face up on the table or because they are in your own hand) and that 2 of them are aces. What is the probability that the next card dealt will be an ace?

   c. Based on a comparison of the results in parts (a) and (b), explain why card counting (counting the number of cards of each type that have appeared on the table) can improve your odds of winning at blackjack.

25. Lotto Psychology. Imagine a game of Lotto in which you choose six numbers between 1 and 42. Each week lottery officials announce the six winning numbers. If all six of your numbers match the six winning numbers (in any order), then you win big bucks.

   a. How many different sets of six numbers are possible?

   b. What are your chances of choosing the six winning numbers?

   c. Are your chances of winning any better with the numbers 16, 22, 3, 40, 33, 11 than with the numbers 1, 2, 3, 4, 5, 6? Is any set of six numbers more likely than any other? Why or why not?

26. Project: State Lotteries. Research the history and status of lotteries in the United States.

   a. How many states now have lotteries? Is there a lottery in your state?

   b. What is the per capita spending on lotteries in the United States? Can you find a state-by-state breakdown in per capita spending?

   c. Describe at least three different lottery games. Choose games from your state, if it has a lottery. Otherwise, choose games from a state of your choice. For each game, find or calculate the odds of winning prizes of various amounts.

   d. Research any general trends in who spends money on lotteries. For example, is per capita lottery spending different among people in different income groups?

   e. Some people believe that lotteries encourage compulsive gambling, but others disagree. Research and describe the evidence on both sides. What do *you* think?

27. Project: Floods. Research a recent episode of extreme flooding somewhere in the world. Describe the flooding and its effects on both people and the environment. For whatever flooding you choose to study, be sure to answer the following questions.

   a. What is the historical probability of a flood of this magnitude in the region where the flood occurred?

   b. Is there any evidence that human activity has increased the natural probability of flooding? Explain.

COMBINING PROBABILITIES

The mathematical study of probability is linked to the gambling of a French con man known as the Chevalier de Méré. In 1654, the Chevalier bet other gamblers even money that he could roll at least one 6 in four rolls of a standard six-sided die. The Chevalier made quite a profit on this bet, despite having calculated his chance of winning incorrectly!

His faulty reasoning went as follows: The probability of rolling a 6 on one roll of a die is $\frac{1}{6}$, so he assumed that his chances would be four times as great in four rolls, or $\frac{4}{6} = 0.67$. In fact, the chance of rolling a 6 within four tries is only 0.52 (as we will show shortly). Nevertheless, because a probability of 0.52 means winning more often than losing, the Chevalier profited over the long run.

By the Way ·············
Chevalier was a title of low-ranking nobility in France. The Chevalier de Méré was a man by the name of Antoine Gombaud.

TIME-OUT TO THINK: The Chevalier guessed that the chance of rolling a 6 within four rolls is $\frac{4}{6}$. Extending his reasoning, the chance of rolling a 6 within six rolls would be $\frac{6}{6} = 1$. Explain why the chance of rolling a 6 within six rolls is *not* 1, and how you could have used this fact to convince the Chevalier that his reasoning was faulty.

As gamblers caught on to this "con game," they were no longer willing to play. The Chevalier therefore proposed a new game in which he rolled *two* dice and bet that he could throw a *double 6* within 24 rolls. He knew that there are $6 \times 6 = 36$ ways for two dice to land (see Unit 8B, Example 5). Because only one of the 36 possible outcomes is a double 6, the probability of a double 6 on any one roll is $\frac{1}{36}$.

The Chevalier then used his faulty reasoning to guess that, if he rolled the pair of dice 24 times, his odds of rolling at least one double 6 would be $\frac{24}{36} = 0.67$. Alas, the actual probability is about 0.49 (as we will see). Because this is *less* than 50%, he began to lose money. For help, he turned to mathematician Blaise Pascal who, in turn, began a correspondence about the mathematics of probability with the mathematician Pierre de Fermat.

The Chevalier had correctly determined the probabilities of *individual* events, such as the $\frac{1}{36}$ probability of getting a double 6 on a *single* roll of two dice. His failure was in combining the individual probabilities to determine the odds of winning on multiple tries. In this unit, we discuss the correct methods of combining probabilities.

JOINT PROBABILITIES

Suppose that we toss two dice and want to know the probability that *both* will come up 4. One way to find the probability is to consider the tossing of the two dice as one *individual* event—a single toss of two dice. Then we can follow the three-step procedure for finding *a priori* probabilities, as we did in Unit 8B. In this case, there are 36 ways for the dice to land and only one way for both of them to show 4, so the probability of a double 4 is $\frac{1}{36}$. That is,

$$P(\text{double } 4) = \frac{1}{36}.$$

Alternatively, we can consider the landing of the two dice as two separate events. For each die (each event), the probability of a 4 is $\frac{1}{6}$. We find the **joint probability** that both events occur together by multiplying the individual probabilities:

$$P(\text{double } 4) = \underbrace{\frac{1}{6}}_{\text{event}} \times \underbrace{\frac{1}{6}}_{\text{event}} = \frac{1}{36}$$

This agrees with the result found by considering the tossing of the two dice as a single event.

The advantage of the multiplication technique is that it can be easily extended to other situations in which we want to know the probability that two or more events will occur together, such as the probability of getting heads ten times in a row on a coin toss or of having a baby *and* getting a pay raise in the same year. There is an important subtlety, however: We must distinguish between events that are *independent* of each other and events that are *dependent* on other events. Let's investigate each case.

Independent Events

The repeated roll of a single die produces **independent events** because the outcome of one roll does *not* affect the outcome of others. Similarly, coin tosses are independent events because the result of one coin toss does not affect what happens when we toss a second coin. Whenever events are independent, we can calculate the joint probability of two or more events simply by multiplying their probabilities.

Joint Probability of Independent Events

Consider two independent events, A and B, that have individual probabilities $P(A)$ and $P(B)$. The joint probability that A and B occur together is

$$P(A \text{ and } B) = P(A) \times P(B).$$

This principle can be extended to any number of independent events. For example, the joint probability of A, B, and a third independent event C is

$$P(A \text{ and } B \text{ and } C) = P(A) \times P(B) \times P(C).$$

TIME-OUT TO THINK: Mathematically speaking, is there any difference between rolling one die three times and rolling three dice once? Explain.

EXAMPLE 1   *Three Coins*

Suppose that you toss three coins. What is the probability of getting three tails?

Solution: Because coin tosses are independent events, we multiply the probability of tails on each individual coin:

$$P(\text{tails on coin 1 } \textbf{\textit{and}} \text{ tails on coin 2 } \textbf{\textit{and}} \text{ tails on coin 3})$$

$$= \underbrace{P(\text{tails})}_{\text{coin 1}} \times \underbrace{P(\text{tails})}_{\text{coin 2}} \times \underbrace{P(\text{tails})}_{\text{coin 3}} = \frac{1}{2} \times \frac{1}{2} \times \frac{1}{2} = \frac{1}{8}.$$

The probability that three tossed coins all land tails is $\frac{1}{8}$. This *a priori* probability is based on the assumption that all three coins are fair. ∎

EXAMPLE 2   *Drawing Cards with Replacement*

Suppose that you draw a card from a deck, replace the card, reshuffle the deck, and draw another card. What is the probability that both draws will be clubs? If you continue to replace the card and shuffle the deck after each draw, what is the probability of drawing a club three times in a row?

Solution: Because you replaced the first card and reshuffled the deck, the outcome of the second draw doesn't depend on the first draw. Thus the two events are *independent*. We therefore find the joint probability by multiplying the individual probabilities.

Because there are equal numbers of cards of each of the four suits (hearts, spades, clubs, and diamonds), the probability of a club on any single draw is $\frac{1}{4}$. The probability of drawing two clubs on two draws, assuming replacement and reshuffling, is

$$P(2 \text{ clubs}) = P(\text{club } \textbf{\textit{and}} \text{ club})$$
$$= P(\text{club}) \times P(\text{club})$$
$$= \frac{1}{4} \times \frac{1}{4} = \frac{1}{16} = 0.0625.$$

The probability of drawing three straight clubs (replacing the card each time) is

$$P(3 \text{ clubs}) = P(\text{club } \textbf{\textit{and}} \text{ club } \textbf{\textit{and}} \text{ club})$$
$$= P(\text{club}) \times P(\text{club}) \times P(\text{club})$$
$$= \frac{1}{4} \times \frac{1}{4} \times \frac{1}{4} = \frac{1}{64} = 0.015625.$$

The chance of getting two clubs in a row is 1 in 16, or about 6%. The chance of getting three clubs in a row is 1 in 64, or just over 1.5%. ∎

EXAMPLE 3   *Consecutive 100-Year Floods*

What is the probability that a "100-year flood" will strike a city in two consecutive years? If that actually happens, what conclusions might you draw?

Solution: A "100-year flood" describes an event whose empirical probability in any year is $\frac{1}{100}$, or 0.01. Assuming that a flood in one year does not affect the likelihood of a flood in another year, 100-year floods in different years are independent events. Thus the probability of a 100-year flood in two consecutive years is

$$P(\text{100-year flood in 2 consecutive years})$$
$$= \underbrace{P(\text{100-year flood})}_{\text{year 1}} \times \underbrace{P(\text{100-year flood})}_{\text{year 2}} = 0.01 \times 0.01 = 0.0001.$$

That is, the probability of a 100-year flood occurring in two consecutive years is about 1 in 10,000, or 0.01%. While events with a probability of 1 in 10,000 can certainly occur, we should remember that this calculation comes from *assuming* that the chance of a flood is 1% in any year. If a "100-year flood" were to occur in two consecutive years, we might wonder whether the 1% probability of a single flood was correct. ■

> **TIME-OUT TO THINK:**   Suppose that the probability of the 100-year flood is based on data showing five such floods in the past 500 years. Describe several reasons why this empirical probability of 1% might not reflect the *true* probability of such a flood occurring next year.

Dependent Events

Suppose that you draw candy from a box that initially contains 5 chocolates and 5 caramels. Clearly, the probability of getting a piece of chocolate on your first try is

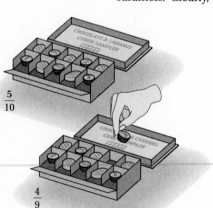

$\frac{5}{10}$

$\frac{4}{9}$

$\frac{5}{10}$, or $\frac{1}{2}$. Now, suppose that you get a chocolate on the first try and quickly eat it. What is the probability of getting another chocolate on your second try?

Because you've already eaten one of the chocolates, the box now contains only 9 pieces of candy, of which 4 are chocolate. Thus, the probability of getting a chocolate on the second try is 4 out of 9, or $\frac{4}{9}$. This is not the same as the $\frac{1}{2}$ probability on the first try because your first draw changed the content of the candy box. In other words, the probability of the second event *is* affected by the outcome of the first event. We therefore say that these are **dependent events**.

Calculating the joint probability of dependent events still involves multiplying the individual probabilities, but we must take into account how prior events affect subsequent ones. In the case of the candy box, we find the probability of getting two chocolates in a row by multiplying the $\frac{1}{2}$ probability on the first try by the $\frac{4}{9}$ probability on the second try:

$$\frac{1}{2} \times \frac{4}{9} = \frac{2}{9}$$

That is, the probability of drawing two chocolates in a row is $\frac{2}{9}$. We can generalize this example to any set of dependent events.

> **Joint Probability of Dependent Events**
>
> Consider two dependent events, A and B. The joint probability that A and B occur together is
>
> $$P(A \textbf{ and } B) = P(A) \times P(B \text{ given } A).$$
>
> We write $P(B \text{ given } A)$, read as "the probability of event B given the occurrence of event A," to show that event B is *dependent* on the outcome of event A. This principle can be extended to any number of individual events. For example, the joint probability of dependent events A, B, and C is
>
> $$P(A \textbf{ and } B \textbf{ and } C) = P(A) \times P(B \text{ given } A) \times P(C \text{ given } A \textbf{ and } B).$$

EXAMPLE 4   *Drawing Cards Without Replacement*

Suppose that you draw a card from a standard deck and then draw a second card *without* replacing the first. What is the probability that both draws will be clubs? What is the probability of drawing three clubs in a row?

Solution: These events are *dependent* since your selection of the first card prevents that card from being chosen again on the second draw. The probability of a club on the first draw is $\frac{13}{52} = \frac{1}{4}$ because a deck of 52 cards contains 13 clubs. However, only 51 cards remain after the first draw. If the first draw was a club, only 12 clubs are left. Thus the probability of a club on the second draw, given a club on the first draw, is $\frac{12}{51}$. The joint probability is

$$P(\text{club and club}) = \underbrace{P(\text{club})}_{\text{first draw}} \times \underbrace{P(\text{club given club on first draw})}_{\text{second draw}}$$

$$= \frac{13}{52} \times \frac{12}{51}$$

$$= 0.0588.$$

The probability of drawing two clubs in a row without replacement is a little less than 6%, which is slightly less than the probability of drawing two clubs with replacement (see Example 2).

If the first two draws are both clubs, there will be 11 clubs left out of a total of 50 remaining cards. Thus the probability of getting a club on the third draw, given clubs on the first two draws, is $\frac{11}{50}$. The joint probability is

$$P(3 \text{ clubs}) = \underbrace{P(\text{club})}_{\text{first draw}} \times \underbrace{P(\text{club given club on first draw})}_{\text{second draw}} \times \underbrace{P(\text{club given clubs on first 2 draws})}_{\text{third draw}}$$

$$= \frac{13}{52} \times \frac{12}{51} \times \frac{11}{50} = 0.013.$$

The probability of drawing three clubs in a row is just over 1%.   ∎

> **TIME-OUT TO THINK:**   Compare the probabilities for getting three clubs in a row when the cards are replaced and reshuffled (Example 2) to when the cards are not replaced in the deck (Example 4). Explain the differences.

EITHER/OR PROBABILITIES

Suppose that we want to know the probability that *either* of two events will occur, rather than the joint probability that both events occur. In that case, we are looking for an **either/or probability**, such as the probability of drawing *either* an ace *or* a king from a deck of cards, or of losing your home *either* to a fire *or* to a hurricane.

Mutually Exclusive Events

A coin can land *either* heads *or* tails, but it can't land both heads *and* tails at the same time. When one event (heads) precludes another (tails) from occurring at the same time, they are said to be **mutually exclusive events**. We can represent mutually exclusive events with a Venn diagram (see Unit 1C) in which the circles do not overlap. For example, we show the possibilities of heads and tails in a coin toss as nonoverlapping circles because a coin cannot land both heads and tails at the same time (Figure 8.6). We find the either/or probability of two mutually exclusive events by adding their individual probabilities.

FIGURE 8.6

A coin cannot land heads and tails at the same time. Heads and tails are mutually exclusive events.

Either/Or Probability for Mutually Exclusive Events

Consider two mutually exclusive events, A and B. The probability that either A *or* B occurs is

$$P(A \textit{ or } B) = P(A) + P(B).$$

This principle can be extended to any number of mutually exclusive events. For example, the probability that either event A, event B, or event C occurs is

$$P(A \textit{ or } B \textit{ or } C) = P(A) + P(B) + P(C),$$

as long as *A*, *B*, and *C* are all mutually exclusive events.

EXAMPLE 5   *Either/Or Dice Probability*

Suppose that you roll a single, six-sided die. What is the probability of getting either a 2 or a 3?

Solution: The outcomes of a 2 or a 3 are mutually exclusive because a single die can yield only one result. The probability of a 2 is $\frac{1}{6}$, and the probability of a 3 also is $\frac{1}{6}$. Therefore the combined probability is

$$P(2 \text{ or } 3) = P(2) + P(3) = \frac{1}{6} + \frac{1}{6} = \frac{2}{6} = \frac{1}{3}.$$

The probability of rolling either a 2 or a 3 is $\frac{1}{3}$. ∎

EXAMPLE 6   *Either/Or Card Probability*

What is the probability of drawing a *face* card (jack, queen, or king) from a standard deck of cards?

Solution: A standard deck of 52 cards contains 4 jacks, 4 queens, and 4 kings. Thus the probability of drawing a jack is $\frac{4}{52}$, or $\frac{1}{13}$. The probability of drawing a queen is also $\frac{1}{13}$, as is the probability of drawing a king. These three events—drawing a jack, queen, or king— are mutually exclusive because each event eliminates the others. For example, drawing a king means that drawing a queen or a jack did not happen. Therefore the combined probability is

$$P(\text{jack } \textit{or } \text{queen } \textit{or } \text{king}) = P(\text{jack}) + P(\text{queen}) + P(\text{king})$$

$$= \frac{1}{13} + \frac{1}{13} + \frac{1}{13} = \frac{3}{13} = 0.23.$$

There is about a 23% chance of drawing a face card. ∎

Not Mutually Exclusive Events

Suppose that you want to know the probability that a card drawn from a deck will be either a queen or a club. The probability of drawing a queen is $\frac{4}{52}$ because there are 4 queens in the deck of 52 cards, and the probability of drawing a club is $\frac{13}{52}$ because there are 13 clubs in the deck of 52 cards. The sum of these individual probabilities is

$$\frac{4}{52} + \frac{13}{52} = \frac{17}{52}.$$

However, this is *not* the probability of drawing a queen or a club! We can see why by drawing a Venn diagram with one circle showing cards that are queens and another showing cards that are clubs (see Figure 8.7 on the following page).

queen *and* club

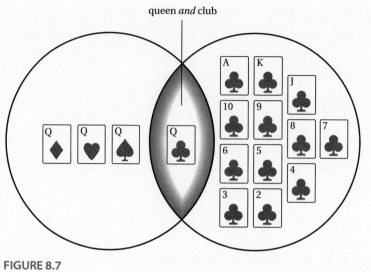

FIGURE 8.7
One card is both a queen and a club. Drawing a queen and drawing a club are not mutually exclusive events.

The nonoverlapping region of the queen circle contains the 3 cards in the deck that are queens but not clubs. The nonoverlapping region of the club circle contains the 12 cards in the deck that are clubs but not queens. The overlap region contains the one card that is both a queen and a club: the queen of clubs. The Venn diagram shows clearly that there are 16 cards that are either a queen or club. Because there are 52 cards in a deck, we conclude that

$$P(\text{drawing a queen } \textbf{\textit{or}} \text{ a club}) = \frac{16}{52} = \frac{4}{13}.$$

Why don't we get the correct probability by adding the individual probabilities? The problem is that, as the Venn diagram shows, a single card *can* be a queen and a club at the same time. This card, the queen of clubs, gets counted twice when we add the individual probabilities: once as a queen and once as a club. The probability of drawing this single card from a deck of 52 cards is  $\frac{1}{52}$. Thus we can correct the error we make in adding the individual probabilities by subtracting this double-counted probability:

$$P(\text{queen } \textbf{\textit{or}} \text{ club}) = P(\text{queen}) + P(\text{club}) - P(\text{queen } \textbf{\textit{and}} \text{ club})$$

$$= \frac{4}{52} + \frac{13}{52} - \frac{1}{52}$$

$$= \frac{16}{52} = \frac{4}{13}$$

We say that drawing a queen and drawing a club are **not mutually exclusive events** because both *can* occur at the same time. Generalizing the procedure we used to find the probability of drawing either a queen or a club, we find the following rule.

Either/Or Probability for Not Mutually Exclusive Events

Assume events A and B are not mutually exclusive. The probability that either A or B occurs is

$$P(A \textbf{ or } B) = P(A) + P(B) - P(A \textbf{ and } B).$$

In this formula, the term $P(A) + P(B)$ is the probability of event A or event B occurring *if* they were mutually exclusive. The last term, $P(A \textbf{ and } B)$, corrects for the case in which events A and B both occur together.

EXAMPLE 7   *Democrats and Women*

Suppose that eight people are in a room: two Democratic men, two Republican men, two Democratic women, and two Republican women. If you select one person at random, what is the probability that the person will be *either* a woman *or* a Democrat?

Solution: Four of the eight people are Democrats, so the probability of selecting a Democrat is $\frac{4}{8}$, or $\frac{1}{2}$. Similarly, four of the eight people are women, so the probability of selecting a woman is also $\frac{1}{2}$. Two of the eight people are both Democrats and women: the two Democratic women. Therefore the probability of selecting someone who is both a Democrat and a woman is $\frac{2}{8}$, or $\frac{1}{4}$. Overall, the probability of selecting *either* a woman *or* a Democrat is

$$P(\text{woman}) \text{ or Democrat} = P(\text{woman}) + P(\text{Democrat}) - P(\text{woman and Democrat})$$

$$= \frac{1}{2} + \frac{1}{2} - \frac{1}{4}$$

$$= \frac{3}{4}.$$

Note that, had we simply added the individual probabilities of a Democrat or woman, we would have counted the two Democratic women twice: once each as a Democrat and once each as a woman. The subtraction corrects for this overcounting. ■

Table 8.2   Summary of Combining Probabilities

Joint Probability: Independent Events	Joint Probability: Dependent Events	Either/Or Probability: Mutually Exclusive Events	Either/or Probability: Not Mutually Exclusive Events
$P(A \textbf{ and } B) = P(A) \times P(B)$	$P(A \textbf{ and } B) = P(A) \times P(B \text{ given } A)$	$P(A \textbf{ or } B) = P(A) + P(B)$	$P(A \textbf{ or } B) = P(A) + P(B) - P(A \textbf{ and } B)$

THE *AT LEAST ONCE* RULE

Suppose that you toss four coins. What is the probability that *at least one* lands heads? There are four different ways to get at least one head: you could get exactly one head, two heads, three heads, or four heads. Thus one way to solve this problem is to add the individual probabilities:

$$P(\text{at least one H in 4 tosses}) = P(1\text{ H}) + P(2\text{ H}) + P(3\text{ H}) + P(4\text{ H})$$

However, there is a simpler approach. Remember that the sum of *all* the individual probabilities, including the probability of *zero* heads, is 1. That is,

$$P(\text{at least 1 head in 4 tosses}) = 1 - P(0\text{ heads in 4 tosses}).$$

This fact makes it much easier to solve the problem. Because the probability of not getting heads in one coin toss is $\frac{1}{2}$, the probability of not getting heads in four coin tosses (a joint probability of four independent events) is:

$$P(0\text{ heads in 4 tosses}) = P(\text{not head}) \times P(\text{not head}) \times P(\text{not head}) \times P(\text{not head})$$
$$= \frac{1}{2} \times \frac{1}{2} \times \frac{1}{2} \times \frac{1}{2}$$
$$= \left(\frac{1}{2}\right)^4 = \frac{1}{16}$$

Therefore the probability of at least one head in four coin tosses is

$$P(\text{at least 1 head}) = 1 - P(0\text{ heads}) = 1 - \frac{1}{16} = \frac{15}{16}.$$

We can generalize this example to any question of the form, "What is the probability of *at least one* particular outcome in a series of independent events?"

The *At Least Once* Rule:

Suppose that the probability of an event A occurring in one trial is $P(A)$. Assuming that all trials are independent, the probability that event A occurs at least once in n trials is

$$P(A \text{ at least once in } n \text{ trials}) = 1 - P(A \text{ does not occur in } n \text{ trials})$$
$$= 1 - [P(\text{not } A)]^n.$$

EXAMPLE 8   *Twenty-first Century Flood*

Consider a region that is prone to 100-year floods (floods that have a 1% chance of occurring in any particular year). What are the chances that the region will experience at least one 100-year flood during the twenty-first century?

Solution: We begin by calculating the odds that a 100-year flood does *not* occur during the twenty-first century. Because a 100-year flood has a 0.01 chance of occurring in any particular year, there is a 0.99 chance that a 100-year flood will *not* occur in a particular year. The probability of no flood in 100 consecutive years is

$$P(\text{no flood in 100 years}) = [P(\text{no flood in one year})]^{100} = (0.99)^{100} = 0.37.$$

The probability that at least one flood *will* occur during the 100 years is

$$P(\text{at least one flood}) = 1 - P(\text{no floods}) = 1 - 0.37 = 0.63.$$

There is a 63% chance that a 100-year flood will occur at least once in the 100 years of the twenty-first century. ∎

> **TIME-OUT TO THINK:**   Does the result in Example 8 surprise you? Suppose that the region did not flood any time during the twentieth century. Does that make a flood during the twenty-first century more likely? Why or why not? Does it suggest that the data on which the probability of flooding was based were flawed or misinterpreted? Explain.

EXAMPLE 9   *Lottery Chance*

Suppose that you purchase 10 lottery tickets for $1 each, and suppose that the probability of winning ($2 or more) on a single ticket is 1 in 10. What is the probability that you will have at least one winning ticket ($2 or more) among the 10 tickets?

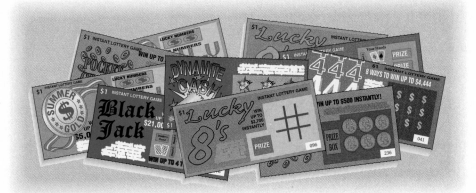

Solution: We begin by finding the probability that *none* of the 10 tickets is a winner. The chance of winning on a single ticket is $\frac{1}{10}$, so the chance of losing is $\frac{9}{10}$, or 0.9. Therefore the chance of losing on all 10 tickets is

$$[P(\text{not winning on one ticket})]^{10} = (0.9)^{10} = 0.35.$$

The probability of getting at least one winner is

$$P(\text{at least one winner}) = 1 - P(\text{no winners}) = 1 - 0.35 = 0.65.$$

The probability of *at least* one winning ticket when you buy 10 lottery tickets is about 65%. In other words, if you spend $10 on lottery tickets, you stand a better than 1 in 3 chance (35%) that you will not have a winning ticket and lose all $10. ∎

Lottery: a tax on people who are bad at math.

—*MESSAGE CIRCULATED ON THE INTERNET*

TIME-OUT TO THINK: Suppose that a lottery states that the probability of winning "$2 or more" is 1 in 10. How do you think the chance of winning $2 compares to the chance of winning more than $2? Why?

Return to the Chevalier

We are now ready to return to the story of the Chevalier de Méré, and see why he won in his first game and lost in the second. Recall that, in the first game, the Chevalier bet that he could roll at least one 6 in four rolls of a die. He guessed that his probability of winning would be $\frac{4}{6}$, but we can find the correct probability with the *at least once* rule.

We know that the probability of rolling a 6 on a single roll is $\frac{1}{6}$, so the probability of *not* rolling a 6 is $\frac{5}{6}$. Therefore the probability of *not* rolling a 6 on all four rolls is

$$\left(\frac{5}{6}\right)^4 = 0.48.$$

The probability of rolling a 6 at least once in four rolls is

$$P(\text{at least one 6 in four rolls}) = 1 - P(\text{no 6s in 4 rolls}) = 1 - 0.48 = 0.52.$$

The Chevalier had a 52% chance of winning in his first game, which is better than even odds. Thus, even though he had guessed the odds incorrectly, he was still likely to come out ahead if he played the game many times.

In the Chevalier's second game, he sought to roll a double 6 within 24 tries. The probability of rolling a double 6 on a single roll is $\frac{1}{36}$, so the probability of *not* rolling a double 6 is $\frac{35}{36}$. Therefore the probability of not rolling a double 6 in all 24 rolls is

$$\left(\frac{35}{36}\right)^{24} = 0.51.$$

The probability of at least one double six in 24 rolls is

$$P(\text{at least one double 6 in 24 tries}) = 1 - P(\text{no double 6s in 24 rolls}) = 1 - 0.51 = 0.49.$$

The Chevalier's probability of winning on this latter bet was about 49%. Although this is only slightly less than 50%, it caused him to lose more often than he won.

TIME-OUT TO THINK: Comment on how the difference between 52% and 49% can be the difference between rags and riches in gambling.

REVIEW QUESTIONS

1. How did the gambling habits of the Chevalier de Méré help launch the mathematical study of probability?

2. Describe the conditions under which we might be interested in a *joint probability*. Give an example.

3. Explain how you can tell whether joint events are *independent* or *dependent*. Give an example of each case.

4. Explain the formula for the joint probability of *independent* events, and give an example of its use.

5. Explain the formula for the joint probability of *dependent* events, and give an example of its use.

6. Describe the conditions under which we might be interested in an *either/or probability*. Give an example in which the events are *mutually exclusive* and an example in which the events are *not* mutually exclusive. How can we use Venn diagrams to illustrate the difference between mutually exclusive and not mutually exclusive events?

7. Explain the formula for the either/or probability of *mutually exclusive* events, and give an example of its use.

8. Explain the formula for the either/or probability of *not mutually exclusive* events, and give an example of its use.

9. What is the *at least once rule*? Give an example of its use.

10. Explain how using the *at least once rule* allows us to calculate the correct chance of winning in the two games of the Chevalier de Méré.

PROBLEMS

1. **Card Probabilities with Replacement.** Suppose that you draw cards from a deck, replacing each card after it is drawn and shuffling the deck.

   a. What is the probability of drawing an ace on the first try?

   b. What is the probability that both of your first two draws will be aces?

   c. What is the probability that your first four draws will be all aces?

   d. What is the probability that your first five draws will be all aces?

2. **Card Probabilities with Replacement.** Suppose that you draw cards from a deck, replacing each card after it is drawn and shuffling the deck.

   a. What is the probability of drawing a king on the first try?

   b. What is the probability that both of your first two draws will be kings?

   c. What is the probability that your first six draws will be all kings?

   d. What is the probability that your first ten draws will be all kings?

3. **Joint Probabilities.** Use the rules for joint probabilities to answer each question. Be sure to decide whether the events are dependent or independent.

   a. If you throw two dice, what is the probability of getting a 6 on both?

   b. If you toss a (fair) coin six times, what is the probability of getting all heads?

   c. Assume that the odds of getting a winning lottery ticket are 1 in 10. What is the probability of purchasing three winning tickets in a row?

   d. Suppose that you draw cards from a deck, but after each draw you replace the card and shuffle the deck. What is the probability of drawing an ace on the first try and a king on the second?

4. **Joint Probabilities.** Use the rules for joint probabilities to answer each question. Be sure to decide whether the events are dependent or independent.

   a. If you throw two dice, what is the probability of getting a 5 on both?

   b. If you toss a (fair) coin ten times, what is the probability of getting all tails?

   c. Assume that the odds of getting a winning lottery ticket are 1 in 8. What is the probability of purchasing four winning tickets in a row?

   d. Suppose that a weather forecast gives the chance of rain as 20% for each of the next four days. What is the probability that it will rain on all four days?

5. **Drawing Candy.** Suppose that you draw candy from a bag that initially contains 20 pieces of candy: 12 chocolate and 8 caramel. Suppose that after drawing the first piece of candy, you put it back in the bag. In that case, what is the probability of drawing two pieces of chocolate in a row? Compare this result to the probability of drawing two chocolates in a row in the case where you eat the first-drawn candy.

6. **Drawing Marbles.** Suppose that you draw marbles from a bag that initially contains 20 red marbles and 10 blue marbles. Suppose that after drawing the first marble, you put it back in the bag. In that case, what is the probability of drawing two red marbles in a row? Compare this result to the case where you keep the first-drawn marble.

7. **Card Probabilities Without Replacement.** Suppose that you draw cards from a deck *without replacement*.

   a. What is the probability of drawing an ace on the first try?

   b. What is the probability that both of your first two draws will be aces?

   c. What is the probability that your first four draws will be all aces?

   d. Suppose that you draw five cards. What is the probability that they will be, *in order*, the ace of spades, king of spades, queen of spades, jack of spades, and 10 of spades?

8. **Card Probabilities Without Replacement.** Suppose that you draw cards from a deck, *without replacement*.

   a. What is the probability of drawing a jack on the first try?

   b. What is the probability that both of your first two draws will be jacks?

   c. What is the probability that your first four draws will be all jacks?

   d. Suppose that you draw five cards. What is the probability that they will be, *in order*, the ace of clubs, 2 of clubs, 3 of clubs, 4 of clubs, and 5 of clubs?

9. **Either/Or Probabilities.** Use the rules for either/or probabilities to answer each question. Be sure to decide whether the events are mutually exclusive. Draw a Venn diagram to illustrate the problem.

   a. What is the probability, on a single draw, of drawing either a 2 or a 4 from a deck of 52 cards?

   b. When rolling a single die, what is the probability that you will roll a 1 or a 2?

   c. When rolling a pair of dice, what is the probability that the sum will be either 7 or 8?

   d. A group of people is comprised of 25 Democratic men, 25 Republican men, 25 Democratic women, and 25 Republican women. If you assign each person in the group a number and then randomly select one of these numbers, what is the probability that the person selected will be a woman *or* a Democrat?

   e. When drawing from a deck of cards, what is the probability that you will draw either a king or a heart?

   f. When drawing from a deck of cards, what is the probability that you will draw a king, an ace, or a diamond?

10. **Either/Or Probabilities.** Use the rules for either/or probabilities to answer each question. Be sure to decide whether the events are mutually exclusive. Draw a Venn diagram to illustrate the problem.

    a. What is the probability, on a single draw, of drawing either an ace or a king from a deck of 52 cards?

    b. When rolling a single die, what is the probability that you will roll a 5 or a 6?

    c. When rolling a pair of dice, what is the probability that the sum will be either 2, 3, or 4?

    d. A group of people is comprised of 25 English men, 25 English women, 25 American men, and 25 American women. If you assign each person in the group a number and then randomly select one of these numbers, what is the probability that the person selected will be a woman *or* an American?

e. When drawing from a deck of cards, what is the probability that you will draw either a king or a spade?

f. When drawing from a deck of cards, what is the probability that you will draw a king, a queen, or a heart?

Game Spinner. *For Problems 11 and 12 refer to the figure below of a spinner that you might find with a board game. Assume that when the arrow is spun it is equally likely to stop in any of the six sectors.*

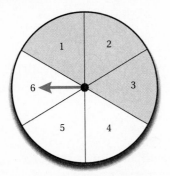

11. What is the probability that the spinner will stop on

a. 6?　　　　　　**b.** white?

c. 1 or 3?　　　　**d.** 1 or blue?

e. 1 or white?

12. What is the probability that the spinner will stop on

a. 1?　　　　　　**b.** blue?

c. 4 or 6?　　　　**d.** 6 or white?

e. 5 or blue?

13. At Least Once Rule.

a. What is the probability that in rolling a single die five times you will roll at least one 6?

b. What is the probability that in tossing a fair coin three times, you will get at least one head?

c. What is the probability that in drawing a card from a standard deck five times (with replacement) you will draw at least one diamond?

d. What is the probability that in drawing a card from a standard deck ten times (with replacement) you will draw at least one ace?

14. At Least Once Rule.

a. What is the probability that in rolling a single die four times you will roll at least one 1?

b. What is the probability that in tossing a fair coin four times, you will get at least one tail?

c. What is the probability that in drawing a card from a standard deck four times (with replacement) you will draw at least one club?

d. What is the probability that in drawing a card from a standard deck twenty times (with replacement) you will draw at least one king?

15. Better Bet for the Chevalier. Suppose that the Chevalier de Méré had bet that he could roll a double 6 within 25 rolls, rather than 24.

a. In that case, what would have been his probability of winning?

b. Had he made this bet, would he still have lost over time? Do you think that he would still have called the mathematician Pascal for help? Explain.

16. Lottery Odds. The probability of a $2 winner in a particular state lottery is 1 in 10, the probability of a $5 winner is 1 in 50, and the probability of a $10 winner is 1 in 500.

a. What is the probability of getting a $2, $5, or $10 winner? Compare to the probability of getting only a $2 winner.

b. If you buy 50 lottery tickets, what is the probability that you will get at least one $5 winner?

c. If you buy 500 lottery tickets, what is the probability that you will get at least one $10 winner?

17. Miami Hurricanes. Studies of the Florida Everglades show that, historically, the Miami region is hit by a hurricane about every 40 years.

a. Based on the historical record, what is the empirical probability that Miami will be hit by a hurricane next year?

b. What is the probability that Miami will be hit by hurricanes in two consecutive years?

c. What is the probability that Miami will be hit by at least one hurricane in the next 10 years?

d. Suppose that global warming causes hurricanes to become twice as common in the future as they have been in the recent past. How would that alter your answers to parts (a), (b), and (c)?

18. **AIDS Among College Students.** Suppose that 2% of the students at a particular college are infected with HIV.

a. If a student has six sexual partners in each of four years of college, what are the odds that at least one of these partners is infected with HIV?

b. If a student has sex, on average, with a new partner each month, what is the probability that at least one partner encountered during a four-year period is infected with HIV?

19. **Poker Hands.** Suppose that you are playing five-card poker with a standard deck of cards and you receive 5 cards from the initial deal.

a. How many different hands are possible?

b. A "royal flush" means that your 5 cards are the ace, king, queen, jack, and 10—all of the same suit. What is the probability of getting a royal flush of spades?

c. What is the probability of getting a royal flush of any suit?

d. (*Challenge Problem*) What is the probability of getting four aces in a five-card hand? What is the probability of getting four of a kind? (Hint: You need to know only two facts–the total number of possible five-card hands, and the number of different hands that have four aces, or four of a kind.)

20. **Finding Mutant Cells.** Biologists often are interested in studying mutant cells. To do so, they may subject a group of cells to radiation that causes many of the cells to mutate "artificially." If a biologist irradiates 1000 cells and, after exposure, the fraction of mutant cells is 0.005, how many cells must be examined to give a 95% chance of finding at least one mutant cell?

21. **Termite Genetics.** Geneticists sometimes study relatively simple organisms, such as termites, to make their work easier. A particular gene in termites has two variants, designated *A* and *a*. The *A* variant is much more common, occurring in 90% of termites, whereas the *a* variant occurs in only 10% of termites. Each termite carries a pair

of these genes; that is, each one has *AA*, *Aa*, or *aa* genes. The termite receives one gene from its mother and one gene from its father. In a large population of termites, approximately what fraction of termites carry the *AA* genes? the *Aa* genes? the *aa* genes?

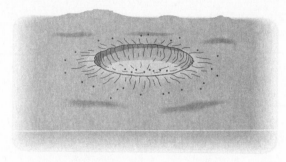

22. **Major Bash.** The Earth is bombarded continually by interplanetary debris, ranging from dustlike particles to large meteorites. When large meteorites hit the earth, noticeable impact craters are created. The probability of a meteorite 100 meters in diameter hitting the Earth is approximately 1 in 10,000 per year.

a. What is the probability that a 100-meter meteorite will *not* hit the Earth in a given year?

b. What is the probability that at least one 100-meter meteorite will hit the Earth during the next 50 years?

c. After how many years does the probability that at least one 100-meter meteorite hits the Earth reach 90%?

d. Assume that such a meteorite could do as much damage as a large nuclear bomb. Is it worth spending large sums of money to try to prevent such impacts? Defend your opinion.

23. **Consecutive 500-Year Floods.** Ordinarily, you might feel relatively safe living in an area that, according to historical and geological records, floods only once every 500 years. Thus, if a flood destroys your home you might well rebuild, imagining that another flood during your lifetime is unlikely. Suppose, however, that your area floods in *two* consecutive years. Should you just chalk it up to bad luck, or should you look for another explanation? Explain your reasoning. Based on your answer, would you rebuild your house a second time? Why or why not?

24. **Probable Weather.** Weather forecasts often state probabilities. Consider a forecast calling for a 20% chance of snow on a given day.

   a. Suppose that the same forecast is made for five consecutive days. What is the probability of snow on at least one of those days?

   b. Write a general formula for the probability of snow on at least one day during a period of n days if each day has a 20% chance of snow.

   c. Based on the formula obtained in part (b), calculate the number of "20% snow" days that must be forecast to be (i) 90% sure of snow on at least one of those days and (ii) 99% sure of snow on at least one of those days. (Hint: You will need to take the logarithm of both sides.)

   d. (*Challenge Problem*) What is the probability of snow on exactly two of the five days?

THE LAW OF AVERAGES

If you toss a fair coin once, the probability of getting either heads or tails is the same; you cannot make a precise prediction about which way the coin will fall. If you toss the coin 100 times, you still cannot make a precise prediction about how many heads will occur. However, you can reasonably expect to get heads *close to* 50% of the time. If you toss the coin 1000 times, you can expect to get heads even closer to 50% of the time. In fact, the more times you toss the coin, the closer to exactly 50% of the outcomes should be heads. This principle is called *the law of averages*.

The Law of Averages

Consider a probability experiment in which the probability of success in a single trial is a fraction P. Suppose that the single trial of this experiment is repeated many times, and that the outcome of each trial is independent of the others. The larger the number of trials, the more likely it is that the overall fraction of successes will be close to the probability P of success in a single trial.

Mathematical Note:
The law of averages is sometimes called *the law of large numbers*.

EXAMPLE 1   *Roulette*

A roulette wheel (Figure 8.8) has 18 black numbers, 18 red numbers, and the numbers 0 and 00 in green.

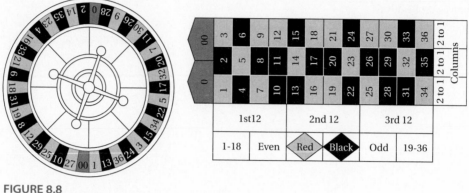

FIGURE 8.8

a)   What is the *a priori* probability of a red number on any spin?

b)   If you spin the wheel three times, how many times should you expect a red number?

c)   If patrons in a casino spin the wheel 100,000 times, how many times should they expect a red number?

Solution:

a)   The *a priori* probability of a red number on any spin is

$$\frac{\text{number of slots that are red}}{\text{total number of slots on wheel}} = \frac{18}{38} = 0.47.$$

b)   Although the probability of getting red on any spin is 47%, there is no way to predict the actual outcome on any single spin. Similarly, with only three spins, you cannot make any reasonable prediction about how many of the spins will come up red.

c)   The law of averages tells us that as the game is played more and more times, the percentage of times that the wheel comes up red should get closer to the 47% probability. Thus, in 100,000 tries, the wheel should come up red close to 47% of the time, or about 47,000 times. ■

THE GAMBLER'S FALLACY

Suppose that you play a game in which you bet $1 on heads each time a coin is tossed (meaning you win $1 if a head appears and lose $1 if a tail appears). In the first 100 coin tosses, heads comes up 44 times and tails comes up 56 times. Thus you win $44 and lose $56, for a net loss of $12. If you continue to play until the coin has been tossed 1000 times, the law of averages says that the percentage of heads is likely to get closer to 50%. Does this mean that are you likely to recover your losses?

In the first 100 tosses, the coin landed heads only 44% of the time. Let's suppose that, after 1000 tosses, the coin has landed heads 48% of the time. This would be consistent with the law of averages, since the percentage grows closer to 50%. In 1000 tosses, 48% heads means 480 heads, and $1000 - 480 = 520$ tails. Thus you would win $480 and lose $520, giving you a net loss of $40.

Note that your net loss *grew* from $12 to $40, even though the percentage of heads grew closer to 50%. This example illustrates that the law of averages can be obeyed even while your losses increase. Unfortunately, many people assume the opposite: that if you're on the losing side of a game, you should keep playing to recover your losses. This mistaken belief is often called **the gambler's fallacy**.

EXAMPLE 2   *The Gambler's Fallacy in Roulette*

Consider the game of roulette as in Example 1. Suppose that you bet on red, and lose four of the first five times you play. Should you expect to recover your losses if you keep playing? Why or why not?

Solution: The probability of getting red on any spin is 47%. Thus if you make this bet a large number of times, on average you should expect to win about 47% of the time—which means you should expect to lose about 53% of the time. Thus, on average, the more you play the more you should expect to lose!

Each spin of the roulette wheel is independent of other spins, so the fact that you have lost on four of your first five tries cannot affect the outcomes of future bets. Because your bet always has a greater probability of losing than of winning, playing more is likely to put you even further into the hole. ■

> **TIME-OUT TO THINK:**   Compulsive gamblers often rationalize their continued play by claiming that their luck is bound to change and that they are "due" for a winning streak. Explain how such claims reflect the gambler's fallacy. Do you think that ignorance of the gambler's fallacy plays any role in explaining why some people become compulsive gamblers? Why or why not?

EXPECTED VALUES

Suppose that an insurance company sells a special type of medical insurance in which it promises to pay you $100,000 in the event that you must quit your job due to serious illness. Based on data from past claims, the empirical probability that a policyholder will make a claim for loss of job is 1 in 500. Should the insurance company expect to earn a profit if they sell the policies for $250 each?

If the company sells only a few policies, the profit or loss is unpredictable. For example, selling 100 policies for $250 each would generate revenue of

$$100 \times \$250 = \$25,000.$$

If none of the 100 policyholders filed a claim, the company would make a tidy profit. On the other hand, if even one policyholder is entitled to a $100,000 claim, the company would face a huge loss.

However, if the company sells a large number of policies, the law of averages tells us that the actual number of claims should be very close to that predicted by probability. For example, if the company sells 1 million policies, it should expect that the number of policyholders making the $100,000 claim will be close to

$$\underbrace{1{,}000{,}000}_{\substack{\text{number} \\ \text{of policies}}} \times \underbrace{\frac{1}{500}}_{\substack{\text{probability of} \\ \text{\$100,000 claim}}} = 2000.$$

Paying these claims will cost

$$2000 \times \$100{,}000 = \$200 \text{ million,}$$

or an *average* of $200 for each of the 1 million policies. Thus if the policies sell for $250 each, the company should expect to earn an average of $250 − $200 = $50 per policy, making a profit of $50 million on sales of 1 million policies.

We can find this same answer with a more formal procedure that will allow us to generalize to other situations. Note that this insurance example involves only two types of *event*: (1) a person buys a policy, and (2) a person files a claim. We say that each of these two events has a particular *value* and *probability* for the company:

- In the event that a person buys a policy, the value to the company is the $250 price of the policy. The probability of this event is 1 because everyone who buys a policy pays the $250.

- In the event that a person files a claim, the value to the company is − $100,000; it is negative because the company loses $100,000 in this case. The probability of this event is $\frac{1}{500}$.

If we multiply the value of each event by its probability, and add the results for each of the events, we find the average or **expected value** of each insurance policy:

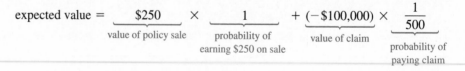

$$\text{expected value} = \underbrace{\$250}_{\substack{\text{value of policy sale}}} \times \underbrace{1}_{\substack{\text{probability of} \\ \text{earning \$250 on sale}}} + \underbrace{(-\$100{,}000)}_{\substack{\text{value of claim}}} \times \underbrace{\frac{1}{500}}_{\substack{\text{probability of} \\ \text{paying claim}}}$$

$$= \$250 + (-\$200) = \$50$$

This expected profit of $50 per policy is the same answer we found earlier. Keep in mind, however, that this expected value is based on applying the law of averages. Thus the company should expect to earn this amount per policy only if it sells a large enough number of policies so that the law of averages comes into play.

> **TIME-OUT TO THINK:** Should the company expect to earn $50 per policy if it sells 500 policies? How about if it sells 500,000 policies? Explain. Comment on why the "expected value" should *not* always be expected.

We can generalize this result to find the expected value in any situation or experiment that involves probability.

Expected Value

Consider a situation or experiment that involves two events, each with its own value and probability. The **expected value** for the situation or experiment is:

$$\text{expected value} = \begin{pmatrix}\text{value of}\\ \text{Event 1}\end{pmatrix} \times \begin{pmatrix}\text{probability of}\\ \text{Event 1}\end{pmatrix} + \begin{pmatrix}\text{value of}\\ \text{Event 2}\end{pmatrix} \times \begin{pmatrix}\text{probability of}\\ \text{Event 2}\end{pmatrix}$$

This formula can be extended to any number of events. For example, if a situation involved three events, the formula becomes:

$$\text{expected value} = \begin{pmatrix}\text{value of}\\ \text{Event 1}\end{pmatrix} \times \begin{pmatrix}\text{probability of}\\ \text{Event 1}\end{pmatrix} + \begin{pmatrix}\text{value of}\\ \text{Event 2}\end{pmatrix} \times \begin{pmatrix}\text{probability of}\\ \text{Event 2}\end{pmatrix}$$

$$+ \begin{pmatrix}\text{value of}\\ \text{Event 3}\end{pmatrix} \times \begin{pmatrix}\text{probability of}\\ \text{Event 3}\end{pmatrix}$$

Note that the expected value should be expected only if the events occur a large number of times so that the law of averages comes into play.

EXAMPLE 3   *A Good Bet?*

Suppose that someone offers you 10 to 1 odds that you will not get three heads when you toss three coins. That is, you lose $1 if you fail to get three heads but gain $10 if you succeed. Is this a good bet for you?

Solution: This game is structured so that there are only two relevant events when you toss the three coins: (1) you get three heads, and (2) you do not get three heads. The value of Event 1, in which you win, is $10. The probability of this event (three heads in three coin tosses) is $\left(\frac{1}{2}\right)^3 = \frac{1}{8}$ (see Unit 8B). The value of event 2, in which you lose, is $-\$1$; it is negative because you lose. The probability of this event is $1 - \frac{1}{8} = \frac{7}{8}$ (because the sum of the probabilities of getting three heads and *not* getting three heads must be 1). Thus the expected value of this game for you is:

$$\text{expected value} = \begin{pmatrix}\text{value of}\\ \text{Event 1}\end{pmatrix} \times \begin{pmatrix}\text{probability of}\\ \text{Event 1}\end{pmatrix} + \begin{pmatrix}\text{value of}\\ \text{Event 2}\end{pmatrix} \times \begin{pmatrix}\text{probability of}\\ \text{Event 2}\end{pmatrix}$$

$$= \$10 \times \frac{1}{8} + (-\$1) \times \frac{7}{8}$$

$$= \$1.25 - \$0.875 = \$0.375$$

Because the expected value of this game is positive, you can expect to have an overall gain if you play it many times—enough times so that the law of averages becomes important.   ∎

TIME-OUT TO THINK: Consider the game in Example 3. If you play this game three times, should you expect to come out ahead? Explain. How about if you play it 50 times? 1000 times?

EXAMPLE 4   *Insurance and Expected Value*

Suppose that an insurance company sells a catastrophic medical insurance policy for a price of $700. Based on data from past claims, the company has calculated the following empirical probabilities.

- An average of 1 in 500 policyholders will file a claim of $100,000.
- An average of 1 in 200 policyholders will file a claim of $30,000.
- An average of 1 in 50 policyholders will file a claim of $10,000.

What is the expected value to the company for each policy sold?

Solution: We begin by making a table of all the "events" associated with this insurance policy. Note that revenues for the company are positive and costs are negative.

event	value	probability	value × probability
sale of policy for $700	$700	1	$700 × 1 = $700
claim of $100,000	−$100,000	$\frac{1}{500}$	−$100,000 × $\frac{1}{500}$ = −$200
claim of $30,000	−$30,000	$\frac{1}{200}$	−$30,000 × $\frac{1}{200}$ = −$150
claim of $10,000	−$10,000	$\frac{1}{50}$	−$10,000 × $\frac{1}{50}$ = −$200

There are four events in this situation, so we find the expected value by adding the four products *value × probability*:

$$\text{Expected earnings} = \underbrace{\$700}_{\substack{\text{value} \times \\ \text{probability for} \\ \text{policy sales}}} + \underbrace{-\$200}_{\substack{\text{value} \times \\ \text{probability for} \\ \$100,000 \\ \text{claims}}} + \underbrace{-\$150}_{\substack{\text{value} \times \\ \text{probability for} \\ \$30,000 \text{ claims}}} + \underbrace{-\$200}_{\substack{\text{value} \times \\ \text{probability for} \\ \$10,000 \text{ claims}}} = \$150$$

On average, the company should expect to earn $150 per policy. However, the company should expect the *actual* earnings to be close to this average only if it sells many thousands of policies. ∎

EXAMPLE 5   *Lottery Expectations*

Suppose that lottery tickets cost $1 each and have the following probabilities of winning: 1 in 5 for a free ticket (worth $1), 1 in 20 for a $2 winner, 1 in 100 for a $5 winner, 1 in 500 for a $10 dollar winner, 1 in 100,000 for a $1000 winner, 1 in 1 million for a $10,000 winner, and 1 in 10 million for a $1 million winner. What is the expected value of a lottery ticket? Discuss the implications.

Solution: Again, the easiest way to proceed is to make a table of all the relevant events with their values and probabilities. Because we are calculating the expected value of a lottery ticket to *you*, the value of a ticket purchase is negative because it costs you money, while the values of all potential winnings are positive.

event	*value*	*probability*	*value × probability*
ticket purchase	−$1	1	$-\$1 \times 1 = -\$1$
win free ticket	$1	$\frac{1}{5}$	$\$1 \times \frac{1}{5} = \$0.20$
win $2	$2	$\frac{1}{20}$	$\$2 \times \frac{1}{20} = \$0.10$
win $5	$5	$\frac{1}{100}$	$\$5 \times \frac{1}{100} = \$0.05$
win $10	$10	$\frac{1}{500}$	$\$10 \times \frac{1}{500} = \$0.02$
win $1000	$1000	$\frac{1}{100,000}$	$\$1000 \times \frac{1}{100,000} = \$0.01$
win $10,000	$10,000	$\frac{1}{1,000,000}$	$\$10,000 \times \frac{1}{1,000,000} = \$0.01$
win $1 million	$1,000,000	$\frac{1}{10,000,000}$	$\$1,000,000 \times \frac{1}{10,000,000} = \$0.10$
sum of last column:			**−$0.51**

The expected value is the sum of all the products *value × probability*; as shown in the table, this sum is −$0.51. In other words, averaged over many tickets, you will lose 51¢ for each lottery ticket that you buy. If you buy, say, 1000 tickets, you should expect to *lose* about

$$1000 \times \$0.51 = \$510.$$

■

By the Way ············
The first legalized lottery to be held in the United States in the twentieth century was initiated by the state of New Hampshire in 1963. By 1990, 42 states and the District of Columbia sponsored lotteries with a gross revenue of over $20 billion.

TIME-OUT TO THINK: Many states use lottery revenues to keep taxes lower than they would be otherwise. Do you think that this is a good economic policy? Is it a good social policy? Note that research shows that lottery players tend to have lower than average incomes. Defend your opinions.

DILBERT® by Scott Adams

DILBERT reprinted by permission of United Features Syndicate, Inc.

THE HOUSE EDGE

A casino makes money because games are set up so that the expected earnings of patrons are negative (that is, losses). Because the casino earns whatever patrons lose, the casino's earnings are positive. The amount that the casino, or *house*, can expect to earn per dollar bet is called the **house edge**. That is, the house edge is the expected value *to the casino* of a particular bet.

The house edge varies from game to game, and on betting strategies as well. It tends to be greatest in games where big winnings are possible, such as slot machines. It tends to be least in games where strategy can improve your odds, such as blackjack.

EXAMPLE 6   *The House Edge in Roulette*

The game of roulette is usually set up so that betting on red is a 1 to 1 bet; that is, you win the same amount of money as you bet if red comes up. Betting on a single number is a 35 to 1 bet; that is, you win 35 times as much as you bet if your number comes up. What is the house edge in each of these two cases? If patrons wager $1 million on such bets, how much should the casino expect to earn?

Solution: We can find the house edge by looking at the *casino's* expected earnings when a patron bets on red. The probability of red is $\frac{18}{38}$ (see Example 1), so the probability of *not red* is $\frac{20}{38}$. Let's call the *value* of red for the casino -1, indicating that the casino pays the amount of the patron's bet when red comes up. The value of *not red* for the casino is 1, because it wins the patron's bet if red does not come up. The house edge when patrons bet on red is

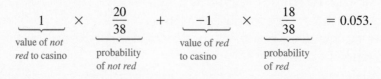

$$\underbrace{1}_{\substack{\text{value of } not \\ red \text{ to casino}}} \times \underbrace{\frac{20}{38}}_{\substack{\text{probability} \\ \text{of } not \text{ red}}} + \underbrace{-1}_{\substack{\text{value of } red \\ \text{to casino}}} \times \underbrace{\frac{18}{38}}_{\substack{\text{probability} \\ \text{of } red}} = 0.053.$$

or 5.3¢ per dollar gambled. In other words, if patrons bet $1 million on red, the casino can expect to earn about

$$\$1 \text{ million} \times 5.3\% = \$53,000.$$

The probability of winning a bet on a single number is $\frac{1}{38}$ because there are 38 numbers on the roulette wheel. The 35 to 1 payoff offered by the casino on this bet means the casino pays the patron 35 times the amount of the bet; thus its value to the casino is -35. The probability that the bet does not win is $\frac{37}{38}$; the casino gains the patron's bet in this case, so its value is 1. Thus the house edge when patrons bet on a single number is

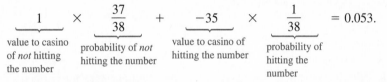

$$\underbrace{1}_{\substack{\text{value to casino} \\ \text{of } \textit{not} \text{ hitting} \\ \text{the number}}} \times \underbrace{\frac{37}{38}}_{\substack{\text{probability of } \textit{not} \\ \text{hitting the number}}} + \underbrace{-35}_{\substack{\text{value to casino of} \\ \text{hitting the number}}} \times \underbrace{\frac{1}{38}}_{\substack{\text{probability of} \\ \text{hitting the} \\ \text{number}}} = 0.053.$$

This is the same house edge as for the red bet. Again, if patrons bet $1 million on single numbers, the house can expect to win $53,000. ∎

> **TIME-OUT TO THINK:** Suppose that instead of betting on one number at a time in roulette, you bet on several numbers at once; for example, suppose that you bet on 3, 8, 12, and 24 in the same spin of the wheel. Does that change the house edge? Why or why not?

REVIEW QUESTIONS

1. In your own words, clearly explain the *law of averages*. Be sure to address both what it *does* mean, and what it *does not* mean.

2. What is the *gambler's fallacy*? Explain how the gambler's fallacy represents a misunderstanding of the law of averages.

3. What is an *expected value*? Explain why we should *not* expect the expected value when an experiment occurs once, but should expect it if the experiment is repeated many times.

4. Explain how expected values are applicable to insurance rates.

5. What do we mean by the *house edge*? Explain how a relatively small house edge can virtually guarantee a huge profit for a casino.

PROBLEMS

• •

1. **Roulette.** A roulette wheel has 18 black numbers, 18 red numbers, and the numbers 0 and 00 in green.

   a. What is the *a priori* probability of a black number on any spin?

   b. If you spin the wheel three times, how many times should you expect a black number? Explain.

   c. If patrons in a casino spin the wheel 100,000 times, how many times should they expect a black number? Explain.

2. **Roulette.** A roulette wheel has 18 black numbers, 18 red numbers, and the numbers 0 and 00 in green.

   a. What is the *a priori* probability of getting 0 or 00 on any spin?

   b. If you spin the wheel three times, how many times should you expect to get 0 or 00? Explain.

   c. If patrons in a casino spin the wheel 100,000 times, how many times should they expect a 0 or 00? Explain.

3. **Dice Rolling.**

   a. If you roll a single fair die once, what is the probability of rolling a 6?

   b. If you roll the die 12 times, how many 6s should you expect to see? Explain.

   c. If you roll the die 1000 times, how many 6s should you expect to see? Explain.

4. **Gambler's Fallacy.** You have lost several games in a row at a casino, and are now deep in a hole.

   a. Does the law of averages suggest that you will now start winning if you continue to bet? Why or why not?

   b. What would the gambler's fallacy "predict" in this situation? Explain why this prediction is incorrect.

5. **Gambler's Fallacy with Coin Tossing.** Suppose that you play a game in which you bet $1 on heads each time a coin is tossed; that is, you win $1 if a head appears and lose $1 if a tail appears. In the first 100 coin tosses, heads comes up 45 times and tails comes up 55 times.

   a. What percentage of the time has heads come up in the first 100 tosses? What is your net gain or loss at this point?

   b. Suppose that you continue to play until there have been 300 tosses, and at that point heads has come up 47% of the time. Is this change in the percentage since the first 100 tosses consistent with what you would expect from the law of averages? Explain. What is your net gain or loss at this point?

   c. How many heads would you have to see in the next 100 tosses, so that you would break even after 400 tosses? Is this likely to occur?

   d. Suppose that, after the first 100 tosses, you decide to keep playing because you are "due" for a winning streak. Explain how this belief would illustrate the gambler's fallacy. Explain.

6. **Gambler's Fallacy with Coin Tossing.** Suppose that you play a game in which you bet $1 on tails each time a coin is tossed; that is, you win $1 if a tail appears and lose $1 if a head appears. In the first 100 coin tosses, tails comes up 47 times and heads comes up 53 times.

   a. What percentage of the time has tails come up in the first 100 tosses? What is your net gain or loss at this point?

   b. Suppose that you continue to play until there have been 1000 tosses, and at that point tails has come up 49% of the time. Is this change in the percentage since the first 100 tosses consistent with what you would expect from the law of averages? Explain. What is your net gain or loss at this point?

   c. How many tails would you have to see in the next 100 tosses, so that you would break even after 1100 tosses? Is this likely to occur?

   d. Suppose that, after the first 100 tosses, you decide to keep playing because you are "due" for a winning streak. Explain how this belief would illustrate the gambler's fallacy. Explain.

7. **Behind in Coin Tossing: Can You Catch Up?** Suppose that you have tossed a fair coin 100 times, getting 38 heads and 62 tails, which is 24 more tails than heads.

   a. Explain why, on your next toss, the difference in the number of heads and tails is as likely to grow to 25 as it is to shrink to 23.

b. Extend your explanation from part (a) to explain why, if you toss the coin 1000 more times, the final difference in the number of heads and tails is as likely to be larger than 24 as it is to be smaller than 24.

c. Suppose that you continue to toss the coin after your first 100 tosses. Explain why the following statement is true: If you stop at any random time, you always are more likely to have fewer heads than tails, in total.

d. Suppose that you are betting on heads with each coin toss. After the first 100 tosses, you are well on the losing side (having lost the bet 62 times while winning only 38 times). Explain why, if you continue to bet, you will most likely remain on the losing side. How is this answer related to the gambler's fallacy?

8. **Should You Play?** Suppose that someone offers you 20 to 1 odds that you will not get four heads when you toss four coins. That is, you lose $1 if you fail to get four heads but win $20 if you succeed. What is the expected value of this bet to you? Should you expect to get the expected value if you play once? What if you play 100 times? Explain.

9. **Should You Play?** Suppose that someone offers you 8 to 1 odds that you will not get a double number (for example, two 1s or two 2s) when you roll two dice. That is, you lose $1 if you fail to get a double number but win $8 if you succeed. What is the expected value of this bet to you? Should you expect to get the expected value if you play once? What if you play 100 times? Explain.

10. **Expected Value and Insurance.** An insurance company sells a homeowner's insurance policy with an annual premium of $1000. Based on data from past claims, the company has calculated the following empirical probabilities.
   - An average of 1 in 200 policyholders will file a claim of $50,000.
   - An average of 1 in 100 policyholders will file a claim of $20,000.
   - An average of 1 in 50 policyholders will file a claim of $5,000.

   What is the expected value to the company for each policy sold?

11. **Expected Value and Insurance.** An automobile insurance company sells an insurance policy with an annual premium of $200. Based on data from past claims, the company has calculated the following empirical probabilities.
   - An average of 1 in 50 policyholders will file a claim of $2000.
   - An average of 1 in 20 policyholders will file a claim of $1000.
   - An average of 1 in 10 policyholders will file a claim of $500.

   What is the expected value to the company for each policy sold?

12. **Lottery Expectations.** Suppose that you buy lottery tickets for $2 each with the following probabilities of winning: 1 in 10 for a $2 winner, 1 in 50 for a $5 winner, 1 in 100 for a $10 winner, 1 in 100,000 for a $1000 winner, 1 in 1 million for a $10,000 winner, and 1 in 10 million for a $1 million winner. What is the expected value of a lottery ticket?

13. **Lottery Expectations.** Suppose that you buy lottery tickets for $1 each with the following probabilities of winning: 1 in 20 for a $2 winner, 1 in 100 for a $5 winner, 1 in 500 for a $10 winner, 1 in 10,000 for a $1000 winner, 1 in 100,000 for a $10,000 winner, and 1 in 10 million for a $1 million winner. What is the expected value of a lottery ticket?

14. **Dice Rolling Expectations.** You are betting on a dice rolling game. The payoffs for each possible roll are shown in the following table. For example, if you roll a 1 you lose $1 and if you roll a 2 you win $2. Suppose you play this game many times. What are your expected earnings or losses?

Roll	1	2	3	4	5	6
Payoff	−$1	+$2	−$3	+$4	−$5	+$6

15. **Dice Rolling Expectations.** You are betting on a dice rolling game. The payoffs for each possible roll are shown in the following table. For example, if you roll a 1 you win $1 and if you roll a 2 you lose $2. Suppose you play this game many times. What are your expected earnings or losses?

Roll	1	2	3	4	5	6
Payoff	+$1	−$2	+$3	−$4	+$5	−$6

16. **House Edge in Roulette.** The probability of winning when you bet on a single number in roulette is 1 in 38, but a winning bet pays $35 for each $1 bet. Suppose that you bet $1 each on the numbers 3, 7, and 29. What is your probability of winning? What is the expected value of this bet for you? What is the house edge? (Hint: If one of your numbers does come up, you still lose the $2 you bet on the other two numbers.)

17. **House Edge in Roulette.** The probability of winning when you bet on a single number in roulette is 1 in 38, but a winning bet pays $35 for each $1 bet. Suppose that you bet $1 each on the numbers 4, 8, 16, and 32. What is your probability of winning? What is the expected value of this bet for you? What is the house edge? (Hint: If one of your numbers does come up, you still lose the $3 you bet on the other three numbers.)

18. **House Edge in Blackjack.** Suppose that in a large casino the house wins on its blackjack tables with a probability of 51.5%. All bets at blackjack are 1 to 1: if you win, you gain the amount you bet and if you lose, you lose the amount you bet.

   a. What is the expected value to you of a single game? What is the house edge?

   b. If you played 100 games of blackjack in an evening, betting $1 on each hand, how much should you expect to win or lose? Explain.

   c. If you played 50 games of blackjack in an evening, betting $5 on each hand, how much should you expect to win or lose? Explain.

   d. If patrons bet $350,000 on blackjack in one evening, how much should the casino expect to earn? Explain.

19. **House Edge in Blackjack.** Suppose that in a large casino the house wins on its blackjack tables with a probability of 50.7%. All bets at blackjack are 1 to 1: if you win, you gain the amount you bet and if you lose, you lose the amount you bet.

   a. What is the expected value to you of a single game? What is the house edge?

   b. If you played 100 games of blackjack in an evening, betting $1 on each hand, how much should you expect to win or lose? Explain.

   c. If you played 100 games of blackjack in an evening, betting $5 on each hand, how much should you expect to win or lose? Explain.

   d. If patrons bet $1,000,000 on blackjack in one evening, how much should the casino expect to earn? Explain.

20. **Profitable Casino.** Suppose that, averaged over all games and all bets being played, a casino has a house edge of 0.07, or 7¢ per dollar gambled. Further suppose that a total of $100 million is wagered in the casino over the course of a year. What are the casino's profits for the year? Explain.

21. **Profitable Casino.** Suppose that, averaged over all games and all bets being played, a casino has a house edge of 0.085, or 8.5¢ per dollar gambled. Further suppose that a total of $1 billion is wagered in the casino over the course of a year. What are the casino's profits for the year? Explain.

22. **Mail Sweepstakes.** You receive a notice in the mail saying that you are eligible to win $1 million in a sweepstakes, simply by filling out and returning the enclosed card. As it turns out, you are one of 10 million people who return the card, which makes your odds of winning 1 in 10 million.

   a. Assume that you spend 32¢ to mail the card back to the sweepstakes. Calculate your expected earnings (or losses) in sending back the card.

   b. Suppose that you decide to purchase three magazine subscriptions, which also are offered to you as an option when you return the card. Later, you learn that you could have purchased the same subscriptions through a student discount service for $7 less than you paid. What are your expected earnings now?

c. Suppose that, on average, each of the 10 million entrants spends an extra $7 on subscriptions. After paying out the $1 million prize, how much profit will the sweepstakes company have earned?

23. **Where Do Lottery Winnings Go?** *Most* people lose *most* of the money that they spend on the lottery. Nevertheless, lottery officials typically advertise that they return half of the money received on tickets to the public in the form of cash prizes. If they are telling the truth, how can most people lose most of the money they spend on the lottery?

24. **Project: The Morality of Gambling.** In a 1995 speech, Republican Senator and presidential candidate Richard Lugar attacked state-supported lotteries and the trend toward increasing legalization of gambling. Specifically, he said: "The spread of gambling is a measure of the moral erosion taking place in our country. . . . It says that if you play enough, you can hit the jackpot and be freed from the discipline of self-support through a job or the long commitment to ongoing education." Write a short essay explaining what you think Senator Lugar meant and describing whether you agree with his assessment of gambling.

PROBABILITY, RISK, AND COINCIDENCE

UNIT 8E

Most of the examples in this chapter have used coin tosses, rolls of dice, cards, or gambling to show how probabilities are calculated. However, probability has a far deeper role in our lives. In this unit, we examine this role briefly by considering how probability comes into play when we analyze risks and when we experience things that may seem like amazing coincidences.

RISK ANALYSIS AND DECISION MAKING

Imagine that a smooth-talking salesman with a reputation for honesty comes to you with a new product:

> *"I can't reveal the details, but you will love this product! It will improve your life in many ways! There is just one problem—it will eventually kill everyone who uses it. Will you buy one?"*

Not likely! After all, could any product be so great that you would die for it? A few weeks later, the salesman shows up again:

> *"No one was buying, so we've made some improvements. Your chance of being killed by the product is now only 1 in 10. Interested?"*

Despite the improvement, most people still would send the salesman home, and wait for his inevitable return:

> *"Okay, this time the product is really ready. You'll love it. We've made it so safe, that it would take some 18 years for it to kill as many people as live in San Francisco. It can be yours, for a mere $20,000!"*

You may be surprised to realize that most Americans eventually jump at this offer. The product is, after all, the automobile. Roughly 40,000 Americans are killed each year in auto accidents; thus the equivalent of the population of San Francisco (735,000 in 1995) is killed in auto accidents in about 18 years. The majority of adult Americans own an automobile, and a new car today typically costs around $20,000.

Figure 8.9 shows a time-series graph of automobile deaths in recent years.

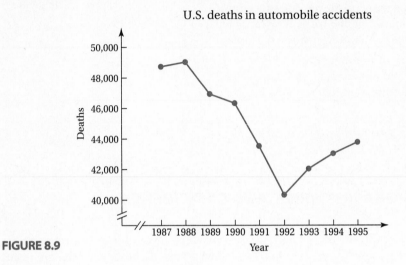

FIGURE 8.9

EXAMPLE 1   *Probability of Death by Auto Accident*

What is the probability that a particular person will die in an auto accident in any single year? What is the probability that the person will die in an auto accident sometime in the next 50 years? Discuss the results.

Solution: We find the empirical probability of death in an auto accident by dividing the 40,000 annual deaths by the 270 million population of the United States (1998).

$$\frac{40,000}{270,000,000} = 1.5 \times 10^{-4} = \frac{1.5}{10,000}$$

That is, about 1.5 out of every 10,000 people, or 15 out of every 100,000 people, dies in an auto accident each year.

To find the probability that a person dies in an auto accident within 50 years, we can use the *at least once* rule. We first find the probability of *not* dying in an auto accident in 50 years. The probability of not dying in a single year is

$$1 - 1.5 \times 10^{-4} = 0.99985.$$

Therefore the probability of not dying in an auto accident in 50 consecutive years is

$$0.99985^{50} = 0.9925,$$

so the probability of dying in an auto accident within 50 years is

$$1 - 0.9925 = 7.5 \times 10^{-3}.$$

That is, about 7.5 out of every 1000 people can be expected to die in auto accidents over a 50-year period. If you attend a college with 10,000 students, about 75 of your schoolmates will be killed in auto accidents within the next 50 years—assuming the probability of death in auto accidents remains the same.   ∎

TIME-OUT TO THINK: Given the risk of death in an automobile and the price of a typical car, how much extra would you be willing to spend to buy a car with better safety features? Explain.

EXAMPLE 2   *Airline vs. Automobile Safety*

Although the number of deaths in airplane accidents varies significantly from year to year, a "typical" year has roughly 500 deaths among roughly 1 billion (10^9) passengers taking trips. (If a flight makes stops, each segment counts as a separate trip, and 100 passengers on a flight counts as 100 separate trips.) The total number of *passenger miles*—the total number of miles flown by all passengers on all trips—is around 1 trillion (10^{12}).

a)   What is the empirical risk of death from a single airplane trip?

b)   What is the empirical risk of death per passenger mile?

c)   Making appropriate estimates, compare the risk of flying to the risk of driving per trip and per mile.

Solution:

a)   The empirical risk of death per trip is the number of deaths divided by the number of trips taken by all passengers:

$$\frac{500 \text{ deaths}}{10^9 \text{ trips}} = 5 \times 10^{-7} \text{ deaths/trip,}$$

or about 1 death in every 2 million trips.

b)   The number of *passenger miles* is the total number of miles flown by passengers, which is about 1 trillion (10^{12}). Thus the empirical probability of death per passenger mile is

$$\frac{500 \text{ deaths}}{10^{12} \text{ passenger miles}} = 5 \times 10^{-10} \text{ deaths/passenger mile.}$$

c)   In Example 1, we found that the probability of being killed in an auto accident is about 0.00015 *per year*. If we assume that the average person takes about three automobile trips per day, or about 1000 trips per year, then the probability of death per trip in an automobile is

$$\frac{0.00015/\text{yr}}{1000 \text{ trips/yr}} = 1.5 \times 10^{-7}/\text{trip.}$$

• Web • Watch •
The U.S. government publishes statistics about transportation safety on the Web.

The average person drives about 15,000 miles per year, so the probability of death per mile driven in a car is

$$\frac{0.00015/\text{yr}}{15,000 \text{ miles/yr}} = 1 \times 10^{-8}/\text{mile}.$$

We can now compare the risks in the airplane and the automobile:

$$\frac{\text{airline risk of death per trip}}{\text{automobile risk of death per trip}} = \frac{5 \times 10^{-7}/\text{trip}}{1.5 \times 10^{-7}/\text{trip}} = 3.3$$

$$\frac{\text{airline risk of death per mile}}{\text{automobile risk of death per mile}} = \frac{5 \times 10^{-10}/\text{mile}}{1 \times 10^{-8}/\text{mile}} = 0.05 = \frac{1}{20}$$

The risk of death *per trip* is about 3 times higher in an airplane than in a car. However, the risk *per mile* is 20 times higher in an automobile. ■

> **TIME-OUT TO THINK:**   Given the risks found in Example 2, would you say that it is safer to fly or to drive? Defend your opinion.

VITAL STATISTICS

• Web • Watch •
Find more vital statistics on this book's web site.

Vital statistics is the name given to all data that concern births and deaths of citizens. Among their many uses, vital statistics are used by insurance companies to assess the risks of clients and to set the rates on insurance policies. Table 8.2 shows a brief table of vital statistics for various causes of death.

Table 8.2	Total Deaths by Cause in the United States for 1995		
Cause	**Total Deaths**	**Cause**	**Total Deaths**
Heart Disease	737,300	Falls	12,600
Cancer	538,500	Poison	10,000
Stroke	157,600	Drowning	4500
Pneumonia/Flu	84,000	Fire	4100
Auto accidents	44,000	Choking on food	2800
Homicide	20,000	Firearms (Accidental)	1400

EXAMPLE 3   *Interpreting the Vital Statistics Table*

a)   Compare the overall probabilities of death by stroke and by drowning. Assume a U.S. population of 260 million for 1995.

b)   What is the death rate due to heart disease in deaths per 100,000 people?

c)   What is the probability of dying by *either* heart disease *or* cancer?

Solution:

a) The empirical probability of death by any particular cause is the number of deaths attributed to that cause divided by the total population of 260 million.

$$P(\text{death by stroke}) = \frac{157{,}600}{260{,}000{,}000} = 0.00061$$

$$P(\text{death by drowning}) = \frac{4500}{260{,}000{,}000} = 0.000017$$

About six people per 10,000 died by stroke, which is about 0.06%. In contrast, only 17 people per *million*, or about 0.0017%, died by drowning. Thus, overall, the probability of death by stroke is

$$\frac{0.061}{0.0017} \approx 36$$

times greater than the probability of death by drowning.

b) Table 8.2 shows 737,300 deaths due to heart disease, so this is the number of deaths *per 260 million people*. We can also write this death rate as

$$\frac{737{,}300 \text{ deaths}}{260{,}000{,}000 \text{ people}} = 0.0028 \frac{\text{deaths}}{\text{person}}$$

Therefore the death rate per 100,000 people is

$$0.0028 \frac{\text{deaths}}{\text{person}} \times 100{,}000 \text{ people} = 280 \text{ deaths per } 100{,}000 \text{ people.}$$

c) If we assume that death by heart disease and cancer are mutually exclusive, the probability of dying by either one is the sum of the individual probabilities. As in part (a), we find that

$$P(\text{death by heart disease}) = \frac{737{,}300}{260{,}000{,}000} = 2.8 \times 10^{-3} = 0.28\%$$

$$P(\text{death by cancer}) = \frac{538{,}500}{260{,}000{,}000} = 2.1 \times 10^{-3} = 0.21\%.$$

The probability of dying by *either* heart disease *or* cancer is

$$0.28\% + 0.21\% = 0.49\%$$

or about 1 in 200. ∎

TIME-OUT TO THINK: Table 8.2 gives only total deaths, but death rates for various causes certainly differ for different age groups. For example, do you think it is true among teenagers that the probability of death by stroke is 40 times greater than the probability of death by drowning? Do you think that the death rate among 30-year-olds by heart disease or cancer is 1 in 200? Explain.

MORTALITY AND LIFE EXPECTANCY

The graphs in Figure 8.10 give two different pictures of death rates by age in the United States. Figure 8.10(a) shows the actual death rate (deaths per 1000) as it changes with age. Note that there is a significant risk of death near birth, after which the death rate drops to very low levels. At about 15 years of age, the death rate begins a continual rise.

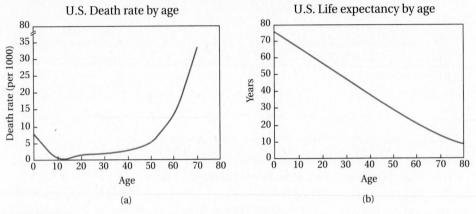

FIGURE 8.10

By the Way ....

In 1997, the highest life expectancies in the world were in Australia (79.6 years) and Canada (79.3 years).

Figure 8.10(b) shows the **life expectancy** of Americans as it changes with age. That is, it shows *how much longer* a person of a given age can expect to live on average. At birth, the life expectancy for Americans is about 75.8 years (72.5 years for males and 78.9 years for females; 1995 data); this means the average age of death is 75.8 years. Careful study of the graph reveals that the age of expected death actually *increases* with age. For example, the graph shows that the life expectancy at age 20 is about 57 years, which means the expected age of death is

$$20 + 57 = 77$$

years. At age 60, the life expectancy is about 21 years, which means the expected age of death is

$$60 + 21 = 81$$

years. In other words, the average American lives to be 75.8, but on average, an American who makes it to 60 will live to be 81.

TIME-OUT TO THINK: The life expectancy of 75.8 years at birth is based on the ages at which Americans die today. Do you think that the average age of death of today's infants really will be 75.8 years? Why or why not?

EXAMPLE 4   *Reading Mortality Graphs*

a)   Approximately what is the death rate for 25- to 35-year-olds? If the population in this age category is 41 million, how many 25- to 35-year-olds can be expected to die in a year?

b) At what average age will today's 80-year-olds die?

c) Suppose that a life insurance company insures 5000 40-year-old people. The cost of the premium is $200 per year and the death benefit is $50,000. How much can the company expect to gain (or lose) in a year?

Solution:

a) Figure 8.10(a) shows that the death rate for people in the 25- to 35-year range is about 2 deaths per 1000 people, or 0.002 deaths per person. Among 41 million people in this age range, the number of deaths in a year will be about

$$41{,}000{,}000 \times 0.002 = 82{,}000.$$

b) Figure 8.10(b) shows that the life expectancy of an 80-year-old person is about eight years. Thus the average 80-year-old will die at age 88.

c) Figure 8.10(a) shows that the death rate for 40-year-olds is about 3 deaths per 1000 or 0.003 deaths per person. Therefore, among the 5000 people insured by the company, about

$$5000 \times 0.003 = 15$$

people can be expected to die in a year. The 5000 people pay

$$5000 \times \$200 = \$1{,}000{,}000$$

in premiums. The total amount paid in death benefits to the families of the 15 people who die is

$$15 \times \$50{,}000 = \$750{,}000.$$

Thus the company's expected profits are

$$\$1{,}000{,}000 - \$750{,}000 = \$250{,}000. \qquad \blacksquare$$

PROBABILITY AND COINCIDENCE

Across the United States, a few dozen people win multi-million dollar prizes in lotteries each year. Thus the probability that *someone* will win a big prize in the next year is virtually certain. Yet the probability that *you* will win the big prize is exceedingly small. It may seem that you are unlucky compared to the mysterious *someone*, but in fact this is just a simple illustration of a basic fact about probability:

> Although a *particular* outcome may be highly unlikely, *some* similar outcome may be extremely likely or even certain to occur.

EXAMPLE 5   *Amazing Card Hand*

Suppose you are dealt the following unimpressive hand in poker: 5 of clubs, 7 of hearts, 3 of diamonds, king of spades, jack of hearts. The player to your left is dealt a royal flush (ace, king, queen, jack, 10 of the same suit). What are the probabilities of these two hands? Should you be amazed by your hand?

or

Solution: Recall that there are 2,598,960 possible hands in five-card poker (see Unit 8A, Example 10). Thus the chances of being dealt *your particular* hand are 1 in 2,598,960, or less than 1 in 2 million! The chances of your partner being dealt a royal flush are also 1 in 2,598,960. The only difference is that the game of poker defines the royal flush to be "special." In fact, you were both certain to get *some* hand of five cards, so there is no reason to be surprised—despite the low odds of your particular hands. ∎

Coincidences are Bound to Happen

People who win the lottery often attribute their winning to "good luck" or to some special "system" they used in choosing a lottery number. Yet, someone was bound to win, and it is merely a coincidence that a particular person held the winning entry.

Coincidences come in many different forms. You may be surprised to find two people in your dinner party with the same birthday, or to find that a friend's mother ran into your mother on a trip to China. You might flip a coin and get heads 10 times in a row, or have a dream that seems to predict an event that later occurs. In any of these cases, it is tempting to chalk it up to something more mysterious than a mere coincidence. But the laws of probability dictate that many coincidences are bound to happen, even though the particular form of a coincidence is unpredictable.

Calvin and Hobbes   by Bill Watterson

EXAMPLE 6   *A Particular Birthday Coincidence*

Suppose there are 25 students in your class. What is the probability that at least one person in the class has the same birthday as yours?

Solution: This is an *at least once* question, so we begin by recognizing that, with 365 days in a year, the probability that any particular student has *your* birthday is $\frac{1}{365}$; the probability that a particular student does *not* have your birthday is $\frac{364}{365}$. Thus the probability that *none* of the 24 other students in the class has your birthday is

Mathematical Note:
We are ignoring leap years when we assume 365 days in a year. Thus the probability in Example 6 does not apply if your birthday happens to be February 29.

$$\left(\frac{364}{365}\right)^{24} = 0.94.$$

The chance that someone in the class *does* have your birthday is

$$1 - 0.94 = 0.06,$$

or about 6%, or about 1 in 16.   ■

EXAMPLE 7   *Some Birthday Coincidence*

What is the probability that at least *some* pair of students in a class of 25 have the *same* birthday? Compare this result to the result from Example 6.

Solution: As usual, calculating this "at least one" probability first requires finding the probability that *no* two people have the same birthday. We begin by considering just two students from the class. The first student has a birthday on 1 of the 365 days in a year. Thus the probability that the second student has a different birthday is

$$\frac{\text{number of possible } \textit{different} \text{ birthdays}}{\text{total number of possible birthdays}} = \frac{364}{365}.$$

Now consider a third student. The probability that all three students have different birthdays is a *dependent* probability: the probability that the third student has a different birthday *given that* the first two students have different birthdays.

$$P(3 \text{ different birthdays}) = P\left(\begin{array}{c}\text{third student has}\\ \text{different birthday}\end{array}\right) \times P\left(\begin{array}{c}\text{first 2 students have}\\ \text{different birthday}\end{array}\right)$$

If the first two students have different birthdays, 2 of the 365 days in the year are "taken," leaving a probability of $\frac{363}{365}$ that the third student has a different birthday. Thus the probability that all three have different birthdays is

$$P(3 \text{ different birthdays}) = \frac{363}{365} \times \frac{364}{365}.$$

Similarly, if the first three students all have different birthdays, the probability that a fourth student has a different birthday from any of the first three is $\frac{362}{365}$. And so on. The probability that all 25 students have different birthdays is the following:

$$\frac{364}{365} \times \frac{363}{365} \times \frac{362}{365} \times \frac{361}{365} \times \frac{360}{365} \times \frac{359}{365} \times \frac{358}{365} \times \frac{357}{365} \times \frac{356}{365} \times \frac{355}{365} \times \frac{354}{365} \times \frac{353}{365}$$

$$\times \frac{352}{365} \times \frac{351}{365} \times \frac{350}{365} \times \frac{349}{365} \times \frac{348}{365} \times \frac{347}{365} \times \frac{346}{365} \times \frac{345}{365} \times \frac{344}{365} \times \frac{343}{365} \times \frac{342}{365} \times \frac{341}{365}.$$

The easiest way to find this product is to do the numerator and denominator separately; after a lot of button pushing on your calculator, you'll find that it is

$$\frac{1.35 \times 10^{61}}{365^{24}} = 0.43.$$

Mathematical Note:
You should recognize that this product can also be written as

$$\frac{364!/340!}{365^{24}}.$$

However, most calculators cannot handle such large factorials.

There is a 0.43 chance that no two students in the class of 25 have the same birthday. The chance that at least two students *do* have the same birthday is

$$1 - 0.43 = 0.57.$$

In other words, there's a 57% chance that at least two of the 25 students have the same birthday! Note that this is a much higher probability than the 6% chance that someone in the class has *your* birthday (see Example 6).

The table below shows the results of the above calculation for different numbers of people.

Number of people	10	20	30	40	50	60
P(at least two birthdays the same)	0.12	0.41	0.71	0.89	0.97	0.99

Note that the probability of finding two people with the same birthday increases rapidly with number. With a group of 40 people, the probability is nearly 90%, and with a group of 60 people, the probability is near certainty. ■

TIME-OUT TO THINK: Suppose you have a dinner party with 25 guests. Should you be surprised if two people learn that they have the same birthday? Should you be surprised if three people drive the same kind of car? Explain.

EXAMPLE 8   *Hot Hand at the Craps Table*

The popular casino game of craps involves rolling dice. Suppose that you are playing craps and suddenly find yourself with a "hot hand:" You roll winners on 10 consecutive bets. Is your hand really "hot?" Should you increase your bet because you are on a hot streak? Assume that you are making bets with a 48.6% chance of winning on a single play (the best odds available in craps).

Solution: First, let's consider the chance of winning 10 times in a row. With a chance of 48.6%, or 0.486, on a single play, the chance of 10 straight wins is

$$0.486^{10} = 0.000735,$$

or about 1 in 1360. Getting a streak with such a low probability might make you think that you really are "hot." However, look around the casino: If it is a large casino, several hundred people may be playing at the craps tables, and there may be *tens of thousands* of individual rolls each night. Thus it is almost inevitable that someone will have a "hot streak" of 10 straight wins; in fact, even longer hot streaks are likely during any given night. Your apparent "hot hand" is a mere coincidence, and your odds of winning on your next bet are still only 0.486. To increase your bet based on a mere coincidence would be foolish, to say the least! ■

"Luck is a mighty queer thing. All you know about it for certain is that it's bound to change."

—Bret Harte, The Outcasts of Poker Flat

REVIEW QUESTIONS

1. Explain how the risk of death by automobiles is calculated.

2. Briefly describe how we can compare risks of different things, and why such comparisons are useful.

3. What are *vital statistics*? How are they used?

4. Explain what we mean by *life expectancy*. Explain the use of mortality and life expectancy data.

5. In your own words, explain the meaning of the following statement: Although a *particular* outcome may be highly unlikely, *some* similar outcome may be extremely likely or even certain to occur.

6. Explain how it is possible that the odds of finding someone in a class of 25 with *your* birthday can be quite low, while the odds of finding two people with the same birthday can be quite high.

7. Explain when and why low probability *streaks* are to be expected.

PROBLEMS

1. **Automobile Risk.** Assume that automobile deaths rose to 45,000 in 1998, and the U.S. population is 270 million.

   a. What was the probability of dying in an automobile accident in 1998? Express your answers both as a percentage and in terms of deaths per 10,000 people. Compare the risk in the two years.

   b. Using 1998 figures, what is the probability of dying in an automobile accident over a period of 10 years?

   c. Using 1998 figures, what is the probability of dying in an automobile accident over a period of 50 years?

2. **U.S. Airline Safety.** According to the National Transportation Safety Board (NTSB), in 1996 the major U.S. commercial carriers reported 38 accidents, resulting in 380 fatalities (including passengers, crew, and people on the ground). These accidents occurred among a total of 8.5 million departures (i.e., flight segments) and 13.7 million flight hours.

   a. Find the fatality rate in terms of fatalities per 100,000 departures.

   b. Find the fatality rate in terms of fatalities per 100,000 flight hours.

   c. Suppose a person flew on 25 commercial departures during 1996. What was this person's empirical probability of dying in an airline accident?

   d. Suppose a person flew 200 flight hours during 1996. What was this person's empirical probability of dying in an airline accident?

3. **Noncommercial Aviation Safety.** The National Transportation Safety Board (NTSB) reported that in 1996 there were 631 fatalities involving noncommercial and nonmilitary planes (private, corporate, and government). These flights accounted for about 23.7 million flight hours. Find the fatality rate in terms of fatalities per 100,000 flight hours. Compare this fatality rate to that for major commercial carriers (see Problem 2 above).

4. **Overall U.S. Aviation Safety.** For all nonmilitary aviation, the NTSB statistics showed 2040 accidents and 1070 fatalities in 1996. In 1995, the numbers were 2175 accidents and 962 fatalities. Assume a U.S. population of 265 million in 1995 and 1996.

   a. Compare the aviation death rates per 100,000 Americans in 1995 and 1996.

   b. Based on these statistics, can you draw any conclusions about whether aviation safety is improving? Why or why not?

5. **High/Low U.S. Birth Rates.** The highest and lowest birth *rates* in the United States in 1995 were in Utah and Maine, respectively. Utah reported 39,577 births, with a birth rate of 20 births per 1000 people. Maine reported 13,896 births, with a birth rate of 11.2 births per 1000 people. Use this data to answer the following questions.

   a. How many people were born per day in Maine? in Utah?

   b. What was the approximate 1995 population of Maine? of Utah?

6. **High/Low U.S. Death Numbers.** In 1996, there were 232,266 deaths in California, which was the highest number in the United States. The lowest number of deaths in any state was for Alaska, with 2562 deaths. The populations of California and Alaska were approximately 30 million and 550,000, respectively.

   a. Compute and compare the 1996 death *rates* for California and Alaska in deaths per 1000.

   b. Based on the fact that California and Alaska had the highest and lowest death numbers, respectively, does it follow that California and Alaska had the highest and lowest death *rates*? Why or why not?

7. **U.S. Birth and Death Rates.** In 1997, the U.S. population was about 270 million. The overall birth rate was 14.7 births per 1000, and the overall death rate was 8.7 deaths per 1000.

   a. Approximately how many births were there in the United States in 1997?

   b. About how many deaths were there in the United States in 1997?

   c. Based on birth and deaths alone (i.e., not counting immigration), about how much did U.S. population rise during 1997?

   d. According to the U.S. Census Bureau, the U.S. population actually increased by 2.4 million during 1997. Based on this fact and your results from part (c), how many people immigrated to the United States during 1997? What fraction of the overall population growth was due to immigration?

8. **Interpreting the Vital Statistics Table.** Use Table 8.2 to answer the following questions. Assume a U.S. population of 260 million for 1995.

   a. Compute and compare the overall probabilities of death by pneumonia/flu and by cancer.

   b. What was the death rate due to cancer in deaths per 100,000 people?

   c. What was the probability of dying by *either* stroke *or* cancer?

9. **Interpreting the Vital Statistics Table.** Use Table 8.2 to answer the following questions. Assume a U.S. population of 260 million for 1995.

   a. Compute and compare the overall probabilities of death by homicide and by automobile accident.

   b. What was the death rate due to homicide in deaths per 100,000 people?

   c. What was the probability of dying by *either* homicide *or* auto accident?

10. **Mortality Rates.** Use the graphs in Figure 8.10 to answer the following questions.

   a. Estimate the death rate for those 50–55 years of age.

   b. Assuming that there were about 13.6 million people 50–55 years of age, how many in this bracket could be expected to die in a year?

   c. To what age could the average 50-year-old expect to live?

   d. Suppose that a life insurance company insures 1 million 50-year-old people. The cost of the premium is $200 per year and the death benefit is $50,000. What is the expected profit or loss for the insurance company?

11. **Birthday Coincidences.** Suppose that 15 students are in your class.

   a. Calculate the probability that someone in the class has the same birthday as yours.

   b. Calculate the probability that at least two people in the class have the same birthday, though not necessarily *your* birthday.

12. **Birthday Coincidences.** Suppose that 23 students are in your class.

    a. Calculate the probability that someone in the class has the same birthday as yours.

    b. Calculate the probability that at least two people in the class have the same birthday, though not necessarily *your* birthday.

13. **Hot Streaks.** Suppose that 2000 people are all playing a game that has a 48% chance of winning.

    a. What is the probability of winning five games in a row?

    b. What is the probability of winning ten games in a row?

    c. On average how many of the 2000 people could be expected to have a hot streak of five games?

    d. On average how many of the 2000 people could be expected to have a hot streak of ten games?

14. **Coin Streaks.** Toss a coin 100 times and record your results (heads or tails) in order. What was your longest streak of consecutive heads or tails? Calculate the probability of this streak by itself. That is, if your longest streak was 4 heads, what is the probability of four heads in four tosses? Should you be surprised that you got such a streak? Explain.

15. **Random Patterns.** Toss a coin 100 times and write the 100 results (H or T) in order in an array with 10 rows and 10 columns. Do the results look truly random? Do you see any apparent patterns? Explain.

16. **Random Number Generators.** Many calculators have a key that is designed to generate a random number when pressed. Find the random number key on a calculator. Either through experimentation, or by consulting the calculator manual, determine what kind of number is generated (e.g., is it always between 0 and 1?). Press the random number button at least 100 times and record your results. What is the average value of these random numbers? Do any apparent patterns emerge in these numbers? Explain why *some* patterns (but not *particular* patterns) are to be expected, even if the numbers are truly random. Describe a way in which, if you pressed it millions of times, you determine whether the random number button is generating truly random numbers.

17. **Joe DiMaggio's Record.** One of the longest standing records in sports is the 56-game hitting streak (in baseball) of Joe DiMaggio. Assume that a player on a long "hot streak" is batting about .400, which is about the best that anyone ever hits over a period of 50 or more games. (A batting average of .400 means the batter gets a hit 40% of the time. Typically, only a handful of players each year hit that well for a period as long as 56 games.)

    a. Suppose that a player gets to bat about four times per game. What is the probability that a player batting .400 will get at least one hit in the four at-bats?

    b. Use the result in part (a) to calculate the probability of a .400 hitter getting a hit in 56 consecutive games.

    c. Suppose that, instead of batting .400, a player has a more ordinary average of .300. In that case, what is the probability of the player's getting a hit in 56 consecutive games? Again, assume four at-bats per game.

    d. Considering the results in parts (b) and (c), and that baseball has been played for about 100 years, are you surprised that *someone* set a record in which he got a hit in 56 consecutive games? Explain clearly. Do you think that DiMaggio's record ever will be broken? Why or why not?

18. **Lottery Psychology.** Imagine a lottery that will offer a prize of $1 million dollars to just one person. In order to offer the prize, the lottery must collect $2 million in revenue.

    a. Suppose that you are part of a crowd of 10,000 attending a college basketball game, at which a lottery with a $1 million prize is being offered. Assuming that all 10,000 people entered the lottery, how much would each entry have to cost in order to generate $2 million in revenue? Do you think it is likely that people would actually pay this much for the lottery? Explain.

b. Real lotteries do in fact give about half their revenues as prizes, and per capita lottery spending in the United States is considerably more than $200 per year. Given that most people would not enter the lottery in part (a), why do you think so many people spend an equivalent amount on lottery tickets at stores? What lessons does this example teach about lottery psychology? Present your answer in a one-page essay.

19. **Automobiles and Vietnam.** Advances in automobile safety (e.g., seat belts and better bumpers) have lowered mortality in auto accidents significantly. During the 1960s and 1970s, approximately 50,000 people were killed in traffic accidents each year.

   a. The Vietnam War saw Americans being killed in combat over a period of about a decade spanning most of the 1960s and the early 1970s. Approximately 50,000 Americans were killed in the Vietnam War; their names are listed on the Vietnam Memorial in Washington, D.C. Compare the number of Americans killed in the Vietnam War with the number killed at home in auto accidents during the same period of time.

   b. The Vietnam War spawned a vast protest movement intended to get the United States out of Vietnam. No similar protest movement has ever been launched against the automobile. Write one- or two-page essay on why the public judges some risks to be "acceptable," even when they may be far more dangerous than other risks deemed "unacceptable." Support your position with further examples of relatively low-risk activities that have generated organized protest in contrast to high-risk activities that go unchallenged.

20. **Project: Let's Make a Deal.** In September 1990, controversy about a probability question erupted over an item in the "Ask Marilyn" column in *Parade* magazine (carried by many newspapers across the country). Her column answered a question loosely based on a TV game show called "Let's Make a Deal." The question was:

Suppose that you're on a game show and you're given the choice of three doors: Behind one door is a car; behind the others, goats. You pick a door, say No. 1, and the host, who knows what's behind the doors, opens another door, say No. 3, which has a goat. He then says to you, "Do you want to change your pick to door No. 2?" Is it to your advantage to switch your choice?

Marilyn answered that the probability of winning was higher if the contestant switched. This answer generated a huge number of letters, including a few from mathematicians, claiming that she was wrong.

   a. Marilyn answered with the following logic. When you first pick door No. 1, the chance that you picked the one with the car is 1/3. The probability that you chose a door with a goat is 2/3. When the host opens door No. 3 to reveal a goat, it does not change the 1/3 probability that you picked the right door in the first place. Thus, as only one other door remains, the probability that it contains the car is 2/3. Briefly discuss this logic. Do you agree with it?

   b. Another way to evaluate the problem is by analyzing all the possibilities. The prizes could be arranged behind the three doors in three possible ways:

	Door No. 1	Door No. 2	Door No. 3
Case 1:	Car	Goat	Goat
Case 2:	Goat	Car	Goat
Case 3:	Goat	Goat	Car

Assume that you choose door No. 1. Clearly, your initial probability of winning the car is 1/3. Now, suppose that the host opens *one of the two* remaining doors to reveal a goat: In case 1, it does not matter which of the two doors he opens because both have goats behind them; in cases 2 or 3, he must use his knowledge of where the car is located in deciding which of the two doors to open. Analyze your probability of winning if you hold to door No. 1 versus switching to the one the host did not open. Can you show that you have a 2/3 chance of winning by switching?

   c. If you still are not convinced that Marilyn was right, here is a third way to analyze the problem. Suppose that there are 100 doors instead of 3, but that only one door has a car behind it; the other 99 doors have goats behind them. After you pick a door, the host opens 98 of the remaining 99 doors to reveal goats. Explain why, at this point, you should switch to the one remaining door. Explain why this similar problem shows that switching in the original problem also is best.

d. This problem differs from the real "Let's Make a Deal" show in one crucial respect: We evaluated only a *single* instance of the game. In the real game, the host did not always open a second door and give the contestant the opportunity to switch. Martin Gardner, a popular writer on mathematical subjects, wrote: "If the host is malevolent, he may open another door only when it's to his advantage to let the player switch, and the probability of being right by switching could be as low as zero." Explain Gardner's statement.

e. Find relevant back issues of *Parade* magazine and read the columns about this controversy. Write a one- or two- page essay summarizing how the controversy unfolded.

21. **Project: Tabloid Predictions.** At the beginning of the year (or end of the previous year) many tabloids and magazines publish a set of predictions for the coming year made by psychics, astrologers, and other "seers." Find a set of these predictions from last year.

  a. Describe each prediction, and state whether it came true.

  b. Does the overall rate at which the predictions came true seem consistent with probability? Explain.

22. **Project: Birthday Intuition.** Ask a group of your friends the following two questions, and keep track of their answers.

  Question 1: How many people must be in a group to be sure that at least two people have the same birthday?

  Question 2: How many people must be in a group to have a 50-50 chance that two people have the same birthday?

  a. What is the correct answer to Question 1? Explain.

  b. What fraction of your friends answered Question 1 correctly? Are you surprised? Why or why not?

  c. Show that the correct answer to Question 2 is 23.

  d. Did any of your friends answer Question 2 correctly? How do the answers given by most of your friends compare to the correct answer?

  e. Based on your results, can you make any general statements about "natural" intuition for probability?

CHAPTER 8
SUMMARY

$\mathcal{I}$n this chapter we explored the mathematics of probability and how it influences our lives. Key lessons to remember from this chapter include the following.

- Because our intuition often fails us in guessing probabilities, we should calculate, or carefully estimate, probabilities whenever possible. With practice, we can improve our intuition for probabilities, and learn to avoid pitfalls such as the gambler's fallacy.

- There are three different approaches to determining probabilities. *A priori* methods rely on counting successes and outcomes. Empirical methods use experiments or observations. Subjective methods are not mathematical and are prone to errors.

- Probability does not allow predictions in a particular instance, such as flipping a coin. However, as described by the law of averages, probability can give precise predictions when large numbers of trials or events are involved.

- Probability plays a role in many practical situations, from gambling at a casino to evaluating everyday risks. Learning to calculate probabilities properly therefore can have many practical benefits.

SUGGESTED READING

Can You Win?, M. Orkin (New York: W. H. Freeman, 1991).

Probabilities in Everyday Life, J. McGervey (New York: Ivy Books, 1989).

What the Odds Are, L. Krantz (New York: Harper Perennial, 1992).

Lady Luck: The Theory of Probability, W. Weaver (Garden City, NJ: Anchor Books, 1963).

The Broken Dice and Other Mathematical Tales of Chance, I. Ekeland (Chicago: University of Chicago Press, 1993).

The Demon-Haunted World, C. Sagan (New York: Ballantine Books, 1997).

Innumeracy, J. A. Paulos (New York: Hill and Wang, 1988).

Chapter 9
PUTTING STATISTICS TO WORK

*I*n Chapter 2 we explored statistics as they arise in news reports, in surveys and opinion polls, in graphics, and in attempts to establish cause and effect. Having now developed a variety of mathematical skills, we are ready to explore the quantitative use of statistics. With the tools in this chapter, you will have a much better understanding of the remarkable power of statistics.

I have a great subject [statistics] to write upon, but feel keenly my literary incapacity to make it easily intelligible without sacrificing accuracy and thoroughness.
SIR FRANCIS GALTON, ENGLISH SCIENTIST (1822–1911)

He uses statistics as a drunken man uses lampposts—for support rather than illumination.
ANDREW LANG, SCOTTISH AUTHOR (1844–1912)

*P*rogressing through this book, we've seen many examples in which conclusions are based on some type of statistical data. We've also seen how data can be uncertain, and how different people can interpret data in different ways. Now we are ready to go "behind the scenes" to explore the actual processing and analysis of data.

In general terms, there are two steps involved in data analysis. The first is often called **exploratory data analysis**. It consists of "getting a feel for the data" and summarizing it in compact ways. The second step is called **inferential statistics** because it allows us to actually draw conclusions from the data and state how confident we are about those conclusions.

UNIT 9A

CHARACTERIZING DATA

The first step in exploratory data analysis is characterizing a data set. We begin by investigating how to characterize a data set that involves a single variable. For example, if you survey 25 of your classmates and record their eye color, you have a data set concerning the variable *eye color*. A good way to begin the analysis of such a data set is by recording the number of times that each value appears; that is, the number of students with blue, brown, green, or hazel eyes. We can then display this data with a table or a graph (Figure 9.1).

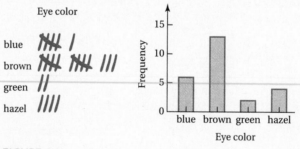

FIGURE 9.1

FREQUENCY TABLES

Imagine that a theater manager takes an exit survey of people who have just seen a new film. Each person is asked to rate the film on a five-point scale:

(poor) 1 2 3 4 5 (excellent)

Suppose that 30 people responded with the following ratings:

1 3 3 2 3 4 3 4 2 3 5 4 3 4 3 5 3 5 4 2 5 2 4 2 1 1 2 5 3 2

This string of numbers is called the **raw data** from the survey. The easiest way to organize the data is to record the **frequency**, which is the number of times that each category appears; the result is called a **frequency table** (Table 9.1).

Table 9.1	Film ratings		
Rating	Frequency	Relative Frequency	Cumulative Frequency
1	3	$\frac{3}{30} = 10.0\%$	3
2	7	$\frac{7}{30} = 23.3\%$	$7 + 3 = 10$
3	9	$\frac{9}{30} = 30.0\%$	$7 + 3 + 9 = 19$
4	6	$\frac{6}{30} = 20.0\%$	$7 + 3 + 9 + 6 = 25$
5	5	$\frac{5}{30} = 16.7\%$	$7 + 3 + 9 + 6 + 5 = 30$
Total	30	$1 = 100\%$	30

Two other useful ways of describing data also are tabulated in Table 9.1.

- The **relative frequency** is the fraction or percentage of times that each response appears. For example, 3 of the 30 people gave the film a rating of 1, so the relative frequency for this response is $\frac{3}{30}$, or 10.0%.

- The **cumulative frequency** is the number of responses in a particular category *and all preceding* categories. For example, the cumulative frequency in row 3 is 19 because 19 people gave the film a rating of 1, 2, or 3.

When the data variable has distinct categories (in this case, five different ratings), we can display the data in a **dotplot** (Figure 9.2a) by making a dot for each response in the appropriate category. Alternatively, we can display the frequencies with a *bar graph* (Figure 9.2b). (See Unit 2C for a review of bar graphs.) Note that these graphs show the scale for both frequency (left scale) and relative frequency (right scale).

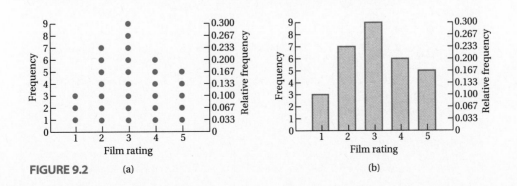

FIGURE 9.2   (a)   (b)

Binning Data

The frequency table in the film rating example was small because only five ratings were used. However, if the data set has many categories or a continuous range of categories, then another method must be used. Consider the data below showing the 1996 per capita income for each of the 50 states.

Alabama	$20,055	Louisiana	$19,824	Ohio	$23,537
Alaska	$24,558	Maine	$20,826	Oklahoma	$19,350
Arizona	$20,989	Maryland	$27,221	Oregon	$22,668
Arkansas	$18,928	Massachusetts	$29,439	Pennsylvania	$24,668
California	$25,144	Michigan	$24,810	Rhode Island	$24,765
Colorado	$25,084	Minnesota	$25,580	South Carolina	$19,755
Connecticut	$33,189	Mississippi	$17,471	South Dakota	$21,516
Delaware	$27,622	Missouri	$22,864	Tennessee	$21,764
Florida	$24,104	Montana	$19,047	Texas	$22,045
Georgia	$22,709	Nebraska	$23,047	Utah	$19,156
Hawaii	$25,159	Nevada	$25,451	Vermont	$22,124
Idaho	$19,539	New Hampshire	$26,520	Virginia	$24,925
Illinois	$26,598	New Jersey	$31,053	Washington	$24,838
Indiana	$22,440	New Mexico	$18,770	West Virginia	$18,444
Iowa	$22,560	New York	$28,782	Wisconsin	$23,269
Kansas	$23,281	North Carolina	$22,010	Wyoming	$21,245
Kentucky	$19,687	North Dakota	$20,710	United States	$24,231

The 50 numbers in this set range from $17,471 to $33,189, and no two numbers are alike. Thus a frequency table for this raw data would not be very interesting, as all 50 entries would have a frequency of 1.

A better way to organize this data for a frequency table is to group, or **bin**, the data. For example, we could create bins that are $1000 wide, such as from $16,500 to $17,499, from $17,500 to $18,499, and so on. Note that, by ending each bin at a "499," we ensure that each data value can fall into only one bin centered on a multiple of $1000. This process is equivalent to rounding the data to the nearest $1000. Only one state has a per capita income that falls into the first bin: Mississippi ($17,471). Table 9.2 shows the frequency table that results when we continue in a similar manner to find the number of states that have incomes in each $1000-wide bin.

Table 9.2   1996 state per capita income

Per Capita Income	Frequency	Relative Frequency	Cumulative Frequency
$17,000 ($16,500–$17,499)	1	$\frac{1}{50} = 2\%$	1
$18,000 ($17,500–$18,499)	1	$\frac{1}{50} = 2\%$	2
$19,000 ($18,500–$19,499)	5	$\frac{5}{50} = 10\%$	7

Per Capita Income	Frequency	Relative Frequency	Cumulative Frequency
$20,000 ($19,500–$20,499)	5	$\frac{5}{50} = 10\%$	12
$21,000 ($20,500–$21,499)	4	$\frac{4}{50} = 8\%$	16
$22,000 ($21,500–$22,499)	6	$\frac{6}{50} = 12\%$	22
$23,000 ($22,500–$23,499)	7	$\frac{7}{50} = 14\%$	29
$24,000 ($23,500–$24,499)	2	$\frac{2}{50} = 4\%$	31
$25,000 ($24,500–$25,499)	10	$\frac{10}{50} = 20\%$	41
$26,000 ($25,500–$26,499)	1	$\frac{1}{50} = 2\%$	42
$27,000 ($26,500–$27,499)	3	$\frac{3}{50} = 6\%$	45
$28,000 ($27,500–$28,499)	1	$\frac{1}{50} = 2\%$	46
$29,000 ($28,500–$29,499)	2	$\frac{2}{50} = 4\%$	48
$30,000 ($29,500–$30,499)	0	0	48
$31,000 ($30,500–$31,499)	1	$\frac{1}{50} = 2\%$	49
$32,000 ($31,500–$32,499)	0	0	49
$33,000 ($32,500–$33,499)	1	$\frac{1}{50} = 2\%$	50
Total	**50**	**1 = 100%**	**50**

(U.S. Department of Commerce, Bureau of Economic Analysis, 1996.)

Having constructed the frequency table for the data, it is now easy to make a *histogram* to display the data (see Unit 2C). Figure 9.3 shows the histogram for the per capita income data with bin sizes of both $1000 (as given in the frequency table above) and $2000.

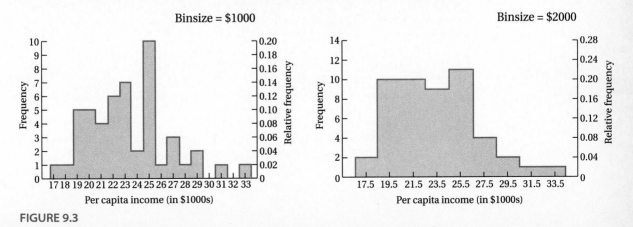

FIGURE 9.3

> **TIME-OUT TO THINK:**   Look through the raw data, and confirm that the values shown in Figure 9.3 with bins that are $2000 wide are correct. What would happen if we made a histogram with bins that are $5000 wide? What are the advantages and disadvantages of placing the data in different-sized bins?

EXAMPLE 1   *Binned Exam Scores*

Consider the following set of 20 scores from a 100-point exam. Determine appropriate bins, and make a frequency table.

76, 80, 78, 76, 94, 75, 98, 77, 84, 88, 81, 72, 91, 72, 74, 86, 79, 88, 72, 75

Solution: The scores range from 72 to 98. Of the many ways to bin the data, one is to group data in 5-point bins. For example, the first bin would represent scores from 95–99, the second bin would represent scores from 90–94, and so on. (Note that there is no overlap between bins.) After we have identified the bins, we simply count the frequency (the number of scores) for each bin. For example, only one score is in bin 95–99 (the high score of 98) and two scores are in bin 90–94 (the scores of 91 and 94). To find the relative frequencies, we divide each frequency by the total number of scores. For example, because there is a total of 20 exam scores, the relative frequency of the 95–99 bin is $1 \div 20 = 0.05$. We find cumulative frequencies by adding all the scores at or above a certain level. For example, the cumulative frequency for bin 90–94 is 3 for the three scores of 90 or above. All the frequencies, relative frequencies, and cumulative frequencies for this binning scheme are shown in Table 9.3.

Beware! To interpret cumulative frequency it is essential to note whether the data are listed in decreasing or increasing order. In Table 9.2, the cumulative frequency gives the number of data points at or *below* a certain per capita income level. In Example 1, the cumulative frequency gives the number of data points at or *above* a certain exam score.

Table 9.3	Exam scores		
Scores	**Frequency**	**Relative Frequency**	**Cumulative Frequency**
95–99	1	0.05 = 5%	1
90–94	2	0.10 = 10%	3
85–89	3	0.15 = 15%	6
80–84	3	0.15 = 15%	9
75–79	7	0.35 = 35%	16
70–74	4	0.20 = 20%	20
Total	**20**	**1.00 = 100%**	**20**

The binned data can be summarized in a histogram (see Figure 9.4 on the next page). Note that the frequency (left scale) and relative frequency (right scale) are both shown.

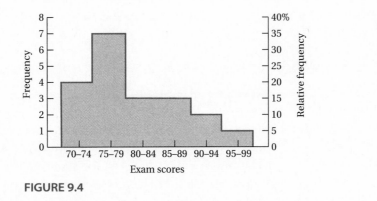

FIGURE 9.4 ∎

MEAN, MEDIAN, AND MODE

Tables and graphs are useful ways of displaying data, but sometimes we may want more compact statistical measures. Let's begin with the familiar term *average*. Statistically speaking, "average" can mean any one of three different things: *mean*, *median*, or *mode*. Each of these terms describes, in different ways, the location of the center of a data set. For example, suppose that 27 students take an exam and, in increasing order, the resulting scores are

$$47, 52, 56, 57, 61, 65, 66, 69, 70, 71, 71, 72, 73, 75,$$
$$77, 77, 77, 78, 81, 82, 85, 87, 87, 91, 93, 96, 97.$$

The **mean** is what most people usually think of as an average. We find it by adding the scores together and then dividing by the total number of scores.

$$\text{mean} = \frac{\left(\begin{array}{c} 47 + 52 + 56 + 57 + 61 + 65 + 66 + 69 + 70 + 71 + 71 + 72 + 73 + 75 + \\ 77 + 77 + 77 + 78 + 81 + 82 + 85 + 87 + 87 + 91 + 93 + 96 + 97 \end{array}\right)}{27}$$

$$= \frac{2013}{27}$$

$$= 74.6$$

Mathematical Note: If you study mathematics further, you will probably use the Greek letter sigma (Σ) to represent sums. For example, we could write Σ (data values) to represent the sum of all the data values in a data set. Note that, if there are n values in a data set, the mean is

$$\text{mean} = \frac{\Sigma \ (\text{data values})}{n}.$$

The **median** is the middle score in the data set. In this case, the median score is 75 because 13 scores are below 75 and 13 scores are above 75. The **mode** is the most common score in a data set. In this case, the mode is 77 because three students had this score and no other score occurred more than twice.

Averages: Mean, Median, and Mode

- The *mean* of a data set is calculated by the formula

$$\text{mean} = \frac{\text{sum of all values}}{\text{total number of values}}.$$

- The *median* is the middle score in the data set. Note that there will be two "middle" values if a data set has an even number of data points; if the two middle values are different, the median lies halfway between them.

- The *mode* is the most common score in a data set. A data set may have more than one mode, or no mode.

EXAMPLE 2   *Average Confusion*

Suppose that a city newspaper surveys wages at local supermarkets, and reports that the average wage of employees is $9.50 per hour. The 15 employees of Stella's Grocery Store immediately request a pay raise, claiming that they work as hard as employees at other stores but their average wage is only $8.10. The store manager quickly rejects their request, telling them that they are *overpaid* because their average wage, in fact, is $9.64. Give an example to show how the manager and the employees might have calculated the average differently.

Solution: Suppose that the hourly wages of the 15 employees, in ascending order, are as follows:

$$\$7.00, \$7.00, \$7.25, \$7.25, \$7.50, \$7.80, \$7.80, \$8.10,$$
$$\$8.10, \$8.45, \$8.45, \$8.45, \$12.15, \$16.80, \$22.50$$

The employees used the median of this set of wages, which is $8.10 (seven employees earn less than $8.10 and seven earn at least as much). However, the manager instead calculated the mean:

$$\text{mean} = \frac{\left(\begin{array}{l}\$7.00 + \$7.00 + \$7.25 + \$7.25 + \$7.50 + \$7.80 + \$7.80 + \$8.10 + \\ \$8.10 + \$8.45 + \$8.45 + \$8.45 + \$12.15 + \$16.80 + \$22.50\end{array}\right)}{15} = \$9.64$$

This example illustrates the confusion that can arise because the mean, median, and mode can all legitimately be called the *average*. It also shows that the choice of which quantity to use as the average is often based on what is advantageous to a person's or group's self-interest. ∎

TIME-OUT TO THINK: In the above example, using the mean is favorable to the manager. Using the median is favorable to the employees. Why is the mean higher than the median? Based strictly on the data, which do you think is the fairest measure of the center of the data set? Explain.

SHAPE OF THE DISTRIBUTION

We can better understand the distinctions between mean, median, and mode by investigating the *shape* of various data sets. First, note that the mean and median are uniquely determined for any data set; that is, a data set can have only *one* value for the mean and *one* value for the median. However, a data set may have more than one mode. Figure 9.5(a) shows a **single-peaked** (or *unimodal*) distribution of exam scores; that is, a distribution with only one mode. In contrast, Figure 9.5(b) shows a **bimodal** distribution with modes at both bin 65–69 and bin 80–84.

> **Mathematical Note:**
> A distribution is considered bimodal if it has two peaks, even if the two peaks are not of the same height.

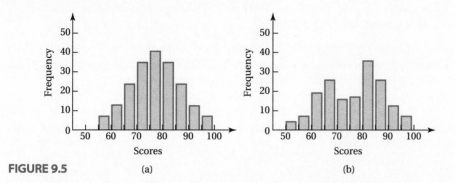

FIGURE 9.5   (a)   (b)

Frequency distributions for three more exam data sets are shown in Figure 9.6, this time with *line charts* (see Unit 2C). The graph in Figure 9.6(a) is **symmetric**; that is, the graph has the same shape, but reversed, on both sides of its peak. Note that the mean, median, and mode are all the same for a symmetric distribution of data.

In contrast, Figure 9.6(b) shows a data distribution in which most of the scores are relatively low, which means the median also is relatively low. However, a few students scored far above the median, and these high scores pull the mean to a higher value than the median. Scores that "stick out" at one end of the distribution are referred to as **outliers**. When most of the outliers lie on the high end of a distribution, as they do in Figure 9.6(b), the distribution is said to be **positively skewed**.

Figure 9.6(c) also has outliers, but this time they are mostly on the low end of the distribution. These low values pull the mean lower than the median, and we say that this distribution is **negatively skewed**.

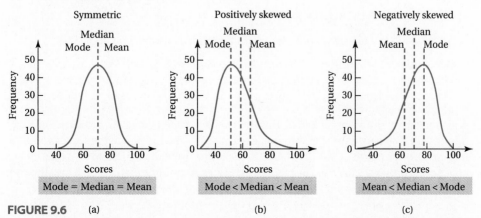

FIGURE 9.6   (a)   (b)   (c)

> **TIME-OUT TO THINK:**   Look back at Example 2. Which values of the employee hourly wages are outliers? How does the presence of these outliers explain why the median and mean are not the same?

EXAMPLE 3   *Distribution of Family Income*

Consider family income for all families in the United States. The median family income in the United States in 1995 was about $36,000. Is this distribution symmetric, positively skewed, or negatively skewed? Explain.

Solution: By definition, half of all families earn less than the median income of $36,000, and half earn more. However, no one can earn less than $0, so the distribution must come to an end at $0 on the low side. In contrast, there is virtually no upper limit on income on the high side, so there are many outliers with very high incomes and the distribution is *positively skewed*. These relatively few wealthy families pull the mean income to a value considerably higher than the median income. Figure 9.7 shows a rough sketch of the distribution of family incomes.

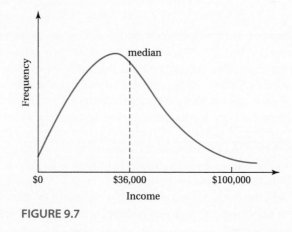

FIGURE 9.7   ■

By the Way ··············

The *Gini index* measures income inequality or the gap between low and high incomes in the United States. From 1947 to 1968 the Gini index decreased. Since 1968 the index has increased steadily. As of 1995, the wealthiest 5% of Americans earn about 21% of the total income in the United States.

THE FIVE-NUMBER SUMMARY

Consider again the 27 exam scores, now with three special scores highlighted:

$$47, 52, 56, 57, 61, 65, 66, 69, 70, 71, 71, 72, 73, 75,$$

$$77, 77, 77, 78, 81, 82, 85, 87, 87, 91, 93, 96, 97$$

The highlighted score of 75 is the median, as we found earlier. Note that the highlighted 66 is the median of the scores *below* 75; this score of 66 is called the **lower quartile**. Similarly, the score of 85 is called the **upper quartile** because it is the median of the scores *above* 75. In general, one-quarter of the data lies *at or above* the upper quartile, and one-quarter lies *at or below* the lower quartile. The median describes the middle of the overall distribution, and the quartiles tell us something about the spread or **dispersion** of the data.

A common way to provide a quick characterization of a distribution is with a **five-number summary**: the median, the upper and lower quartiles, and the high and low extremes. For the exam scores, the five-number summary is:

- median = 75
- upper quartile = 85
- lower quartile = 66
- high score = 97
- low score = 47

A simple way to display the five-number summary is with a **box plot** (sometimes called a *box and whisker* plot). We draw a box around the middle two quartiles, make a line through the box for the median, and then draw two "whiskers" from the box to cover the range. The box plot for the exam scores is shown in Figure 9.8.

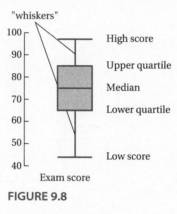

FIGURE 9.8

EXAMPLE 4   *Five-Number Summary*

Consider the following two sets of twenty 100-meter running times (in seconds):

> *Set 1:* 9.92, 9.97, 9.99, 10.01, 10.06, 10.07, 10.08, 10.10, 10.13, 10.13, 10.14, 10.15, 10.17, 10.17, 10.18, 10.21, 10.24, 10.26, 10.31, 10.38

> *Set 2:* 9.89, 9.90, 9.98, 10.05, 10.35, 10.41, 10.54, 10.76, 10.93, 10.98, 11.05, 11.21, 11.30, 11.46, 11.55, 11.76, 11.81, 11.85, 11.87, 12.00

Using five-number summaries and box plots, compare the two data sets.

Solution: The two data sets are already in ascending order, making it easy to construct the five-number summary. Each has 20 data points, so the median lies halfway between the tenth and eleventh times; the lower quartile is the median of the *lower half* of the times, which lies halfway between the fifth and sixth times; and the upper quartile is the median of the *upper half* of the times, which lies halfway between the fifteenth and sixteenth times. Thus the five-number summaries for the two data sets are as shown on the following page.

By the Way ············
Quartiles are not the only way to summarize data. For example, data are sometimes grouped in fifths, giving *quintiles*.

Beware! The upper and lower quartiles are usually found with the data in *ascending* order. Thus, in the example of running times, the lower quartile is located among the *faster* times because these times are listed first in ascending order.

First Set of Times	Second Set of Times
• median = 10.135 s	• median = 11.015 s
• upper quartile = 10.195 s	• upper quartile = 11.655 s
• lower quartile = 10.065 s	• lower quartile = 10.38 s
• fastest time = 9.92 s	• fastest time = 9.89 s
• slowest time = 10.38 s	• slowest time = 12.00 s

Figure 9.9 shows the box plots for the two data sets. Note that the box plots make it easy to see some key features of the data sets. For example, it is immediately clear that the second data set has a higher median time, as well as a greater spread of times.

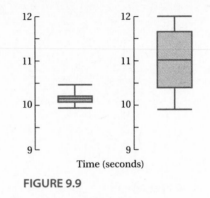

FIGURE 9.9

VARIANCE AND STANDARD DEVIATION

Another way to describe the spread or dispersion of a data set is with statistical measures called the *variance* and the *standard deviation*. To illustrate the calculation and meaning of these measures, consider two sets of 10 exam scores:

Set 1: 45, 55, 63, 72, 77, 79, 81, 84, 88, 97

Set 2: 45, 65, 71, 72, 74, 75, 76, 78, 88, 97

We begin by calculating the *mean* of each data set.

$$\text{mean (Set 1)} = \frac{45 + 55 + 63 + 72 + 77 + 79 + 81 + 84 + 88 + 97}{10} = 74.1$$

$$\text{mean (Set 2)} = \frac{45 + 65 + 71 + 72 + 74 + 75 + 76 + 78 + 88 + 97}{10} = 74.1$$

Both data sets have the same mean, and both also have the same lowest and highest scores (45 and 97, respectively). Nevertheless, histograms of the two data sets show that the first set is more spread out and the second set is more peaked (Figure 9.10).

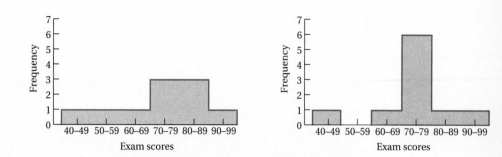

FIGURE 9.10

We can quantify the "spread" in a data set by looking at the **deviation from the mean** (or just **deviation**) for each of its data points. For example, the high score is 97, so its deviation from the mean is

$$\text{deviation} = \text{score} - \text{mean} = 97 - 74.1 = 22.9.$$

Table 9.4 shows the 10 scores for each data set in ascending order, along with the deviation for each score and the *square* of the deviation. The bottom row of the table shows the total for each column. Note the following key features of the table.

- For a particular score, the deviation from the mean is positive if it lies above the mean and negative if it lies below the mean.

- The *sum* of the deviations is always *zero*. Thus this sum cannot tell us anything about the dispersion in the data set.

- When we *square* the deviations, all the values become positive. Note also that squaring the deviations has a magnifying effect: points with large deviations have *very* large squared deviations.

- The sum of the squared deviations is larger for Set 1, which we know has a larger dispersion (see Figure 9.10). We therefore conclude that this sum tells us something about the dispersion.

Table 9.4	Deviations and squared deviations for exam scores				
	Set 1			**Set 2**	
Score	**Deviation (Score − Mean)**	**(Deviation)²**	**Score**	**Deviation (Score − Mean)**	**(Deviation)²**
45	$45 - 74.1 = -29.1$	$(-29.1)^2 = 846.81$	45	$45 - 74.1 = -29.1$	$(-29.1)^2 = 846.81$
55	$55 - 74.1 = -19.1$	$(-19.1)^2 = 364.81$	65	$65 - 74.1 = -9.1$	$(-9.1)^2 = 82.81$
63	$63 - 74.1 = -11.1$	$(-11.1)^2 = 123.21$	71	$71 - 74.1 = -3.1$	$(-3.1)^2 = 9.61$
72	$72 - 74.1 = -2.1$	$(-2.1)^2 = 4.41$	72	$72 - 74.1 = -2.1$	$(-2.1)^2 = 4.41$
77	$77 - 74.1 = 2.9$	$(2.9)^2 = 8.41$	74	$74 - 74.1 = -0.1$	$(-0.1)^2 = 0.01$
79	$79 - 74.1 = 4.9$	$(4.9)^2 = 24.01$	75	$75 - 74.1 = 0.9$	$(0.9)^2 = 0.81$
81	$81 - 74.1 = 6.9$	$(6.9)^2 = 47.61$	76	$76 - 74.1 = 1.9$	$(1.9)^2 = 3.61$
84	$84 - 74.1 = 9.9$	$(9.9)^2 = 98.01$	78	$78 - 74.1 = 3.9$	$(3.9)^2 = 15.21$
88	$88 - 74.1 = 13.9$	$(13.9)^2 = 193.21$	88	$88 - 74.1 = 13.9$	$(13.9)^2 = 193.21$
97	$97 - 74.1 = 22.9$	$(22.9)^2 = 524.41$	97	$97 - 74.1 = 22.9$	$(22.9)^2 = 524.41$
741	**Sum = 0**	**Sum = 2234.9**	**741**	**Sum = 0**	**Sum = 1680.9**

In fact, the variance and standard deviation are both based on the sum of the squared deviations, and are defined as follows.

The **variance** is the sum of the squared deviations divided by the total number of data points minus 1; that is

$$\text{variance} = \frac{\text{sum of squared deviations}}{n - 1}$$

where n is the total number of data points in the data set.

The **standard deviation** is the *square root* of the *variance*:

$$\text{standard deviation} = \sqrt{\text{variance}} = \sqrt{\frac{\text{sum of squared deviations}}{n - 1}}$$

Thus, for the 10 scores in Set 1, the variance and standard deviation are

$$\text{variance} = \frac{2234.9}{10 - 1} = \frac{2234.9}{9} = 248.3 \quad \text{and}$$

$$\text{standard deviation} = \sqrt{248.3} = 15.8.$$

For Set 2, we find

$$\text{variance} = \frac{1680.9}{10 - 1} = \frac{1680.9}{9} = 186.8 \quad \text{and}$$

$$\text{standard deviation} = \sqrt{186.8} = 13.7.$$

Note that both the variance and standard deviation are smaller for Set 2, just as we expect since this data has a smaller dispersion.

Mathematical Note: The reason for dividing by $n - 1$, rather than n, is a bit technical but has to do with the fact that statisticians have found it gives a better measure of dispersion. Note that if n is large, n and $n - 1$ are nearly equal, and we get about the same results either way. The use of $n - 1$ in the formulas assumes that we are dealing with actual data (for example, a sample), and not with inference to a population.

In general, the variance is a very large number compared to the actual exam scores because it uses the *squares* of the deviations. The square root in the standard deviation formula "makes up" for squaring all the deviations in the variance formula; it also gives the standard deviation the same units as the data. For this reason, the standard deviation provides a better description of the average spread of the scores.

TIME-OUT TO THINK: Look carefully at the individual scores and their deviations in the two data sets (Figure 9.10). Does the standard deviation of 15.8 seem like a reasonable "average" for the deviations in Set 1? How about the standard deviation of 13.7 for Set 2? Explain.

EXAMPLE 5   *Calculating Standard Deviation*

Two sets of five students take a quiz, with the following results:

> Group 1 scores: 19, 20, 21, 22, 23
> Group 2 scores: 12, 16, 19, 28, 30

Find the mean and standard deviation for each group and compare these statistics for the two groups.

Solution: First, we find the mean for each group.

$$\text{Group 1 mean} = \frac{19 + 20 + 21 + 22 + 23}{5} = \frac{105}{5} = 21$$

$$\text{Group 2 mean} = \frac{12 + 16 + 19 + 28 + 30}{5} = \frac{105}{5} = 21$$

Note that the mean is the same for both groups, even though the data sets are very different. To find the standard deviations, we calculate the deviations and squared deviations for each data point as shown in Table 9.5.

Table 9.5   Calculation of the deviations for two sets of exam scores

Group 1			Group 2		
Score	Deviation	(Deviation)^2	Score	Deviation	(Deviation)^2
19	$19 - 21 = -2$	$(-2)^2 = 4$	12	$12 - 21 = -9$	$(-9)^2 = 81$
20	$20 - 21 = -1$	$(-1)^2 = 1$	16	$16 - 21 = -5$	$(-5)^2 = 25$
21	$21 - 21 = 0$	$0^2 = 0$	19	$19 - 21 = -2$	$(-2)^2 = 4$
22	$22 - 21 = 1$	$1^2 = 1$	28	$28 - 21 = 7$	$7^2 = 49$
23	$23 - 21 = 2$	$2^2 = 4$	30	$30 - 21 = 9$	$9^2 = 81$
		Sum = 10			Sum = 240

Each group has $n = 5$ data points, so the standard deviations are:

$$\text{Group 1:}\quad \text{standard deviation} = \sqrt{\frac{10}{5 - 1}} = \sqrt{2.5} = 1.6$$

$$\text{Group 2:}\quad \text{standard deviation} = \sqrt{\frac{240}{5 - 1}} = \sqrt{60} = 7.7$$

Although both groups of scores have the same mean, the standard deviation of Group 2 is much larger because its scores are more spread out.   ∎

Standard Deviation on a Calculator or Computer

There are many strategies for finding the standard deviation, but they depend on the type of calculator or computer software that you have available.

- If you have a very basic calculator, you'll need to construct a table of the deviations and squared deviations (as above), then find the sum of the squared deviations and use it in the standard deviation formula.

- If you have a calculator that has a memory, you can compute the individual deviations, square them, and add them into the memory (often with a key denoted (M+)). The memory will keep a cumulative sum of the squared deviations. When you've finished computing all the squared deviations, recall this sum from memory and use it in the standard deviation formula.

- Many scientific calculators, as well as most software packages for statistics, offer built-in statistical operations. These operations allow you to enter data into a list. Once the list is made, pushing appropriate buttons will automatically give you the standard deviation, as well as many other statistics such as mean, median, mode, and variance.

REVIEW QUESTIONS

1. What is a *frequency table*? Give an example of how one is constructed.

2. Explain the difference between *frequency, relative frequency*, and *cumulative frequency*.

3. What do we mean by *binning* of data? Give an example of how binning is useful.

4. Define and distinguish between *mean, median*, and *mode*.

5. How do we characterize the shape of a data distribution? Contrast *single-peaked* and *bimodal* distributions. Contrast *symmetric, positively skewed*, and *negatively skewed* distributions.

6. What are *outliers*? How do they affect the relationship between mean and median in a data set?

7. What are the five numbers in the *five-number summary*? Explain the calculation and meaning of each. How does a *box plot* represent a five-number summary?

8. What is a *deviation from the mean*? Explain how the deviations are used to compute the *variance* and *standard deviation* of a data set.

9. How is standard deviation related to the spread, or *dispersion*, of a data set?

PROMBLEMS

1. **Term Paper Results.** The following student grades were given on a term paper in a small class:

A+, A, A−, B+, B+, B, B, B, B−, B−, C+,
C, C, C, C−, C−, C−, D+, D, F, F, F

   a. Sort the data into five bins: A, B, C, D, F (ignoring + and − grades). Then make a frequency table for the binned data. Include columns for relative frequency and cumulative frequency.

   b. Make a pie chart of the binned data. On each slice of the pie indicate its percentage of the total.

   c. Make a bar graph of the binned data.

2. **Exam Scores.** The results of an exam given to 25 students are as follows:

67, 54, 89, 67, 98, 68, 88, 78, 87, 82, 91, 80,
73, 59, 85, 86, 76, 99, 38, 75, 84, 87, 77, 69, 90

   a. Sort the scores into bins of 10 (that is, 90–99, 80–89, etc.), and make a frequency table for the binned data. Include columns for relative frequency and cumulative frequency.

   b. Make a pie chart of the binned data. On each slice of the pie indicate its percentage of the total.

   c. Make a bar graph of the binned data.

3. **Exploring Tables.** The following table shows the times (rounded to the nearest 0.1 s) for the 10 fastest runners in the semifinals of a men's 100-meter dash. Only a few of the blanks are filled in. Based on your understanding of frequency tables, complete the remaining entries in the table.

Time	Frequency	Relative Frequency	Cumulative Frequency
10.0			
10.1	3		4
10.2		0.2	
10.3			
Total			**10**

4. **Exploring Tables.** The following table shows the weights of ten gymnasts rounded to the nearest five pounds. Only a few of the blanks are filled in. Based on your understanding of frequency tables, complete the remaining entries in the table.

Weight	Frequency	Relative Frequency	Cumulative Frequency
85			
90		0.2	
95	3		6
100			
Total			**10**

5. **Mean, Median, and Mode.** The scores on a homework assignment for 10 people are

5, 5, 6, 7, 7, 9, 9, 9, 9, 10.

   a. What is the *mean* of the set?

   b. What is the *median* of the set?

   c. What is the *mode* of the set?

6. **Mean, Median, and Mode.** The scores on a homework assignment for 10 people are

15, 15, 20, 20, 25, 25, 25, 25, 30, 40.

   a. What is the *mean* of the set?

   b. What is the *median* of the set?

   c. What is the *mode* of the set?

7. **Commuter Statistics.** For ten days a commuter records her travel time to work (in minutes).

22.1, 23.4, 23.8, 24.1, 25.3, 26.3, 26.3, 26.5, 26.8, 27.0

   a. Calculate the mean for this set of times.

   b. What is the median for this set of times?

   c. Give a five-number summary for this set of data.

   d. Make a box plot for the data.

   e. Calculate the standard deviation for this set of times.

   f. Look again at the data. Does the standard deviation from part (e) make sense? Explain.

8. Racing Statistics. Consider the following set of 100-meter times (in seconds).

10.5, 10.3, 11.1, 10.8, 10.9, 11.1, 11.1, 11.3, 11.0, 11.1

a. Calculate the mean for this set of times.

b. What is the median for this set of times?

c. Give a five-number summary for this set of data.

d. Make a box plot for the data.

e. Calculate the standard deviation for this set of times.

f. Look again at the data. Does the standard deviation from part (e) make sense? Explain.

9. Deviations in Height. Suppose you measure the heights of seven students to the nearest centimeter. Your results are

172, 145, 166, 178, 190, 188, 169.

a. Give a five-number summary for this set of data.

b. Make a box plot for the data.

c. Calculate the mean for the data. Give your answer to four significant digits.

d. Calculate the standard deviation for this set of data.

e. Briefly explain why the standard deviation you found makes sense for this set of data.

10. Comparing Standard Deviations. Two heats of a 100-meter dash are run with six runners in each heat, with the following results (in seconds).

Heat 1:   9.98, 10.01, 10.05, 10.09, 10.10, 10.15

Heat 2:   9.91, 9.95, 10.14, 10.23, 10.30, 10.35

Find the mean and standard deviation for each heat, and interpret the differences.

11. Air Force Cadets. Suppose that you are evaluating a new group of 50 pilot candidates for the Air Force. You perform a hand-eye reflex test, in which you time how fast (to a precision of 0.1 s) each candidate can react by pressing a buzzer in response to a flashing light. The following table lists the results in terms of frequency.

Reflex Time	Frequency
1.1	5
1.2	10
1.3	20
1.4	8
1.5	4
1.6	2
1.7	1
Total	**50**

a. Add columns to this table for the relative frequency and cumulative frequency.

b. Construct a histogram for this data set.

c. Construct a line chart for the relative frequency data.

d. Construct a line chart for the cumulative frequency data.

e. Calculate the mean for the data (to three significant digits).

f. What is the median for the data?

g. What is the mode for the data?

h. Compute the standard deviation for the data.

i. Would you say that this distribution is symmetric, positively skewed, or negatively skewed? Why?

12. **Kids and TV.** You've studied the habits of a sample of 25 fourth-grade children, finding the number of hours of television, to the nearest hour, watched by these children each weekday:

$$5, 8, 4, 1, 2, 3, 0, 2, 6, 4, 4, 4,$$
$$5, 1, 6, 6, 2, 4, 3, 4, 0, 2, 4, 4, 3$$

a. Construct a frequency table with columns for the frequency, relative frequency, and cumulative frequency.

b. Construct a histogram for this data set.

c. Construct a line chart for this data set.

d. Calculate the mean for the data (to two significant digits).

e. What is the median for the data?

f. What is the mode for the data?

g. Would you say that this distribution is symmetric, positively skewed, or negatively skewed? Why?

h. Give a five-number summary for this set of data.

i. Make a box plot for the data.

13. **Income and Degrees.** A 1993 survey collected the following income data for U.S. residents by educational level.

Educational Level	Median Income
No high school diploma	$9,000
High school graduate	$11,000
Bachelor's degree	$27,000
Master's degree	$35,000
Professional degree	$56,000
Doctoral degree	$48,000

a. Speculate on how these data were gathered and identify some potential sampling errors.

b. Why do you think that *median* salary is displayed? How might the figures change if *mean* annual salary were displayed?

14. **Cumulative Frequency Line Chart.** The cumulative frequency line chart shown in the next column gives the results of a hypothetical 100-meter race. Study the line chart and answer each question. Explain your answers.

a. How many runners were timed in the race?

b. How many runners ran 10.0 s or *faster*?

c. How many runners ran 10.4 s?

d. What was the median time for this set of data?

e. What was the mode for this set of data?

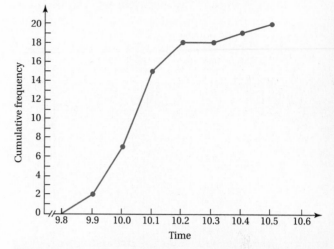

Skewness. *For the distributions in Problems 15–19, state whether you expect it to be symmetric, positively skewed, or negatively skewed and briefly explain why.*

15. An exam (with a maximum possible score of 100) in which most students do very well, say, in the 80s and 90s, but a few students do very poorly.

16. An exam (with a maximum possible score of 100) in which most students do very poorly, say, in the 30s and 40s, but a few students do very well.

17. The speeds of cars on a freeway.

18. The weights of newborn babies.

19. Amount of pizza, per week, consumed by college students.

20. **U.S. Family Income.** To study family income in the United States, you choose an unbiased sample of 200 families. You summarize the results of your sample:
   - The mean family income was $20,000.
   - The median family income was $18,000.
   - The standard deviation was $5000.
   - The extremes of family incomes were a low of $2300 and a high of $862,000.

a. In your sample, about how many families earned less than $18,000? Explain.

b. Would you characterize this distribution as symmetric, positively skewed, or negatively skewed? Why?

c. Based on what you know about U.S. family incomes, how well do you think your sample statistics compare to the true population parameters? Explain.

21. **The Shape of Data.** The histograms below show the distribution of data for (i) time between eruptions of Old Faithful Geyser in Yellowstone National Park, (ii) failure time of computer chips, and (iii) weights of rugby players. The number of data points is N.

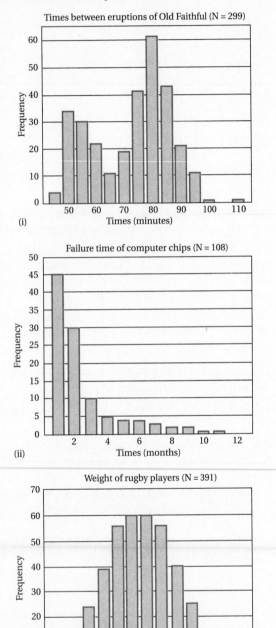

Times between eruptions of Old Faithful (N = 299)

(i)

Failure time of computer chips (N = 108)

(ii)

Weight of rugby players (N = 391)

(iii)

(*Handbook of Small Data Sets* by Hand, et. al., 1994). Answer the following questions about each data set, estimating the data values from the histograms.

a. Give the five-number summary for each data set.

b. Describe the shape of the distribution and explain why each distribution has that shape.

22. **Fruit Fly Genetics.** Genetic theory predicts that the four eye colors—red (R); pink (P); brown (B); white (W)—in mated fruit flies will occur in the ratio 9R:3P:3B:1W. You conduct an experiment in which you check the eye colors of 32 fruit fly offspring and observe:

R, R, P, R, B, W, R, R, R, R, R, P, P, R, R, P,
P, B, B, B, R, R, W, R, B, P, B, P, R, R, P, R

a. Make a frequency table from these data for the four different eye colors. Include the relative frequency, but not the cumulative frequency, in your table.

b. Construct a pie chart for the data in your frequency table. Label each slice with the appropriate percentage.

c. Compare the results from the experimental data with the results predicted from theory. For example, did red occur more often, less often, or exactly as often as predicted by theory? What about brown and pink?

d. Does the fact that your results do not agree perfectly with the theory mean that the theory is invalid? Why or why not?

23. **Project: Randomized Sampling.** Suppose you want to survey people on a sensitive issue such as personal income taxes. In order to get complete and honest responses, the people in the survey should know that their responses will be confidential; that is, it should be impossible to match the responses with each individual. Here is a way to insure that confidentiality.

Ask each respondent to flip a coin. Then give the following instructions:

- If the coin comes up HEADS, then answer the DECOY question: *were you born on an even day of the month?*
- If the coin comes up TAILS, then answer the REAL question: *have you ever failed to report any income on your tax return?*

Note that the response of each person cannot be determined: even if you received a response with a name on it, there is no way to tell whether that person answered the decoy question or the real question.

a. Suppose the surveys are returned and you find that there are 64 YES responses and 36 NO responses. How do you estimate the percentage of people who have failed to report income on their tax returns? (Hint: Because of the coin flip, you can expect that close to 50% of the people answered the decoy question while the rest answered the real question. Furthermore, because about half the days of the month have an even number, the responses to the decoy question should be split almost evenly between YES and NO.)

b. What other decoy questions could you use?

c. Create a survey question and use the randomized sampling technique in your class. Describe how you chose your real question and your decoy question, and summarize your findings.

LINEAR REGRESSION MODELS

In the previous unit, we worked with data sets involving a single variable, such as test scores or family income. However, it is quite common to collect data on two or more variables. When data sets involve two or more variables, the first step in exploratory data analysis usually is to search for correlations between variables. If a correlation is found, we can begin to search for cause and effect relationships. Keep in mind that, as discussed in Unit 2E, a correlation does not necessarily imply causality. It also is possible for a correlation between two variables to be a coincidence, or for the correlated variables to have a common underlying cause. In this unit, we examine how to search for correlations, and how to use correlations to model relationships between two variables.

SCATTERPLOTS AND CORRELATION

Suppose that you record the height and weight of 12 randomly chosen classmates. You will have a data set consisting of 12 *pairs* of numbers: one height and one weight for each person. Therefore we can make a *scatterplot* (see Unit 2C) of the data by plotting each person's height on the horizontal axis and weight on the vertical axis. Suppose that the result is the scatterplot in Figure 9.11. Note that the points show a definite pattern: shorter people have lower weights and taller people have higher weights. We therefore say that there is a *correlation* between the heights and weights in this data set.

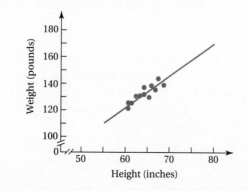

FIGURE 9.11

Recall that only two data points are needed to determine a unique straight line. Generally, when there are *more* than two data points, a single straight line will not pass through all of them. For example, the height-weight data points in Figure 9.11 do not all fall on a straight line. However, the straight line shown in the figure seems to "fit" the data quite well; that is, all the data points lie close to the straight line. The straight line that *best fits* a set of data is called the **best-fit line**.

TIME-OUT TO THINK: Try drawing a few other straight lines, each with a different slope, through the data in Figure 9.11. Do any of your other lines appear to fit the data as well as the line shown? Can you convince yourself that the line shown is the best-fit line? Explain.

Mathematical Note:
More technically, the vertical distance of each point from the line represents a *deviation*, so we can calculate a *variance* by summing the squares of all these distances. The best-fit line is the one that minimizes the variance.

How do we know whether a line is really the best-fit line? The best-fit line is the particular line in which the spread of the data points around the line is the smallest. If we draw a line through a set of data points on a scatterplot, we can measure the vertical distance between each data point and the line. The best-fit line minimizes the sum of all these distances. A mathematical technique called **linear regression** can be used to find the equation of the best-fit line. (The best-fit line is sometimes called a *regression line*.) In fact, many calculators have built-in functions for linear regression, making it easy to find the equation of the best-fit line. However, in this book we will draw best-fit lines "by eye:" simply look at a data set and draw the line that *appears* to minimize the distances of the data points.

THE CORRELATION COEFFICIENT

When data points cluster tightly near the best-fit line, as they do in Figure 9.11, the correlation between the two variables is strong. More precisely, statisticians describe the strength of a correlation with a number known as a **correlation coefficient**, usually denoted by the letter r. The correlation coefficient is defined so that:

- If $r = 1$, the two variables are *perfectly* and *positively* correlated. All the data points lie precisely on the best-fit line, and an increase in one variable means an increase in the other (Figure 9.12a).

- If r is between 0 and 1, the two variables are *positively* correlated, but not perfectly. That is, the data points do not all lie precisely on the best-fit line, but an increase in one variable *tends* to mean an increase in the other. The stronger this tendency, the closer the correlation coefficient is to 1. Figure 9.12(b) shows a data set with $r = 0.5$.

- If $r = 0$, the two variables are completely *uncorrelated*: there is no linear relationship between them (Figure 9.12c).

- If r is between 0 and -1, the two variables are *negatively* correlated, but not perfectly. That is, an increase in one variable *tends* to mean a *decrease* in the other. The stronger this tendency, the closer the correlation coefficient is to -1. Figure 9.12(d) shows a data set with $r = -0.5$.

- If $r = -1$, the two variables are *perfectly* and *negatively* correlated. All the data points lie precisely on the best-fit line, and an increase in one variable means a *decrease* in the other (Figure 9.12e).

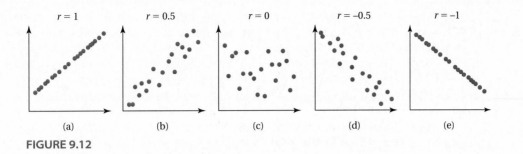

FIGURE 9.12

EXAMPLE 1   *Interpreting Correlation Coefficients*

Suppose that a biological study collects field data on the rate of cricket chirping at different outside temperatures. The researchers report that they found a correlation coefficient of $r = -0.25$ between the rate of chirping and the temperature. They then claim that "there is evidence to suggest that the rate of chirping decreases linearly with the temperature." Do you agree with their claim? Explain.

Solution: The correlation coefficient of $r = -0.25$ means that as the temperature increases, the rate of chirping tends to decrease. However, a correlation coefficient of $r = -0.25$ means a fairly *weak* correlation between the variables. Thus the researchers are correct in saying "there is evidence" for the correlation, but they've left out the fact that the evidence is pretty weak. ∎

Visual Linear Regression

Table 9.6 shows changes over several decades in the men's and women's world records for the mile run. The two data sets (men and women) are plotted in Figure 9.13, along with the best-fit lines for each data set.

Table 9.6 World records in the mile run (Minutes: Seconds)			
Date	**Women's Record**	**Date**	**Men's Record**
1967	4:37	1942	4:06
1971	4:35	1945	4:01
1973	4:29	1954	3:59
1979	4:22	1958	3:55
1981	4:21	1964	3:54
1985	4:17	1975	3:51
1989	4:15	1981	3:48
1996	4:12	1993	3:44

Data courtesy of Hal Bateman, *USA Track and Field.*

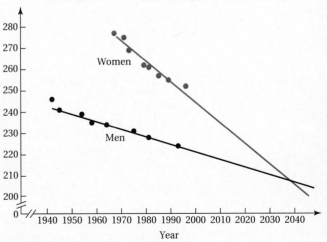

FIGURE 9.13

Note that the data points do not fall exactly on the best-fit lines for the two sets of data. However, they are close, so the best-fit lines represent reasonable models for the changes in the world records. Because the linear models fit the data so well, we might use them to make predictions about world records in the future. For example, the men's and women's lines intersect in the year 2038. That is, according to these models, the women's and men's mile records will be equal in 2038, and the women's record will be faster in the years beyond.

Clearly, the models must eventually break down: If you extend both lines even farther into the future, they make the impossible prediction that the mile records eventually become zero! As with many linear models, these have only a limited range of validity. Where do they stop being valid? That's difficult to say; close examination of the data suggests that the linear model for the men's record may already be breaking down in the 1990s. However, the women's record may continue to decrease linearly. Will it continue to decrease beyond the men's record? Only time will tell.

> **TIME-OUT TO THINK:**   Imagine that, in the year 2040, we make a similar graph of men's and women's world records for the mile. Do you think that a linear model will still appear to be a good fit for the data? Explain. Do you think that the women's world record will have equaled the men's world record? Why or why not?

EXAMPLE 2   *Visual Linear Regression*

Figure 9.14(a) shows height and weight data for 10 men (black dots) and 10 women (blue dots). Find the best linear fit for each set "by eye," and find an equation for each best-fit line. Use the equation for the best-fit line to the men's data to predict the weight of a 6′ 10″ man. Does the prediction seem reasonable?

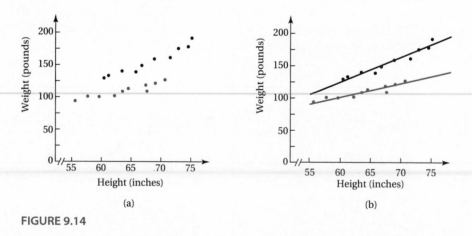

(a)   (b)

FIGURE 9.14

Solution: If you draw the best-fit lines by visually trying to minimize the distances between the data points and the lines, you should come up with something very similar to the lines shown in Figure 9.14(b). (These are the actual best-fit lines for the data, calculated with linear regression techniques.)

Each line represents a linear model for a (*height, weight*) function. We can find the equation of each line using the techniques discussed in Unit 6B. We begin by measuring the *rise* and *run* of each best-fit line: you should use a ruler to confirm that the men's line rises 3.8 pounds along the vertical axis for each 1 inch along the horizontal axis, so its slope is about 3.8 pounds per inch. Letting h stand for height in inches and w for weight in pounds, the equation for this line is

$$w = 3.8 \times h + \text{initial value of } h.$$

Now we can choose any point on the line, such as (60 in., 123 lb). Substituting these values for (h, w) in our equation, we find

$$123 = (3.8 \times 60) + \text{initial value of } h.$$

Solving for the initial value of h gives

$$\text{initial value of } h = 123 - (3.8 \times 60) = -105.$$

Thus the equation for the men's line is

Men: $w = 3.8 \times h - 105,$

where h is measured in inches and w is measured in pounds.

You should confirm that a similar procedure results in the following equation for the women's line:

Women: $w = 2.4 \times h - 37$

Now, let's use the equation for the men's line to predict the weight of a 6'10" man. A height of 6'10" is equivalent to 82 inches, so we substitute $h = 82$ in. in the equation for the men's line:

$$w = 3.8 \times h - 105 = 3.8 \times 82 - 105 = 206.6$$

This linear model predicts that the weight of a 6'10" man will be about 207 pounds. In fact, this weight is quite low: most men of that height weigh considerably more than 207 pounds. The problem is that none of the data points in our model involved men of such great height. This example illustrates the danger of using a linear model to predict values that fall outside the range of the data set on which the model is based. ∎

TIME-OUT TO THINK: The data in Figure 9.14 show that heights and weights have a strong linear correlation over relatively small ranges of height. Do you think that they also have a strong linear correlation over larger ranges of height? Why or why not?

EFFECTS OF OUTLIERS

Perhaps the greatest difficulty in searching for correlations is caused by outliers—individual points that are conspicuously separated from the rest. Let's return to the data shown in Figure 9.11, but imagine that a new student appears in class who is 6 feet (72 inches) tall

By the Way ············
In 1877, Sir Francis Galton conducted a study in which he showed that the heights of children with tall parents are *less* far above average than the heights of the parents; he therefore said that the heights of the children *regress* toward the average, and called the line in his data describing this relationship a *regression line*. This is the origin of the term *regression* as it is used in statistics.

and weighs 160 pounds. The new height-weight data for the class is shown in Figure 9.15(a). The point representing the new student is an outlier because it lies far above and to the right of the other points. However, this outlier falls very close to the best-fit line of the original data set, so there is little change in the line or the correlation coefficient. Indeed, this added point makes the correlation between the heights and weights look even stronger.

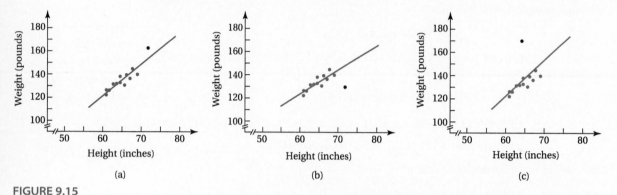

FIGURE 9.15

Figure 9.15(b) shows data for the original class with a new student who is 6 feet tall and weighs only 130 pounds. The outlier representing this new student pulls the entire best-fit line down. The slope of the new line is much shallower than that of the original line, and the correlation now appears to be much weaker.

Figure 9.15(c) shows data for the original class and a new student member who is 5′4″ (64 inches) tall and weighs 170 pounds. The data point representing this new student is an outlier far above the original data set. This outlier gives the best-fit line a greater slope, and again makes the correlation appear weaker.

Thus even one outlier can have a significant effect on the correlation, which in turn affects the predictive power of the data. The treatment of outliers is both important and subtle. If an outlier is a genuine data point, it must be included and allowed to alter the best-fit line accordingly. In fact, if the outliers in either Figure 9.15(b) or (c) are genuine points, they might lead you to question the validity of assuming a linear correlation between height and weight in the class. However, if you suspect that an outlier is the result of an error in measurement or in recording the data, then it should be removed.

"Common wisdom among statisticians is that at least 5% of all data points are corrupted, either when they are initially recorded or when they are entered into the computer."

—*JESSICA UTTS*

> **TIME-OUT TO THINK:**   Consider the outliers in Figures 9.15(b) and (c). Do you think that these could be genuine data points? If they are genuine, do you think they invalidate a conclusion that any linear correlation exists between height and weight? Explain.

NONLINEAR REGRESSION

Throughout this unit, we tried to fit straight lines to data points. However, if the relationship between two variables is not linear, a straight-line fit will not work. Figure 9.16 shows two hypothetical data sets between variables called *x* and *y* in which the data points clearly cannot be fit with a straight line. In each case, the graphs instead show *best-fit curves*.

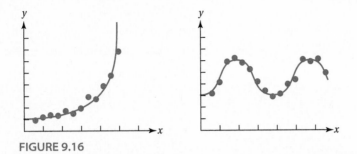

FIGURE 9.16

Techniques for finding best-fit curves go by the name of **nonlinear regression**, and are similar in principle to linear regression. However, nonlinear regression can be much more complex in practice: while there is only one best-fit line to any set of data, there may be many best-fit curves, each with a different shape. Thus nonlinear regression usually involves making some assumptions about what shape a curve *should* have, before seeking the best-fit curve.

REVIEW QUESTIONS

1. What is a scatterplot? Give an example of how you would make one.

2. What is the purpose of the process of *linear regression*? What do we mean by a *best-fit line*?

3. Explain the meaning of a *correlation coefficient*. Be specific in the interpretation of different values of *r*.

4. Explain how outliers can affect a best-fit line. Under what conditions should outliers be trusted?

5. What is *nonlinear regression*? When is it useful?

PROBLEMS

1. **Examples of Correlation.** Give an example of two variables that are positively correlated, in which both variables change with respect to time.

2. **Examples of Correlation.** Give an example of two variables that are negatively correlated, in which both variables change with respect to time.

3. **Examples of Correlation.** Give an example of two variables that are positively correlated, neither of which changes with respect to time.

4. **Examples of Correlation.** Give an example of two variables that are negatively correlated, neither of which changes with respect to time.

5. **Television Time.** The table below shows the average daily TV viewing time per household over a 40-year period.

Year	1954	1964	1974	1984	1994
Viewing hours	4.6	5.3	6.1	7.0	7.2

   a. Comment on how these figures might have been determined and how reliable they might be.

   b. Make a histogram and a line chart of the data.

   c. Draw a straight line on the histogram of part (b) that appears to fit the data. Does the line provide a good fit?

d. Based on the linear fit of part (c), what do you project the average TV viewing time will be in 2004? Do you think this prediction is reasonable?

6. Speed Limits and Death Rates. The following table gives the death rates and speed limits in 10 countries.

Country	Death Rate (per 100 million vehicle miles)	Speed Limit (miles per hour)
Norway	3.0	55
United States	3.3	55
Finland	3.4	55
Britain	3.5	70
Denmark	4.1	55
Canada	4.3	60
Japan	4.7	55
Australia	4.9	65
Netherlands	5.1	60
Italy	6.1	75

(D. J. Rivkin, *New York Times*, November 25, 1986)

a. In what order are the countries in this table listed? Why do you think this order was chosen?

b. Make a scatterplot of these data. Based on the 10 data points, are speed limits and highway death rates strongly correlated? Explain.

c. Are there any outliers in this plot? If the outlier(s) is (are) removed, does the strength of the correlation change?

d. The source article's title was "Fifty-five mph speed limit is no safety guarantee." Based on the data, do you agree? Explain.

7. Diving Reflex. A well-known way to slow the heartbeat is to immerse a subject's face in cold water (known as the diving reflex). A study involving ten children was designed to measure the diving reflex. The childrens' faces were immersed in water of different temperatures, and the reduction in heart rate (compared to the normal heart rate) of each child was measured. The results that were obtained are shown in the table at the top of the next column.

Temperature (°F)	Reduction in Heart Rate (beats/min)
68	2
65	5
70	1
62	10
60	9
55	13
58	10
65	3
69	4
63	6

(*Statistics by Example* by T. Sincich, 1990)

a. Make a scatterplot of the data.

b. Draw what appears to be the best-fit line through the data.

c. Comment on the degree of correlation of the data. Is the correlation positive or negative?

d. Estimate the correlation coefficient of the data. Based on these data, would you say that the diving reflex is a real effect? Explain.

8. Bowling Scores. A (hypothetical) study was designed to determine how practice time affects bowling averages. The following data were collected, giving hours per week of practice time and bowling averages (maximum possible is 300).

Practice Time (hours/week)	Bowling Average (out of 300)
3	102
6	120
8	140
12	189
9	141
2	100
14	193
7	137
6	115
10	178

a. Make a scatterplot of the data.

b. Draw what appears to be the best-fit line through the data.

c. Comment on the degree of correlation of the data. Is the correlation positive or negative?

d. Estimate the correlation coefficient of the data. Based on these data, would you say that a person's bowling average could be improved by additional practice time? Explain.

9. Women in the Labor Force. The following data show how the number of women in the U.S. labor force has changed since 1900. (Data for 1950–1990 include only women over age 16.)

Year	Number of Female Workers (thousands)	Percentage of Female Population Working	Percentage of Total Labor Force
1900	5319	18.8	18.3
1910	7445	21.5	19.9
1920	8637	21.4	20.4
1930	10,752	22.0	22.0
1940	12,845	25.4	24.3
1950	18,389	33.9	29.6
1960	23,240	37.7	33.4
1970	31,543	43.3	38.1
1980	45,487	51.5	42.5
1990	56,554	57.5	45.3
1995	60,944	58.9	46.1

(1997 *Information Please Almanac*)

a. Use a time-series diagram to show how each variable (total number, percentage of female population, and percentage of total labor force) changed over time.

b. Imagine drawing a best-fit line for each of the variables shown. Describe whether these lines would fit the data well. Do you think you could use these lines to predict the number of women in the work force in 2010? Why or why not?

c. Assume that the numbers of men and women in the entire population are equal. What can you conclude about the percentage of men that work? Explain.

d. Draw a pie chart for the 1995 data that has four regions: the percentage of the overall population made up of working women, nonworking women, working men, and nonworking men.

Correlation Coefficients. *For the scatterplots of Problems 10–15, state whether the correlation between the two variables is positive or negative and strong or weak. Then give an estimate of the correlation coefficient, explaining your reasoning.*

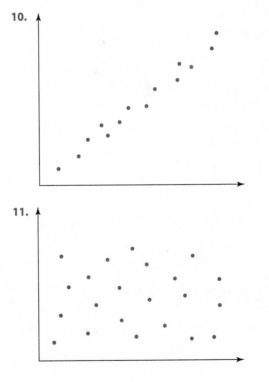

10.

11.

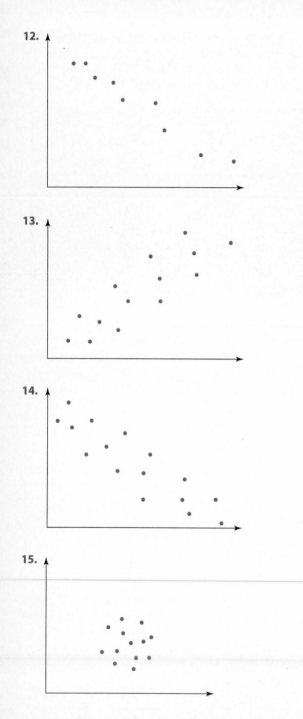

12.

13.

14.

15.

16. Education and Salary. The table at the top of the next column shows the mean earnings of men and women in terms of their education.

Education (years after high school)	Earnings, Men	Earnings, Women	Earnings, Both Sexes
No HS diploma	$16,745	$9790	$14,013
0 (HS diploma only)	$26,333	$15,970	$21,431
2 (Associate's degree)	$33,881	$22,429	$27,780
4 (Bachelor's degree)	$46,111	$26,841	$36,980
6 (Master's degree)	$58,301	$34,911	$47,609
8 (Doctorate)	$71,016	$47,733	$64,550
Overall	$33,251	$19,414	$26,792

(1998 *Information Please Almanac,* 1996 data)

  a. Make a scatterplot of education vs. earnings for all three categories shown. (For simplicity; use −2 years as the "years after high school" for no HS diploma.)

  b. Make a visual best-fit line for the three categories.

  c. Can you make any general statements about a correlation between educational level and earnings that apply to all three categories? Explain.

  d. What can you conclude about the differences between the men's and the women's earnings? Explain.

17. Project: World Record Marathon Times. Locate (perhaps in an almanac) the successive world record marathon times for both men and women, with data going back as far as possible. Record your data both in a table and on a scatterplot, and draw best-fit lines for the men's data and the women's data. Based on your results, discuss whether the data can be modeled with linear functions, and whether these functions can be used to predict future world records. Do you think that the world record for women may surpass the world record for men sometime in the future? Explain.

18. Project: Collecting Data for Linear Regression. Collect data for two variables that are plausibly related in a linear way. For example, you might collect data for the relations (*weight, shoe size*) or (*height, weight*) from your classmates. Or within your community, you might collect data for the variables (*time in years, population*) or (*time in years, sales tax revenue*). Then carry out the following analysis on your data set.

a. Plot the data with appropriate axes and labels.

b. Draw a best-fit line for the data set, and determine the equation for the line. Discuss whether this linear model gives a valid representation for the data set.

c. Describe several ways that your data set and analysis might be useful.

STATISTICAL INFERENCE

UNIT 9 C

In the previous two units, we looked at exploratory methods that allow us to get a feel for a data set. These methods produce *sample statistics*—numbers such as the median, mean, and standard deviation—that characterize a sample taken from a larger population. But, as we discussed in Unit 2A, the goal of a statistical study usually is to learn something about the *population parameters* that characterize an entire population. One of the keys to inferring population parameters from sample statistics is the use of something called the *normal distribution*. In this chapter, we first investigate the normal distribution, and then see how it can be used to help us make predictions about an entire population when we have data from only a sample.

THE NORMAL DISTRIBUTION

In Unit 9A we discussed *symmetric distributions* of a single variable. A very special and common type of symmetric distribution has the shape of a bell; in statistics, this type of distribution is called a **normal distribution** (or a **bell-shaped curve**). All normal distributions have the same characteristic bell shape, and differ only in their mean and standard deviation. Real data can only approximate a perfect normal distribution, so we usually think of the normal curve overlaying a histogram or line chart of the real data. Figure 9.17 shows two such normal distributions. Each one represents a frequency distribution for heights of women who belong to a particular club. Both distributions have the same mean, but the broader one has a greater standard deviation.

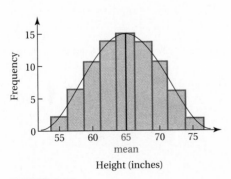
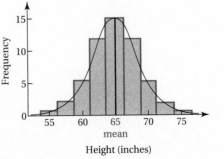

FIGURE 9.17

The normal distribution is extremely important because it occurs so often. Many human attributes, when sampled randomly, take the form of a normal distribution. For example, most women tend to be fairly average in height, which creates the peak of the distribution near the mean. Moving away from the average of the distribution, we find fewer and fewer women who are extremely short or extremely tall, which produces a *tailing off* on either side of the mean.

In fact, any quantity that is the result of *many* factors is likely to follow a normal distribution. For example, the height and weight of an adult are the result of many genetic and environmental factors. Scores on SAT tests or IQ tests tend to be normally distributed because each test score is the sum of the results from many individual test questions. Sports statistics, such as batting averages, tend to be normally distributed because they involve many people with many different levels of skill.

> **TIME-OUT TO THINK:** Would you expect grade point averages at your school to be normally distributed? Why or why not?

The Standard Deviation in Normal Distributions

Suppose we have a data set for a single variable such as height. As we have seen, we can make a frequency distribution for the data and compute their standard deviation. The standard deviation, usually denoted by σ (the Greek letter *sigma*), has a special interpretation with the normal distribution. This interpretation can be summarized with the **68-95-99.7 Rule** (Figure 9.18).

Mathematical Note:
On a more technical level, adding and subtracting the standard deviation from the mean locates the two points on either side of the mean at which the normal curve is the steepest.

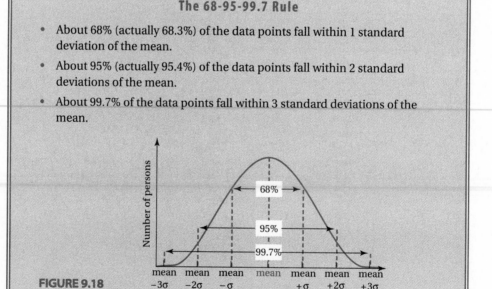

The 68-95-99.7 Rule

- About 68% (actually 68.3%) of the data points fall within 1 standard deviation of the mean.
- About 95% (actually 95.4%) of the data points fall within 2 standard deviations of the mean.
- About 99.7% of the data points fall within 3 standard deviations of the mean.

FIGURE 9.18

EXAMPLE 1   *SAT Scores*

Each test that makes up the SAT is designed so that its scores have a mean of 500 and a standard deviation of 100. Interpret this statement.

Solution: From the 68-95-99.7 rule, about 68% of students score within 1 standard deviation (100 points) of the mean (500 points), which means they score between 400 and 600. About 95% of the students score within 2 standard deviations (200 points) of the mean, or between 300 and 700. And about 99.7% of the students score within 3 standard deviations (300 points) of the mean, or between 200 and 800. The overall distribution is shown in Figure 9.19.

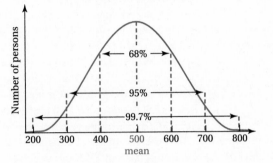

Distribution of SAT scores

FIGURE 9.19

EXAMPLE 2   *One Side of the Normal Curve*

Suppose that 1000 students take an exam. The scores are normally distributed with a mean of 75 and a standard deviation of 7 (Figure 9.20). About how many students scored above 89?

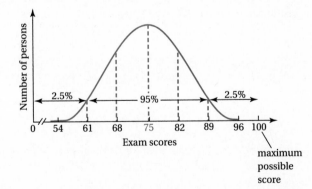

FIGURE 9.20

Solution: A score of 89 is 14 points above the mean of 75. Because the standard deviation is 7, this score also is 2 standard deviations above the mean. The 68-95-99.7 rule tells us that about 95% of the scores are *within* 2 standard deviations of the mean, so about 5% of the scores are *farther* than 2 standard deviations from the mean. Half of this 5%, or 2.5%, are scores more than 2 standard deviations *below* the mean; the other 2.5% are more than 2 standard deviations *above* the mean. Thus about 2.5% of 1000 students, or 25 students, scored above 89.

EXAMPLE 3   *Normal Auto Prices*

Because the price of an automobile involves many variables (for example, dealer preparation charges, taxes, and bargaining), different people tend to pay different prices for the same model of car. Suppose that a survey shows that the prices paid for a particular new car model are normally distributed with a mean of $16,000 and a standard deviation of $400 (Figure 9.21). Assume that 100,000 cars of this model are sold. How many people paid between $15,600 and $16,400? How many paid less than $15,600? How many paid more than $16,800?

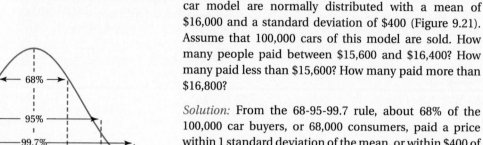

FIGURE 9.21

Solution: From the 68-95-99.7 rule, about 68% of the 100,000 car buyers, or 68,000 consumers, paid a price within 1 standard deviation of the mean, or within $400 of $16,000 in this case. Thus about 68,000 people paid between $15,600 and $16,400. The remaining 32,000 car buyers paid *either* less than $15,600 or more than $16,400. Because a normal distribution is symmetric, half of these 32,000 people, or 16,000 consumers, paid less than $15,600. To determine how many paid more than $16,800, we recognize that this price is 2 standard deviations (2 × $400 = $800) above the mean price of $16,000. Following the reasoning used in Example 2, about 2.5% of the 100,000 customers, or 2,500 people, paid more than $16,800. ◼

Z-SCORES AND PERCENTILES

You are probably familiar with the idea of **percentiles** from standardized tests. If 35% of the test takers score below you, you are in the 35th percentile. Similarly, if 85% of the test takers score below you, you are in the 85th percentile. The percentile ranking is simply the percentage of the group with scores *below* a given score.

> **TIME-OUT TO THINK:**   Explain why there is no 100th percentile.

In a normal distribution, the percentile of a particular data point depends on what we call its **z-score**, or *standardized score*, which describes how many standard deviations it lies above or below the mean. For example,

- the z-score of a data point located *at* the mean is 0, because it is 0 standard deviations from the mean.

- the z-score of a data point 1 standard deviation *above* the mean is 1.

- the z-score of a data point 2 standard deviations *below* the mean is −2.

Generalizing, we find the following rule.

Computing z-scores

The number of standard deviations of a particular point from the mean is called its *z-score*. We can compute the *z*-score with the following formula.

$$z\text{–score} = \frac{\text{value of data point} - \text{mean}}{\text{standard deviation}}$$

Note that data points below the mean have a negative *z*-score and data points above the mean have a positive *z*-score.

EXAMPLE 4   *Finding z-scores*

Suppose that a normal distribution has a mean of 100 and a standard deviation of 10. Find the *z*-scores for data points at 85, 100, and 125.

Solution: We can calculate the *z*-scores for the data points by using the *z*-score formula with a mean of 100 and standard deviation of 10.

$$z\text{–score for 85:} \qquad \frac{85 - 100}{10} = -1.5$$

$$z\text{–score for 100:} \qquad \frac{100 - 100}{10} = 0$$

$$z\text{–score for 125:} \qquad \frac{125 - 100}{10} = 2.5$$

Figure 9.22 shows the locations of these points on a graph, in terms of both *z*-scores and actual data values. We can interpret these z-scores as follows: 85 is 1.5 standard deviations below the mean, 100 is equal to the mean, and 125 is 2.5 standard deviations above the mean.

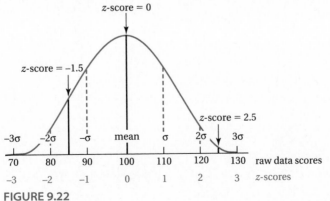

FIGURE 9.22

EXAMPLE 5   *Car Price z-scores*

Consider the car price distribution shown in Figure 9.21. What is the z-score for a price of $15,500? for a price of $17,700? Interpret the results in terms of the standard deviation.

Solution: The mean price shown in Figure 9.21 is $16,000, and the standard deviation is $400. Thus the z-scores for the two prices are:

$$z\text{-score for } \$15,500: \quad \frac{\$15,500 - \$16,000}{\$400} = -1.25$$

$$z\text{-score for } \$17,700: \quad \frac{\$17,700 - \$16,000}{\$400} = 4.25$$

In other words, a price of $15,500 is 1.25 standard deviations *below* the mean price paid for the car. A price of $17,700 is 4.25 standard deviations *above* the mean price paid for the car. ∎

Using a z-score Table

The relation between z-scores and percentiles is best summarized by a table, such as Table 9.7. Once we calculate the z-score of a data point, we simply look up the corresponding percentile in the table.

Table 9.7 *z-scores and percentiles*

z-score	Percentile	z-score	Percentile	z-score	Percentile	z-score	Percentile
−4.0	0.00	−1.0	15.87	0.0	50.00	1.1	86.43
−3.5	0.02	−0.95	17.11	0.05	51.99	1.2	88.49
−3.0	0.13	−0.90	18.41	0.10	53.98	1.3	90.32
−2.9	0.19	−0.85	19.77	0.15	55.96	1.4	91.92
−2.8	0.26	−0.80	21.19	0.20	57.93	1.5	93.32
−2.7	0.35	−0.75	22.66	0.25	59.87	1.6	94.52
−2.6	0.47	−0.70	24.20	0.30	61.79	1.7	95.54
−2.5	0.62	−0.65	25.78	0.35	63.68	1.8	96.41
−2.4	0.82	−0.60	27.43	0.40	65.54	1.9	97.13
−2.3	1.07	−0.55	29.12	0.45	67.36	2.0	97.72
−2.2	1.39	−0.50	30.85	0.50	69.15	2.1	98.21
−2.1	1.79	−0.45	32.64	0.55	70.88	2.2	98.61
−2.0	2.28	−0.40	34.46	0.60	72.57	2.3	98.93
−1.9	2.87	−0.35	36.32	0.65	74.22	2.4	99.18
−1.8	3.59	−0.30	38.21	0.70	75.80	2.5	99.38
−1.7	4.46	−0.25	40.13	0.75	77.34	2.6	99.53
−1.6	5.48	−0.20	42.07	0.80	78.81	2.7	99.65
−1.5	6.68	−0.15	44.04	0.85	80.23	2.8	99.74
−1.4	8.08	−0.10	46.02	0.90	81.59	2.9	99.81
−1.3	9.68	−0.05	48.01	0.95	82.89	3.0	99.87
−1.2	11.51	0.0	50.00	1.0	84.13	3.5	99.98
−1.1	13.57	—	—	—	—	4.0	100.00

EXAMPLE 6   *Cholesterol Levels*

Suppose that the cholesterol levels of American men are normally distributed with a mean of 200 and a standard deviation of 15. If your cholesterol level is 190, in what percentile are you located? What cholesterol level corresponds to the 90th percentile, at which treatment may be necessary?

Solution: The *z*-score for a cholesterol level of 190 is

$$z\text{-score} = \frac{\text{value of data point} - \text{mean}}{\text{standard deviation}} = \frac{190 - 200}{15} = -0.667.$$

Table 9.7 shows that a *z*-score of −0.65 corresponds to the 25.78th percentile, and a *z*-score of −0.7 corresponds to the 24.20th percentile. Because a *z*-score of −0.67 lies between these two values, we conclude that it corresponds to a value somewhere in the 24th or 25th percentile.

Table 9.7 also tells us that percentile 90.3 corresponds to a *z*-score of 1.3. Thus the 90th percentile is about 1.3 standard deviations above the mean. Given that the mean cholesterol level is 200 and the standard deviation is 15, a cholesterol level 1.3 standard deviations above the mean is

$$200 + (1.3 \times 15) = 219.5.$$

The 90th percentile corresponds to a cholesterol level of 219.5.   ■

EXAMPLE 7   *SAT Scores*

Suppose that you score in the 85th percentile on an SAT exam, while your best friend scores in the 47th percentile. Assuming that the mean on the exam is 500 and the standard deviation is 100, approximately what were your actual scores?

Solution: From Table 9.7, the 85th percentile corresponds to a *z*-score between 1.0 (percentile 84.13) and 1.1 (percentile 86.43). This means that you scored between 1 and 1.1 standard deviations, or between 100 and 110 points, above the mean of 500. Thus your score was between 600 and 610. The 47th percentile falls between a *z*-score of −0.1 (percentile 46.02) and −0.05 (percentile 48.01), which means your friend scored between 0.1 and 0.05 standard deviations, or between 10 and 5 points, below the mean of 500. Thus your friend's score was between 490 and 495.   ■

MARGIN OF ERROR AND CONFIDENCE INTERVALS

In Unit 2A, we discussed how surveys and opinion polls often quote a margin of error. This allows researchers to extrapolate results from a relatively small sample to a large population. With an understanding of the normal distribution, we can look in more detail at how margins of error are calculated. We will consider the most common situation in which the goal of the survey is to estimate a population *fraction* (or percentage); for example, the fraction of a population that supports a candidate or the fraction of a population that responds favorably to a drug.

Taking Many Samples

Imagine that you want to know what percentage of the 400 people employed at a local company is happy with their jobs. You decide to survey all 400 people, and each Y(es) or N(o) below is one person's answer to the question "are you happy with your job?"

```
YNYYNYNNYNYYNNNYYYYNYNYNYYYNYYNYYNNNYYYNYNYNYNYYYY
YYNYYNYYNNNYYYNYNYNYNYYYYYNYYNYNNYNYYNNNYYYYNYNYNY
NYYYYNYNYNYYYNYYNYYNNNYYYNYNYNYNYYYYYNYYNYNNYNYYNN
NNNYYYNYNYNYNYYYYYNYYNYNYNNYNYYNNNYYYYNYNYNYYYNYYNYY
NYYNYNNYNYYNNNYYYNNNYYYNYNYNYNYYYYYYNYNYNYNYYYNYYNYY
NYYYNYYNYYYYNNNYYYNNNYYYNYNYNYYNYNNYNYNYNYNYYYYYYN
YNYNYYYYYYNYNYNYYNYNNYNYYNNNYYYNNNYYYNYNYNNYYYNYYNYY
YYYNYNYNYYYNYYNYNNYNYYNNNYYYNNNYYYNYNYNYNYNYYYNYYNYY
```

You could, of course, count the number of Ys in the above list. You'd find that 240 of the 400 responses were Y(es), so the *exact* percentage of employees happy with their jobs is

$$P = \frac{240}{400} = 60\%.$$

This percentage, $P = 60\%$, is a *population parameter*. It reflects the true percentage of the *population* of 400 employees that is happy with their jobs.

However, suppose instead that you choose to count the responses of only 25 people. Imagine that you draw 25 responses from the above list at random and generate the following *sample*.

YNYYNYYNYNYYNYYYNNYYYNNYY

Note that the percentage of Ys in this list is

$$p = \frac{16}{25} = 64\%.$$

This percentage, $p = 64\%$, is a *sample statistic* because it reflects the percentage only among a *sample* drawn from the entire population. (Note that we use capital P for the population parameter and lower case p for the sample statistic.)

Now, suppose that you repeat the process of drawing a random sample of 25 responses hundreds of times, and calculate the sample percentage p each time. Because the true percentage of Y(es) responses in the population is 60%, you might expect to find that the percentage of Ys in many of your samples also is close to 60%. Relatively few samples would have percentages far from 60%. In fact, if you make a histogram of how many samples resulted in each different percentage p from 0% through 100%, it would turn out to be a *normal distribution* like that shown in Figure 9.23. The fact that the sample statistics form a normal distribution can be proven mathematically, and is part of an important mathematical theorem called the *central limit theorem*.

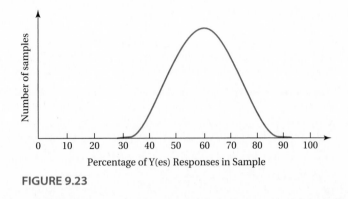

Number of samples

Percentage of Y(es) Responses in Sample

FIGURE 9.23

> **TIME-OUT TO THINK:**   Take a random sample (perhaps by closing your eyes and pointing) of 25 responses from the table of 400 Ys and Ns above. What percentage of Ys do you choose? Where does the result of your sample appear in Figure 9.23?

Note that the *mean* of the distribution in Figure 9.23 is the actual population parameter of 60%, and is the most likely result from a sample. Moreover, we can use the properties of the normal distribution to estimate the *probability* that any single sample reflects the true population parameter.

> **Mathematical Note:**   This result from the central limit theorem holds only under these conditions: (1) the actual population parameter must have a single, clearly defined value; (2) the samples must be chosen randomly; (3) the sample size must be relatively large— large enough so that each possible response (e.g., Y or N) is likely to occur at least five times; and (4) the population must be much larger than the sample size.

According to the central limit theorem, the standard deviation of a distribution of samples such as that shown in Figure 9.23 is approximately

$$\frac{1}{2\sqrt{n}},$$

where n is the size of each sample. Because Figure 9.23 is a distribution based on samples of $n = 25$ responses, its standard deviation is approximately

$$\frac{1}{2\sqrt{25}} = \frac{1}{2 \times 5} = 0.1 = 10\%.$$

We can now apply the 68-95-99.7 rule. If we took many samples of $n = 25$ responses, we'd find that

- about 68% of the samples have a percentage of Ys within 10% (one standard deviation) of the true population percentage $P = 60\%$ (that is, between 50% and 70%).

- about 95% of the samples have a percentage of Ys within 20% (two standard deviations) of the true population percentage $P = 60\%$ (that is, between 40% and 80%).

- about 99.7% of the samples have a percentage of Ys within 30% (three standard deviations) of the true population percentage $P = 60\%$ (that is, between 40% and 90%).

In other words, if we draw a *single* random sample, there is a 68% chance that it will lie within 10% of the mean, a 95% chance that it will lie within 20% of the mean, and a 99.7% chance that it will lie within 30% of the mean. This is the key to understanding sampling.

The *margin of error* in a survey is usually defined to be *two* standard deviations, or

$$2 \times \frac{1}{2\sqrt{n}} = \frac{1}{\sqrt{n}}$$

for a sample of size n. In other words, when we measure a sample statistic with a random sample from a large population, there is a 95% likelihood that it lies within the margin of error of the true population parameter. The margin of error is therefore said to define a *95% confidence interval,* as we discussed in Unit 2A.

EXAMPLE 8   *Checking the Margin of Error*

A poll is conducted to determine the fraction of voters that favors a new nonsmoking initiative. The poll claims to have randomly surveyed 1017 likely voters, and to have a margin of error of ±3.1%. Are these figures consistent?

Solution: The poll seeks to estimate a population fraction: the fraction of voters in the population that support the initiative. Thus, the margin of error formula given above applies in this situation. Assuming that the sample was chosen randomly, the margin of error in a poll with $n = 1017$ is approximately

$$\frac{1}{\sqrt{1017}} = 0.031 = 3.1\%.$$

The margin of error quoted by the polling organization is correct for a sample size of 1017.   ∎

EXAMPLE 9   *Election Polls*

On the same day, three polling organizations survey voters on their preferences for the next election. Poll 1 surveys 500 people and finds that 49% plan to vote for Smith. Poll 2 surveys 1500 people and finds that 47.5% plan to vote for Smith. Poll 3 surveys 3000 people and finds that 47% plan to vote for Smith. Find the margin of error for each survey. If 50% of the vote is needed to win the election, is Smith likely to win? Are the three surveys consistent?

Solution: Again, we are estimating a population fraction: the percentage of people who support Smith. The margins of error for surveys of 500, 1500, and 3000 people are given below.

Poll #1 ($n = 500$ people):   Margin of error $= \dfrac{1}{\sqrt{500}} = 0.045 = 4.5\%$

Poll #2 ($n = 1500$ people):   Margin of error $= \dfrac{1}{\sqrt{1500}} = 0.026 = 2.6\%$

Poll #3 ($n = 3000$ people):   Margin of error $= \dfrac{1}{\sqrt{3000}} = 0.018 = 1.8\%$

By adding and subtracting the margin of error in each survey, we find the following 95% confidence intervals.

Poll #1: Smith's percentage will be 49% ± 4.5%, or between 44.5% and 53.5%.

Poll #2: Smith's percentage will be 47.5% ± 2.6%, or between 44.9% and 50.1%.

Poll #3: Smith's percentage will be 47% ± 1.8%, or between 45.2% and 48.8%.

Neither Poll #1 nor Poll #2 can be used to predict the outcome of the election because their confidence intervals include values on both sides of 50%. However, Poll #3 gives a 95% probability that Smith will receive between 45.2% and 48.8% of the vote. Based on this result, Smith probably will not receive the 50% needed to win this election. Note also that an outcome showing Smith receiving less than 48.8% of the vote falls within the 95% confidence limits of the first two polls. Thus all three surveys are consistent with the prediction that Smith will receive less than 48.8% of the vote. ∎

EXAMPLE 10   *Unemployment Data*

The U.S. unemployment rate is calculated from a monthly survey of 60,000 households conducted by the Bureau of Labor Statistics. Suppose that the unemployment rate for a particular month is reported as 5.5%. What is the 95% confidence interval for this result?

Solution: The survey concerns the percentage of people who are unemployed. Thus, with $n = 60,000$ people in a sample, the margin of error is about

$$\frac{1}{\sqrt{60,000}} = 0.004 = 0.4\%.$$

The 95% confidence interval is 5.5% ± 0.4%, or from 5.1% to 5.9%. ∎

TIME-OUT TO THINK: Suppose that the reported unemployment rate falls from 5.9% in April to 5.8% in May. Should you be confident that a smaller percentage of the population is unemployed? Why or why not?

EXAMPLE 11   *TV Nielsen Ratings*

Nielsen Media Research has conducted surveys of American radio and TV viewing prefer-ences since 1936. As of 1997, the organization uses random samples of 5000 households to determine the top TV programs during a given week. Suppose that, during a week in March, the NCAA basketball tournament is found to have a 65.0% share of the view-ing audience. What is the margin of error in this figure and what is the 95% confidence interval?

Solution: Once again we are estimating a population percentage. Thus, with $n = 5000$ households in the sample, the margin of error is

$$\frac{1}{\sqrt{5000}} = 0.014 = 1.4\%.$$

The 95% confidence interval is 65% ± 1.4%. Thus, with 95% confidence, between 63.6% and 66.4% of the viewing audience was watching the NCAA basketball tournament. ∎

REVIEW QUESTIONS

1. What is the *normal distribution*? How is it related to a *bell-shaped curve*?

2. Under what conditions should we expect data to follow a normal distribution? Give a few examples of data sets that follow a normal distribution.

3. Describe the *68-95-99.7 rule*, and how it is used.

4. What is a *percentile*? What is a *z-score*? How are they related? Describe the use of Table 9.7.

5. Explain the meaning of the *margin of error* in a survey in terms of a distribution of many samples.

6. How does the margin of error in a survey depend on the sample size?

7. How is the margin of error related to the idea of a 95% confidence interval?

PROBLEMS

1. **Normal Distributions.** Assume that a set of test scores is distributed according to the normal distribution with a mean of 60 and a standard deviation of 10.

   a. What percentage of the scores lies between 50 and 70?

   b. What percentage of the scores lies between 60 and 70?

   c. What percentage of the scores lies between 50 and 60?

   d. What percentage of the scores lies between 40 and 80?

   e. What percentage of the scores lies between 30 and 90?

   f. What percentage of the scores lies between 30 and 60?

2. **SAT Scores.** The SAT is designed so the scores are nor-mally distributed with a mean of 500 and a standard deviation of 100.

   a. What percentage of the scores falls between 400 and 600?

   b. What percentage of the scores falls between 300 and 700?

   c. What percentage of the scores falls between 500 and 700?

   d. In what percentile is a score of 600?   a score of 300?

3. **Graphing the Normal Distribution.** Make a careful sketch by hand of a normal distribution with a mean of 10 and a standard deviation of 2.

   a. On this graph, indicate (with vertical lines or shading) the region under the curve that is one standard deviation on either side of the mean. What percentage of the observations lies in this region?

   b. Indicate the region under the curve that is two standard deviations on either side of the mean. What percentage of the observations lies in this region?

4. **Graphing the Normal Distribution**. Make a careful sketch by hand of a normal distribution with a mean of 20 and a standard deviation of 4.

   a. On this graph, indicate (with vertical lines or shading) the region under the curve that is one standard deviation on either side of the mean. What percentage of the observations lies in this region?

   b. Indicate the region under the curve that is two standard deviations on either side of the mean. What percentage of the observations lies in this region?

5. **Raw Data to Z-Scores.** Suppose a data set has a mean of 50 and a standard deviation of 8. Convert the following data values to z-scores.

   a. 50   b. 45
   c. 60   d. 40
   e. 65   f. 42
   g. 58   h. 75

6. **Raw Data to Z-Scores.** Suppose a data set has a mean of 80 and a standard deviation of 5. Convert the following data values to z-scores.

   a. 80   b. 85
   c. 75   d. 90
   e. 92   f. 78
   g. 101   h. 72

7. **Z-Scores and Percentiles.** Use Table 9.7 to answer the following questions.

   a. What is the z-score of a data value that is 1 standard deviation above the mean? In what percentile is that data value?

   b. What is the z-score of a data value that is 1.5 standard deviations below the mean? In what percentile is that data value?

   c. What is the z-score of a data value that is 2 standard deviations below the mean? In what percentile is that data value?

   d. What is the z-score of a data value that is 1.5 standard deviations above the mean? In what percentile is that data value?

   e. How many standard deviations above the mean is a data value in the 94th percentile?

   f. How many standard deviations below the mean is a data value in the 6th percentile?

8. **Z-Scores and Percentiles.** Use Table 9.7 to answer the following questions.

   a. What is the z-score of a data value that is 0.5 standard deviation above the mean? In what percentile is that data value?

   b. What is the z-score of a data value that is 1 standard deviation below the mean? In what percentile is that data value?

   c. What is the z-score of a data value that is 2 standard deviations above the mean? In what percentile is that data value?

   d. What is the z-score of a data value that is 1.9 standard deviations above the mean? In what percentile is that data value?

   e. How many standard deviations above the mean is a data value in the 85th percentile?

   f. How many standard deviations below the mean is a data value in the 10th percentile?

9. **GRE Scores.** Suppose that the scores on the Graduate Record Exam (GRE) are normally distributed, with a mean of 497 and a standard deviation of 115.

   a. If a graduate school requires a GRE score of 650 for admission, to what percentile does this correspond?

   b. If a graduate school requires a GRE score in the 95th percentile for admission, to what actual score does this correspond?

10. **Textbook Survey.** Suppose that, at a university with 25,000 students, the amount of money spent by students on books is normally distributed. The mean book amount per semester per student is $150, with a standard deviation of $25. About how many of the 25,000 students spend *less* than $100 on books each semester? Explain.

11. **Close to Normal Income Distributions.** Assume that the income distribution for assembly line workers at a large manufacturing plant is nearly normal with a mean of $30,000 and a standard deviation of $6000. Using Table 9.7, estimate the percentage of workers that earned between $27,000 and $33,000.

12. **Stanford-Binet IQ Test Scores.** Assume that IQ test scores are normally distributed with a mean of 100 and a standard deviation of 15. Using Table 9.7, determine the range of IQ scores for the middle 50% of the population. (That is, the scores in percentiles 25 through 75.)

13. **Heart Rates.** Suppose you do a survey on the resting heart rate of 200 classmates. After plotting the data on a histogram, you see that the distribution is roughly normal with a mean and median very close to 70 (beats per minute). You calculate the standard deviation and find that it is 15 (beats per minute).

    a. About how many classmates have a resting heart rate between 55 and 85?

    b. About how many classmates have a resting heart rate between 40 and 100?

    c. About how many classmates have a resting heart rate between 50 and 90?

    d. If you have a resting heart rate of 45, in what percentile are you?

    e. A friend has a resting heart rate of 110. In what percentile is she?

14. **Sample and Population Percentages.** Suppose that, in a suburb of 12,345 people, a total of 6523 people actually moved to the suburb within the last five years. You survey 500 people, and find that 245 of the people in your sample moved to the suburb in the last five years. What is the population percentage of people who moved to the city in the last five years? What is the sample percentage of people who moved to the city in the last five years? Does your sample appear to be representative of the population? Discuss.

15. **Sample and Population Percentages.** Suppose that, in a school with 1348 students, 305 students are left-handed. You survey 100 students, and find that 23 of the students in your sample are left-handed. What is the population percentage of left-handed students? What is the sample percentage of left-handed students? Does your sample appear to be representative of the population? Discuss.

16. **Estimating Population Percentages.** You select a random sample of 150 people at a convention attended by 1608 people. Within your sample, you find that 73 people have traveled from abroad. Based on this sample statistic, estimate how many people at the convention traveled from abroad.

17. **Estimating Population Percentages.** A random sample of 1320 people is selected from the 4500 people attending a soccer game. Within the sample, 103 people are supporting the visiting team. Based on this sample statistic, estimate how many people at the game support the visiting team.

18. **Distribution of Sample Percentages.** Suppose that 34% of the students at a large university are from out-of-state. Imagine that you select many random samples of 400 students from this university, and determine the percentage of out-of-state students in each sample. If you calculate the mean and standard deviation of the percentages found in the many samples, what values should you expect? Why? Make a sketch of the distribution of sample percentages that you would expect to find from the many samples.

19. **Distribution of Sample Percentages.** Suppose that you work at a large company with several thousand employees, 45% of whom are women. Imagine that you select many random samples of 900 employees, and

compute the percentage of women in each sample. If you calculate the mean and standard deviation of the percentages found in the many samples, what values should you expect? Why? Make a sketch of the distribution of sample percentages that you would expect to find from the many samples.

20. **Election Predictions.** In a random sample of 1600 people from a large city, it is found that 900 support the current mayor in the upcoming election. Based on this sample, would you claim that the mayor will win a majority of the votes? Explain.

21. **Election Predictions.** In a random sample of 2500 people from a large city, it is found that 1300 support the current mayor in the upcoming election. Based on this sample, would you claim that the mayor will win a majority of the votes? Explain.

22. **Unemployment Survey.** The U.S. Department of Labor's Bureau of Labor Statistics estimates the unemployment rate in the United States monthly by surveying 60,000 households.

   a. The unemployment rate generally is stated each month to the nearest tenth of a percent. For the margin of error with a sample size of 60,000, is this precision reasonable? Explain.

   b. Suppose that the number of households surveyed were increased by a factor of four (to 240,000). By how much would the margin of error change?

   c. Suppose that the number of households surveyed were decreased by a factor of four (to 15,000). By how much would the margin of error change?

23. **Nielsen TV Ratings.** Nielsen Media Research uses samples of 5000 households to rank TV shows. Suppose Nielsen reports that NFL Monday Night Football had 35% of the TV audience. What is the 95% confidence interval for this result?

24. **Opinion Poll.** A poll finds that 54% of the population approves of the job that the president is doing; the poll has a margin of error of 4%.

   a. What is the 95% confidence range on the true population percentage that approves of the president's performance?

   b. According to this poll, what is the probability that *less* than half the population approves of the president's performance?

   c. What was the size of the sample for this poll?

25. **Project: Sample Percentages Experiment.** When all 120 people at a party are asked whether they are single (S) or married (M), the responses are as follows:

   S M S S S S M S M M M S M M S S M M M M S S S M
   S M S S S S M S M M S S M S M M M S S S M M M S
   S M S S S S S S M S M S M S M S M S S S M M M M
   S S M M M M S S M S M S M M M S S S M S M S M
   S M M M S M S S S S M S M S S S S M S M S M S S

   a. Describe a way (for example, drawing tags labeled with S or M from a hat) to randomly sample 16 of the people at the party.

   b. Select a random sample of 16 people using the method in part (a) and compute the percentage of married people in the sample.

   c. Repeat part (b) 15 times and record the sample percentage for each sample. What is the mean of your sample proportions?

   d. Make a histogram of the 15 sample proportions. Does it have the shape you should expect? Explain.

   e. Based on your 15 samples, what would you predict as the fraction of people at the party that are married? Does your prediction agree with the actual population fraction (which you can find by counting the responses in the complete list)?

26. **Project: Sample Means Experiment.** We studied the problem of estimating a population proportion (fraction) from sample proportions. Another common problem is to estimate a population mean (average) from sample means. As shown below (adapted from *Seeing Through Statistics* by J. M. Utts, 1996), the two problems have some similarities. Suppose the weights of 50 people at a party are:

   140 150 120 141 139 138 144 160 145 135 110
   159 155 145 149 139 164 160 124 158 102 140
   174 138 133 153 150 151 149 148 128 129 168
   164 140 143 151 142 148 149 152 153 129 132
   135 155 149 149 144 162

The goal is to estimate the mean of the weights using the means of small samples.

   a. Describe a way (for example, drawing tags labeled with the individual weights from a hat) to randomly sample nine of the people at the party.

   b. Select a random sample of nine people using the method in part (a) and compute the mean of the weights in the sample.

c. Repeat part (b) 20 times and record the sample mean for each sample. What is the mean of your sample means? Explain.

d. Make a histogram of the 20 sample means. Does it have the shape you should expect? Explain.

e. Based on your 20 samples, what would you predict as the mean weight of people at the party? Does your prediction agree with the actual mean weight for the population (which you can find by computing the mean weight from the complete list)?

UNIT 9D

SAMPLE ISSUES IN STATISTICAL RESEARCH

The best way to understand statistics in everyday life is by examining real studies. In this Unit, we discuss a few interesting issues that can arise in statistical studies. These issues were chosen both because they show how statistical methods are used and because they illustrate the care that must be taken in interpreting statistical research.

TRENDS IN THE SAT

The SAT has been taken by many college-bound high school students since 1941. There are two parts to the SAT: the verbal test and the math test, each with a maximum of 800 points. The mean score of the 11,000 students who took each part in 1941 was scaled to 500. Each year since 1941, the test has had a few common items with the previous year's test. The College Entrance Examination Board (which designs the test) claims that it can make valid comparisons from year to year by analyzing the results of the common items that appear on tests in different years. In other words, it claims that a score of 500 represents the same level of achievement on the test, no matter what year the test is taken.

Because scores from year to year are supposed to be comparable, SAT scores have been widely used to assess the general state of American education. Figure 9.24 shows the average scores on the verbal and math parts of the SAT between 1972 and 1995.

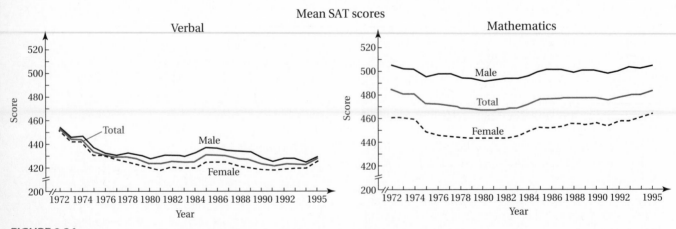

FIGURE 9.24
Verbal and Mathematics SAT scores, 1972–1995.

If the scores from year to year are truly comparable, it would appear that students taking the SAT today do not perform as well as students a few decades ago—particularly in verbal skills. Common sense raises two questions that must be answered to assess the validity of SAT scores as a measure of trends in American education.

1. Are trends among the *sample* of high school students who take the SAT representative of trends for the *population* of all high school students?

2. Even with a short section of common questions that link one year to the next, the test changes every year. Can test scores from one year legitimately be compared to scores in other years?

Unfortunately, neither question can be answered with a clear *yes*. For example, many people argue that much of the long-term decline in SAT scores is the result of changes in the makeup of students who take the test. Figure 9.25 shows that only about a third of high school graduates took the SAT in the 1970s, whereas over 40% of high school graduates were taking the SAT by the 1990s. If these samples represent the top tier of high school students, the decline in score averages might simply reflect the fact that a greater range of student abilities is represented in later samples. Because of the change in the sample makeup, inference from sample trends to population trends is extremely difficult.

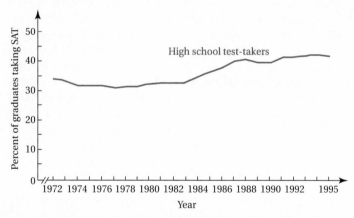

FIGURE 9.25

Percentage of high school graduates taking the SAT, 1972–1995.

The second question is even more difficult. For example, even if some test questions stay the same from one year to the next, different material may be emphasized in high schools—thereby changing the likelihood of students answering the same questions correctly from one year to the next.

Further complications arise when the test undergoes major changes. For example, the 1994 test was the first to allow the use of calculators on the mathematics section, additional time was allotted for the entire test, and an entire section of prior verbal SATs was deleted (a section on antonyms). Note also, in Figure 9.24, that scores in mathematics seem to have jumped significantly in both 1994 and 1995 relative to earlier years, and verbal scores rose significantly in 1995. Clearly, the change in the structure of the test makes it difficult to determine whether these rises in average scores reflect improvements in education.

The search for trends in SAT scores became even more difficult in 1996. The average scores on the SAT had fallen from around 500 in the 1960s to the mid-400s by the 1990s. In 1996, the College Board "recentered" SAT scores so that the average score on both the verbal and math sections once again became 500. The net effect of this change was to raise everyone's SAT scores significantly. For example, a score of 500 in 1997 might be the equivalent of only a 480 in 1995. Clearly, this change makes it far more difficult to do year-to-year comparisons of SAT scores.

> **TIME-OUT TO THINK:**   Do you think that the recentering of SAT scores was a good idea? Why or why not?

Average SAT scores are used not only to make comparisons over time, but also to compare results between different states or different school systems. However, this type of comparison also can be difficult. The first two columns of Table 9.8 show how several selected states ranked nationally in terms of average SAT scores. For example, Alabama's rank of 14 means it had the fourteenth highest average SAT score among the 50 states. However, as shown in the third column, different states had widely different percentages of high school students taking the SAT: the range in the table goes from 4% for Mississippi and Utah to 88% for Connecticut. If the students taking the SAT in any particular state represent top-tier students, then the meaning of the average SAT score is very different from state to state. In that case, for example, Utah's average score really represents the average score among the top 4% of Utah high school students—which could not be expected to represent the average of *all* Utah high school students. In contrast, Connecticut's average score would be a fairly good measure of its average high school students, since the vast majority of students in Connecticut take the SAT. The final column of Table 9.8 shows an "adjusted ranking" based on the percentage of test takers in each state.

Table 9.8   Ranking of states by SAT scores

State	Rank	Percentage of High School Students Taking SAT	Adjusted Rank
Alabama	14	9%	37
Arkansas	18	6%	47
Colorado	23	28%	8
Connecticut	33	88%	14
Mississippi	16	4%	50
New Hampshire	27	78%	3
Utah	4	4%	42

(*New York Times*, 3/27/97)

MEDICAL RESEARCH AND THE PLACEBO EFFECT

In Unit 2A, we discussed how control groups can be created in statistical experiments by using a *placebo*. This practice is particularly common in testing the effectiveness of new drugs or therapies. A placebo is something that looks and feels like the drug being tested but lacks its active ingredients. In a *double-blind experiment*, neither patients nor researchers know who receives the placebo and who receives the real drug until the end of the study.

Surprisingly, the placebo often proves to be a highly effective treatment itself! Indeed, some studies have shown that as many as 30–40% of patients receiving a placebo show improvement, at least temporarily. This **placebo effect** is well documented, but not well understood.

The placebo effect was identified at least 400 years ago in an essay written by Michel de Montaigne, a great French Renaissance philosopher. The placebo effect has been documented in countless studies. However, the results are often difficult to interpret. For example, in a recent study of the drug Proscar, which is designed to shrink enlarged prostate glands, patients on the placebo showed no improvement; that is, their prostate glands continued to enlarge. However, these patients reported feeling better, demonstrating that the placebo had a strong psychological effect even though the physical effect of the placebo was nonexistent.

> *A strong imagination brings on the [placebo effect]....*
> *Everyone feels its impact, but some are knocked over by it.*
> *... [Doctors] know that there are men on whom the mere*
> *sight of medicine is operative.*
> —*Michel de Montaigne* (1533–1592)

In other studies, the way in which the drug and placebo are administered has been explored. If either the drug or the placebo is given to patients by a doctor with a positive or encouraging attitude, there can be more improvement than if they are administered in a skeptical way. Similarly, patients who sign a consent form saying that they *might* receive a placebo tend to show more improvement than patients who know nothing about the use of a placebo.

> *With proper treatment, a cold can be cured in a week. Left to itself, it may linger for seven days.*
> —*A Medical Folk Saying*

It is important to understand the placebo effect for at least two reasons. First, it causes problems in research. After all, if up to one-half of patients receiving the placebo report improvement, how can researchers determine whether the real treatment makes a difference? In the terminology of the drug-testing industry, the drug itself has a *specific effect*. All other factors—placebos, the role of the doctor, and the attitude of the patient—are called *nonspecific effects*. Some statistical studies try to separate the specific effects from the nonspecific effects and explore the interaction between them.

The second reason why it is important to understand the placebo effect is that it may be an important part of the healing process itself. It appears that because most people trying new treatments *believe* or *hope* that they will be beneficial, the placebo effect is likely to be present. If this is true, then nonspecific effects, such as placebos and positive attitudes, will be an increasingly important part of the treatments that we use for disease.

> **TIME-OUT TO THINK:**   Suppose that the effectiveness of some new treatment turns out to be entirely the result of the placebo effect. Should the treatment still be used? Should it be used only under certain circumstances? Defend your opinion.

EXAMPLE 1   *Drug-Placebo Data*

Suppose that two new drugs, one a strong antibiotic and the other an herbal medicine, are tested to determine their effectiveness in alleviating cold symptoms. One-third of the patients in the study were given the antibiotic, one-third received the herbal medicine, and one-third were given a placebo. The table below shows the percentage of patients who showed significant improvement with each treatment. Comment on the results.

Antibiotic	Herbal	Placebo
48%	32%	36%

Solution: Fewer than half the patients showed significant improvement with any treatment, which is reason to question the effectiveness of any of the treatments. However, the antibiotic group did significantly better than the placebo group, suggesting that the antibiotic was more effective than no treatment at all. The herbal remedy was no more effective than the placebo (in fact, less effective) suggesting that it has no specific effects. ∎

STATISTICAL SIGNIFICANCE

Look again at Example 1 above. At first glance, you might be tempted to conclude that the antibiotic represents an effective treatment while the herbal remedy does not. But suppose that the study were to be repeated with a different sample of people. Would you expect to see the same results?

By now, you realize that any statistical study always involves some degree of randomness. We definitely should *not* expect to get precisely the same results if we repeat the study. Just as we can assign a margin of error and level of confidence to a result obtained by an opinion poll, statisticians have developed ways to assess the **statistical significance** of a result from a study. A result is said to have high statistical significance if it is very *unlikely* to have occurred by chance.

There are many methods for assessing statistical significance, and you will surely study some of them if you take a course in statistics. Here, we describe just a few of the basic ideas. Consider a study in which the effectiveness of vitamin C is compared to the effectiveness of a placebo at preventing colds. Everyone in the study takes a daily pill, which may be either vitamin C or the placebo, for a period of 10 weeks. The question "did a person get a cold during the 10 weeks?" has two possible answers: *yes* or *no*. Similarly, the question "what treatment was used?" has two possible answers: vitamin C or the placebo.

Table 9.9 summarizes the results of the study. Note that it has a row for each of the two possible treatments (vitamin C or placebo) and a column for each of the two possible responses (cold or no cold), along with a row and a column showing totals. For example, the table shows that 45 people who received the vitamin C caught a cold. This type of table is called a **contingency table** because all possibilities, or contingencies, of the study are considered.

Table 9.9	Contingency Table for the Vitamin C versus Placebo Study		
	Cold	No Cold	Total
Vitamin C	45	60	105
Placebo	75	40	115
Total	120	100	220

Because the purpose of the study is to assess the effectiveness of vitamin C at preventing colds, the results can be explained in one of two ways:

- the **hypothesis**: vitamin C is more effective than a placebo at preventing colds.
- the **null hypothesis**: vitamin C has no more effect on colds than the placebo.

The basic idea behind assessing statistical significance is to test whether the study data are sufficient to decide between the hypothesis and the null hypothesis. Note that 120 out of 220 people, or 54.5%, got a cold. If the null hypothesis is true, then the treatments made no difference to the results, so we should expect the percentages of people getting colds in both the vitamin C and placebo groups to be close to this overall 54.5%. If the hypothesis is true, then the percentage of people getting colds should be much smaller among the vitamin C group than the placebo group.

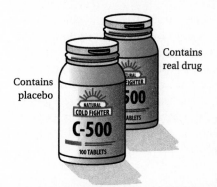

Contains real drug

Contains placebo

The actual results show that the percentage of people getting colds in the vitamin C group was 45 out of 105, or 42.9%, which is 54.5 − 42.9 = 11.6 percentage points lower than predicted by the null hypothesis. The percentage of people getting colds in the placebo group was 75 out of 115, or 65.2%, which is 65.2 − 54.5 = 10.7 percentage points higher than predicted by the null hypothesis. Thus the question of statistical significance boils down to this: could these differences have occurred by chance, or do they suggest that the vitamin C really has an effect?

Statisticians answer this question by calculating the probability that the differences could have come about by chance, and using this probability to define a level of confidence in the results of the study. (This calculation technique goes by the name of a *chi-squared test*.) For example, if the probability of the differences occurring by chance is found to be 5%, then there is only a 5% chance that the null hypothesis is true and a 95% probability that the vitamin C *did* have an effect. We would therefore say that the study suggests that vitamin C helps prevent colds, with results significant at a 95% level of confidence.

We will not cover the details of such probability calculations in this book, but it should be clear that the level of significance will be affected by two key factors in the study:

1. The magnitude of the differences between the actual results and the results expected by the null hypothesis. The larger the differences, the less likely that they occurred by chance.

2. The size of the sample. The larger the number of people in the sample, the less likely that any difference could have occurred by chance.

In our vitamin C example, the results summarized in Table 9.9 turn out to be significant at about the 95% level. Thus, from this study alone, we could claim 95% confidence that vitamin C helps to prevent colds. However, it is very important to remember that 95% confidence is not a sure thing. Moreover, this level of confidence assumes that the study was conducted properly in all stages, and no biases were introduced. For example, a vitamin C versus placebo study should be *double-blind*, in which neither the patients nor researchers know which treatment is given (see Unit 2A). If some patients were able to figure out whether they received vitamin C or the placebo, the study would then be biased and the conclusions unreliable—no matter what level of confidence the calculations suggest.

TIME-OUT TO THINK: Look back at Example 1 in this unit, and suppose that the groups receiving the different treatments consisted of 25 people each. Would you conclude that the antibiotic works and the herbal remedy does not? How would your conclusions change if each group consisted of 10,000 people? Would your conclusions change if you learned that the study was not double-blind? Explain.

THE GENETICS OF INTELLIGENCE

In the 1910s, psychologist Henry Goddard did intelligence testing on immigrants arriving at Ellis Island and concluded that 80% of Jews, Italians, and Hungarians and 90% of Russians were "feeble-minded." As a result, hundreds were deported each year. In Germany, the Nazis latched onto similar ideas when they claimed that they were genetically superior to other people. Jews, blacks, gypsies, homosexuals, and other groups were labeled "subhuman" and chosen for extermination. The Nazis deliberately killed more than 13 million civilians, including 6 million Jews—more than two-thirds of Europe's prewar Jewish population. These historical misuses of intelligence testing probably explain why no other subject in statistics is more widely debated.

Measuring Intelligence: IQ

The best known measure of intelligence, the *intelligence quotient* or IQ, was first developed by the French psychologist Alfred Binet (1857–1911). Binet originally defined the IQ to represent a child's "mental age," which he measured with his IQ test, divided by the child's actual age and converted to a percentage. For example, if a 5-year-old scored as well as an average 6-year-old on the test, Binet said that the 5-year-old had a mental age of 6 and an IQ of $\frac{6}{5} \times 100\% = 120$ Today, the IQ test is defined so that the average person has an IQ of 100, and scores are normally distributed with a standard deviation of 15 points.

EXAMPLE 2   *IQ Percentiles*

What is the range in which 99.7% of IQ scores fall? What IQ score must you have to be in the 98th percentile of IQ scores?

Solution: From the 68-95-99.7 rule, the range in which 99.7% of IQ scores fall is within 3 standard deviations of the mean, or within $3 \times 15 = 45$ points of 100. Thus 99.7% of IQ scores fall between 55 and 145. Table 9.7 in Unit 9C shows that the 98th percentile corresponds to a *z*-score of about 2.05, or 2.05 standard deviations above the mean. Thus, the 98th percentile lies

$$2.05 \times 15 = 30.75$$

points above the mean of 100, or at an IQ of 130.75.   ■

Is Intelligence Hereditary?

Two primary questions lie at the heart of any statistical attempt to demonstrate a hereditary link to intelligence. First, does the IQ truly measure intelligence? Second, is intelligence an innate quality or one that can be manipulated through education?

Neither question has a clear answer. Many psychologists argue that IQ is a valid measure of intelligence, but others believe that a single number can never adequately describe something as complex as human intelligence.

The second question is even more complex. For example, research has shown that *physical* connections between neurons in the brains of young children develop in response to external stimuli almost from the moment of birth and perhaps even before. Because surroundings can affect physical brain development, separating genetic and environmental effects may well be impossible even when testing the intelligence of very young children.

Perhaps the only clear way to distinguish the effects of heredity and environment is through the study of identical twins separated at birth. Such twins have the same genetic makeup but are raised in different environments. Unfortunately for scientists, identical twins separated at birth are rare. Further, the history of this research is badly tainted by a famous case of scientific fraud. In the 1950s, British psychologist Cyril Burt claimed he studied more than 50 pairs of identical twins separated at birth, finding that the twins had nearly identical IQ scores despite the different environments in which they were raised. Burt therefore claimed that IQ is hereditary, and further claimed that different races have different innate levels of intelligence. However, Burt's work was later shown to be fraudulent; indeed, he most likely invented the data reported in his studies!

An interesting note on the question of heredity and intelligence comes from two studies published nearly simultaneously in 1997. A study published by British and Swedish scientists (*Science*, June 1997) claimed to find that heredity is responsible for *at least half* of the difference in scores on tests that measure various mental skills. An American study (*Nature*, July 1997) combined the results of 212 earlier studies and concluded that heredity accounts for *less than half* of the factors that determine IQ. Thus, two studies, published only one month apart, reached contradictory conclusions. Clearly, the debate over how much of intelligence is innate is unlikely to be resolved soon. Moreover, in light of the long and troubled history of this debate, you probably should study new claims about nature versus nurture in intelligence very carefully.

By the Way ············
Alfred Binet developed IQ tests for the purpose of identifying children who needed special help in school and then providing them with that help. He warned against taking his tests as a measure of innate intelligence.

By the Way ············
An excellent summary of how Cyril Burt's fraud was detected, along with a summary of the sad history of attempts to link intelligence with race, is provided in *The Mismeasure of Man*, by Stephen Jay Gould (New York: W. W. Norton and Company, 1981).

REVIEW QUESTIONS ●

1. List several reasons why it may be difficult to compare SAT scores from one year to another. Why is it difficult to compare average SAT scores between different states?

2. What is the *placebo effect*? Explain how it can make it difficult to determine whether a drug has any *specific effects*.

3. What do we mean by statistical significance? Briefly explain how we can contrast results expected by a *hypothesis* and by a *null hypothesis* to determine the statistical significance of the conclusions of a study.

4. How is IQ defined? List a few ways that intelligence tests have been abused in the past, and a few reasons why it is difficult to determine how much of intelligence is inherited.

PROBLEMS ●

1. **Recentered SAT Verbal Scores.** Figure 9.24 shows that, prior to recentering in 1995, the mean SAT verbal score was about 430. Assume that the mean score after recentering was 500, and that the standard deviation remained 100 points both before and after recentering.

   a. Suppose that a student scored 500 on the SAT verbal test prior to recentering. What is the z-score for this value? Use Table 9.7 to determine the student's percentile.

   b. Suppose that another student scored 500 on the SAT verbal test after recentering. What is the z-score and percentile for this student?

   c. Briefly comment on the difficulty of comparing the scores of the students described in parts (a) and (b).

2. **Recentered SAT Mathematics Scores.** Figure 9.24 shows that, prior to recentering in 1995, the mean SAT math score was about 480. Assume that the mean score after recentering was 500, and that the standard deviation remained 100 points both before and after recentering.

   a. Suppose that a student scored 500 on the SAT mathematics test prior to recentering. What is the z-score for this value? Use Table 9.7 to determine the student's percentile.

   b. Suppose that another student scored 500 on the SAT mathematics test after recentering. What is the z-score and percentile for this student?

   c. Briefly comment on the difficulty of comparing the scores of the students described in parts (a) and (b).

3. **SAT Research.** Look for recent articles that discuss the validity of the SAT as a measure of improvements in overall education in the United States. Summarize your findings and your own opinion on this issue in a one- to two-page essay.

4. **Statistical Significance in the News.** Look for a recent newspaper article on a statistical study in which the idea of *statistical significance* is used in the article. Write a one-page summary of the study and the result that is considered to be statistically significant. Also include a brief discussion of whether you believe the result, given its statistical significance.

5. **Contingency Table for Grades and Study.** A study of 400 students in freshman biology was designed to determine if grades could be raised from the first semester to the second semester by attending a special study session during the second semester. The results are summarized in the following contingency table.

	Improvement	No improvement	Total
Study session	145	55	200
No study session	75	125	200
Total	220	180	400

   a. Construct a hypothesis and a null hypothesis for this study.

   b. Briefly describe the results expected in this study if the *null hypothesis* were true.

c. Briefly describe the results expected in this study if the *hypothesis* were true.

d. In terms of percentages, how do the actual results compare to the results expected according to the null hypothesis?

e. In your opinion, would you consider these results to be statistically significant? Why or why not? Be sure to discuss how the magnitude of the differences and the sample size affects your opinion.

6. **Contingency Table for Flu Treatments**. A study of 500 people was designed to determine whether flu shots were effective in preventing flu during the winter months. The results are summarized in the following contingency table.

	Flu	No flu	Total
Flu shots	100	200	300
No flu shots	100	100	200
Total	200	300	500

a. Construct a hypothesis and a null hypothesis for this study.

b. Briefly describe the results expected in this study if the *null hypothesis* were true.

c. Briefly describe the results expected in this study if the *hypothesis* were true.

d. In terms of percentages, how do the actual results compare to the results expected according to the null hypothesis?

e. In your opinion, would you consider these results to be statistically significant? Why or why not? Be sure to discuss how the magnitude of the differences and the sample size affects your opinion.

7. **Drinking and Driving Court Case**. A case that eventually went to the Supreme Court concerned whether there should be different ages at which young men and young women can buy beer. At issue was whether young men are more of a risk than young women when it comes to drinking and driving. A roadside survey determined how many young men and young women who were stopped had been drinking in the last two hours. The results are given in the following contingency table.

Drank Alcohol in Last 2 Hours?			
	Yes	No	Total
Men	77	404	481
Women	16	122	138
Total	93	526	619

(*Seeing Through Statistics,* by Jessica Utts, 1996, and *Statistical Reasoning in Law and Public Policy,* by J. L. Gastwirth, 1988.)

a. Construct a hypothesis and a null hypothesis for this study.

b. Briefly describe the results expected in this study if the *null hypothesis* were true.

c. Briefly describe the results expected in this study if the *hypothesis* were true.

d. In terms of percentages, how do the actual results compare to the results expected according to the null hypothesis?

e. In your opinion, would you consider these results to be statistically significant? Why or why not? Be sure to discuss how the magnitude of the differences and the sample size affects your opinion.

f. The Supreme Court concluded that the statistical significance of this data was not enough to support different drinking ages for men and women. Does this finding surprise you? Why or why not? Present your answer in two to three paragraphs.

8. **IQ Superstitions?** Steven Jay Gould, author of *The Mismeasure of Man,* offers the following suggestion about the interpretation of IQ scores. Explain what you think Gould means. Do you agree with him? Why or why not? Present your answers in a one-page essay.

A certain skepticism about what there is to IQ besides being good at certain sorts of tests may make us less superstitious about its importance.

9. **Project: Overhead Power Lines and Cancer.** Hundreds of scientific and statistical studies have been done in the last 20 years to determine whether high-voltage overhead power lines increase the incidence of cancer among those living nearby. A summary study reported in the

Journal of the American Medical Association (July 1997), based on many previous studies, concludes that there is no significant link between power lines and cancer. Find this report and as many previous studies as possible. Write a two-page paper that summarizes the work that has been done and the current state of the issue.

10. **Project: Nature, Nurture, and IQ.** The questions of whether and how much heredity (nature) and environment (nurture) determine IQ are far from being resolved. Seek out recent articles concerning this controversy, and write a two-page paper that both summarizes the current state of the issue and gives your own views.

CHAPTER 9
SUMMARY

ata and statistics are everywhere, so an understanding of statistical methods is crucial to understanding society. Despite the complexity of many statistical techniques and problems, the basic ideas and techniques can be understood.

- The purpose of exploratory data analysis is to characterize the data. This is done with the use of frequency tables, sample statistics (such as the mean, median, mode, standard deviation), five-number summaries, and box plots.

- When a data set consists of two variables, the technique of linear regression determines the degree of correlation between the variables and how well a straight line fits the data points.

- The purpose of statistical inference is to make predictions about a large population based on data gathered from a relatively small sample of the population.

- The normal distribution lies at the heart of statistics and describes many data sets. Using the 68-95-99.7 rule, z-scores, and percentiles, it is possible to make precise statements about data sets described by the normal distribution.

- The results of statistical studies appear everyday in countless sources. The techniques of this chapter and Chapter 2 will allow you to read and interpret these studies with a critical eye.

SUGGESTED READING

Seeing Through Statistics, J. M. Utts (Belmont, CA: Duxbury Press, 1996).

Statistics by Example, T. Sincich (San Francisco, CA: Dellen Publishing Co., 1990).

Statistics: Concepts and Controversies, D. Moore (New York, NY: W. H. Freeman, 1991).

Statistical Methods for Psychology, D. C. Howell (Belmont, CA: Duxbury Press, 1992).

Statistical Reasoning in Law and Public Policy, J. L. Gastwirth (New York: Academic Press, 1988).

Statistics for Business and Economics, H. Kohler (New York: HarperCollins, 1994).

The Mismeasure of Man, S. J. Gould (New York: W. W. Norton and Company, 1981).

The Mismeasure of Woman, C. Tavris (New York: Touchstone Books, 1993).

The Bell Curve Debate, R. Jacoby and N. Glauberman (eds) (New York: Times Books, 1995).

Chapter 10

MATHEMATICS AND THE ARTS

*I*n our daily lives, mathematics and the fine arts often are treated as though they are utterly different: one rigorous and exact, the other creative and boundless. But this distinction is an illusion. In fact, mathematics and the fine arts are inextricably linked, and both have developed together through history.

[Mathematics] seems to stand for all that is practical, poetry for all that is visionary, but in the kingdom of the imagination you will find them close akin, and they should go together as a precious heritage to every youth.
FLORENCE MILNER, *SCHOOL REVIEW,* 1898

I never had a taste for . . . geometry except in so far as [it] could serve as a means of arriving at some sort of knowledge . . . for the good and convenience of life.
GIRARD DESARGUES (1593–1662),
FRENCH ARCHITECT AND ENGINEER

*T*he connections between art and mathematics are threaded through history to the present day. The ties are most evident in architecture and design: the Great Pyramids in Egypt, the Eiffel Tower in France, and New York City's Chrysler Building were all built with the aid of mathematics. The ties between mathematics and painting or sculpture are less tangible. But as we will see, mathematics has supported art in profound ways over the centuries. Conversely, mathematics has come to be viewed by many as an art in its own right.

Of course, Beethoven's Eroica Symphony or Michelangelo's ceiling of the Sistine Chapel can be reduced to an adequate mathematical formula. There is, however, one little catch. The only person capable of making such an all-inclusive analysis must be able to feel the emotions that Beethoven and Michelangelo felt, to think in sound and paint as they did.

—*J. Murray Barbour*, music historian

In this chapter, we explore some of the connections between mathematics and art. We begin with a brief overview of some key ideas from geometry that underlie much of art and architecture, and then explore several of the most direct connections between mathematics, art, and music.

UNIT 10A

FUNDAMENTALS OF GEOMETRY

The word **geometry** literally means *earth measure.* Many ancient cultures developed geometric ideas to survey the boundaries of flood basins around agricultural fields and to establish patterns of planetary and star motion. However, geometry was always more than just a practical science, as demonstrated by the artistic use of geometric shapes and patterns in cave paintings, pottery decorations, and ancient architecture.

The Greek mathematician Euclid (c. 325–270 B.C.) summarized Greek knowledge of geometry in a 13-volume textbook called *Elements.* The geometry described in Euclid's work, now called **Euclidean geometry,** is the familiar geometry of lines, angles, and planes.

By the Way ·············
Euclid's *Elements* was the primary textbook for geometry throughout the Western world for almost 2000 years. Until recently, it was the second-most reproduced book of all time (after the Bible) and, by almost any measure, the most successful textbook in history.

> **TIME-OUT TO THINK:**   Euclid worked at a university called the Museum, so named because it honored the Muses—the patron goddesses of the sciences and the fine arts. Thus the Greeks thought that the sciences and fine arts went hand-in-hand. Do the sciences and the arts still seem so clearly linked today? Why or why not?

POINTS, LINES, AND PLANES

Geometric objects, such as points, lines, and planes, represent *idealizations* that do not exist in the real world (Figure 10.1). A geometric **point** is imagined to have zero size. No real object has zero size, but many real phenomena approximate geometric points. Stars, for example, appear as points of light in the night sky.

A mathematical point is the most indivisible and unique thing that art can present.

—JOHN DONNE (1573–1631)

FIGURE 10.1
Representations of a point, a line, and a plane.

A geometric **line** is formed by connecting two points along the shortest possible path. A line is imagined to extend infinitely in length and to have no thickness. Because no physical object is infinite in length, we usually work with **line segments**, or pieces of a line. Long straight sections of lane dividers on a highway make good approximations to line segments.

A geometric **plane** is a perfectly flat surface that extends infinitely in length and width but has no thickness. A sheet of paper, a smooth tabletop, and the surface of a chalkboard are everyday approximations to segments of planes.

TIME-OUT TO THINK: Describe at least three additional everyday realizations of points, line segments, and segments of planes. How does each real object compare to its geometric counterpart?

Dimension

The **dimension** of an object can be thought of as the number of independent directions in which you could move if you were on the object. If you were a geometric prisoner confined to a point, you would have no place to go, so a point has zero dimensions. A line is one-dimensional because, if you walk on a line, you can move in only one direction. (Note that forward and backward count as the same direction, but one positive and the other negative.) In a plane, you can move in two independent directions, such as north/south and east/west (or any combination of those two directions). Thus a plane is two-dimensional. In a three-dimensional **space,** such as the world around us, you can move in three independent directions: north/south, east/west, and up/down.

Euclid alone has looked on Beauty bare ...

—EDNA ST. VINCENT MILLAY, SONNET

Another way to think about dimension is by the number of **coordinates** required to locate a point (Figure 10.2). A line is one-dimensional because it requires only one coordinate, such as x, to locate a point. A plane is two-dimensional because it requires two coordinates, such as x and y, to locate a point. Three coordinates, such as x, y, and z, are needed to locate a point in three-dimensional space.

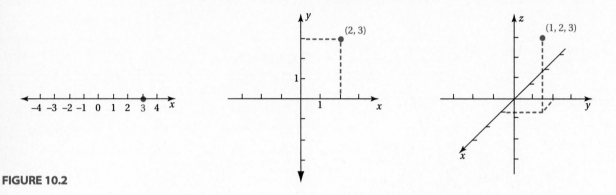

FIGURE 10.2

Angles

The intersection of two lines or line segments forms an **angle.** The point of intersection is called the **vertex.** Figure 10.3(a) shows an arbitrary angle with its vertex at point A, so we call it *angle A,* denoted as $\angle A$. The most common way to measure angles is in *degrees* (°), derived from the ancient base-60 numeral system of the Babylonians. By definition, a full circle encompasses an angle of 360°, so an angle of 1° represents $\dfrac{1}{360}$ of a circle (Figure 10.3b). To measure an angle, we imagine its vertex as the center of a circle. Figure 10.3(c) shows that $\angle A$ **subtends,** or cuts, $\dfrac{1}{12}$ of a circle. Thus $\angle A$ measures $\dfrac{1}{12} \times 360°$, or 30°.

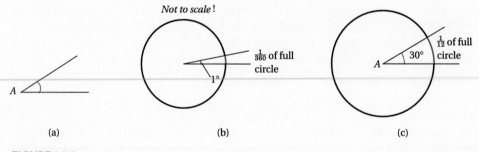

(a)   (b)   (c)

FIGURE 10.3

Mathematical Note:

Angles are also sometimes measured in *radians,* in which case the angle subtended by a full circle is defined to be 2π radians. From this definition, we find that $1° = \dfrac{2\pi}{360}$ radians, or 1 radian $= \dfrac{360°}{2\pi} \approx 57.8°$.

Some angles have special names, as shown in Figure 10.4.

- A **right angle** measures 90°.
- A **straight angle** is formed by a straight line and measures 180°.
- An **acute angle** is any angle whose measure is less than 90°.
- An **obtuse angle** is any angle whose measure is between 90° and 180°.

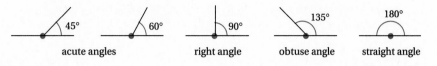

| acute angles | right angle | obtuse angle | straight angle |

FIGURE 10.4

TIME-OUT TO THINK: Draw another acute angle and another obtuse angle. How is the term *acute* in an acute illness related to its meaning in an acute angle? If we say that someone is being obtuse, does the term *obtuse* bear any relation to its meaning in an obtuse angle? Explain.

Two lines or line segments in a plane that meet in a right angle are said to be **perpendicular** (Figure 10.5a). Two lines or line segments in a plane that are the same distance apart at all points are said to be **parallel** (Figure 10.5b). Parallel lines in a plane can never meet.

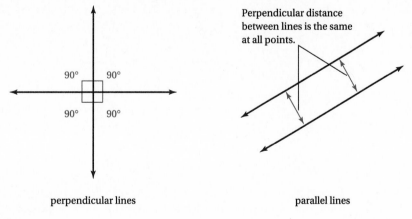

| perpendicular lines | parallel lines |

FIGURE 10.5

EXAMPLE 1   *Angles*

Find the angles that subtend the following:
a) a semicircle (half a circle)   b) a quarter circle
c) an eighth of a circle   d) a hundredth of a circle.

Solution:

a) An angle subtending a semicircle is $\frac{1}{2} \times 360° = 180°$.

b) An angle subtending a quarter circle is $\frac{1}{4} \times 360° = 90°$.

c) An angle subtending an eighth of a circle is $\frac{1}{8} \times 360° = 45°$.

d) An angle subtending a hundredth of a circle is $\frac{1}{100} \times 360° = 3.6°$. ■

PLANE GEOMETRY

Plane geometry involves problems that can be solved with the geometry of two-dimensional planes. Here, we examine only problems involving *circles* and *polygons*. We can make a circle by using a string and a pencil to trace a curve that is a constant distance, or **radius,** from a fixed point, or *center* (Figure 10.6). The **diameter** of a circle is the distance across the circle passing through its center. Thus the diameter is twice the radius of the circle.

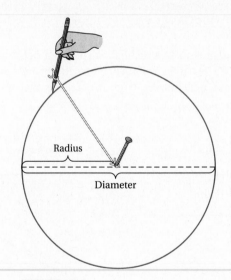

Radius

Diameter

FIGURE 10.6

By the Way
The root *poly* comes from a Greek word for *many,* and *polygon* means a many-sided figure.

Any closed shape in the plane made from straight line segments is a **polygon** (Figure 10.7). A **regular polygon** is a polygon in which all the sides have the same length and all interior angles are equal. Table 10.1 shows several common regular polygons and their names.

FIGURE 10.7
Examples of polygons.

Table 10.1	A Few Regular Polygons				
Sides	Name	Picture	Sides	Name	Picture
3	Equilateral triangle		6	Regular hexagon	
4	Square		8	Regular octagon	
5	Regular pentagon		10	Regular decagon	

Perimeter

The perimeter of a plane shape is simply the length of its boundary. For a polygon, we can find the perimeter by adding the lengths of each individual edge. The perimeter of a circle is called the **circumference.** We can measure the circumference of a circle by carefully laying a piece of string around it, then measuring the length of the string. We can also find the circumference of a circle with a simple formula:

$$\text{circumference of circle} = \pi \times \text{diameter} = 2 \times \pi \times \text{radius}$$

EXAMPLE 2   *Interior Design*

A window consists of a 4-foot by 6-foot rectangle capped by a semicircle (Figure 10.8). How much trim is needed to go around the window?

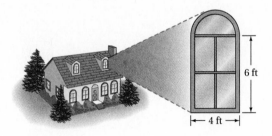

FIGURE 10.8

Solution: The trim around the window must cover the 4-foot base of the rectangle, two 6-foot sides, and the semicircular cap. The straight edges have a total length of

$$4 \text{ ft} + 6 \text{ ft} + 6 \text{ ft} = 16 \text{ ft.}$$

The perimeter of the semicircular cap is half of the circumference of a full circle with a diameter of 4 feet, or

$$\frac{1}{2} \times \pi \times 4 \text{ ft} \approx \frac{1}{2} \times 3.14 \times 4 \text{ ft.} \approx 6.3 \text{ ft.}$$

Thus the total length of trim needed for the window is about

$$16 \text{ ft} + 6.3 \text{ ft} = 22.3 \text{ ft.}$$

Areas

As discussed in Unit 3A, we can measure the area enclosed by any shape by counting the number of small squares that would fit within it (see Figure 3.1). We can also find areas for many shapes with simple formulas. For example, the area of a circle is:

$$\text{area of circle} = \pi \times \text{radius}^2 = \pi \times r^2$$

Another familiar area formula is that of a rectangle: length × width. Sometimes, it is convenient to think of the length and width as the base b and the height h (Figure 10.9a). In that case, a diagonal cut through the rectangle produces two right triangles, and makes the area of each right triangle clear:

$$\text{area of triangle} = \frac{1}{2} \times \text{base} \times \text{height} = \frac{1}{2} \times b \times h$$

In fact, this area formula holds for all triangles (Figure 10.9b).

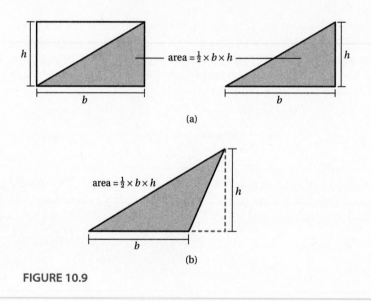

FIGURE 10.9

Similarly, we can find the area formula for a **parallelogram**—a four-sided polygon in which the opposite sides are parallel—by visualizing how we can transform it into a rectangle (Figure 10.10). We then see that its area formula is the same as that of a rectangle:

$$\text{area of parallelogram} = \text{base} \times \text{height} = b \times h$$

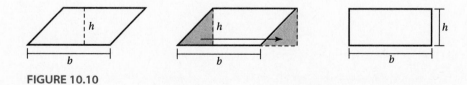

FIGURE 10.10

Many other areas can be computed from these basic formulas. For example, we can find the area of any *quadrilateral* (a four-sided polygon) by dividing it into two triangles and adding the areas of the two triangles (Figure 10.11).

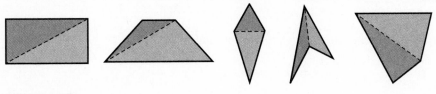

FIGURE 10.11

EXAMPLE 3   *Building Stairs*

You have built a stairway in a new house and want to cover the space beneath the stairs with plywood (Figure 10.12). The stairway is 12 feet along its base and rises at height of 12 feet. What is the area to be covered?

Solution: The area to be covered is triangular, with base and height both equal to 12 feet. Thus the area of this triangle is

$$\text{area} = \frac{1}{2} \times b \times h$$
$$= \frac{1}{2} \times (12 \text{ ft}) \times (12 \text{ ft}) = 72 \text{ ft}^2.$$

The area to be covered is 72 square feet.

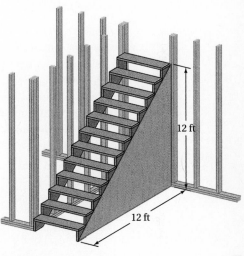

FIGURE 10.12

EXAMPLE 4   *City Park*

A city park is to be developed on an open block bounded by two sets of parallel streets (Figure 10.13). The streets along the block are each 55 yards long and the perpendicular distance between the streets is 39 yards. How much sod should be purchased to cover the entire park in grass?

Solution: The city park is a parallelogram with a base of 55 yards and height of 39 yards. Thus the area of the park is

$$\text{area} = b \times h = 55 \text{ yd} \times 39 \text{ yd} = 2145 \text{ yd}^2.$$

The city will need to purchase 2145 square yards of sod.

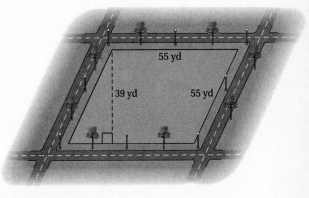

FIGURE 10.13

EXAMPLE 5 *Interior Design*

Look back at the window shown in Figure 10.8. How much glass is needed for the window?

Solution: The total area of the window is the area of the 4-ft × 6-ft rectangle plus the area of a semicircle with a diameter of 4 feet, or a radius of 2 feet. The area of the rectangle is

$$4 \text{ ft} \times 6 \text{ ft} = 24 \text{ ft}^2.$$

The area of a semicircle is half of the area of a full circle, or

$$\frac{1}{2} \times \pi \times (2 \text{ ft})^2 \approx 6.3 \text{ ft}^2.$$

Thus the total amount of glass needed for the window is about $24 \text{ ft}^2 + 6.3 \text{ ft}^2 = 30.3 \text{ ft}^2$. ∎

SOLID GEOMETRY

A **solid** is a closed three-dimensional object, such as a box or a sphere, and *solid geometry* involves problems in three-dimensional space. Two of the most important properties of a solid are its *volume* and its *surface area*. We can always measure volume by counting how many small cubes it would take to fill a solid (see Figure 3.3). Surface area is the total area of all the external faces of a solid. We can measure a solid's surface area by covering it with a piece of cloth and then measuring the area of the cloth. Some solids have simple formulas for volume and surface area. Table 10.2 shows the names of several common solids, along with their volume and surface area formulas.

Table 10.2 Volume and Surface Area Formulas			
Solid Figure	**Picture**	**Volume**	**Surface Area**
Rectangular prism (box)		$V = l \times w \times h$	$A = 2(l \times w + l \times h + w \times h)$
Cube		$V = l^3$	$A = 6 \times l^2$
Right circular cylinder		$V = \pi \times r^2 \times h$	$A = 2\pi \times r^2 + 2\pi \times r \times h$
Sphere		$V = \frac{4}{3} \times \pi \times r^3$	$A = 4\pi \times r^2$

EXAMPLE 6 *Water Reservoir*

A water reservoir has a rectangular base that measures 30 meters by 40 meters, and vertical walls 15 meters high. At the beginning of the summer, the reservoir was filled to capacity. At the end of the summer, the water depth was 4 meters. How much water was used?

Solution: The reservoir has the shape of a rectangular prism, so the volume of water in the reservoir is its length times its width times the depth of water it contains. When filled at the beginning of the summer, the volume of water was

$$30 \text{ m} \times 40 \text{ m} \times 15 \text{ m} = 18,000 \text{ m}^3.$$

At the end of the summer, the amount of water remaining was

$$30 \text{ m} \times 40 \text{ m} \times 4 \text{ m} = 4800 \text{ m}^3.$$

Therefore the amount of water used was $18,000 \text{ m}^3 - 4800 \text{ m}^3 = 13,200 \text{ m}^3$. ■

$\mathcal{T}$HINKING ABOUT . . .

Plato, Geometry and Atlantis

Plato (427–347 B.C.) was one of the many ancient Greeks who emphasized the study of geometry and sought to find geometric patterns in nature. He believed that the heavens must exhibit perfect geometric form, and therefore argued that the Sun, Moon, planets, and stars must move in perfect circles—an idea that was not proven false until the early 1600s. He also believed that the heavenly bodies were held in place by crystalline spheres, which is why the phrase the *music of the spheres* came to mean heavenly music or other things of perfection.

Another of Plato's ideas about geometry and the universe involved the five *perfect solids*—the only possible solid shapes in which all the faces are regular polygons (Figure 10.14). Plato believed that four of the perfect solids represented the four elements that the Greeks thought made up the universe: earth, water, fire, and air. The dodecahedron, he believed, represented the universe as a whole.

Plato presented his ideas in a series of written dialogues. In the dialogue *Timaeus* in which he discussed the role of the perfect solids in the universe, he also invented a moralistic tale about a fictitious land he called *Atlantis*. Interestingly, while Plato's ideas about the universe were abandoned long ago, millions of people today believe that Atlantis really existed. Plato's fiction, in the end, has more adherents than the ideas in which he firmly believed. Commenting on this irony, the popular author Isaac Asimov wrote:

If there is a Valhalla for philosophers, Plato must be sitting there in endless chagrin, thinking of how many foolish thousands, in all the centuries since his time . . . who have never read his dialogues or absorbed a sentence of his serious teachings . . . believed with all their hearts in the reality of Atlantis.

tetrahedron (4 triangular faces)   cube (6 square faces)   octahedron (8 triangular faces)   dodecahedron (12 pentagonal faces)   icosahedron (20 triangular faces)

FIGURE 10.14
The five perfect solids.

EXAMPLE 7 *Comparing Volumes*

Which holds more soup: a can with a diameter of 3 inches and a height of 4 inches or a can with a diameter of 4 inches and a height of 3 inches?

Solution: Recall that radius $= \frac{1}{2}$ diameter. Soup cans have the shape of right circular cylinders, so the volumes of the two cans are:

Can 1: $V = \pi \times r^2 \times h = \pi \times (1.5 \text{ in.})^2 \times 4 \text{ in.} = 28.27 \text{ in.}^3$

Can 2: $V = \pi \times r^2 \times h = \pi \times (2 \text{ in.})^2 \times 3 \text{ in.} = 37.70 \text{ in.}^3$

The second can, with the larger radius but shorter height, has the larger capacity. ∎

EXAMPLE 8 *Size of the Earth*

The radius of the Earth is about 6400 km. What is the total volume of the Earth? What is the total surface area of the Earth?

Solution: The Earth is spherical in shape, so we use the volume and surface area formulas for a sphere with $r = 6400$ km.

$$V = \frac{4}{3} \times \pi \times r^3$$

$$= \frac{4}{3} \times \pi \times (6400 \text{ km})^3 = 1.1 \times 10^{12} \text{ km}^3$$

$$A = 4\pi \times r^2$$

$$= 4\pi \times (6400 \text{ km})^2 = 5.1 \times 10^8 \text{ km}^2$$

The volume of the Earth is about 1.1 trillion cubic kilometers, and the surface area of the Earth is about 510 million square kilometers. ∎

SCALING LAWS

The ideas of surface area and volume are particularly useful when we investigate *scaling*: how one measurement is affected when we change another one by some *scale factor* (see Unit 4C). The most direct application of scaling laws is to scale models. Suppose that we make an engineering scale model of a car on a scale of 1 to 10 (Figure 10.15). That is, the actual car will be 10 times as long, 10 times as wide, and 10 times as tall as the model car. Given this scale, we might ask: How will the surface area and volume of the actual car compare to the surface area and volume of the model?

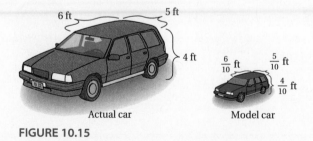

FIGURE 10.15

Consider the area of the car roof, which is its length times its width. Because the length and width of the actual car roof will each be 10 times their sizes in the model, we can calculate the area of the actual car roof from the area of the model car roof as follows:

$$\text{actual roof area} = \text{actual roof length} \times \text{actual roof width}$$
$$= (10 \times \text{model roof length}) \times (10 \times \text{model roof width})$$
$$= 10^2 \times \text{model roof length} \times \text{model roof width}$$
$$= 10^2 \times \text{model roof area}$$

The end result shows that the actual roof area is greater than the model roof area by the *square* of the scale factor, or by $10^2 = 100$ in this case.

We can do a similar calculation for the volume by considering, for example, the box-shaped (rectangular prism) passenger compartment.

$$\text{actual volume} = \text{actual length} \times \text{actual width} \times \text{actual height}$$
$$= (10 \times \text{model length}) \times (10 \times \text{model width}) \times (10 \times \text{model height})$$
$$= 10^3 \times (\text{model length}) \times (\text{model width}) \times (\text{model height})$$
$$= 10^3 \times \text{model volume}$$

Thus the actual car volume is greater than the model car volume by the *cube* of the scale factor, or by $10^3 = 1000$ in this case.

Generalizing our results, we find:

- *Areas* always scale with the *square* of a scale factor.
- *Volumes* always scale with the *cube* of a scale factor.

EXAMPLE 9   *Doubling Your Size*

Suppose that, magically, your size suddenly doubled; that is, your height, width, and depth doubled. For example, if you were 5 feet tall before, you now are 10 feet tall. (a) By what factor has your waist size increased? (b) With your sudden growth, you will need to replace your wardrobe. How much more material will be required for your new set of clothes? (c) By what factor has your weight changed?

Solution:

a) Waist size is measured around your waist, and thus is like a perimeter. Therefore your waist size simply doubles, just like your other *linear* dimensions of height, width, and depth. If you had a 30-inch waist before, it is now a 60-inch waist.

b) Clothing covers surface *area* and therefore scales as the *square* of the scale factor. The scale factor by which you have grown is 2 (doubling), so your surface area has grown by a factor of $2^2 = 4$. If your shirt size used 2 square yards of material before, it now uses 8 square yards of material.

c) Your weight depends on your *volume*, which scales as the *cube* of the scale factor. Your new volume, and new weight, are therefore $2^3 = 8$ times their old values. If your old weight was 100 pounds, your new weight is 800 pounds.   ∎

The Surface Area to Volume Ratio

Another important scaling idea concerns the *relative* scaling of areas and volumes. We define the **surface area to volume ratio** for any object as its surface area divided by its volume:

$$\text{surface area to volume ratio} = \frac{\text{surface area}}{\text{volume}}$$

Note that, because surface area scales with the square of a scale factor and volume scales with the cube of the scale factor, the surface area to volume ratio must scale with the reciprocal of the scale factor:

$$\text{scaling of surface area to volume ratio} = \frac{(\text{scale factor})^2}{(\text{scale factor})^3} = \frac{1}{\text{scale factor}}$$

Thus when an object is "scaled up," its surface area to volume ratio *decreases*. When an object is "scaled down," its surface area to volume ratio *increases*. That is:

- Larger objects have *smaller* surface area to volume ratios than similarly proportioned small objects.
- Smaller objects have *larger* surface area to volume ratios than similarly proportioned large objects.

EXAMPLE 10   *Chilled Drink*

Suppose that you have a few ice cubes and you want to cool a drink quickly. Should you crush the ice before you put it into your drink? Why or why not?

Solution. A drink is cooled by contact between the liquid and the ice surface. Thus the greater the surface area of the ice, the more rapidly the drink will cool. Because smaller objects have *larger* surface area to volume ratios, the crushed ice will have more total surface area than the same volume of ice cubes. The crushed ice therefore cools the drink much more quickly.   ■

REVIEW QUESTIONS

1. What do we mean by *Euclidean geometry*?

2. Give a geometric definition for each of the following: point, line, line segment, plane, plane segment, space. Give an example of an everyday object that can represent each of these geometrical objects.

3. What do we mean by *dimension*? How is dimension related to the number of *coordinates* needed to locate a point?

4. Define an *angle* in geometric terms. What is the *vertex*? What do we mean when we say an angle *subtends* some

portion of a circle? Distinguish between right angles, straight angles, acute angles, and obtuse angles.

5. What is plane geometry? What does it mean for lines to be *perpendicular* or *parallel* in a plane?

6. What is a polygon? How do we measure the *perimeter* of a polygon? Describe how we calculate the areas of a few simple polygons.

7. Describe how we construct a circle. What are the formulas for the circumference and area of a circle?

8. What is solid geometry? Describe how we calculate the volumes and surface areas of a few simple solids.

9. What are the *scaling laws* for area and volume? Explain.

10. What is a *surface area to volume ratio*? How does this ratio change if we make an object bigger? smaller?

PROBLEMS

1. **Geometry Around You.** Look around your room. Briefly describe at least three realizations each of (a) points, (b) lines, and (c) planes.

2. **Dimension.** Examine a closed book.

   a. How many dimensions are needed to describe the book? Explain.

   b. How many dimensions describe the surface (cover) of the book? Explain.

   c. How many dimensions describe an edge of the book? Explain.

   d. Describe some aspect of the book that represents zero dimensions.

3. **Angles and Circles.** Find the angles that subtend:

   a. $\frac{1}{3}$ circle

   b. $\frac{1}{12}$ circle

   c. $\frac{1}{20}$ circle

4. **Angles and Circles.** Find the angles that subtend:

   a. $\frac{1}{30}$ circle

   b. $\frac{1}{90}$ circle

   c. $\frac{3}{4}$ circle

5. **Fractions of Circles.** Find the fraction of a circle subtended by an angle of:

   a. $1°$

   b. $4°$

   c. $15°$

   d. $30°$

6. **Fractions of Circles.** Find the fraction of a circle subtended by an angle of:

   a. $60°$

   b. $90°$

   c. $180°$

   d. $235°$

7. **Angle Practice.** Find each of the unknown angles. Explain your reasoning.

   a. 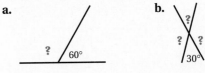   b.

8. **Angle Practice.** Find each of the unknown angles. Explain your reasoning.

   a. 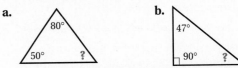   b.

9. **Perpendicular and Parallel.** Suppose that you mark a single point on a line that lies in a fixed plane. Can you draw any other lines in that plane that pass through the point and are *perpendicular* to the original line? Can you draw any other lines that pass through the point and are *parallel* to the original line? Explain.

10. **Perpendicular and Parallel.** Suppose that you draw two parallel lines in a plane. If a third line is perpendicular to one of the two parallel lines, is it necessarily perpendicular to the other as well? Explain.

11. **Circle Practice.** Find the circumference and area for each circle described.

    a. A circle with a radius of 6 meters

    b. A circle with a radius of 4 kilometers

    c. A circle with a diameter of 25 centimeters

12. **Circle Practice.** Find the circumference and area for each circle described.

    a. A circle with a radius of 16 meters

    b. A circle with a diameter of 9 millimeters

    c. A circle with a diameter of 0.5 kilometers

13. **Distance Measurement.** The following city map shows the bus stop nearest to a movie theater. The city blocks are square and are 40 meters long.

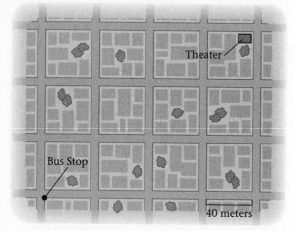

a. What is the shortest *direct* path (as the crow flies) from the bus stop to the theater? (Hint: You'll need to use the Pythagorean Theorem.)

b. What is the shortest path along the sidewalk?

c. Draw two possible routes on the figure that both have the shortest length along sidewalks.

14. **Building Stairs.** Refer to Figure 10.12 showing the area to be covered with plywood under a set of stairs. Suppose that instead of being 12 feet tall, the stairs will rise at a steeper angle and be 14 feet tall. What is the area to be covered in that case?

15. **No Calculation Required.** The end views of two different barns are shown in the following figure. Without calculating, which end has the greater surface area? Explain how you know.

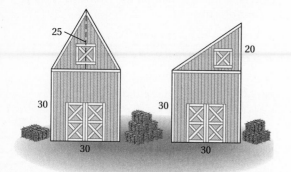

16. **City Park.** The following figure shows a city park in the shape of a parallelogram with a rectangular playground in its center. If all but the playground is covered with grass, what area is covered by grass?

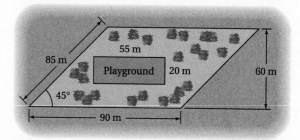

17. **Back Yard.** The following figure shows the layout of a backyard that is to be seeded with grass except for the patio and flower garden. What area is to be seeded with grass?

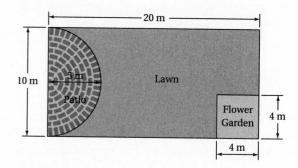

Solid Problems. *Solve Problems 18–24 using the formulas from solid geometry.*

18. A competition swimming pool is 50 meters long, 25 meters wide, and has an average depth of 2 meters. How much water does the pool hold?

19. A large convention center measures 40 meters by 60 meters in floor area, with a ceiling 10 meters high. How much air does it hold, in cubic meters? In liters?

20. How many cubical dice, 3 cm on a side, can be packed into a box that measures 30 cm by 60 cm by 90 cm?

21. A heat duct in the college library has a circular cross section with a radius of 10 inches and a length of 25 feet. What is the volume of the duct and how much paint (in square feet) is needed to paint the duct?

22. Your company manufactures oil drums that have a radius of 0.25 meter and a height of 1 meter. The material used

for the tops and bottoms of the drums costs $7.25 per square meter and the material used for the sides costs $4.50 per square meter. What is the capacity of a single drum? What is the cost of materials for a single drum?

23. A lacrosse ball has a diameter of 4.5 inches. What is its volume and surface area?

24. Three tennis balls fit perfectly when stacked in a cylindrical can. Without calculating, state which is greater: the circumference of the can or the height of the can. Explain your reasoning.

25. **Architectural Model.** Suppose that you build an architectural model of a new concert hall using a scale of 1 to 25.

   a. How will the height of the actual concert hall compare to the height of the scale model?

   b. How will the surface area of the concert hall compare to the surface area of the scale model?

   c. How will the volume of the actual concert hall compare to the volume of the scale model?

26. **Architectural Model.** Suppose that you build an architectural model of a new office complex using a scale of 1 to 50.

   a. How will the height of the actual office complex compare to the height of the scale model?

   b. How will the amount of paint needed for the exterior of the actual office complex compare to the amount of paint needed for the scale model?

   c. Suppose that you wanted to fill both the scale model office complex and the actual office complex with marbles. How many more marbles would be required for the actual building than for the model?

27. **Tripling Your Size.** Suppose that you magically tripled in size; that is, your height, width, and depth tripled.

   a. By what factor has your arm length increased?

   b. By what factor has your waist size increased?

   c. How much more material will be required for your new set of clothes?

   d. By what factor has your weight increased?

28. **Comparing People.** Consider a person, let's call him or her Sam, who is 10% taller than you but proportioned in exactly the same way. (That is, Sam looks like a larger version of you.)

   a. How tall are you? How tall is Sam?

   b. What size is your waist? What size is Sam's waist?

   c. How much do you weigh? How much does Sam weigh?

29. **Squirrels or People?** Squirrels and humans are both mammals that maintain a warm body temperature through metabolism that takes place in the body *volume*. Mammals must constantly generate internal heat to replace the heat that they lose through the *surface area* of their skin.

   a. In general terms, how does the surface area to volume ratio of a squirrel compare to that of a human being?

   b. Which animal must maintain a higher rate of metabolism to replace the heat they lose through their skin: squirrels or humans? Based on your answer, which animal would you expect to eat more food in proportion to its body weight each day? Explain.

30. **Earth and Moon.** Both the Moon and the Earth are thought to have formed with similar internal temperatures about 4.6 billion years ago. Both worlds gradually lose this internal heat to space as the heat passes out through their surfaces.

   a. The diameter of the Earth is about four times greater than the diameter of the Moon. How does the surface area to volume ratio of the Earth compare to that of the Moon.

   b. Based on your answer to part (a), which would you expect to have a hotter interior today: the Earth or the Moon? Why?

   c. Use your answer to part (b) to explain why the Earth remains volcanically active today, while the Moon has no active volcanoes.

31. **Optimizing Area.** You have 132 meters of fence that you want to use to enclose a corral on a ranch. By experimenting with different shapes, determine what shape gives the corral the maximum possible *area*. What is the area of the corral in that case? Explain your reasoning.

32. Optimizing Boxes. You design boxes for a moving company that must have square bottoms and a volume of 1 cubic meter. Material for the boxes costs 10 ¢/m². The boxes currently in use measure 0.5 m by 0.5 m by 4 m.

  a. Confirm that the volume of the current boxes is 1 cubic meter. How much do the materials for producing one of the current boxes cost?

  b. By experimenting with other possible dimensions, find an alternative box shape that requires less total material. Can you find the most efficient possible box design?

  c. How much will your new box save, per box? Suppose that the moving company produces 1 million boxes per year. How much would the company save in total with your new design? Also discuss any drawbacks to changing the dimensions of the box.

33. Optimal Container Design. You are designing wooden crates (rectangular prisms) that must have a volume of 2 cubic meters. The wood used to make the crates costs $12 per square meter. Using trial and error, what is the most economical design for the crates, and how much will each crate cost to manufacture? Explain how you determine the most economical design.

34. Soda Can Design. Standard soft drink cans hold 12 ounces, or 355 milliliters, of soda. Thus their volume is 355 cm³. For this problem, assume that soda cans must be right circular cylinders.

  a. The cost of materials for a can depends on its surface area. Through trial and error with cans of different sizes, find the dimensions of a 12-ounce can that has the lowest cost for materials. Explain how you arrive at your answer.

  b. Compare the dimensions of your can from part (a) to the dimensions of a real soda can from a vending machine or store. Suggest some reasons why real soda cans might not have the dimensions that minimize their use of material.

35. Design of the Human Lung. The human lung has approximately 300 million nearly spherical air sacs (alveoli), each with a diameter of about 1/3 millimeter. The key feature of the air sacs is their surface area because on their surfaces gas is exchanged between the blood stream and the air.

  a. What is the total surface area of the air sacs? What is the total volume of the air sacs?

  b. Suppose that a single sphere were made that had the same volume as the total volume of the air sacs. What is the radius and surface area of such a sphere? How does this surface area compare to that of the air sacs?

  c. If a single sphere had the same surface area as the total surface area of the air sacs, what would be its radius?

  d. Based on your results, comment on the design of the human lung.

36. Automobile Engine Capacity. The size of a car engine is often stated as the total volume of its cylinders.

  a. American car manufacturers often state engine sizes in cubic inches. Suppose that a six-cylinder car has cylinders with a radius of 2.22 inches and a height of 3.25 inches. What is the engine size?

  b. Foreign car manufacturers often state engine sizes in liters. Compare the engine size of the car in part (a) to a foreign car with a 2.2-liter engine.

  c. Look up the number of cylinders and the engine size for your car (if you don't own a car, choose a car that you'd like to own). Estimate the dimensions (radius and height) of the cylinders in your car. Explain your work.

37. The Chunnel. The world's longest tunnel is the English Channel Tunnel, or "Chunnel," connecting Dover, England, and Calais, France. The Chunnel consists of three separate, adjacent tunnels. Its length is approximately 50 km. Each of the three tunnels is shaped like a half-cylinder 4 meters high. How much earth (volume) was removed to build the Chunnel?

38. The Trucker's Dilemma. It is a dark and stormy night. Through the beat of the wipers on your truck you see a rickety country bridge marked "Load Limit 40 Tons." You know your truck, *White Lightning*, like the back of your hand: it weighs 16.3 tons. Trouble is, you are carrying a cylindrical steel water tank that is full of water. The empty weight of the tank is printed on its side: 1750 pounds. But what about the water? Fortunately, the Massey-Fergusson trucker's almanac in your glove box tells you that every cubic inch of water weighs 0.03613 pounds. So you dash out into the rain with a tape measure to find the dimensions of the tank: length = 22 ft, diameter = 6 ft 6 in. Back in the truck, dripping and calculating, do you risk crossing? Explain.

39. Melting Ice Caps. Some scientists are concerned that global warming could eventually cause ice on the Antarctic continent to melt. As this melting ice flows into the ocean, sea level would rise. The ice on the Antarctic continent covers a surface area of about 1.3×10^7 km², with an average depth of about 2 km. (Note: Even if global warming does cause such melting of ice in Antarctica, it probably would take several centuries to melt *all* the ice.)

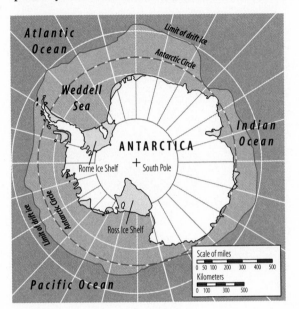

a. Calculate the total volume of ice on the Antarctic continent. Explain your calculation.

b. Because ice is less dense than water, melting the Antarctic ice would add a volume of water to the ocean that is about $\frac{5}{6}$ the volume of the ice. How much water would be added to the ocean if all the Antarctic ice melts, in cubic kilometers?

c. The total surface area of the Earth that is covered by oceans is about 3.4×10^8 km². Based on this fact and your answer to part (b), estimate the rise in sea level if all the Antarctic ice melts. Give your answer in kilometers, meters, and feet. Explain your calculation.

d. In one or two paragraphs, discuss the human impact of a sea level rise such as that found in part (c).

40. Project: The Geometry of Ancient Cultures. Although the work of the ancient Greeks became the basis of modern geometry, many ancient cultures made extensive use of geometry. Research the use of geometry in an ancient culture of your choice. Some possible areas of focus: (1) study the use of geometry in ancient Chinese art and architecture; (2) investigate the geometry and purpose of Stonehenge; (3) compare and contrast the geometry of the Egyptian pyramids to those of Central America; (4) study the geometry and possible astronomical orientations of Anasazi buildings and communities; or (5) research the use of geometry in the ancient African empire of Aksum (in modern-day Ethiopia).

41. Project: The Great Pyramids of Egypt. Egypt's Old Kingdom began in about 2700 B.C. and lasted for 550 years. During that time at least six pyramids were built as monuments to both the life and afterlife of the pharaohs. These pyramids remain among the largest and most impressive structures constructed by any civilization. The building of the pyramids required a mastery of art, architecture, engineering, and social organization at a level unknown before that time. The collective effort required to complete the pyramids transformed Egypt into the first nation-state in the world. Of the six pyramids, the best known are those on the Giza plateau outside Cairo, and the largest of those is the Great Pyramid built by Pharaoh Khufu (or Cheops to the Greeks) in about 2550 B.C. With a square base of 756 feet on a side and a height of 481 feet, the pyramid is laced with tunnels, shafts, corridors and chambers, all leading to and from the deeply concealed king's burial chamber. The stones used to build the pyramids were transported, often hundreds of miles, with sand sledges and river barges, by a labor force of 100,000, as estimated by the Greek historian Herodotus. Historical records suggest that the Great Pyramid was completed in approximately 25 years.

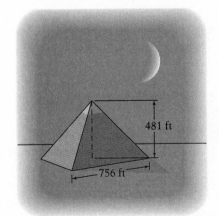

a. To appreciate the size of the Great Pyramid of Khufu, compare its height to the length of a football field (which is 100 yards long).

b. The volume of a pyramid is given by the formula $V = \frac{1}{3} \times$ area of base $\times$ height. Use this formula to estimate the volume of the Great Pyramid. State your answer in both cubic feet and cubic yards.

c. The average size of a limestone block in the Great Pyramid is 1.5 cubic yards. How many blocks were used to construct this pyramid?

d. A modern research team, led by Mark Lehner of the University of Chicago, estimated that the use of winding ramps to lift the stones and desert clay and water for lubrication would allow placing one stone every 2.5 minutes. If the pyramid workers labored 12 hours a day, 365 days a year, how long would it have taken to build the Great Pyramid? How does this estimate compare with historical records? Why did Lehner's research team conclude that the Great Pyramid could have been completed with only 10,000 laborers, rather than the 100,000 laborers estimated by Herodotus?

e. Constructed in 1889 for the Paris Exposition, the Eiffel Tower is a 980-foot iron lattice structure supported on four arching legs. The legs of the tower are at the corners of a square with sides of length 120 feet. If the Eiffel Tower were a solid pyramid, how would its volume compare to the volume of the Great Pyramid?

UNIT 10B

MATHEMATICS AND MUSIC

The roots of mathematics and music are entwined in antiquity. Pythagoras (c. 500 B.C.) claimed that "all nature consists of harmony arising out of number." He imagined that the planets circled the Earth on invisible heavenly spheres, obeying specific numeric laws and emitting the ethereal sounds known as the "music of the spheres." Thus he saw a direct connection between geometry and music. Over a thousand years later, the standard curriculum in medieval universities was the *quadrivium* (Latin for *crossroads*) consisting of arithmetic, geometry, music, and astronomy.

By the Way ··············

Evidence of music is found in nearly all ancient cultures. Indeed, based on evidence dating back 30,000 years, some archaeologists suspect that music may predate both history and speech.

SOUND AND MUSIC

Any vibrating object produces sound. The vibrations produce a **wave** (much like a water wave) that propagates through the surrounding air in all directions. When such a wave impinges on the miraculously designed organ called the ear, it is perceived as sound. Of course, some sounds, such as speech and screeching tires, do not qualify as music. Most musical sounds are made by vibrating strings (violins, cellos, guitars, and pianos), vibrating reeds (clarinets, oboes, and some organ pipes), or vibrating columns of air (other organ pipes, horns, and flutes).

One of the most basic qualities of sound is **pitch.** For example, a tuba has a "lower" pitch than a flute, and a violin has a "higher" pitch than a bass guitar. To understand the origin of pitch, find a taut string (a guitar string works best, but a stretched rubber band will do). When you pluck the string, it produces a sound with a certain pitch. Next, use your finger to hold the midpoint of the string in place, and pluck either half of the string. Note that a higher-pitched sound is produced, demonstrating an ancient musical principle discovered by the Greeks: *The shorter the string, the higher the pitch.*

But what *is* pitch? It took many centuries to arrive at a precise understanding of pitch; it is attributed to both Galileo (1564–1642) and the French monk Marin Mersenne (1588–1648). Plucking the fully taut string causes it to vibrate up and down along its length (Figure 10.16a). For example, if the string vibrates up and down 100 times per second, its **frequency** is said to be 100 **cycles per second** (**cps**); each cycle corresponds to one vibration up *and* down. When you pluck only half the string (by holding down its middle), the resulting wave is *half* as long as the first wave (Figure 10.16b), and its frequency therefore is twice as much, or 200 cps. Thus pitch is related to the frequency of the vibrating string: the higher the frequency, the higher the pitch.

For any particular string, plucking its full length causes it to vibrate at its **fundamental frequency.** Plucking half its length generates a wave with twice the fundamental frequency, called the **first harmonic** (or **overtone**). Halving the length of the string again (Figure 10.16c), doubles the frequency again so that the resulting tone has *four* times the fundamental frequency.

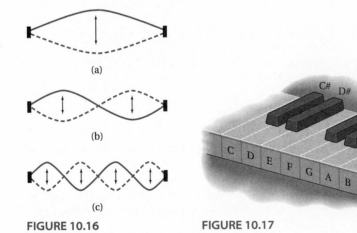

FIGURE 10.16   **FIGURE 10.17**

Now we come to another discovery of the ancient Greeks: doubling the frequency of a tone raises the pitch by an **octave**—perhaps the most appealing and natural combination of notes in music. The piano keyboard (Figure 10.17) is helpful here. An octave is the interval between, say, middle C and the next higher C. For example, middle C on the piano has a frequency of 260 cps, the C above middle C has a frequency of 2×260 cps = 520 cps, and the next higher C has a frequency of 2×520 cps = 1040 cps. Similarly, the C below middle C has a frequency of about $\frac{1}{2} \times 260$ cps = 130 cps.

Music is the universal language of mankind.
—HENRY WADSWORTH LONGFELLOW

> **TIME-OUT TO THINK:**   The note middle A (above middle C) has a frequency of about 440 cps. What is the frequency of the A an octave higher? What is the frequency of the A an octave below middle A?

SCALES

The musical tones that span an octave comprise a **scale.** The Greeks invented the seven-note (or diatonic) scale that corresponds to the white keys on the piano. In the seventeenth century, Johann Sebastian Bach adopted a 12-tone scale for his keyboard music, which corresponds to both the white and black keys on a modern piano. Through Bach's music, the 12-tone scale spread throughout Europe, making it a foundation of Western music. Many other scales are possible: For example, three-tone scales are common in African music, scales with more than 12 tones occur in Asian music, and 19-tone scales are used in contemporary music.

On the 12-tone scale, the frequency separating each tone is called a **half-step**, corresponding to consecutive notes on the piano keyboard. For example, E and F are separated by a half-step, as are F and F# (read "F sharp"). In each half-step, the frequency increases by some *multiplicative* factor; let's call it *f*. Thus the frequency of the note C is the frequency of B times the factor *f*, the frequency of B is the frequency of A times the factor *f*, and so on. The frequencies of the notes across the entire scale are related as follows:

$$C \xrightarrow{f} C\# \xrightarrow{f} D \xrightarrow{f} D\# \xrightarrow{f} E \xrightarrow{f} F \xrightarrow{f} F\# \xrightarrow{f} G \xrightarrow{f} G\# \xrightarrow{f} A \xrightarrow{f} A\# \xrightarrow{f} B \xrightarrow{f} C$$

Because an octave corresponds to an increase in frequency of a factor of 2, the factor *f* must have the property:

$$\underbrace{f \times f \times f \times f \times f \times f \times f \times f \times f \times f \times f \times f}_{12 \text{ times}} = f^{12} = 2$$

Thus, *f* must be the *twelfth root* of two, or $f \approx 1.05946$. You may want to use your calculator to verify that $f^{12} \approx (1.05946)^{12} \approx 2$.

We can now calculate the frequency of every note of a 12-tone scale. Starting from middle C, with its frequency of 260 cps, we multiply by $f \approx 1.05946$ to find that the frequency of C# is 260 cps $\times$ 1.05946 = 275 cps. Multiplying again by $f \approx 1.05946$ gives the frequency of D as 275 cps $\times$ 1.05946 = 292 cps. Continuing in this way generates Table 10.3, shown on the next page.

Column 4 of Table 10.3 shows that a few tones have simple ratios of frequency to middle C. For example, the frequency of G is approximately $\frac{3}{2}$ times the frequency of middle C (musicians call this interval a *fifth*), and the frequency of F is approximately $\frac{4}{3}$ times the frequency of middle C (musicians call this interval a *fourth*). According to many musicians, the most pleasing combinations of notes, called **consonant tones,** are those whose frequencies have a simple ratio. Referring to consonant tones, the Chinese philosopher Confucius observed that small numbers are the source of perfection in music.

Human speech is like a cracked kettle on which we tap crude rhythms for bears to dance to, while we long to make music that will melt the stars.

—GUSTAVE FLAUBERT

Table 10.3	Frequencies of Notes in the Octave Above Middle C		
Note	**Frequency (cps)**	**Ratio to Frequency of Preceding Note**	**Ratio to Frequency of Middle C**
C	260		$1.00000 = 1$
C#	275	1.05946	1.05946
D (second)	292	1.05946	1.12246
D#	309	1.05946	1.18921
E (third)	328	1.05946	$1.25992 \approx \dfrac{5}{4}$
F (fourth)	347	1.05946	$1.33484 \approx \dfrac{4}{3}$
F#	368	1.05946	1.41421
G (fifth)	390	1.05946	$1.49831 \approx \dfrac{3}{2}$
G#	413	1.05946	1.58740
A (sixth)	437	1.05946	$1.68179 \approx \dfrac{5}{3}$
A#	463	1.05946	1.78180
B (seventh)	491	1.05946	1.88775
C (octave)	520	1.05946	$2.00000 = 2$

$By\ the\ Way$ ············

All entries in Column 3 are the same because the same factor, $f \approx 1.05946$, separates every pair of notes. The parenthetical terms in Column 1 are names used by musicians to describe intervals between the note shown and middle C.

EXAMPLE 1   *The Dilemma of Temperament*

Because the ratios of frequencies in Table 10.3 are not exactly ratios of whole numbers, tuners of musical instruments have the problem of *temperament*, which can be demonstrated as follows. Start at middle C with a frequency of 260 cps. Using the whole number ratios in Table 10.3, what is the frequency if you raise C by a sixth to A, raise A by a fourth to D, lower D by a fifth to G, and lower G by a fifth to C? Having returned to the same note, have you also returned to the same frequency?

Solution: According to Table 10.3, raising a note by a sixth increases its frequency by a factor of $\dfrac{5}{3}$; thus the frequency of A above middle C is $\dfrac{5}{3} \times 260$ cps = 433.33 cps. Raising this note by a fourth increases its frequency by $\dfrac{4}{3}$; thus D has a frequency of $\dfrac{4}{3} \times 433.33$ cps = 577.77 cps. Lowering D by a fifth $\left(\text{a factor of } \dfrac{2}{3}\right)$ to G gives a frequency of $\dfrac{2}{3} \times 577.77 =$ 385.18 cps. Finally, lowering G by another fifth $\left(\text{a factor of } \dfrac{2}{3}\right)$ puts us back to middle C, with a frequency of $\dfrac{2}{3} \times 385.18$ cps = 256.79. Note that, by using whole number ratios, we have not returned to the proper frequency of 260 cps for middle C. The problem is that the whole number ratios are not exact. That is, $\dfrac{5}{3} \times \dfrac{4}{3} \times \dfrac{2}{3} \times \dfrac{2}{3} = \dfrac{80}{81}$ is close to, but not *exactly* 1. ∎

Musical Scales as Exponential Growth

The increase in frequencies in a scale is just another example of exponential growth. Each successive frequency is $f \approx 1.05946$ times more, or approximately 5.9% more than the previous frequency. In other words, the frequencies increase at a fixed relative growth rate. Thus we can use the exponential growth law to find any frequency on the scale. Suppose we start at a frequency that we call Q_0. Then the frequency, Q, of the note n half-steps higher is given by the exponential growth law (see Unit 7C):

$$Q = Q_0 \times 1.05946^n$$

EXAMPLE 2   *Exponential Growth on Musical Scales*

Use the exponential growth law to find the frequency of the note one fifth above middle C, the note one octave and one fifth above middle C, and the note two octaves and one fifth above middle C.

Solution: We let the frequency of middle C be the initial value for the scale; that is, we set $Q_0 = 260$ cps. Table 10.3 shows that the note a fifth above middle C is G, which is seven half-steps above middle C. Thus we let $n = 7$ in the exponential law, and find that the frequency of G is

$$Q = Q_0 \times 1.05946^7 = 390 \text{ cps.}$$

The note one octave and a fifth above middle C is $12 + 7 = 19$ half-steps above middle C. Letting $n = 19$, the frequency of this note is

$$Q = Q_0 \times 1.05946^{19} = 779 \text{ cps.}$$

The note two octaves and a fifth above middle C is $(2 \times 12) + 7 = 31$ half-steps above middle C. Letting $n = 31$, the frequency of this note is

$$Q = Q_0 \times 1.05946^{31} = 1558 \text{ cps.} \qquad \blacksquare$$

FROM TONES TO MUSIC

Although the simple frequency ratios of "pure" tones are the building blocks of music, the sounds of music are far richer and more complex. For example, a plucked violin string does much more than produce the simple frequency of its vibration. The string motion is transferred through the bridge of the violin to its top, and the ribs transfer those vibrations to the back of the instrument. With the top and back of the violin in oscillation, the entire instrument acts as a resonating chamber, which excites and amplifies the higher harmonics of the original tone.

Similar principles generate rich and complex sounds in all instruments. Figure 10.18 (top) shows a typical sound wave that might be produced by an instrument. Note that it isn't a simple wave like those pictured in Figure 10.16; instead it consists of a combination of simple waves that are the harmonics of the fundamental. In fact, the complex wave is the *sum* of the other simple waves shown. The fact that a musical sound can be expressed as a sum of simple harmonics is surely the deepest connection between mathematics and music.

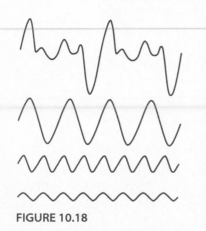

FIGURE 10.18

The French mathematician Jean Baptiste Joseph Fourier first enunciated this principle in about 1810; it was one of the most profound discoveries in mathematics.

Although mathematics helps in understanding music, many mysteries remain. For example, in about 1700, an Italian craftsman known as Stradivarius made what are still considered to be the finest violins and cellos ever produced. Despite years of study by mathematicians and scientists, no one has succeeded in reproducing the unique sounds of a Stradivarius instrument.

Stradivarius was essentially a craftsman of science, one with considerable, demonstrable knowledge of mathematics and acoustical physics.

—THOMAS LEVENSON
IN MEASURE FOR
MEASURE: A MUSICAL
HISTORY OF SCIENCE
(PP. 207–208)

THE DIGITAL AGE

When we talk about sound waves and imagine music to consist of waves, we are working with the **analog** picture of music. Until about 20 years ago, all musical recordings (phonograph cylinders, records, and tape recordings) were based on the analog picture of music. Storing music in the analog mode requires literally storing sound waves; for example, on records, the grooves in the vinyl surface are etched with the shape of the original musical sound wave. If you have listened to analog recordings, you know that this shape can easily become distorted or damaged.

Most recent musical recordings use a **digital** picture of music, in which the sound waves are represented by lists of *numbers.* Modern computer technology generally prevents undesired changes to numbers, making digital storage of music much more reliable than analog storage. Today, the most common medium for storing digital music is the **compact disc.** Computers convert music into numbers which are stored on compact discs, and computer chips in your compact disc player convert the numbers back into sound.

Another advantage of digital technology is that it provides easy and endless ways to "process" music. Through techniques of **filtering** and **digital signal processing,** the sounds of a musical recording can be modified with computers. For example, extraneous sounds (background noises or "hiss") can be detected and removed. Changing the music by amplifying certain frequencies or attenuating others also is possible. Moreover, once digital music can be modified, it is a short step to creating music. Instruments called **synthesizers** can create and imitate a tremendous variety of sounds without strings, brass tubes, or reeds—they do it digitally! In the digital age, the dividing line between mathematics and music all but vanishes.

REVIEW QUESTIONS

1. What is pitch? How is it related to the frequency of a musical note?

2. Define *fundamental frequency, first harmonic,* and *octave.* Why are these concepts important in music?

3. What is a 12-tone scale? How are the frequencies of the notes on a 12-tone scale related to one another?

4. Explain how the notes of the scale are generated by exponential growth.

5. How do the wave forms of real musical sounds differ from the wave forms of simple tones? How are they related?

6. What is the difference between an analog and a digital recording of music? What are the advantages of digital recording?

PROBLEMS

1. **Octaves.** Starting with a tone having a frequency of 110 cycles per second, find the frequencies of the tones that are one, two, three, and four octaves higher.

2. **Octaves.** Starting with a tone having a frequency of 880 cycles per second, find the frequencies of the tones that are one, two, three, and four octaves lower.

3. **Notes of a Scale.** Find the frequencies of the 12 notes of the scale that start at the A above middle C, which has a frequency of 437 cycles per second.

4. **Notes of a Scale.** Find the frequencies of the 12 notes of the scale that start at the G above middle C, which has a frequency of 390 cycles per second.

5. **Exponential Growth and Scales.** Starting at middle C with a frequency of 260 cps, find the frequency of the following notes:
   a. five half-steps above middle C
   b. a fifth above middle C
   c. an octave and a fourth above middle C
   d. 36 half-steps above middle C
   e. four octaves and three half-steps above middle C

6. **Exponential Growth and Scales.** Starting at middle A with a frequency of 437 cps, find the frequency of the following notes:
   a. six half-steps above middle A
   b. a third above middle A
   c. an octave and a fourth above middle A
   d. 25 half-steps above middle A
   e. two octaves and two half-steps above middle A

7. **Exponential Decay and Scales.** What is the frequency of the note five half-steps *below* middle A (which has a frequency of 437 cps)? eight half-steps *below* middle A?

8. **The Dilemma of Temperament.** Start at middle A with a frequency of 437 cps. Using the whole number ratios in Table 10.3, what is the frequency if you raise A by a fifth to E? What is the frequency if you raise E by a fifth to B? What is the frequency if you lower B by a sixth to D? What is the frequency if you lower D by a fourth to A? Having returned to the same note, have you also returned to the same frequency? Explain.

9. **Circle of Fifths.** The circle of fifths is generated by starting at a particular musical note and stepping upward by intervals of a fifth (seven half-steps). For example, starting at middle C, a circle of fifths includes the notes $C \rightarrow G \rightarrow D' \rightarrow A' \rightarrow E'' \rightarrow B'' \rightarrow \ldots$, where each (') denotes a higher octave. Eventually the circle comes back to C several octaves higher.
   a. Show that the frequency of a tone increases by a factor of $2^{7/12} = 1.498$ if it is raised by a fifth. (Hint: Recall that each half-step corresponds to an increase in frequency by a factor of $f = 1.05946$.)
   b. By what factor does the frequency of a tone increase if it is raised by two fifths?
   c. Starting with middle C at a frequency of 260 cycles per second, find the frequencies of the other notes in the circle of fifths.
   d. How many notes are required for the circle of fifths to return to a C? How many octaves are covered by a complete circle of fifths?

e. What is the ratio of frequencies of the C at the beginning of the circle and the C at the end of the circle?

f. A circle of *fourths* is generated by starting at any note and stepping upward by intervals of a fourth (five half-steps). By what factor is the frequency of a tone increased if it is raised by a fourth? How many steps are needed to complete the entire circle of fourths? How many octaves are covered in a complete circle of fourths?

10. Rhythm and Mathematics. In this unit we focused on musical sounds, but rhythm and mathematics are also closely related. For example, in "4/4 time," there are four *quarter notes* in a measure. If two quarter notes have the duration of a *half note*, how many half notes are in one measure? If two *eighth notes* have the duration of a quarter note, how many eighth notes are in one measure? If two *sixteenth notes* have the duration of an eighth note, how many sixteenth notes are in one measure?

11. Project: Experimenting with Waves. Have a friend hold both ends of a cut rubber band so that it is stretched tight. Pluck the entire rubber band, look for the waves on the rubber band, and listen to the pitch of the tone produced; you are hearing the fundamental frequency.

a. Have your friend maintain the same tension in the rubber band. Pinch the rubber band at its midpoint and pluck either half of the rubber band. How does the pitch of the resulting tone change? Explain why the pitch changes, in terms of frequency.

b. Experiment further by pinching the rubber band at various points (keeping the length of the entire rubber band fixed) and plucking on either side of the pinch point. Describe your experiments and the resulting sounds.

PERSPECTIVE AND SYMMETRY

UNIT 10C

We are now ready to turn our attention to some of the connections between mathematics and the visual arts, such as painting, sculpture, and architecture. At least three aspects of the visual arts relate directly to mathematics: *perspective, symmetry*, and *proportion*. In this unit we see how Renaissance mathematicians and artists discovered perspective almost simultaneously, and then we explore the notion of symmetry that runs deeply through all of the arts. In the next unit, we will focus on the concept of proportion and some of its many applications in mathematics, the arts, and nature.

People of all cultures have used geometrical ideas and patterns in their artwork. In the western world, the ancient Greeks developed strong ties between the arts and mathematics because both endeavors were central to their view of the world. Much of the Greek outlook was lost during the Middle Ages, but the Renaissance brought at least two new developments that made mathematics an essential tool of artists. First, there was a renewed interest in natural scenes, which brought a need to paint with realism. Second, many of the artists of the day also worked as engineers and architects.

The desire to paint landscapes with three-dimensional realism brought Renaissance painters face-to-face with the matter of perspective. In their attempts to capture depth and volume on a two-dimensional canvas, these artists made a science of painting. The painters Brunelleschi (1377–1446) and Alberti (1404–1472) are generally credited with developing, in about 1430, a system of perspective that involved geometrical thinking. Alberti's principle that a painting "is a section of a projection" lies at the heart of drawing with perspective.

We can get an idea of the complexity of painting with perspective with an example. Suppose you want to paint a simple view looking down a hallway with a checkerboard tile floor. The situation is shown in Figure 10.19, in which we see a side view of the artist's eye, the canvas, and the hallway. Note that the front of the hallway is closer to the artist than the back. We also see four lines labeled L_1, L_2, L_3, and L_4. The two side walls of the hallway intersect the floor and the ceiling along these four lines. These lines are important because they are parallel to each other in the scene and perpendicular to the canvas (or the extension of the canvas). If such parallel lines are treated properly, the resulting painting will have good perspective.

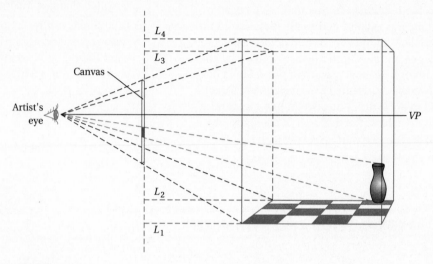

FIGURE 10.19

Let's now look at the scene as the artist sees and paints it. The artist looks down the hallway with the point of view shown in Figure 10.20.

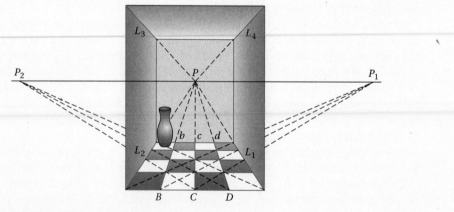

FIGURE 10.20
Adapted from M. Kline, *Mathematics in Western Culture.*

The lines L_1, L_2, L_3, and L_4, which are parallel in the actual scene, are no longer parallel in the painting. In fact, they all meet at a single point, labeled P, which is called the **principal vanishing point**. Thus we have the first principle of perspective discovered by the Renaissance painters:

All lines that are parallel in the real scene and perpendicular to the canvas must intersect at the principal vanishing point of the painting.

Note that other lines that are parallel to lines L_1, L_2, L_3, and L_4, such as the lines going straight down the hallway along the floor tiles, also meet at the principal vanishing point. For example, the line connecting the points B and b intersects P, as does the line connecting C and c, and the line connecting D and d.

What happens to lines that are parallel in the actual scene, but not perpendicular to the canvas, such as the diagonal lines along the floor tiles? If you study Figure 10.20 carefully, you'll see that such sets of lines intersect in their own vanishing points, all on the horizontal line passing through the principal vanishing point; this line is called the **horizon line**. For example, the right-slanting diagonals of the floor tiles all are parallel in the actual scene. But, in the painting, these lines meet at the vanishing point labeled P_1 on the horizon line. Similarly, the left-slanting diagonals meet at a vanishing point called P_2 on the horizon line. In fact, all sets of lines that are mutually parallel in the real scene must meet at their own vanishing point on the horizon line.

> **TIME-OUT TO THINK:**   Imagine looking along a set of long parallel lines that stretches far into the distance, such as a set of train tracks or a set of telephone lines. The lines will appear to your eyes to get closer as you look into the distance. If you were painting a picture of the scene, where would you put the principal vanishing point? Why?

Leonardo da Vinci (1452–1519) contributed greatly to the science of perspective. His broad interests in not only the arts, but also engineering, science, and mathematics, led him to say "let no one who is not a mathematician read my works." We can see da Vinci's mastery of perspective in many of his paintings. If you study *The Last Supper* (see Figure 10.21 in color insert), you will notice several parallel lines in the actual scene intersecting at the principal vanishing point of the painting. Note that the vanishing point is directly behind the central figure of Christ.

The German artist Albrecht Dürer (1471–1528) further developed the science of perspective. Near the end of his life, he wrote a widely read book that stressed the use of geometry and encouraged artists to paint according to mathematical principles. Figure 10.22 is one of Dürer's woodcuts, in which he shows an artist using his principles of perspective. A string from a point on the lute is attached to the wall at

FIGURE 10.22

the point corresponding to the artist's eye. At the point that the string passes through the frame, a point is placed on the canvas. As the string is moved to different points on the lute, a drawing of the lute is created on the canvas—a drawing in perfect perspective as shown in the woodcut.

The artist Jan Vredeman de Vries (1527–1604) summarized much of the science of perspective in a book he published in 1604. Figure 10.23 shows a sketch from this book that illustrates how thoroughly perspective can be analyzed. It also shows that the analysis of perspective in even a simple drawing can be quite complex.

FIGURE 10.23

TIME-OUT TO THINK: Identify some of the vanishing points in the sketch shown in Figure 10.23. What parallel lines from the real scene are converging at each vanishing point?

Curiously, perspective drawing was occasionally abused in the name of art. The painting *False Perspective* (see Figure 10.24 in color insert) by the English artist William Hogarth (1697–1764) deliberately reminds us of how essential perspective is in art. Note where the fishing line of the near man lands, and how the woman in the window appears to be lighting the pipe of a man on a distant hill. The familiar work of Maurits C. Escher (1898–1972) also confounds us with its use and abuse of perspective. His drawing *Belvedere* (see Figure 10.25) illustrates a good use of perspective, and yet the pillars of the structure in the figure are cleverly drawn to show impossible positions.

FIGURE 10.25

SYMMETRY

The term *symmetry* has many meanings in both mathematics and in everyday use. Sometimes it refers to a kind of balance. For example, da Vinci's painting *The Last Supper* (see Figure 10.21 in color insert) is symmetrical because the disciples are grouped in four groups of three, with two groups on either side of the central figure of Christ. A human body is symmetrical because a vertical line drawn through the head and navel divides the body into two (nearly) identical parts (Figure 10.26).

Symmetry can also refer to the repetition of patterns in an object. Symmetries in this sense can be found in art from around the world. Native American pottery is often decorated with simple borders that used repeating patterns. Similar symmetries can be found in African, Moslem and Moorish art as shown in the strip pattern of Figure 10.27.

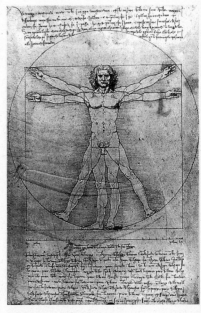

FIGURE 10.26
Leornardo da Vinci's sketch showing the symmetry of the human body.

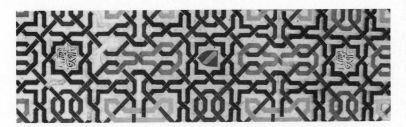

FIGURE 10.27

In mathematics, the word symmetry describes an operation that leaves something unchanged. For example, a circle still looks the same if it is rotated about its center, say by 90°. A square still looks the same if it is flipped across one of its diagonals. There are many such mathematical symmetries, many of them quite subtle. However, three symmetries are easy to identify:

- **Reflection symmetry**: an object remains unchanged when reflected across a straight line. For example, the letter A has reflection symmetry about a vertical line, while the letter H has reflection symmetry about a vertical and a horizontal line.

- **Rotation symmetry**: an object remains unchanged when rotated through some angle about a point. A circle, such as the letter O, has rotation symmetry, as does the letter S.

By the Way ············
In mathematics and physics, *translating* an object means moving it in a straight line, without rotating it.

- **Translation symmetry:** a pattern remains the same when shifted, say to the right or left. The pattern ...XXX... (with the X's continuing in both directions) has translation symmetry because it still looks the same if we shift it to the left or to the right.

(a)   (b)

FIGURE 10.28

EXAMPLE 1   *Finding Symmetries*

Identify the types of symmetry found in each star in Figure 10.28.

Solution:

a) The five-pointed star has five lines about which it can be flipped (reflected) without changing its appearance, so it has five reflection symmetries (Figure 10.29a). Because it has five vertices that all look the same, it can be rotated by $\frac{1}{5}$ of a full circle, or $\frac{360°}{5} = 72°$, and still look the same. Similarly, its appearance remains unchanged if it is rotated by $2 \times 72° = 144°$, $3 \times 72° = 216°$, or $4 \times 72° = 288°$. Thus this star has four rotational symmetries.

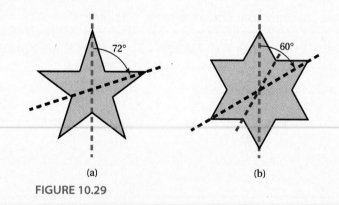

(a)   (b)

FIGURE 10.29

b) The six-pointed star has six reflection lines about which it can be flipped (reflected) without changing its appearance, so it has six reflection symmetries; three of them are shown in Figure 10.29b. Because of its six equivalent vertices, it has rotation symmetry when rotated by $\frac{1}{6}$ of a full circle, or $\frac{360°}{6} = 60°$. It also has symmetry if rotated by $2 \times 60° = 120°$, $3 \times 60° = 180°$, $4 \times 60° = 240°$, or $5 \times 60° = 300°$. Thus this star has five rotational symmetries. ■

Symmetries in Painting

Gustave Dore's (1832–1883) engraving *The Vision of the Empyrean* offers a dramatic illustration of rotation symmetry (Figure 10.30). This grand image of the cosmos can be rotated by many different angles and, at least on a large scale, appears much the same.

FIGURE 10.30

Sometimes, it is the *departures* from symmetry that make art effective. The twentieth-century work *Supernovae* (Figure 10.31) by the Hungarian painter Victor Vasarely (1908–) may have started as a symmetric arrangement of circles and squares. But the gradual deviations from that pattern make a powerful visual effect.

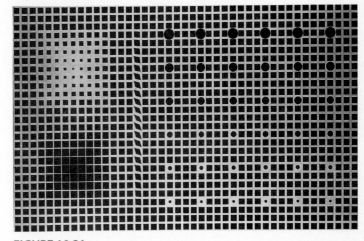

FIGURE 10.31

Given the strong ties between mathematics and art, you may not be surprised that mathematical algorithms (recipes) can generate art on computers. Figure 10.32 (see color insert) shows an intricate Persian rug design generated on a computer. The algorithm can be varied to give an endless variety of patterns and symmetries.

TILINGS

By the Way ············

The word *tessellation* comes from a Latin word for the small tiles used in mosaics.

A form of art called *tilings* (or *tessellations*) involves covering a flat area, such as a floor, with geometrical shapes. Tilings usually involve regular or symmetric patterns. Tilings are found in ancient Roman mosaics, stained glass windows, and the elaborate courtyards of Arab mosques—as well as in many modern kitchens and bathrooms.

More precisely, a **tiling** is an arrangement of *polygons* (see Unit 10A) that interlock perfectly with no overlapping. The simplest tilings use just one type of regular polygon. Figure 10.33 shows three such tilings made with equilateral triangles, squares, and regular hexagons, respectively. Note that there are no gaps or overlaps between the polygons in any of the three cases. In each case, the tiling is made by translating (shifting) the same basic polygon in various directions. Thus these tilings have translation symmetry.

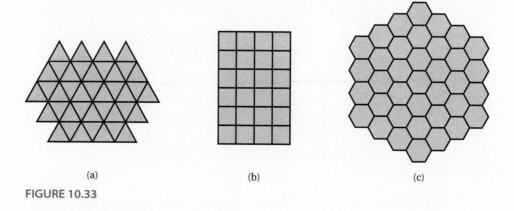

(a)   (b)   (c)

FIGURE 10.33

What happens if you try to make a tiling with, say, regular pentagons? If you try, you'll find that it simply does not work. If you measure the interior angles of a regular pentagon, you'll find that each measures 108°. Thus, as Figure 10.34 shows, the angle that remains when three regular pentagons are placed next to each other is too small to fit another regular pentagon. In fact, a mathematical theorem states that tilings with a single regular polygon are possible only with equilateral triangles, squares, and hexagons.

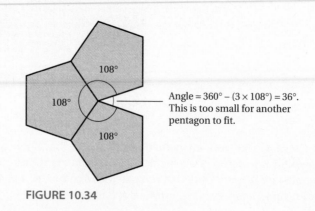

Angle = 360° − (3 × 108°) = 36°.
This is too small for another pentagon to fit.

FIGURE 10.34

TIME-OUT TO THINK: Try to draw a tiling made from regular octagons. Does it work? Why or why not?

Of course, more tilings are possible if we remove the restriction of using only a single type of regular polygon. For example, by allowing more than one different regular polygon, but still requiring that the arrangement of polygons look the same around each vertex (intersection point), it is possible to make the eight different tiling patterns shown in Figure 10.35 (see color insert).

TIME-OUT TO THINK: Verify that each of the tilings in Figure 10.35 uses only regular polygons. How many different regular polygons are used in each of these tilings? Verify that the same arrangement of polygons appears around each vertex. (Note: look carefully at the polygons; there are no circles in this figure.)

Tilings that use *irregular* polygons (those with sides of different lengths) are endless in number. As an example, suppose we start with an arbitrary triangle that has no special properties (other than three sides). The easiest way to tile a region with this triangle is by translating it parallel to two of its sides as shown in Figure 10.36.

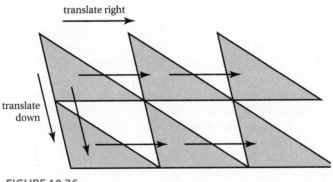

FIGURE 10.36

We shift the original triangle to the right so that the new triangle just touches the original triangle in a single point. We also shift the original triangle down so that the new triangle just touches the original triangle in a single point. Then we repeat these right/left and up/down translations as many times as we like. The gaps created in this process are themselves triangles that interlock perfectly with the translated triangles to create a tiling. Figure 10.37 on the following page shows another example, this time created by first reflecting an arbitrary triangle to produce a wing-shaped object, and then translating this object up/down and right/left.

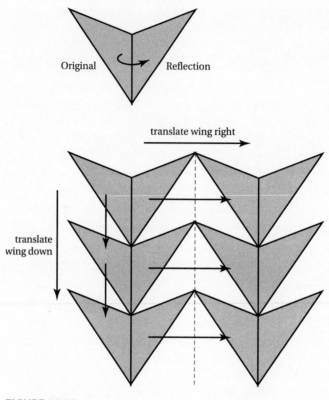

FIGURE 10.37

All of the tilings discussed so far are called *periodic tilings* because they have a pattern that is repeated throughout the tiling. In recent decades, mathematicians have begun to explore tilings that are *aperiodic*, meaning that they do not have a pattern that repeats throughout the entire tiling. Figure 10.38 (see color insert) shows an aperiodic tiling created by British mathematician Roger Penrose, a leader in research on tilings. If you look at the center of the figure, there appears to be a fivefold symmetry (a rotational symmetry that you would find in a pentagon). However, if the figure were extended indefinitely in all directions, the same pattern would never be repeated.

Tilings can be beautiful and practical for such things as floors and ceilings. However, recent research also shows that tilings may be very important in nature. In particular, many molecules and crystals apparently have patterns and symmetries that can be best understood with the same mathematics used to study tilings in art.

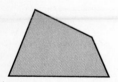

EXAMPLE 2   *Quadrilateral Tiling*

Create a tiling by translating the quadrilateral shown at the left. As you translate the quadrilateral, make sure that the gaps left behind have the same quadrilateral shape.

Solution: We can find the solution by trial and error, translating the quadrilateral in different directions until we have correctly shaped gaps. Figure 10.39 shows the solution.

Translate along both diagonals of the quadrilateral.

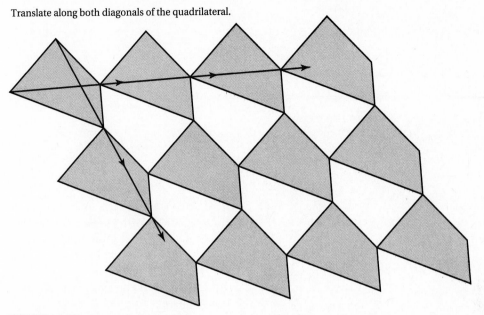

FIGURE 10.39

Note that the translations are in two directions: the directions of the two diagonals of the quadrilateral. The gaps between the translated quadrilaterals are themselves quadrilaterals that interlock perfectly to complete the tiling. ∎

REVIEW QUESTIONS

1. Describe in your own words what is meant by *perspective* and *symmetry*.

2. Describe how the principal vanishing point in a picture is determined.

3. What is the horizon line? How is it important in a painting showing perspective?

4. Briefly describe and distinguish between *reflection symmetry, rotation symmetry,* and *translation symmetry*. Draw a simple picture that shows each type of symmetry.

5. What is a *tiling*? Draw a simple example.

6. Briefly explain why there are only three possible tiling patterns that consist of a single, regular polygon. What are the three patterns?

7. Briefly explain why infinitely more tilings are possible if we remove the restriction of using regular polygons.

8. What is the difference between periodic and aperiodic tilings?

PROBLEMS

1. Vanishing Points. Consider the simple drawing below of a road and a telephone pole.

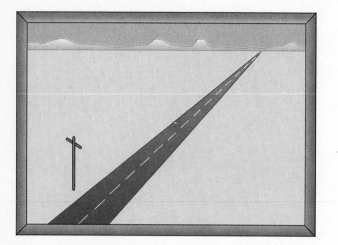

   a. Locate a vanishing point for the drawing. Is it the principal vanishing point?

   b. Draw three more telephone poles, receding into the distance, with proper perspective.

2. Correct Perspective. Consider the two boxes shown below. Which one is drawn with proper perspective relative to a single vanishing point? Explain.

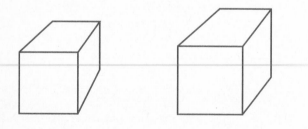

3. Drawing With Perspective. Use the square, circle, and triangle at the top right and make them into three-dimensional solid objects: a box, a cylinder, and a triangular prism, respectively. The given objects should be used as the front faces of the three-dimensional objects and all figures should be drawn with correct perspective relative to the given vanishing point *P*.

• *P*

4. Drawing MATH With Perspective. Use the letters MATH below and make them into three-dimensional solid letters. The given letters should be used as the front faces of the three-dimensional letters and all letters should be drawn with correct perspective relative to the given vanishing point *P*.

• *P*

MATH

5. Proportion and Perspective. The drawing below shows two poles drawn with correct perspective relative to a single vanishing point. As you can check, the first pole is 2 cm tall in the drawing and the second pole is 2 cm away from the first pole in the drawing.

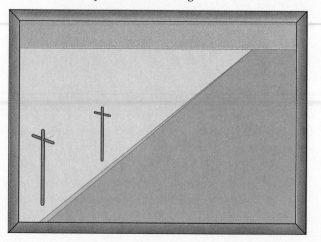

a. Draw three more identical poles that are equally spaced in the drawing.

b. What are the heights of these three new poles in the drawing?

c. In the actual scene, would these five poles be equally spaced? Explain.

6. **Two Vanishing Points.** The figure below shows a road receding into the distance. Draw a second road in the direction of the arrow that intersects the first road. Be sure that the vanishing points of the two roads lie on a horizontal horizon line.

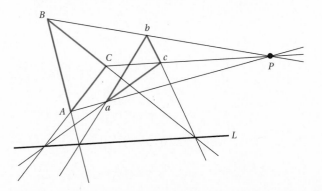

7. **Desargues' Theorem.** An early theorem of projective geometry, proved by French architect and engineer Girard Desargues (1593–1662), says that if two triangles (see below) are drawn so that the straight lines joining corresponding vertices (*Aa, Bb,* and *Cc*) all meet in a point *P* (corresponding to a vanishing point), then the corresponding sides (*AC* and *ac, AB* and *ab, BC* and *bc*), if extended, will meet in three points that all lie on the same line *L*.

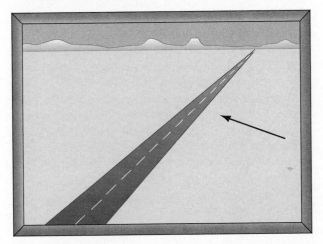

Draw two triangles of your own in such a way that the conditions of Desargues' Theorem are satisfied. Verify that the conclusions of the theorem are true.

8. **Symmetry in Letters.** Find all of the capital letters of the alphabet that have

a. right/left reflection symmetry (such as A);

b. top/bottom reflection symmetry (such as H);

c. both right/left and top/bottom reflection symmetry;

d. a rotational symmetry.

9. **Star Symmetries.**

a. How many reflection symmetries does a four-pointed star have? How many rotational symmetries does a four-pointed star have?

b. How many reflection symmetries does a seven-pointed star have? How many rotational symmetries does a seven-pointed star have?

10. **Symmetries of Geometric Figures.**

a. Draw an equilateral triangle (all three sides have equal length). How many degrees can the triangle be rotated about its center so that it remains unchanged in appearance? (There are several correct answers.)

b. Draw a square (all four sides have equal length). How many degrees can the square be rotated about its center so that it remains unchanged in appearance? (There are several correct answers.)

c. Draw a regular pentagon (all five sides have equal length). How many degrees can the pentagon be rotated about its center so that it remains unchanged in appearance? (There are several correct answers.)

d. Can you see a pattern in parts (a), (b), and (c)? How many degrees can a regular *n*-gon be rotated about its center so that it remains unchanged in appearance? How many different angles answer this question for an *n*-gon?

Identifying Symmetries. *Identify all of the symmetries in the figures of Problems 11–14.*

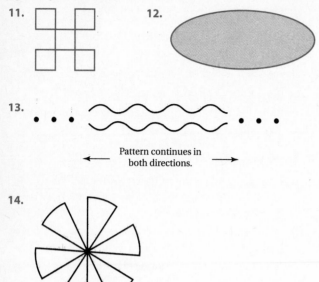

11.

12.

13.

Pattern continues in
both directions.

14.

Tilings from Translating Triangles. *Make a tiling from each triangle in Problems 15 and 16 using translations only, as in Figure 10.36.*

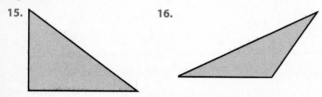

15.

16.

Tilings from Translating and Reflecting Triangles. *For Problems 17 and 18, make a tiling from the given triangle using translations and reflections, as in Figure 10.37.*

17. the triangle in Problem 15

18. the triangle in Problem 16

Tilings from Quadrilaterals. *Make a tiling from the quadrilaterals in Problems 19 and 20 using translations, as in Figure 10.39.*

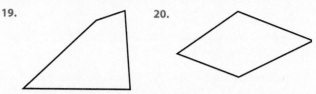

19.

20.

21. **Why Quadrilateral Tilings Work.** Consider the tiling with quadrilaterals shown in Figure 10.39 of the text. Look at any of the points at which four quadrilaterals meet in the tiling. Call such a point P. Given that the sum of the inside angles of any quadrilateral is 360°, show that the sum of the angle around the point P is also 360°, thus proving that the quadrilaterals interlock perfectly.

22. **Tiling with a Rhombus.** A rhombus is a quadrilateral in which all four sides have the same length and opposite sides are parallel. Show how a tiling can be made from a rhombus using (a) translations only (as in Figure 10.36) and (b) an initial reflection and then translations (as in Figure 10.37).

23. **Project: Symmetry and Proportion in Art.** Find one piece of pre-twentieth-century art and one piece of twentieth-century art that you like. Use as many ideas from this unit as possible (involving perspective and symmetry) to write a two- to three-page analysis and comparison of these two pieces of art.

PROPORTION AND THE GOLDEN RATIO

In Unit 10C, we studied how mathematics enters into art through the ideas of symmetry and perspective. In this unit, we turn our attention to the third major mathematical idea involved with art: proportion. As we will see, matters of proportion underlie not only the art created by humans, but the "art" we find in nature as well.

The importance of *proportion* was expressed well by astronomer Johannes Kepler (1571–1630):

> *Geometry has two great treasures: one is the theorem of Pythagoras; the other, the division of a line into extreme and mean ratio. The first we may compare to gold; the second we may name a precious jewel.*

Kepler's statement about *the division of a line into extreme and mean ratio* describes one of the oldest principles of proportion, which dates back to the time of Pythagoras (c. 500 B.C.) when scholars asked the following question: How can a line segment be divided into two pieces that have the most visual appeal and balance?

Surprisingly, although this was a question of beauty, there seemed to be general agreement on the answer. Suppose that a line segment is divided into two pieces as shown in Figure 10.40. We will call the length of the long piece L and the length of the short piece 1. The Greeks claimed that the most visually pleasing division of the line had the property that the ratio of the length of the long piece to the length of the short piece is the same as the ratio of the length of the entire line segment to the length of the long piece. That is:

$$\frac{L}{1} = \frac{L + 1}{L}$$

$$\underset{\text{FIGURE 10.40}}{\underline{\qquad L \qquad \bullet \quad 1 \quad}}$$

> *The senses delight in things duly proportioned.*
>
> —St. Thomas Aquinas (*1225–1274*)

This statement of proportion can be solved (see Problem 8) to find that the length of the long piece has the special value, symbolized by the Greek letter ϕ (pronounced *fie* or *fee*), which is

$$\phi = \frac{1 + \sqrt{5}}{2} = 1.61803\ldots.$$

The number ϕ is more commonly called the **golden ratio** or **golden section**; it is also sometimes referred to as the *divine proportion*. Note that, because of the $\sqrt{5}$ term, the golden ratio is an irrational number. Its value is approximately 1.6, or $\frac{8}{5}$.

Thus, as Figure 10.41 shows, for any line segment that is divided in two pieces according to the golden ratio, the ratio of the long piece to the short piece is

$$\frac{x}{y} = \phi = \frac{1.61803\ldots}{1} \approx \frac{8}{5}.$$

$$\underset{\text{FIGURE 10.41}}{\underline{\overset{1.61803\ldots \qquad\qquad 1}{} \underset{x \qquad\qquad\quad y}{\bullet}}}$$

By the Way ....
The Greek letter ϕ is the first letter in the Greek spelling of the name Phydias, a Greek sculptor who may have used the golden ratio in his work.

EXAMPLE 1   *Calculating a Golden Ratio*

Suppose that the line segment at the right is divided in the golden ratio. If the length of the piece labeled x is 5 cm, how long is the entire line segment?

Solution: We are given that the line segment is divided in the golden ratio, which means that

$$\frac{x}{y} = \phi.$$

We can solve for *y* by multiplying both sides by *y* and dividing both sides by ϕ.

$$\frac{x}{y} = \phi \qquad \Rightarrow \qquad y = \frac{x}{\phi}$$

We are given that *x* = 5 cm, so we substitute this value and an approximate value for ϕ to find *y*:

$$y = \frac{5 \text{ cm}}{\phi} \approx \frac{5 \text{ cm}}{1.6} \approx 3.1 \text{ cm}$$

The entire segment is made up of both *x* and *y*, so its total length is approximately

$$5 \text{ cm} + 3.1 \text{ cm} = 8.1 \text{ cm.} \qquad \blacksquare$$

THE GOLDEN RATIO IN ART HISTORY

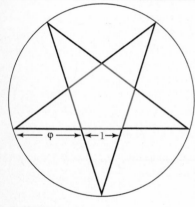

FIGURE 10.42

Although the ancient Greeks generally struggled with the notion of irrational numbers, they embraced the golden ratio and used it in their philosophy and their art. Furthermore, they discovered that it reappeared as the answer to many other geometrical questions. For example, the *pentagram* (Figure 10.42) was the seal of the mystical Pythagorean Brotherhood (see Unit 4A). This five-pointed star inscribed in a circle produces a pentagon at the center. The golden ratio occurs in at least ten different ways in the pentagram. For example, if the length of the sides of the pentagon is 1, then the lengths of the arms of the star are ϕ.

TIME-OUT TO THINK: Using a ruler to measure, find at least one other place in the pentagram of Figure 10.42 where the ratio of the lengths of two line segments is the golden ratio.

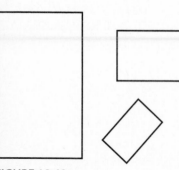

FIGURE 10.43

From the golden ratio it is a short step to the other famous Greek expression of proportion, the **golden rectangle**: a rectangle whose long side is ϕ times longer than its short side. A golden rectangle can be of any size, but its sides must have a ratio of $\phi \approx \frac{8}{5}$. Figure 10.43 shows three golden rectangles.

The golden rectangle had both practical and mystical importance to the Greeks. It became a cornerstone of their philosophy of *aesthetics*— the study of beauty. There is considerable speculation about the uses of the golden rectangle in art and architecture in ancient times. For example, it is widely claimed that many of the great monuments of antiquity, such as the Pyramids in Egypt, were designed in accordance with the golden rectangle. And, whether by design or by chance, the proportions of the Parthenon (in Athens, Greece) match those of the golden rectangle very closely.

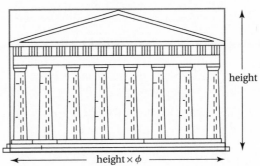

By the Way ....
The Parthenon was completed in about 430 B.C. as a temple to Athena Parthenos, the Warrior Maiden. It stands on the Acropolis (which means *the uppermost city*), about 500 feet above Athens.

FIGURE 10.44

The golden rectangle also appears in many more recent works of art and architecture. The book *De Divina Proportione*, illustrated by Leonardo da Vinci in 1509, is filled with references to and uses of ϕ. Indeed, da Vinci's unfinished painting *St. Jerome* (see Figure 10.45 in color insert) places the central figure inside of an imaginary golden rectangle. More recently, the French impressionist painter Georges Seurat is said to have used the golden ratio on every canvas. The abstract geometric paintings of the twentieth-century Dutch painter Piet Mondrian (see Figure 10.46 in color insert) are filled with golden rectangles.

Today, the golden rectangle appears in many everyday items. For example, photographs, note cards, cereal boxes, posters, and windows often have the proportions of the golden rectangle. But the question remains as to whether the golden rectangle is really more pleasing. In the late nineteenth century, the German psychologist Gustav Fechner (1801–1887) studied the question statistically. He showed several rectangles with various length-to-width ratios to hundreds of people and recorded their choices for the most and least visually pleasing rectangles. The results as given in Table 10.4 show that almost 75% of the participants chose rectangles that were very close to the golden rectangle.

By the Way ....
Despite evidence of the appeal of the golden rectangle and ratio, there are theories to the contrary. A 1992 study by George Markowsky discredits the claim that the golden ratio was used in art and architecture, attributing them to coincidence and bad science. He also claims that statistical studies of peoples' preferences, such as Fechner's research, are not conclusive (see problem 15).

Table 10.4 Fechner's data.		
Length-to-Width Ratio	Most Pleasing Rectangle (Percentage response)	Least Pleasing Rectangle (Percentage response)
1.00	3.0	27.8
1.20	0.2	19.7
1.25	2.0	9.4
1.33	2.5	2.5
1.45	7.7	1.2
1.50	20.6	0.4
$\phi \approx$ **1.62**	**35.0**	**0.0**
1.75	20.0	0.8
2.00	7.5	2.5
2.50	1.5	35.7

TIME-OUT TO THINK: Do you think that the golden rectangle is visually more pleasing than other rectangles? Explain.

EXAMPLE 1   *Household Golden Ratios*

The following household items were found to have the dimensions shown. Which items come the closest to having the proportions of the golden ratio?

- Standard sheet of paper: 8.5 in. × 11 in.
- 8 × 10 picture frame: 8 in. × 10 in.
- 35 mm slide: 35 mm × 23 mm

Solution: The ratio of the sides of a standard sheet of paper is $\frac{11}{8.5} = 1.29$, which is 20% less than the golden ratio. The ratio of the sides of a standard picture frame is $\frac{10}{8} = 1.25$, which is 22% less than the golden ratio. A careful measurement of a 35 mm slide shows that it is 35 mm × 23 mm. The ratio of its sides is $\frac{35}{23} = 1.52$, which is 5% less than the golden ratio. Of the three objects, the 35 mm slide comes closest to being a golden rectangle.   ■

THE GOLDEN RATIO IN NATURE

The golden ratio also appears to be common in the "artwork" of nature. One striking example arises by creating a spiral from golden rectangles. We begin by dividing a golden rectangle to make a square on its left side, as shown in Figure 10.47(a). If you measure the sides of the remaining, smaller rectangle to the right of the square, you'll see that it is a smaller golden rectangle.

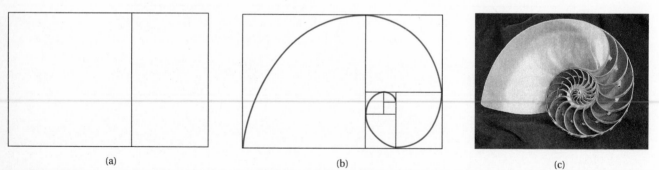

(a)   (b)   (c)

FIGURE 10.47

We now repeat this splitting process on the second golden rectangle, this time making the square on the top (instead of the left) of the golden rectangle. This split makes a third, even smaller, golden rectangle. Continuing to split each new golden rectangle in this manner generates the result shown in Figure 10.47(b). Now, we connect opposite corners of all the squares with a nice smooth curve. The result is a continuous curve called a **logarithmic spiral** (or *equiangular spiral*). This spiral very closely matches the spiral shape of the beautiful chambered nautilus shell (Figure 10.47c).

Another intriguing connection between the golden ratio and nature comes from a problem in population biology, first posed by a mathematician known as Fibonacci in 1202. Fibonacci's problem essentially asked the following question about the reproduction of rabbits.

Suppose that a pair of baby rabbits takes one month to mature into adults, then produces a new pair of baby rabbits the following month and in each subsequent month. Further suppose that each newly born pair of rabbits matures and gives birth to additional pairs with the same reproductive pattern. If no rabbits die, how many pairs of rabbits are in the population at the beginning of each month?

Figure 10.48 shows the solution to this problem for the first six months. Note that the number of pairs of rabbits at the beginning of each month forms a sequence of numbers that begins 1, 1, 2, 3, 5, 8. Fibonacci found that, by computing the numbers of pairs of rabbits in each month, the numbers in this sequence continue to grow with the following pattern:

$$1, 1, 2, 3, 5, 8, 13, 21, 34, 55, \ldots$$

This sequence of numbers is known as the **Fibonacci sequence**.

By the Way ············
Fibonacci, also known as Leonardo of Pisa, is credited with popularizing the use of Hindu-Arabic numerals in Europe. His book called *Liber Abaci* ("Book of the Abacus"), published in 1202, explained their use and the importance of the number zero.

Beginning of:	Month 1	Month 2	Month 3	Month 4	Month 5	Month 6
Population (pairs):	1	1	2	3	5	8

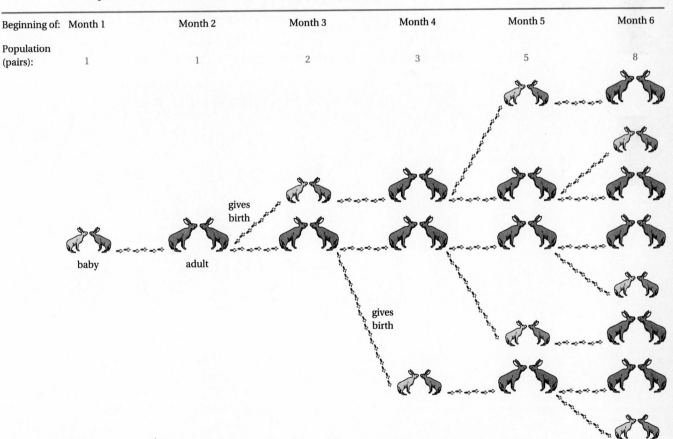

FIGURE 10.48

Let's call the nth Fibonacci number F_n, so we have $F_1 = 1$, $F_2 = 1$, $F_3 = 2$, $F_4 = 3$, and so forth. The most basic property of the Fibonacci sequence is that the next number in the sequence is the sum of the previous two numbers. For example, note that

$$F_3 = F_2 + F_1 = 1 + 1 = 2 \quad \text{and} \quad F_4 = F_3 + F_2 = 2 + 1 = 3.$$

We can express this rule in general as

$$F_{n+1} = F_n + F_{n-1}.$$

> **TIME-OUT TO THINK:**   Confirm that the above rule works for Fibonacci numbers F_3 through F_{10}. Use the rule to determine the eleventh Fibonacci number (F_{11}).

The connection between the Fibonacci numbers and the golden ratio becomes clear when we compute the ratios of successive Fibonacci numbers.

EXAMPLE 3   *Ratios of Fibonacci numbers*

Compute the ratios of successive Fibonacci numbers for the first 18 numbers. Do these ratios approach a single number? If so, what is it?

Solution: The ratios of the first 18 Fibonacci numbers are shown in Table 10.5.

Table 10.5 Ratios of successive Fibonacci numbers	
$F_3/F_2 = 2/1 = 2.0$	$F_{11}/F_{10} = 89/55 = 1.618182$
$F_4/F_3 = 3/2 = 1.5$	$F_{12}/F_{11} = 144/89 = 1.617978$
$F_5/F_4 = 5/3 = 1.667$	$F_{13}/F_{12} = 233/144 = 1.618056$
$F_6/F_5 = 8/5 = 1.600$	$F_{14}/F_{13} = 377/233 = 1.618026$
$F_7/F_6 = 13/8 = 1.625$	$F_{15}/F_{14} = 610/377 = 1.618037$
$F_8/F_7 = 21/13 = 1.6154$	$F_{16}/F_{15} = 987/610 = 1.618033$
$F_9/F_8 = 34/21 = 1.61905$	$F_{17}/F_{16} = 1597/987 = 1.618034$
$F_{10}/F_9 = 55/34 = 1.61765$	$F_{18}/F_{17} = 2584/1597 = 1.618034$

Note that, as we go further out in the sequence, the ratios of successive Fibonacci numbers get closer and closer to the golden ratio $\phi = 1.61803. \ldots$ ∎

The Fibonacci Sequence in Nature

There are many examples of the Fibonacci sequence in both art and nature. The heads of sunflowers and daisies consist of a clockwise spiral superimposed on a counterclockwise spiral (both of which are logarithmic spirals), as shown in Figure 10.49. The number of individual florets in each of these intertwined spirals is a Fibonacci number; for example, 21 and 34, or 34 and 55. Biologists have also observed that the number of petals on many common flowers is a Fibonacci number (for example, iris have 3 petals, primroses have 5 petals, ragworts have 13 petals and daisies have 34 petals). The arrangement of leaves on the stem of many plants also exhibits the Fibonacci sequence. And the spiraling Fibonacci numbers can also be identified on pine cones and pineapples.

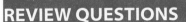

(a)   (b)

FIGURE 10.49

REVIEW QUESTIONS

1. Explain the *golden ratio* in terms of proportions of line segments.

2. How is a golden rectangle formed?

3. What evidence suggests that the golden ratio and golden rectangle hold particular beauty?

4. What is a *logarithmic spiral*? How is it formed from a golden rectangle?

5. What is the Fibonacci sequence?

6. What is the connection between the Fibonacci sequence and the golden ratio? Give some examples of the Fibonacci sequence in nature.

PROBLEMS

1. **Golden Ratio.** Draw a line segment 4 inches long. Now subdivide it according to the golden ratio. Verify your work by computing (a) the ratio of the length of the whole segment to the long segment length and (b) the ratio of the long segment length to the short segment length.

2. **Golden Rectangles.** Measure the sides of each rectangle below, and compute the ratio of the long side to the short side for each rectangle. Which one is a golden rectangle?

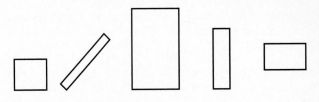

Dimensions of Golden Rectangles. Problems 3–6 give the length of one side of a golden rectangle. Find the length of the other side. Notice that the other side could be either longer or shorter than the given side. Use the approximation $\phi \approx 1.62$ for your work.

3. 1.5 inches

4. 2.5 meters

5. 6.4 kilometers

6. 0.12 centimeters

7. Everyday Golden Rectangles. Find at least three everyday objects with rectangular shapes (for example, billboards, cereal boxes, windows). In each case, measure the side lengths and calculate the ratios. Are any of these objects golden rectangles? Explain.

8. Finding ϕ. The property that defines the golden ratio is

$$\frac{L}{1} = \frac{L+1}{L}.$$

a. Show that, by multiplying both sides by L and rearranging, this equation becomes

$$L^2 - L - 1 = 0.$$

Confirm that substituting the value of ϕ for L satisfies this equation.

b. The quadratic formula states that, for any equation of the form $ax^2 + bx + c = 1$, the solutions are given by

$$x = \frac{-b + \sqrt{b^2 - 4ac}}{2a} \quad \text{and}$$

$$x = \frac{-b - \sqrt{b^2 - 4ac}}{2a}.$$

Use the quadratic formula to solve for L in the formula for the golden ratio. Show that one of the roots is ϕ.

9. Properties of ϕ.

a. Enter $\phi = (1 + \sqrt{5})/2$ into your calculator. Show that the number $1/\phi = \phi - 1$.

b. Now compute ϕ^2. How is this number related to ϕ?

10. Logarithmic Spirals. Draw a rectangle that is 10 cm on a side. Follow the procedure described in the text for subdividing the rectangle until you can draw a logarithmic spiral. Do all work carefully, and show your measurements with your work.

11. The Lucas Sequence. A sequence called the *Lucas sequence* is closely related to the Fibonacci sequence. The Lucas sequence begins with the numbers $L_1 = 1$ and $L_2 = 3$ and then uses the same relation $L_{n+1} = L_n + L_{n-1}$ to generate $L_3, L_4, \dots$.

a. Generate the first 10 Lucas numbers.

b. Compute the ratio of successive Lucas numbers L_2/L_1, L_3/L_2, L_4/L_3, and so on. Can you determine if these ratios approach a single number? What number is it?

12. Graphing Fechner's Data. Consider Gustav Fechner's data shown in Table 10.4. Make a histogram that displays the responses for both the most pleasing and least pleasing rectangle proportions.

13. Project: The Golden Navel. An old theory claims that, on average, the ratio of the height of a person to the height of his/her navel is the golden ratio. Collect "navel ratio data" from as many people as possible. Graph the ratios in a histogram, find the average ratio over your entire sample, and discuss the outcome. Do your data support the theory?

14. Project: Mozart and the Golden Ratio. In a paper called "The Golden Section and the Piano Sonatas of Mozart" (*Mathematics Magazine*, Vol. 68, No. 4, 1995), John Putz gives the lengths (in measures) of the first part (*a*) and the second part (*b*) of the movements of the 19 Mozart piano sonatas. A few of the data are given below.

a = Length of first part	b = Length of second part
38	62
28	46
56	102
56	88
24	36
77	113
40	69
46	60
15	18
39	63
53	67

a. The ratio of the length of the whole movement to the length of the longer segment is $\frac{a+b}{b}$. Add a third column to the table and compute this ratio for the given data.

b. Make a histogram of the ratios that you computed in part (a) (choosing an appropriate bin size). Comment on how well they are approximated by ϕ.

c. Read the article by John Putz. Do you believe that Mozart composed with the golden ratio in mind?

15. Project: Debunking the Golden Ratio. Find the article "Misconceptions about the Golden Ratio" by George Markowsky (*College Mathematics Journal*, Vol. 23, No. 1, 1992). Choose at least one of the misconceptions that Markowsky discusses, summarize it, and then discuss whether you find his argument convincing. Discuss your opinion of whether the golden ratio has been consciously used by artists and architects in their work.

· ·

FRACTAL GEOMETRY

UNIT 10E

In Units 10C and 10D, we studied how the visual arts are linked with the familiar Euclidean geometry of points, lines, and planes. We also saw several ways in which geometry was linked to "art" in nature, in forms such as shells and the arrangements of flower petals. However, mathematicians have recently developed a new type of geometry, called **fractal geometry**, that has an even greater application to natural forms. Fractal geometry has proven so good at imitating nature that it is now used in art and film to create imaginary yet realistic landscapes. Figure 10.50 (see color insert) shows such an imaginary landscape, generated entirely on a computer with the mathematics of fractal geometry.

WHAT ARE FRACTALS?

We can investigate fractals by envisioning measurements made with "rulers" of different lengths, such as 100 meters, 1 meter, 1 millimeter, and so on. Each length laid out by a ruler is called an **element** (of length), and we must follow one important rule: we will count only *whole numbers* (not fractions) of elements. For example, if we find that 5.6 rulers of 100-meter length fit along some object, then we will count only five elements that are 100 meters in length. We would therefore find that the approximate total length of the object is

$$5 \times 100 \text{ meters} = 500 \text{ meters}.$$

More generally, the approximate length of any object is

$$\text{total length} \approx \text{number of elements} \times \text{length of each element}.$$

By the Way ·············

Purchased by New York City in 1856, Central Park was designed by Frederick Law Olmsted and Calvert Vaux. It was one of the first public parks laid out by landscape architects. Olmsted also worked to preserve such natural areas as Yosemite National Park in California.

Measuring the Perimeter of Central Park

Imagine that you are asked to measure the perimeter of Central Park in New York City, which was deliberately laid out in the shape of a rectangle (Figure 10.51). Suppose that you start with 100-meter rulers. You begin at one corner of the park, and lay the rulers end-to-end around the edges of the park. Because you are allowed to count only whole numbers of rulers, your measurement will be somewhat short of the actual perimeter unless the 100-meter rulers fit the park perimeter perfectly.

You can get a better approximation of the park perimeter by using a shorter ruler. For example, if you lay 10-meter rulers end-to-end, you'll get within 10 meters of the actual length of each side. You could do better yet with an even shorter ruler.

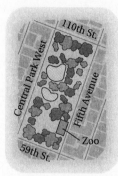

FIGURE 10.51

Figure 10.52 shows how your estimates of the park perimeter will change with the length of the ruler. A very long ruler underestimates the actual length because of the whole number rule. But rulers with a length of less than about 1 meter yield nearly identical measurements. This result should not be surprising. After all, shorter rulers simply give you better estimates of the exact perimeter.

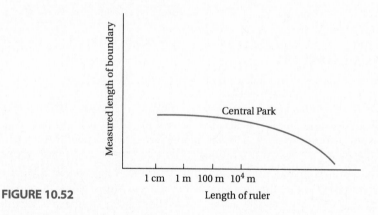

FIGURE 10.52

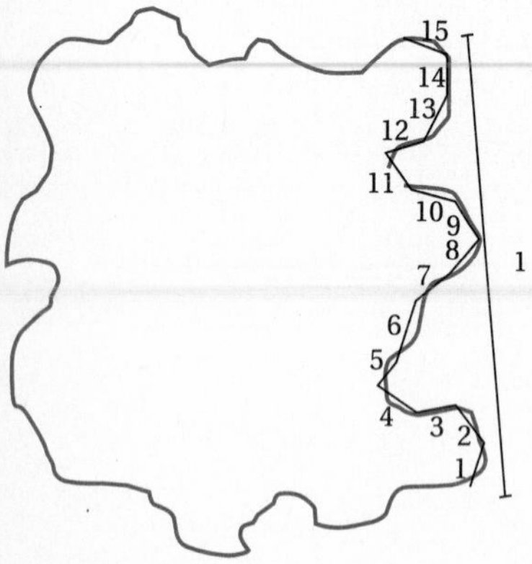

Measuring an Island Coastline

Next, imagine that you are asked to measure the perimeter, or coastline, of an *island*. To avoid problems with tides and waves, imagine that it is winter and the water around the island is frozen. Thus your task is to measure the length of the coastline defined by the ice-land boundary.

Again, you begin by using a 100-meter ruler, laying it end-to-end around the island. The 100-meter ruler (the single long line segment in Figure 10.53) will adequately measure large-scale features such as bays and estuaries, but will miss features such as promontories and inlets that are less than 100 meters across. Switching to a 10-meter ruler will allow you to follow many features that were missed by the 100-meter ruler. As a result, you'll measure a *longer* perimeter (the short line segments in Figure 10.53).

FIGURE 10.53

The 10-meter ruler is still too long to measure all the features along the coastline, so you'll find an even greater perimeter if you switch to a 1-meter ruler. In fact, as you use shorter and shorter rulers to measure the coastline, you'll get longer and longer estimates of the perimeter. Figure 10.54 shows how the measured length of the coastline grows larger when measured with shorter rulers. Whereas the perimeter of Central Park is clearly defined simply by using a small enough ruler, we cannot agree on the "true" length of the coastline because the length depends on the length of the ruler used for measurement.

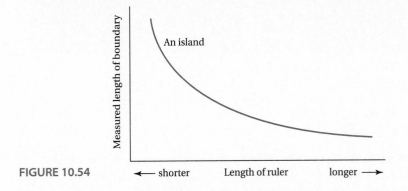

FIGURE 10.54

Rectangles and Coastlines Under Magnification

Imagine viewing a piece of the rectangular perimeter of Central Park under a magnifying glass. No new details would appear; it is still a straight-line segment (Figure 10.55a). In contrast, if you view a piece of the coastline under a magnifying glass, you will see details that were not visible without magnification (Figure 10.55b).

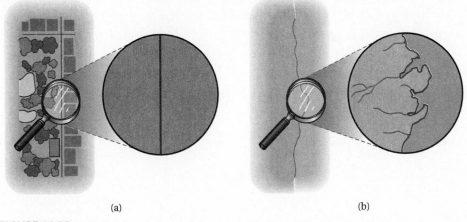

(a)   (b)

FIGURE 10.55

Objects like the coastline that continually reveal new features at smaller scales are called **fractals**. Many natural objects are thus fractals. For example, coral and mountain ranges both reveal more and more features when looked at under greater magnification, and hence are fractals (see Figure 10.56 on the following page).

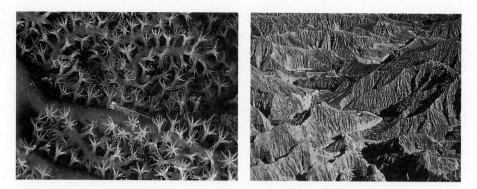

coral   mountains

FIGURE 10.56

FRACTAL DIMENSION

The boundary of Central Park is one-dimensional because a single number locates any point. For example, if you tell people to meet 375 meters from the park's northwest corner, going clockwise, they'll know exactly where to go.

In contrast, if you tell people to meet 375 meters along the coastline from a particular point on the island, different people will end up in different places depending on the length of the ruler they use to measure the distance. Thus the coastline is *not* an ordinary one-dimensional object. But neither is it a two-dimensional object that has area. Instead, we say that the coastline has a **fractal dimension** that falls "in-between" the ordinary dimensions. The fractal dimension of the coastline lies *between* one and two, indicating that it has some properties that are best thought of in terms of length (one dimension) and others that are more like area (two dimensions).

A New Definition of Dimension

If we measure a 1-inch line segment with a 1-inch ruler, we find *one* 1-inch element of length along it (Figure 10.57). Using a ruler that is smaller by a *factor* of 2, or $\frac{1}{2}$ inch in length, we find *two* elements of length along the 1-inch line segment. And, if we choose a ruler that is smaller by a *factor* of 4, or $\frac{1}{4}$ inch in length, we find *four* elements of length along the 1-inch line segment.

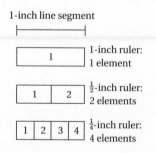

FIGURE 10.57

Let's use R to represent the *reduction factor* in the length of the ruler, and N to represent the *factor* by which the *number of elements* increases. We can now restate our results for the line segment:

- Reducing the ruler length by a *reduction factor R = 2* (to 1/2 inch) leads to an increase in the *number of elements* by a factor N = 2.

- Reducing the ruler length by a *reduction factor R = 10* (to 1/10 inch) leads to an increase in the *number of elements* by a factor N = 10.

Generalizing for the line segment: decreasing the ruler length by a *reduction factor R* leads to an increase in the *number of elements* by a factor N = R.

We can use a similar process to determine the area of a square by counting the number of *area elements* that fit within it (Figure 10.58). Using a ruler that has the same 1-inch length as a side of the square, we can fit a single area element in the square. If we reduce the ruler length by a factor R = 2, to $\frac{1}{2}$ inch, we fit N = 4 times as many elements in the square. Reducing the ruler length by R = 4, to $\frac{1}{4}$ inch, fits N = 16 times as many elements in the square. Generalizing for the square: Reducing the ruler length by a reduction factor R increases the number of area elements by a factor $N = R^2$.

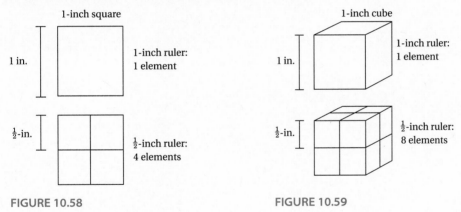

FIGURE 10.58   **FIGURE 10.59**

Finally, we can measure the volume of a cube by counting the number of *volume elements* that fit within it (Figure 10.59). A ruler of the same 1-inch length as a side of the cube makes a single volume element that fits in the cube. This time, reducing the ruler length by a factor R = 2, to $\frac{1}{2}$ inch, allows us to fit N = 8 times as many volume elements in the cube. Reducing the ruler length by R = 4, to $\frac{1}{4}$ inch, fits N = 64 times as many elements in the cube. Generalizing for the cube: Reducing the ruler length by a factor R increases the number of volume elements by a factor $N = R^3$.

Let's summarize our results:

- For a one-dimensional object (e.g., a line segment), we found $N = R^1$.

- For a two-dimensional object (e.g., a square), we found $N = R^2$.

- For a three-dimensional object (e.g., a cube), we found $N = R^3$.

Note that, in each case, the dimension of the object shows up as a power. We can now generalize this new definition of dimension.

The **fractal dimension** of an object is defined as a number D such that

$$N = R^D,$$

where N is the factor by which the number of elements increases when we shorten a ruler by a reduction factor R.

EXAMPLE 1   *Finding a Fractal Dimension*

Suppose that you measure an object in which every time you reduce the length of your ruler by a factor of 3, the number of elements you measure increases by a factor of 4. What would be the fractal dimension of this object?

Solution: We are given that reducing the ruler length by a factor $R = 3$ increases the number of elements by a factor $N = 4$. Thus we are looking for a fractal dimension D such that

$$4 = 3^D.$$

Because D appears as an exponent, we can "bring it down" by taking the logarithm of both sides of the equation.

$$\log_{10} 4 = \log_{10} 3^D$$

Using the rule that $\log_{10} a^x = x \times \log_{10} a$,   the equation becomes

$$\log_{10} 4 = D \times \log_{10} 3.$$

Now we divide both sides by $\log_{10} 3$ to find

$$\frac{\log_{10} 4}{\log_{10} 3} = \frac{D \times \log_{10} 3}{\log_{10} 3} \quad \Rightarrow \quad D = \frac{\log_{10} 4}{\log_{10} 3} \approx 1.2619.$$

The *fractal dimension* of this object is about 1.2619.   ∎

$\mathcal{By\ the\ Way}$ ....
The snowflake curve sometimes is called a *Koch curve,* after Helga von Koch who first described it in 1906.

The Snowflake Curve

What kind of object could have a fractal dimension like the one we calculated in Example 1? Let's consider a special object called a **snowflake curve**, generated by a drawing process that begins with a straight-line segment. As shown at the bottom of Figure 10.60, we designate the starting line segment L_0. We then generate L_1 with the following three steps.

1. Divide the line segment L_0 into three equal pieces.
2. Remove the middle piece.
3. Replace the middle piece with two segments of the same length arranged as two sides of an equilateral triangle.

Note that L_1 consists of four line segments and that each is $\frac{1}{3}$ the length of L_0 (because L_0 was divided into three equal pieces).

Next, we repeat the three steps on *each* of the four segments of L_1. The result is L_2, which has 16 line segments, each $\frac{1}{9}$ the length of L_0.

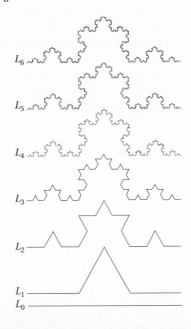

FIGURE 10.60

TIME-OUT TO THINK: Count the segments shown in Figure 10.60 to confirm that L_2 has 16 line segments. Measure to confirm that each is $\frac{1}{9}$ the length of L_0. Why are the segments $\frac{1}{9}$ the length of L_0? How long are the segments of L_3?

Repeating the three-step process on each segment of the current figure generates L_3, L_4, L_5, and so forth. If we could repeat this process an *infinite* number of times, the snowflake curve, denoted L_∞, would be the ultimate result. Thus any figure that we actually draw, whether L_6 or $L_{1,000,000}$, is only an approximation to the true snowflake curve.

Now, imagine measuring the length of the complete snowflake curve, L_∞. A ruler with the length of L_0 would simply lay across the base of the snowflake curve, missing all fine detail and measuring only the straight-line distance between the endpoints. This ruler would yield only *one* element along the snowflake curve. Reducing the ruler length by a factor $R = 3$ would make it the same length as each of the four segments of L_1; thus it would find four elements along the snowflake curve, or $N = 4$ times more elements than the first ruler. Reducing the ruler length by another factor $R = 3$ makes it $\frac{1}{9}$ the length of L_0, or the length of the 16 segments of L_2. Thus the ruler finds 16 elements along the snowflake curve, or $N = 4$ times more elements than the previous ruler.

In general, every time we reduce the ruler length by another factor $R = 3$, we find $N = 4$ times more elements. Thus the fractal dimension of the snowflake curve is a D such that

$$4 = 3^D.$$

As we found in Example 1, this means that the snowflake curve has fractal dimension $D = 1.2619$.

What does it mean to have a fractal dimension of 1.2619? The fact that the fractal dimension is greater than 1 means that the snowflake curve has more "substance" than an ordinary one-dimensional object. In a sense, the snowflake curve begins to fill the part of the plane in which it lies. The closer the fractal dimension of an object is to 1, the more closely it resembles a collection of line segments. The closer the fractal dimension is to 2, the more closely it comes to filling a part of a plane.

EXAMPLE 2   *How Long Is a Snowflake Curve?*

How much longer is L_1 than L_0? How much longer is L_2 than L_0? Generalize your results, and discuss the length of a complete snowflake curve.

Solution: From Figure 10.60, we see that L_1 consists of four line segments and that each is $\frac{1}{3}$ the length of L_0. Thus L_1 is $\frac{4}{3}$ times the length of L_0, or

$$\text{Length of } L_1 = \frac{4}{3} \times (\text{length of } L_0).$$

L_2 has 16 line segments, each $\frac{1}{9}$ the length of L_0. Thus its length is $\frac{16}{9}$ times the length of L_0, or

$$\text{Length of } L_2 = \frac{16}{9} \times (\text{length of } L_0) = \left(\frac{4}{3}\right)^2 \times (\text{length of } L_0).$$

Generalizing, we can say that

$$\text{Length of } L_n = \left(\frac{4}{3}\right)^n \times (\text{length of } L_0).$$

Because $\left(\frac{4}{3}\right)^n$ grows without bound as n gets larger, we conclude that the complete snowflake curve, L_∞, must be infinitely long! ■

The Snowflake Island

The **snowflake island** is a *region* (island) bounded by three snowflake curves. The process of drawing the snowflake island begins with an equilateral triangle (Figure 10.61). Then we convert *each* of the three sides of the triangle into a snowflake curve, L_∞. We cannot draw the complete snowflake island because it would require an infinite number of steps, but Figure 10.61 shows the results in which the sides are L_1, L_2, and L_6. Note that, because the *coastline* of the snowflake island consists of snowflake curves, the fractal dimension of the coastline is the same 1.2619 as that of the snowflake curve.

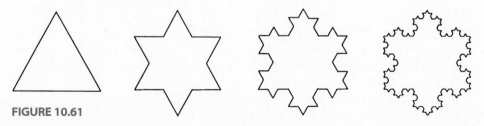

FIGURE 10.61

The snowflake island illustrates some truly extraordinary properties. In Example 2, we found that a single snowflake curve is infinitely long. Thus the coastline of the snowflake island also must be infinitely long, since it is made of three snowflake curves. However, Figure 10.61 shows that the island's *area* is clearly contained within the bounds of the page. Thus we have the intriguing result that a snowflake island is an object with a *finite area* and an *infinitely long* boundary.

Real Coastlines and Borders

Note that each of the four pieces of L_2 in Figure 10.60 looks exactly like L_1, except they are smaller. Similarly, L_3 consists of four pieces that each look like L_2. In fact, if we magnify *any* piece of the snowflake curve L_∞, it will look exactly like one of the earlier curves, L_0, L_1, L_2, . . . , used in its generation. Because the snowflake curve looks similar to itself when examined at different scales, we say that it is a **self-similar** fractal. The snowflake curve's *self-similarity* is due to a repeated application of a simple set of rules.

Natural objects, such as real coastlines, also reveal new details under higher magnification. However, unlike the case with self-similar fractals, a magnified view of a natural object isn't likely to look *exactly* like the original object. Nevertheless, if we have data about how measured lengths change with different "ruler" sizes, we can still assign a fractal dimension to a natural object.

The first significant data concerning the fractal dimensions of natural objects was collected by L. F. Richardson in about 1960. Richardson's data represented measurements and estimates of the lengths of various coastlines and international borders measured by "rulers" of varying sizes. His data suggest that most coastlines have a fractal dimension of about $D = 1.25$—very close to the fractal dimension of a snowflake curve.

EXAMPLE 3   *The Fractal Border of Spain and Portugal*

Portugal claims that its international border with Spain is 987 kilometers in length. Spain claims that the border is 1214 kilometers in length. However, the two countries agree on the location of the border. How is this possible?

Solution: The border follows a variety of natural objects including rivers and mountain ranges. It is therefore a fractal, with more and more detail revealed on closer and closer examination. Like the snowflake curve, the border length will be *longer* if it is measured with a shorter "ruler." Thus it is quite possible for Spain and Portugal to agree on the border's location but disagree on its length: It simply means that they measured the border with different length "rulers." Spain claims a longer border length, so it must have used the shorter ruler for the measurement.   ■

THE FASCINATING VARIETY OF FRACTALS

The process of repeating a rule over and over to generate a self-similar fractal is called **iteration**. Different sets of rules can produce a fascinating variety of self-similar fractals. Let's look at just a few.

Consider a fractal generated from a line segment to which the following rule is applied repeatedly: *Delete the middle third of each line segment of the current figure* (Figure 10.62). With each iteration, the line segments become shorter until eventually the line turns to dust. The limit (after infinitely many iterations) is a fractal called the **Cantor set**. Because this ephemeral structure results from *diminishing* a one-dimensional line segment, its fractal dimension is *less* than 1.

FIGURE 10.62

Another interesting fractal, called the **Sierpinski triangle**, is produced by starting with a solid black equilateral triangle and iterating with the following rule: *For each black triangle in the current figure, connect the midpoints of the sides and remove the resulting inner triangle* (Figure 10.63). In the complete Sierpinski triangle, which would require *infinitely* many iterations to produce, every remaining black triangle would be infinitesimally small. Therefore, the total area of the black regions in the complete Sierpinski triangle is zero! The fractal dimension of the Sierpinski triangle is between 1 and 2; it is less than 2 because "material" has been removed from the initial two-dimensional triangle.

FIGURE 10.63

FIGURE 10.64

A closely related object is the **Sierpinski sponge** (Figure 10.64). It is generated by starting with a solid cube and iterating with the rule: *Divide each cube of the current object into 27 subcubes, and remove the central subcube and the center cube of each face.* The resulting object has a fractal dimension between 2 and 3; it has less than a full three dimensions because material has been removed from the space occupied by the sponge.

In fact, an infinite variety of fractals can be generated by iteration. Some are remarkably beautiful, such as the famous **Mandelbrot set** (see Figure 10.65 in color insert). This fractal is so popular that you probably have seen it before on T-shirts or calendars, or in other books.

All the self-similar fractals we have considered so far are created by repeating the *exact* same set of rules in each iteration. An alternative approach, called **random iteration,** is to introduce slight, random variations in every iteration. The resulting fractals therefore are *not* precisely self-similar, but they will be close. Such fractals often appear remarkably realistic. For example, Barnsley's fern (see Figure 10.66 in color insert) is a fractal produced by random iteration, and it looks very much like a real fern.

The fact that fractals so successfully replicate natural forms suggests an intriguing possibility: Perhaps *nature* produces the many diverse forms that we see around us through simple rules that are applied repeatedly and with a hint of randomness. Because of this observation and because modern computers allow iterations to be carried out thousands or millions of times, fractal geometry surely will remain an active field of research for decades to come.

• Web • Watch •
Check out many more fractals at the book web site.

REVIEW QUESTIONS ....

1. What is a *fractal*? Explain why measuring a fractal with a shorter ruler leads to a longer measurement.

2. Why do *fractal dimensions* fall in-between the ordinary dimensions of 0, 1, 2, and 3?

3. Explain the meaning of the factors R and N used in calculating fractal dimensions.

4. What is the *snowflake curve*? Explain why we cannot actually draw it, but can only draw partial representations of it.

5. What is the *snowflake island*? Explain how it can have an infinitely long coastline, yet have a finite area.

6. What do we mean by a *self-similar* fractal? How is a self-similar fractal, like the snowflake curve, similar to a real coastline? How is it different?

7. Briefly describe what we mean by the process of *iteration* in generating fractals. Describe the generation of the *Cantor set*, the *Sierpinski triangle*, and the *Sierpinski sponge*. Describe the fractal dimensions of each.

8. What is *random iteration*? Why do objects generated by random iteration make scientists think that fractals are important in understanding nature?

PROBLEMS ....

1. **Ordinary Dimensions for Ordinary Objects.**

   a. Suppose that you want to measure the length of the sidewalk in front of your house. Describe a thought process by which you can conclude that $N = R$ for the sidewalk and hence that its fractal dimension is the same as its ordinary dimension of 1.

   b. Suppose that you want to measure the area of your living room floor, which is square-shaped. Describe a thought process by which you can conclude that $N = R^2$ for the living room and hence that its fractal dimension is the same as its ordinary dimension of 2.

   c. Suppose that you want to measure the volume of a cubical swimming pool. Describe a thought process by which you can conclude that $N = R^3$ for the pool and hence that its fractal dimension is the same as its ordinary dimension of 3.

2. **Fractal Dimensions for Fractal Objects.**

   a. Suppose that you are measuring the length of the stream frontage along a piece of mountain property. You begin with a 15-meter ruler and find just one element along the length of the stream frontage. When you switch to a 1.5-meter ruler, you are able to trace finer details of the stream edge and you find 20 elements along its length. Switching to a 15-centimeter ruler, you find 400 elements along the stream frontage. Based on these measurements, what is the fractal dimension of the stream frontage?

   b. Suppose that you are measuring the area of a very unusual square leaf with many holes, perhaps from hungry insects, in a fractal pattern (e.g., similar to the Sierpinski triangle, Figure 10.63). You begin with a 10-cm ruler, and find that it lies over the entire square, making just one element. When you switch to a 5-cm ruler, you are better able to cover areas of leaf while skipping areas of "holes" and you find three area elements. You switch to a 2.5-cm ruler and find nine area elements. Based on these measurements, what is the fractal dimension of the leaf? Explain *why* the fractal dimension is less than 2.

   c. Suppose that you are measuring the volume of a cube cut from a large rock that contains many cavities forming a fractal pattern. Beginning with a 10-m ruler, you find just one volume element. Smaller rulers allow you to ignore cavities, gauging only the volume of rock material. With a 5-m ruler you find six volume elements. With a 2.5-m ruler you find 36 volume elements. Based on these measurements, what is the fractal dimension of the rock? Explain why a fractal dimension between 2 and 3 is reasonable. (Ignore the practical difficulties of making this measurement caused by the fact that you cannot see through a rock to find all its holes!)

3. **Ordinary and Fractal Dimensions.** Find the dimension of each object and state whether or not it is a fractal.

   a. When measuring the length of an object and reducing the length of your ruler by a factor of 3, the number of length elements increases by a factor of 3.

   b. When measuring the area of an object and reducing the length of your ruler by a factor of 3, the number of area elements increases by a factor of 9.

   c. When measuring the volume of an object and reducing the length of your ruler by a factor of 3, the number of volume elements increases by a factor of 27.

   d. When measuring the length of an object and reducing the length of your ruler by a factor of 3, the number of length elements increases by a factor of 4.

   e. When measuring the area of an object and reducing the length of your ruler by a factor of 3, the number of area elements increases by a factor of 12.

   f. When measuring the volume of an object and reducing the length of your ruler by a factor of 3, the number of volume elements increases by a factor of 36.

4. **Ordinary and Fractal Dimensions.** Find the dimension of the object and state whether or not it is a fractal.

   a. When measuring the length of an object and reducing the length of your ruler by a factor of 4, the number of length elements increases by a factor of 4.

   b. When measuring the area of an object and reducing the length of your ruler by a factor of 4, the number of area elements increases by a factor of 16.

   c. When measuring the volume of an object and reducing the length of your ruler by a factor of 4, the number of volume elements increases by a factor of 64.

   d. When measuring the length of an object and reducing the length of your ruler by a factor of 4, the number of length elements increases by a factor of 6.

   e. When measuring the area of an object and reducing the length of your ruler by a factor of 4, the number of area elements increases by a factor of 24.

   f. When measuring the volume of an object and reducing the length of your ruler by a factor of 4, the number of volume elements increases by a factor of 80.

5. **Fractal Patterns in Nature.** Describe at least five natural objects that exhibit fractal patterns. In each case, explain the structure that makes the pattern a fractal, and estimate its fractal dimension.

6. **The Quadric Koch Curve and Quadric Koch Island.** To draw the *quadric Koch curve* (one of many variations of the snowflake curve), one begins with a horizontal line segment and applies the following rule: Divide each line segment into four equal pieces; replace the second piece with three line segments of equal length that make the shape of a square above the original piece; and replace the third piece with three line segments making a square below the original piece. The quadric Koch curve would result from infinite applications of the rule; the first three stages of the construction are shown in the following figure.

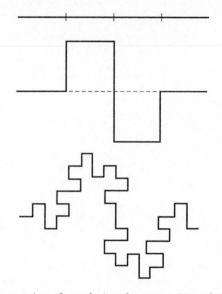

   a. Determine the relation between N and R for the quadric Koch curve.

   b. What is the fractal dimension of the quadric Koch curve? Can you draw any conclusions about the length of the quadric Koch curve? Explain.

   c. The *quadric Koch island* is constructed by beginning with a square. Then each of the four sides of the square is replaced with a quadric Koch curve. Explain why the total area of the quadric Koch island is the same as the area of the original square. How long is the coastline of the quadric Koch island?

7. **The Cantor Set.** Recall that the Cantor set is formed by starting with a line segment and then successively removing the middle one-third of each segment in the current figure (Figure 10.62). If a ruler the length of the original line segment is used, it detects one element in the Cantor set because it can't "see" details smaller than itself. If the ruler is reduced in size by a factor of $R = 3$, it will find two elements (only solid pieces of line, not holes, are measured). If the ruler is reduced in size by a factor of $R = 9$, how many elements does it find? Based on these results, what is the fractal dimension of the Cantor set? Explain why this number is less than 1.

8. **Natural Fractals Through Branching.** One way that natural objects reveal fractal patterns is by branching. For example, the intricate structure of the human lung, the web of capillaries in a muscle, the branches or roots of a tree, or the successive division of streams in a river delta all involve branching at different spatial scales. Explain why structures formed by branching resemble self-similar fractals. Further, explain why fractal geometry rather than ordinary geometry leads to a greater understanding of such structures.

9. **Fractal Dimension from Measurements.** An ambitious and patient crew of surveyors has used various rulers to measure the length of the coastline of Dragon Island. The table at top right gives the measured length of the island, L, and the length of the ruler used, r.

r meters	100	10	1	0.1	0.01	0.001
L meters	315	1256	5,000	19,905	79,244	315,479

a. Make a second table with the entries $\log_{10} r$ and $\log_{10} L$.

b. Graph these data ($\log_{10} r$, $\log_{10} L$) on a set of axes. Connect the data points.

c. If the graph of the data is close to a straight line, it is an indication that the coastline is a self-similar fractal. Does the coastline of Dragon Island appear to be a self-similar fractal?

d. What is the approximate slope of the line through the data? Call the slope s. The fractal dimension of the coastline is $D = 1 - s$. What is the fractal dimension of Dragon Island?

10. **Project: Fractal Research.** Many popular accounts of fractals are available and there are many impressive Internet sites devoted to fractals. Locate at least two articles or web sites related to fractals and use them to write a two- to three-page paper on either a specific use of fractals or a technique for generating fractals.

CHAPTER 10
SUMMARY

*I*n this chapter we have seen the deep connections between mathematics and the arts, beginning with the ancient patterns of Euclidean geometry and progressing through the new science of fractal geometry. The next time you examine a piece of art or music, you might keep in mind the following key ideas.

- The principles of plane and solid geometry were used in many cultures, and formalized by the ancient Greek mathematician Euclid.

- Musical tones, such as those that make up scales, have precise mathematical relationships to each other. In fact, exponential growth laws can be used to describe the frequencies of musical tones.

- You can gain a greater appreciation of art and architecture by understanding the mathematical principles of perspective, symmetry, and proportion. Indeed, many artists of the Renaissance and later ages used these mathematical ideas deliberately in their work.

- Proportion shows up in both art and nature. At least some evidence suggests that the golden ratio plays a particularly important role in both.

- Euclidean geometry has served art and science well for more than 2000 years. Today, however, the recently developed mathematics of fractal geometry is opening entirely new avenues for studying patterns in nature and developing computer-generated art.

SUGGESTED READING

Fractals: Form, Chance, and Dimension, Benoit Mandelbrot (San Francisco: W. H. Freeman and Company, 1977).

Gödel, Escher, Bach: An Eternal Golden Braid, D. R. Hofstadter (New York: Basic Books, 1979).

Mathematics, David Bergamai, ed. (Alexandria, VA: Time-Life Books, 1980).

Mathematics, The Science of Patterns: The Search for Order in Life, Mind, and the Universe, K. Devlin (New York: Scientific American Library, 1994).

Measure for Measure: A Musical History of Science, T. Levenson (New York: Simon and Schuster, 1994).

Nature's Numbers, I. Stewart (New York: Basic Books, 1995).

The Divine Proportion: A Study in Mathematical Beauty, H. E. Huntley (New York: Dover Books, 1970).

The Golden Mean, C. F. Linn (New York: Doubleday, 1974).

The Mathematical Tourist, I. Peterson (New York: W. H. Freeman, 1988).

The Science of Fractal Images, H. O. Peitgen and D. Saupe, eds. (New York: Springer-Verlag, 1988).

FIGURE 10.21
The Last Supper by
Leonardo da Vinci
(1452–1519).
See page 563.

FIGURE 10.24
False Perspective by
William Hogarth (1697–1764).
See page 564.

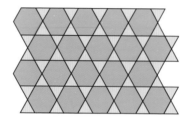

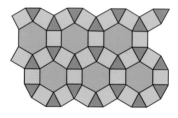

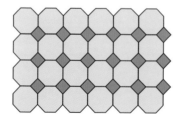

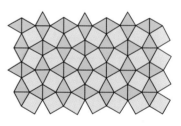

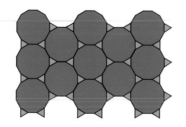

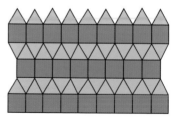

FIGURE 10.35
Eight tilings, each having
more than one type of
regular polygon.
See page 569.

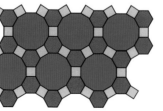

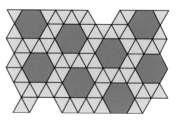

FIGURE 10.38
A Penrose tiling by
Roger Penrose.
See page 570.

FIGURE 10.45
St. Jerome by Leonardo da Vinci (1452–1519).
See page 577.

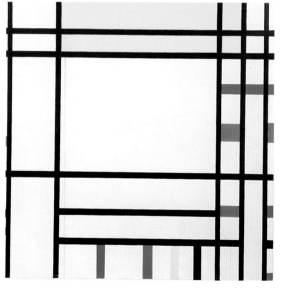

FIGURE 10.46
Place de la Concorde by Piet Mondrian (1872–1944).
See page 577.

FIGURE 10.50
Untitled fractal landscape
by Anne Burns.
See page 583.

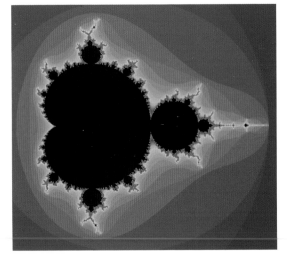

FIGURE 10.65
The Mandelbrot set generated by Robert L.
Devaney. See page 592.

FIGURE 10.66
Two views of Barnsley's fern by Michael Barnsley. One view is a
magnification of the other. See page 592.

Chapter 11

DISCRETE MATHEMATICS IN BUSINESS AND SOCIETY

*D*iscrete mathematics is the mathematics of individual objects, data points, choices, or decisions. In its simplest form, it is little more than counting or arithmetic. However, in the past 50 years, discrete mathematics has found a wide variety of applications in business and society. In this chapter, we investigate how discrete mathematics is applied to large networks (such as communications or computer networks), to making large projects more efficient, and to decision making in democracies.

The wrinkles progress among themselves in a phalanx—beautiful under networks of foam, and fade breathlessly while the sea rustles in and out of the seaweed.
MARIANNE MOORE (1887–1972), "A GRAVE"

You could not step twice into the same river; for other waters are ever flowing on to you.
HERACLITUS (C. 540–480 B.C.)

*By convention there is
color, by convention
sweetness, by conven-
tion bitterness, but in
reality there are
atoms and space.*

—DEMOCRITUS,
C. 460–400 B.C.

FIGURE 11.1

We usually think of a tabletop as a smooth, or **continuous,** surface. But the table actually is a collection of individual atoms. When we consider its atoms, we are thinking of the table as a **discrete** collection of distinct particles.

Clocks provide another illustration of the distinction between the discrete and the continuous (Figure 11.1). On a *continuous* (or *analog*) clock, the hands sweep out smooth circles around the clock face, giving the impression that they pass through *every instant* of time. In contrast, a *discrete* (or *digital*) clock shows the time ticking away one second at a time, giving the impression that time moves ahead in short steps.

The fundamental distinction between the continuous and the discrete also is reflected in mathematics. When a variable can take on *any* value in some interval of numbers, we are dealing with continuous mathematics. For example, time always moves ahead smoothly so *time* generally appears as a continuous variable in mathematical functions. When a variable can take on only *isolated* (usually integer) values, we are dealing with discrete mathematics. For example, the number of people in a room can only be a natural number and not a fraction, so *number of people* is a discrete variable when it appears in a mathematical function. We have already dealt with discrete mathematics in probability (Chapter 8), where we count individual possibilities, and in statistics (Chapters 2 and 9), where we deal with individual data points. We are now ready to explore a few of the important applications of discrete mathematics in business and society.

UNIT 11A

NETWORK ANALYSIS

By the Way ·············

In mathematics, network analysis is sometimes called *graph theory.*

A **network** is a collection of points that are interconnected in some way. The telephone system is a network in which each individual telephone represents a point, and any telephone can be connected to any other through the phone lines. The Internet is a network that connects individual computers together. These networks are extremely complex, as they must allow any two telephones or computers to connect on demand; that is, whenever a user wants to make a connection.

Fortunately, the principles involved in **network analysis** are relatively simple and easily visualized. Let's begin in the same way that the mathematics of network analysis began, in the early eighteenth century in the Prussian town of Königsberg (now Kaliningrad, Russia) on the Pregel River.

THE BRIDGES OF KÖNIGSBERG

Königsberg had seven bridges straddling the Pregel River (Figure 11.2a). A popular pastime was to try to find a path that started and ended at the same point and crossed each bridge *exactly once.* Most citizens of Königsberg believed that such a tour was impossible, but no one was sure. The problem came to the attention of the Swiss mathematician Leonhard Euler (pronounced *oiler).*

By the Way ············

Leonhard Euler (1707–1783) published more than 700 books and papers, making fundamental contributions to mathematics, astronomy, music, hydraulics, optics, and mechanics.

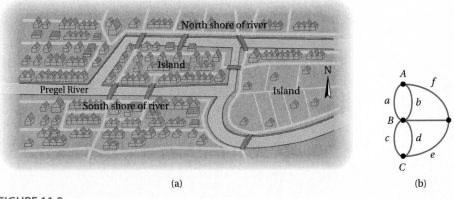

(a)   (b)

FIGURE 11.2

Euler recognized that he could approach the problem by using the simplified representation shown in Figure 11.2(b).

- Each of the seven bridges is represented by a line or curve, called an **edge**, labeled with a lowercase letter (*a, b, c, . . .).*

- Each of the four "pieces" of land to which bridges connect—the two islands, the north shore, and the south shore—is represented by a dot, or **vertex**, labeled with an uppercase letter (*A, B, C, . . .).*

This *network* looks rather different from the actual river and bridges, but it captures all the essential aspects of the problem. For example, the three bridges that land on the north side of the river all connect to the same vertex *A* because the problem concerns a path across *bridges,* not across land. Similarly, the problem concerns only crossing the bridges, so the lengths and shapes of the edges that represent bridges are not precise or to scale. They are drawn in a way that makes the picture easy to see. The problem of finding a path that starts and ends at the same point and crosses all the bridges exactly once is now equivalent to that of finding a path through the network that traverses each edge exactly once, while starting and ending at the same vertex.

> **TIME-OUT TO THINK:**   Which vertex in Figure 11.2(b) represents the land on the south side of the river? Which one represents the west island? the east island? Match each of the bridges in Figure 11.2(a) with its corresponding edge in Figure 11.2(b).

In other applications, network vertices might represent power stations, railway terminals, jobs on an assembly line, or computers, while edges represent connections between them. Networks don't have to represent physical connections. For example, a network could have vertices representing businesses and edges representing economic ties between them.

EXAMPLE 1   *Intranet*

Figure 11.3 shows the layout of computers and cables in a small office *intranet* (a network within an office). Draw a network to represent this intranet.

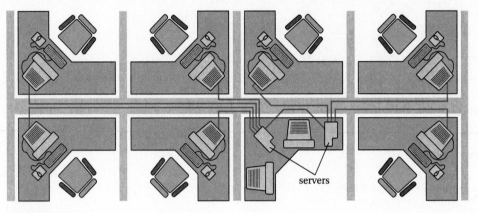

FIGURE 11.3

Solution: Each of the computers becomes a vertex in the network, labeled with an uppercase letter. Each of the cables becomes an edge, labeled with a lowercase letter. Figure 11.4 shows one way of drawing the resulting network.

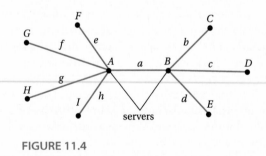

FIGURE 11.4

EXAMPLE 2   *Geographic Network*

Consider a map of southwestern Europe (Figure 11.5). Draw a network in which vertices represent countries and edges represent the relationship "share a common border." What could such a network be used for?

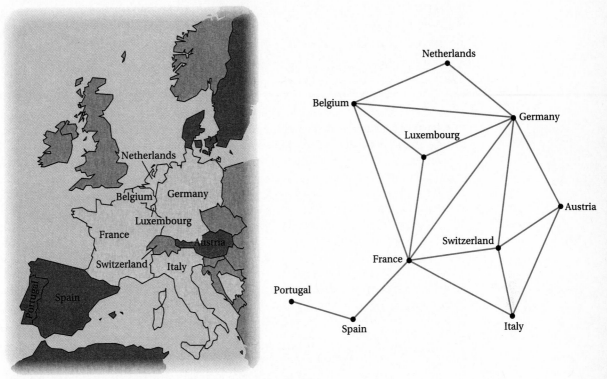

FIGURE 11.5

FIGURE 11.6

Solution: The network is shown in Figure 11.6. The network makes it much easier to see which countries share borders. For example, Luxembourg's borders with Belgium, Germany, and France are clear. One possible use of this network would be to plot trade routes that minimize border crossings.   ∎

Euler Circuits

A path through a network that starts and ends at the same point and traverses every edge exactly once is called an **Euler circuit**. Thus, the Königsberg bridge problem comes down to the question of whether its network has an Euler circuit.

Before we tackle the Königsberg bridge network, let's look for Euler circuits in some simpler networks. It's easy to find an Euler circuit in Figure 11.7(a): start from any vertex, and go around the edges in either a clockwise or counterclockwise direction. We will have started and ended at the same point, and traversed every edge exactly once. We can also find an Euler circuit for Figure 11.7(b): for example, start at vertex E, and follow the edges in the order $a \to b \to c \to d \to e \to f \to g$. This path takes us back to vertex E, and we will have traversed every edge exactly once.

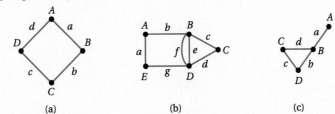

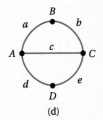

FIGURE 11.7   (a)   (b)   (c)   (d)

TIME-OUT TO THINK: Draw an Euler circuit on each of Figure 11.7(a) and (b) by putting arrows on the appropriate edges. Can you find an Euler circuit for Figure 11.7(b) that traverses the edges in an order besides $a \rightarrow b \rightarrow c \rightarrow d \rightarrow e \rightarrow f \rightarrow g$?

Now consider the network in Figure 11.7(c). We cannot find an Euler circuit through this network because, no matter where we start, we have to go *out and back* along edge *a* in order to traverse it at all. Thus there is no way to traverse all the edges *once* while starting and ending at the same point. A similar problem prevents us from finding an Euler circuit for Figure 11.7(d): we can draw an Euler circuit around the lower or upper *half* of the network, but no matter where we start, we cannot get back to our starting point without either missing an edge or traversing an edge twice.

Is there a general rule that tells us when an Euler circuit exists? Note that all the vertices in Figure 11.7(a) and (b)—the networks that have Euler circuits—have either two or four edges connected to them. For example, vertex *B* in Figure 11.7(b) is connected to the four edges *b*, *c*, *e*, and *f*. In contrast, vertex *A* in Figure 11.7(c) has only one edge connected to it, and vertices *A* and *C* in Figure 11.7(d) have three edges connected to them. Apparently, the key to whether a network has an Euler circuit lies in whether its vertices have an even or an odd number of connected edges.

Test for an Euler Circuit

An Euler circuit exists for a network only if all of the vertices have an even number of edges.

EXAMPLE 3   *The Optimal Mail Route*

An ingenious mail carrier is loath to work any harder than necessary. Her assignment is to deliver mail along the shaded sidewalks of the city blocks shown in Figure 11.8. Can she park her truck in one place and do the job without walking any of the sidewalks more than once? If so, how?

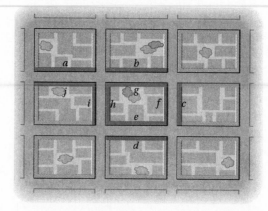

FIGURE 11.8

■ = delivery route
■ = not on delivery route

Solution: The mail carrier's question essentially asks whether there is an Euler **circuit for** her route. We begin by transforming the map into a network. We represent each intersection as a vertex, and each shaded sidewalk along which she must deliver mail as an edge (Figure 11.9). Note that we show distinct edges for either side of each street since she must walk each side separately to deliver the mail.

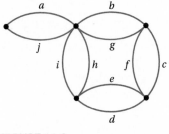

FIGURE 11.9

Every vertex in this network has an even number of edges. Therefore an Euler circuit exists and her desired route is possible. One of several possible Euler circuits is the path $a \to b \to c \to d \to \cdots \to i \to j$. That is, by parking her truck at the leftmost intersection and delivering mail along the path of the Euler circuit, she can complete her route with the minimum possible effort. ◼

Answering the Bridges of Königsberg Problem

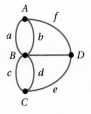

Is it possible to cross each of the seven bridges of Königsberg exactly once while starting and ending in the same place? Take another look at the bridge network (Figure 11.2) shown again to the right. Note that vertices A, C, and D each have three edges, and vertex B has five edges. All of these vertices fail to meet the requirement of having an even number of edges. Therefore this network has no Euler circuit, so the people of Königsberg were searching in vain for a path that crossed each bridge once and returned to its starting point. This may seem an anticlimactic end to the bridges of Königsberg story, but Euler's work on the problem started a broad field of mathematical research.

Finding Euler Circuits

If a network has an Euler circuit, how can we find it? One approach is by trial and error. But there is a more systematic method, or *algorithm*, that we will call the **burning bridges rule.** It is particularly useful when networks are complex.

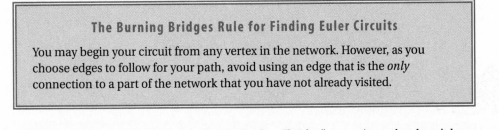

The Burning Bridges Rule for Finding Euler Circuits

You may begin your circuit from any vertex in the network. However, as you choose edges to follow for your path, avoid using an edge that is the *only* connection to a part of the network that you have not already visited.

In other words, never use an edge that is the last "bridge" to territory that hasn't been visited.

EXAMPLE 4   *Applying the Burning Bridges Rule*

Find an Euler circuit for the network shown in Figure 11.10.

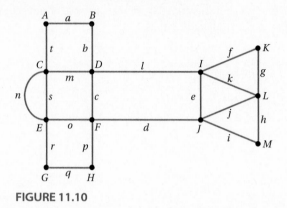

FIGURE 11.10

Solution: First, we should confirm that the network has an Euler circuit. All the vertices have either two or four edges. Because these numbers are even, the network has an Euler circuit.

Now we can find the Euler circuit by applying the *burning bridges rule.* Starting arbitrarily at vertex A, we begin tracing the edges in the order $a \rightarrow b \rightarrow c \rightarrow d \rightarrow e$. At most vertices we have several choices for a path, but so far the choices have been unimportant. Having arrived at vertex I, we have a choice of turning left and using edge l or going right along edge f or edge k. Using edge l violates the burning bridges rule because it is the only remaining link to edges f through k—edges that have not yet been covered. Therefore we must proceed along either edge f or k. Once we choose f or k, we can proceed in several different ways, one of which is shown in Figure 11.11.

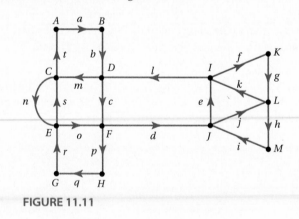

FIGURE 11.11   ■

FIGURE 11.12

NETWORK TERMINOLOGY

Before we consider other network applications, a bit of terminology will be useful. A network that forms a closed ring, such as the network in Figure 11.12 is called a **cycle.** Note that each vertex in a cycle is connected to *exactly two* other vertices.

Closely related to a cycle is a **circuit:** a path *within* a network that begins and ends at the same vertex. The network in Figure 11.13 contains many distinct circuits, such as the path connecting the vertices $A \to B \to C \to A$, or the path connecting the vertices $A \to G \to F \to D \to B \to A$. Note that neither of these circuits is an *Euler circuit* because they do not traverse *every* edge once.

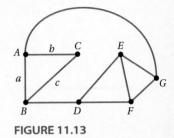

FIGURE 11.13

> **TIME-OUT TO THINK:**   Find at least four other circuits in Figure 11.13. Does this network have an Euler circuit? Why or why not?

The two networks in Figure 11.14 are said to be **complete** because each vertex is connected to *every* other vertex. A complete network always includes many circuits. For example, a path around the outer edges forms a circuit.

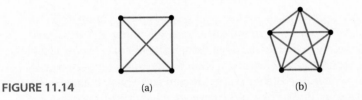

FIGURE 11.14   (a)   (b)

A **tree** is a network in which all the vertices are connected, but in which there are no cycles within the network (Figure 11.15). Among their many uses, trees can be used to represent genealogies and family trees.

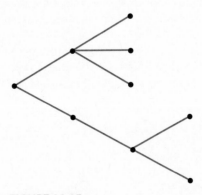

FIGURE 11.15

The **order** of any network is its number of vertices. For example, the two complete networks in Figure 11.14 have order 4 and order 5, respectively; the tree network in Figure 11.15 has order 9. Each vertex in a network may be described further by its **degree:** the number of edges connected to it. For example, each of the vertices in a cycle (Figure 11.12) has degree 2; all the vertices in the network of Figure 11.14(b) have degree 4.

EXAMPLE 5   *Network Classification*

Describe each network shown in Figure 11.16. Identify the order of the network and degree of each of its vertices.

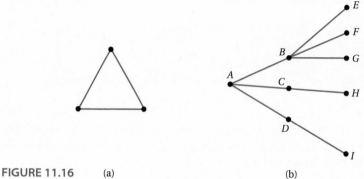

FIGURE 11.16   (a)   (b)

Solution:

a)   This network is a *cycle* because it forms a closed ring. It is also *complete* because each vertex is connected to both of the other vertices. It has *order* 3 because it has 3 vertices. Each vertex has *degree* 2 because it is connected to the other 2 vertices.

b)   This network is a *tree* because all the vertices are connected but it contains no cycles. It has order 9 because it has 9 vertices. Vertex *A* has degree 3 because it is connected to 3 other vertices. Vertex *B* has degree 4. Vertices *C* and *D* have degree 2, and the remaining vertices have degree 1.   ■

MINIMUM COST NETWORKS: CONNECTING ALL THE TOWNS

One very practical application of networks is in minimizing the cost when places or objects are connected together. For example, imagine that the local telephone company needs to connect the seven towns of Sunshine County, shown in Figure 11.17(a), with new high-speed telephone lines. Figure 11.17(b) shows a network representation of Sunshine County with distances between towns indicated on the edges; note that no edges are shown between towns where lakes or mountains would make it very difficult to string wires.

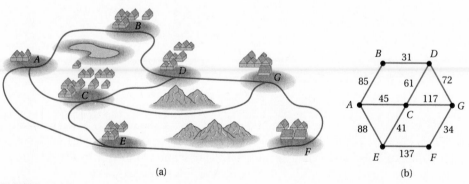

(a)   (b)

FIGURE 11.17

Let's assume that the cost of putting up new lines is proportional to the distance. In that case, the numbers on the edges also represent the cost of connections between towns. In general, numbers on edges are called *costs* (or *weights*), even when they don't have anything to do with money.

The telephone company's objective is to find a set of edges in the original network that

- either directly or indirectly links every town to every other town, and
- has the minimum possible cost (or length).

A set of edges that meets the first criterion, linking all the towns, is called a **spanning network.** Two of the many possible spanning networks for Sunshine County are shown with bold edges in Figure 11.18. The total cost of a spanning network is the sum of the individual costs on its edges. Note that the total cost of the spanning network in Figure 11.18(b) is substantially lower than the total cost of the one in Figure 11.18(a). In fact, if you tried every possible spanning network, you'd find that the one shown in Figure 11.18(b) has the lowest cost of all of them; that is, it is the **minimum cost spanning network** for this problem.

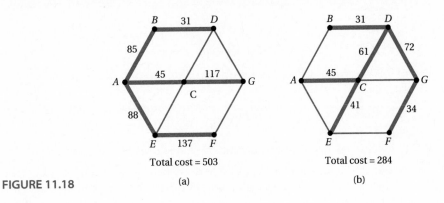

FIGURE 11.18   (a)   (b)

EXAMPLE 6   *Spanning Network Costs*

Find another spanning network for Sunshine County and calculate its total cost. Compare it to the total cost of the spanning networks shown in Figure 11.18.

Solution: The network shown in Figure 11.19 is another spanning network because it connects all the towns. We find its total cost by adding the costs of its edges.

$$85 + 31 + 88 + 41 + 137 + 34 = 416$$

This spanning network has a lower total cost than the one in Figure 11.18(a), but is still substantially higher than the minimum cost network in Figure 11.18(b).

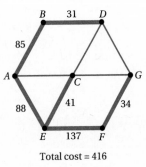

Total cost = 416

FIGURE 11.19   ■

Finding Minimum Cost Networks

Fortunately, there are ways to find the minimum cost spanning network without trying every possibility. One simple procedure is called **Kruskal's algorithm.**

Kruskal's Algorithm for Finding Minimum Cost Networks

Step 1.   Make a list of the edges from the least expensive to the most expensive.

Step 2.   Begin with the least expensive edge; make it bold to indicate it is part of the minimum cost spanning network. Continue to select edges in order of increasing cost until every vertex is connected, either directly or indirectly, to every other vertex.

Step 3.   If any cycles have been created within the spanning network, remove the most expensive edge of each cycle. The final result is the minimum cost spanning network.

FIGURE 11.20

EXAMPLE 7   *Minimum Cost Power Lines*

Figure 11.20 shows a network of towns (vertices) and possible power-line connections (edges). The cost of building a power line (in $ millions) along each edge also is shown. How should the power lines be built to minimize the total cost? What is the total cost?

Solution:

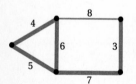

FIGURE 11.21

Step 1:   The edges, in order of increasing cost, have costs of 3, 4, 5, 6, 7, and 8.

Step 2:   We start by making the edge with cost 3 bold. Then we continue to bold the edges for costs of 4, 5, 6, and 7, as shown at the left. All the vertices are now connected (Figure 11.21).

Step 3:   The procedure created a circuit connecting the edges with costs 4, 5, and 6. We remove the edge with cost 6 because it is the most expensive edge in the circuit. The final result is shown in Figure 11.22.

This is the minimum cost spanning network. Its total cost is

$$3 + 4 + 5 + 7 = 19.$$

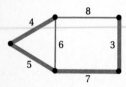

FIGURE 11.22

Because costs in this problem represent millions of dollars, the cost of building this power-line network will be $19 million.   ∎

REVIEW QUESTIONS

1. Explain the distinction between continuous and discrete mathematics.

2. What is a *network*? What is *network analysis*?

3. Briefly describe the problem of the bridges of Königsberg. What was its eventual resolution found by Euler?

4. Explain how a network can be used to represent the Königsberg bridges problem. What do the *edges* and *vertices* represent? Give some examples of what edges and vertices can represent in other networks.

5. What is an Euler circuit? How can we tell whether one exists for a network?

6. Describe the *burning bridges rule* for finding Euler circuits, and give an example of its application.

7. Define and draw an example for each of the following types of network: a *cycle* of order 3; a *complete network* of order 4; a *tree* of order 7.

8. What is a *circuit*? What makes an *Euler circuit* special?

9. How do we define the *degree* of a vertex? What must the degree of the vertices be in order for a network to have an Euler circuit?

10. What do we mean by a *spanning network*? How do we find the cost of a spanning network? What is a *minimum cost spanning network*?

11. Describe *Kruskal's algorithm*, and give an example of its use.

PROBLEMS

1. **Discrete or Continuous?**

   a. As you wait at a bus stop, is the arrival of buses discrete or continuous? Explain.

   b. Suppose that you are skydiving. Does your altitude as you fall vary discretely or continuously? Explain.

   c. Identify three activities in your daily life that represent discrete processes.

2. **Discrete or Continuous?**

   a. Consider the possible results of a census of the number of people in your hometown. Are the possible results discrete or continuous? Explain.

   b. The balance in your bank account earns interest with daily compounding. Does your balance change in a discrete or continuous way? Explain.

   c. Identify three activities in your daily life that represent continuous processes.

3. **Network Models.** Describe at least three examples from your own experience or from the news that could be modeled with networks. In each case, explain what the vertices and edges in the network would represent.

4. **Practical Uses of Euler Circuits.** Example 3 in the text illustrates a practical use of Euler circuits: Mail carriers *do* park their trucks in one place and seek efficient delivery routes that return them to their trucks. Describe, in words, at least three other practical situations in which Euler circuits are useful.

5. **City Streets.** Consider the map of streets and intersections shown.

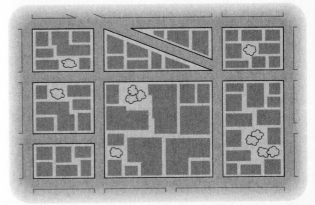

   a. Let edges represent the streets and let vertices represent the intersections and draw the network that results.

   b. Use the same city map and assume that each street has a sidewalk on both sides of it. Draw the network that results when edges are used to represent sidewalks instead of streets.

6. Layout of an Art Gallery. The floor plan of a small art gallery is shown below. Draw a network that has each of the gallery rooms as vertices and doors between neighboring rooms as edges.

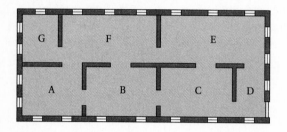

7. Friendships and International Trade. Among the members of a fourth-grade class, Amy trades baseball cards with Beth, Cate, and Daniel; and Beth trades with Daniel and Cate. Draw a network that represents these trading relations. Edges in the network represent the relationship "trades with." Based on this example, explain how networks might also be used to describe international trade relations.

8. Euler Circuits. The town of Sleepy Waters is entwined around the Tranquility River with eight bridges and three islands as shown.

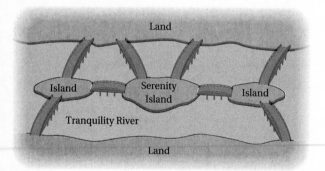

a. How many vertices are needed to draw a network representing the town? How many edges are needed?

b. Draw a network that represents the town.

c. Is it possible to begin on Serenity Island and walk in a closed loop that crosses each bridge exactly once and returns to the starting point? Explain.

9. Euler Circuits. Baytown is located on the coast with three islands nearby. Ferry boats can be taken between the town and the islands along the routes shown in the figure at the top of the next column.

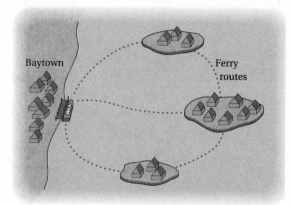

a. How many vertices are needed to draw Baytown and the islands? How many edges are needed to draw the ferry routes? Explain.

b. Draw a network that represents the town, islands, and ferry routes.

c. Is it possible to begin at Baytown and make a closed loop that uses each ferry boat exactly once and returns to Baytown? Explain.

10. Reading Meters. A gas meter reader must visit all the houses on the streets of the map shown.

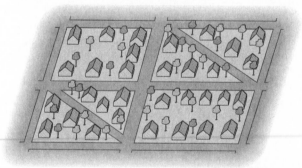

a. Draw a network for this map in which each vertex represents an intersection and each edge represents a street that the meter reader must walk.

b. Does this network have an Euler circuit? If so, draw it. If not, why not?

c. Explain why the Euler circuit will be useful to the meter reader.

11. Checking Parking Meters. An attendant must make hourly inspections of parking meters along the sidewalks (shaded strips) on the following map. Convert the map to a network in which edges represent sidewalks and vertices represent intersections. Can you find a circuit beginning and ending at City Hall that covers each sidewalk exactly once? If so, show it on your network.

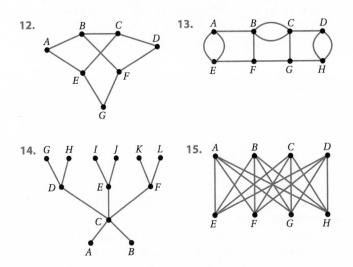

■ = inspection route
□ = not on inspection route

Network Terminology. *For the networks shown in Problems 12–17, answer the following questions.*

  a. What is the order of the network?

  b. What is the degree of each vertex of the network?

  c. Does the network have a special form (such as a tree, a cycle, or a complete network)? If so, what?

  d. Does the network have an Euler circuit? Explain.

  e. In those cases in which an Euler circuit does exist, apply the burning bridges rule to find an Euler circuit. In each case explain the steps of the method and note the vertices at which critical decisions must be made.

12.

13.

14.

15.

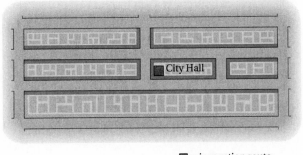

16.

17.

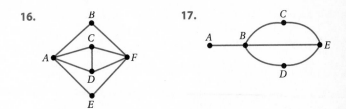

18. Drawing Networks.

  a. Draw a network of order 6 whose vertices all have an odd degree.

  b. Draw a tree of order 7 with at least one vertex with an even degree.

  c. Draw a network of order 5 with no vertices having an odd degree.

19. Drawing Networks.

  a. Draw a network of order 5 whose vertices all have an even degree.

  b. Draw a network of order 6 in which all vertices have degree 2.

  c. Draw a network of order 8 in which all vertices have degree 3.

20. Cycle Networks. Draw a *cycle* of order 5. How many edges does it have? How many edges does a cycle of order 6 have? Find a general rule that gives the number of edges in a cycle with n vertices.

21. Complete Networks. Draw complete networks of order 5 and order 6. How many edges does each network have? Show that the number of edges in these networks satisfies the following rule: a complete network with n vertices has $n \times (n - 1)/2$ edges.

22. Neighboring States. Find a map of Tennessee and its immediate neighbors. Letting vertices represent states and edges represent the relationship "is a neighbor of," draw a network showing all connections among Tennessee and its neighbors. What is the order of the network? What is the degree of the vertex corresponding to Tennessee?

23. Neighboring States. Find a map of Missouri and its immediate neighbors. Letting vertices represent states and edges represent the relationship "is a neighbor of," draw a network showing all connections among Missouri and its neighbors. What is the order of the network? What is the degree of the vertex corresponding to Missouri?

24. Family Trees. A family tree is a network in which each vertex is a person and each edge represents a parent-child relationship.

   a. Sebastian and Emily are the parents of three children: Paul, Victor, and Alice. Draw a family tree showing only these five people.

   b. Victor is married to Melinda, and they have two children: Debra and Katy. Add Melinda and the two children to the family tree from part (a).

   c. Make a family tree for your own family that includes all your siblings, your parents and their siblings, and your grandparents.

25. Soccer Tournament. In the three weeks that remain in the season, the five teams of the Skyline Soccer League (A, B, C, D, and E) must play the games indicated in the following table. (An X means the two teams play, and a blank means they do not play.) Draw a network that shows the teams as vertices and relationships "plays against" as edges. What is the order of the network you have drawn? What is the degree of each vertex of the network? Is the network complete? Explain.

—	A	B	C	D	E
A		×	×		×
B	×			×	
C	×			×	×
D		×	×		×
E	×		×	×	

26. Soccer Schedule. As a wise coach who knows some network theory, you have drawn the following network to show the games remaining in the final four weeks of the soccer league. The vertices represent teams, and the edges represent the relationship "plays against." Convert the network to a table, similar to the table in Problem 25, that shows all the games to be played.

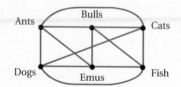

27. Spanning Networks. Consider the network and the three spanning trees labeled I, II, and III shown in the next column. Find the cost of each of the three spanning trees. Which of the three spanning trees has the minimum cost?

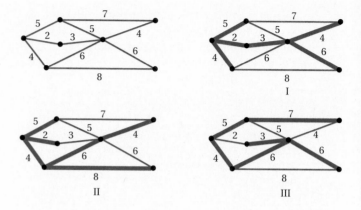

Kruskal's Algorithm. *Use Kruskal's algorithm to find the minimum cost spanning tree of the networks shown in Problems 28 and 29.*

28.   **29.**

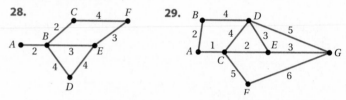

30. Community Planning. A mountain community is being planned and provision must be made for water. The following map shows the building sites (the vertices of the network) and the feasible routes for water pipelines (the edges of the network). The weights on the vertices give the distances between sites in kilometers. A single well can supply all the houses and be drilled on any of the housing sites. Assume that the cost of supplying water to all the homes depends on the total length of pipe used. Find the minimum cost network that supplies water to all the houses.

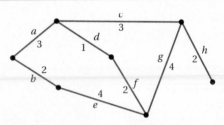

31. Local Area Networks. If every workstation in a computer network is connected to at least one other workstation, every user can communicate with every other user. Consider the layout shown on the next page, in which vertices represent workstations and edges represent feasible connections. The weight on each edge gives the cost of the wire needed for that link. Find the minimum cost network that achieves full connectivity between all the computers.

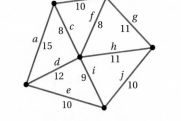

32. **Traversable Networks.** For any network, an Euler circuit must traverse all the edges *and* return to the starting vertex. Sometimes a path traverses all the edges but doesn't return to the starting vertex. A network with this property is said to be *traversable*. The rule that determines whether a network is traversable is:

If a network has two vertices with an odd number of edges and all other vertices have an even number of edges, it is traversable but does not have an Euler circuit. If all the vertices have an even number of edges, the network is traversable and it has an Euler circuit.

  a. Explain why a traversable network can have two vertices with an odd degree.

  b. Consider the networks shown in Problems 12–17. For each one, determine whether it is traversable. If it is, show a traversing path.

THE TRAVELING SALESMAN PROBLEM

Imagine that you work as a traveling salesperson. Your job requires you to make presentations about your products to companies located in five cities. How do you find a route that allows you to visit all the cities once and return home with the minimum amount of travel?

This is an example of a famous problem in mathematics called the **traveling salesman problem.** We can represent this problem as a network by letting vertices represent the cities. Edges, labeled with distances, represent possible routes between pairs of cities (Figure 11.23).

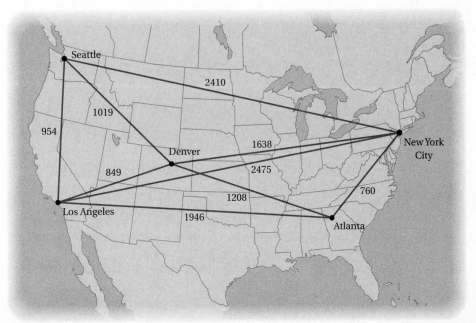

FIGURE 11.23

The solution to the traveling salesman problem is the shortest path that starts and ends in the same place and visits each city exactly once. That is, it is the shortest *circuit* (closed path) through a network that visits all the *vertices* exactly once.

In this unit, we will investigate the traveling salesman problem. As you might expect, we will find that it has many important applications that go far beyond traveling salesmen.

HAMILTONIAN CIRCUITS

A circuit that passes through every *vertex* of a network exactly once and returns to the starting vertex is called a **Hamiltonian circuit,** after the nineteenth-century Irish mathematician William Rowan Hamilton (1805–1865). The networks in Figure 11.24(a) and (b) have Hamiltonian circuits on the paths indicated with arrows. Note that, while Hamiltonian circuits pass through every *vertex,* they do not necessarily traverse every *edge.* Many networks do not have Hamiltonian circuits. For example, the network in Figure 11.24(c) does not have a Hamiltonian circuit because it is impossible to find a circuit that passes through all vertices without passing through at least one vertex twice.

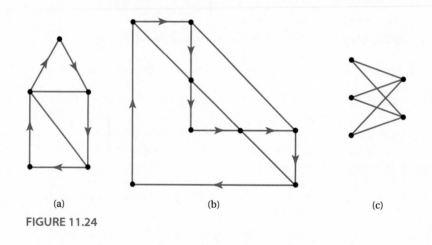

(a)   (b)   (c)

FIGURE 11.24

EXAMPLE 1   *Hamiltonian Circuit for Sunshine County*

Find at least two Hamiltonian circuits for the Sunshine County network shown in Figure 11.17(b). Also find one circuit that visits all the vertices but is *not* a Hamiltonian circuit. (The costs on the edges are not important in finding the Hamiltonian circuits.)

Solution: Figure 11.25(a) and (b) on the next page show two Hamiltonian circuits for the Sunshine County network. In each case, the circuits can start at any vertex; they must end at that same vertex and must visit every other vertex exactly once along the way. The circuit in Figure 11.25(c) also visits all the vertices. However, it is *not* a Hamiltonian circuit because it visits vertex *C* more than once.

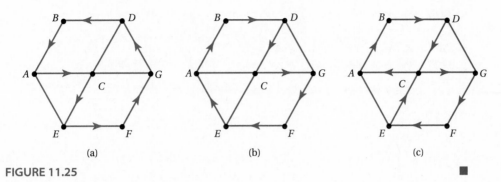

FIGURE 11.25 ■

Finding Hamiltonian Circuits

In Unit 11A, we found a test for whether a network has an Euler circuit and a method for finding an Euler circuit. Unfortunately, finding *Hamiltonian* circuits is much more difficult. In fact, there is no known rule for *efficiently* determining whether a Hamiltonian circuit even exists in a network, except in a few special cases. For example, a *cycle* always has a Hamiltonian circuit (Figure 11.26a), as does a *complete* network (Figure 11.26b). A *tree* never has a Hamiltonian circuit because it contains no circuits at all (Figure 11.26c).

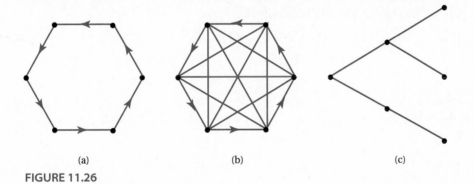

FIGURE 11.26

Aside from such special cases, trial and error is the only way to find Hamiltonian circuits. Moreover, some networks have many Hamiltonian circuits within them. For example, a *complete* network of order 4 (that is, 4 vertices) has three different Hamiltonian circuits (Figure 11.27).

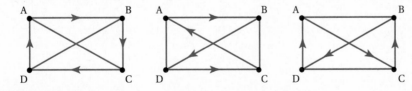

FIGURE 11.27

You might imagine that complete networks of higher order would have more Hamiltonian circuits—and they do. In fact, a simple formula tells us the number of Hamiltonian circuits in a complete network.

Hamiltonian Circuits in Complete Networks

The number of Hamiltonian circuits in a complete network of order n is

$$\frac{(n-1)!}{2}.$$

Note that this formula says that a complete network of order $n = 4$ has

$$\frac{(4-1)!}{2} = \frac{3!}{2} = \frac{3 \times 2 \times 1}{2} = 3$$

Hamiltonian circuits, just as we found in Figure 11.27.

EXAMPLE 2   *Complete Network of Order 5*

How many Hamiltonian circuits are there in a complete network of order 5? Draw such a network, and identify all of its Hamiltonian circuits.

Solution: A complete network of order $n = 5$ has

$$\frac{(5-1)!}{2} = \frac{4!}{2} = \frac{4 \times 3 \times 2 \times 1}{2} = 12$$

Hamiltonian circuits. A complete network of order 5 has every vertex connected to every other vertex. The 12 Hamiltonian circuits are shown in Figure 11.28.

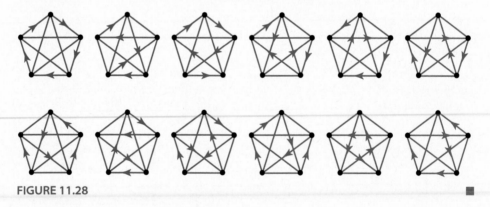

FIGURE 11.28

SOLVING TRAVELING SALESMAN PROBLEMS

The solution to a traveling salesman problem is the shortest path that starts and ends in the same place and visits each city once. Thus the solution is not just *any* Hamiltonian circuit, but the *particular* Hamiltonian circuit that is the shortest one in a network.

For example, suppose that you're planning a summer vacation during which you want to visit five national parks. Let's make the problem slightly more exotic by assuming that you will rent a private airplane at one park and do all your traveling between parks by air. If you have to pay for the plane according to the number of miles flown, what is the lowest cost path you can take? Although sales are not involved in this problem, it is mathematically equivalent to the traveling salesman problem: You must begin and end at the same place (you must return the rented airplane), and you seek the shortest path that visits all five parks.

Table 11.1 lists the five national parks and the distances between them. We can represent this data with a network in which vertices represent the national parks and edges represent the air routes between them (Figure 11.29). The edges are labeled with distances in miles, but the network does not need to be drawn to scale.

Table 11.1	Air Miles Between National Parks				
—	Bryce	Canyonlands	Capitol Reef	Grand Canyon	Zion
Bryce	—	136	69	108	52
Canyonlands	136	—	75	202	188
Capitol Reef	69	75	—	151	123
Grand Canyon	108	202	151	—	101
Zion	52	188	123	101	—

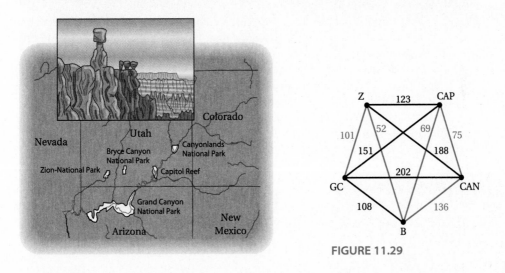

FIGURE 11.29

EXAMPLE 3   *Hamiltonian Circuits in the National Park Network*

How many Hamiltonian circuits are there in the national park network shown in Figure 11.29? Two possible circuits are shown in Figure 11.30 on the following page. Describe each in words, and find its total length.

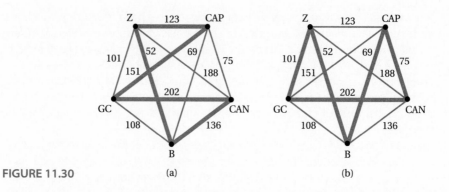

FIGURE 11.30

(a)   (b)

Solution: This is a *complete* network of order 5, since it is possible to fly from any national park to any of the four others. Hence it has 12 Hamiltonian circuits (see Example 2). If we start from Bryce, the circuit in Figure 11.30(a) shows a trip taken in the order Canyonlands, Grand Canyon, Capitol Reef, Zion, and back to Bryce (or the reverse order). Its total length is

$$136 \text{ mi} + 202 \text{ mi} + 151 \text{ mi} + 123 \text{ mi} + 52 \text{ mi} = 664 \text{ mi}.$$

The circuit in Figure 11.30(b) goes from Bryce to Capitol Reef, Canyonlands, Grand Canyon, Zion, and back to Bryce (or the reverse order). Its total length is

$$69 \text{ mi} + 75 \text{ mi} + 202 \text{ mi} + 101 \text{ mi} + 52 \text{ mi} = 499 \text{ mi}. \qquad \blacksquare$$

An Explosion of Possibilities

Example 3 above shows that the circuit in Figure 11.30(b) is substantially shorter than the circuit in Figure 11.30(a). But is it the *shortest* circuit for the national park problem? The only way to know for sure is to calculate the length of *all* 12 Hamiltonian circuits for this network. If you do so, you'll find that the circuit in Figure 11.30(b) is indeed the shortest one. Thus it is the solution to this particular version of the traveling salesman problem because it is the lowest cost trip among the five national parks.

Unfortunately, finding and checking all the Hamiltonian circuits for a problem is worse than just tedious. In fact, the number of Hamiltonian circuits in many real problems is so great that even the fastest imaginable computers could never check all of them.

EXAMPLE 4   *Airline Planning for 15 Cities*

The CEO of your company wants to take the company jet on a 15-city tour, beginning and ending at the company headquarters in Chicago (one of the 15 cities). How many possible routes must be checked to find the shortest route among the 15 cities? Suppose that you have a computer that can find and measure 1 million Hamiltonian circuits per second. How much computer time will it take to find the shortest route?

Solution: In principle, it is possible to fly between any pair of cities. Thus a network representing this problem is a *complete* network because every vertex representing a city is connected to every other. For a complete network of order $n=15$, the number of Hamiltonian circuits is

$$\frac{(n-1)!}{2} = \frac{(15-1)!}{2} = \frac{14!}{2} = 4.4 \times 10^{10},$$

or 44 billion. If your computer can find and measure 1 million of these circuits each second, it will take

$$\frac{4.4 \times 10^{10} \text{ circuits}}{10^6 \text{ circuits/s}} = 44,000 \text{ s},$$

or

$$44,000 \text{ s} \times \frac{1 \text{ min}}{60 \text{ s}} \times \frac{1 \text{ hr}}{60 \text{ min}} = 12 \text{ hr}$$

to check them all. That is, it will take about half a day of computer time to check all the Hamiltonian circuits and find the shortest route.   ■

EXAMPLE 5   *Airline Planning for 30 Cities*

Just as you finished planning the 15-city tour in Example 5, the CEO informs you that she actually needs to visit 30 cities. Is it possible to find the shortest route by checking all the possibilities? Explain.

Solution: In this case, the problem involves a complete network of order $n = 30$. Thus it has

$$\frac{(n-1)!}{2} = \frac{(30-1)!}{2} = \frac{29!}{2} = 4.4 \times 10^{30}$$

possible Hamiltonian circuits. Checking 1 million circuits per second, it would take

$$\frac{4.4 \times 10^{30} \text{ circuits}}{10^6 \text{ circuits/s}} = 4.4 \times 10^{24} \text{ s},$$

or

$$4.4 \times 10^{24} \text{ s} \times \frac{1 \text{ min}}{60 \text{ s}} \times \frac{1 \text{ hr}}{60 \text{ min}} \times \frac{1 \text{ day}}{24 \text{ hr}} \times \frac{1 \text{ yr}}{365 \text{ days}} = 1.4 \times 10^{17} \text{ yr}$$

to check every circuit and find the shortest route. Note that $10^{17} = 10^7 \times 10^{10}$. Thus 10^{17} years is some 10 million (10^7) times longer than the current age of the universe of about 10 billion (10^{10}) years. Clearly, this is not a viable way to plan the CEO's trip.   ■

By the Way ············

In 1994, computer scientist Leonard Adelman (University of California) chose a network problem for the first-ever test of a computer using DNA molecules rather than electric charge to represent data. Theoretically, such DNA computers could be a billion times faster than current supercomputers—in which case Example 5 could be solved in only 100 million years or so!

Nearly Optimal Solutions to Traveling Salesman Problems

No known method besides trial and error is guaranteed to find the shortest Hamiltonian circuit in an arbitrary network. Unfortunately, as Example 5 above shows, checking all the possible Hamiltonian circuits is impossible in many real problems. However, there are some methods that usually (but not always) find nearly optimal solutions; that is, solutions that find circuits with a total length *close to* that of the shortest circuit. One such method that can be used for complete networks is called the **nearest neighbor method.**

> ### The Nearest Neighbor Method
>
> To find a nearly optimal solution to a traveling salesman problem, begin at any vertex. From this starting point, take the shortest path to a neighboring vertex. Proceed to the nearest vertex that has not yet been visited. Continue this process of visiting "nearest neighbors" until the circuit is complete.

EXAMPLE 6   *Nearest Neighbor Method for the National Park Network*

Apply the nearest neighbor method to the national parks network (Figure 11.29), beginning at Bryce. Does this solution seem to be "nearly optimal?"

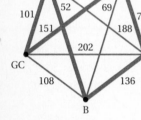

FIGURE 11.31

Solution: The *nearest neighbor* to Bryce is Zion, 52 miles away. From Zion, the nearest neighbor not already visited is Grand Canyon, 101 miles away. From Grand Canyon, the nearest unvisited neighbor is Capitol Reef, 151 miles away. Finally, from Capitol Reef we have only one place left to visit: Canyonlands, 75 miles away. We then return to Bryce. The circuit is shown in Figure 11.31. Its total length is

$$52 \text{ mi} + 101 \text{ mi} + 151 \text{ mi} + 75 \text{ mi} + 136 \text{ mi} = 515 \text{ mi}.$$

This is slightly longer than the 499 mile distance of the shortest circuit for this network (Figure 11.30b). However, it is longer only by a factor of about

$$\frac{515 \text{ mi}}{499 \text{ mi}} = 1.03,$$

or about 3% longer. Thus it seems reasonable to say that this solution is *nearly* optimal, because its cost is only 3% higher than the cost of the optimal solution. ∎

CASE STUDY   *Computer Circuit Board*

A computer circuit board may require thousands of small thin holes to be drilled at precise locations so that computer components can be inserted. The holes are drilled by a special machine, or robot (Figure 11.32a) on the next page. In principle, the robot could start drilling holes at any location on the board, but it must finish back at its starting point so that it is ready for the next board. In other words, the robot's task is a *traveling salesman problem*: it must start and end in the same place while drilling at the locations of every required hole. Finding the shortest possible path is important because the drilling machine is expensive to operate, and a shorter path allows the machine to drill more boards each day.

   A network representing this problem would have a vertex for each of the thousands of holes that must be drilled. Moreover, the network would be *complete* because, in principle, the robot can go from any one hole location to any other. Example 5 showed that we could never test all the Hamiltonian circuits in a complete network with just 30 vertices, and a complete network with thousands of vertices is a far more complex problem. Fortunately, the problem can be approached with methods for finding nearly optimal solutions, such as the one shown in Figure 11.32(b) for a circuit board with 3038 holes.

FIGURE 11.32   (a)   (b)   ■

OPERATIONS RESEARCH

Applications of the traveling salesman problem are almost endless. They include scheduling of public maintenance vehicles, such as snowplows and street cleaners; finding optimal layouts for sewer, power, and water lines; designing telephone switching devices for routing long distance calls; laying out highways and railroads; designing distribution strategies for merchandise; devising efficient airline routes and schedules; and much more.

All of these applications of the traveling salesman problem fall under a general branch of research called **operations research** (or *management science*). Operations research seeks strategies for carrying out complex tasks in organized and efficient ways. The increasing globalization of travel and commerce makes operations research ever more important to governments and businesses, as even relatively small improvements in efficiency can have a big impact on costs, revenues, or profits.

By the Way ············
Operations research developed into a true science in the years following World War II. Its growth has gone hand-in-hand with the increasing power of computers, which allow new optimization strategies to be tested and refined.

EXAMPLE 7   *Snowplow Efficiency*

Consider a major city with hundreds of miles of roads that must be plowed after a snowstorm. Suppose that the snowplows move at a speed of 10 miles per hour, and that their total operating costs are $150 per hour. Imagine that the standard route followed by the snowplow drivers involves a total of 1600 miles of driving. A computer program developed through operations research helps city planners discover a new route that involves only 1330 miles (about 17% shorter). How much will the city save after a snowstorm?

Solution: The new route is shorter than the old one by

$$1600 \text{ mi} - 1330 \text{ mi} = 270 \text{ mi}.$$

At a speed of 10 miles per hour, the time saved for the snowplows will be

$$\frac{270 \text{ mi}}{10 \text{ mi/hr}} = 27 \text{ hr}.$$

The total savings will be $27 \text{ hr} \times \dfrac{\$150}{\text{hr}} = \$4050.$

The new route will save the city more than $4000 every time it must send out the snowplows. In a major storm, in which the snowplows must clear the roads several times, the savings could easily be tens of thousands of dollars.   ■

---

*T*HINKING ABOUT . . .

Recent Successes in Operations Research

Operations research is a major factor in the increasing efficiency of modern business. Below are two quotes concerning recent successes of operations research (from *Interfaces* 24:1, Jan.-Feb. 1994).

"Delta Airlines flies over 2500 domestic flight legs every day, using about 450 aircraft from 10 different fleets. The fleet assignment problem is to match aircraft to flight legs so that seats are filled with paying passengers. Recent advances in [operations research] and computer hardware make it possible to solve optimization problems of this scope for the first time. Delta is the first airline to solve [this problem] to completion . . . [The

solution] is expected to save Delta Airlines $300 million over the next three years."

—FROM "COLDSTART: FLEET ASSIGNMENT AT DELTA AIRLINES"

BY R. SUBRAMANIAN ET AL.

"North Carolina uses [operations research] to produce a pupil transportation funding process that encourages operational efficiency and reduces expenditures. The new process has led to changes in bus routes and schedules, adjustments in school start and stop times, and reductions in the inventory of buses. Between 1990 and 1993, the state waved $25.2 million in captial costs and $27.9 million in operating costs, and it expects saving to increase."

—FROM "IMPROVING PUPIL TRANSPORTATION"

BY T. SEXTON ET AL.

REVIEW QUESTIONS

1. Give a simple example of the *traveling salesman problem* in its traditional form. List a few types of problems that are mathematically equivalent to traveling salesman problems, but that don't involve traveling salesmen.

2. What is a *Hamiltonian circuit*? How does it differ from an Euler circuit?

3. How many Hamiltonian circuits can be drawn in a complete network?

4. How is the solution to a traveling salesman problem related to Hamiltonian circuits? Why are traveling salesman problems so difficult to solve?

5. What do we mean by a *nearly optimal* solution to a traveling salesman problem? Describe the *nearest neighbor method* for finding a nearly optimal solution. Give an example of its use.

6. What is *operations research*? Why is it important?

PROBLEMS

1. **Number of Hamiltonian Circuits.**

   a. How many different Hamiltonian circuits are there in a complete network of order $n = 8$?

   b. How many different Hamiltonian circuits are there in a complete network of order $n = 18$?

   c. Suppose that a computer could check Hamiltonian circuits at a rate of one per second. How long would it take to check all the circuits in each of the networks in parts (a) and (b)?

2. **Number of Hamiltonian Circuits.**

   a. How many different Hamiltonian circuits are there in a complete network of order $n = 10$?

   b. How many different Hamiltonian circuits are there in a complete network of order $n = 20$?

   c. Suppose that a computer could check Hamiltonian circuits at a rate of one per second. How long would it take to check all the circuits in each of the networks in parts (a) and (b)?

Practice with Hamiltonian Circuits. *For Problems 3–6, consider the networks shown. In each case, follow the edges according to the arrows, and determine whether that path forms a Hamiltonian circuit. Justify your reasoning and explain your answer carefully. If a path does not form a Hamiltonian circuit, can you find a different path that does?*

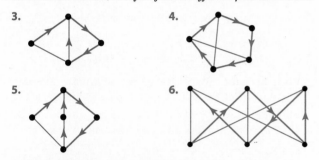

3.

4.

5.

6.

7. **Traveling the National Parks.** Suppose that you are taking a trip to Bryce, Canyonlands, Capitol Reef, and Grand Canyon National Parks. Use the data in Table 11.1 to answer the following questions.

   a. Draw a network that shows the four national parks and the distances between them.

   b. What is the total length of a trip that visits the parks in the order of Bryce, Canyonlands, Capitol Reef, Grand Canyon, and back to Bryce?

   c. Use the nearest neighbor method and start at Grand Canyon to find a circuit that passes through the three other parks and returns to Grand Canyon. Describe your work.

   d. Repeat part (c), but start and end at Bryce.

   e. Is either of the circuits in parts (c) or (d) the shortest possible circuit? Explain.

8. **The Mail Must Go Through.** The table at the top right shows the distances between pairs of towns in a rural county.

—	Avila	Barila	Camilla	Dorrito	Eldorado
Avila	—	12	9	5	7
Barila	12	—	10	2	6
Camilla	9	10	—	15	8
Dorrito	5	2	15	—	9
Eldorado	7	6	8	9	—

   a. Make a network showing the five towns and the distances between them.

   b. As a mail truckdriver, you must visit each town once every weekday. Find the shortest route that visits each town exactly once and returns to the starting town. Explain the methods that you use.

9. **A Traveling Salesperson Problem.** The network in the following figure shows the flight times (in hours) between cities that a salesperson must visit. Starting once from each city in the network, apply the nearest neighbor method to find a circuit that visits each city exactly once. Do you get different circuits for different starting cities? Can you conclude whether any of your circuits represent the shortest possible circuit? Explain.

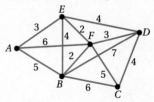

10. **Five-City Vacation.** A section of a road atlas looks like the following figure showing distances between various pairs of the cities St. Louis, Chicago, Kansas City, Memphis, and Louisville. For the connections shown, find the shortest circuit that passes through each city exactly once and starts and finishes in the same city. Some connections between cities are missing. Would a shorter circuit be possible if all connections between the cities were available?

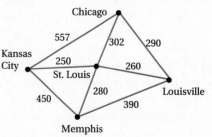

11. **Car Shuttles.** Each week Abe, Barbara, Carl, or Dolores hosts a bridge game at his or her house. Everyone is eager to save gasoline, so the host always picks up and drops off the other three players. The location of each player's house and the distances between them are shown. What path should Abe use to pick up his guests to minimize the distance traveled? Would Barbara, Carl, or Dolores use a different route? Explain.

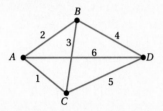

12. **Taxi Routing.** A taxi driver based at a train station (the black square) must deliver five riders to five different hotels. The locations of the hotels are shown and the travel times (in minutes) are indicated on the edges. Find the route that minimizes the total travel time for the taxi, beginning and ending at the train station. Assume that no time is spent waiting at the hotels.

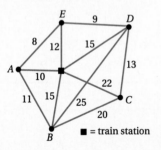

13. **Overnight Delivery.** An overnight delivery company must be able to deliver packages by air between any pair of 35 cities. The cost of flying its planes is $15,000 per hour. Its current schedule requires a total of 890 hours of flight time each day (among its fleet of planes). Suppose that a more efficient set of routes can be found that reduces the total flight requirement to 860 hours per day. How much money will the company save each day? Each year?

14. **Interstate Trucking.** A computer manufacturer uses trucks to deliver computers throughout the United States. Its fleet of trucks currently drives an average of 75,000 miles per day, at a cost of approximately $0.95 per mile. Suppose that an operations research project shows

that it can accomplish the same deliveries, but can reduce its total truck driving by 2%. How much will the company save each year?

15. **Distribution of Rental Cars.** The following figure shows the four distribution sites for a rental car company. At one particular time two sites have an excess of cars and two sites have a shortage of cars, as shown. Each of the sites with an oversupply is connected to each of the sites with an undersupply by a route that has a certain cost per car (shown as a number on the edge). Proceed by trial and error to find the best (least expensive) scheme for redistributing the rental cars.

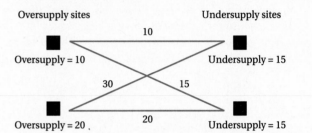

16. **Project: Your Own Traveling Salesman Problem.** Consider a day on which you must run errands to at least four different places. Locate each of the places you must visit on a street map, and highlight your normal route between your starting point (e.g., home or work) and these places. Draw a network that represents each place you must visit as a vertex, and represent possible routes between places with edges. Label the distances between places on the edges. Find the total distance along your normal route. Then apply the nearest neighbor method to find another circuit that visits all the places. Is this new route more efficient than your old route? Can you find an even shorter route? Explain.

17. **Project: Operations Research.** Find an example in which a business or government has used operations research to attain a significant reduction in costs or improvement in efficiency. In one or two pages, summarize how the operations research was conducted and the results that were achieved.

PROJECT DESIGN

Another type of problem that relies on networks is planning a project with many stages or tasks. Some of the tasks may be performed simultaneously, whereas others have a definite precedence (that is, one must precede another). Problems like this arise in everything from planning space flights to designing shopping centers. Finding an efficient way to schedule the various tasks can have a huge impact on the cost and time for a large project.

A HOUSE BUILDING PROJECT

We can investigate the principles of project design problems by considering a relatively small project: building a house. Figure 11.33 shows a simplified **flowchart** for the various tasks in building a house. The flowchart is a network in which the *edges* represent tasks and the *vertices* represent completion points for the tasks. Time flows from left to right through this network, specifying the precedence of the various tasks.

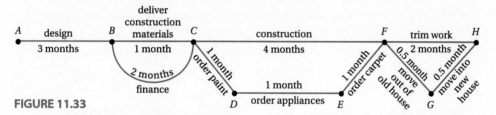

FIGURE 11.33

To interpret this network, note that:

- Each edge is labeled with the number of months needed to complete the task. For example, the label on the *finance* edge indicates that two months are required to obtain financing for the house.

- In some phases of the project, only one task can be undertaken at a time. For example, only the *design* task can be carried out between vertices *A* and *B*.

- During other phases, two or more tasks can be carried out concurrently. For example, between points *C* and *F*, *construction* can take place at the same time that *paint*, *appliances*, and *carpet* are ordered.

EXAMPLE 1   *Interpreting the Project Design Network*

Interpret the *construction, order paint, order appliances,* and *order carpet* edges (tasks) in the house building network shown in Figure 11.33.

Solution: The *construction* edge indicates that the project plan allows four months for construction. The *order paint, order appliances,* and *order carpet* edges show that the plan allows one month for each of these tasks. The fact that these edges connect the same two vertices (*C* and *F*) as the *construction* edge means that the paint, appliances, and carpet should be ordered *during* the period of construction. It also means that all these tasks must be completed before the project can continue after construction. The fact that only three months are needed for ordering these supplies, while four months are needed for construction, means that a month can be spared in ordering these supplies without delaying the project schedule. ∎

Limiting Tasks and Critical Path

In parts of the network where two or more tasks take place simultaneously, the project cannot continue until *all* the simultaneous tasks are completed. For example, the project cannot reach point *F* until *all* the tasks connecting vertices *C* and *F* are completed. The task that requires the most time between stages in the project is called the **limiting task** for that period. For example, *construction* is the limiting task between stages *C* and *F* because it takes four months, while *order paint, order appliances,* and *order carpet* require only three months combined.

The minimum time required to complete the project is dictated by the limiting tasks. We can represent this minimum time as a path through the network that includes all the limiting tasks. This is called the **critical path** for the network, because the tasks along this path must be completed on time if the project is to be completed in the minimum possible time.

> The minimum completion time for the project is the length of the *critical path* through the network. It is the path that includes all the limiting tasks, and hence has the *largest* total time of any path through the network.

EXAMPLE 2   *Limiting Tasks*

What is the limiting task between stages *B* and *C* in Figure 11.33? How about between stages *F* and *H*?

Solution: Between stages *B* and *C*, the tasks *deliver construction materials* and *finance* take place simultaneously. *Finance* is the limiting task because it takes two months, while *deliver construction materials* takes only one month.

Between stages *F* and *H*, *trim work* takes place simultaneously with *moving out of the old house* and *moving into the new house. Trim work* is the limiting task because it takes two months, while the moves take one month combined.   ∎

EXAMPLE 3   *Finding the Critical Path*

Find the critical path for the house building project in Figure 11.33.

Solution: We start at point *A*. The design phase takes us to *B* in three months. The limiting task between *B* and *C* is *finance*, so we follow the *finance* edge to *C*. Between *C* and *F*, *construction* is the limiting task, so we follow the *construction* edge to *F*. We complete the critical path by following the *trim work* edge to *H*, since *trim work* is the limiting task between *F* and *H*. The critical path is shown in Figure 11.34.

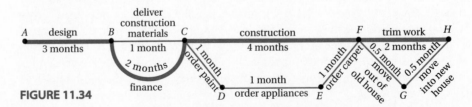

FIGURE 11.34

The total time along the critical path, which is the minimum time required for the project, is

$$3 \text{ mo} + 2 \text{ mo} + 4 \text{ mo} + 2 \text{ mo} = 11 \text{ mo.} \qquad \blacksquare$$

> **TIME-OUT TO THINK:** Find the total time along the path formed by edges *design, deliver construction materials, construction,* and *trim work.* Explain why, although this path adds up to a shorter time than the critical path, it is *not* the minimum completion time for the project.

IS THE PROJECT ON SCHEDULE?

We can investigate the scheduling requirements in more depth by exploring the conditions under which the project remains on schedule, and the conditions under which it falls behind.

Earliest Start and Finish Times

The **earliest start time** (**EST**) for a task is the earliest time since the beginning of the project at which the task can start (assuming the project is on schedule). For example, the earliest start time for the *finance* task in Figure 11.33 is three months because it cannot start until the *design* task is complete.

The **earliest finish time** (**EFT**) for a task is the earliest time since the beginning of the project at which the task may be completed (assuming the project is on schedule). For example, the earliest finish time for the *finance* task is five months: the three months before it can start and the two months required for its completion. The following general rules translate these definitions into the language of the network.

> - The earliest start time (EST) of a task leaving a particular vertex is the *largest* of the earliest finish times of the tasks entering that vertex.
> - The earliest finish time (EFT) of a task is the earliest start time of that task *plus* the time required for the task. That is,
>
> $$\text{EFT} = \text{EST} + \text{time for task.}$$

EXAMPLE 4   *Finding Earliest Start and Finish Times*

Find the earliest start and finish times for the *deliver construction materials* and *construction* tasks in Figure 11.33.

Solution: The *deliver construction materials* task leaves from vertex *B*. The only task entering vertex *B* is *design*, which has an earliest finish time of three months. Thus the earliest start time for *deliver construction materials* is three months. Because *deliver construction materials* takes one month to complete, its earliest finish time is

$$\text{EFT} = \underbrace{3 \text{ months}}_{\text{EST}} + \underbrace{1 \text{ month}}_{\substack{\text{time required} \\ \text{for } task}} = 4 \text{ months.}$$

The *construction* task leaves from vertex *C,* which is entered by two tasks: *deliver construction materials* and *finance.* Construction cannot begin until both of those tasks are completed. Because *finance* is the limiting task entering vertex *C,* its earliest *finish* time of five months is also the earliest *start* time for construction. Construction requires four months, so its earliest finish time is four months after its earliest start time, or

$$5 \text{ months} + 4 \text{ months} = 9 \text{ months}.$$

All of these values, and more, are summarized in Table 11.2 on page 629 which shows earliest start times [EST] in Column 2 and earliest finish times [EFT] in Column 4. ■

Latest Start and Finish Times

The **latest start time** (**LST**) for a task is the latest time at which it can start *without delaying* the overall project. Similarly, the **latest finish time** (**LFT**) of a task is the time by which it must be completed to avoid delays. Again, we can express these ideas with two rules.

- The latest finish time (LFT) of a task *entering* a particular vertex is the *smallest* of the latest start times of the tasks *leaving* that vertex.
- The latest start time (LST) of a task is the latest finish time of that task *minus* the time required for the task. That is,

$$LST = LFT - \text{time for task.}$$

EXAMPLE 5 *Finding Latest Start and Finish Times*

Find the latest start and finish times for the *trim work, move into new house, move out of old house,* and *order carpet* tasks.

Solution: The minimum time required to complete the entire project is 11 months (see Example 3). Therefore 11 months is the latest finish time for the *trim work* task, since this task enters the final vertex. Because *trim work* takes 2 months to complete, its latest start time is 2 months prior to its latest finish time, or

$$11 \text{ months} - 2 \text{ months} = 9 \text{ months}.$$

The task *move into new house* also enters the final vertex, so it also has a latest finish time of 11 months. Because this task requires only 0.5 months to complete, its latest start time is

$$11 \text{ months} - 0.5 \text{ month} = 10.5 \text{ months}.$$

The task *move out of old house* enters vertex *G.* The only task leaving that vertex is *move into new house.* Thus the latest start time of 10.5 months for *move into new house* is the latest finish time for *move out of old house.* Because *move out of old house* requires 0.5 month, its latest start time is

$$10.5 \text{ months} - 0.5 \text{ month} = 10 \text{ months}.$$

The *order carpet* task enters vertex *F,* which has two tasks leaving it: *move out of old house* and *trim work*. Thus *order carpet* must be completed before both those other tasks have begun. Because *trim work* has the smaller latest start time of 9 months, this is the latest finish time for *order carpet*. We subtract the 1 month required for *order carpet* to find its latest start time of

$$9 \text{ months} - 1 \text{ month} = 8 \text{ months}.$$

All of the latest start times [LST] and latest finish times [LFT] are shown in Table 11.2.   ■

Table 11.2	Scheduling Demands for House Building Project					
Task	EST	LST	EFT	LFT	Slack Time	Critical Path?
design	0	0	3	3	0	Yes
deliver construction materials	3	4	4	5	1	No
finance	3	3	5	5	0	Yes
construction	5	5	9	9	0	Yes
order paint	5	6	6	7	1	No
order appliances	6	7	7	8	1	No
order carpet	7	8	8	9	1	No
trim work	9	9	11	11	0	Yes
move out of old house	9	10	9.5	10.5	1	No
move into new house	9.5	10.5	10	11	1	No

EXAMPLE 6   *Project Delay*

Suppose that the *order carpet* task doesn't begin until 9.5 months after the project begins. What is the effect on the project?

Solution: In Example 5, we found that the latest start time for *order carpet* is 8 months. In other words, this is the latest that *order carpet* can begin without delaying the project. If *order carpet* actually begins 9.5 months after the project begins, then the entire project is

$$9.5 \text{ months} - 8 \text{ months} = 1.5 \text{ months}$$

behind schedule.   ■

Slack Time

In Example 4 we found that the earliest start time for *deliver construction materials* is three months, and its latest start time is four months. In other words, this task cannot begin earlier than three months into the project, but there is no harm to the overall schedule if it begins as much as four months into the project. We say that *deliver construction materials* has a **slack time** of one month. Note that this one month of slack time also is the difference between the earliest and latest finish times for *deliver construction materials*. In general,

$$\text{slack time} = \text{LST} - \text{EST} = \text{LFT} - \text{EFT}.$$

Table 11.2 summarizes the scheduling demands for each of the tasks in the house building project. Note that tasks on the critical path have no slack time because they must be completed on schedule if the overall project is to remain on schedule. Here is another hint: to compute EST and LST, it's easiest to start at the *beginning* of the project and work forward; to compute EFT and LFT, it's easiest to start at the *end* of the project and work backward.

EXAMPLE 7   *Impact of Slack Time*

The task *move out of old house* is supposed to take 0.5 month. Suppose that, unexpectedly, it takes a full month. What is the impact on the overall schedule? Next, suppose that *construction* takes five months to complete, rather than the expected four months. What is the impact on the overall schedule?

Solution: The task *move out of old house* has one month of slack time. Therefore an "overrun" of 0.5 month for this task has no impact on the overall schedule—as long as everything else remains on time.

The *construction* task is on the critical path, and therefore has no slack time. A one-month "overrun" on this task will delay the entire project by one month. ■

I call on this nation to commit itself to achieving the goal, before this decade is out, of landing a man on the Moon and returning him safely to Earth.

—*President John F. Kennedy, May 1961 (The goal was accomplished with Apollo 11 in July 1969.)*

APPLICATIONS OF SCHEDULING PROBLEMS

The house building example illustrates the basic principles of a **scheduling problem,** and the method we used to analyze it is called the **critical path method** (CPM). In practice, scheduling problems can be far more complex than our simple example. Imagine the problem of scheduling all the various tasks in building a new freeway overpass (where traffic must be rerouted during certain phases of construction), in making sure a big-budget movie is ready for release by the holiday season, or in coordinating a military response to a terrorist attack.

The house building example has one key ingredient that is not always present in real scheduling problems: it contained good estimates of the time required for each task. Knowing these times allowed us to find the critical path—which is necessary for CPM analysis. When task times are uncertain, as is the case when a project has never been tried before, a variation on CPM called the *project evaluation and review technique* (PERT) must be used. PERT was developed in the 1950s and 1960s to address the scheduling problems in the newly developing space program, including the Apollo project Moon landings. Today, large computer programs are used to run CPM or PERT-based analyses of nearly all complex projects. Indeed, it would be difficult to overestimate the importance of CPM and PERT to the success of major projects undertaken today by businesses and governments.

Although we have only touched on CPM analysis—and haven't done any problems using PERT—remember that both are extensions of the network ideas discussed in this chapter. It's a long way from the bridges of Königsberg to the Moon, but the road is paved with the mathematics of networks.

EXAMPLE 8   *Major Project Delays*

Suppose that a new, multi-level freeway overpass is being built in an area with heavy daily traffic. The project has been contracted to a team that has total labor costs of $500,000 per day. In addition, the traffic detours required during the project are costing local businesses an estimated $200,000 per day in lost revenues and extra driving time. The project was budgeted to be completed in two months. Suppose instead that unexpected delays make it take four months. What is the total cost of the delays?

Solution: A two-month delay is about 60 days. Because the labor rate is $500,000 per day, this results in a budget overrun of

$$\frac{\$500,000}{\text{day}} \times 60 \text{ days} = \$30 \text{ million.}$$

Moreover, it costs local businesses

$$\frac{\$200,000}{\text{day}} \times 60 \text{ days} = \$12 \text{ million.}$$

Overall, this two-month delay costs some $42 million!   ■

REVIEW QUESTIONS

1. Describe the meaning of the edges and vertices in the house building project example.

2. What is a *limiting task?* How are limiting tasks related to the critical path for the project?

3. Explain the meaning and calculation of each of the following: *earliest start time, earliest finish time, latest start time,* and *latest finish time.*

4. What is *slack time?* Why do tasks on the critical path always have zero slack time?

5. Explain what we mean by CPM and PERT. How are they important?

PROBLEMS

Scheduling a Paint Job. *For Problems 1–11, refer to the schedule for a painting job shown in the figure below. The estimated time for the completion of each task (in hours) is shown on the corresponding edge of the network.*

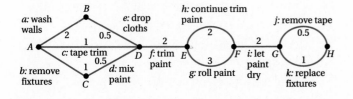

1. Which tasks are on the critical path between *A* and *E*?

2. Which tasks are on the critical path between *E* and *H*?

3. Give the tasks through which the critical path for the entire project passes.

4. What is the length of the critical path for the entire project?

5. What are the EST and LST for the tasks *d: mix paint* and *f: trim paint?*

6. What are the EST and LST for the tasks *g: roll paint* and *j: remove tape?*

7. What are the EFT and LFT for the tasks *d: mix paint* and *f: trim paint*?

8. What are the EFT and LFT for the tasks *g: roll paint* and *j: remove tape*?

9. What is the effect of a one-hour delay in task *c: tape trim*?

10. What is the effect of a one-hour delay in task *g: roll paint*?

11. Make a table similar to Table 11.2 showing the EST, LST, EFT, LFT, and slack times for each task of the project.

Building a Hotel. *Problems 12–22 refer to the scheduling network shown in the figure for building a new hotel. Completion times, in months, for each task are shown on the edges.*

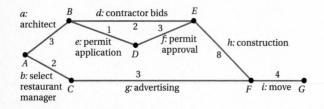

12. Which tasks are on the critical path between *A* and *E*?

13. Which tasks are on the critical path between *B* and *G*?

14. Give the tasks through which the critical path for the entire project passes.

15. What is the length of the critical path for the entire project?

16. What are the EST and LST for the tasks *e: permit application* and *g: advertising*?

17. What are the EST and LST for the tasks *d: contractor bids* and *h: construction*?

18. What are the EFT and LFT for the tasks *e: permit application* and *g: advertising*?

19. What are the EFT and LFT for the tasks *d: contractor bids* and *h: construction*?

20. What is the effect of a one-month delay in task *d: contractor bids*?

21. What is the effect of a one-month delay in task *e: permit application*?

22. Make a table similar to Table 11.2 showing the EST, LST, EFT, LFT, and slack times for each task of the project.

Doing the Laundry. *The table at the top right gives the critical times (in minutes) for the job of washing two loads of laundry (one person). Refer to this table in Problems 23–27.*

Task	EST	LST	EFT	LFT	Slack Time
a: sort colors and whites	0	0	15	15	
b: wash colors	15	15	45	45	
c: remove stains from whites	15	25	35	45	
d: dry colors	45	45	90	90	
e: wash whites	45	60	75	90	
f: fold colors	90	120	105	135	
g: dry whites	90	90	135	135	
h: fold whites	135	135	150	150	

23. Fill in the slack time column in the table.

24. Which tasks are on the critical path?

25. How much time is needed for the entire project?

26. Which pairs of tasks take place at the same time? Which tasks must be done alone?

27. Use the table to make a network for the project.

Cooking a Pasta Dinner. *The following table gives the critical times (in minutes) for the job of cooking a pasta dinner (alone). Refer to this table in Problems 28–32.*

Task	EST	LST	EFT	LFT	Slack Time
a: boil water	0	0	5	5	
b: wash vegetables	0	1	4	5	
c: cook pasta	5	5	17	17	
d: heat up sauce	5	7	15	17	
e: make salad	5	8	14	17	
f: set table	17	17	22	22	

28. Fill in the slack time column in the table.

29. Which tasks are on the critical path?

30. How much time is needed for the entire project?

31. Which pairs of tasks take place at the same time? Which tasks must be done alone?

32. Use the table to make a network for the project.

33. **Project: Scheduling and Construction.** Locate the site of a residential or commercial construction project in your area and try to meet a contractor or supervisor. Is there a master plan for the construction project that has deadlines for individual stages of the project? Has the construction company used CPM (critical path method) to develop a schedule for the project? Report on the techniques that are being used and comment on how you could improve the scheduling of the project with what you have learned in this chapter.

34. **Project: Mars Mission.** The idea of sending humans to Mars has been discussed seriously ever since the Apollo Moon landings. At present, the most likely scenario for a Mars mission involves a multinational project with a journey of several months in each direction and a stay of a year or more on the surface by 8 to 12 astronauts. Discuss some of the tasks that should appear in a CPM or PERT analysis of such a mission, as well as the uncertainties in both the completion times and budgets for those tasks.

VOTING: DOES THE MAJORITY ALWAYS RULE?

**UNIT
11D**

When the American democracy in which we live was proposed over 200 years ago, it was a radical idea. No one even knew if it would survive. Today democracy has emerged as the most stable and enduring form of government. The right to vote is one of the fundamental rights and responsibilities of citizens in a democracy. Voting may seem like a straightforward business that is far removed from mathematics. But there are some genuine mathematical issues associated with voting, and some unexpected dilemmas.

VOTING WITH TWO CHOICES

The simplest type of voting occurs when there are only two choices. For two choices, the most common way of deciding the vote is by **majority rule**: The candidate receiving more than 50% of the votes wins. While we take this rule for granted, it has three properties that are worth noting.

- Every vote has the same *weight*. That is, no one person's vote counts more or less than any other person's.

- There is *symmetry* between the candidates: if all the votes were reversed, the loser would become the winner.

- If a vote for the loser were changed to a vote for the winner, the outcome of the election would not be changed.

We can illustrate these properties with a simple example. Suppose that Sharon and Robert are running for captain of the debate team. Sharon wins with 10 votes to Robert's 9 votes. The first property, that all votes have the same weight, means we care only about the vote totals—not the names of the people who cast the individual votes. The second property tells us that if we reverse all the votes, so that Sharon receives 9 votes and Robert receives 10, then Robert becomes the winner. The third property tells us that if a vote for Robert (the loser) is changed to a vote for Sharon (the winner), Sharon still wins: this change would raise Sharon's vote total to 11 and reduce Robert's to 8.

No one pretends that democracy is perfect or all-wise. Indeed it has been said that democracy is the worst form of government except for all those others that have been tried from time to time.

—WINSTON CHURCHILL

By the Way ············
In 1952, Kenneth May proved that majority rule is the *only* voting system for two candidates that satisfies all three properties.

> **TIME-OUT TO THINK:** Suppose that a vote for the *winner* is changed to a vote for the loser. Can the outcome of the election change in that case? Explain.

U.S. Presidential Elections

By the Way ············

When you cast a ballot for president, you actually are casting votes for your state's *electors*. The electors cast their votes a few weeks later.

U.S. presidential elections involve a twist on majority rule even when only two major candidates are involved. The **popular vote** in a presidential election reflects the total number of votes received by each candidate. However, presidents are elected by **electoral votes.** As mandated by the U.S. Constitution, each state gets as many electoral votes as the state has members of Congress. Thus every state gets at least three electoral votes because every state has two senators and at least one representative. In most states, the electoral votes are determined by a winner-take-all system: *All* the electoral votes of that state go to the candidate with the most popular votes. But the winner of the presidential election is chosen by majority rule among the electoral votes.

EXAMPLE 1   *1960 Presidential Election*

The 1960 U.S. presidential election had the following results.

Candidate	Popular vote	Electoral vote
John F. Kennedy	34,226,731	303
Richard M. Nixon	34,108,157	219

Assuming that all votes were cast for either of the two candidates, contrast the outcomes of the popular and electoral votes.

By the Way ············

The winner of the popular vote has lost the electoral vote twice in U.S. history. In 1876, Samuel J. Tilden won the popular vote over Rutherford B. Hayes, but Hayes became President because he won the electoral vote. In 1888, Benjamin Harrison won the popular vote over Grover Cleveland, but lost to Cleveland in the electoral count.

Solution: Kennedy's fraction of the popular vote was

$$\frac{34,226,731}{34,226,731 + 34,108,157} = \frac{34,226,731}{68,334,888} = 0.5009 = 50.09\%.$$

Thus he won the popular vote, but barely! Kennedy's fraction of the electoral vote was

$$\frac{303}{303 + 219} = \frac{303}{522} = 0.5805 = 58.05\%,$$

a far more comfortable margin than his popular victory. (Note: The actual percentages were slightly different than calculated in this problem because of votes received by several minor party candidates.) ∎

> **TIME-OUT TO THINK:** Do you think that it would be fairer if presidential elections were decided by popular vote rather than electoral vote? Why or why not?

Variations on Majority Rule

U.S. senators are generally allowed to speak about a legislative bill for as long as they wish before it is brought to a vote. If a senator opposes a particular bill, but fears that it will gain a majority in a vote, the senator may choose to speak continuously (at least during the hours that the issue is under consideration)—and thereby prevent the vote from ever taking place. This technique for preventing a vote on a bill is called a **filibuster**. The filibuster can be ended only by a vote of 3/5, or 60%, of the senators. This situation, in which 60% rather than 50% of the votes are required to end a filibuster, illustrates that majority rule is not always followed strictly.

Whenever a candidate or issue must receive *more* than a majority of the vote to win—such as 60% of the vote, 75% of the vote, or a unanimous vote—we say that a **super majority** is required. Criminal trials provide one of the most common cases in which a super majority is required: all states require a super majority vote of the jury to reach a verdict, and many states require that this vote be unanimous. A jury that cannot reach this super majority or unanimous agreement is called a *hung jury*. When a jury is hung, the judge generally declares a mistrial, and the case must be tried again or dropped.

The U.S. Constitution requires super majority votes for many specific issues. For example, an international treaty can be ratified only by a 2/3 super majority vote of the Senate. Amending the Constitution first requires a 2/3 super majority vote on the amendment in both the House and the Senate, and then the amendment must also be approved by 3/4 of the states.

Another variation on majority rule occurs with **veto** power. For example, a bill proposed in the U.S. Congress normally becomes law if it receives a majority vote in both the House and the Senate. However, if the president vetoes the bill, it can become law only if it then receives a 2/3 super majority vote in both the House and the Senate to override the veto. The courts can also effectively veto a popular vote. For example, even if a proposition in a state election receives a huge majority of the popular vote, it will not become law if the courts declare that the proposition violates the U.S. Constitution.

By the Way ....
Senate rules do not specify what a senator can say about an issue. During some filibusters, senators have read novels aloud in order to keep speaking. In 1964, the Civil Rights Act was passed by Congress only after 75 days of filibuster by southern senators.

By the Way ....
The U.S. Constitution also specifies that it may be amended if 2/3 of the states call for "a convention for proposing amendments." However, such a call for a convention has never occurred.

EXAMPLE 2   *Majority Rule?*

Evaluate the outcome in each of the following cases.

a) 59 of the 100 senators in the U.S. Senate favor a new bill on campaign finance reform. The other 41 senators are adamantly opposed and start a filibuster. Will the bill pass?

b) A criminal conviction in a particular state requires a vote by 3/4 of the jury members. On a nine-member jury, seven jurors vote to convict. Is the defendant convicted?

c) A proposed amendment to the U.S. Constitution has passed both the House and Senate with more than the required 2/3 super majority. Each state holds a vote on the amendment. Overall, 75% of the public supports the amendment, but it garners less than a majority vote in 14 of the 50 states. Is the Constitution amended?

d) A bill limiting the powers of the president has the support of 73 out of 100 senators and 270 out of 435 members of the House of Representatives. But the president promises to veto the bill if it is passed. Will it become law?

Solution:

a) The filibuster can be ended only by a vote of 3/5 of the Senate, or 60 out of the 100 senators. Thus the 59 senators in favor of the bill cannot stop the filibuster. The bill will not become law.

b) Seven out of nine jurors represents a super majority of $\frac{7}{9} = 77.8\%$. This percentage is more than the required 3/4 (75%), so the defendant is convicted.

c) The amendment must be approved by 3/4, or 75%, of the states. But the 14 states that do not support the amendment represent $\frac{14}{50} = 28\%$ of the 50 states. Thus only $100\% - 28\% = 72\%$ of the states voted to ratify the amendment, which is not enough for it to become part of the Constitution.

d) The bill has the support of $\frac{73}{100} = 73\%$ of the Senate and $\frac{270}{435} = 62\%$ of the House. But overriding a presidential veto requires a super majority vote by $\frac{2}{3} = 66.7\%$ in both the House and the Senate. The 62% support in the House is not enough to override the veto, so the bill will not become law.   ■

VOTING WITH THREE OR MORE CHOICES

Imagine that the results in a three-candidate election for governor of a certain state are as follows:

Candidate	*Percentage of Vote*
Smith	32%
Jones	33%
Wilson	35%

None of the three candidates has received a majority of the vote, so who should become the governor?

One way to decide this election is by looking to see who received the most votes, called a **plurality** of the vote. Wilson has the greatest percentage, and thus becomes governor if we decide this election by plurality. An alternative way to decide the election is to hold a **runoff** between the top two vote getters. In that case, another election must be held between Jones and Wilson.

Note that the plurality method for deciding the election is simpler (and cheaper) because it does not require holding a second election. However, it can lead to a situation in which the majority of the electorate opposes the new governor. For example, suppose that all of Smith's supporters prefer Jones to Wilson. In that case, the combined 65% of the electorate that voted for Smith and Jones would rather see Jones elected governor than Wilson. By the plurality method, Wilson would become governor despite being opposed by 65% of the voters. By the runoff method, Jones would easily defeat Wilson to become governor.

EXAMPLE 3   *1992 Presidential Election*

The three major candidates in the 1992 U.S. presidential election received the numbers of popular and electoral votes shown on the next page. Analyze the outcome. Is it possible that Bush could have won if Perot had not run?

By the Way

A U.S. presidential candidate must receive a *majority* of the electoral vote to become president—not just a plurality. If no candidate receives an electoral majority, the House of Representatives chooses the president. This has happened twice in U.S. history: with the elections of Thomas Jefferson in 1801 and John Quincy Adams in 1825.

	Popular Vote	Popular Percentage	Electoral Vote	Electoral Percentage
Clinton	44,909,889	43.3%	370	68.8%
Bush	39,104,545	37.7%	168	31.2%
Perot	19,742,267	19.0%	0	0%
Total	103,756,701	100%	538	100%

Solution: Clinton won a *plurality* of the popular vote, but there was no majority winner in the popular vote. However, Clinton won a large *majority* in the electoral vote. If Perot had not run, the 19% of the votes that he received would presumably have been divided in some way between Clinton and Bush. Note that Bush needed an additional 12.3% of the vote to reach a 50% total (37.7% + 12.3% = 50%). Thus if $\frac{12.3}{19} \approx 65\%$ of the Perot voters preferred Bush to Clinton, Bush could have won the race in Perot's absence.   ■

TIME-OUT TO THINK: Do you think that a runoff is more fair than choosing the winner by plurality in a three-way race? Why or why not?

Preference Schedules

So far, we have examined situations in which each voter is allowed to vote for only a single candidate. However, there are many cases in which voters are instead asked to rank candidates in order of preference. This type of election is used for sports polls, sports awards (such as the Heisman Trophy), consumer surveys, beauty pageants, and the Academy Awards for motion pictures.

Imagine a club of 55 people that holds an election among five candidates for president. For simplicity, let's call the candidates A, B, C, D, and E. Each ballot asks the voter to rank these candidates in order of preference (Figure 11.35).

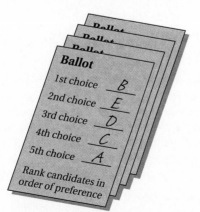

FIGURE 11.35
Sample ballot for the club election.

Mathematical Note:
With five candi-
dates in an election,
there are 5! = 120
possible orders in
which they can be
ranked. This
example uses only
6 of these 120
possible orders.
These results were
chosen to give a
particularly
interesting analysis,
and have been used
by many voting
theorists.

Table 11.3 shows a special type of table, called a **preference schedule**, that summarizes the results of the election. Each column shows a different ranking among the candidates, and the number at the bottom of the column indicates the number of people who chose that ranking. For example, the first column shows that 18 voters ranked the candidates in the order A, D, E, C, B. The second column shows that 12 voters ranked the candidates in the order B, E, D, C, A. The remaining columns show, respectively, that 10 voters chose the rank order C, B, E, D, A; 9 voters chose the order D, C, E, B, A; 4 voters chose the E, B, D, C, A; and 2 voters chose the order E, C, D, B, A.

Table 11.3	Preference Schedule for the Club Election					
First	A	B	C	D	E	E
Second	D	E	B	C	B	C
Third	E	D	E	E	D	D
Fourth	C	C	D	B	C	B
Fifth	B	A	A	A	A	A
	18	12	10	9	4	2

Look carefully at the results in Table 11.3. Can you decide which candidate should be declared the winner? You may be surprised to learn that a winner cannot be declared without at least some controversy. Let's investigate.

Choosing a Winner by Plurality
One method for selecting the winner is to count who obtained the most *first-place* votes. That is, which candidate received a *plurality* in this election? Counting the number of first-place votes for each candidate in Table 11.3 reveals the following results.

- A received 18 first-place votes.
- B received 12 first-place votes.
- C received 10 first-place votes.
- D received 9 first-place votes.
- E received $4 + 2 = 6$ first-place votes.

The supporters of Candidate A can rightfully argue that A has received a plurality and should be declared the winner.

Choosing a Winner by a Runoff Between Top Two
"Not so fast!" yell the supporters of Candidate B. They suggest that a runoff should be held between A and B because these two candidates had the most first-place votes. Table 11.3 shows that, although 18 people voted for A in first place, the other 37 people all ranked A in last place. These 37 people will vote for B over A in a runoff. Thus B wins the runoff easily, by a vote of 37 to 18. Supporters of Candidate B can now proclaim victory for their candidate.

Choosing a Winner by Sequential Runoffs
Now the supporters of Candidate C chime in. They claim that a single runoff is unfair because it ignores rankings below the first two. They suggest a series of runoffs, in which

the candidate with the fewest first-place votes is eliminated at each stage, continuing until someone claims a majority. Candidate E received the fewest first-place votes in the original election, and therefore is eliminated from the first runoff. Because we have a preference schedule showing the rankings of all voters, we can see how this runoff will turn out without actually holding a second election.

The first column of Table 11.3 shows that 18 people ranked E in third place, followed by C and B. With E eliminated, these 18 people will now rank C in third place and B in fourth place. Similarly, the second column shows that 12 people ranked E in second place, followed by D, C, and A. Eliminating E means they will now move D up to second place, C to third, and A to fourth. Most importantly, the 6 voters who originally ranked E first (last two columns of Table 11.3), now must move their second choices up to first. By revising Table 11.3 with the changes caused by eliminating Candidate E , we find the following results for the first runoff.

First	A	B	C	D	B	C
Second	D	D	B	C	D	D
Third	C	C	D	B	C	B
Fourth	B	A	A	A	A	A
	18	12	10	9	4	2

Candidate A still has the most first-place votes, but these 18 votes are much less than a majority of the 55 votes. Therefore we need a second runoff. The candidate with the fewest first-place votes is now D (with 9 votes), so D is eliminated from this runoff. The 9 people who previously had D in first place (Column 4) must now move their other choices up, awarding their first-place votes to C. Proceeding as before to adjust all the rankings with Candidate D eliminated, we find the following new results after the second runoff.

First	A	B	C	C	B	C
Second	C	C	B	B	C	B
Third	B	A	A	A	A	A
	18	12	10	9	4	2

Candidate C is now the leader with $10 + 9 + 2 = 21$ votes. However, this total is still short of a majority out of 55, so we need one final runoff. This time B has the fewest first-place votes ($12 + 4 = 16$) and is eliminated. Note that all 16 people who previously ranked B first had ranked C in second place. Thus C gets all these votes in the runoff, raising C's total to $21 + 16 = 37$ votes, or $\frac{37}{55} = 67.3\%$ of the total.

This method for choosing a winner through a series of runoffs that eliminates the candidate with the fewest first-place votes at each stage is called the **sequential runoff** method. By this method, Candidate C is declared the winner. Note that, because we have a preference schedule showing how all voters rank all the candidates, we do not actually need to hold multiple elections to determine the results of a sequential runoff.

Choosing a Winner by a Point System

Now it is the turn of Candidate D's supporters, who argue that the winner should be selected by a **point system.** Because there are five candidates, first-place votes are worth 5 points, second-place votes are worth 4 points, and so on, down to 1 point for fifth-place votes. We can calculate the total number of points won by each candidate by multiplying the number of votes in each column of Table 11.3 by the assigned point value and adding these products together.

A gets $(18 \times 5) + (12 \times 1) + (10 \times 1) + (9 \times 1) + (4 \times 1) + (2 \times 1) = 127$ points

B gets $(18 \times 1) + (12 \times 5) + (10 \times 4) + (9 \times 2) + (4 \times 4) + (2 \times 2) = 156$ points

C gets $(18 \times 2) + (12 \times 2) + (10 \times 5) + (9 \times 4) + (4 \times 2) + (2 \times 4) = 162$ points

D gets $(18 \times 4) + (12 \times 3) + (10 \times 2) + (9 \times 5) + (4 \times 3) + (2 \times 3) = 191$ points

E gets $(18 \times 3) + (12 \times 4) + (10 \times 3) + (9 \times 3) + (4 \times 5) + (2 \times 5) = 189$ points

Candidate D is now the winner, by virtue of having the largest number of points.

A system that assigns points for every ranking down the line, as was done in this example, is called a **Borda count.** A standard Borda count assigns one point for last place and an additional point for each higher place. Many common ranking systems use modified Borda counts. For example, polls that rank college football teams usually use a system in which voters rank their top 25 teams in order. A first-place ranking is worth 25 points, second-place is worth 24 points, and so on, to 1 point for twenty-fifth place. At a meet between two college swim teams, the winner may be decided by a point system that awards 4 points for first place, 2 points for second place, and 1 point for third place.

By the Way ············
The Borda count is named for French mathematician and astronomer Jean-Charles de Borda (1733–1799). Among his interests were fluid mechanics, instruments for navigation, and politics.

> **TIME-OUT TO THINK:** Look for an example in which a point system is used to determine a winner, such as a football poll or a sports competition. How are points assigned to the different places? Would the outcome be the same if the winner had instead been chosen by a plurality (counting first place only)?

EXAMPLE 4   *Majority Loser*

Suppose that the seven sportswriters of Seldom County rank the county's three women's volleyball teams according to the following preference schedule:

First	A	C	B
Second	B	B	C
Third	C	A	A
	4	1	2

Determine the winner by plurality and by a Borda count. Discuss the results.

Solution: By plurality, Team A is the winner, having received 4 out of the 7 first-place votes (which is not just a plurality, but also a majority). For the Borda count, we assign 3 points for a first-place vote, 2 points for a second-place vote, and 1 point for a third-place vote, and find the total number of points.

$$\text{Team A gets } (4 \times 3) + (1 \times 1) + (2 \times 1) = 15 \text{ points}$$
$$\text{Team B gets } (4 \times 2) + (1 \times 2) + (2 \times 3) = 16 \text{ points}$$
$$\text{Team C gets } (4 \times 1) + (1 \times 3) + (2 \times 2) = 11 \text{ points}$$

So Team B is the winner by the Borda count method.

Although the majority of the sportswriters voted for Team A as the best team, Team B gets ranked first by the Borda count. The fact that the Borda count winner can be different from the majority winner is a well-known shortcoming of the method. ∎

Choosing a Winner by Pairwise Comparisons

So far, we have seen from the preference schedule in Table 11.3 that each of the four candidates A, B, C, and D can be declared the winner depending on which counting method for the votes is used. But what about Candidate E?

Candidate E's supporters point out an important new fact about the election rankings. Suppose, they say, that the vote had been only between E and A, without the other candidates. Column 1 of Table 11.3 has A ranked higher than E, so these 18 voters would choose A over E. However, E is ranked higher than A in all the other columns, so E would get the remaining 37 votes (out of the 55 total) in the contest against A. That is, E beats A by 37 to 18.

Now suppose that the vote had been only between E and B. Note that E is ranked higher than B in Columns 1, 4, 5, and 6, so E gets $18 + 9 + 4 + 2 = 33$ votes. The remaining 22 votes go to B. Thus E beats B by 33 to 22. A similar comparison of the outcome of a race between only E and C shows that E is the winner by 36 to 19, and E comes out ahead in a contest with D by 28 to 27.

> **TIME-OUT TO THINK:**   By studying Table 11.3, confirm the outcomes given above for the two-way races between E and each other candidate.

Because Candidate E beats every other candidate in one-on-one contests, E's supporters now declare victory in the election. This method of deciding an election by analyzing the outcome of one-on-one contests is called the **method of pairwise comparisons.** More generally, the winner by this method is the candidate who wins the most one-on-one contests.

The method of pairwise comparisons is also called the *Condorcet method,* because it was invented by French mathematician and political leader Marie Jean Antoine Nicholas de Caritat, Marquis de Condorcet (1743–1794). Condorcet, as he is usually called, invented this method because he thought it would lead to fairer elections. However, this method suffers a major drawback: It does not always produce a clear winner. (Condorcet was aware of this problem.)

By the Way ············
Condorcet did pioneering work in probability and calculus before becoming a leader of the French Revolution. He argued forcefully for equal rights for women, for universal free education, and against capital punishment. In 1794, as extremists took control of the revolution, he was arrested because of his aristocratic background. He died in prison the next day, in a death labeled a suicide by his captors.

EXAMPLE 5   *Pairwise Paradox*

Consider the following simple three-candidate preference schedule.

First	A	C	B
Second	B	A	C
Third	C	B	A
	14	12	10

Can you find a winner through pairwise comparisons? Explain.

Solution: With three candidates, three pairwise comparisons are possible: A versus B, B versus C, and A versus C. Columns 1 and 2 show A ranked ahead of B, but B is ahead of A in Column 3. Thus A beats B in a one-on-one contest by 26 to 10. In the B versus C contest, B is ranked ahead of C in Columns 1 and 3, so B beats C by 24 to 12. Similarly, A is ranked ahead of C only in Column 1, so C beats A by 22 to 14. Summarizing, we have the following results for the pairwise comparisons:

- A beats B
- B beats C
- C beats A

Note that, because A beats B and B beats C, the first two results suggest that A should also beat C. However, the third result shows that A actually loses to C. Thus the results are paradoxical, and there is no way to establish a clear winner.   ■

> **TIME-OUT TO THINK:**   Does Example 5 have a plurality winner? Does it have a winner by a Borda count? Explain.

SUMMARY: CHOOSING A WINNER IS NOT SO EASY

Table 11.3 showed a fairly simple set of election results for a contest with five candidates. Yet, by analyzing these results in different ways, we found that every candidate could reasonably claim to be the winner! Other election results may not be so ambiguous, but the point should be clear: Different people can reasonably disagree about who should be declared the winner in an election with more than two candidates. In fact, as we will see in the next unit, a mathematical theorem *proves* that there is *no absolutely fair* way of deciding elections among more than two candidates.

REVIEW QUESTIONS •

1. What is *majority rule*? When can it definitively decide an election?

2. Contrast the *popular vote* and *electoral vote* in a presidential election.

3. What is a *filibuster*? What percentage of the vote is required to end one?

4. What is a *super majority*? Give several examples in which a super majority is required to decide a vote.

5. What is a *veto*? How does a veto affect the idea of majority rule?

6. Describe how a three-way election can be decided either by plurality or by runoff. Will both methods necessarily give the same results? Explain.

7. What is a *preference schedule*? Give an example of how to make one.

8. Using the preference schedule of Table 11.3, explain how the vote can be decided in five different ways depending on whether we look for the winner by plurality, single runoff, sequential runoffs, a point system, or pairwise comparisons.

PROBLEMS •

1. **1876 Presidential Election.** In the 1876 presidential election, Samuel J. Tilden received 4,285,992 votes to Rutherford B. Hayes' 4,033,786 votes. However, Hayes received 185 electoral votes to Tilden's 184. Assume that all votes were cast for one of these two candidates. Contrast the outcomes of the popular and electoral votes in terms of percentages. Who became president?

2. **1888 Presidential Election.** In the 1888 presidential election, Benjamin Harrison received 5,540,309 votes to Grover Cleveland's 5,439,853 votes. However, Cleveland received 233 electoral votes to Harrison's 168. Assume that all votes were cast for one of these two candidates. Contrast the outcomes of the popular and electoral votes in terms of percentages. Who became president?

Other Close Presidential Elections. *The following table shows the results of three other close presidential elections. Use these data in Problems 3–5. In each case, assume that all votes were cast for one of the two candidates listed.*

Year	Major Candidates	Electoral Vote	Popular Vote
1880	James A. Garfield	214	4,449,053
	Winfield S. Hancock	155	4,442,035
1916	Woodrow Wilson	277	9,129,606
	Charles E. Hughes	254	8,538,221
1976	Jimmy Carter	297	40,830,763
	Gerald R. Ford	240	39,147,973

3. Contrast the outcomes of the popular and electoral votes in terms of percentages for the 1880 election.

4. Contrast the outcomes of the popular and electoral votes in terms of percentages for the 1916 election.

5. Contrast the outcomes of the popular and electoral votes in terms of percentages for the 1976 election.

6. Super Majorities.

   a. Of the 100 senators in the U.S. Senate, 57 favor a new bill on health care reform. The opposing senators start a filibuster. Is the bill likely to pass?

   b. A criminal conviction in a particular state requires a vote by 2/3 of the jury members. On a 12-member jury, 7 jurors vote to convict. Will the defendant be convicted?

   c. A proposed amendment to the U.S. Constitution has passed both the House and Senate with more than the required 2/3 super majority. Each state holds a vote on the amendment and it receives a majority vote in all but 15 of the 50 states. Is the Constitution amended?

   d. A tax increase bill has the support of 70 out of 100 senators and 260 out of 435 members of the House of Representatives. The president promises to veto the bill if it is passed. Is it likely to become law?

7. Super Majorities.

   a. According to the by-laws of a corporation, a 2/3 vote of the shareholders is needed to approve a merger. Of the 15,890 shareholders voting on a certain merger, 10,580 approve of the merger. Will the merger happen?

   b. A criminal conviction in a particular state requires a vote by 3/4 of the jury members. On a 15-member jury, 11 jurors vote to convict. Will the defendant be convicted?

   c. A proposed amendment to the U.S. Constitution has passed both the House and Senate with more than the required 2/3 super majority. Each state holds a vote on the amendment and it receives a majority vote in 32 of the 50 states. Is the Constitution amended?

   d. A tax increase bill has the support of 68 out of 100 senators and 292 out of 435 members of the House of Representatives. The president promises to veto the bill if it is passed. Is it likely to become law?

8. 1912 Presidential Election. The 1912 U.S. presidential election featured three major candidates, with the vote split as follows.

Candidate	Popular Votes	Electoral Votes
Woodrow Wilson	6,286,214	435
Theodore Roosevelt	4,126,020	88
William Taft	3,483,922	8

For this problem, assume that all votes were cast for one of these three candidates. (Actually, other candidates split 7.6% of the total vote.)

   a. Calculate each candidate's percentage of the popular vote. Who won a plurality? Did any candidate win a majority?

   b. Calculate each candidate's percentage of the electoral vote. Who won a plurality? Did any candidate win a majority?

   c. Suppose that Taft had dropped out of the election. Is it possible that Roosevelt would have won the popular vote? Could Roosevelt have become president in that case? Explain.

   d. Suppose that Roosevelt had dropped out of the election. Is it possible that Taft would have won the popular vote? Could Taft have become president in that case? Explain.

9. Three-Candidate Elections. Consider an election in which the votes were cast as follows.

Candidate	Percentage of Vote
Able	32%
Best	40%
Crown	28%

   a. Who won a plurality? Does any candidate have a majority? Explain.

   b. What percentage of Crown's votes would Able need to win a runoff election?

10. Three-Candidate Elections. Consider an election in which the votes were cast as follows.

Candidate	Percentage of Vote
Davis	21%
Earnest	38%
Fillipo	41%

   a. Who won a plurality? Does any candidate have a majority? Explain.

   b. What percentage of Davis' votes would Earnest need to win a runoff election?

11. **Three-Candidate Elections**. Consider an election in which the votes were cast as follows.

Candidate	Number of Votes
Grand	130
Height	150
Inviglio	185

a. Who won a plurality? Does any candidate have a majority? Explain.

b. How many of Grand's votes would Height need to win the runoff election?

12. **Three-Candidate Elections**. Consider an election in which the votes were cast as follows.

Candidate	Number of Votes
Joker	225
King	360
Lord	285

a. Who won a plurality? Does any candidate have a majority? Explain.

b. How many of Joker's votes would Lord need to win the runoff election?

13. **Ballots to Schedule**. Ten voters are asked to rank three brands of soda: A, B, and C. The ten voters turn in the following ballots showing their preferences in order:

ABC   CBA   CAB   ACB   ABC
CAB   ABC   ABC   CAB   ACB

Make a preference schedule for these ballots.

14. **Ballots to Schedule**. Twelve voters are asked to rank four brands of candy bar: A, B, C, and D. The twelve voters turn in the following ballots showing their preferences in order:

ABCD   CBAD   CADB   ACDB   DABC   ABCD
CADB   ABCD   ABCD   CADB   DABC   ACDB

Make a preference schedule for these ballots.

Preference Schedules. *For the preference schedules in Problems 15–19, answer the following questions.*

a. How many votes were cast in the survey?

b. Find the plurality winner. Did the plurality winner also receive a majority? Explain.

c. Find the winner by a runoff of the top two candidates.

d. Find the winner of a sequential runoff.

e. Find the winner by a Borda count.

f. Find the winner, if any, by the method of pairwise comparisons.

g. Summarize the results of the various methods of determining a winner. Based on these results, is there a clear winner? If so, why? If not, which candidate would be selected as the winner, and why?

15.

First	B	D	C	A	D	C
Second	D	A	D	D	A	A
Third	C	C	A	C	B	B
Fourth	A	B	B	B	C	D
	20	15	10	8	7	6

16.

First	B	D	D	C	E
Second	A	B	B	A	A
Third	C	A	E	B	D
Fourth	D	C	C	D	B
Fifth	E	E	A	E	C
	9	7	6	4	3

17.

First	A	A	B	B	C	C
Second	B	C	A	C	A	B
Third	C	B	C	A	B	A
	30	5	20	5	10	30

18.

First	A	B	D
Second	B	A	C
Third	C	D	B
Fourth	D	C	A
	10	10	10

19.

First	E	B	D
Second	D	C	A
Third	A	E	B
Fourth	B	A	C
Fifth	C	D	E
	40	30	20

20. Condorcet Winner. If a candidate wins *all* head-to-head (two-way) races with other candidates, that candidate is called the *Condorcet winner*. A Condorcet winner automatically wins by the pairwise comparison (Condorcet) method. Consider the following preference schedule for a four-candidate election. Is there a Condorcet winner? Explain.

First	B	B	A	A
Second	A	A	C	D
Third	C	D	D	C
Fourth	D	C	B	B
	30	30	30	20

21. Pairwise Comparison Question. Suppose 15 voters rank 4 candidates. How many pairs of candidates must be examined to carry out the pairwise comparison method? How many pairs of candidates must be examined to carry out the pairwise comparison method with five candidates?

22. Borda Question. In a preference schedule with 5 candidates and 25 voters, what is the total number of points awarded to the candidates using the usual Borda count weights?

23. Borda Question. Suppose 25 voters rank 4 candidates, A, B, C, and D. Using the usual Borda count weights, A receives 60 points, B receives 50 points, and C receives 10 points. Is it possible that D has won this election by the Borda count? Explain.

24. Pairwise Paradox. Consider the following preference schedule. Can you find a winner through pairwise comparisons? Explain.

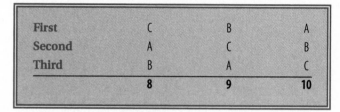

First	C	B	A
Second	A	C	B
Third	B	A	C
	8	9	10

25. Project: College Sports Poll. Find a recent college football or basketball poll ranking the top 25 teams. Describe how the rankings were determined. Is the team with the most first-place votes also the top-ranked team? Explain.

26. Project: Academy Awards. The election process for the Academy Awards (for films) involves several stages and several different voting methods. Investigate the full election procedure for the Academy Awards. Describe the procedure and comment on its fairness.

THEORY OF VOTING

In Unit 11D, we saw that different methods of counting votes can lead to different results. In fact, mathematicians, economists, and political scientists have discovered many other surprises about voting. In this unit, we investigate just a few of these surprises that arise in voting.

WHICH METHOD IS MOST FAIR?

When there are only two choices in an election, the clear winner is the choice that gets a majority of the votes. However, when there are three or more choices, choosing a winner can be much more difficult. In Unit 11D, we studied five different methods for deciding an election:

- *Plurality:* The candidate with the most first-place votes wins.
- *Top-two runoff:* The winner is decided by a runoff between the two candidates that receive the most first-place votes.
- *Sequential runoffs:* The winner is chosen by a series of runoffs in which the candidate with the fewest first-place votes is eliminated at each stage.
- *Point system* (Borda count): The winner is chosen by adding up points with a system where last place is worth 1 point, second-to-last is worth 2 points, and so on to first place.
- *Pairwise comparisons:* The election is decided by analyzing the outcomes of one-on-one contests between each pair of candidates.

Sometimes, all these methods will give the same winner. Other times, different methods may give different winners. In extreme cases, the five methods can produce five different winners! In fact, there are other possible methods for counting votes besides the ones listed here. Thus, the key question in deciding the outcome becomes: Which method for deciding the winner is the most fair?

CRITERIA OF FAIRNESS

Before we can answer the question of which method is the most fair, we must define what we mean by "fairness." Judgments of fairness are necessarily very subjective. Nevertheless, mathematicians and political scientists have come up with four basic criteria that must be met for a voting system to be considered fair. These **fairness criteria** are the following:

- **Criterion 1:** If a candidate receives a majority of the first-place votes, that candidate should be the winner.
- **Criterion 2:** If a candidate is favored over every other candidate in pairwise races, then that candidate should be declared a winner.
- **Criterion 3:** Suppose that Candidate X is declared the winner of an election, and then a second election is held. If some voters rank X even higher in the second election (without changing the order of other candidates), then X should also win the second election.

By the Way ·············
Voting theorists refer to these four criteria as, respectively: the *majority criterion;* the *Condorcet criterion;* the *monotonicity criterion;* and the *independence of irrelevant alternatives criterion.*

- **Criterion 4:** Suppose that Candidate X is declared the winner of an election, and then a second election is held. If voters do not change their preferences, but one or more of the losing candidates drops out, then X should also win the second election.

Because the election in Table 11.3 (Unit 11D) gave five different winners by the five different voting methods we've discussed, you may already be guessing that none of these methods can always satisfy all four criteria. However, to be sure that these criteria are clear, let's examine a few more examples in which we test each criterion.

EXAMPLE 1   *An Unfair Plurality*

Consider the following preference schedule. Suppose that the winner is chosen by plurality. Does this method satisfy the four fairness criteria?

First	A	B	C
Second	B	C	B
Third	C	A	A
	5	4	2

Solution: Let's begin by interpreting the table. Column 1 shows that five voters prefer A to B or C. Column 2 shows that four voters prefer B to C or A, and Column 3 shows that two voters prefer C to B or A. Thus A has the most first-place votes, and is the plurality winner. Now let's apply the four fairness criteria to this result.

Criterion 1: Because no candidate received a majority of first-place votes in this election, the first criterion does not apply.

Criterion 2: To check this criterion, we need to examine pairwise races between the candidates. There are three pairs to consider: A versus B, A versus C, and B versus C. Here are the results:

- A is ranked ahead of B in Column 1 (5 votes) but behind in Columns 2 and 3 (4 + 2 = 6 votes). Thus B beats A by 6 to 5.

- A is ranked ahead of C in Column 1 (5 votes) but behind in Columns 2 and 3 (4 + 2 = 6 votes). Thus C beats A by 6 to 5.

- B is ranked ahead of C in Columns 1 and 2 (5 + 4 = 9 votes) but behind in Column 3 (2 votes). Thus B beats C by 9 to 2.

Note that B beats both A and C in pairwise contests. Thus, by Criterion 2, B should be the winner. Thus the plurality choice of A as the winner is unfair because A would lose to B in a head-to-head contest.

Criterion 3: We check this criterion by imagining a second election in which some voters move A higher in their rankings, but do not change the order in which they have B and C ranked. Column 1 cannot change because A already is ranked first. Moving A higher in Columns 2 or 3 may give A even more first-place votes, but cannot reduce A's number of first-place votes. Thus A would still win a plurality, and Criterion 3 is satisfied.

Criterion 4: This time we imagine a second election in which one of the losers drops out. Suppose that C were to drop out. Then B would move up to first place in Column 3, picking up the 2 first-place votes in this column. Because B already has 4 first-place votes in Column 2, B would then have 6 first-place votes, and would win the election by 6 to 5 over A. Thus the plurality method is unfair in this case because A would lose to B if C were to drop out of the running.

In summary, the plurality choice of A violates two of the four fairness criteria in this election. The basic problem is clear if we look back at the criteria that are violated: most voters apparently prefer candidate B over candidate A, but choosing a winner by plurality makes A the winner.   ■

EXAMPLE 2   *An Unfair Runoff*

Consider the following preference schedule. Suppose that a winner is chosen by a runoff. Are the four fairness criteria satisfied?

First	C	A	C	B
Second	B	C	A	A
Third	A	B	B	C
	9	**13**	**5**	**11**

Solution: First, note that A has 13 first-place votes (Column 2), B has 11 first-place votes (Column 4), and C has $9 + 5 = 14$ first-place votes (Columns 1 and 3). Thus B has the fewest first-place votes and is eliminated from the runoff. Because the 11 voters who chose B ranked A second, A will pick up all these first-place votes in the runoff, and win over C by 24 to 14. Thus A is the winner of this election in a runoff. Now let's evaluate the fairness criteria.

Criterion 1: No candidate received a majority in the original election, so this criterion does not apply.

Criterion 2: If you examine the pairwise matchups in this election, you'll find that B beats A by 20 to 18, A beats C by 24 to 14, and C beats B by 27 to 11. Note that these results are paradoxical: the fact that voters prefer B to A and A to C suggests that B should beat C, but instead C beats B. Thus no candidate is a clear winner in pairwise comparisons, so the second criterion does not apply.

Criterion 3: This criterion asks us to imagine a second election in which the winner of the first election, A, picks up some additional votes. Suppose that the five voters who chose the order in Column 3 decide to move A up to first place, meaning that they now rank the candidates in the order A, C, B (note that this order is the same as that in Column 2). Then the new results are as follows:

First	C	A	A	B
Second	B	C	C	A
Third	A	B	B	C
	9	**13**	**5**	**11**

Now we must examine what happens if we choose a winner of this second election by the runoff method. The first-place results are now 18 for A, 11 for B, and 9 for C. Thus C has the fewest first-place votes and is eliminated from the runoff. The 9 people who voted for C had B ranked second, so their votes will go to B in the runoff, giving B a total of $11 + 9 = 20$ votes in the runoff. Thus B will beat A by 20 to 18. Notice what has happened: A won the original election, then picked up some additional votes in the second election. But this gain caused A to lose the second election to B!

Criterion 4: This criterion asks us to imagine a second election in which one or more of the losers of the first election drops out. Candidate B was eliminated in the original runoff election, so the results cannot be affected if B drops out. However, if C drops out instead, B picks up the 9 first-place votes from Column 1, and A picks up the 5 first-place votes from Column 3. Then A would have $13 + 5 = 18$ first-place votes and B would have $9 + 11 = 20$ first-place votes. This would change the winner from A to C, so Criterion 4 is violated.

In summary, this election violates Criteria 3 and 4. The unfairness of the former violation is particularly striking: It certainly wouldn't be fair if A is hurt by gaining votes—but that is exactly what happened in this case. ■

EXAMPLE 3   *A Fair Election*

Consider the following preference schedule. Suppose that the winner is chosen by plurality. Does this choice satisfy the four fairness criteria?

First	A	B	C
Second	B	C	B
Third	C	A	A
	10	4	2

Solution: The winner by the plurality method is A, with 10 votes. Because this plurality is also a majority of the 16 votes cast, the first criterion is satisfied. Examining pairwise matchups shows that A beats B by 10 to 6, and also beats C by 10 to 6. Thus A is also the pairwise winner, and the second criterion is satisfied. The third criterion also is satisfied because, if we imagine a second election in which A picks up additional votes, it will not reduce A's majority. Finally, the fourth criterion asks what happens in a second election in which one or more of the losers drops out. Eliminating either B or C cannot reduce A's majority, so A still wins. In summary, this particular election satisfies all four fairness criteria. ■

> **TIME-OUT TO THINK:** Example 1 shows an election decided by plurality that *does not* meet all four fairness criteria. Example 3 shows an election decided by plurality that *does* meet the criteria. Note also that the plurality is a *majority* in Example 3. Is it generally true that an election decided by a majority will be fair? Explain.

Arrow's Impossibility Theorem

Suppose that we continue to test the fairness criteria on many different elections decided by any of the five methods we have discussed. Sometimes, as in Example 3 above, we will find that all four criteria are satisfied and we can declare the election to be fair. Other times, as in Examples 1 and 2, the results will violate one or more of the fairness criteria.

Despite the fact that some elections are clearly fair and others are not, we can find some general rules. For example, elections decided by plurality always satisfy Fairness Criteria 1 and 3, but can sometimes violate Criteria 2 and 4. A similar analysis of the other voting methods we have discussed gives the summary shown in Table 11.4.

Table 11.4	Fairness Criteria and Voting Systems (Y = criterion always holds, N = criterion may be violated)				
Criterion	Plurality	Top-Two Runoff	Sequential Runoff	Borda Count	Pairwise Comparison
1	Y	Y	Y	N	Y
2	N	N	N	N	Y
3	Y	N	N	Y	Y
4	N	N	N	N	N

Table 11.4 carries the disconcerting message that none of the five voting systems will always give a fair outcome. During the two centuries following the American and French revolutions, political theorists spent a great deal of effort trying to come up with a better voting system—one that would always give fair results. Unfortunately, their quest was futile, because a perfect voting system will never be found. In 1952, economist Kenneth Arrow proved mathematically that it is *impossible* to find a voting system that will always satisfy all four fairness criteria. This result is one of the landmark applications of mathematics to social theory, and is now known as **Arrow's Impossibility Theorem.**

By the Way ··············
Kenneth Arrow received the 1972 Nobel Prize in Economics for his mathematical analysis of voting systems that led him to discover the Impossibility Theorem.

TIME-OUT TO THINK: Arrow's Impossibility Theorem tells us that no voting system can be perfect. However, some systems may still be superior to others. Among the voting systems we have discussed, which do you think would be the best to use for electing a president? Defend your opinion.

APPROVAL VOTING

Democratic voting systems have traditionally been based on the principle of *one person, one vote*. However, in light of the fact that no voting system can be perfect, some political theorists have proposed alternative methods of voting. Arrow's Impossibility Theorem assures us that none of these methods can be perfect either, but a new method *might* give fair results more often than traditional methods.

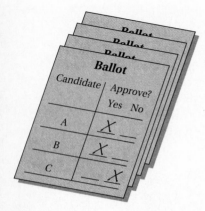

FIGURE 11.36

One such alternative voting system, called **approval voting,** asks voters simply to specify whether they approve or disapprove of each candidate. They may approve of as many candidates as they like, and the candidate with the most approval votes wins. As an example, consider a three-candidate race for governor between Candidates A, B, and C. If we use approval voting, the ballot would look something like Figure 11.36.

Suppose that Candidates A and B happen to be closely aligned politically, while Candidate C is on the opposite end of the political spectrum. In that case, voter opinions about the candidates might actually be as follows:

- 32% want A as their first choice, but would also approve of B.
- 32% want B as their first choice, but would also approve of A.
- 1% want A as their first choice, and approve of neither B nor C.
- 35% want C as their first choice, and approve of neither A nor B.

On the approval ballot, we will find that:

A is approved by $32\% + 32\% + 1\% = 65\%$.

B is approved by $32\% + 32\% = 64\%$.

C is approved by 35%.

Thus, by approval voting, A becomes the new governor and C is clearly in last place. However, if this election had been decided by plurality, C would have been elected by virtue of the most first-place votes, with 35%.

> **TIME-OUT TO THINK:**   Do you think that approval voting is a better way to decide this election than plurality? Do you think it is better in general? Defend your opinion.

EXAMPLE 4   *A Drawback to Approval Voting*

Suppose that the opinions of voters in a particular three-way election are as follows:

- 26% want A as their first choice, but would also approve of B.
- 25% want A as their first choice, and approve of neither B nor C.
- 15% want B as their first choice, and approve of neither A nor C.
- 18% want C as their first choice, but would also approve of B.
- 16% want C as their first choice, and approve of neither A nor B.

Contrast the results if the election is decided by approval voting and by plurality.

Solution: By approval voting, we find the following results:

A is approved by $26\% + 25\% = 51\%$.

B is approved by $26\% + 15\% + 18\% = 59\%$.

C is approved by $18\% + 16\% = 34\%$.

Thus B is the winner by approval voting. However, if we count only first choices, we find the following outcome by the plurality method:

A is the first choice for $26\% + 25\% = 51\%$.

B is the first choice for 15%.

C is the first choice for $18\% + 16\% = 34\%$.

Thus a *majority* of voters want A as their first choice for governor. However, because a larger majority finds B "acceptable," B wins in the approval voting.   ■

> **TIME-OUT TO THINK:**   Example 4 shows that while approval voting ensures that the winning candidate is acceptable to the largest number of people, another candidate might be the first choice of the majority. Do you think this is a serious drawback to approval voting? Why or why not?

VOTING POWER

So far, we have taken for granted that every voter has the same amount of power to influence an election as every other voter. However, this is not necessarily the case. For example, shareholders in a corporation generally are given votes in proportion to the number of shares they own. A shareholder with 10 shares gets 10 votes, and a shareholder with 1000 shares gets 1000 votes. Thus every voter at a shareholder meeting is *not* equally empowered to influence the outcome of an election.

A similar situation can arise in politics when voters form groups, or **coalitions**, that agree to vote the same way on a particular issue. Such coalition-building can drastically affect the power of individual voters to influence the outcome of a vote. Several techniques have been devised for measuring the effective power wielded by different voters when all voters do not have the same power. Although we will not discuss the details of these techniques here, we can give a brief taste of the type of situation that can arise. Suppose, for example, that the 100 members of the U.S. Senate happened to be divided as follows:

49 Democrats

49 Republicans

2 reformists

Moreover, suppose that all 49 Democrats favor a particular bill, while all 49 Republicans are opposed. The 2 reformists may not care which way the election goes but, if they vote together, their two votes will decide the outcome of the election. Both the Democrats and Republicans are likely to work hard to woo the votes of the reformists, perhaps by agreeing to support other bills that the reformists favor but which would not otherwise pass. Thus the power of the 2 reformists to influence decision making is much greater than their two votes out of 100 would seem to imply.

EXAMPLE 5   *Missing the Big Vote*

A small corporation has four shareholders. The 10,000 shares in this corporation are divided among the shareholders as follows:

Shareholder A owns 2,650 shares (26.5% of the company).

Shareholder B owns 2,550 shares (25.5% of the company).

Shareholder C owns 2,500 shares (25% of the company).

Shareholder D owns 2,300 shares (23% of the company).

The corporation has scheduled a key vote about buying out another company. Each shareholder's vote is counted in proportion to the number of shares the person owns. Suppose that Shareholder D misses the vote. Does it matter?

Solution: At first, it would seem disastrous for a major shareholder to miss the chance to influence the future of the company. However, note that any two of the three largest shareholders can form a majority by voting together:

A and B: 26.5% + 25.5% = 52%

A and C: 26.5% + 25% = 51.5%

B and C: 25.5% + 25% = 50.5%

In contrast, Shareholder D cannot be part of a majority unless two other shareholders vote the same way. But, in that case, the other two shareholders already determine the outcome by themselves. Thus Shareholder D has *no power* to affect the election outcome.   ■

CONCLUSIONS: LESSONS ABOUT DEMOCRACY

Georgia Congressional Districts

In this unit and the previous one, we have examined the surprising complexity of democratic voting. In some cases, we found that the choice of a voting system can be just as important to determining the outcome of an election as the actual preferences of voters.

In fact, there are many more subtleties to democratic politics than we have investigated here. For example, members of the U.S. House of Representatives are elected by voters in *districts* within each state. The U.S. Constitution leaves it to the states to decide how to draw the borders of the districts. The problem of creating these borders is usually called **apportionment** (because voters must be apportioned, or divided, among the different districts). As you might imagine, districts can be drawn in an enormous number of different ways, and drawing districts differently can lead to different people being elected. Again, mathematical analysis shows that there is no single best solution to apportionment problems. As a result, apportionment problems often are the subject of intense political debate, and sometimes end up being decided by the courts.

In summary, mathematical analysis offers many important lessons about democracy. Perhaps the most important one is this: Democracy is much more difficult in practice than it is in principle, which is why it is so important to be politically and mathematically aware if you hope to influence the future of your town, state, country, or world.

REVIEW QUESTIONS

1. Briefly summarize each of the four criteria of fairness. For each one, given an example that describes why an election would be unfair if the criterion were violated.

2. What is *Arrow's impossibility theorem*? Summarize its meaning and its importance.

3. What is *approval voting*? How is it different from the traditional idea of one person, one vote?

4. Give an example in which different voters may wield different amounts of power in an election.

5. What is a *coalition*? How can coalition-building influence power?

6. What is *apportionment*? Briefly describe why apportionment problems are complex.

PROBLEMS

1. **Plurality and Criterion 1.** Explain in words why the plurality method always satisfies Fairness Criterion 1.

2. **Runoff Methods and Criterion 1.** Explain in words why both the top-two runoff and the sequential runoff methods always satisfy Fairness Criterion 1.

3. **Point System and Criterion 1.** Consider the preference table below for three candidates. Which candidate wins by the point system (Borda count)? Is Fairness Criterion 1 satisfied? Explain.

First	A	B
Second	B	C
Third	C	A
	3	2

4. **Point System and Criterion 1.** Devise your own preference schedule with three candidates, three rankings, and seven voters in which the point system (Borda count) violates Fairness Criterion 1. Explain your work.

5. **Pairwise Comparison and Criterion 1.** Explain in words why the pairwise comparison method always satisfies Fairness Criterion 1.

6. **Plurality and Criterion 2.** Consider the preference table shown below. Which candidate is the plurality winner. Does this choice satisfy Fairness Criterion 2? Explain.

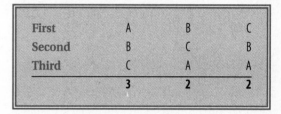

First	A	B	C
Second	B	C	B
Third	C	A	A
	3	2	2

7. **Plurality and Criterion 2.** Devise a preference schedule with three candidates (A, B, and C) and 11 voters in which C is the plurality winner and yet, A beats C and B beats C in head-to-head races. Explain your work.

8. **Pairwise Comparison and Criterion 2.** Explain in words why the pairwise comparison method always satisfies Fairness Criterion 2.

9. **Sequential Runoff and Criterion 2.** Suppose the sequential runoff method is used on the following preference schedule. Is Fairness Criterion 2 satisfied? Explain.

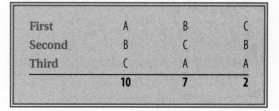

First	A	B	C
Second	B	C	B
Third	C	A	A
	10	7	2

10. **Sequential Runoff and Criterion 2.** Suppose the sequential runoff method is used on the following preference schedule. Is Fairness Criterion 2 satisfied? Explain.

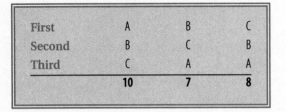

First	A	B	C
Second	B	C	B
Third	C	A	A
	10	7	8

11. **Sequential Runoff and Criterion 2.** Devise a preference schedule with three candidates and three rankings such that the sequential runoff method (top-two runoff) violates Fairness Criterion 2. Explain your work.

12. **Point System and Criterion 2.** Suppose the point system (Borda count) is used on the following preference schedule. Is Fairness Criterion 2 violated? Explain.

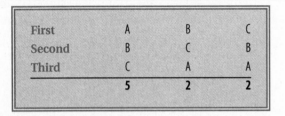

First	A	B	C
Second	B	C	B
Third	C	A	A
	5	2	2

13. **Point System and Criterion 2.** Devise a preference schedule with four candidates such that the point system violates Fairness Criterion 2.

14. **Plurality and Criterion 3.** Explain in words why the plurality method always satisfies Fairness Criterion 3.

15. **Sequential Runoff and Criterion 3.** Suppose the sequential runoff method is used on the following preference schedule. Is Fairness Criterion 3 satisfied? Explain.

First	A	B	A	C
Second	B	C	C	A
Third	C	A	B	B
	7	8	4	10

16. **Point System and Criterion 3.** Explain in words why the point system (Borda count) always satisfies Fairness Criterion 3.

17. **Pairwise Comparison and Criterion 3.** Explain in words why the pairwise comparison method always satisfies Fairness Criterion 3.

18. **Plurality and Criterion 4.** Suppose the plurality method is used on the following preference schedule. Is Fairness Criterion 4 satisfied? Explain.

First	A	B	C
Second	B	C	B
Third	C	A	A
	6	2	5

19. **Plurality and Criterion 4.** Devise a preference schedule with three candidates and nine votes such that the plurality method violates Fairness Criterion 4. Explain your work.

20. **Sequential Runoff and Criterion 4.** Suppose the sequential runoff method is used on the following preference schedule. Is Fairness Criterion 4 satisfied? Explain.

First	A	B	C
Second	C	C	B
Third	B	A	A
	8	6	3

21. **Sequential Runoff and Criterion 4.** Devise a preference schedule with three candidates such that the sequential runoff method violates Fairness Criterion 4. Explain your work.

22. **Point System and Criterion 4.** Suppose the point system (Borda count) method is used on the following preference schedule. Is Fairness Criterion 4 satisfied? Explain.

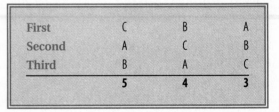

First	C	B	A
Second	A	C	B
Third	B	A	C
	5	4	3

23. Point System and Criterion 4. Devise a preference schedule with three candidates such that the point system (Borda count) method violates Fairness Criterion 4. Explain your work.

24. Pairwise Comparison and Criterion 4. Suppose the pairwise comparison method is used on the following preference schedule. Is Fairness Criterion 4 satisfied? Explain.

First	A	A	E	C	D
Second	E	C	B	B	B
Third	C	D	A	A	A
Fourth	D	E	C	D	E
Fifth	B	B	D	E	C
	1	1	1	1	1

25. Pairwise Comparison and Criterion 4. Devise a preference schedule with four candidates such that the pairwise comparison method violates Fairness Criterion 4.

Fairness Criteria. *In Problems 26–30, consider the following preference schedule for four candidates.*

First	A	B	C	D
Second	D	A	B	C
Third	B	D	A	A
Fourth	C	C	D	B
	16	10	8	7

26. Suppose the winner is decided by plurality. Analyze whether this choice satisfies the four fairness criteria.

27. Suppose the winner is decided by a top-two runoff. Analyze whether this choice satisfies the four fairness criteria.

28. Suppose the winner is decided by sequential runoffs. Analyze whether this choice satisfies the four fairness criteria.

29. Suppose the winner is decided by a Borda count (point system). Analyze whether this choice satisfies the four fairness criteria.

30. Suppose the winner is decided by pairwise comparisons. Analyze whether this choice satisfies the four fairness criteria.

Fairness Criteria. *In Problems 31–35, consider the following preference schedule for five candidates (Table 11.3 of Unit 11D).*

First	A	B	C	D	E	E
Second	D	E	B	C	B	C
Third	E	D	E	E	D	D
Fourth	C	C	D	B	C	B
Fifth	B	A	A	A	A	A
	18	12	10	9	4	2

31. Suppose the winner is decided by plurality. Analyze whether this choice satisfies the four fairness criteria.

32. Suppose the winner is decided by a top-two runoff. Analyze whether this choice satisfies the four fairness criteria.

33. Suppose the winner is decided by sequential runoffs. Analyze whether this choice satisfies the four fairness criteria.

34. Suppose the winner is decided by a Borda count (point system). Analyze whether this choice satisfies the four fairness criteria.

35. Suppose the winner is decided by pairwise comparisons. Analyze whether this choice satisfies the four fairness criteria.

36. Approval Voting. Suppose that Candidates A and B have moderate political positions, while Candidate C is relatively conservative. Voter opinions about the candidates are as follows:
- 30% want A as their first choice, but would also approve of B.
- 29% want B as their first choice, but would also approve of A.
- 1% want A as their first choice, and approve of neither B nor C.
- 40% want C as their first choice, and approve of neither A nor B.

a. If all voters could vote only for their first choice, which candidate would win by plurality?

b. Which candidate wins by an approval vote?

37. **Approval Voting.** Suppose that Candidates A and B have moderate political positions, while Candidate C is quite liberal. Voter opinions about the candidates are as follows:

    - 28% want A as their first choice, but would also approve of B.
    - 29% want B as their first choice, but would also approve of A.
    - 1% want B as their first choice, and approve of neither A nor C.
    - 42% want C as their first choice, and approve of neither A nor B.

    a. If all voters could vote only for their first choice, which candidate would win by plurality?

    b. Which candidate wins by an approval vote?

38. **Swing Votes.** Suppose that the Senate has the following breakdown in party representation: 49 Democrats, 49 Republicans, and 2 Independents. Further, suppose that all Democrats and Republicans vote along party lines. Assuming that a majority is required to pass a bill, explain why the 2 Independents, despite holding only 2% of the Senate seats, effectively hold power equal to that of either of the large parties.

39. **Power Voting.** Imagine that a small company has four shareholders who hold 26%, 26%, 25%, and 23% of the company's stock. Assume that votes are assigned in proportion to share holding (e.g., if there are a total of 100 votes, the four people get 26, 26, 25, and 23 votes, respectively). Also assume that decisions are made by strict majority vote. Explain why, although each individual holds roughly one-fourth of the company's stock, the individual with 23% holds *no* effective power in voting.

40. **Project: Political Coalitions.** Israel is one of several countries in which two large parties and several smaller parties are represented in the Parliament. The smaller parties hold great power because of their ability to form winning coalitions. Investigate the current alignment of political parties in the Israeli (or any other) government. Discuss the majority coalitions that are theoretically and practically possible.

CHAPTER 11
SUMMARY

$\mathcal{I}$n this chapter, we investigated several applications of discrete mathematics to business and society. We found that the mathematics of networks is playing an increasingly large role in business. We also found that mathematical analysis of voting yields many surprising results. Keep in mind the following key lessons from this chapter.

- Networks are increasingly a part of everyone's life, with direct applications of networks to interconnected systems such as telephone lines, electrical power grids, and the Internet.

- Network analysis also allows us to study complex management problems that can have dramatic effects on business efficiency and on the bottom line.

- The discrete mathematics of voting yields many important results that have a direct impact on the working of our democracy.

- Amazingly, a mathematical theorem proves that no perfect voting system is possible. Democracy, at its root, must involve give and take, and compromise.

SUGGESTED READING

Introduction to Management Science, B. Taylor (Englewood Cliffs, NJ: Prentice Hall, 1995).

Graphs, Models, and Finite Mathematics, J. Malkevitch and W. Meyer (Englewood Cliffs, NJ: Prentice Hall, 1974).

Social Choice and Individual Values, K. Arrow (New York: John Wiley and Sons, 1963).

"The Choice of Voting Systems," R. Niemi and W. Riker, *Scientific American* (234, June 1976, pp. 21–27).

"The Symmetry and Complexity of Elections," Donald Saari, *Complexity* (Vol. 2, January/February 1997).

Theory of Voting, R. Farquharson (New Haven: Yale University Press, 1969).

Topics in the Theory of Voting, P. Straffin, UMAP Expository Monograph (Boston: Birkhauser, 1980).

Chapter 12

THE POWER OF NUMBERS: A FEW MORE TOPICS

We have covered a lot of ground in the previous eleven chapters, but we have only glimpsed the remarkable power that mathematics gives us for understanding the world. Fortunately, with the skills you have developed, you can learn much more on your own. In this final chapter, we explore just a few more useful topics that illustrate the breadth of mathematics.

In a sense, we are all mathematicians—and superb ones. It makes no difference what you do. Your real forte lies in navigating the complexities of social networks, weighing passions against histories, calculating reactions, and generally managing a system of information that, when all laid out, would boggle a computer. But if all this is true, why haven't you noticed this ability by now?
A. K. DEWDNEY, IN *200% OF NOTHING* (P. 147)

Mathematics is a way of thinking that can help make muddy relationships clear. It is a language that allows us to translate the complexity of the world into manageable patterns. In a sense, it works like turning off the houselights in a theater the better to see a movie. Certainly something is lost when the lights go down; you can no longer see the faces of those around you or the inlaid patterns on the ceiling. But you gain a far better view of the subject at hand.
K. C. COLE, IN *THE UNIVERSE AND THE TEACUP* (P. 6)

athematics is a language that helps us solve problems involving numbers. The utility of mathematics stems from the way it enables us to use a few *unifying* ideas to address many seemingly different questions. By tapping the unifying power of mathematics, we can cross boundaries between subject areas and develop a truly interdisciplinary view of the real world. In this chapter, we study a few more topics that demonstrate the practical utility of mathematics.

BALANCING THE FEDERAL BUDGET

A national debt, if it is not excessive, will be to us a national blessing.

—ALEXANDER HAMILTON, 1781

Our first topic in this chapter is one that has dominated much of the political landscape for the past 30 years: balancing the federal budget. In the late 1990s, the budget is nearing balance for the first time in this period, but keeping it balanced in the future still poses a major challenge. In addition, we are still faced with the difficulty of reducing the overall debt that has accumulated over the past three decades.

In theory, the federal government works like a small business or your personal finances. Both have **receipts** (or *income*), and **outlays** (or *expenses*). **Net income** is defined as the difference between receipts and outlays.

$$\text{net income} = \text{receipts} - \text{outlays}$$

There can be no freedom or beauty about a home life that depends on borrowing and debt.

—HENRIK IBSEN, 1879

When receipts exceed outlays, the net income is positive and the business has a **surplus** (profits). When outlays exceed receipts, the net income is negative and the business has a **deficit**.

A deficit must be covered either with reserves (accumulated savings from prior surpluses) or by taking out a loan. If a loan is used to finance the deficit, the borrower must pay interest to the lender. The total amount of money owed by the business or government, which may be the accumulation of deficits from many years, is called the **debt**.

Between 1955 and 1997, the U.S. government had only four surplus years—and not one since 1970 (Figure 12.1). During this time, the federal debt grew to nearly $5.5 trillion.

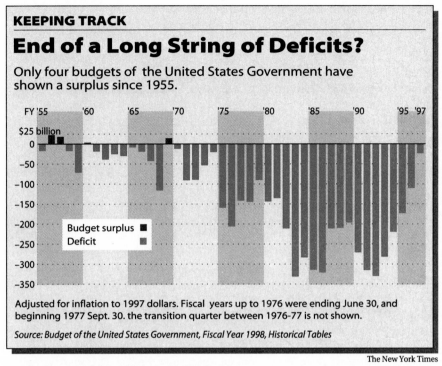

KEEPING TRACK

End of a Long String of Deficits?

Only four budgets of the United States Government have shown a surplus since 1955.

Adjusted for inflation to 1997 dollars. Fiscal years up to 1976 were ending June 30, and beginning 1977 Sept. 30. the transition quarter between 1976-77 is not shown.

Source: Budget of the United States Government, Fiscal Year 1998, Historical Tables

The New York Times

FIGURE 12.1   Copyright © 1997 by the New York Times Company. Reprinted with permission.

A SMALL BUSINESS ANALOGY

Before studying the federal budget, let's investigate the simpler books of an imaginary company with not-so-imaginary problems. Table 12.1 shows four years of budgets for the Wonderful Widget Company, which started with a clean slate at the beginning of 1995.

Table 12.1	Budget Summary for the Wonderful Widget Company (all amounts in thousands of dollars)			
	1995	1996	1997	1998
Total Receipts	$854	$908	$950	$990
Outlays				
Operating	$525	$550	$600	$600
Employee Benefits	200	220	250	250
Security	275	300	320	300
Interest on debt	0	12	26	47
Total Outlays	1000	1082	1196	1197
Surplus/Deficit	−146	−174	−246	−207
Debt (accumulated)	−146	−320	−566	−773

The first column shows that, during 1995, the company had total receipts of $854,000 and total outlays of $1,000,000. Thus the company's net income was

$$\$854{,}000 - \$1{,}000{,}000 = -\$146{,}000.$$

The negative net income tells us that the company had a first-year *deficit* of $146,000. To cover this deficit (so that, for example, checks written for outlays didn't bounce), the company had to borrow money, ending the year $146,000 in debt. Note that the table shows both deficit and debt as negative numbers because they represent money owed to someone else.

In 1996, receipts increased to $908,000, while outlays increased to $1,082,000. Among the 1996 outlays was a $12,000 interest payment on the first-year debt of $146,000. Thus the deficit for 1996 was

$$\$908{,}000 - \$1{,}082{,}000 = -\$174{,}000.$$

The company therefore had to borrow $174,000 to cover this deficit. Further, it had no money with which to pay off the debt from 1995. Thus the total debt at the end of 1996 was

$$\$146{,}000 + \$174{,}000 = \$320{,}000.$$

Here is the key point: Because the company failed to balance its budget in 1996, its total debt continued to grow. As a result, its interest payment in 1997 increased to $26,000.

After another deficit year in 1997, the company's owners decided on a change of strategy in 1998. Relative to 1997, they froze operating expenses and employee benefits and actually *cut* security expenses. Note, however, that the interest payment rose substantially because the total debt increased in 1997. Despite the attempts to curtail outlays and despite another increase in receipts in 1998, the company still ran a deficit and the total debt continued to grow.

TIME-OUT TO THINK: Suppose that you are a loan officer for a bank in 1999, when the Wonderful Widget Company comes asking for further loans to cover its increasing debt. Would you agree to lend it the money? If so, would you attach any special conditions to the loan? Explain.

THE FEDERAL BUDGET

• Web • Watch •
The complete federal budget, as well as several shorter summaries, are available on the Web.

The Widget Company example shows that a succession of annual deficits leads to a rising debt. The increasing interest payments on that debt, in turn, make it even easier to run deficits in the future. The Widget Company story is a mild version of what has happened to the U.S. federal budget.

Table 12.2 provides a five-year summary of the federal budget, including data for several important categories of outlays. Figure 12.2 shows relative shares of *all* income and outlays for 1996 as pie charts. Note that the wedge labeled "borrowing" in the revenue pie chart represents the *net deficit* shown in Table 12.2. That is, like the Widget Company, the government must cover its deficit by borrowing money. The government borrows money from "the public" by selling Treasury notes, bills, and bonds to investors (see Unit 5E).

Table 12.2	U.S. Federal Budget Summary, 1993–1997 (all amounts in billions of dollars)				
	1993	**1994**	**1995**	**1996**	**1997**
Total Receipts	1,154	1,259	1,352	1,453	1,579
Total Outlays	1,409	1,462	1,516	1,560	1,601
Net Deficit (borrowed from public)	−255	−203	−164	−107	−22
Outlays—Selected Categories					
Social Security	305	320	336	350	367
Federal Employee Retirement Benefits	60	63	66	68	72
Medicare	131	145	160	174	190
Health Care (excluding Medicare)	87	94	102	107	113
National Defense	291	282	272	266	270
Veterans Benefits and Services	36	38	38	37	39
Housing and Food Assistance to Poor	57	61	65	65	64
International Affairs	17	17	16	13	15
Educational and training	37	31	38	36	36
Net Interest on the Debt	199	203	232	241	244
Gross Federal Debt	−4,351	−4,644	−4,921	−5,182	−5,370

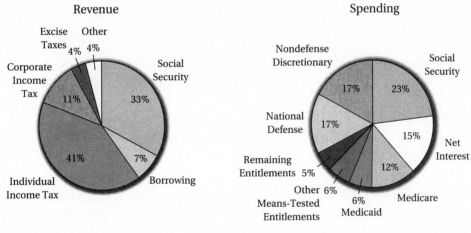

FIGURE 12.2

Pie charts from Federal Budget Guide 1998.

TIME-OUT TO THINK: Which categories in Table 12.2 are rising most dramatically? Based on the trends in spending, predict how the pie chart for outlays will look in 2005 compared to the one shown in Figure 12.2. Explain your reasoning.

EXAMPLE 1   *Interest on the Debt*

Table 12.2 shows that the debt at the end of 1995 was about $4900 billion ($4.9 trillion), and the interest paid on the debt in 1996 was $241 billion. What interest rate does this represent? Suppose that the interest rate had been one percentage point higher. How would the interest payment have been affected?

Solution: The interest rate is the interest paid divided by the debt.

$$\text{interest rate} = \frac{\$241 \text{ billion}}{\$4900 \text{ billion}} = 0.049 = 4.9\%$$

The government paid interest on the debt at a rate of about 4.9%. If the rate had been one percentage point higher, or 5.9%, the interest payment would have been

$$\text{interest payment} = \text{debt} \times \text{interest rate} = \$4900 \text{ billion} \times 5.9\% = \$289 \text{ billion}.$$

Note that this is $48 billion higher than the actual interest payment of $241 billion. That is, a rise in interest rates of just one percentage point could increase the interest payment—and hence the next year's deficit—by nearly $50 billion! ∎

> **Mathematical Note:**
> This calculation assumed that the debt remained fixed at $4.9 trillion throughout 1996. In reality, the debt continued to grow during the year.

Strange Numbers: Net Deficit and Gross Debt

Take another look at Table 12.2. Note that the gross *debt* in 1994 was higher than that in 1993 by

$$\$4{,}644 \text{ billion} - \$4{,}351 \text{ billion} = \$293 \text{ billion}.$$

However, the 1994 net *deficit* was only $203 billion. How did the debt rise by an amount greater than the deficit?

The answer lies in a subtlety having to do primarily with **trust funds**—special accounts designed to meet future obligations. For example, if the government collects more in Social Security taxes than it pays out in Social Security benefits, the difference becomes part of the Social Security trust fund.

To see how the trust fund affects the deficit and debt numbers, imagine that in a particular year:

- Total government receipts are $1,380 billion, of which $380 billion comes from Social Security taxes; and

- Total government outlays are $1,580 billion, of which $340 billion is paid out in Social Security benefits.

The **net deficit** for the year is the difference between outlays and receipts:

$$\text{net deficit} = \$1{,}380 \text{ billion} - \$1{,}580 \text{ billion} = -\$200 \text{ billion}$$

The government must cover this net deficit by borrowing $200 billion from the public. The **net debt** is the total amount of money that the government has borrowed from the public. If the net debt were $3000 billion at the end of the previous year, the $200 billion net deficit would increase the net debt to $3200 billion.

Note, however, that Social Security recorded a surplus of

$$\$380 \text{ billion} - \$340 \text{ billion} = \$40 \text{ billion}.$$

By law, this $40 billion must be added to the Social Security Trust Fund. But there's a problem: the government already spent it! Therefore the government places "IOUs" in the trust fund that promise to reimburse the trust fund for the $40 billion, plus interest, at some later date. In effect, the government has borrowed $40 billion from future recipients of Social Security benefits—which means you!

The **gross deficit** for the year is the sum of the net deficit and the money borrowed from the Social Security Trust Fund, or

$$\$200 \text{ billion} + \$40 \text{ billion} = \$240 \text{ billion}.$$

The **gross debt** is the total amount of money that the government has borrowed from both the public and the trust fund; thus the gross debt for the year rises by $240 billion.

In summary, the year-to-year debt in Table 12.2 increases by more than the year-to-year deficit because the table shows the *net* deficit and the *gross* debt. Although this may seem a bit like comparing apples and oranges, it is the way the government routinely quotes its numbers. There's at least some justification for this practice: the *net* deficit accurately reflects the annual differences in outlays and receipts, while the *gross* debt accurately reflects the amount that the government is obligated to repay someday.

By the Way ··········
This example assumes that the Social Security surplus is the *only* difference between the net and gross deficits. In reality, several other trust funds, and a few other items, also contribute to this difference.

> **TIME-OUT TO THINK:**   Suppose that the government balances its annual budget so that it has no *net* deficit. Could the gross debt still increase? Explain.

Balancing the Budget

Depending on whose opinion you hear, the federal deficit and debt are anything from a minor problem to a major crisis. In any case, balancing the budget involves two related problems. First is the deficit, which arises from the annual failure to balance receipts and outlays. Second is the debt. The government cannot begin to pay off the debt until it eliminates the deficit and achieves a surplus.

For the purposes of the federal budget, it is useful to categorize spending in three broad categories:

- **Mandatory expenses** are expenses that *must* be paid. The primary mandatory expense is interest on the debt. If this were not paid, the government would default on its obligations and the entire economy might be thrown into turmoil.

- **Entitlements** are payments to individuals, such as payments through Medicare and Social Security. They are called *entitlements* because the law says that certain individuals are *entitled* to these benefits. Entitlement spending can be changed only if the law changes.

- All other spending is **discretionary**, including spending for national defense, education, international aid, and scientific research. Discretionary spending is set by Congress, and approved by the president, on a year-to-year basis.

By the Way ··········
Entitlements sometimes are categorized as a mandatory expense because the law mandates that they be paid.

Careful study of Table 12.2 shows why deficits will be difficult to control in the future. Note that, from 1993 to 1996, receipts increased from $1154 billion to $1453 billion, or by

$$\$1453 \text{ billion} - \$1154 \text{ billion} = \$299 \text{ billion}.$$

Meanwhile, outlays increased from $1409 billion to $1560 billion, representing a rise of only

$$\$1560 \text{ billion} - \$1409 \text{ billion} = \$151 \text{ billion}.$$

By the Way ·············

In 1997, the gross deficit was $101 billion or $79 billion greater than the net deficit of $22 billion.

In other words, the dramatic decrease in the deficit between 1993 and 1996 came about largely due to increased tax revenues—a direct result of a strong economy.

Assuming this trend would continue, in 1998 the government forecast several years of *net surplus* beginning in 1999. Unfortunately, even if this "balanced budget" is achieved, longer term problems remain. Most importantly, entitlement spending is expected to grow dramatically as retiring "baby boomers" begin collecting Social Security and Medicare benefits. Moreover, a zero *net* deficit is not the same as a zero *gross* deficit. Under even the most optimistic predictions, the gross debt will continue to increase, and hence the mandatory spending for interest on the debt will increase as well.

The rise in mandatory and entitlement spending threatens to cause a ballooning deficit in the future. Mathematical models of the budget that extrapolate current trends show huge deficits beginning sometime after 2010. Ultimately, maintaining balanced budgets will require either cuts in entitlement spending or large tax increases—both of which are politically difficult to achieve.

TIME-OUT TO THINK: Given that about half of all Americans receive entitlement benefits of some type, explain why it would be difficult for politicians to cut these benefits. Do you think they *should* be cut? Defend your opinion.

EXAMPLE 2   *A Budget Forecast*

In 1996, total federal spending for mandatory programs and entitlements was about $902 billion, and discretionary spending was about $658 billion. Total receipts were $1453 billion. Suppose that spending on mandatory programs and entitlements rises by 7% per year, while receipts rise by 3% per year. Forecast the spending on mandatory programs and entitlements and receipts in 2016. Comment on the impact on the deficit and on discretionary spending.

Solution: Because we are given a starting number and a percentage growth rate, we can calculate a future value by using the exponential growth law from Unit 7C. This equation takes the form

$$Q = Q_0 \times (1 + r)^t,$$

where Q is the value of the quantity after time t, Q_0 is the starting value of the quantity, and r is the fractional growth rate.

To forecast the mandatory and entitlement spending, we set Q_0 = $902 billion for the 1996 spending, t = 20 years from 1996 to 2016, and r = 0.07 for the 7% growth rate. The spending in 2016 is

$$Q = \$902 \text{ billion} \times (1 + 0.07)^{20} = \$3490 \text{ billion}.$$

For the receipts, we set Q_0 = $1453 billion in 1996, t = 20 years from 1996 to 2016, and r = 0.03 for the 3% growth rate. The receipts in 2016 are

$$Q = \$1453 \text{ billion} \times (1 + 0.03)^{20} = \$2624 \text{ billion}.$$

Note that mandatory and entitlement spending alone will be larger than receipts by

$$\$3490 \text{ billion} - \$2624 \text{ billion} = \$866 \text{ billion}.$$

Even if discretionary spending were held steady at its 1996 level of $658 billion, the deficit for 2016 would be

$$\$866 \text{ billion} + \$658 \text{ billion} = \$1524 \text{ billion}.$$

In other words, this model predicts a deficit of more than $1.5 trillion by 2016! Clearly, we will have to find some way to ensure either that spending does not rise at such a great rate or that receipts increase at a much greater rate than the rates used in this model. ■

Retiring the Debt

Even if the budget is truly balanced (zero *gross* deficit), the nation will still be saddled with the *debt*. As we saw in Unit 4C, the 1997 gross debt already represented more than $20,000 for every person in the United States. Given the difficulties of even reaching a zero gross deficit in the near future, it is unlikely that the debt will drop significantly in the next couple decades, let alone be retired. Indeed, few economists believe that the debt will be retired within the lifetime of anyone living today.

As we saw in Example 2, we can forecast future deficits or surpluses by making assumptions about the growth rates of receipts and outlays. We can then estimate how and when the *debt* can be retired by calculating how it increases or decreases each year. Table 12.3 shows the outcome of such calculations using a model in which receipts and outlays change by fixed percentages each year; this model is based on receipts, outlays, and debt at the end of 1997. Although this model is oversimplified, it does give a general idea of the challenge of retiring the debt. For example, the model shows that, if we wish to have the debt retired by 2025, the federal budget would require a steady decline in outlays of 1% per year at the same time that receipts rise by 1% per year.

Table 12.3	Year of Zero Debt			
	Receipts +1%	Receipts +2%	Receipts +3%	Receipts +4%
Outlays −2%	2019	2014	2011	2010
Outlays −1%	2025	2017	2013	2011
Outlays 0%	2045	2022	2016	2012
Outlays +1%	Never	2036	2020	2015
Outlays +2%	Never	Never	2031	2019

> **TIME-OUT TO THINK:** Briefly explain why it would be politically difficult for the government to sustain cuts in outlays of 1% per year at the same time that revenue grows by 1% per year. Do you think any other entries in Table 12.3 are more realistic? Explain.

EXAMPLE 3   *Retiring the Debt with a National Lottery*

Suppose that, through some political miracle, the gross deficit is held to zero indefinitely. To retire the gross debt, the government decides to have a national lottery. If this lottery produces a net income of $50 million per week, and all of this money goes toward retiring the debt, when will the debt be retired? Use the 1997 gross debt of $5.4 trillion.

> • Web • Watch •
> Learn more about efforts to control the federal budget from the *Concord Coalition* Web site.

Solution: With $50 million ($5 \times 10^7$) per week going toward paying off the $5.4 trillion ($5.4 \times 10^{12}$) debt, the debt would be retired in

$$\frac{\$5.4 \times 10^{12}}{\$5 \times 10^7/\text{week}} = 108{,}000 \text{ weeks.}$$

There are 52 weeks in a year, so 108,000 weeks is equivalent to

$$108{,}000 \text{ weeks} \times \frac{1 \text{ year}}{52 \text{ weeks}} = 2077 \text{ years.}$$

With $50 million per week going toward debt reduction, it would take more than 2000 years to retire the debt!   ∎

REVIEW QUESTIONS

1. Define receipts, outlays, net income, surplus, and deficit as they apply to budgets.

2. Explain why years of deficits make it more difficult to balance a budget in the future.

3. How does the government cover its deficits? How large have deficits been in recent years?

4. What is a *trust fund*, such as the Social Security trust fund?

5. What is the difference between the *net deficit* and the *gross deficit*? Which is larger? Which is usually quoted in reports about the federal budget? Which will be eliminated if we have a "balanced budget?"

6. What is the difference between the *net debt* and the *gross debt*? Which is larger? Which is usually quoted in reports about the federal budget?

7. Differentiate between *mandatory expenses, entitlements*, and *discretionary spending*. Discuss the difficulties in cutting expenses in each category.

8. Discuss the prospects of retiring the federal debt in your lifetime.

PROBLEMS

1. **The Wonderful Widget Company Future.** Extending the budget summary of the Widget Company (Table 12.1), assume that for 1999: total receipts are $1,050,000; operating expenses are $600,000; employee benefits are $200,000; and security costs are $250,000.

   a. Based on the accumulated debt at the end of 1998, calculate the 1999 interest payment. Assume an interest rate of 8.2%.

   b. Calculate the total outlays for 1999 and the year-end surplus or deficit, and the year-end accumulated debt.

   c. Based on the accumulated debt at the end of 1999, calculate the 2000 interest payment, again assuming an 8.2% interest rate.

   d. Assume that in 2000 the Widget Company has receipts of $1,100,000, holds operating costs and employee benefits to their 1999 levels, and spends no money on security. Calculate the total outlays for 2000, the year-end surplus or deficit, and the year-end accumulated debt.

   e. Imagine that you are the CEO (Chief Executive Officer) of the Wonderful Widget Company at the end of 2000. Write a three-paragraph statement to shareholders about the company's future prospects.

2. **The Wonderful Widget Company Future.** Extending the budget summary of the Widget Company (Table 12.1), assume that for 1999: total receipts are $975,000; operating expenses are $850,000; employee benefits are $290,000; and security costs are $210,000.

a. Based on the accumulated debt at the end of 1998, calculate the 1999 interest payment. Assume an interest rate of 8.2%.

b. Calculate the total outlays for 1999 and the year-end surplus or deficit, and the year-end accumulated debt.

c. Based on the accumulated debt at the end of 1999, calculate the 2000 interest payment, again assuming an 8.2% interest rate.

d. Assume that in 2000 the Widget Company has receipts of $1,050,000, holds operating costs and employee benefits to their 1999 levels, and spends no money on security. Calculate the total outlays for 2000, the year-end surplus or deficit, and the year-end accumulated debt.

e. Imagine that you are the CEO (Chief Executive Officer) of the Wonderful Widget Company at the end of 2000. Write a three-paragraph statement to shareholders about the company's future prospects.

3. **Analysis of the Federal Budget.** Consider the federal budget summary given in Table 12.2. For 1997, calculate the percentages of total outlays that went to each of the following: interest on the debt; defense spending; Social Security; and international affairs. Comment on these percentages; do they surprise you?

4. **Analysis of the Federal Budget.** Consider the federal budget summary given in Table 12.2. For 1996, calculate the percentages of total outlays that went to each of the following: interest on the debt; defense spending; Social Security; housing and food assistance to the poor; and international affairs. Comment on these percentages; do they surprise you?

5. **Interest on the Debt.** Using the data in Table 12.2, calculate the interest rate paid on the 1996 debt in 1997. Suppose that the interest rate had been one percentage point higher. How would the interest payment have been affected? What if the interest rate had been one percentage point lower?

6. **Interest on the Debt.** Using the data in Table 12.2, calculate the interest rate paid on the 1997 debt in 1998. Assume the interest payment in 1998 was $250 billion. Suppose that the interest rate had been 0.5 percentage point higher. How would the interest payment have been affected? What if the interest rate had been 0.5 percentage point lower?

7. **Net Deficit and Gross Debt.** Using the data in Table 12.2, calculate the change in the gross debt from 1995 to 1996. How does this number compare to the net deficit for 1996? Explain why the two numbers are not the same.

8. **Net Deficit and Gross Debt.** Using the data in Table 12.2, calculate the change in the gross debt from 1996 to 1997. How does this number compare to the net deficit for 1997? Explain why the two numbers are not the same.

9. **Budget Forecast.** In 1996, total federal spending for mandatory programs and entitlements was $902 billion, and discretionary spending was about $658 billion. Total receipts were $1453 billion. Suppose that spending on mandatory programs and entitlements rises by 6% per year, while receipts rise by 4% per year. Forecast the spending on mandatory programs and entitlements in 2016, and receipts in 2016. Contrast your results with the results from Example 2, and comment.

10. **Budget Forecast.** In 1996, total federal spending for mandatory programs and entitlements was $902 billion, and discretionary spending was about $658 billion. Total receipts were $1453 billion. Suppose that spending on mandatory programs and entitlements rises by 8% per year, while receipts rise by 3% per year. Forecast the spending on mandatory programs and entitlements in 2016, and receipts in 2016. Contrast your results with the results from Example 2, and comment.

11. **Per Capita Debt.** If half of all Americans work, what is the 1997 federal debt per laborer? Assume a 1997 U.S. population of 270 million.

12. **National lottery.** Assume that the federal government could raise $70 million per week in a national lottery and direct those funds to reducing the federal debt. If the government balances the budget each year (so the debt does not increase), how long would it take to retire the debt using lottery proceeds? Explain.

13. **Year of Zero Debt.** Study Table 12.3.

a. If receipts increase at an annual rate of 2% and outlays are held constant, when (according to this model) will the debt be eliminated? Are these reasonable assumptions? Why or why not?

b. If receipts increase at an annual rate of 3% and outlays decrease at an annual rate of 1%, when (according to this model) will the debt be eliminated? Comment on these assumptions.

c. Explain briefly why the debt will never be eliminated if receipts increase by 2% and outlays increase by 3%. What would happen in this case?

14. **Web Project: Balanced Budget?** In early 1998, projections made by the White House Office of Management and Budget predicted a balanced budget in 1999, with surpluses continuing for at least several years thereafter. Using information available on the Web, find out how projections have changed since that time. Does it still seem likely that surpluses will continue for several years? Explain.

15. **Web Project: Social Security Trust Fund.** Using information available on the Web, research the current status of the Social Security Trust Fund and potential future problems in paying out benefits. For example, when is the fund projected to start paying out more than it is taking in each year? Given that the trust fund surplus has actually been spent to support other government programs, how will the government make good on its promises to future social security recipients? Write a two- to three-page report that summarizes your findings.

16. **Project: Entitlement Spending.** The following table shows the distribution of entitlement payments to U.S. families, broken down by family income and family type (data from the Congressional Budget Office, early 1995).

Family Description	Percentage of Families Receiving Benefits	Average Benefit
All families	49%	$10,320
By income:		
$0–$29,999	58%	$9,950
$30,000–$99,000	37%	$11,710
$100,000 or more	31%	$15,220
By type:		
With children	39%	$8,200
Elderly	98%	$13,970
All other	32%	$6,930

a. Note that 58% of low-income families ($0–$29,999 income) receive an average benefit of $9950. Suppose that the total amount of these benefits were distributed equally among *all* low-income families (rather than 58%). In that case, what would the average benefit be for these families? Explain your work.

b. Do a calculation similar to that in part (a) to find the average benefit among *all* high-income families (more than $100,000). Compare this to your result in part (a).

c. Suppose that, based on your results in parts (a) and (b), someone claims that the government spends almost as much money on entitlements to high-income families as to low-income families. Is this statement supported by the data? Why or why not? (*Hint:* Does the table give any information about the *number* of families in each income category?)

d. In 1995, entitlements comprised approximately 55% of the total federal budget of $1.4 trillion. How much money was spent on entitlements?

e. In this table, "elderly" means people over age 65. Estimate (or find through research) the fraction of the U.S. population in this category. Then estimate the total amount of money spent on benefits for the elderly. Explain your estimates and uncertainties. Based on your estimate, what fraction of 1995 entitlement spending went to the elderly? What fraction of the overall 1995 federal budget?

f. Note that nearly half of all U.S. families receive entitlement benefits. Identify any benefits that *you* personally receive (don't forget to include any federal education grants or student loans). If you receive any benefits, compare the total amount you receive each year to the total amount you pay in taxes. Overall, are you supporting the government or is the government supporting you? For additional research, survey your class or a group of friends and identify all of the entitlement benefits received.

g. Based on the information in this problem, and other information you have gathered about entitlements received by yourself and friends, discuss the political difficulties of cutting entitlement spending.

· ·

ENERGY: OUR FUTURE DEPENDS ON IT

We pay energy bills to the power companies, we use energy from gasoline to run our cars, and we argue about whether nuclear energy is a sensible alternative to fossil fuels. On a personal level, we often talk about how energetic we feel on a particular day. But what *is* energy?

Broadly speaking, energy is what makes matter move or heat up. We need energy from food to keep our hearts beating, to maintain our 37°C (98.6°F) body temperatures, and to walk or run. A car needs energy to move the pistons in its engine, which in turn spin the wheels. A light bulb needs energy to generate the heat that makes its filament glow. Making sure that we will have enough energy to keep our civilization running is one of the most challenging issues of our time.

BASIC CONCEPTS OF ENERGY

Energy comes in three basic forms:

- Whenever matter is moving it has energy of motion, or **kinetic energy**. Falling rocks, the moving blades on an electric mixer, a car driving down the highway, and the molecules moving in the air around us are examples of objects with kinetic energy.

- **Potential energy** is energy that is stored; it has the *potential* to make something move if it is released. For example, a rock perched on a ledge has *gravitational potential energy* because it will fall if it slips off the edge. Gasoline contains *chemical potential energy*, which a car engine converts to the kinetic energy of the moving car. Power companies supply *electrical potential energy* that we use to run appliances.

- Light carries energy, called **radiative energy**. For example, the energy of light activates the molecules in your eyes that enable you to see; it also fuels the process of *photosynthesis* in plants.

By the Way ············
The word *kinetic* comes from a Greek word meaning *motion*.

Conservation of Energy

When we talk about *conserving* energy, we usually mean using less energy so that we can preserve energy resources such as oil. Scientifically, however, the term *conservation of energy* has a different meaning:

> ### The Law of Conservation of Energy
>
> Energy can neither be created nor destroyed; it can only change from one form to another.

For example, a moving car has kinetic energy. This kinetic energy comes from the release of chemical potential energy stored in the gasoline. Gasoline comes from oil, which is made from the remains of plants that lived long ago. These plants built up chemical potential energy in their stems and leaves through photosynthesis, which uses the radiative energy of sunlight. Thus the energy that runs our cars actually came from the Sun!

Mass-Energy and Nuclear Power

By the Way ············

The amount of energy, E, stored in an amount of mass, m, is given by Einstein's famous formula $E = mc^2$, where c is the speed of light.

Where did the Sun get its energy? It turns out, as first discovered by Einstein in 1905, that mass itself is a form of potential energy, often called **mass-energy**. In other words, any little bit of mass can be turned into other forms of energy—at least in principle. In practice, there are two basic ways to extract mass-energy for other purposes:

- **Nuclear fusion** is the process of fusing light chemical elements into heavier ones. For example, four hydrogen nuclei can be fused to make one helium nucleus. A little less than 1% of the mass "disappears" in this process, releasing energy that can appear as heat, light, or motion.
- **Nuclear fission** is the process of splitting heavy elements, such as uranium or plutonium, into lighter ones. Like fusion, this process releases a little less than 1% of the mass-energy stored in the matter.

The Sun is made up mostly of hydrogen, and produces energy through nuclear fusion that occurs deep in its core. Thus the energy of sunlight actually comes from mass-energy stored in the hydrogen inside the Sun. On Earth, nuclear power plants generate energy through nuclear *fission*.

Energy Versus Power

The words *energy* and *power* are often used together, but they are not quite the same thing. Technically, power is the rate at which energy is used. You probably are familiar with **Calories** as a unit of energy. A typical adult uses about 2000 Calories of energy each day. Because power is a rate of energy usage, power has units such as Calories per hour or Calories per minute. In general,

$$\text{power} = \frac{\text{energy}}{\text{time}}.$$

EXAMPLE 1   *Pedal Power*

By the Way ············

One calorie (lower case "c") is the amount of energy needed to raise the temperature of one gram of water by 1°C. A food Calorie (capital "C") is 1000 calories.

Suppose that you ride a stationary bicycle for exercise. One day, you ride for an hour and burn 500 Calories of energy. The next day, you ride for only half an hour, but burn the same 500 Calories. Compare your power on the bike for the two days.

Solution: You used the same 500 Calories of energy each day. The first day you used this energy in one hour, so your power was

$$\text{power} = \frac{500 \text{ Cal}}{1 \text{ hr}} = 500 \text{ Cal/hr}.$$

The second day you used the 500 Calories in half an hour, so your power was

$$\text{power} = \frac{500\ \text{Cal}}{\frac{1}{2}\ \text{hr}} = 1000\ \text{Cal/hr}.$$

In other words, you rode the bike with more power on the second day, which enabled you to use the same amount of energy in less time. ■

UNITS OF ENERGY AND POWER

Just as there are many different units for measuring height, such as inches, feet, and meters, there are many alternatives to Calories for measuring energy. If you look closely at an electric bill, you'll find that the power company probably charges you for electrical energy in units called *kilowatt-hours*. If you purchase a gas appliance, its energy requirements may be labeled in *British Thermal Units*, or BTUs. Internationally, and in science, the favored unit of energy is the **joule**.

The standard international unit of power is the familiar **watt,** defined as one joule per second, or

$$1\ \text{watt} = 1\ \frac{\text{joule}}{\text{s}}.$$

For example, a 100-watt light bulb consumes 100 joules of energy each second. In 10 seconds, it consumes 1000 joules of energy. A 50-watt light bulb consumes the same 1000 joules of energy in 20 seconds.

Electric Bills: Kilowatt-Hours

Take a look at a home electric bill. It probably lists the amount of electricity consumed in units of **kilowatt-hours.**

1 kilowatt-hour = 1 kilowatt × 1 hour = 1000 watts × 1 hour

We can convert this unit to joules by remembering that a watt is a joule per second:

$$1\ \text{kilowatt-hour} = 10^3\ \text{watts} \times 1\ \text{hr}$$

$$= 10^3\ \frac{\text{joules}}{\text{s}} \times 1\ \text{hr} \times 60\ \frac{\text{min}}{\text{hr}} \times 60\ \frac{\text{s}}{\text{min}}$$

$$= 3.6 \times 10^6\ \text{joules}$$

Thus a kilowatt-hour is just another name for 3.6 million joules.

By the Way ·············
Technically, a joule is equivalent to *1 kilogram-square meter per square second;* that is,

$$1\ \text{joule} = 1\ \frac{\text{kg} \times \text{m}^2}{\text{s}^2}.$$

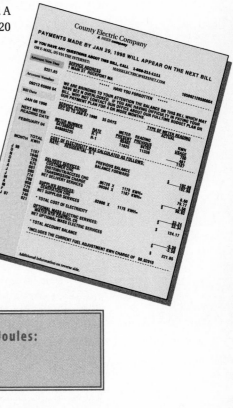

Conversion Factor Between Kilowatt-Hours and Joules:

$$1\ \text{kilowatt-hr} = 3.6 \times 10^6\ \text{joules}$$

EXAMPLE 2   *Operating Cost of a Light Bulb*

Suppose that your utility company charges 6¢ per kilowatt-hour of energy. How much does it cost to keep a 100-watt light bulb on for a week?

Solution: A 100-watt light bulb consumes energy at the rate of 100 watts, or 100 joules per second. In a week it uses

$$100 \, \frac{\text{joules}}{\text{s}} \times 1 \text{ week} \times 7 \, \frac{\text{days}}{\text{wk}} \times 24 \, \frac{\text{hr}}{\text{day}} \times 60 \, \frac{\text{min}}{\text{hr}} \times 60 \, \frac{\text{s}}{\text{min}} = 6 \times 10^7 \text{ joules}.$$

Converting to kilowatt-hours, we find:

$$6 \times 10^7 \text{ joules} \times \underbrace{\frac{1 \text{ kilowatt-hour}}{3.6 \times 10^6 \text{ joules}}}_{\text{conversion factor}} \approx 17 \text{ kilowatt-hours}.$$

Thus the light bulb will use 17 kilowatt-hours of energy in a week. Therefore the cost is

$$17 \text{ kilowatt-hr} \times \frac{\$0.06}{1 \text{ kilowatt-hr}} = \$1.02. \qquad \blacksquare$$

TIME-OUT TO THINK: How much does a typical 100-watt light bulb cost? (Remember that light bulbs usually are sold in packages of two or four bulbs.) Suppose that you could buy an energy efficient light bulb that costs twice as much, but provides the same light using half as much power. Would it be worth it?

Food Energy: Calories

Our bodies expend energy that we obtain from the chemical potential energy stored in food. Traditionally, the energy that we can extract from food is measured in Calories. We can convert between Calories and joules using the following conversion factor.

Conversion Factor Between Calories and Joules:

1 Calorie = 4184 joules

The number of Calories that you need to consume each day depends on many factors. For example, when you exercise you need to consume more Calories, and you may need more food to maintain your body heat on a cold day than on a warm day.

TIME-OUT TO THINK: If you run at the same pace, do you burn more calories on a hot day or a cold day? Explain.

EXAMPLE 3   *Burning Calories Through Running*

Vigorous running requires about 15 Calories per minute for an athlete. How much energy is spent in an hour of running? Suppose that you were running with such an energy expenditure on a treadmill that converted the kinetic energy of your motion into electricity. Could you supply enough power to light a 100-watt light bulb? Explain.

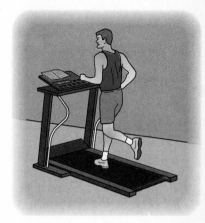

Solution: An hour of running will use

$$15 \frac{\text{Cal}}{\text{min}} \times 60 \text{ min} \approx 900 \text{ Cal.}$$

Converting to joules, we find that this is equivalent to

$$900 \text{ Cal} \times 4184 \frac{\text{joules}}{\text{Cal}} = 4 \times 10^6 \text{ joules.}$$

An hour of running consumes roughly 4 million joules of energy, which is slightly more than one kilowatt-hour of energy (3.6×10^6 joules). Thus your power consumption during the hour of running is more than one kilowatt, or 1000 watts. Not all this power goes to the motion of your legs: some goes to your heart beat, your breathing, and your general metabolism. Nevertheless, because one kilowatt is ten times as much power as that required for a 100-watt light bulb, you will be able to keep the light bulb shining as you run on the treadmill. ■

ENERGY COMPARISONS

Most of us are quite familiar with units such as meters or miles, and can form a "mental picture" of 5 meters or 10 miles. However, most of us are not very familiar with joules, which can make energy calculations difficult to interpret. In general, the best way to understand energy calculations is by making comparisons with energies that are familiar. Table 12.4 lists the energies of various items for the purposes of comparison.

Table 12.4 Energy Comparisons.	
Item	**Energy (joules)**
Average daytime solar energy striking Earth, per m^2 per second	1×10^3
Energy released by metabolism of 1 average candy bar	1×10^6
Energy needed for 1 hour of walking (adult)	1×10^6
Kinetic energy of average car traveling at 60 mi/hr	1×10^6
Energy needed for 1 hour of running (adult)	4×10^6
Daily energy needs of average adult	1×10^7
Energy released by burning 1 liter oil	1.2×10^7
Daily electrical energy used in an average home	5×10^7
Energy released by fission of 1 kg of uranium-235	5.6×10^{13}
Energy released by fusion of hydrogen contained in 1 liter of water	7×10^{13}
Energy released by 1-megaton H-bomb	5×10^{15}
Energy released by major earthquake (magnitude 8)	2.5×10^{16}
Kinetic energy in the winds of a major hurricane	10^{17}
U.S. annual energy consumption	10^{20}
Estimated energy in world fossil fuel reserves	3×10^{22}
Annual energy generation of Sun	10^{34}

EXAMPLE 4   *U.S. Energy Needs*

Based on Table 12.4, approximately how many liters of oil would be needed to supply all U.S. energy needs for a year?

Solution: The U.S. annual energy consumption is about 10^{20} joules. Because 1 liter of oil yields 1.2×10^7 joules, the amount of oil needed would be

$$\frac{10^{20} \text{ joules}}{1.2 \times 10^7 \text{ joules/liter}} = 10^{20} \text{ joules} \times \frac{1 \text{ liter}}{1.2 \times 10^7 \text{ joules}} = 8 \times 10^{12} \text{ liters.}$$

The annual energy consumption of the United States is equivalent to the energy of burning about 8 trillion liters of oil, or roughly 2 trillion gallons of oil. This much oil would occupy a volume of nearly 8 cubic kilometers, which is equivalent to a cube-shaped reservoir about 2 kilometers (1.2 miles) on a side.   ∎

EXAMPLE 5   *Fusion Power*

No one has yet succeeded in creating a commercially viable way to produce energy through nuclear fusion. However, suppose that at some time in the future we are able to build fusion power plants that are safe and cost-efficient. If we could extract all the hydrogen from water and use it for fusion, how much water would we need each day to meet U.S. energy needs?

Solution: We divide the U.S. annual energy consumption by the amount of energy available through fusion of 1 liter of water (from Table 12.4):

$$\frac{10^{20} \text{ joules}}{7 \times 10^{13} \text{ joules/liter}} = 1.4 \times 10^6 \text{ liters}$$

That is, we would need about 1.4 million liters of water *per year*; dividing by 365 days in a year, this becomes

$$\frac{1.4 \times 10^6 \text{ liters/yr}}{365 \text{ days/yr}} \approx 4000 \text{ liters/day.}$$

In other words, all the energy needs of the United States could be met if we could generate fusion power from the hydrogen contained in just 4000 liters of water per day. If we divide by the number of minutes in a day, we find that 4000 liters per day is about the same as

$$\frac{4000 \text{ liters/day}}{24 \text{ hr/day} \times 60 \text{ min/hr}} = 3 \text{ liters/min,}$$

which is a little less than 1 gallon per minute.   ∎

By the Way ············
Scientists working on fusion power actually use a rare form of hydrogen called *deuterium*. (Ordinary hydrogen nuclei contain just a single proton; deuterium contains a proton and a neutron.) About 1 in 50,000 hydrogen atoms is deuterium, so the flow rate needed to power the United States would be 50,000 times higher than calculated in this example—but still only about the flow rate of a small creek.

> **TIME-OUT TO THINK:** If you invented a fusion power plant, could you meet the energy needs of the *entire* United States with the water flowing from your kitchen sink? Discuss. (*Hint:* You can measure the flow rate of water from your kitchen sink by timing how long it takes to fill a 1 liter pitcher.)

EXAMPLE 6   *Solar Energy*

A typical household uses an average of about 1000 watts of power (including lights, appliances, and hot water). Suppose that solar cells are able to make use of about 10% of the energy coming from the Sun. How much of the roof would need to be covered with solar cells to power the house?

Solution: Table 12.4 shows that solar energy provides an average of about 1000 joules per square meter per second, or 1000 watts per square meter, in the daytime. Thus if solar cells used all this energy, it would take just 1 square meter of solar cells to power the house (in the daytime). If solar cells used only 10% of the solar energy, then the house would need 10 square meters of solar cells. Of course, more will be needed to collect energy for nights and cloudy days, and a way to store this energy will also be needed.   ■

REVIEW QUESTIONS

1. What is *energy*? Describe each of the three basic forms of energy.

2. What is the law of conservation of energy? Give an example of its application.

3. What is *mass-energy*, and what does it have to do with *nuclear energy*?

4. What is the difference between nuclear fusion and nuclear fission? Which one is used in today's nuclear power plants?

5. What is the difference between *energy* and *power*? What are the standard units for power?

6. List at least three common units of energy. Under what circumstances do the different units tend to be used?

PROBLEMS

1. **Exercise Power.** Suppose that you burn 800 Calories while playing a basketball game for an hour.

   a. What is your average power during the game? Give your answer in Calories per hour.

   b. Convert your answer from part (a) to watts. Suppose that this power could be harnessed somehow. Is it enough to keep a 100-watt light bulb shining? Explain.

   c. What is the total energy you use, in joules, during the basketball game?

   d. Using data from Table 12.4, determine how many candy bars you would need to replace the energy you expend during the game. Explain your work.

2. **Exercise Power.** Suppose that you burn 500 Calories while doing an aerobics class for 45 minutes.

   a. What is your average power during the class? Give your answer in Calories per hour.

   b. Convert your answer from part (a) to watts. Suppose that this power could be harnessed somehow. Is it enough to keep a 100-watt light bulb shining? Explain.

   c. What is the total energy you use, in joules, during the class?

   d. Using data from Table 12.4, determine how many candy bars you would need to replace the energy you expend during the class. Explain your work.

3. **Operating Cost of a Refrigerator.** Suppose that your refrigerator uses an average power of 350 watts, and your utility company charges 6¢ per kilowatt-hour of energy. How much does it cost to run your refrigerator each year? Explain.

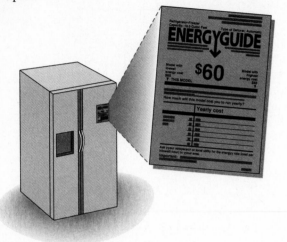

4. **Operating Cost of a Hair Dryer.** Suppose that you have an 1800-watt hair dryer, which you use for an average of 10 minutes per day. If your utility company charges 7¢ per kilowatt-hour of energy, how much does it cost to run the hair dryer each day? each year?

5. **Electric Bill.** Your electric bill states that you used 1250 kilowatt-hours of energy in the past month.

   a. How many joules of energy did you use?

   b. How many liters of oil would be needed to provide this energy? how many gallons? how many barrels? (1 barrel of oil = 42 gallons of oil)

   c. What is your average *power* use, in watts?

6. **Electric Bill.** Your electric bill states that you used 970 kilowatt-hours of energy in the past month.

   a. How many joules of energy did you use?

   b. How many liters of oil would be needed to provide this energy? how many gallons? how many barrels? (1 barrel of oil = 42 gallons of oil)

   c. What is your average *power* use, in watts?

7. **Fission Power.** Suppose that we could generate all the energy needed in the United States by nuclear fission of uranium-235 (U-235). How much uranium would we need each year? Explain.

8. **The Fusion Home.** Suppose that you invented a home fusion device that could generate energy by fusing the hydrogen found in ordinary tap water.

   a. How much tap water would you need to generate the daily electrical energy used by a typical home? Explain your assumptions.

   b. Suppose that you received a contract to supply electrical energy for a city with 100,000 homes. How much water would you need each day to supply your fusion reactor? How much water would you need each year? Explain your assumptions.

9. **Energy Comparisons.** Use the data in Table 12.4 to answer each of the following questions.

   a. Compare the energy of a 1-megaton hydrogen bomb to the energy in the winds of a major hurricane.

   b. How many candy bars do you need to fuel a 1-hour run?

   c. Suppose that we could somehow capture *all* the energy released by the Sun for just one second. Would this energy be enough to supply U.S. energy needs for a year? Explain.

10. **Energy Comparisons.** Use the data in Table 12.4 to answer each of the following questions.

    a. Compare the estimated energy in world fossil fuel reserves to the annual U.S. energy requirement.

    b. Compare the energy released by a 1-megaton nuclear bomb to the energy released by an earthquake of magnitude 8.

    c. How large an area of solar collectors would be needed to supply solar energy to an average home? Explain your work and your assumptions.

11. **Furnace Power in BTUs.** In the United States, the energy that comes from burning kerosene or natural gas is often measured in *British Thermal Units* (BTUs): 1 BTU = 1055 joules. Suppose that you have a furnace rated at 1000 BTUs per hour. What is its power in watts?

12. **Human Wattage.** The average person requires about 2500 food Calories per day.

    a. Convert this quantity to joules and kilowatt-hours.

    b. What *power* (in watts) does the average human body run at? Compare your answer to the wattage used by some familiar appliance.

c. Estimate the total amount of energy needed from food by an average person over a year.

d. Suppose that, instead of getting energy from food, you were able to live by burning oil. How much oil would you use each year?

e. U.S. annual energy consumption is about 10^{20} joules. What is the *per capita* energy consumption? Compare this value to the energy needed from food alone.

13. A Power Plant. Suppose a new power plant can generate a gigawatt (billion watts) of power.

a. How much energy, in kilowatt-hours, can it generate each month?

b. If the average home uses 1000 kilowatt-hours per month, how many homes can this power plant supply with energy?

c. If the power plant generates its energy from oil, how many *barrels* of oil (1 barrel = 42 gallons) does it require each month? each day?

14. Nuclear Power Plants. Operating at full capacity, the Fort St. Vrain Nuclear Power Station in Platteville, Colorado, can generate 330 megawatts of power.

a. If the Fort St. Vrain station operated at full capacity for a month, how much energy would it produce? Give your answer in both kilowatt-hours and joules.

b. If a typical household uses 1000 kilowatt-hours of electricity each month, how many households could have their energy needs met by the Fort St. Vrain station?

c. Calculate the total weight, in kilograms, of U-235 needed to provide one month's worth of power for the number of households in parts (a) and (b).

15. Energy from Junk Mail. The flow of junk mail through the average person's mailbox seems endless. Most of it goes directly into the trash; a small percentage is recycled. Suppose, instead, that junk mail were burned to make energy. Burning 1 gram of paper releases 2×10^4 joules of energy.

a. Estimate the total mass of junk mail, in grams, received in U.S. mailboxes each year. Clearly explain your assumptions and associated uncertainties.

b. Estimate the total amount of energy, in joules, that could be produced from America's annual supply of junk mail.

c. Use your answer from part (b) to estimate the average power, in watts, that could be supplied by burning junk mail. Compare this quantity to the power output of a 1 gigawatt power station.

d. Total U.S. electrical power production is about 400 gigawatts. What fraction of U.S. electricity needs could be produced by junk mail? Is junk mail a realistic source of energy? Explain.

16. The Power of Photovoltaics. Photovoltaic cells convert sunlight directly into electricity. In direct sunlight, each square meter receives about 1000 watts of solar power. That is, if a panel of photovoltaic cells were 100% efficient and had an area of 1 square meter, it would generate about 1000 joules of energy each second.

a. Suppose that a 1 m^2 panel of photovoltaic cells has an efficiency of 12%. Further, suppose that it receives an average of six hours of sunlight per day. How much energy would it produce during those six hours? Give your answer in both kilowatt-hours and joules.

b. Suppose that the energy produced by the panel in part (a) can be stored in batteries so that it can be released at a uniform rate day and night (24 hours a day). What is the *average power* that can be supplied by the solar panel?

c. A typical U.S. household requires 1 kilowatt of power. Under the assumptions of parts (a) and (b), how large an array of photovoltaic cells would be needed to meet this demand? Would it fit on the roof of a typical single-family home?

d. Under the assumptions of parts (a) and (b), how large an array of photovoltaic cells would be needed to supply the total U.S. electricity usage of 400 gigawatts of power?

e. The land area of the continental United States is approximately 10 million km^2. If all U.S. electricity were supplied by photovoltaic cells, what fraction of the land area would we need to cover with photovoltaic arrays? Comment on the environmental impact of such a system, and contrast with the environmental impact of current electrical energy sources (e.g., fossil fuels, nuclear energy, and hydroelectric energy).

17. Wood for Energy? A total of about 180,000 terawatt-years of solar energy reaches the Earth's surface each year, of which 0.06% is used by plants in photosynthesis. Of the energy used in photosynthesis, 1% is stored in plant matter (e.g., wood). (*Hint:* 1 terawatt-year = 1 terawatt $\times$ 1 year; 1 terawatt = 10^{12} watts.)

   a. Calculate the total amount of energy stored in plants each year over the entire Earth. Give your answer first in terawatt-years, then convert to joules.

   b. Suppose that power stations generated electricity by burning plant matter. If *all* the energy stored in plants each year could be converted to electricity, what average power would be possible?

   c. Total world power use for electricity is on the order of 10 terawatts. Based on your answer to part (b), calculate the fraction of world power that could be supplied during one year if we burned *all* the living plants on Earth.

   d. Considering your results from parts (b) and (c), can you draw any conclusions about why humans depend on *fossil* fuels, such as oil and coal, which are the remains of plants that died long ago? Explain.

18. Nuclear Materials. Natural uranium ore consists primarily of uranium-238 (U-238) with only small amounts of uranium-235 (U-235). Mined uranium can be *enriched* by discarding much of the U-238 so that the remaining material has a larger fraction of U-235; producing 1 ton of enriched uranium requires about 8 tons of natural uranium. A typical 1-gigawatt nuclear power plant uses about 30 metric tons of enriched uranium each year.

   a. The world's existing nuclear power plants, combined, generate about 400 gigawatts of power (1990 estimate). How much natural uranium (in tons) is needed to supply the world's nuclear reactors each year?

   b. World uranium reserves (i.e., the amount of uranium available for future mining) are estimated to be about 6 million tons. Estimate how long these reserves will last. Explain your assumptions and uncertainties.

19. Nuclear Fission Bomb. Consider a nuclear fission bomb in which the energy is produced by fission of one kilogram of U-235.

   a. How much energy is released by this bomb? (Consult Table 12.4.)

   b. The explosion of a ton of TNT releases 5×10^9 joules. How many tons of TNT would be needed to create an explosion of equivalent energy to that of the fission bomb?

   c. How does this fission bomb compare in energy to a 1-megaton hydrogen bomb?

   d. How does this fission bomb compare in energy to the roughly 20-kiloton bomb (equivalent to 20 kilotons of TNT) that destroyed Hiroshima in 1945?

   e. Suppose that a terrorist got hold of 1 kilogram of U-235, and used it to make an atomic bomb. Assume that the rest of the bomb weighs about 500 pounds and hence is transportable by car, boat, or airplane. Comment on the damage that such a bomb could do.

20. The Hibernating Bear. Mammals such as the Alaskan brown bear slow their metabolic rate during hibernation. During a six-month hibernation, a brown bear metabolizes about 100 kilograms of fat from a typical body mass of 450 kilograms. Metabolizing animal fat yields about 3.8×10^7 joules per kilogram of mass.

   a. Calculate the total amount of energy used by a brown bear during hibernation. Then calculate the average metabolic rate of the bear, in watts, during hibernation.

   b. Biologists have found that the normal (nonhibernating) resting metabolic rate of mammals is roughly proportional to the $\frac{3}{4}$ power of body weight. For example, if one mammal weighs four times as much as another, its metabolic rate is $4^{3/4} = 2.8$ times higher. Given that the resting metabolic rate of a 75-kg human is about 75 watts, estimate the resting metabolic rate of the brown bear.

   c. Based on your results from parts (a) and (b), compare the metabolic rate of the brown bear during hibernation and nonhibernation.

   d. In animals that enter "true" hibernation (e.g., many rodents), their hibernating metabolic rate is less than half their normal metabolic rate. In light of this fact, are brown bears true hibernators? Is it safe to walk into the den of a hibernating brown bear? Explain.

21. Project: Compact Fluorescent Light Bulbs. Compact fluorescent light bulbs fit most existing light sockets and produce light with less power than ordinary (incandescent) light bulbs. The average life of a compact fluorescent bulb is estimated to be about 10,000 hours, whereas the average life of an incandescent bulb is about 1000 hours.

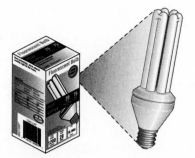

a. Suppose that you replace a 100-watt incandescent light bulb with a 27-watt compact fluorescent bulb that supplies the same lighting intensity. Over its 10,000-hour life, how much energy (kilowatt-hours) is saved with the new bulb? If the energy is generated by the combustion of oil, how much oil will be saved?

b. Determine the cost of electricity per kilowatt-hour in your area. Based on this cost, how much will you save on your electric bill over 10,000 hours by switching from a 100-watt incandescent bulb to a 27-watt compact fluorescent bulb?

c. Find the purchase prices of both 100-watt incandescent bulbs and 27-watt compact fluorescent bulbs. By accounting for the purchase price, the lifetimes of the bulbs, and their energy costs, calculate the *total* cost of each type of bulb for 10,000 hours of lighting. Which is more cost effective?

22. Project: Personal Energy Audit. Do a thorough electrical energy audit of your home, apartment, or dormitory. Although there are several ways to approach this problem, you might try the following.

• Determine the power requirement, in watts, of all electrical appliances.

• Estimate the number of hours that each appliance is used during a typical month.

• Calculate the total energy, in kilowatt-hours, used by each appliance for a month.

• Sum all these amounts to get your total monthly electricity use, in kilowatt-hours.

a. What is your estimated daily, monthly, and annual use of energy for electricity? If you live in a home or apartment where you have an electric meter, compare your daily estimate to a direct daily meter reading. Compare your monthly estimate to the amount shown on your monthly electricity bill. Discuss any discrepancies.

b. Find the cost of electricity (per kilowatt-hour) in your area. Based on your findings in part (a), calculate your average daily, monthly, and annual costs for electricity.

c. Burning 1 kilogram of coal yields about 1.6×10^9 joules of energy. Assume that your local power plant generates energy from coal and that it delivers electricity with an efficiency of about 30% (that is, 30% of the energy released from the coal ends up as electricity at your residence). How much coal is needed to supply your daily, monthly, and annual electrical needs?

d. Find out how electricity actually is generated in your area. Discuss the use of resources and environmental impacts associated with your personal electricity consumption.

· ·

DENSITY AND CONCENTRATION

UNIT 12C

In ancient Syracuse, King Hieron once received a new crown from a goldsmith. The crown was supposed to be pure gold, but Hieron suspected the goldsmith might have mixed in some silver. Hieron turned for help to Archimedes (c. 287–212 B.C.), one of the greatest scientists of the ancient world.

Archimedes was at a loss until, one day, he noticed the water overflowing as he stepped into his bath. He suddenly realized that the volume of water overflowing was equal to the volume of his body immersed in the bath. Herein lay the solution: Silver is *less dense* than gold, so he merely needed to immerse the crown and an equal weight of pure gold into water. If the water level rose the same amount in both cases, the crown was made

By the Way · · · · · · · · · · · ·
The ancient town of Syracuse is located in Sicily on the coast of the Ionian Sea.

of pure gold; if the crown caused the water to rise more than the pure gold, it contained silver. Thrilled at this flash of insight, Archimedes supposedly ran naked through the streets of Syracuse shouting "Eureka!"—a Greek word meaning "I have found it." For the goldsmith, alas, the news wasn't so great. The crown was partly silver, and the king had him executed.

THE CONCEPT OF DENSITY

At the heart of Archimedes' discovery lies the concept of **density**, which describes how tightly something is packed. In the case of materials, we measure density as mass per unit volume. That is,

$$\text{density} = \frac{\text{mass}}{\text{volume}}.$$

The most common metric units of density are *kilograms per cubic meter* (kg/m^3) or *grams per cubic centimeter* (g/cm^3).

A useful guide for putting densities in perspective is the density of water, which is approximately 1 gram per cubic centimeter, or 1000 kilograms per cubic meter:

$$\text{density of water} = 1\,\frac{\text{g}}{\text{cm}^3} = 1000\,\frac{\text{kg}}{\text{m}^3}$$

EXAMPLE 1   *Pebble Density*

A 40-gram pebble has a volume of 10 cubic centimeters. What is its density? Will it sink or float in water?

Solution: We find the density by dividing the pebble's mass by its volume:

$$\frac{40\,\text{g}}{10\,\text{cm}^3} = 4\,\frac{\text{g}}{\text{cm}^3}$$

The density of the pebble is 4 grams per cubic centimeter. Because the pebble is more dense than water, the pebble will sink in water.   ■

> **TIME-OUT TO THINK:** Explain why you can float better in a swimming pool if your lungs are filled with air than when you fully exhale.

EXAMPLE 2   *Solving for Mass*

Suppose that you have a 20-liter bucket of plaster with a density of 2 grams per cubic centimeter. What is its mass?

Solution: We can find the mass from the volume and density by multiplying both sides of the density equation by volume:

$$\text{density} \times \text{volume} = \frac{\text{mass}}{\text{volume}} \times \text{volume} \quad \Rightarrow \quad \text{mass} = \text{density} \times \text{volume}$$

Because 1 liter is 1000 cm³, the volume of 20 liters is equivalent to 20,000 cm³. Substituting this volume and the given density, we find that

$$\text{mass} = 2 \, \frac{\text{g}}{\text{cm}^3} \times 20,\!000 \text{ cm}^3 = 40,\!000 \text{ g}.$$

The bucket of plaster weighs 40,000 grams, or 40 kilograms.

Other Uses of Density

The concept of density may be applied to other quantities such as population or information. In most cases, the appropriate units for the density will be clear from the context.

EXAMPLE 3   *Population Density*

Manhattan Island has a population of about 1.5 million people living in an area of about 57 km². What is its population density? If there were no high-rise apartments, how much space would be available per person?

Solution: The context of population density suggests units of people per unit area. We divide the population by the given area to find that

$$\text{population density} = \frac{1.5 \times 10^6 \text{ people}}{57 \text{ km}^2} = 26,\!000 \, \frac{\text{people}}{\text{km}^2}.$$

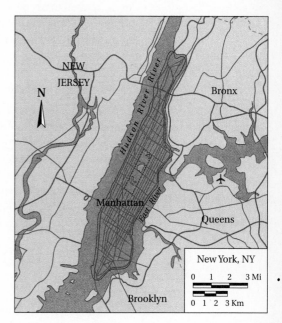

The average population density of Manhattan is about 26,000 people per square kilometer. If there were no high rises, each resident would have

$$\frac{1 \text{ km}^2}{26,\!000 \text{ people}} = 3.8 \times 10^{-5} \, \frac{\text{km}^2}{\text{person}}$$

of land. We can interpret this number better by converting from square kilometers to square meters:

$$3.8 \times 10^{-5} \, \frac{\text{km}^2}{\text{person}} \times \left(\frac{1000 \text{ m}}{1 \text{ km}} \right)^2 = 3.8 \times 10^{-5} \, \frac{\text{km}^2}{\text{person}} \times \frac{10^6 \text{ m}^2}{1 \text{ km}^2} = 38 \, \frac{\text{m}^2}{\text{person}}$$

There would be only 38 square meters of land per person, which is equivalent to a room about 6 meters, or 20 feet, on a side. And this does not include any space for roads, schools, or other common properties. Clearly, Manhattan Island could not fit so many residents without high rises.

EXAMPLE 4   *Information Density*

A standard compact disc stores information with tiny variations on its surface. Each variation represents one *bit* of information, and eight bits comprise a *byte*. Standard discs hold 650 megabytes of information on a surface area of about 90 square centimeters. What is the information density on a standard disc?

By the Way ··········
Newer, digital videodiscs (DVD) hold about 10 *gigabytes,* or 10,000 megabytes—roughly 15 times as much information as a standard compact disc.

Solution: The context in this case suggests units for *information density* of megabytes per unit area. Substituting the given values, we find that the information density of a standard compact disc is

$$\frac{650 \text{ megabytes}}{90 \text{ cm}^2} = 7 \frac{\text{megabytes}}{\text{cm}^2}.$$

Each square centimeter holds about seven megabytes of information. A megabyte is equivalent to about 500 pages of plain text, so a square centimeter of compact disc could store the equivalent text of seven 500-page books. ■

THE CONCEPT OF CONCENTRATION

Closely related to the concept of density is the concept of **concentration**, which describes the amount of one substance mixed with another. Like density, the concept of concentration can be applied to many different things, with units chosen appropriately for the context.

EXAMPLE 5   *Blood-Alcohol Concentration*

By the Way ··········
Brain functions begin to be impaired when the blood-alcohol concentration reaches 0.05%, and a blood-alcohol concentration of 0.5% is usually fatal.

Alcohol from beer, wine, and other liquors is absorbed into the bloodstream. The **blood-alcohol concentration** measures the percentage of a person's total volume of blood that consists of alcohol. In most states, a blood-alcohol concentration of 0.1% can warrant a DWI (driving while intoxicated) arrest. An average-sized person has about 5 liters of blood. How much beer does an average person need to consume before becoming legally intoxicated? Assume that the concentration of alcohol in the beer is 6%; that is, alcohol represents 6% of the total volume of the beer. Also assume that the alcohol is immediately absorbed into the bloodstream.

Solution: We are given that a concentration of 0.1% alcohol in an average person's 5 liters of blood will make that person legally intoxicated. This amount of alcohol in the blood is

$$0.1\% \times 5 \text{ liters} = 0.001 \times 5 \text{ liters} = 0.005 \text{ liters}.$$

Next, we need to determine how much beer a person must drink to get 0.005 liters of alcohol. Using the 6% alcohol concentration in the beer, we have that

$$\text{amount of alcohol} = 0.06 \times \text{amount of beer}.$$

Dividing both sides by 0.06 we find that

$$\text{amount of beer} = \frac{\text{amount of alcohol}}{0.06} = \frac{0.005 \text{ liters}}{0.06} = 0.08 \text{ liters}.$$

Thus, for an average person, legal intoxication requires consuming less than 0.1 liter of beer, which is about three ounces. In other words, less than half a glass of beer contains enough alcohol to put an average person at the legal intoxication limit!

However, our assumption that the alcohol is absorbed into the bloodstream immediately is not completely realistic. For example, most people don't drink their beer all at once, and food in the stomach slows the absorption of alcohol into the bloodstream. Moreover, metabolic processes (primarily in the liver) remove alcohol from the bloodstream at a rate of about 10–15 milliliters per hour. Thus the 5 milliliters of alcohol in 3 ounces of beer will be metabolized within about half an hour. ■

> **TIME-OUT TO THINK:** Suppose that you drink two 12-ounce bottles of beer. Given the rate of alcohol metabolism, how long should you wait before attempting to drive?

EXAMPLE 6   *Contaminated Water*

A sample of water that flows from a nineteenth-century mine is found to be contaminated with lead chloride (a lead salt) at a concentration of 0.1 gram per liter. How much lead chloride is contained in 500 liters of water? Should you be concerned about drinking from this water supply?

Solution: The given units for the concentration are a mass of the *contaminant* (lead chloride) per unit volume of water. That is,

$$\text{lead chloride concentration} = \frac{\text{mass of dissolved lead chloride}}{\text{volume of water}} = \frac{0.1 \text{ g}}{\ell}.$$

The amount of lead chloride contained in 500 liters of water is

$$500 \ \ell \times \frac{0.1 \text{ g}}{\ell} = 50 \text{ g}.$$

Lead is a toxic substance linked to many health problems, including mental retardation in children, so avoiding this water supply would be a good idea! ∎

By the Way ·············
The EPA standard for safe water requires levels of lead chloride below 0.00015 grams per liter—more than 600 times lower than the concentration in this example.

EXAMPLE 7   *Parts Per Million*

The U.S. Environmental Protection Agency (EPA) sets health-based standards for concentrations of various pollutants in the air. For example, the EPA standard for carbon monoxide is 9 *parts per million* (ppm). That is, air in which the carbon monoxide concentration is greater than 9 ppm is considered unhealthy. Express this concentration as a percentage.

Solution: The units of *parts per million* express a fraction: the number of particles of one substance mixed in with a million particles of another. In this case, the EPA standard calls for no more than 9 molecules of carbon monoxide for each 1 million molecules of air. We can express this concentration as a fraction by writing

$$\frac{9 \text{ molecules CO}}{1{,}000{,}000 \text{ molecules air}} = 0.000009 \text{ molecule CO/molecule air}.$$

As a percentage, this concentration is

$$0.000009 \times 100\% = 0.0009\%.$$

That is, carbon monoxide is considered to be at dangerous levels when it makes up only 0.0009% of the air. ∎

• Web • Watch •
You can find EPA safety standards for many pollutants and toxic substances at the EPA Web site.

> **TIME-OUT TO THINK:** If you live in an area with severe pollution, the local media may report concentrations of various pollutants. Check your newspaper and local TV news. Do they ever report on pollution? What units of concentration do they use?

REVIEW QUESTIONS ●●

1. What do we mean by *density*? Give several examples of density used in different contexts.

2. What do we mean by *concentration*? Give several examples of concentration used in different contexts.

3. What is blood-alcohol concentration? How much alcohol does it take to be legally intoxicated?

4. What do units of *parts per million* mean?

PROBLEMS ●●

Calculating Densities. *For Problems 1 and 2, find the average density of the following objects in grams per cubic centimeter.*

1. **a.** A rock with a volume of 15 cm³ and a mass of 0.25 kg.

   b. A box with a volume of 7 liters and a mass of 200 grams.

2. **a.** A sphere with a radius of 10 cm and a mass of 2 kg.

   b. A cube with side length of 6 in. and a weight of 10 lb.

Granite and Iron. *For problems 3 and 4, the density of granite is about 2.7 grams per cm³ and the density of iron is about 7.9 grams per cm³.*

3. **a.** What is the weight, in kilograms, of a granite slab that measures 1 m × 1 m on its face and is 2 cm thick? What is its weight in pounds?

   b. How much would the slab in part (a) weigh if it were made of iron instead of granite?

4. **a.** Compare the volumes of two rocks, one made of granite and one made of iron, that both weigh 1 kg.

   b. When iron is molten (liquid), its density is only 7.0 grams per cm³. If you start with 1 m³ of solid iron, what is its volume when it is molten?

5. **Population Density.**

   a. New Jersey and Wyoming have areas of 7419 and 970,000 square miles, respectively, and populations of 7.7 million and 450,000, respectively. Calculate and compare their population densities.

   b. Find the population and the area of your hometown. Calculate its population density.

6. **Population Density.**

   a. The land area of the United States is about 3.5 million square miles, and the population is about 250 million people. What is the average population density?

   b. Find the land area and population of China. Compare its average population density to that of the United States.

7. **DVD Density.** A digital videodisc (DVD) holds about 10 gigabytes of information in an area about the same as that of a compact disc (90 cm²).

   a. Calculate the information density on the DVD.

   b. The text in a typical 500-page book contains about 1 megabyte of information. How many 500-page books could be stored on 1 cm² of the DVD? How many such books could be stored on the entire disc?

8. **Compact Disc Density.** Suppose that a future compact disc holds 250 gigabytes of information on a surface area of 90 cm².

   a. Calculate the information density on the disc.

   b. The text in a typical 500-page book contains about 1 megabyte of information. How many 500-page books could be stored on 1 cm² of the disc? How many such books could be stored on the entire disc?

9. **Blood-Alcohol Concentration: Wine.** Humans generally have 70 cm³ of blood *per kilogram* of weight.

   a. About how many liters of blood are in *your* body?

   b. Suppose that you drink wine containing 13% alcohol by volume. How much would you have to drink to reach a blood-alcohol concentration of 0.1%?

   c. How much wine would you have to drink to reach a near-fatal limit of 0.5% blood-alcohol concentration?

d. Suppose that you drank enough to raise your blood-alcohol concentration to 0.2%, but then stopped. If your body metabolizes alcohol at the rate of 1 milliliter per hour, how long would it take to reduce your blood-alcohol concentration to below the legal limit of 0.1%?

10. **Blood-Alcohol Concentration: Whiskey.** Humans generally have 70 cm^3 of blood *per kilogram* of weight.

 a. About how many liters of blood are in *your* body?

 b. Suppose that you drink whiskey containing 40% alcohol by volume. How much would you have to drink to reach a blood-alcohol concentration of 0.1%?

 c. How much whiskey would you have to drink to reach a near-fatal limit of 0.5% blood-alcohol concentration?

 d. Suppose that you drank enough to raise your blood-alcohol concentration to 0.3%, but then stopped. If your body metabolizes alcohol at the rate of 1 milliliter per hour, how long would it take to reduce your blood-alcohol concentration to below the legal limit of 0.1%?

11. **Lead Chloride Contamination.** Water flowing from an abandoned mine is contaminated with lead chloride at a concentration of 0.0002 grams per milliliter. How much lead chloride is contained in 1500 liters of water? Should you be concerned about drinking this water? Why or why not?

12. **Particulate Pollution.** A sample of air taken during rush hour in downtown Denver on a December day exceeds the EPA standard for particulate pollution of 120 micrograms per cubic meter by 20%. What is the particulate concentration in micrograms per cubic meter? At this concentration, how many grams of particulate matter would be contained in a room that measures 5 meters in length, 3 meters in width, and 2 meters in height?

13. **Gaseous Pollution.** The EPA standard for carbon monoxide is 9 ppm (parts per million), and its standard for ozone is 120 ppb (parts per billion). On a day in which both gases are at their maximum EPA levels, what is the ratio of carbon monoxide molecules to ozone molecules? Does this ratio depend on the sample size? Explain.

14. **Air Pollution Reduction.** Through a variety of conservation, emissions, and public transportation initiatives, a city is able to reduce its average winter carbon monoxide levels by 0.3 ppm per year. If the worst year was 1990 with average winter levels of 6 ppm, and the 0.3 ppm per year reduction can be sustained, what will the levels be in 2005? Do you think the assumptions in this problem are realistic? Explain.

15. **Plutonium Release.** Fires occurred at the Rocky Flats nuclear weapons plant (west of Denver, Colorado) in 1957 and 1969, releasing as much as 100 pounds of plutonium (in the form of plutonium dioxide) into the atmosphere. Assume that at Rocky Flats the plutonium concentration in the air was 1.5 micrograms per cubic meter ($\mu g/m^3$), and that as the plutonium was carried downwind its concentration decreased by 0.5 $\mu g/m^3$ with each mile from the plant. How far downwind were concentrations within the EPA's "safe" level of 0.5 $\mu g/m^3$? Did the plutonium pose any danger to workers in downtown Denver, about five miles downwind of Rocky Flats? Explain.

16. **Mining Gold Ore.** The *grade*, or *purity*, of ore (a mixture of precious metal and surrounding rock found in a mine) describes the concentration of the precious metal in the surrounding rock. For example, if gold ore has a grade of 0.3 ounce per ton, 0.3 ounce of gold is found in each ton of the ore. Suppose that a mine is producing gold ore with an average grade of 0.3 ounce per ton, and that pure gold can be sold for $350 per ounce.

Uses of Gold

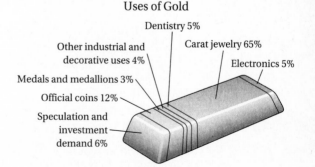

 a. How many tons of ore must be processed to produce $1000 worth of gold?

 b. Suppose that the cost of processing the ore is $20 per ton. What is the net profit on each ton of ore?

c. Suppose that, after the highest grade ore has been removed, the remaining ore in the mine has an average grade of 0.1 ounce per ton. Further, because this ore is more difficult to extract, the cost of processing it is $40 per ton. Can this ore be mined at a profit? Explain.

17. **Stellar Corpses: White Dwarfs and Neutron Stars.** A few billion years from now, after exhausting its nuclear engines, the Sun will become a type of remnant star called a *white dwarf*. It will still have nearly the same mass (about 2×10^{30} kg) as the Sun today, but its radius will be only about that of the Earth (about 6400 km).

   a. Calculate the average density of the white dwarf in units of kilograms per cubic centimeter.

   b. What is the mass of a teaspoon of material from the white dwarf? (*Hint:* A teaspoon is about 4 cubic centimeters.) Compare this mass to the mass of something familiar (e.g., a person, a car, a tank).

   c. A neutron star is a type of stellar remnant compressed to even greater densities than a white dwarf. Suppose that a neutron star has a mass that is 1.4 times the mass of the Sun but a radius of only 10 kilometers.

What is its density? Compare the mass of 1 cubic centimeter of neutron star material to the total mass of Mt. Everest (about 5×10^{10} kg).

18. **Web Project: Pollution Progress.** Using data available on the Web, investigate the average concentrations of various pollutants in a major city of your choice. Find the EPA standards for each pollutant, and find some of the hazards associated with exposure to each pollutant. Track how the levels of pollution in this city have changed over the past 20 years. Based on your findings, it is likely that pollution in this city will get better or worse over the next decade? Summarize your findings and your conclusions in a one- to two-page report.

19. **Project: Alcohol Poisoning.** Research recent cases in which teenagers or young adults died from alcohol poisoning (that is, they drank so much alcohol so fast that their blood-alcohol concentration exceeded the fatal limit). What did the individuals drink, and how quickly? Do you think they were aware of the dangers from this drinking? Summarize your findings, and write a series of recommendations on how such deaths might be prevented in the future.

UNIT 12D

LOGARITHMIC SCALES: EARTHQUAKES, SOUNDS, AND ACIDS

You've probably heard the strength of an earthquake described by its *magnitude*, the loudness of a sound described in *decibels*, or the acidity of a household cleanser described by its *pH*. These three methods of measurement share something in common: All three are **logarithmic scales**, or scales that involve logarithms (see Unit 6C). Because logarithms are ways of expressing powers, logarithmic scales are useful whenever a quantity varies over a very wide range of numbers.

THE MAGNITUDE SCALE FOR EARTHQUAKES

Earthquakes are a fact of life for much of the world's population (Figure 12.3). In the United States, California and Alaska are most prone to earthquakes, although earthquakes can strike almost anywhere. Most earthquakes are so minor that they can hardly be felt, but severe earthquakes can kill tens of thousands of people. Table 12.5 lists the frequencies of earthquakes of various strength according to standard categories defined by geologists.

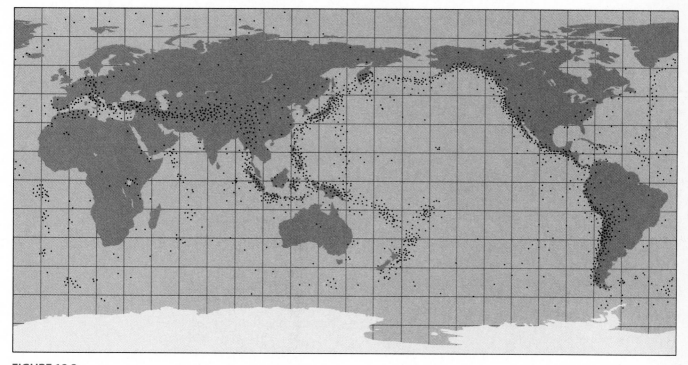

FIGURE 12.3

Dots represent earthquake locations over a twenty-year period.

Table 12.5	Earthquake Categories and Their Frequency	
Category	Magnitude	Approximate Number per Year (Worldwide Average Since 1900)
Great	8 and up	1
Major	7–8	18
Strong	6–7	120
Moderate	5–6	800
Light	4–5	6000
Minor	3–4	50,000
Very minor	less than 3	magnitude 2–3: 1000 per day
		magnitude 1–2: 8000 per day

Scientists describe the strength of an earthquake using the earthquake **magnitude scale**. The magnitude of an earthquake is related to the total energy released by the earthquake as shown on the following page.

By the Way ············

The original magnitude scale was created by Charles Richter in 1935. This *Richter scale* measured only the up and down motion of the ground during a quake. Magnitude 0 was defined as the smallest detectable quake, and each increase of 1 magnitude corresponded to a factor of 10 increase in ground motion.

The Earthquake Magnitude Scale

An earthquake of magnitude M releases an amount of energy E, measured in *joules*, that can be calculated by the following formula:

$$\log_{10} E = 4.4 + 1.5M \qquad \text{or, equivalently,} \qquad E = (2.5 \times 10^4) \times 10^{1.5M}.$$

This scale is *logarithmic* because the magnitude is related to the logarithm of the energy released.

Earthquakes of the same magnitude may cause vastly different amounts of damage depending on *how* their energy is released. The energy released from earthquakes is transmitted through the ground by *seismic waves*, some of which travel along the surface of the Earth and others which travel through its interior. In general, only the surface waves cause damage because they are the only ones that make the land move up and down. Thus a moderate earthquake that releases most of its energy in surface waves may do more damage than a strong earthquake that releases most of its energy in interior waves. The degree of damage is also influenced by the type of surface bedrock that transmits the seismic waves.

An earthquake itself rarely causes any human death. Instead, deaths occur when buildings collapse on people, when fires are ignited by broken gas lines, when landslides occur, or when tsunamis are caused by the earthquake. Table 12.6 lists some of the major earthquakes of the past 100 years. Note that the greatest numbers of deaths are not necessarily from the largest quakes, but rather from quakes in regions where people cannot afford the high cost of construction designed to withstand moderate earthquakes.

Table 12.6   Notable Earthquakes of the Past 100 Years

Date	Location	Magnitude	Estimated Deaths
April 1906	San Francisco	8.3	700
December 1908	Sicily	7.5	160,000
December 1920	Gansu, China	8.6	100,000
September 1923	Sagami Bay, Japan	8.3	100,000
February 1931	New Zealand	7.9	255
August 1950	Assam, India	8.4	30,000
February 1960	Morocco	5.8	12,000
December 1972	Nicaragua	6.2	10,000
July 1976	Tangshan, China	7.9	250,000–700,000
September 1985	Mexico City	8.1	7000
December 1988	Armenia	6.9	25,000–45,000
October 1989	San Francisco	7.1	90
June 1990	Iran	7.7	40,000
January 1994	Los Angeles	6.7	61
January 1995	Kobe, Japan	6.9	5500
May 1997	Iran	7.5	2000
May 1998	Afghanistan	6.9	4000–5000

TIME-OUT TO THINK: Surface waves from earthquakes make the ground roll up and down like ripples moving outward on a pond. Given this type of motion, suggest a few ways that buildings could be designed to withstand earthquakes. Do you think it is possible to make a building that could withstand *any* earthquake? Why or why not?

EXAMPLE 1   *The Meaning of 1 Magnitude*

How much energy is represented by an increase of 1 magnitude on the earthquake scale?

Solution: Because we are comparing *energies*, it's easiest to work with the version of the magnitude formula that gives the energy, E:

$$E = (2.5 \times 10^4) \times 10^{1.5M}$$

If two earthquakes have magnitudes M_1 and M_2, the ratio of their corresponding energies E_1 and E_2 is

$$\frac{E_1}{E_2} = \frac{(2.5 \times 10^4) \times 10^{1.5M_1}}{(2.5 \times 10^4) \times 10^{1.5M_2}} = \frac{10^{1.5M_1}}{10^{1.5M_2}} = 10^{1.5(M_1 - M_2)}.$$

A magnitude difference of 1 means that $M_1 - M_2 = 1$. Therefore the energy ratio is

$$\frac{E_1}{E_2} = 10^{1.5 \times 1} = 10^{1.5} = 32.$$

> **A Brief Review:**
> Recall the rule for dividing powers of 10:
> $$\frac{10^x}{10^y} = 10^{x-y}$$

Thus, each magnitude on the earthquake scale corresponds to an increase in total energy by a factor of about 32. For example, a magnitude 5 earthquake releases about 32 times as much energy as a magnitude 4 earthquake, and a magnitude 8.5 earthquake releases about $32 \times 32 = 1024$ times as much energy as a magnitude 6.5 earthquake.   ∎

EXAMPLE 2   *Comparing Disasters*

Compare the estimated energies released by the 1906 and 1989 San Francisco earthquakes.

Solution: We can use the energy ratio formula from Example 1 with the 1906 earthquake as Quake 1 and the 1989 earthquake as Quake 2. We find that

$$\frac{E_{1906}}{E_{1989}} = 10^{1.5(M_{1906} - M_{1989})} = 10^{1.5(8.3 - 7.1)} = 10^{1.8} = 63.$$

The 1906 quake released more than 60 times as much energy as the 1989 quake.   ∎

MEASURING SOUNDS IN DECIBELS

Another commonly used logarithmic scale measures the loudness of a sound. The **decibel scale** actually compares sounds to one another. It is defined so that a sound of 0 **decibels**, abbreviated **dB**, represents the softest sound audible to the human ear. The loudness of any other sound in decibels can then be found from the formula on the following page.

By the Way ··········
A decibel is 1/10 of a *bel*, a unit named for Alexander Graham Bell (1847–1922). Bell's mother was deaf, and his father was a pioneer in teaching speech to the deaf. Bell became a professor of vocal physiology at Boston University and married a deaf pupil. He patented his most famous invention, the telephone, in 1876.

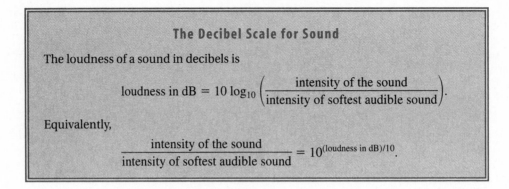

The Decibel Scale for Sound

The loudness of a sound in decibels is

$$\text{loudness in dB} = 10 \log_{10}\left(\frac{\text{intensity of the sound}}{\text{intensity of softest audible sound}}\right).$$

Equivalently,

$$\frac{\text{intensity of the sound}}{\text{intensity of softest audible sound}} = 10^{(\text{loudness in dB})/10}.$$

Table 12.7 lists the approximate loudness of some common sounds. In absolute terms, the intensity of the softest audible sound is about 10^{-12} watts per square meter.

Table 12.7	Typical Sounds in Decibels	
Decibels	**Times Louder Than Softest Audible Sound**	**Example**
140	10^{14}	jet at 30 meters
120	10^{12}	strong risk of damage for human ear
100	10^{10}	siren at 30 meters
90	10^{9}	threshold of pain for human ear
80	10^{8}	busy street traffic
60	10^{6}	ordinary conversation
40	10^{4}	background noise in average home
20	10^{2}	whisper
10	10^{1}	rustle of leaves
0	1	threshold of human hearing
-10	0.1	inaudible sound

EXAMPLE 3   *Computing Decibels*

Suppose that a sound is 100 times as intense as the softest audible sound. What is its loudness, in decibels?

Solution: We are looking for the loudness in decibels, so we use the first form of the decibel scale formula:

$$\text{loudness in dB} = 10 \log_{10}\left(\frac{\text{intensity of the sound}}{\text{intensity of the softest audible sound}}\right)$$

The ratio in parentheses is 100 because we are given that the sound is 100 times as intense as the softest audible sound. Thus we find that

$$\text{loudness in dB} = 10 \log_{10}(100) = 10 \times 2 = 20 \text{ dB}.$$

A Brief Review:

Recall that
$\log_{10}100 = 2$
because
$10^2 = 100$.

A sound that is 100 times more intense than the softest possible sound has a loudness of 20 dB. Table 12.7 shows that this is the sound of a whisper. ∎

Comparing Sounds

We can compare the loudness of any two sounds by working with the decibel scale formula. If we multiply both sides of the decibel scale equation (second form) by the *intensity of the softest audible sound*, we find that

$$\text{intensity of a sound} = 10^{(\text{sound in dB})/10} \times (\text{intensity of softest audible sound}).$$

Now we can compare the intensities of two sounds, Sound 1 and Sound 2, by dividing their intensities. We find that

$$\frac{\text{intensity of Sound 1}}{\text{intensity of Sound 2}} = \frac{10^{(\text{Sound 1 in dB})/10} \times (\text{intensity of softest audible sound})}{10^{(\text{Sound 2 in dB})/10} \times (\text{intensity of softest audible sound})},$$

or

$$\frac{\text{intensity of Sound 1}}{\text{intensity of Sound 2}} = \frac{10^{(\text{Sound 1 in dB})/10}}{10^{(\text{Sound 2 in dB})/10}}.$$

EXAMPLE 4   *Sound Comparison*

How much more intense is a 57-dB sound than a 23-dB sound?

Solution: We can use the above formula with the 57-dB sound as Sound 1 and the 23-dB sound as Sound 2. We find that

$$\frac{\text{intensity of 57-dB sound}}{\text{intensity of 23-dB sound}} = \frac{10^{57/10}}{10^{23/10}} = \frac{10^{5.7}}{10^{2.3}} = 10^{3.4} = 2512.$$

A sound of 57 dB is about 2500 times more intense than a sound of 23 dB. ∎

Sound and Distance: An Inverse Square Law

As sound travels through the air, it spreads out over larger and larger *areas*. Because area scales as the *square* of a scale factor (see Unit 10A), sound spreads out over an area related to the distance squared (Figure 12.4). As a result, the amount of sound energy striking your eardrums goes *down* by the square of the distance. For example, doubling your distance from a source of sound causes the intensity of the sound you hear to *decrease* by a factor of $2^2 = 4$; tripling the distance from the source reduces the intensity by a factor of $3^2 = 9$; and increasing the distance by a factor of 10 reduces the intensity by a factor of $10^2 = 100$.

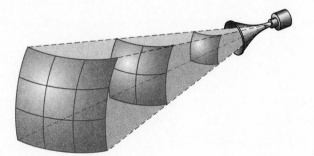

FIGURE 12.4

By the Way ··········
What we perceive as sound actually is tiny pressure changes in the air. Sound travels through the air as a wave of such pressure changes, called a *sound wave*. When a sound wave strikes the eardrum, its energy causes the eardrum to move in response. The brain analyzes the motions of the eardrum, perceiving sound.

In general, we can say that the intensity of a sound at Distance 1 is related to its intensity at Distance 2 by

$$\frac{\text{sound intensity at Distance 1}}{\text{sound intensity at Distance 2}} = \left(\frac{\text{Distance 2}}{\text{Distance 1}}\right)^2.$$

Because Distances 1 and 2 are inverted in the fraction on the right compared to the one on the left, and because the distances are squared, we say that this formula is an **inverse square law**. Many other quantities that vary with distance, such as the brightness of a light and the strength of gravity, also follow inverse square laws.

EXAMPLE 5   *Sound Advice*

How far should you be from a jet to avoid a strong risk of damage to your ear?

Solution: According to Table 12.7, the sound from a jet at a distance of 30 meters is 140 dB, while 120 dB is the level of sound that poses a strong risk of ear damage. Comparing these two sounds, we find that

$$\frac{\text{intensity of 140-dB sound}}{\text{intensity of 120-dB sound}} = \frac{10^{140/10}}{10^{120/10}} = \frac{10^{14}}{10^{12}} = 10^2 = 100.$$

That is, the sound of a jet at 30 meters is 100 times more intense than the sound needed to cause damage to your ear. Thus, to be safe, you should be far enough from the jet that its sound intensity is 100 times less than it is at 30 meters. Because sound intensity follows an inverse square law with distance, moving 10 times farther away makes the intensity $10^2 = 100$ times weaker. Thus you should be 10×30 m $= 300$ m from the jet to be safe.   ∎

THE pH SCALE FOR ACIDITY

By the Way ....
The word *acid* comes from the Latin *acidus,* meaning sour. Fruits taste sour because they are acidic. Bases are substances that can neutralize the action of acids. Common antacid tablets (for upset stomachs) are basic and work by neutralizing stomach acid.

If you check the labels of many household products, including cleansers, drain openers, and shampoo, you will see that they state a quantity called the **pH**. The pH is used by chemists to classify substances as **neutral**, **acidic**, or **basic** (also called *alkaline*). By definition:

- Pure water is *neutral* and has a pH of 7.
- *Acids* have a pH lower than 7.
- *Bases* have a pH higher than 7.

Table 12.8 gives a few typical pH values.

Table 12.8	Typical pH Values			
Solution	**pH**		**Solution**	**pH**
pure water	7		drinking water	6.5–8
stomach acid	2–3		baking soda	8.4
vinegar	3		household ammonia	10
lemon juice	2		drain opener	10–12

TIME-OUT TO THINK: Check around your house or apartment for labels that state a pH. Are the substances acids or bases?

Chemically, acidity is related to the concentration of positively charged hydrogen ions. The ions themselves are denoted H^+ for hydrogen with a positive charge. The concentration of the hydrogen ions is denoted $[H^+]$, and usually measured in units of moles per liter. A **mole** is *Avogadro's number* of particles, or about 6×10^{23} particles.

By the Way ············
Neutral hydrogen atoms consist of a single proton and a single electron. H^+ ions are positive because they are missing their electron, and hence are simply protons.

> ### The pH Scale
>
> The pH scale is defined by the equivalent formulas
> $$pH = -\log_{10}[H^+] \qquad \text{or} \qquad [H^+] = 10^{-pH},$$
> where $[H^+]$ is the hydrogen ion concentration in moles per liter.

EXAMPLE 6   *Finding pH*

What is the pH of a solution with a hydrogen ion concentration of 10^{-12} mole per liter?

Solution: Using the first version of the formula for the pH, we find that
$$pH = -\log_{10}[H^+] = -\log_{10}10^{-12} = -(-12) = 12.$$
A solution with a hydrogen ion concentration of 10^{-12} mole per liter has pH 12. Because this pH is well above 7, it is a strong base. ∎

A Brief Review:
Recall that
$$\log_{10}10^x = x.$$

Acid Rain

Normal raindrops are mildly acidic, with a pH slightly under 6. However, the burning of fossil fuels releases sulfur or nitrogen that can form sulfuric or nitric acids in the air. These acids can make raindrops far more acidic than normal, creating the problem known as **acid rain**. Acid rain in the northeastern United States and *acid fog* in the Los Angeles area have been observed with pH as low as 2—the same acidity as pure lemon juice!

Acid rain can kill trees and other plants, doing serious damage to forests. Many forests in the northeastern United States and southeastern Canada have been damaged by acid rain. Acid rain can also "kill" lakes by making the water so acidic that nothing can survive. Thousands of lakes in the northeastern United States and southeastern Canada are "dead." Surprisingly, you can often recognize a dead lake by its exceptionally clear water—living organisms tend to make the water cloudy.

By the Way ············
Acid rain in the northeastern United States and Canada is caused primarily by emissions from coal-burning power plants and industries; the problem is particularly bad when coal with a high sulfur content is burned. The acid fog of Los Angeles probably comes from automobile emissions.

EXAMPLE 7   *Acid Rain Versus Normal Rain*

How much more acidic is acid rain with a pH of 2 than ordinary rain with a pH of 6?

Solution: We can answer this question by comparing the hydrogen ion concentrations in the acid rain and the ordinary rain. For the acid rain with pH 2, the hydrogen ion concentration is

$$[H^+] = 10^{-pH} = 10^{-2}.$$

For the ordinary rain with pH 6, the hydrogen ion concentration is

$$[H^+] = 10^{-pH} = 10^{-6}.$$

Thus the hydrogen ion concentration in the acid rain is

$$\frac{10^{-2}}{10^{-6}} = 10^{-2-(-6)} = 10^4$$

times greater than that in the ordinary rain. That is, the acid rain is 10,000 times more acidic than the ordinary rain.   ■

REVIEW QUESTIONS

1. What is the *magnitude scale* for earthquakes? Why do we say that it is *logarithmic*?

2. How much energy is represented by an increase of 1 magnitude on the earthquake scale?

3. What is the *decibel scale*? Describe how it is defined.

4. How does sound intensity vary with distance? Why do we call it an *inverse square law*?

5. What is pH? What pH values define an acid, a base, or a neutral substance?

6. What is acid rain? Why is it a serious environmental problem?

PROBLEMS

Thinking in Logarithmic Scales. *For Problems 1 and 2, briefly describe, in words, the effects you would expect in each given situation.*

1. **a.** An earthquake of magnitude 2.8 strikes the Los Angeles area.

   b. You have your ear against a new speaker when it emits a sound with an intensity of 160 dB.

   c. A young child (too young to know better) finds and drinks from an open bottle of drain opener with pH 12.

2. **a.** An earthquake of magnitude 8.5 strikes the New York City area.

   b. Your friend is calling you from across the street in New York City, with a shout that registers 90 dB. Traffic is heavy, and several emergency vehicles are passing by with sirens.

   c. A jet plane flies overhead at an altitude of 1 kilometer.

   d. A forest situated a few hundred miles from a coal-burning industrial area is subjected regularly to acid rain, with pH 4, for many years.

Earthquake Magnitudes. *For Problems 3 and 4, use the formula for the magnitude scale for earthquakes and data from Table 12.6, as needed, to answer each of the following questions.*

3. **a.** How much energy, in joules, is released by an earthquake of magnitude 5?

   b. How much energy, in joules, was released by the 1985 earthquake in Mexico City?

   c. How much energy, in joules, was released by the 1960 earthquake in Morocco?

4. **a.** How much more energy is released by an earthquake of magnitude 8 than one of magnitude 6?

b. Compare the energy of a magnitude 7 earthquake to that released by a 1-megaton nuclear bomb (5×10^{15} joules).

c. What magnitude of earthquake would release an energy equivalent to that of a 1-megaton nuclear bomb? Which would be more destructive? Why?

5. LA and China Earthquakes. Compare the energies of the 1994 Los Angeles earthquake that killed 61 people and the 1976 Tangshan earthquake that killed on the order of half a million. Do you think that an earthquake of magnitude 7.9 could cause a similar number of deaths in Los Angeles? Why or why not?

6. The Decibel Scale. Answer each of the following questions. As necessary, refer to Table 12.7.

a. How much louder is the sound of busy street traffic than the softest audible sound?

b. What is the loudness, in decibels, of a sound 45 million times louder than the softest audible sound?

c. How much louder (more intense) is a 35-dB sound than a 10-dB sound?

d. How much louder (more intense) is an 85-dB sound than a 15-dB sound?

e. Suppose that a sound is 100 times louder (more intense) than a whisper. What is its loudness in decibels?

7. The Decibel Scale. Answer each of the following questions. As necessary, refer to Table 12.7.

a. How much louder is the sound of a siren at 30 meters than the softest audible sound?

b. What is the loudness, in decibels, of a sound 18 trillion times louder than the softest audible sound?

c. What is the loudness, in decibels, of a sound 1000 times softer (less intense) than the softest audible sound?

d. How much louder (more intense) is a 125-dB sound than a 95-dB sound?

e. Suppose that a sound is 2 million times louder (more intense) than the threshold of pain for the human ear. What is its intensity in decibels? How does this sound compare to that of a jet at 30 meters?

8. Sound Intensity. In absolute terms, the intensity of the softest audible sound is about 10^{-12} watts per square meter. Calculate the absolute intensity, in watts per square meter, for each of the following sounds.

a. A jet at 30 meters   **b.** A whisper

c. A sound of 62 decibels   **d.** A sound of 160 decibels

9. Sound and Distance. Use the formula that relates the intensity of a sound to distance to answer each of the following questions.

a. The decibel level for busy street traffic in Table 12.7 is based on the assumption that you stand very close to the noise source—say, 1 meter from the street. If your house is 100 meters from a busy street, how loud will the street noise be, in decibels?

b. At a distance of 10 meters from the speakers at a concert the sound level is 135 dB. How far away should you be sitting to avoid the risk of damage to your hearing?

c. Imagine that you are a spy at a restaurant. The conversation you want to hear is taking place in a booth across the room, about 8 meters away. The people speak in soft voices so that, to each other, they hear voices of about 20 dB (assume that they sit about 1 meter apart). How loud is the sound of their voices when it reaches your table? If you have a miniature amplifier in your ear and want to hear their voices at 60 dB, by what factor must their voices be amplified?

10. Variation in Sound with Distance. Suppose that a siren is placed 0.1 meter from your ear.

a. How many times louder will the sound you hear be than that of a siren at 30 meters (see Table 12.7)?

b. How loud will the siren next to your ear sound, in decibels?

c. How likely is it that this siren would cause damage to your eardrum? Explain.

11. The pH Scale. Consider the following questions about the pH scale.

a. If the pH of a solution increases by 1 (e.g., from 4 to 5 or from 7 to 8), how much of a change does that represent in the hydrogen ion concentration? Does the increase in pH make the solution more acidic or more basic?

b. What is the hydrogen ion concentration of a solution with pH 8.5?

c. What is the pH of a solution with a hydrogen ion concentration of 0.1 mole per liter? Is this solution an acid or a base?

12. **The pH Scale.** Consider the following questions about the pH scale.

  a. If the pH of a solution increases by 1.5 (e.g., from 4 to 5.5 or from 7 to 8.5), how much of a change does that represent in the hydrogen ion concentration?

  b. What is the hydrogen ion concentration of a solution with pH 3.5?

  c. What is the pH of a solution with a hydrogen ion concentration of 10^{-9} mole per liter? Is this solution an acid or a base?

13. **Toxic Dumping in Acidified Lakes.** Consider a situation in which acid rain has heavily polluted a lake to a level of pH 4. An unscrupulous chemical company dumps some acid into the lake illegally. Assume that the lake contains 100 million gallons of water and that the company dumps 100,000 gallons of acid with pH 2.

  a. What is the hydrogen ion concentration $[H^+]$ of the lake polluted by the acid rain alone?

  b. Suppose that the unpolluted lake, without acid rain, would have pH 7. If the lake were then polluted by the company acid alone (no acid rain), what hydrogen ion concentration $[H^+]$ would it have?

  c. What is the concentration $[H^+]$ after the company dumps the acid into the acid rain-polluted lake (pH 4)? What is the new pH of the lake?

  d. If the U.S. Environmental Protection Agency can test for changes in pH of only 0.1 or greater, could the company's pollution be detected?

14. **Project: Acid Rain.** Investigate the problem of acid rain in a region where it has been a particular problem, such as the northeastern United States, southeastern Canada, the Black Forest in Germany, eastern Europe, or China. Write a two- to three-page report on your findings. The report should include a description of the acidity of the rain, the source of the acidity, the damage being caused by the acid rain, and the status of efforts to alleviate this damage.

CHAPTER 12

SUMMARY

*I*n this final chapter, we studied a variety of practical mathematical applications. The most important lesson to take from this variety of topics is that you have the ability to understand all of them, and many more. We hope that, in the future, you will see mathematics as a tool you can use to understand *your* world, and to help you live a better life.

SELECTED ANSWERS TO ODD-NUMBERED PROBLEMS

CHAPTER 1

UNIT 1A

13. *False cause*: we cannot conclude that the Soviet Union broke up because of Reagan's actions based solely on the fact that his actions preceded the breakup.

15. *Circular reasoning*: to someone listening to the question, the conclusion is almost inevitable.

17. *Personal attack*: the mother's own smoking is irrelevant to the legitimacy of her advice.

19. *Appeal to ignorance*: an absence of proof for one conclusion (telepathy does not exist) does not prove the opposite conclusion (telepathy does exist).

21. *Personal attack*: the fact that the Senator gets campaign funds from the NRA does not necessarily mean that he or she will only introduce bills that are to the NRA's liking.

23. *False cause*: the mere fact that there are now more overweight people and fewer smokers does not prove that quitting smoking causes overeating.

UNIT 1B

1. This statement is a proposition.

3. This statement is a proposition.

5. This statement is not a proposition.

7. *The Gettysburg Address was not given by George Washington.* The original proposition is false, so its negation is true.

9. *Mark Twain did not write Tom Sawyer.* The original proposition is true, so its negation is false.

11. *New York is the capital of the United States.* The original proposition is true, so its negation is false.

13. *Some snakes are not mammals.* The original proposition is false, so its negation is true.

15. The Senate supports the bill.

17. The Dean supports affirmative action.

19. He believes anti-war demonstrations should be allowed.

21. True. **23.** False. **25.** Exclusive *or*.

27. Probably inclusive, but not clear.

29. Probably exclusive, but not clear.

31. True. **33.** False. **35.** True. **37.** True. **39.** True.

41. False.

43.

p	q	$\sim q$	p or $\sim q$
T	T	F	T
T	F	T	T
F	T	F	F
F	F	T	T

45.

p	q	$\sim p$	$\sim p$ or q
T	T	F	T
T	F	F	F
F	T	T	T
F	F	T	T

47.

p	q	r	p or q	(p or q) and r
T	T	T	T	T
T	T	F	T	F
T	F	T	T	T
T	F	F	T	F
F	T	T	T	T
F	T	F	T	F
F	F	T	F	F
F	F	F	F	F

49.

p	q	r	~r	p or q	(p or q) and ~r
T	T	T	F	T	F
T	T	F	T	T	T
T	F	T	F	T	F
T	F	F	T	T	T
F	T	T	F	T	F
F	T	F	T	T	T
F	F	T	F	F	F

51. a. 42   **c.** 18

53. False.   **55.** True.   **57.** True.   **59.** True.

61. True.

63.

p	q	r	q and r	p ⟹ (q and r)
T	T	T	T	T
T	T	F	F	F
T	F	T	F	F
T	F	F	F	F
F	T	T	T	T
F	T	F	F	T
F	F	T	F	T
F	F	F	F	T

65.

p	q	r	p and q	(p and q) ⟹ r
T	T	T	T	T
T	T	F	T	F
T	F	T	F	T
T	F	F	F	T
F	T	T	F	T
F	T	F	F	T
F	F	T	F	T
F	F	F	F	T

67. The converse is *if Marco lives in the United States, then he lives in Chicago.* The inverse is *if Marco does not live in Chicago, then he does not live in the United States.* The contrapositive is *if Marco does not live in the United States, then he does not live in Chicago.*

69. The converse is *if I got wet, then I went swimming.* The inverse is *if I do not go swimming, then I will not get wet.* The contrapositive is *if I do not get wet, then I did not go swimming.*

71. The converse is *if it's cold-blooded, then it's a reptile.* The inverse is *if it is not a reptile, then it is not cold-blooded.* The contrapositive is *if it is not cold-blooded, then it's not a reptile.*

73. *If a person is a member of Congress, then that person is a lawyer.* Antecedent: *a person is a member of Congress,* Consequent: *that person is a lawyer.* False.

75. *If a person is a musician, then that person can play the saxophone.* Antecedent: *a person is a musician,* Consequent: *that person can play the saxophone.* False.

77. (i) Already in standard form. (ii) S = police officers; P = women.
(iii)

Stated: some police officers here.

79. (i) No Republicans are socialists. (ii) S = Republicans; P = socialists.
(iii)

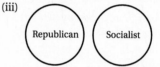

81. (i) No bachelors are married people. (ii) S = bachelors; P = married people.
(iii)

83. (i) Already in standard form. (ii) S = Ronald Reagan; P = great presidents.
(iii)

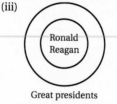

Great presidents

85. a. 28 people.
c. 13 men in the room.
e. 7 Republican men.
g. 6 Republican women.

87. a. 62;   **c.** 8;
e. 38 people.

• • • • • • • • •
UNIT 1C

1. (i) Premise: All islands are tropical lands.

Premise: All tropical lands are lands with jungles.

Conclusion: All islands are lands with jungles.

(ii) Valid.　　　　　　　　　　(iii) Not sound.

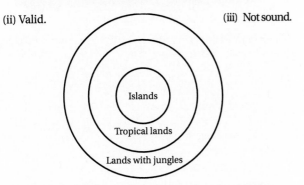

3. (i) Premise: All dairy products are products containing protein.

Premise: No soft drinks are products containing protein.

Conclusion: No soft drinks are dairy products.

(ii) Valid.

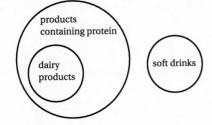

(iii) Sound.

5. (i) Premise: No women are NFL quarterbacks.

Premise: Some NFL quarterbacks are tall people.

Conclusion: Some tall people are not women.

(ii) Valid.

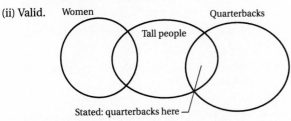

(iii) Not sound.

7. (i) Premise: Some lobbyists are people who work for the oil industry.

Premise: All lobbyists are persuasive people.

Conclusion: Some persuasive people are people who work for the oil industry.

(ii) Valid.

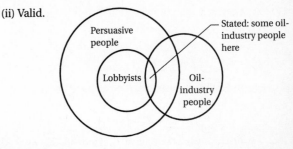

(iii) Sound.

9. (i) Premise: No uninsured person is a person who can get medical treatment.

Premise: Some people are uninsured people.

Conclusion: Some people are not people who can get medical treatment.

(ii) Valid.

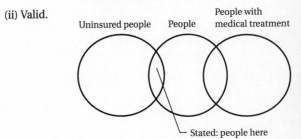

(iii) Not sound.

11. (i) Premise: States in the EST zone are states east of the Mississippi River.

Premise: Maine is in the EST zone.

Conclusion: Maine is a state east of the Mississippi River.

(ii) Valid.

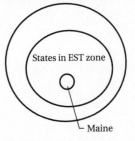

(iii) Not sound.

13. A sound argument must be valid.

15. Example of valid argument with two false premises and a true conclusion:

P: All mammals swim. (false)

P: All fish are mammals. (false)

C: All fish swim. (true)

17. (i) Already in standard form. (ii) Affirming the antecedent. (iii) Valid.

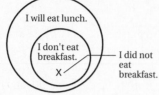

19. (i) Already in standard form. (ii) Denying the antecedent. (iii) Invalid.

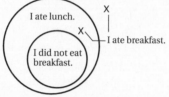

(iv) Denying the antecedent is an invalid argument.

21. (i) Premise: If interest rates decline, then the bond market improves.

Premise: The bond market improved.

Conclusion: Interest rates declined.

(ii) Affirming the consequent.

(iii) Invalid.

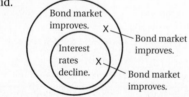

(iv) Affirming the consequent is an invalid argument.

23. (i) Premise: If a person is a nurse, then the person knows CPR.

Premise: Tom is a nurse.

Conclusion: Tom knows CPR.

(ii) Affirming the antecedent.

(iii) Valid.

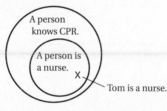

29. (i) Premise: If you shop, then I will make dinner.

Premise: If I make dinner, then you will take out the trash.

Conclusion: If you shop, then you will take out the trash.

(ii) Valid. (iii) Sound if the premises are true.

31. (i) Premise: If we live in the United States, then we have the right to say anything at any time.

Premise: If we have the right to say anything at any time, then we have the right to yell "fire!" in a theater.

Conclusion: If we live in the United States, then we have the right to yell "fire!" in a theater.

(ii) Valid. (iii) Not sound because the first premise is false.

33. (i) Premise: If taxes are cut, the U.S. government will have less revenue.

Premise: If the U.S. government has less revenue, then the deficit will be larger.

Conclusion: If taxes are cut, then the deficit will be larger.

(ii) Valid. (iii) Not sound because the deficit can be affected by spending as well as revenue.

35. (i) Premises are all true. (ii) The three test cases make a strong but not definitive argument. (iii) The conclusion is true.

37. (i) The premise is true (at least in the opinion of most musicians). (ii) The argument seems contrived to show only examples that begin with "B" and is therefore weak. (iii) The conclusion is false.

39. Many different number pairs can be substituted into the expression $a + b = b + a$ and in all cases the equation will be satisfied. This amounts to a strong inductive argument for the truth of this statement.

41. One set of three numbers that does not satisfy the equation suffices to prove that the statement is false in general. For example $2^2 + 3^2 \neq (2 + 3)^2$.

43. Inductive.   **45.** Inductive.

• • • • • • • • •
UNIT 1D

1. One possible way to analyze the argument:

P1: Asbestos is a known cancer-causing substance.

P2: This school contains asbestos.

A1: The school contains asbestos in amounts that are harmful.

A2: It is important to protect the occupants of the school.

A3: The only way to handle buildings with asbestos is to close them.

C: The school should be closed.

$$\frac{P1 + P2 + A1 + A2 + A3}{\downarrow}$$
$$C$$

With the assumed premises, the argument is deductively valid.

3. One possible way to analyze the argument:

A1: If people ride in limousines, then they must be rich.

P1: I saw Jenny in a limousine.

C: Jenny must be rich.

$$\frac{A1 + P1}{\downarrow}$$
$$C$$

With the assumed premise, the argument is deductively valid; without them it is unconvincing.

5. One possible way to analyze the argument:

P1: The U.S. provided weapons to the Afghan rebels.

A1: The Soviet Union lost the war because the Afghan rebels had weapons.

C: The Soviet Union lost the war because the U.S. provided the weapons to the rebels.

$$\frac{A1 + P1}{\downarrow}$$
$$C$$

With the assumed premise, the argument is deductively valid.

7. One possible way to analyze the argument:

P1: A criminal offense occurs every two seconds.

P2: Violent crimes occur every 16 seconds.

P3: Robberies occur every 48 seconds.

I1: There is a crime problem that must be combated.

A1: Increasing the conviction rate and strengthening the police force are the most effective ways to combat crime.

C: We should increase the conviction rate and strengthen the police force

With the intermediate conclusion and assumed premise, the argument is deductively valid. However, the assumed premise is not necessarily true, so the argument is unconvincing.

9. If you go on the trip Option 1 costs $250 less than Option 2. If you decide not to go on the trip, Option 1 costs $75 more than Option 2. Because $250 is about three times as much as $75, you might argue that if the likelihood of canceling is more than three times the likelihood of traveling, then Option 2 is the better choice. Otherwise, take Option 1.

11. The Saturday night stay saves $365 on airfare, but the hotel costs an extra $210 for the 2 nights (Saturday and Sunday) and meals cost an extra $110 for the 2 days, making the total cost quite similar for the two trip options.

13. The round-trip distance is 3000 miles, so it will take 9 trips to earn a free ticket on the frequent flier program. Thus if you make 10 trips, Airline A is cheaper.

CHAPTER 2

UNIT 2A

5. a. The population is all lawns, or perhaps all lawns in similar climate zones. The sample is the lawns of your 30 friends. The population parameter is the percentage of all lawns that do better with More-Grow.

   b. The raw data consist of the 30 responses: 18 for More-Grow and 12 for Go-Green.

   c. The sample statistic is the percentage of lawns that do better with More-Grow.

   d. Your estimate of the population parameter is 60%.

7. The population is all students who will vote in the election. The sample of students who will vote must be selected randomly, perhaps by a random drawing of student ID numbers, so that there is no bias for age, sex, major, residence (on-campus vs. off-campus).

9. The population is all people in this country. Blood types do not change from birth, so age is not a factor. An unbiased sample must be chosen randomly so it fairly represents the sex and racial background of people in the population.

11. The population is all lung cancer victims. The sample would consist of autopsy records of victims randomly selected to represent the ages, sex, living conditions, and life styles of all victims.

13. The population is all people who get colds. The random sample should include both people who do drink 3 cups of herbal tea per day and those who don't.

15. We can be 95% confident that between 59.4% to 66.6% of adults believe their children will have a higher standard of living.

17. We can be 95% confident that the actual percentage of unemployed people is between 5.2% and 6.0%.

19. Observational with case-control study.

21. Observational with no case-control groups.

23. Observational with no case-control groups.

25. Observational study with a survey. It could not use a placebo or be single- or double-blind.

UNIT 2B

1. Unbiased.

3. Unbiased.

5. Possible bias because of the company's financial interest in the outcome.

7. Possible bias because shoppers during selected hours may not be representative of all people.

15. a. 92%. **b.** 35%.
  c. The conclusion that two-thirds of all employers spy on employees does not follow directly, but could be true from the data.

UNIT 2C

1. a. About 6 times more Asians immigrated than Africans.
  b. Since 1 cm in height corresponds to 50,000 immigrants, the height of the bar for the former Soviet Union would be about 1.1 cm and the height of the bar for South America would be about 0.9 cm.

3.

Income by education

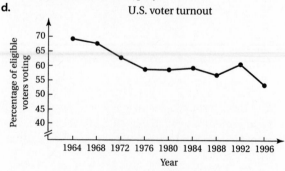

5. a. 8.2 (million) people.
  b. Spanish: 55.4%, French: 5.5%, German: 4.8%, Italian: 4.2%, Chinese: 3.8%, All Others: 26.3%. Spanish: 199°, French: 20°, German: 17°, Italian: 15°, Chinese: 14°, All Others: 95°.

c.

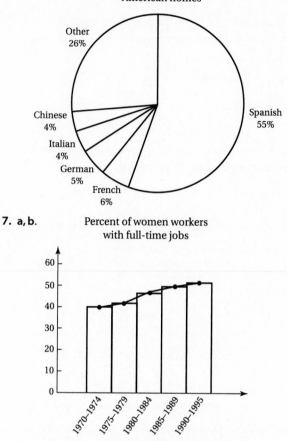

Non-English languages spoken in American homes

7. a, b.

Percent of women workers with full-time jobs

c. There is an upward trend in the data as the percentage of full-time women workers increases in time.

9. a. Voter turnout was lowest in 1996 (54.2% of eligible voters actually voted).
  b. Unemployment was highest in 1976 (7.7%).
  c. Visually there does not seem to be a correlation between voter turnout and unemployment.
  d.

U.S. voter turnout

U.S. unemployment

13. Age and height of adults are uncorrelated.

• • • • • • • • •
UNIT 2D

1. a. No; only the difference between grain production and grain consumption is shown on the graph.
   b. −6 million tons. The bar for this data point would extend *downward* about one-eighth of the way to the −50 mark.

3. a, b.

Year	Net Interest	Defense
1960	8%	52%
1970	7%	42%
1980	8%	22%
1990	15%	24%
1999	14%	15%

c. For 1988:

	Percent of Budget	Amount Spent ($ billions)
Individuals	50%	500
Defense	27%	270
Interest	14%	140
Other	9%	90

d. For 1995:

	Percent of Budget	Amount Spent ($ billions)
Individuals	60%	900
Defense	20%	300
Interest	12%	180
Other	8%	120

5. a. 9:30 P.M., 10:30 P.M., . . . ,4:30 A.M.
   c. Oneonta: At 9:30 P.M., approximately 10 birds; at 11:30 P.M., about 35 birds; at 1:30 A.M., about 60 birds.
   e. Oneonta and Jefferson had the most sightings.

7. a. Tuition did not decrease in either category.
   c. 1988.
   e. $750, tuition was about $10,750.

9. a.

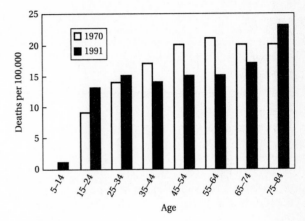

U.S. suicide rate (Deaths per 100,000)

• • • • • • • • •
UNIT 2E

13. Sufficient cause.   **15.** Probabilistic cause.
17. Probabilistic cause.   **19.** Sufficient cause.
21. Probabilistic cause.   **23.** Probabilistic cause.
25. Probabilistic cause.

CHAPTER 3

• • • • • • • • •
UNIT 3A

3. a. There are 1000 cm^3/liter.   **b.** One atmosphere is 14.7 pounds/$in.^2$.   **c.** The Earth's average density is 5.5 g/cm^3.
   d. The rocket accelerated at 7 m/s^2.
   e. The gold content of the ore is 5 oz/ft^3.

5. a. 3,500,000 mi^2; 97,574,400,000,000 ft^2; 2,240,000,000 acres.
   b. 120,000 mi^2; 3,345,408,000,000 ft^2; 76,800,000 acres.
   c. 3.4%

7. a. 150 m^3.   **b.** 3960 $in.^3$.   **c.** 25,000,000 ft^3.

9. a. $1.50.   **b.** 100,000 cm/km.   **c.** 1760 yd/mi.
   d. 1.26 carats.

11. Sorry! You are wrong because your answer has units of lb^2/$. The correct answer is $7.70/lb × 0.11 lb = $0.85.

13. Sorry! The units of your answer are lb/$ not $/lb. You should also be suspicious of the large price difference in your answers. The correct answer is $11/50 lb = $0.22/lb. The price per pound of the 50-pound bag is less than the price per pound of the 1-pound bag.

15. a. 128 ounces per gallon.   **b.** 525,600 minutes in a year.
    c. 290,400 ft/hr or 80.7 ft/s.

17. a. 1 ft^2 = 144 in.^{2}.   **b.** 5500 yd^2 or 49,500 ft^2.
    c. 135 ft^3.

19. $1349.50.

21. a. 1 lire is less that $1.   **b.** $12.24.   **c.** $8.40.
    d. 31,521 lire.   **e.** 686 francs.

23. a. 105 dog years.
    b. 2/7 to 3/7 real years or about 15 weeks to 22 weeks.

25. Approximately 250,000 pages.

27. a. 15,603,840,000 ft^3.   **b.** 1.3%.

UNIT 3B

3. a. Example: 5'10" is the same as 70 inches.
    b. 81 inches or 2.25 yd.   **c.** 1.275 miles, or about 1/5 of a 6.2-mile road race and 1/20 of a 26-mile marathon.
    d. 8 pounds.   **e.** 0.352 oz av.   **f.** 4 liquid pints, 3.44 dry pints.   **g.** 10,752,000 in.^{3}, 322,560 million in.^{3}, 187 million ft^3.

5. a. Smaller by a factor of 1000.   **b.** Smaller by a factor of 1000.   **c.** Smaller by a factor of 100.   **d.** Larger by a factor of 1 billion.

7. a. 32,808 feet.   **b.** 0.8 kilometers or 800 meters.
    c. 75 liters.   **d.** 0.30 cubic inches.   **e.** 68 kilograms.

11. a. −223°C.   **c.** 499,727°C.   **e.** 373 K.

13. a. 2.75 miles, 4418 meters, 4.418 km.   **c.** 31,996 ft, 6.06 mi, 9752.4 m, 9.7524 km.

15. a. $3.18/gallon.   **c.** $0.65/lb.

17. a. 18 karats.   **b.** 8.75 grams.

19. Cullinan diamond: 621,200 milligrams, 621.2 grams, 1.37 lb.

UNIT 3C

1. a. There are 6 solutions (T = 3 axles, C = 2 axles). T = 11 and C = 1; T = 9 and C = 4; T = 7 and C = 7; T = 5 and C = 10; T = 3 and C = 13; T = 1 and C = 16.
    c. There are 5 solutions (B = 4 axles, T = 3 axles, C = 2 axles). B =0, T = 0, and C = 5; B =1, T = 0, and C = 3; B =2, T = 0, and C = 1; B =0, T = 2, and C = 2; B =1, T = 2, and C = 0.

3. 100 km.

5. 63 cm.

9. Hint: What if Reuben were born on December 31?

11. That man is my son.   **13.** She gains $200.

15. Hint: Select one ball from the first barrel, two balls from the second barrel, three balls from the third, and so forth.

21. Alma visited at 8:00 P.M., Bess visited at 9:00 A.M., Cleo visited at 10:00 P.M., and Dina visited at 11:00 A.M.

CHAPTER 4

UNIT 4A

1. a. Cardinal.   **b.** Ordinal.   **c.** Cardinal.
    d. Nominal.

3. a. Rational number.   **b.** Rational number.
    c. Natural number.   **d.** Integer.   **e.** Rational number.   **f.** Rational number.   **g.** Rational number.
    h. Irrational number.

5. a. Negative integer.   **b.** Negative rational number.
    c. Positive irrational number.
    d. In general, an irrational number.

7. a. 0.75. The decimal terminates.
    b. 0.363636 . . . The repeating pattern is 36.
    c. 0.625. The decimal terminates.
    d. 0.16666 . . . The repeating pattern is 6.
    e. 0.571428571428 . . . The repeating pattern is 571428.
    f. 0.55. The decimal terminates.
    g. 0.616666 . . . The repeating pattern is 6.
    h. 0.621621621 . . . The repeating pattern is 621.

9. Natural (counting) number; exact.

11. Rational number; approximate.

13. Rational number; exact.   **15.** 10/x, where $x \neq 0$.

17. a − 12, where a is any number.

19. The pattern is *adding* a set of consecutive integers. The pattern stops after six terms. The full statement is 1 + 2 + 3 + 4 + 5 + 6.

21. The pattern is adding numbers that increase by a factor of 10 each time. The pattern ends and the entire statement reads: 1 + 10 + 100 + 1,000 + 10,000 + 100,000 + 1,000,000.

23. The pattern is adding successive powers of 2. The pattern ends and the entire statement reads: $2^0 + 2^1 + 2^2 + 2^3 + 2^4 + 2^5 + 2^6 + 2^7 + 2^8 + 2^9 + 2^{10}$.

25. The pattern is listing the powers of two. The pattern ends and the entire statement reads: 2, 4, 8, 16, 32, 64, 128, 256, 512, 1024.

27. The pattern is adding numbers that increase as the square of successive integers. The pattern continues forever and with three more terms reads: 1 + 5 + 14 + 30 + 55 + 91 + 140 + 204, ...

29. a. 53.   **b.** 66.   **c.** 121.   **d.** 181.   **e.** 1443.
f. 44.   **g.** IV.   **h.** XXXVII.   **i.** XLI.   **j.** XLIX.
k. CVI.   **l.** CCCXXXIV.

31. a. 1776.   **b.** 1379.

●●●●●●●●●
UNIT 4B

1. a. 22.1%.   **b.** 86.7%.   **c.** 41.0%.

3. a. 766.67.   **b.** 666.67.   **c.** 168.54.

5. a. Absolute change = 5; relative change = 10%.
b. Absolute change = 0.4; relative change = 400%.
c. Absolute change = 1; relative change = 0.1%.

7. a. Absolute change = 45 million;
relative change = 20.45%.
b. Absolute change = −110,000;
relative change = −44%.
c. Absolute change = −$12 billion;
relative change = −4.5%.

9. a. Relative change = 85%.   **b.** Relative change = 205%.
c. Relative change = 65%.   **d.** Relative change = 15%.

11. a. Absolute difference = $0.66; 1 English pound is 66%
larger than $1.
b. Absolute difference = −0.37 mi; 1 kilometer is 37%
smaller than 1 mile.
c. Absolute difference = −13.5 million; New York is 42.7%
less populous than California.

13. a. Brian's income is 4 times larger than Wilson's income.
b. Kathy earns 20% less than Martha.
c. Retail price is 1.4 times the wholesale price.
d. Regular price = $46.29.

15. If A is p% larger than B, then B cannot be p% smaller
than A.

17. $1000.69.

19. The price is *lower* in Longmont because $5 is a *larger*
percentage of the purchase price in Longmont.

21. The company's profits have improved over two years, but
only by 8.75%.

23. a. $1000.   **b.** $250.   **c.** −75%.

25. a. False, 40% × 20% = 8%, not 80%.   **b.** False.
c. True.

27. The fallacy arises because the two groups, Union members
and Democrats, could overlap.

29. a. 18.33%.   **b.** 12.62%.

31. a. $75 billion.   **b.** $417.   **c.** 25,000.   **d.** 16,733.
e. 784,906.

●●●●●●●●●
UNIT 4C

3. a. $5 \times 10^6 = 5,000,000 = 5$ million.
b. $7 \times 10^9 = 7,000,000,000 = 7$ billion.
c. $-2 \times 10^{-2} = -0.02 = -2$ hundredths.
d. $8 \times 10^{11} = 800,000,000,000 = 800$ billion.
e. $1 \times 10^{-7} = 0.0000001 = 1$ ten millionth.
f. $9 \times 10^{-4} = 0.0009 = 9$ ten thousandths.

5. a. 6×10^2.   **b.** 9×10^{-1}.   **c.** 5×10^4.
d. 3×10^{-3}.   **e.** 5×10^{-4}.   **f.** 7×10^{10}.

7. a. 1×10^6.   **b.** 1.5×10^5.   **c.** $4.5 \times 10^0 = 4.5$.
d. 1.8×10^{-1}.   **e.** 5.40×10^8.   **f.** 5.30×10^{25}.

9. a. $2.2 \times 10^{-4} = 0.00022$.
b. $2 \times 10^{-1} = 0.2$.
c. $9.828 \times 10^7 = 98,280,000$.
d. $6.667 \times 10^1 = 66.67$.
e. $3.5 \times 10^4 = 35,000$.
f. $1.501 \times 10^{-10} = 0.0000000001501$.

11. a. $\$5.3 \times 10^{12}$.   **b.** $\$6.32 \times 10^{11}$.   **c.** $\$8 \times 10^{11}$.
d. 5.4×10^9 bytes.

13. a. To a high degree of accuracy, $10^{26} + 10^7$ is equal to 10^{26}.
b. To a high degree of accuracy, $10^{81} - 10^{62}$ is equal to 10^{81}.

15. a. 5×10^9.   **b.** 8.1×10^{30}.   **c.** 7.5×10^{-8}.
d. 2.7×10^{21}.   **e.** $1.5 \times 10^1 = 15$.   **f.** -1.29×10^5.
g. 3.0×10^{-7}.   **h.** 1.96×10^{-2}.

17. a. Approximately 3×10^7; exact value $= 2.8 \times 10^7$ which is
6.7% below the estimate.
b. Approximately $2 \times 10^4 = 20,000$; exact value = 15,384.
c. Approximately 5×10^8; exact value $= 4.95 \times 10^8$ which is
1% below the estimate.
d. Approximately $10^5 = 100,000$; exact value $= 1.2 \times 10^5$.
e. Approximately 2.1×10^6; exact value $= 2.04 \times 10^6$ which
is within 3% of the estimate.
f. Approximately 0.05; exact value = 0.056 which is within
about 12% of the estimate.

19. a. 5 billion is 20 times larger than 250 million.
b. 9.3×10^2 is 30,000 times larger than 3.1×10^{-2}.
c. 10^{-8} is 50,000 times larger than 2×10^{-13}.
d. 3.5×10^{-2} is 500,000 times larger than 7×10^{-8}.
e. 1000 is 1 million times larger than 0.001.
f. 10^{12} is 10^{21} times larger than 10^{-9}.

23. a. A 10-story building is between 100 and 120 feet high. A
football field is 300 feet long—almost three times longer
than the building is high.
b. Assuming the United States is 3200 miles long, the
walker can cover 2 miles per hour and can walk 8 hours
per day, the trip will take 1600 hours = 200 days.

25. a. 1 to 100,000.   **b.** 1 to 15,840.   **c.** 1 to 2,000,000.
d. 1 to 328.

27. 10^{24} atoms, or 100 times more than the number of stars in the observable universe.

29. One billion years is 6.7 m along the timeline. Human history is 0.000067 m = 0.067 mm along the timeline.

31. 160,000 bills per second for $1 bills; 1600 bills per second for $100 bills.

33. **a.** About 50,000 km². **b.** No. **35.** $25 million per year.

37. **a.** 1.3×10^3 gallons per person per day.
   b. 130 gallons of water per day.

· · · · · · · · ·
UNIT 4D

7. **a.** 2365.985, 2366.0, 2370, and 2400.
   b. 322354.090, 322354.1, 322350, 322,400.
   c. 6000.000, 6000.0, 6000, 6000.
   d. 11.333, 11.3, 10, 0.
   e. 578.555, 578.6, 580, 600.
   f. 0.452, 0.5, 0, 0.
   g. −12.100, −12.1, −10, 0.
   h. −850.765, −850.8, −850, −900.
   i. −10,995.624, −10,995.6, −11,000, −11,000.

9. The first scale has better accuracy. The second scale has better precision.

11. **a.** Three significant digits; precise to the nearest person.
   b. Four significant digits; precise to the nearest tenth (0.1) of a liter.
   c. Six significant digits. **d.** Three significant digits.

13. **a.** Four significant digits. **b.** Three significant digits.
   c. Two significant digits; precise to the nearest 0.00001 mm.
   d. One significant digit; precise to the nearest 0.000001 m.

15. **a.** 5×10^5. **b.** 5.0×10^5. **c.** 5.0000×10^5.
   d. 5.0000000×10^5.

17. **a.** Implied precision: nearest ten thousandth of an inch.
   b. Implied precision: every vote counted.
   c. Prices of textbooks are between $70 and $170.
   d. Implied precision: nearest 10 million people.
   e. Implied precision: nearest year.

19. **a.** 44 cm. **b.** 280 kg. **c.** 140 liters. **d.** 1 hour, 15 minutes. **e.** 415.4×10^8 km. **f.** 69 kg.

21. **a.** 2700 m². **b.** 2.0 km/min. **c.** 1600 kg².
   d. 4 g/cm³.

CHAPTER 5

· · · · · · · · ·
UNIT 5A

1.

	You		Your friend	
Year	Interest	Balance	Interest	Balance
0	—	$500	—	$500
1	$25	$525	$25	$525
2	$25	$550	$26.25	$551.25

1. (cont'd)

	You		Your friend	
Year	Interest	Balance	Interest	Balance
3	$25	$575	$27.56	$578.81
4	$25	$600	$28.94	$607.75
5	$25	$625	$30.39	$638.14

3. **a.** $2687.83. **b.** $26,532.98. **c.** $162,822.98.

5. **a.** $1731.08. **b.** $2323.65. **c.** $21,248.27.
   d. $13,623.37.

7. $533.58; APY = 6.7%.

9. $859.60; APY = 7.5%.

11. **a.** APY = 6.77%. **b.** APY = 8.80%. **c.** APY = 6.82%.

13. Over the 10 years, account 1 increases in value by $708 or 70.8%. Account 2 increases in value by $733 or 73.3%.

15. **a.** $1221.40, $2225.54. **b.** $2568.05, $5436.56.
   c. $13,498.59, $33, 201.17.

17. **a.** APY = 4.08%. **b.** APY = 5.13%. **c.** APY = 6.18%.

19.

n	1	4	12	365	500	1000
APY	12.00	12.55	12.68	12.75	12.75	12.75

   b. APY = 12.75%. **e.** $563.75, $911.06.

21. Brian: $2003.72, $3935.36. Celeste: $1843.14, $4205.83.

23. **a.** $4224. **b.** $4106. **c.** $4079. **d.** $4066.

25. For Plan A $13,597. For Plan B $15,526.

29. **a.** $Y = 14.3$ years. **c.** $Y = 68.1$ years.

· · · · · · · · ·
UNIT 5B

1. Accumulated balance = $174,550.39; total amount deposited = $24,000.

3. Accumulated balance = $86,144.21; total amount deposited = $43,200.

5. Yolanda: accumulated balance = $15,528.23; total amount deposited = $12,000. Zach: accumulated balance = $15,093.47; total amount deposited = $12,000.

7. Juan: accumulated balance = $32,775.87; total payments = $24,000. Maria: accumulated balance = $33,736.06; total payments = $25,000.

9. $329.96. **11.** $256.13.

13. 19.52 years; about 58 years. **15.** No; balance = $23,772.17.

17. Monthly payments of $419.36. **19. a.** $24,435.28.

21. **a.** $2543.20. **b.** $12,151.65.

· · · · · · · · ·
UNIT 5C

1. **a.** $40,000. **b.** 7%. **c.** $310. **d.** 20 years.
   e. 240. **f.** $74,400. **g.** $34,400 or 46% will be interest.

3. a. Monthly payments = $241.26.
   b. Monthly payments = $1261.28.
   c. Monthly payments = $1449.17.

5. Monthly payments = $166.07; total payments = $5978.52; 16% of the payments are interest.

7. Monthly payments = $477.83; total payments = $86,009.40; 42% of the payments are interest.

9. a. Monthly payments = $224.93.
   b. Monthly payments = $316.69.
   c. With a 20-year term, total payments = $53,985.60. With a 10-year term, total payments = $38,002.80.

11. a. Monthly payment for Loan 1 = $95.57; monthly payment for Loan 2 = $130.17; monthly payment for Loan 3 = $158.34.
   b. Total payments = $67,444.20.
   c. Monthly payment for the consolidated loan = $325.43.

13. a. Month 1 balance = $1093.00.
   Month 2 balance = $984.40.
   Month 3 balance = $874.17.
   Month 4 balance = $762.28.
   Month 5 balance = $648.71.
   Month 6 balance = $533.44.
   b. The loan will be paid off in the tenth month. The balances in months 7 through 10 are: $416.45, $297.69, $177.16, $54.81.

15. After 8 months the balance is $126.09.

17. Monthly payment for Loan 1 = $308.77; total payments = $11,115.72.
Monthly payment for Loan 2 = $241.79; total payments = $11,605.92.
Monthly payment for Loan 3 = $202.76; total payments = $12,165.60.
Loan 3 is the only option.

19. Monthly payment for Loan 1 = $1061.89; total payments = $191,140.20.
Monthly payment for Loan 2 = $930.36; total payments = $223,286.40.
Monthly payment for Loan 3 = $810.49; total payments = $291,776.40.

21. Loan 1 monthly payments = $587.01; total payout over 30 years is $211,323.60. Loan 2 monthly payments = $559.37; total payout over 30 years is $201,373.20.

23. Loan 1 monthly payments = $545.74; total payout over 30 years is $198,466.40. Loan 2 monthly payments = $696.89; total payout over 30 years is $129,040.20.

25. With monthly payments of $500, you could afford a loan principal of $62,140.93. With a 20% down payment, you can afford a house that costs $77,676.16.

• • • • • • • • •
UNIT 5D

1. Taxable income $26,250.   **3.** Taxable income $24,500.

5. $500.   **7.** $0.   **9.** $280.   **11.** $6708.

13. $19,861.   **15.** $20,326.   **17.** $23,360.

19. a. $13,040.   **b.** $11,594.

21. FICA tax = $2142; federal tax = $2805; overall tax rate = 17.7%.

23. FICA tax = $3427; federal tax = $7225; overall tax rate = 23.1%.

25. FICA tax = $7374; federal tax = $7355; overall tax rate = 30.6%.

27. Pierre's federal tax = $36,686; overall tax rate = 30.6%. Katherine's federal tax = $21,937; overall tax rate = 18.3%.

29. Fred's federal tax = $99,424; overall tax rate = 36.1%. Tamara's federal tax = $52,937; overall tax rate = 19.2%.

31. Each $600 contribution to a tax-deferred savings plan will reduce your take-home pay by $432.

33. Each $800 contribution to a tax-deferred savings plan will reduce your take-home pay by $512.

35. Savings = $1526 per year, $127 per month.

37. a. $60.   **b.** $33.
   c. Effective interest on the home equity loan = $24.

• • • • • • • • •
UNIT 5E

1. a. Not liquid, very safe investment, annual return 6%.
   b. Fairly liquid, relatively low risk, moderate return.
   c. Less liquid than large company stock, risky, return unpredictable.
   d. Poor liquidity, risky, high return.

3. a. Total return = 87.3%; annual return = 13.4%.
   b. Total return = 56.3%; annual return = 2.3%.
   c. Total return = 105.4%; annual return = 7.5%.
   d. Total return = −44.4%; annual return = −17.8%.

5. Small Stocks $1,903,911, Large Stocks $542,412, Bonds $24,223, Treasury Bills $6360.

7. a. MGX.   **b.** $8.63 and $8.38.   **c.** Stock closed at $8.51 two days ago.   **d.** 10,000 shares.
   e. No dividends.   **f.** Earnings per share = $0.42.

13. Lost $1.14 per share including commission.

15. Lost $37.87 per share including commission.

17. a. 6.3%.   **b.** 6.7%.   **c.** 8.9%.

19. a. $892.50.   **b.** $637.00.   **c.** $717.50.   **21.** $540.

CHAPTER 6

●●●●●●●●●
UNIT 6A

3. a. As the weight of the bag of apples increases, the price of the apples also increases.
   b. The price of movies increases as time goes on.
   c. As the price of a product increases, the demand generally decreases.
   d. The strength of the Earth's gravitational force decreases as the distance from the Earth increases.

5.

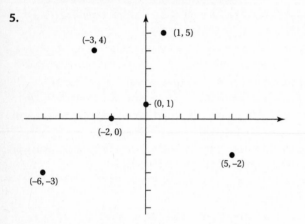

7. a. The independent variable is time, measured in years, and the dependent variable is world population.
   b. Because of the data values given, we can take the domain to be the years between 1950 and 1990. The range is all populations between about 2.5 billion and 6 billion.
   c. The function shows a steadily increasing world population between 1950 and 1990 (and beyond).

9. a. The variables are (*time, temperature*) or (*date, temperature*).
   b. The domain is all days over the course of a year. The range is temperatures between 38° and 85° (although for graphing purposes the range could be made larger).
   c.

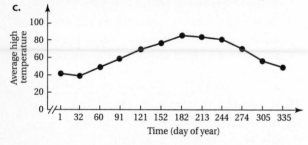

   d. It makes sense to fill in the graph between the data points.

11. a. The variables are (*time, projected population*).
   b. The domain is all years between 1995 and 2030. The range is population values between 258.3 million and 302.0 million.
   c.

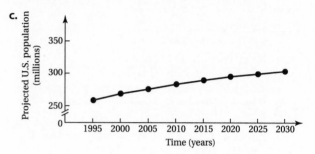

   d. It make sense to fill in the graph between the data points.

13. a. (8000 ft, 23 in.), (17,000 ft, 16 in.), and (25,000 ft, 12 in.).
   b. 5000 ft, 12,000 ft, and 25,000 ft.
   c. Pressure reaches 5 in. Mercury at about 50,000 ft.

●●●●●●●●●
UNIT 6B

1. a. 0.75 inches per hour.   **b.** 1.25 inches per hour.

3. a. Slope $= -62.5$ mi/hr.
   b. Slope $= -27.8$ units/dollar.
   c. Slope $= 0.56°C/°F = 5/9°C$.
   d. Slope $= 0.25$ (mi/gal)/(mi/hr).

5. Rate of change is $= -0.25$ in./hr. In 6.5 hours, the water depth changes by -1.625 in. In 12.5 hours, the water depth changes by -3.125 in.

7. Rate of change $= 0.2$ in./yr. In 4.5 years, the tree will increase in diameter by 0.9 inches. In 20.5 years, the tree will increase in diameter by 4.1 inches.

9. Rate of change $= -2$ cm/hr. In 3.8 hours the length of the candle changes by -7.6 cm. In 4.2 hours the length of the candle changes by -8.4 cm.

11. Rate of change $= 4$ in./hr. In 5.5 hours the snow depth will increase by 22 in. In 7.8 hours the snow depth will increase by 31.2 in.

13. Rate of change $= -2°F/1000$ ft $= -0.002°F/$ft. At 6000 ft above sea level, the change in boiling point is $-12°F$; boiling point $= 200°F$. At 12,000 ft above sea level, the change in boiling point is $-24°F$; boiling point $= 188°F$.

15. a. Initial value $= 120$ chips; rate of change $= 14$ chips per hour.

b.

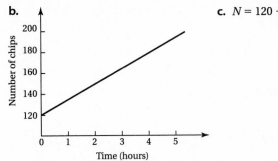

c. $N = 120 + 14t$.

17. a. (*time, diameter*).
    b. Rate of change $= 0.2$ in./yr; initial value $= (0$ yr, 4 in.$)$.

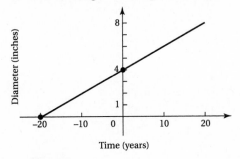

    c. 20 years.
    d. Not too accurate.

19. a. (*time, amount of sugar*).
    b. Rate of change $= -0.1$ grams/day;
        initial value $= (0$ days, 5 grams$)$.

    c. 50 days.

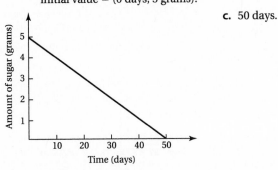

21. a. (*load, maximum speed*).

    c. 100 tons.

    b.

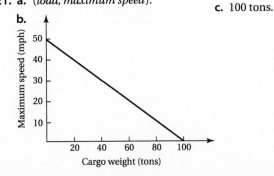

23. a. (*time, price*).   **b.** $p = 25 - 5t$.   **c.** $7.50.
    d. Probably gives a good estimate of prices of memory over
        a short period of time.

25. a. (*time, candle length*).   **b.** $L = 20 - 2t$.
    c. 10 hours.

27. a. (*miles, rental cost*).   **b.** $r = 40 + 0.10m$.
    c. 500 miles.

29. a. (*time, population*).   **b.** $p = 2000 + 200t$.
    c. 8000 people.

31.

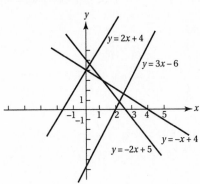

33. 3 years.

35. a. $y = 3x + 1$.

b. $y = 4x + 7$.

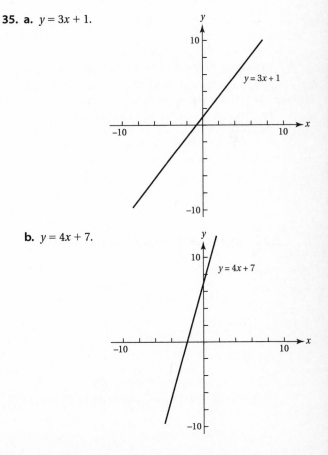

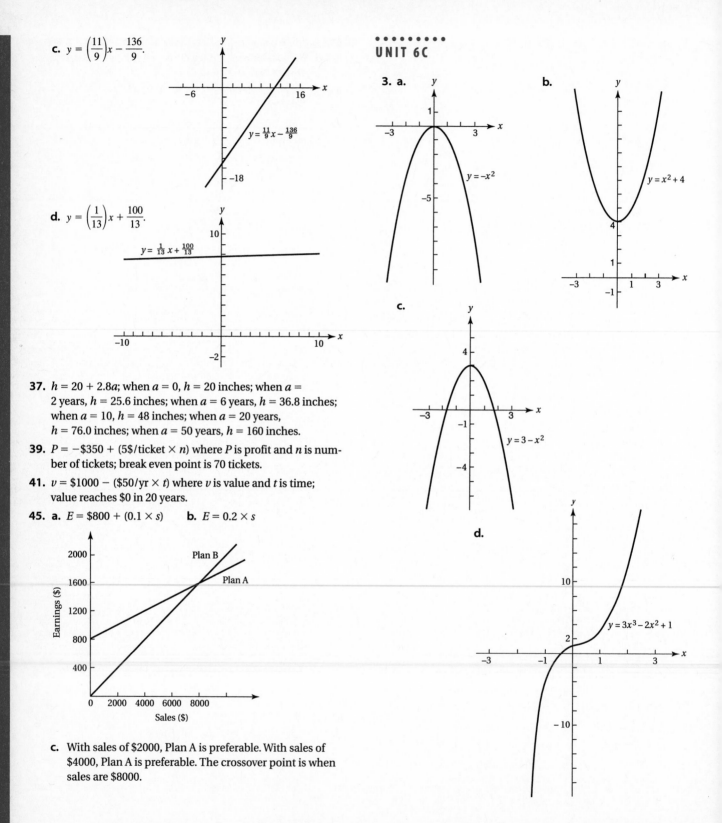

c. $y = \left(\dfrac{11}{9}\right)x - \dfrac{136}{9}$.

$y = \frac{11}{9}x - \frac{136}{9}$

d. $y = \left(\dfrac{1}{13}\right)x + \dfrac{100}{13}$.

$y = \frac{1}{13}x + \frac{100}{13}$

37. $h = 20 + 2.8a$; when $a = 0$, $h = 20$ inches; when $a = 2$ years, $h = 25.6$ inches; when $a = 6$ years, $h = 36.8$ inches; when $a = 10$, $h = 48$ inches; when $a = 20$ years, $h = 76.0$ inches; when $a = 50$ years, $h = 160$ inches.

39. $P = -\$350 + (5\$/\text{ticket} \times n)$ where P is profit and n is number of tickets; break even point is 70 tickets.

41. $v = \$1000 - (\$50/\text{yr} \times t)$ where v is value and t is time; value reaches $0 in 20 years.

45. a. $E = \$800 + (0.1 \times s)$ **b.** $E = 0.2 \times s$

Plan B
Plan A

c. With sales of $2000, Plan A is preferable. With sales of $4000, Plan A is preferable. The crossover point is when sales are $8000.

UNIT 6C

3. a. $y = -x^2$

b. $y = x^2 + 4$

c. $y = 3 - x^2$

d. $y = 3x^3 - 2x^2 + 1$

e.

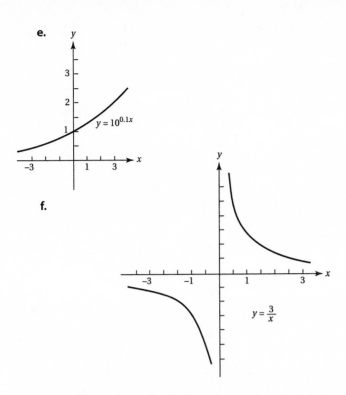

$y = 10^{0.1x}$

f.

$y = \dfrac{3}{x}$

5. a. Time for the 400-mile trip decreases as speed for the trip increases.

b. $t = 400/v$.

c.

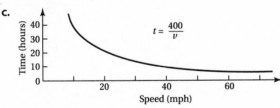

$t = \dfrac{400}{v}$

7. a. $\lambda = c/v$. **b.** 2.8 meters. **c.** 429 meters.

9. a. 0.015 m^3. **b.** 7.5 m. **c.** 0.86 cm.

d. 0.18 m. **e.** 96.84 cm.

11. a. If $p = 0$, then the revenue is $R = 0$.

b. If $p = 8$, then $R = 0$.

c. If $p = 3$, then $R = 75$. If $p = 5$, then $R = 75$.

d. If $p = 4$, then $R = 80$, which is the maximum revenue possible with this model.

13. 78% over 5 years. **15. a.** $N = 2300$. **b.** $t = 12.56$ days.

17. $Y = 4.5$ years. **19.** Doubling time = 8.67 years.

21. $Y = 5.8$ years.

23. a. Monthly payments = \$224.93; total amount paid = \$53,983.

b. $Y = 8.56$ years; total amount paid = \$36,120.

25. a. Monthly payment = \$804.62; total of \$9655 per year.

b. Biweekly payment = \$402.31; total of \$10,460 per year.

c. $Y = 21.9$ years.

d. With monthly payments for 30 years, you pay a total of \$289,440. With biweekly payments for 22 years, you pay a total of \$229,944.

27. $\dfrac{1}{6^3} = 0.0046$; $\dfrac{1}{6^7} = 0.0000036$.

CHAPTER 7

UNIT 7A

3. a. Linear. **b.** Exponential. **c.** Exponential.

d. Linear.

5. Square #20 has $2^{19} = 524,288$ grains; total number of grains = 1,048,575.

7. a. 1.3×10^{12} tons. **b.** Grain on chessboard is 650 times annual harvest.

9. a. At 11:55, 2^{55} bacteria; the bottle is $1/32$ full.

b. At 11:05, 2^5 bacteria; the bottle is $1/2^{55}$ or 2.8×10^{-17} full.

11. 2.5 meters thick over the entire Earth.

13.a.

Year	Population	Year	Population
2000	6×10^9	2550	1.229×10^{13}
2050	1.2×10^{10}	2600	2.458×10^{13}
2100	2.4×10^{10}	2650	4.915×10^{13}
2150	4.8×10^{10}	2700	9.830×10^{13}
2200	9.6×10^{10}	2750	1.966×10^{14}
2250	1.92×10^{11}	2800	3.932×10^{14}
2300	3.84×10^{11}	2850	7.864×10^{14}
2350	7.68×10^{11}	2900	1.573×10^{15}
2400	1.536×10^{12}	2950	3.146×10^{15}
2450	3.072×10^{12}	3000	6.291×10^{15}
2500	6.144×10^{12}		

b. Between 2800 and 2850. **c.** Shortly after 2150.

UNIT 7B

1. a. By a factor of $2^8 = 256$ in 24 hours; by a factor of $2^{56} = 7.2 \times 10^{16}$ in a week.

b. By a factor of $2^3 = 8$ in 30 years; by a factor of $2^5 = 32$ in 50 years.

c. 44 years.

3. a. $1260; $2000.
  b. 44,123; 124,800.
  c. 16,777,216; 4,294,967,296.

5. 7.35 billion; 17.49 billion; 34.98 billion.

7. a. 100 months; 1.09; 1.95.
  b. 44 years; 7 billion; 29 billion; 4.2×10^{16}.
  c. 37 years; 1.21.

9. a. $\frac{1}{4}$ of its size in 70 years; $\frac{1}{16}$ of its size in 140 years.

  b. $\frac{1}{16}$ of its size in 40 years; $\frac{1}{128}$ of its size in 70 years.

  c. $\frac{1}{4}$ of its size in 24 hours; $\frac{1}{8}$ of its size in 36 hours.

11. a. 4.27 gm; 0.17 gm.   **b.** 496 trees; 39 trees.
  c. 35.35 mg; 8.84 mg.

13. a. 35 years; by a factor of 0.91 of its original value.
  b. 21.2 years; by a factor of 0.85 of its original value; by a factor of 0.19 of its original value.
  c. 7 hours; by a factor of 0.37 of its original value; by a factor of 0.09 of its original value.
  d. 11.7 years.   **e.** 140,000 years; 61 kg.

15. a. 4 days; exact.   **b.** 2 days; exact.   **c.** 4 days; exact.
  d. 14 years; good approximation.
  e. 14 years; good approximation.
  f. 14 hours; good approximation.

17. a. True.   **b.** True.   **c.** True.   **d.** False.

19. a. 34.19 years; by a factor of 0.904 of its original value.
  b. 20.58 years; by a factor of 0.845; by a factor of 0.186.
  c. 6.56 hours; by a factor of 0.35; by a factor of 0.079.
  d. 11.16 hours.   **e.** 138,155 years; 60.5 kg.

21. a. Approximate 8.24 years; exact 8.5 years.
  b. Approximate 15.05 years; exact 15.25 years.

. . . .

UNIT 7C

1. a, b.

Year	Linear Population	Exponential Population
1990	100,000	100,000
1991	110,000	110,000
1992	120,000	121,000
1993	130,000	133,100
1994	140,000	146,410
1995	150,000	161,051
1996	160,000	177,156
1997	170,000	194,872
1998	180,000	214,359

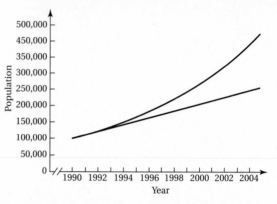

1999	190,000	235,795
2000	200,000	259,374
2001	210,000	285,312
2002	220,000	313,843
2003	230,000	345,227
2004	240,000	379,750
2005	250,000	417,724

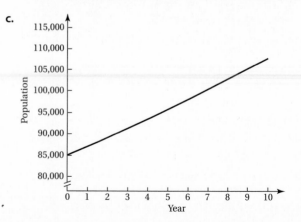

c. Linear City grows with a constant *absolute* growth rate. Exponential City grows with a constant *percentage* growth rate.

3. a. $Q = 85,000 \times (1 + 0.024)^t = 85,000 \times (1.024)^t$.

b.

Year	0	1	2	3
Population	85,000	87,040	89,129	91,268
Year	4	5	6	7
Population	93,458	95,701	97,998	100,350
Year	8	9	10	
Population	102,759	105,225	107,750	

c.

5. a. $Q = 800 \times (1 + 0.03)^t = 800 \times (1.03)^t$.
b.

Year	0	1	2	3	4	5
Homicides	800	824	849	874	900	927
Year	6	7	8	9	10	
Homicides	955	984	1013	1043	1075	

c.

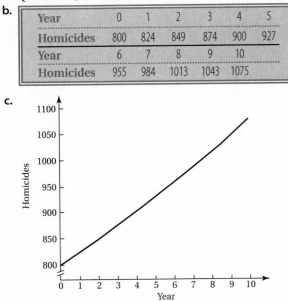

7. a. $Q = 10{,}000 \times (1 - 0.003)^t = 10{,}000 \times (0.997)^t$.
b.

Month	0	1	2	3
Population	10,000	9970	9940	9910
Month	4	5	6	7
Population	9881	9851	9821	9792
Month	8	9	10	
Population	9763	9733	9704	

c.

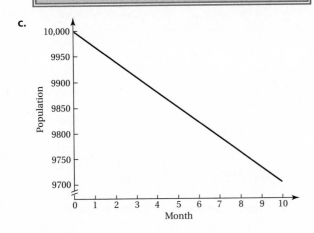

9. a. $Q = \$0.25 \times (1 - 0.1)^t = \$0.25 \times (0.9)^t$.

b. The values in the table are rounded to the nearest cent.

Week	0	1	2	3	4	5
Value	$0.25	$0.23	$0.20	$0.18	$0.16	$0.15
Week	6	7	8	9	10	
Value	$0.13	$0.12	$0.11	$0.10	$0.09	

c.

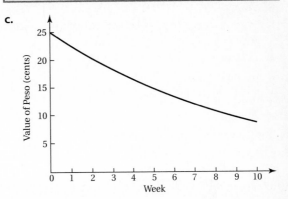

11. a. $Q = \$2000 \times (1 + 0.05)^t = \$2000 \times (1.05)^t$.
b.

Year	0	1	2	3
Monthly Salary	$2000	$2100	$2205	$2315
Year	4	5	6	7
Monthly Salary	$2431	$2553	$2680	$2814
Year	8	9	10	
Monthly Salary	$2955	$3103	$3258	

c.

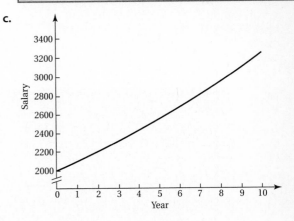

13. In 2010, population = 153,519; in 2100,
population = 2,195,400.

15. a. $126.68.   **b.** $1773.48.
c. $44.34.

17. a. By a factor of 1.074; annual growth rate is $r = 0.074$ or 7.4%.

    b. By a factor of 0.908; annual decay rate is $r = 0.092$ or 9.2%.

19. 33.28 years.

21. a. Between 3 and 4 weeks.

    b. 5 milligrams.

    c. Your cat can graze provided diet remains below 5 cm² of grass.

23. a. 1.055 billion years;   **b.** 3.88 billion years.

25. a. 15.9 mg;   **b.** 120 hours.

•••••••••
UNIT 7D

1. Approximate doubling time = 50 years.

3. a. 1.4% per year;   **b.** 2.2% per year;

    c. 1.6% per year.

5. a. Birth rate decreased by about 25% between 1975 and 1995.

    b. Death rate decreased slightly between 1975 and 1995.

    c. Net growth rates in 1975, 1985, and 1995 were 21.1, 16.9, and 14.7 per 1000, respectively.

7. a. Birth rate increased then decreased slightly between 1975 and 1995.

    b. Death rate was virtually constant between 1975 and 1995.

    c. Net growth rates in 1975, 1985, and 1995 were 5.1, 7.0, and 6.3 per 1000, respectively.

9. 2.4%, 1.2%, 0.3% per year.

11. 400,000

CHAPTER 8

•••••••••
UNIT 8A

1. (i)

	Sedan	Station Wagon	Hatchback
Color 1	Car #1	Car #2	Car #3
Color 2	Car #4	Car #5	Car #6
Color 3	Car #7	Car #8	Car #9
Color 4	Car #10	Car #11	Car #12
Color 5	Car #13	Car #14	Car #15
Color 6	Car #16	Car #17	Car #18
Color 7	Car #19	Car #20	Car #21
Color 8	Car #22	Car #23	Car #24

(ii)

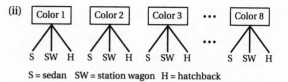

S = sedan   SW = station wagon   H = hatchback

(iii) By the multiplication principle, there are $8 \times 3 = 24$ choices.

3. (i)

	C1	C2	C3	C4	C5	C6	C7	C8
P1	1	2	3	4	5	6	7	8
P2	9	10	11	12	13	14	15	16
P3	17	18	19	20	21	22	23	24
P4	25	26	27	28	29	30	31	32

(ii)

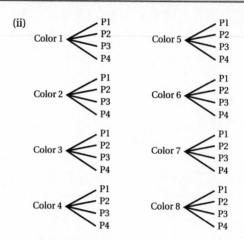

(iii) By the multiplication principle, there are $8 \times 4 = 32$ wallpaper selections.

5. 336 packages.   **7.** 360 schedules.

9. a. 10,000 house numbers (or 9000, if zero is not used for the first digit).

    b. 282,475,249 tunes.   **c.** 11,881,376 words.

    d. 512 different arrangements.

11. a. 5040;   **b.** 840;   **c.** 35;   **d.** 3,991,680;

    e. 495;   **f.** 21.

13. a. 95,040 arrangements.   **b.** 720 passwords.

    c. 479,001,600 schedules.   **d.** 5.74×10^{16} card sequences.

15. a. 252 subcommittees.   **b.** 91,390 hands.

    c. 210 arrangements.   **d.** 125,970 squads.

17. a. $_9C_3 = \dfrac{9!}{6!3!} = \dfrac{9 \times 8 \times 7}{3 \times 2} = 84$ is the number of three-card arrangements that can be made from nine cards.

c. $_4P_2 = \dfrac{4!}{2!} = 4 \times 3 = 12$ is the number of ways to choose a president and vice-president from a four-person board of directors.

e. $_{12}C_8 = \dfrac{12!}{4!8!} = \dfrac{12 \times 11 \times 10 \times 9}{4 \times 3 \times 2} = 495$ is the number of 8-topping pizzas that can be made with 12 toppings to choose from.

19. 7.41×10^{11} batting orders; 2×10^9 years.

21. 28 chimes; combinations.   **23.** 50 three-digit addresses.

25. 1,771,561 passwords with repetition.

27. a. 8,000,000 phone numbers; two million people can be served.
   b. 10,000 telephone numbers within a single exchange; at least two exchanges, assuming 4 people per telephone.

29. a. 100,000 five-digit zip codes.
   b. 2700 people per zip code.
   c. 10^9 nine-digit zip codes.

31. a. 120 ways;   **b.** 28 ways;   **c.** 6 ways.

33. a. 8.1×10^{67};   **b.** 1.5×10^{62} years.

• • • • • • • • •
UNIT 8B

5. $P(2) = \dfrac{4}{52} = \dfrac{1}{13}$.

7. $P(\text{red}) = \dfrac{26}{52} = \dfrac{1}{2}$.

9. $P(\text{no spade}) = \dfrac{39}{52} = \dfrac{3}{4}$.

11. a. $\dfrac{2}{17}$;   **b.** $\dfrac{5}{17}$;   **c.** $\dfrac{10}{17}$;   **d.** $\dfrac{7}{17}$.

13. a. $P(\text{not 2 heads}) = \dfrac{3}{4}$.

b. $P(\text{not odd}) = \dfrac{1}{2}$.

c. $P(\text{not face card}) = \dfrac{10}{13}$.

d. $P(\text{not double}) = \dfrac{5}{6}$.

15. a.

Coin #1	Coin #2	Coin #3	Coin #4
H	H	H	H
H	H	H	T
H	H	T	H
H	H	T	T
H	T	H	H
H	T	H	T
H	T	T	H
H	T	T	T
T	H	H	H
T	H	H	T
T	H	T	H
T	H	T	T
T	T	H	H
T	T	H	T
T	T	T	H
T	T	T	T

b.

Result	4H	3H, 1T	2H, 2T	1H, 3T	4T
# Occurrences	1	4	6	4	1
Probability	$\dfrac{1}{16}$	$\dfrac{1}{4}$	$\dfrac{3}{8}$	$\dfrac{1}{4}$	$\dfrac{1}{16}$

c. $P(\text{3H, 1T}) = \dfrac{1}{4}$.   **d.** $P(\text{not 4T}) = \dfrac{15}{16}$.

17. a. $P(\text{sum of 4}) = \dfrac{1}{12}$.   **b.** $P(\text{not sum of 4}) = \dfrac{11}{12}$.

c. $P(\text{2 and 5}) = \dfrac{1}{18}$.

d. $P(\text{one 3}) = \dfrac{5}{18}$, excluding double 3; $P(\text{one 3}) = \dfrac{11}{36}$; including double 3.

e. You would expect to win $1.61 for every $1 you lose.

21. a. 1 to 2.   **b.** 1 to 3.   **c.** 1 to 3.   **d.** 3 to 10.

23. a. $15.   **b.** $25.

25. a. 5,245,786.   **b.** 1 in 5,245,786.
   c. The chances are the same for any set of 6 numbers.

• • • • • • • • •
UNIT 8C

1. a. $P(A) = 0.077$.   **b.** $P(AA) = 0.0059$.
   c. $P(AAAA) = 0.000035$.   **d.** $P(AAAAA) = 0.000003$.

3. a. $P(6, 6) = 0.028$.   **b.** $P(6H) = 0.016$.
    c. $P(3 \text{ wins}) = 0.001$.   **d.** $P(AK) = 0.0059$.

5. With replacement, $P = 0.36$; without replacement, $P = 0.35$.

7. a. $P(A) = \dfrac{1}{13}$.   **b.** $P(AA) = 0.0045$.
    c. $P(AAAA) = 0.000004$.   **d.** $P = 3 \times 10^{-9}$.

9. a. $P(2 \text{ or } 4) = 0.15$.   **b.** $P(1 \text{ or } 2) = \dfrac{1}{3}$.   **c.** $P = 0.31$.
    d. $P(\text{woman or Democrat}) = \dfrac{3}{4}$.   **e.** $P(\text{king or heart}) = \dfrac{4}{13}$.

    f. $P(\text{ace or king or diamond}) = \dfrac{19}{52}$.

11. a. $P(6) = \dfrac{1}{6}$.   **b.** $P(\text{white}) = \dfrac{1}{2}$.   **c.** $P(1 \text{ or } 3) = \dfrac{1}{3}$.
    d. $P(1 \text{ or blue}) = \dfrac{1}{2}$.   **e.** $P(1 \text{ or white}) = \dfrac{1}{6} + \dfrac{1}{2} = \dfrac{2}{3}$.

13. a. $P(\text{at least one } 6) = 0.60$.   **b.** $P(\text{at least one } H) = \dfrac{7}{8}$.
    c. $P(\text{at least one } D) = 0.76$.   **d.** $P(\text{at least one } A) = 0.55$.

15. a. $P = 0.5055$.   **b.** He would have won over time.

17. a. $P = 0.025$;   **b.** $P = 0.000625$;   **c.** $P = 0.224$.

19. a. $_{52}C_5 = 2{,}598{,}960$ different hands.
    b. $P(\text{royal flush in a given suit}) = 3.8 \times 10^{-7}$.
    c. $P(\text{any royal flush}) = 1.52 \times 10^{-6}$.
    d. $P = 1.8 \times 10^{-5}$; $P = 2.4 \times 10^{-4}$.

21. $P(Aa) = 0.18$.

UNIT 8D

1. a. $P(\text{black}) = 47\%$.
    b. You cannot make a reasonable prediction with only 3 spins.
    c. In 100,000 tries, the wheel should come up red about 47,000 times.

3. a. $P(6) = \dfrac{1}{6} = 16.7\%$.
    b. Probably less than 3 or 4 times, but an accurate prediction is impossible with so few rolls.
    c. Somewhere around 160 to 170 times.

5. a. Net loss of $10.   **b.** Net loss of $18.
    c. You would have to get 59 heads in the next 100 tosses.

7. a. On the next toss you could get a head which makes the difference 23, or equally likely, you could get a tail which makes the difference 25.

    b. On each successive toss, the difference is equally likely to increase as decrease. Therefore, after 1000 tosses, the original difference of 24 is just as likely to be greater than it is to be less.

    c. Once you have fewer heads than tails (say 38 heads vs. 62 tails), the difference between the number of heads and tails is equally likely to increase as decrease, so the deficit of heads is likely to remain.

9. Expected value = $0.50.

11. $60 per policy.

13. −$0.53 per ticket.

15. −$0.17 per game.

17. $P = \dfrac{4}{38}$; expected earnings = −$0.211; house edge = $0.053.

19. a. Expected value = −$0.014; house edge is 1.4 cents per dollar.
    b. Expected loss = $1.40.
    c. Expected loss = $7.
    d. Expected earnings of the casino = $14,000.

21. $85 million.

UNIT 8E

1. a. 1.66×10^{-4}.
    b. $P = 0.00166$.
    c. $P = 0.00827$.

3. 2.66×10^{-5} fatalities per flight hour.

5. a. In Utah there were 108 births per day; in Maine there were 38 births per day.
    b. Utah population = 1,978,850; Maine population = 1,240,714.

7. a. 4.0 million births.
    b. 2.3 million deaths.
    c. Net population gain = 1.7 million people per year.
    d. Population increase due to immigration = 700,000; 29.2% of the population increase was due to immigration.

9. a. $P(\text{death by homicide}) = 7.7 \times 10^{-5}$; $P(\text{death by auto}) = 1.7 \times 10^{-4}$.
    b. 7.7 deaths per 100,000 people.
    c. $P(\text{death by homicide or automobile}) = 2.5 \times 10^{-4}$.

11. a. 4%;   **b.** 25%.

13. a. 0.025.
    b. 0.00065.
    c. 50 people expected to win 5 games in a row.
    d. 1.3 people expected to win 10 games in a row.

17. a. 0.8704.
    b. 0.0004 or 1 in 2500.
    c. 2×10^{-7}.

CHAPTER 9

●●●●●●●●●
UNIT 9A

1. a.

Grade	Frequency	Relative Frequency	Cumulative Frequency
A	3	$\frac{3}{22} = 13.6\%$	3
B	7	$\frac{7}{22} = 31.8\%$	10
C	7	$\frac{7}{22} = 31.8\%$	17
D	2	$\frac{2}{22} = 9.1\%$	19
F	3	$\frac{3}{22} = 13.6\%$	22
Total	22	99.9%	22

b.

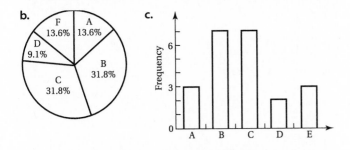

c.

3.

Time	Frequency	Relative Frequency	Cumulative Frequency
10.0	1	0.1	1
10.1	3	0.3	4
10.2	2	0.2	6
10.3	4	0.4	10
Total	10	1.0	10

5. a. Mean = 7.6.
  b. Median is 8 (halfway between 7 and 9).
  c. Mode = 9.

7. a. Mean = 25.16 seconds.
  b. Median is 25.8 seconds (halfway between 25.3 and 26.3).

c. The five-number summary of the scores is the following: the median is 25.8 seconds; the upper quartile is 26.5 seconds; the lower quartile is 23.8 seconds; the fastest time is 22.1 seconds; and the slowest time is 27.0 seconds.

d.

e. Standard deviation = 1.70.

9. a. The five-number summary of the scores is the following: the median is 172 cm; the upper quartile is 188 cm; the lower quartile is 166 cm; the largest height is 190 cm; and the smallest height is 145 cm.

b.

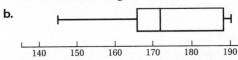

c. Mean = 172.6 cm.   **d.** Standard deviation = 15.2.

11. a.

Reflex Time	Frequency	Rel. Freq.	Cum. Freq.
1.1	5	0.1	5
1.2	10	0.2	15
1.3	20	0.4	35
1.4	8	0.16	43
1.5	4	0.08	47
1.6	2	0.04	49
1.7	1	0.02	50
Total	50	1.0	50

b.

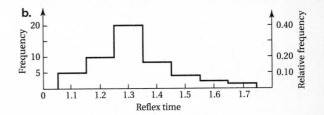

c.

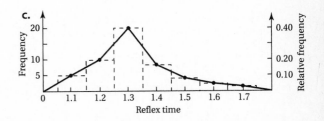

d.

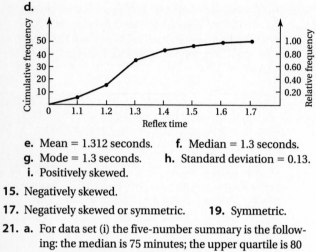

e. Mean = 1.312 seconds.   **f.** Median = 1.3 seconds.
g. Mode = 1.3 seconds.   **h.** Standard deviation = 0.13.
i. Positively skewed.

15. Negatively skewed.

17. Negatively skewed or symmetric.   **19.** Symmetric.

21. a. For data set (i) the five-number summary is the following: the median is 75 minutes; the upper quartile is 80
minutes; the lower quartile is 60 minutes; the longest
time is 110 minutes; and the lowest time is 45 minutes.
b. Data set (i) is bimodal because it has two distinct peaks.

- - - - - - - - -
UNIT 9B

5. b.

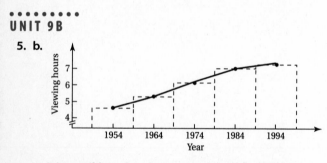

c. It is possible to draw by eye a straight line that fits the
data very well.
d. A projection based on the straight-line fit may be
unreliable. It appears that the increasing trend in
television-viewing time may not continue.

7. a, b.

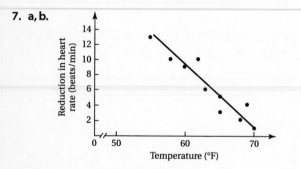

c. Strong negative correlation.
d. $r = -0.90$.

9. a.

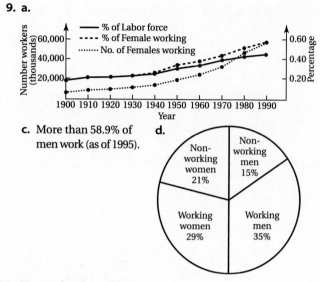

c. More than 58.9% of   **d.**
men work (as of 1995).

11. Uncorrelated; $r = 0.00$.

13. Weak positive correlation; $r = 0.50$.

15. Uncorrelated; $r = 0.00$.

- - - - - - - - -
UNIT 9C

1. a. 68.3%.   **b.** 34%.   **c.** 34%.   **d.** 95%.
    e. 99.7%.   **f.** 49.85%.

3.

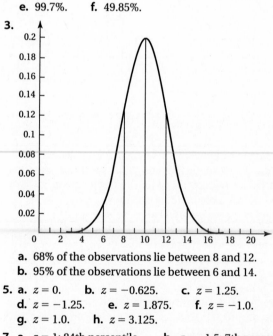

a. 68% of the observations lie between 8 and 12.
b. 95% of the observations lie between 6 and 14.

5. a. $z = 0$.   **b.** $z = -0.625$.   **c.** $z = 1.25$.
    d. $z = -1.25$.   **e.** $z = 1.875$.   **f.** $z = -1.0$.
    g. $z = 1.0$.   **h.** $z = 3.125$.

7. a. $z = 1$; 84th percentile.   **b.** $z = -1.5$; 7th percentile.
    c. $z = -2$; 2nd percentile.   **d.** $z = 1.5$; 93rd percentile.

7. e. 1.6 standard deviations above the mean.
 f. 1.55 standard deviations below the mean.
9. a. 90th percentile.   **b.** Score = 693.
11. 38% of the people earn between $27,000 and $33,000.
13. a. 68% of the classmates.   **b.** 95% of the classmates.
 c. 80% of the classmates.   **d.** 5th percentile.
 e. 99.6th percentile.
15. Population: 22.6%; sample: 23%.
17. About 350 supporters.
19. Mean = 0.45, standard deviation = 0.017.

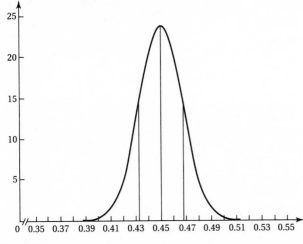

21. Yes.
23. 33.6% to 36.4%.

UNIT 9D

1. a. z-score = 0.7; 76th percentile.
 b. z-score = 0.0; 50th percentile.
5. a. Hypothesis: attending the study session is more likely to lead to an improvement in a student's grade than not attending the study session. Null hypothesis: attending the study session has no more effect on grade improvement than not attending the study session.
 b. 55% of the students will improve their grade.
 c. The percentage of students attending the study session who improve their grade should be significantly greater than 55% and the percentage of students not attending the study session who improve their grade should be significantly less than 55%.
 d. 72.5% of the students attending the study session improved their grade; 37.5% of the students not attending the study session improved their grade.

e. The percentage of students who improved their grade with study sessions is significantly greater than expected by chance. The number of students who improved their grade without study sessions is significantly less than expected by chance. These differences are statistically significant.
7. a. Hypothesis: men are more of a drinking-and-driving risk than women. Null hypothesis: men and women are equal drinking risks.
 b. 15.0% of the people in the study have been drinking in the last two hours.
 c. The percentage of men who have been drinking in the last two hours should be greater than 15% and the percentage of women who have been drinking in the last two hours should be less than 15%.
 d. 16.0% of men were drinking in the last two hours; 11.5% of women were drinking in the last two hours.
 e. The percentages expected by chance are close to the actual percentages for both men and women, suggesting that the differences are not statistically significant.

CHAPTER 10

UNIT 10A

3. a. 120°.   **b.** 30°.   **c.** 18°.
5. a. $\dfrac{1}{360}$.   **b.** $\dfrac{1}{90}$.   **c.** $\dfrac{1}{24}$.   **d.** $\dfrac{1}{12}$.
7. a. 120°;   **b.** 150°.
9. 1 perpendicular line; 0 parallel lines.
11. a. 37.7 m; 113.1 m^2.   **b.** 25.1 km; 50.3 km^2.
 c. 157.1 cm; 1963.5 cm^2.
13. a. 200 meters.   **b.** 7 blocks or 280 meters.
15. Area of the left barn is greater than the area of the right barn.
17. 144.7 square meters.   **19.** 24,000 m^3; 2.4 × 10^7 liters.
21. Volume = 54.50 ft^3; surface area = 130.85 ft^2.
23. Volume = 47.71 in.^{3}; surface area = 63.62 in.^{2}.
25. a. 25 times larger than the height of the model.
 b. 625 times larger than the surface area of the model.
 c. 15,625 times larger than the volume of the model.
27. a. Arm length increases by a factor of 3.
 b. Waist size increases by a factor of 3.
 c. Amount of clothing increases by a factor of 9.
 d. Weight increases by a factor of 27.

29. a. Squirrels have a much higher surface area to volume ratio than humans.

   b. Squirrels lose much more heat for their volume than humans and must maintain a higher rate of metabolism.

31. Maximum possible area = 1387 square meters.

33. Lowest cost design = $114.

35. a. Surface area = 1.05×10^8 mm^2; volume = 5.7×10^6 mm^3.

   b. Radius = 111 mm; surface area = 1.55×10^5 mm^2.

   c. Radius = 2.9×10^3 mm.

37. 0.004 cubic kilometers.

39. a. Volume = 2.6×10^7 km^3.

   b. Volume of water = 2.2×10^7 km^3.

   c. 65 m.

UNIT 10B

1. 220 cps; 440 cps; 880 cps; 1760 cps.

3.

Note	Frequency (cps)	Note	Frequency (cps)
A	437	E	655
A#	463	F	694
B	491	F#	735
C	520	G	779
C#	551	G#	825
D	583	A	874
D#	618		

5. a. 347 cps; **b.** 390 cps; **c.** 694 cps;
   d. 2080 cps; **e.** 4946 cps.

7. 327 cps; 275 cps.

9. a. A factor of $2^{7/12} = 1.498$.

   b. A factor of 2.24.

   c.

Note	Frequency (cps)	Note	Frequency (cps)
C	260	C#	4401
G	389	G#	6593
D	583	D#	9876
A	874	A#	14,794
E	1309	F	22,162
B	1961	C	33,198
F#	2938		

d. 12 notes, 7 octaves. **e.** A factor of 128.

f. A factor of $2^{5/12} = 1.335$; 12 notes; 5 octaves.

UNIT 10C

1. b.

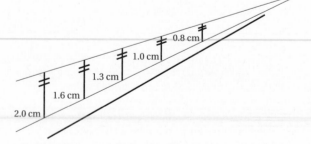

3.

5. a.

   b. 1.3 cm, 1.0 cm, 0.8 cm.

   c. If the poles are equally spaced in the drawing, then they would not be equally spaced in the real scene. Alternatively, if the poles are equally spaced in the real scene, then they would not appear equally spaced in the drawing.

9. a. 4 reflection symmetries, rotational symmetries of 90°, 180°, and 270°.

   b. 7 reflection symmetries, rotational symmetries of 51.4°, 102.8°, 154.2°, 205.6°, 257.0°, 308.4°.

11. Reflection symmetries: it can be reflected across a vertical line through its center, a horizontal line through its center, or either of its diagonals; rotation symmetries: it can be rotated though 90°, 180°, and 270°.

13. Translation symmetry and reflection symmetry.

15.

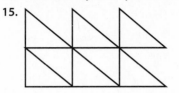

17.

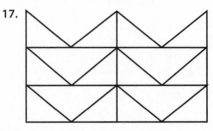

19.

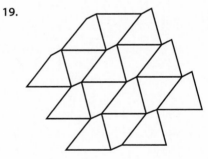

21. The angles around a point P are precisely the angles that appear inside of a single quadrilateral. Thus the angles around P have a sum of 360° and the quadrilaterals around P fit together perfectly.

UNIT 10D

Throughout these solutions, we use a value of $\phi = 1.62$.

3. 2.43 inches or 0.93 inches.

5. 10.37 km or 3.95 km.

9. b. $\phi^2 = \phi + 1$.

11. a.

n	L_n	L_n/L_{n-1}
1	1	–
2	3	3
3	4	1.33
4	7	1.75
5	11	1.57
6	18	1.64
7	29	1.61
8	47	1.62
9	76	1.617
10	123	1.618

b. The ratios approach ϕ.

UNIT 10E

1. a. Start with a meter stick and measure the length of the sidewalk, counting the number of times you lay the stick down on the sidewalk. Now reduce the length of the stick to 0.1 meter—a reduction by a factor of $R = 10$. Repeat the measurement of the sidewalk and count the number of times you lay the stick down on the sidewalk. You should find that the number of elements for this second measuring is $N = 10$ times the number of elements of the first measurement. If this pattern continues, that every time the ruler is decreased in length by a factor of $R = 10$, the number of elements increases by a factor of $N = 10$, then the sidewalk is a one-dimensional object.

3. a. Dimension is 1 and the object is ordinary (non-fractal).
   b. Dimension is 2 and the object is ordinary (non-fractal).
   c. Dimension is 3 and the object is ordinary (non-fractal).
   d. Dimension is 1.26 and the object is a fractal.
   e. Dimension is 2.26 and the object is a fractal.
   f. Dimension is 3.26 and the object is a fractal.

7. Fractal dimension is 0.63.

9. d. Fractal dimension is approximately 1.6.

CHAPTER 11

UNIT 11A

1. a. Discrete;   **b.** Continuous.

5. a.

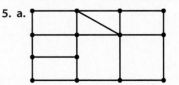

b. If edges represent sidewalks, then each edge in this network will be replaced by two edges, one for each sidewalk.

7. Vertices represent the four people, Amy, Beth, Cate, and Daniel, and edges to represent the relation "trades with."

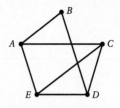

9. a. Four vertices and five edges.   **b.**
   c. No Euler circuit.

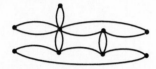

11. There are many Euler circuits.

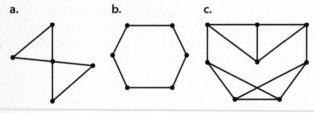

13. a. Order 8.   **b.** Vertices A, D, E, F, G, H have degree 3, vertices B and C have degree 4.   **c.** No special form.
   d. No Euler circuits.

15. a. Order 8.   **b.** All vertices have degree 4.
   c. No special form.   **d.** There are Euler circuits.

17. a. Order 5.   **b.** Vertex A has degree 1, vertices C and D have degree 2, vertex E has degree 3, vertex B has degree 4.
   c. No special form.   **d.** No Euler circuits.

19. Example of networks with the given properties are shown below. They are not unique!

   a.   **b.**   **c.**

21. 10 edges in a complete network of order 5; 15 edges in a complete network of order 6.

23. The order of the network is 9 and the degree of the Missouri vertex is 8.

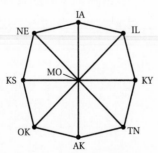

25. Order five. Vertices A, C, D, and E have degree three, vertex B has degree two. The network is not complete.

27. Network I has the shortest length.

29. Minimum cost spanning tree has length 16.

31. The minimum cost is 45 units.

• • • • • • • • •
UNIT 11B

1. a. 2520.   **b.** 1.78×10^{14}.   **c.** 5,644,343 years.

3. Not a Hamiltonian circuit.

5. Not a Hamiltonian circuit.

7. a. The network will look like Figure 11.29 of the text except that the vertex for Zion will be missing along with all of the edges connected to it. The resulting network has four vertices and a total of six edges.
   b. 470 miles.
   c. Grand Canyon, Bryce, Capitol Reef, Canyonlands, Grand Canyon; distance 454 miles.
   d. Same as part (c).
   e. Circuit of (c) and (d) is the shortest.

9. The circuits BCDFEAB and FEABCDF have length 23.

11. The shortest circuit for Abe is ACDBA with length 12.

13. $450,000 per day; $164,250,000 per year.

15. $550.

• • • • • • • • •
UNIT 11C

1. a, e, and f.

3. a, e, f, g, i, and k (vertices A, B, C, D, E, F, G, H).

5. Task d: EST = 1 hour; LST = 2 hours.
   Task f: EST = 2.5 hours; LST = 2.5 hours.

7. Task d: EFT = 1.5 hours; LFT = 2.5 hours.
   Task f: EFT = 4.5 hours; LFT = 4.5 hours.

9. No delay in overall project schedule.

11.

Task	EST	LST	EFT	LFT	Slack
a	0.0	0.0	2.0	2.0	0
b	0.0	1.0	1.0	2.0	1.0
c	0.0	1.5	1.0	2.5	1.5
d	1.0	2.0	1.5	2.5	1.0
e	2.0	2.0	2.5	2.5	0
f	2.5	2.5	4.5	4.5	0
g	4.5	4.5	7.5	7.5	0
h	4.5	5.5	6.5	7.5	1.0
i	7.5	7.5	9.5	9.5	0
j	9.5	10.0	10.0	10.5	0.5
k	9.5	9.5	10.5	10.5	0

13. e, f, h, and i.   **15.** 19 months.

17. Task d: EST = 3 months; LST = 5 months. Task h: EST = 7 months; LST = 7 months.

19. Task d: EFT = 5 months; LFT = 7 months; it is not on the critical path. Task h: EFT = 15 months; LFT = 15 months.

21. A one month delay in this task will delay the project.

23. Slack times: a 0, b 0, c 10, d 0, e 15, f 30, g 0, h 0.

25. 150 minutes.

27.

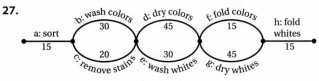

29. a, c, and f.

31. Tasks (a, b) and (c, d, e) take place at the same time. Task f must be done alone.

· · · · · · · · ·
UNIT 11D

1. Tilden narrowly won the popular vote, but Hayes narrowly won the electoral vote and became President.

3. Garfield narrowly won the popular vote and easily won the electoral vote to become President.

5. Carter narrowly won the popular vote and comfortably won the electoral vote to become President.

7. a. No.   **b.** No.   **c.** No.   **d.** Yes.

9. a. Best wins a plurality, but not a majority.
   b. Able needs 64.3% of Crown's votes.

11. a. Inviglio wins a plurality, but not a majority.
   b. Height needs 63.8% of Grand's votes.

13.

First	A	A	C	C
Second	B	C	A	B
Third	C	B	B	A
	4	2	3	1

15. a. 66 votes were cast.
   b. D is the plurality winner (but not by a majority).
   c. D is the winner of the top two runoff.
   d. The winner by sequential runoff is D.
   e. D is the winner by the Borda count.
   f. D wins by the pairwise comparison method.
   g. As the winner by all five methods, candidate D is clearly the winner of the election.

17. a. 100 votes were cast.
   b. C is the plurality winner (but not by a majority).
   c. A is the winner of the top two runoff.
   d. A is the sequential runoff method.
   e. B is the winner by the Borda count.
   f. B wins by the pairwise comparison method.
   g. There is not a clear winner.

19. a. 90 votes were cast.
   b. E is the plurality winner (but not by a majority).
   c. B the winner of the top two runoff.
   d. B the winner of the sequential runoff.
   e. B is the winner by the Borda count.
   f. D is the winner by the pairwise comparison method.
   g. Candidates B and E each win by two methods, so the outcome is debatable.

21. With four candidates, there are 6 pairwise races. With five candidates, there are 10 pairwise races.

23. D has 130 points which is enough to win the election.

· · · · · · · · ·
UNIT 11E

3. B wins by the Borda count; Criterion 1 is violated.

7. One possible answer:

First	B	A	C	C
Second	A	B	A	B
Third	C	C	B	A
	2	4	2	3

C is the plurality winner, but A beats C and B beats C in one-on-one races.

9. Criterion 2 is satisfied.

11. One possible answer:

First	B	A	C
Second	A	C	A
Third	C	B	B
	3	2	4

Candidate C wins in a sequential runoff. But A beats B and C in head-to-head races, violating Criterion 2.

13. One possible answer:

First	A	D	C
Second	B	B	B
Third	C	A	D
Fourth	D	C	A
	4	8	3

15. Criterion 3 is violated.

19. One possible answer:

First	A	B	C
Second	B	A	B
Third	C	C	A
	3	2	4

Candidate C would win by plurality. However, if Candidate A drops out Candidate B wins by plurality.

21. One possible answer:

First	A	B	C
Second	B	A	B
Third	C	C	A
	4	3	5

After Candidate B is eliminated, Candidate A wins the runoff. However, if Candidate C were to drop out, B would win the election. So Criterion 4 is violated.

23. One possible answer:

First	A	B
Second	C	A
Third	B	C
	2	3

A wins by the point system. However, if Candidate C were to drop from the race, then Candidate B would win. Thus Criterion 4 is violated.

25. One possible answer:

First	B	C	D
Second	A	A	C
Third	D	D	B
Fourth	C	B	A
	2	2	2

Pairwise comparisons would lead to a tie between C and D. However, if A were to drop out of the election, then D would win. Criterion 4 is violated.

27. Criterion 1 does not apply. Criterion 2 is satisfied. Criterion 3 is satisfied. Criterion 4 is satisfied.

29. Criterion 1 does not apply. Criterion 2 is satisfied. The point system always satisfies Criterion 3. Criterion 4 is satisfied.

31. Criterion 1 does not apply. Criterion 2 is violated. The plurality method always satisfies Criterion 3. Criterion 4 is violated.

33. Criterion 1 does not apply. Criterion 2 is violated. Criterion 3 is satisfied. Criterion 4 is violated.

35. Criterion 1 does not apply. Criterion 2 is satisfied. The pairwise comparison method always satisfies Criterion 3. Criterion 4 is satisfied.

37. a. C wins by plurality.   **b.** B wins by approval vote.

CHAPTER 12

UNIT 12A

1. a. $63,000.
   b. Total outlays = $1,113,000; deficit = − $63,000; debt = $836,000.
   c. $69,000.
   d. Total outlays = $869,000; surplus = $231,000; debt = $605,000.

3. Debt 15.2%; defense 16.9%; social security 22.9%; international affairs 0.94%.

5. 4.90%; $290 billion; $192 billion.

7. Change in gross debt = $261 billion; net deficit = $107 billion. The gross debt reflects both the net deficit (money borrowed from the public) and money borrowed from various trust funds. In 1996 about $154 billion was borrowed from trust funds.

9. Mandatory spending in 2016 = $2893 billion; receipts in 2016 = $3184 billion.

11. $39,778 per worker.   **13. a.** 2022;   **b.** 2013.

UNIT 12B

1. a. 800 Calories per hour.
   b. 920 watts, enough for nine 100-watt light bulbs.
   c. 3.3×10^6 joules.   **d.** 3.3 candy bars.

3. $184 per year.

5. a. 4.5×10^9 joules.
   b. 375 liters; 99 gallons; 2.4 barrels.   **c.** 1.7 kilowatts.

7. 1.8×10^6 kg.

9. a. The energy in a hurricane is 20 times greater than the energy released in the bomb.
  b. 4 candy bars.
  c. The total energy from the Sun in one second could easily supply the annual U.S. energy needs.

11. 294 watts.

13. a. 7.2×10^8 kilowatt-hr.
  b. 7.2×10^5 homes.
  c. 46,667 barrels per day.

17. a. 1 terawatt-year; 3.2×10^{19} joules.
  b. 1 terawatt.
  c. One-tenth of the world's power needs.

19. a. 5.6×10^{13} joules.
  b. 11,200 tons.
  c. A 1-megaton hydrogen bomb releases 89 times more energy than a 1-kilogram sample of U-235.
  d. The Hiroshima bomb was 1.8 times more powerful than the fission bomb of part (a).

• • • • • • • • •
UNIT 12C

1. a. 0.017 kg/cm^3; **b.** 0.0286 g/cm^3.

3. a. 54 kg = 119 pounds; **b.** 160 kg.

5. a. New Jersey 1038 people per square mile; Wyoming 0.5 people per square mile.

7. a. 0.11 gigabytes/cm^2.
  b. A square centimeter of disk can hold 110 typical 500-page books. Entire disk could store 9,900 such books.

11. 300 grams.

13. In *any* sample of this air, there are on average 75 times as many carbon monoxide molecules as ozone molecules.

15. One mile downwind the levels were 1.0 µg/m^3 and 2 miles downwind the levels were 0.5 µg/m^3 which is the EPA's "safe" level. This means that, within the assumptions of this model, five miles from the plant the levels are safe.

17. a. 1800 kg/cm^3.
  b. The teaspoon of stellar matter will weigh almost eight tons!
  c. 6.7×10^{11} kg—ten times more than Mt. Everest.

• • • • • • • • •
UNIT 12D

1. a. Little effect.
  b. Serious ear damage.
  c. The effect would be serious.

3. a. 7.9×10^{11} joules.
  b. 3.5×10^{16} joules.
  c. 1.3×10^{13} joules.

5. The China earthquake was 64 times stronger than the Los Angeles earthquake.

7. a. 10^{10} times.
  b. 133 dB.
  c. -30 dB.
  d. 1000 times more intense.
  e. 153 dB; louder than a jet at 30 meters.

9. a. 40 dB.
  b. 56 meters away.
  c. 1.76 dB; a factor of 6.3×10^5.

11. a. An *increase* of 1 unit on the pH scale means that a solution becomes 10 times more basic.
  b. 3.2×10^{-9} moles per liter.
  c. pH $=1$. This solution is an acid.

13. a. 10^{-4} moles of hydrogen ions per liter.
  b. 1×10^{-5} moles per liter or a pH of 5.
  c. 1.1×10^{-4} moles per liter or a pH of -3.96.
  d. No.

Credits

Prologue

Page 1, photo of workers of various professions. ©Douglas E. Walker/Masterfile.

Chapter 1

Page 13, photo of U.N. General Assembly, courtesy of the United Nations; Page 21, photo of students in classroom, ©Rhoda Sidney/Stock Boston; Page 57, photo of college graduation, ©Lionel Delevingne/Stock Boston.

Chapter 2

Page 59, photo of business meeting, ©Steve Niedorf/The Image Bank; Page 83, photo of man in library, ©Jeffry W. Myers/Stock Boston; Page 96, photo of multi-lingual sign, ©Susan Van Etten; Page 97, Figure for Problem 10; Reprinted with permission from *Science*, Vol. 276, 11 April 1997, P. 192. Copyright 1997 American Association for the Advancement of Science. Source: Nobel Foundation; Page 120, photo of water testing, ©John Dewaele/Stock Boston; Page 121, photo of ozone layer, courtesy of NASA; Page 122, photo of Los Angeles traffic, ©Spencer Grant/Stock Boston; Page 125, photo of Marie Curie, ©The Granger Collection.

Chapter 3

Page 127, photo of world as a puzzle, ©Michael Simpson, 1995/FPG International; Page 139, photo of map with currency, ©Susan Van Etten; Page 142, photo of four dogs, ©Patricia Hollander/Stock Boston; Page 143, photo of Glen Canyon Dam, ©Don B. Stevenson/The Picture Cube; Page 156, photo of 1500 meter race, ©Bob Daemmrich/Stock Boston.

Chapter 4

Page 173, photo of fossil in amber, ©John Cancalosi/Stock Boston; Page 174, photo of Egyptian hieroglyphs, ©The Granger Collection; Page 178, image of Bosthius and Pythagoras, ©Corbis/Bettmann; Page 192, photo of newspaper reporters, ©Fredrik Bodin/Stock Boston; Page 199, photo of Mount Washington, ©Spencer Grant/Stock Boston; Page 205, satellite photo of River Nile, Courtesy of NASA; Page 213, photo of New York Marathon, ©Corbis/Bettmann; Page 229, photo of recycling newsprint, ©Inga Spence/The Picture Cube.

Chapter 5

Page 233, photo montage of bank options, ©Susan Van Etten; Page 247. photo of Anne Scheiber, ©AP/Wide World; Page 264, photo of wallet with credit cards, ©Susan Van Etten; Page 267, photo of house, ©Susan Van Etten; Page 278, photo of couple working on taxes, ©Jean-Claude Lejeune/Stock Boston; Page 300, photo collage of mutual funds, ©Susan Van Etten.

Chapter 6

Page 307, photo of astronomers, ©Jeff Zaruba/Tony Stone Images; Page 322, photo of hiker, ©Geroge Bellerose/Stock Boston; Page 326, photo of truck on Vermont road, ©Kindra Clineff/The Picture Cube; Page 337, photo of redwood tree cross-section, ©Susan Van Etten; Page 338, photo of swimmer, ©J.D.Sloan/The Picture Cube; Page 354, photo of irrigation circles, ©George W. Gardner/Stock Boston; Page 357, photo of stamps, ©Susan Van Etten.

Chapter 7

Page 359, photo of babies around the world, ©Ed Honowitz/Tony Stone Images; Page 364, photo of petri dishes, ©Susie Fitzhugh/Stock Boston; Page 368, photo of prairie dogs, ©Leonard Lee Rue III/Animals, Animals; Page 373, photo of Russian babies, ©Todd Phillips/The Picture Cube; Page 385, Figure 7.5, *Vital Signs 1997*, Worldwatch Institute, W. W. Norton & Company, New York, page 47. Reprinted with permission.; Page 393, photo of rhinos, ©Susan Van Etten; Page 401, Figure 7.11, Hollingsworth, T. H., *Historical Demography* (Ithaca, NY: Cornell University Press, 1969). Reprinted with permission.

Chapter 8

Page 405, photo of horse race, ©Richard Mackson, 1986/FPG International; Page 414, photo of swim meet, ©Elwin D. Williamson/The Picture Cube; Page 421, photo of graduation, ©Susan Van Etten; Page 425, photo of coin toss, ©Christopher S. Johnson/Stock Boston; Page 434, photo of flood, ©AP/Wide World; Page 449, photo of Miami hurricane, courtesy of The American Red Cross; Page 475, photo of Joe DiMaggio, ©Baseball Hall of Fame/Boston Herald.

INDEX